Physics for Realists

Mechanics

Modern Physics with a Common Sense Grounding

Physics for Realists:

Mechanics

Physics with a Common Sense Grounding

Anthony Rizzi

Press of the Institute for Advanced Physics

Published by IAP Press, Baton Rouge, LA (May 22, 2008)
IAP Press: IAPpress@iapweb.org

 IAP Press is a division of the Institute for Advanced Physics. This book is printed on acid-free paper.

ISBN: 978-0-9816470-0-5
Library of Congress Control Number: 2008926205
Includes appendices, index.
1. Physics-textbooks. I. Rizzi, Anthony II. Title.

QC21.3.H.35 2008
530-dc22

Printed in the United States of America
Rochester, New York

Author

Anthony Rizzi, Ph.D., Director, Institute for Advanced Physics, LA

Reviewers & Contributors

Primary: Joseph Haller, Ph.D., physicist, Institute for Advanced Physics

Benedict Ashley, O.P., Ph.D., philosopher, St. Louis Univ., MO
(special role in the reviewing process)
Vincent Bork, Ph.D., physicist and diocesan priest, VA
Murray Daw, Ph.D., Professor of Physics, Clemson University, SC
Thomas De Koninck, Professor of Philosophy, Universite Laval, Canada
Matthew Green, LC, Professor of Philosophy, Thornwood, NY
(including front and back cover designs)
Stephen Guse, M.A., Captain USN (retired), VA
Clifton Hill, C.S.Sp., Prof. of Physics (retired), Duquesne University, PA
John Keck, Ph.D., physicist, NH (including front cover concept)
Kenneth Klenk, Ph.D., Director, Science Systems Consulting, Inc., SD
Dermott Mullan, Professor of Physics, University of Delaware
Mark Wyman, Ph.D., Research Fellow, Perimeter Institute of
Theoretical Physics, Canada

Technical Editor

Clifton Hill, C.S.Sp., Prof. of Physics (retired), Duquesne University, PA

Equation Editor

Jamie Anson, B.S., Education & Development Projects Manager,
Institute for Advanced Physics, LA

Special thanks to:

IAP's *Foundational Funders* each of whom made this work possible.

Heartfelt thanks to:

Mrs. Susan Rizzi and Giuseppe Rizzi for much work, and Kateri,
Nicolo and Thomasina Rizzi for their special contributions during
the long process of the creating and writing of this book.

Preface

In this book, we cover Newtonian mechanics from a point of view that keeps contact with one's pre-scientific experience. This is what we mean by *Physics for Realists*. It brings out and emphasizes that physics is about the world we see around us. It applies to everything. I remember when I was an undergraduate at MIT in one of the big lecture halls for a freshman electricity and magnetism course. The professor was discussing how electromagnetism is involved with everything around us. It's something the textbook never explicitly said, so it made an impact on us when he said it so starkly. The lone girl in the front row was knitting or sewing while she watched and looked up and said: "even sewing?". And the professor after a slight pause, said yes even sewing and everyone laughed, but she brought home the point very well that it has to do with the world around us and as such is very relevant to everyone's lives.

Physics for Realists (PFR) is for people who want to be firmly grounded in the world of things around them. There is a second related philosophical sense, where *realism* is opposed to *philosophical idealism*, not against wanting to dream great dreams but against the idea that our minds know only themselves and ideas, not things. Of course, science is about real things, so we cannot accept such an opinion, but oftentimes this is not made clear to students by connecting everything back to their immediate sensorial contact with real physical things.

This starting point makes this textbook unique. Indeed, this text distinguishes itself in at least five major ways:

1. Common sense starting point.
2. Unifying practical theme in addition to theoretical theme. The student learns how to make a manned trip to Mars.
3. History that, until now, was not in current physics textbooks. The new history underscores the continuity of physics over the centuries (and in so doing cements the concepts and gives students an important pathway by which to remember them).
4. Special Relativity is included as an advanced chapter. It also is explained starting from common sense and using the basis of Newtonian physics laid in the text up to that point. The chapter completes and further manifests the sense in which this text is *a book of the study of impetus* (momentum).
5. Free download of *Interactive Physics*™ program, with *PFR* specific files, gives students a virtual experimental lab. *Interactive Physics*™ is an award-winning educational software from Design Simulation Technologies.

The wonder and awe that nature elicits through her deep beauty is hampered when textbooks and teachers leap-frog over the student's common sense. Indeed, it is precisely the contact with reality given in everyday life that is the starting point of science. Simple concepts, such as substance and property, which are at the root of common sense, can be understood at a very young age. Indeed, they are understood at a very young age, at a pre-scientific (or infra-scientific) level, though they are seldom explicitly raised by educators, let alone given a clear definition. Science depends on having experienced much, and, especially as children, we do experience much before thinking deeply about it. In this text, we make explicit use of this experience and critically analyze it to bring out the essential truth common sense contains while purging it of spurious error that attaches to it without such rigorous analysis. For instance, inertia is hard for students, because it appears to contradict their common sense. However, an explicit analysis, such as given in this text, shows common sense is right in one regard, but not in another. Knowing where it went wrong is crucial, but so is knowing where it was right. Starting with common sense manifests the firm ground upon which modern science, and all of our knowledge, stands as well as reveals the nature of modern science, what it includes and what it leaves out, showing what we know at the current level, e.g. Newtonian physics of the first part of this text, and what we do not.

The second item, i.e. the unifying practical theme, is also not seen in standard physics texts. As every experimental physicist knows, nature is always much more complex than appears at first glance. Every experimentalist has to learn by experience how to make his experiment work. This is because when we return from theory to experiment we move from abstraction to the specific reality, which means we have to confront all those things we have not incorporated into our theories. A practical theme, which involves, at least, having to think how one might do something, forces one back into this specificity. It forces one to think more deeply about the meaning of the theory.

But, why Mars? Mars has captured the imagination of many for ages and still does. Indeed, America now has a commitment to a manned mission to the red planet. Mars is one of the "wanderers" in the sky, i.e. one of the planets. Newton's discovery was successful because it, for the first time, causally explained the motion of Mars and the other wanderers. Indeed, it was Mars that Kepler used to formulate his second law which was so instrumental in the development of Newtonian mechanics. If there were no Mars, or it did not have its relatively high elliptical orbit, the history of physics might be quite different.

As the student learns new areas of physics, he learns how to solve some part of how one would actually make a trip to Mars. Then, in chapter nine the student learns how to execute the Manned Mission to Mars. In this way, he hones his understanding of theoretical principles, learns some practical thinking, develops his technical problem solving technique, all while he is led to ponder a future with astronauts walking on Mars. As part of this process, we develop a unique Martian calendar that is available as a java applet at: www.iapweb.org/MartianCalendar.

Third, the history written in physics texts almost universally begins with Galileo, giving little credit to key predecessors of Galileo, often maligning both the medieval contribution and the contribution of Aristotle. Though medieval historians now know that this is incorrect at a very important level, few physicists know it, and it has not managed, until now, to reach the physics texts. Knowing this history helps students see the continuity in the evolution of ideas and the dependency that even great thinkers like Newton have, as he was the first to say, on those that came before him. The missing link is the medieval period in which modern science was conceived, finally being born at the end of that period. Seeing this dependency helps students realize their own such dependency and thus to take seriously the need for study. Further, having such knowledge helps the pedagogy, because the history mirrors the path the student must take from common sense grounding to the highly mathematical modern science. Thus, it helps cement his understanding while giving him another pathway by which to remember the material. Skipping over history leads, as is famously said, to repeating its mistakes. In this case, the central one is not properly understanding what we include and what we leave out in our thinking in modern physics. Or, more to the point, to not know what it is we mean and what we do not mean by our theories in modern science and what we know and what we still seek to know.

Next, the fourth numbered item: the topic of Newtonian physics is incomplete in just this way. In it, we, sensibly enough, leave much out so that we can focus on what we know. Newtonian physics remains true in its domain of validity. However, at some point we would like to understand nature more deeply by including more specifics, specifics that we left out up to this point. Special relativity is that next level, covering a larger domain than Newtonian physics. It gives us a concrete example of how this augmentation happens, thus helping the student to get a real glimpse at how modern science moves forward and what changes and what does not. Furthermore, the common sense basis for Newtonian mechanics naturally leads to questions about how one can then proceed to special relativity with its radically different and more profound mathematical structure. This common sense discussion reveals the depth that modern physics has and, again, its meaning in a way that is not found in other approaches. Special relativity is an advanced subject and will probably be left for the advanced students, but teachers should try to say what they can about the subject at a generic level to help the student understand something about the nature of physics and the world that it studies.

The fifth item, the interactive simulations of the Newtonian physics, helps fill out the natural terminus mentioned later in which understanding is fed by contact with the concrete physical world. Real physical experiments always, of course, remain the only way to do this. Still, as mentioned, practical thinking (as opposed to theoretical) precedes experiment, including modeling of the situation under consideration. Computer modeling is often helpful. The *Interactive Physics*™ software allows one to switch on and off various effects (such as gravity, fields or masses), just as we do in our thinking in Newtonian physics, when we consider an effect to be so small as to be negligible. My children started using the program when they were six or seven; they called it circle *game!,* because to them it was fun. And it is. One can make any simple, even some fairly complex, mechanical situation and watch what happens. It's a great way to learn the principles and has its own place in the learning process.

Interactive Simulation

The book gives Interactive Physics files appropriate to the given section in orange font. Often the files will be highlighted with an orange box such as shown here. *Interactive Physics*™ (IP) files and video clips can be found at *www.iapweb.org/interactivephysics*. The IP files require the IP program, while the videos can be played on any computer. To access the directory, you will need your username: *pfr* and password: *m3chanics*. See below for download information.

Though the fifth is the last item listed, there is one other important aspect about the book that bears mentioning. A concentrated effort has been made to include many problems about everyday life, so that students may realize that physics starts with our senses and comes back to them. Again, everything we see around us is the business of physics. Furthermore, physics in the wide sense of the study of the physical world, though it is not all we know, it is the *basis* for all we know. It is in this sense, that physics *is* the base science, and so essential.

We, the author and contributors at the Institute for Advanced Physics, welcome your comments, as we continually strive to improve the usefulness of the text. In particular, send your suggestions or any corrections you may find to us at iappress@iapweb.org. Though much effort has been put into proofing the text, no doubt errata remain. Hence, for your convenience, we will keep a running compilation of errata at iapweb.org/pfrmechanics/errata. These will be incorporated into the next printing of the book.

Also, there is a solutions manual available to instructors. Please email us at iappress@iapweb.org.

Lastly, we plan on writing teacher's manuals for various levels. Please check IAP's website at iapweb.org/pfrmechanics/teachers_manuals

Interactive Physics™

Your version of *Interactive Physics™* software is available as a download at: www.interactivePhysics.com/iap.php

To access this directory you need:
username: physics
password: IPRizzi

Interactive Physics™

The files for this textbook can be found at:
www.iapweb.org/interactivephysics
To access this directory you need:
username: pfr
password: m3chanics

Contents

Detailed outline of each chapter is found on the chapter's first page.
Appendices on the categories, the sciences, Mars and various physical facts of interest are found in the back of the text along with an index

Chapter I

What is Physics?

Introduction

As you know, the study of physics has led to the discovery of black holes, the stretching of the universe, lasers, quasars, quarks and a host of other fascinating things. The material in this course is essential if you want to begin to understand these mysteries. To really understand any of those things, we need to go back to the beginnings, the foundation, and build. These roots are themselves equally fascinating in a different way. They bring to light the deep order in the world around us.

Nothing is more immediately known to us than the physical world, because we know everything through our senses. Since early childhood, we have likely been told many times and certainly have noticed ourselves that we know the world through the things we touch and countless sights, sounds, smells and tastes. We immediately notice that physical things can and do change and in so doing reveal more about the world. A plant growing, for instance, helps us realize it is alive.

In moments of clarity, everything around us evokes awe and wonder. Such wonder makes us ask: How? Why? What? (and other W's: Where? When? etc.). Such questions are the beginning of **physics**, *the study of the physical world.* Physics, in this very basic sense, is the science we ***must*** begin with, because all we know comes via the physical. From physics, one proceeds, as shown in Figure 1-1 , to the other basic physical sciences (chemistry, biology, and their many subdivisions and derived sciences) and then to the applied sciences.[1] Of course, not all of physics is needed to move on, just the foundational principles.

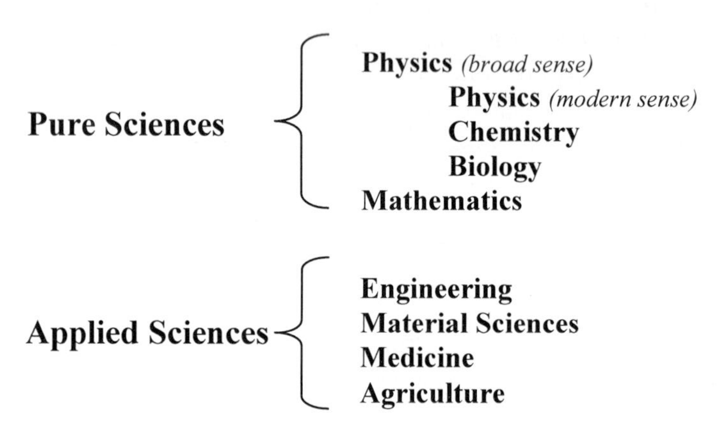

Figure 1-1: ***The pure sciences*** seek knowledge of things ***in order to understand***, whereas the ***applied sciences*** use information from the pure sciences to learn ***how to use things***. We group mathematics as a physical science only because its content *ultimately* comes from the physical world through our senses.

In physics, we seek first to know what things are and how and why they are the way they are. Modern physics has, for the larger part, limited itself to the general inanimate aspects of physical things, leaving the specifications of chemistry, biology and the cognitive ability of animals and man to other disciplines, although properly speaking chemistry and even biology are also part of physics. In a misuse of terminology that partly arises from the need for specialization to probe the deep complexity of the world, the word *physics* is now largely limited to the study of the *fundamental* aspects of the physical world (see Figure 1-1). Despite this usage, at some level, it is recognized by all that physics is needed in order to approach chemistry and biology in a truly scientific way. Indeed, with its grasp of the broad fundamentals, physics has started to explore certain regions of biology.

In any case, we should indeed start developing the science of physics by studying the simplest physical things that are directly accessible to our senses. We want to extract the most general information that we can about the physical world. We can then quickly move to those things more remotely accessible, again razing away the irrelevant, trying to

[1] Appendix II gives more detail and fills in the rest of the sciences.

get to the simple truth, trying, as Einstein once said, to get as simple as possible but no simpler. This latter principle, called Ockham's razor, was articulated by Aristotle (384-322 BC) and later also used by his medieval successors including William of Ockham (1285–1349 AD), whose name became attached to it, though he deviated widely from Aristotle, making fundamental errors.

Looking for the most general principles and starting with the simplest physically accessible things doesn't exclude taking information from more complex things. Indeed, thinking about more complex and/or less accessible things may lead to crucial insights. In short, the approach of starting from the simpler things is only the primary path.

Of course, before we start our rigorous study of physics, we know a lot about the physical world just from living in the world of people, animals, plants and inorganic things. Most of this knowledge is generic and confused; such knowledge is called ***infrascientific*** (from *infra*, meaning below, as in the word infrastructure) and is defined as *knowledge that has not yet been made rigorous by scientific analysis.*

So, ***physics is the rigorous study of the physical world***. Still, we have not yet talked about theory and experiment, the two distinctive components of a truly rigorous understanding of the world. Obviously, the point of experiment is to gather information about some characteristic of reality, and theory is explanation (at some level) or model of the data. The things around us can be understood at a simple level by observation and simple thought. For example, we hear a noise and conclude something made the noise. This is a simple observation and explanation, and only the beginnings of what we mean by experiment and theory in modern physics. Beyond this then, what is theory and experiment and how do they interact? This is best understood by experimenting and theorizing oneself, which you will begin to do in this course. To begin to understand, we must note that modern physics, including its theoretical and experimental side, is really only understandable in light of the distinguishing characteristic of modern physics, mathematics.

The major tool of physics is mathematics. Why? It is a question that is seldom if ever addressed in physics texts, but the great applicability of mathematics has led physicists to call it the *language* of physics and ultimately to talk about the "unreasonable effectiveness of mathematics."[2] For instance the equations of Newton, which we study in this book, have proven valid and useful in the ordinary world of our immediate experience as well as in our solar system , other solar systems and beyond. Those equations were part of the foundation that led to trains and automobiles as well as spaceflight. Furthermore, each piece of technology that permeates our daily lives, from cell phones, TV's and planes to feather-light glasses, superglue and sinks, required the equations of physics in its development. For example, TV's and cell phones use the principles of semiconductor physics which incorporate the equations of quantum mechanics which are also used in understanding the nature of light. Planes and spacecraft require use of the equations of thermodynamics in their engine design, which are the same equations that "govern" a refrigerator. The diverse areas of weather forecasting and plumbing each require the use of the same general equations of hydrodynamics. What indeed, you may now echo, is this eerie "effectiveness of mathematics?"

---

[2] Physicist Eugene Wigner (1902-1995) was a renowned physicist known for his contributions to the foundations of quantum mechanics through the use of symmetry principles. The quoted phrase comes from his 1960 article titled "The Unreasonable Effectiveness of Mathematics in the Natural Sciences."

Shortly, we will break the mold and answer this interesting question. Take special care to digest the answer so as to have it in easy mental reach, because a true appreciation for the answer will only come as you learn and do physics, seeing for yourself the "effectiveness of mathematics." It's often the mathematical side of physics that makes many think physics is too hard. By starting with what we know first in the mathematics and the physics, much of the mental gymnastics that make it seem harder than it is can be avoided.

Though it remains true that learning modern physics requires hard work and persistence, the reward of seeing the beauty of the physical world especially as revealed by the mathematics is well worth it. Physicist Paul Dirac hinted at this when he went so far as to say, "If one is working from the point of view of getting beauty into one's equation, ... one is on a sure line of progress."[3] This, of course, cannot be taken to mean that one doesn't need input from the real physical world, for, as we've said, that's where everything we know gets its base. Let the deep beauty Dirac points to help spur us on to learn physics. The "effectiveness of mathematics" is a sign of its beauty. Though we can see evidence of its effectiveness, we cannot truly appreciate it from the outside; even more so, the full beauty can only be seen from the inside, by someone who knows the physics. In reply to a journalist who asked him to explain mathematical beauty, Dirac hits this point.

> *"Dirac asked the journalist "Do you know mathematics?", and when the journalist replied "No", Dirac said, "Then you can't understand the concept of mathematical beauty"*[4]

Our subject, Newtonian mechanics, is the natural—and the standard—place to start the process of a modern scientific understanding of nature. Newtonian mechanics begins with the study of locomotion of bodies that are accessible to the unaided senses. We analyze locomotion quantitatively to get a detailed understanding of the general aspects of local motion. Through instruments that aid or augment the senses, we later find that this understanding extends beyond the domain immediately accessible to the senses. It extends to bodies that are smaller and larger than we can see, to longer and shorter times, faster and slower speeds than we can see and bodies that are heavier and lighter than those encountered in ordinary experience. In Chapter 10, we will learn some special relativity, which incorporates all of Newtonian mechanics and yet allows us to deal with bodies that move near the speed of light.

Obviously, Newtonian mechanics is, comparatively speaking, close to our sensorial experience, though it is conceptually challenging because of its abstract generality. Again, by pointing out clearly how we get these generalities, including what we leave out in getting there, we will see it is not as difficult as some presentations make it seem. Because of its closeness to sense experience, Newtonian Mechanics (as given in Newton's *Principia*) was the first example of a theory of real generality that looked at the world as mathematical.

But, we jump ahead of ourselves. We need to lay some ground-work that we will use throughout the study of physics. All science must have base principles. We will see that these principles are broad and thus very simple and unrestrictive. From what we said

[3] http://www-groups.dcs.st-and.ac.uk/~history/Quotations/Dirac.html
[4] http://plus.maths.org/issue21/news/dirac/

earlier, those principles must ultimately come from the senses and be verified in the senses. There will be at least two levels of principles, the foundational level and the principles of Newtonian physics.

Physicist Sir Arthur S. Eddington (1882-1944) once remarked that it is ultimately molar physics that matters.[5] Remember, "molar" means something with (about) 10^{23} atoms or more, which is enough atoms to be sensible, i.e. we can have direct sensorial contact with it. So, he implicitly repeats our principle that the study of the physical, *physics*, begins with what we can sense.

Since our main goal in this text is to learn the principles of Newtonian mechanics, it is not necessary to prove all the underlying principles upon which they in turn rest. We can call those principles the *foundational principles of physics*. Of course, those foundational principles that are not immediately evident to the senses can and have been proved by others (see appendix I). Yet, without having to explicitly prove them, we can and will make use of our already implicit knowledge of these principles; after all, we assume them implicitly in everything we think and say. The foundational principles need to be articulated clearly and their meaning understood to avoid confusion. Let's take a look now at some starting points.

How Do We Know Physical Things?

Sense and Abstract Knowledge

We have two different types of knowledge: particular knowledge and abstract (generic or general) knowledge. The knowledge we get directly from our senses, which we call *sensorial knowledge*, gives us particular knowledge. Yet, I don't get this *particular* knowledge or awareness of the coldness of this particular water glass without also getting the knowledge of the *general* reality called coldness, that is, coldness in the abstract, coldness that is not necessarily attached to this glass. Because of the contact with the glass, I now know, if I did not already, the general idea of coldness. This general or abstract knowledge by which I can put many things in one category is called *intellectual knowledge*. By it, I can put all cold things in one category. We will obviously need and use both intellectual and sensorial knowledge. Now, we can move to directly applicable principles.

The Fundamental Principles

As may be expected, the foundational principles are very obvious and simple. Indeed, they are so much so that one might first think it's a waste of time to discuss them. Yet, they are the foundation, the starting point. They must be explicitly acknowledged as such, for despite their obviousness, not a few have denied them, thus implicitly denying all the many physical conclusions which stand on them. As for their simplicity, as Isaac Newton and Albert Einstein point out in another context, simple principles--like all deep principles--have a grandeur that is only appreciated by contemplating and living with them. They are:

1) ***Things exist.***
2) ***Things change.***

[5] The full quote is "Molar physics has the last word in observation for the observer is molar."

More explicitly: Things are something, but can become something else. They are actually something (act), but are potentially something else (potency).

Example*:* An apple is one thing before it is eaten and is another after it is digested and part of one's body.

3) ***Principle of Non-Contradiction:***

Something cannot be and not be at the same time and in the same way.

Example: A shirt cannot be red and green at the same time and in the same way. It could be red and then green *or* it could be part red and part green, but not simultaneously red and green in the same place.

The principle of logic that we either have X or not X derives from this fact. The daisy is blooming. The daisy is not blooming.

4) ***Principle of Causality:***

Something cannot change itself. Something can not give itself something it doesn't have. It can, however, act on (change) other things according to what it is.

5) ***Substances and their properties***:

a. Some things (substances) exist of themselves, while others can only exist as aspects of a substance (call them properties).

b. Obviously, the first are referred to using nouns and the second using adjectives.

Example: "The daisy is yellow." The daisy is the substance, and yellow is a property of that substance. Yellow cannot exist by itself, but only as a property of something(s).

The Categories of Properties

To say more, let's take a glass of ice water. From this drink, we can learn a lot beyond just the general existence of substances and their properties. By way of terminology, we will, for now, call any physical substance a *body*;[6] at this point, there are no requirements on it except that it exists of itself. We will now determine other physical, but generic requirements. Physical things are characterized by the fact that they are changeable.

Holding the glass, one feels its coldness; the glass has a ***quality*** that cools my hand down. Simultaneously, my hand warms up the glass. The glass acts on me and is able to receive my action and vice versa. Both it and I act and receive. Here we have two broad types of properties; call them ***action*** and ***reception***. I also see that these processes indicate a ***relation*** between my hand and the glass. Further, I note that the glass and drink are extended, i.e. have length, width and height; I can feel that the coldness stops where the glass stops; I can feel variations in the temperature that reveal to me that one part is here another part there, i.e. separated from it. This extendedness, this property of having parts, each of which is in contact with the next but outside the next, we call ***quantity***. I also notice that the glass is in a ***place***, in an ***environment***. That is, there are bodies like my hand in direct contact with the glass, which determine what will happen to it; my hand, for instance, warms up the glass. The bodies immediately surrounding a substance then define a property called place. Those bodies are, in turn, surrounded by a larger group of bodies

[6] Later we will use body in a much more indeterminate, vague way; at that point, *body* will used to mean a substance, group of substances or sometimes even a part of a substance.

which determine the property called environment. Now, the glass can be rotated without necessarily changing the surrounding bodies, i.e. without changing the place or environment; we call this property the ***orientation***. Lastly, all this happens during a span of ***time*** and the glass exists at a moment in time.

Hence, we see there are nine different generic types of properties of physical things. (see also Figure 1-2):[7]

The first two types are ***intrinsic*** to the substance:

1) **Quantity** (extension): plane, circle (disc), number, …3-dimensions,
2) **Quality:** figure, color, hardness, pressure, hot, cold, tones, smells, tastes

The remaining categories are ***relative*** to another substance in some way:

3) **Relation**: equality, similarity: two circles of the same *size (extension)*, two shades of the same color; cause and effect
4) **Action**: heating, moving
5) **Reception**: to be heated, to be moved
6) **Place**: a fish in a certain region of water in the tank
7) **Orientation:** the cylinder is rotated by 90 degrees
8) **Environment**: the water around a fish
9) **Time:** now, today, yesterday, a second from now

Note that the categories do not overlap; for example, a quantity cannot be quality; they are *univocal*. The categories cannot exist without each other; we say they are co-relative. The distinction between the first two categories of quality and quantity is especially important. The most characteristic difference between them is that quantity is additive, i.e., two quantities can be put together to get a new quantity, while quality is not additive. For example, no number, even an infinite number, of mediocre mathematicians will equal a great mathematician; or again, no matter how many snowballs one accumulates, they will never be able to generate the temperature needed to cook dinner, for the quality of temperature is not additive. This non-additive nature of quality is also reflected in the fact that bare quantities are said to be equal or not, whereas qualities are said to be similar or not.

Though all real physical things have properties in all nine categories,[7] the categories exist in a substance in an order; that is, there is a dependence of one on another. Now, physical things, which we will also call bodies or material things, are things that can change; they can change from one thing into another and/or their properties can change. Thus, physical things cannot have any quality without quantity, for the quality of one part acts on another part. If I weren't here and the glass there, the glass could not act on my hand. Why so? If, *per impossible*, I think of physical things such as the glass and my hand as without extension-- i.e. without quantity-- they would be both at one point; that is not separated, for that's what we mean by extension. Therefore, one thing would be in whatever state the other is with no way for one to change the other (or even to change at all if there is nothing else outside of them).

---

[7] As discussed in appendix I, because physical things are characterized by being able to change, they must have at least the nine categories of properties.

Indeed, all the categories of properties of physical things require the property of quantity. For instance, try to think of yellow without thinking of it as extended; it's not possible. Yet, we can consider bodies with quantity that only have a minimal unifying quality, such as shape in geometry. We can, for instance, consider a disc without any reference to color and we can think of the number six without thinking of it as relating to a quality such as color. When we do this, we are in the realm of mathematics. We have abstracted all away but the quantity and only a thin presence of quality. In geometry, for instance, we keep the simple quality known as shape, which is the boundary of a quantity.[8] It's so simple many miss it, yet the shape of a disc (circle) is, for instance, obviously *qualitatively* different from that of a ruler (rectangle). By next leaving out (abstracting) everything but the number of parts of a given extended object, for instance by counting the number of parts of a disc, we come to the even more abstract quantitative concept of number.

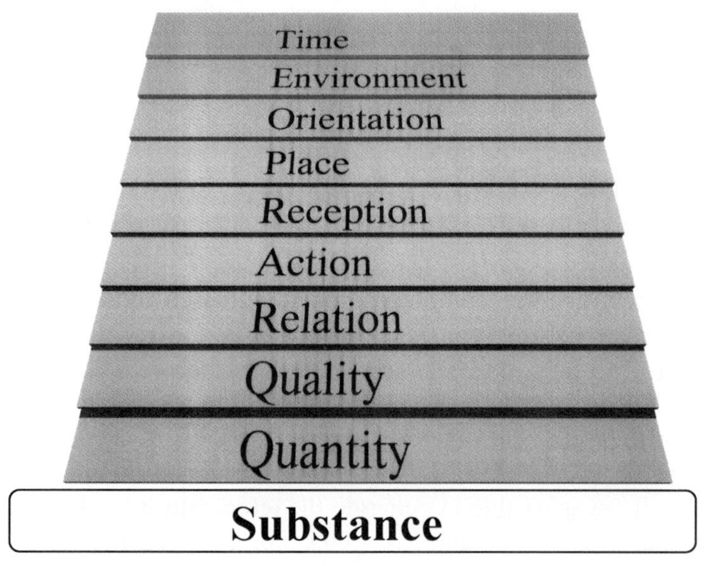

Figure 1-2: In gray are the nine categories of properties that are interrelated or co-relative one with the other. Quantity is the base property, upon which all the others stand. By abstracting, we can separate out quantity and consider it isolated from the rest excepting a thin presence of quality (shape, for instance in geometry); this is what we do in mathematics. The importance of this type of abstraction is indicated in the figure by a large separation between the quantity layer and the quality layer. Of course, the properties presuppose a substance, i.e., that which has properties.

These principles, being first principles, are obviously used implicitly in everything we do. For instance, engineering firms, artificial intelligence researchers and many others implicitly use this understanding in establishing methods to properly sort objects and tasks to facilitate problem solving. For example, the firm *Kepner -Tregoe* has a problem identification algorithm, which is used in many industries, that makes implicit use of the

[8] In arithmetic, we keep the quality of wholeness; for example, three is the "whole" composed of three units.

above foundational principles. At a simpler level, this kind of sorting is what we do in the initial stages of the game *twenty questions*.

To better understand the categories in familiar context, consider how understanding substance and the nine categories of properties could expand the way you play the game of twenty questions. If for instance, you were trying to guess *triangle*, how would you do it? Right at the beginning, the pre-20 question: "Is it animal, vegetable or mineral?" would get a shrug or other unhelpful response. With the above more general understanding of physical things, we would first try to determine whether it is a substance or property. If it is a property, we would then seek to determine into what category it falls. The triangle, for instance, belongs in the subcategory of shape, in the category of quality. If it is a substance, one would proceed to find out what type of substance it is, such as animal or mineral. When you start talking about properties of human beings that relate to thinking or free will, obviously more has to be discussed, but here we are interested in the purely physical aspects.

The Empiriometric Method

A great contribution of the ancient Greeks was to see, beginning with Pythagoras (6th century B.C.), that mathematics can give great insights into the world. Pythagoras was so impressed by this fact that he attached religious significance to his own concept of number.[9] The great philosopher Plato (see box on Aristotle) was to make much of this insight, though he never reduced things purely to mathematics. The great scientist/philosopher Aristotle (see box) uncovered and articulated the first principles of understanding nature. He also pointed out the dividing lines for the sciences. In so doing, he saw that because quantity is the ***first*** property of all physical things (bodies), one can treat them as mathematical, look at them under that light. Greek scholars began this work and made real progress in statics (the study of stationary objects).

Although promising, the Greek initial conception of science died, and it was not until a millennium later that science was robustly conceived in a culture that could successfully give birth to modern science. Namely, in Christian medieval Europe, men such as Bishop Robert Grosseteste (c 1175-1253 AD) emphasized the importance of experiment and mathematics, while Bishop Thomas Bradwardine (c 1290-1349 AD),[10] one of the "Oxford Calculators,"[11] began to apply mathematics to *motion* with some success. Moreover, they and their medieval colleagues strove to understand physical things, not only as mathematical, but as fully physical. They were the critical link in the evolutionary path of understanding that culminated with Galileo and Newton.

Along this path, in the 1500 and 1600's, were also Vieta, Stevin, Descartes, and Wallis who were the first to set out the systemization and symbolization which are

---

[9] The story is told that Pythagoras believed strongly in the absoluteness of numbers (specifically the natural numbers), and could not accept the existence of irrational numbers. According to the story, when Hippasus, one of his disciples, discovered that $\sqrt{2}$ was irrational, i.e. not expressible as a ratio of natural numbers, Pythagoras could not accept the existence of this irrational number. So, he sentenced Hippasus to death by drowning lest someone discover Hippasus proof.

[10] Interestingly, each was Archbishop of Canterbury at one point in his life.

[11] They discovered and proved the mean speed law (see history box in Chapter 2), emphasized the distinction between kinematics (the path of a motion) and dynamics (the study of the effects of forces), and distinguished instantaneous speed from average speed.

characteristic of modern mathematics; the great mathematician Rene Descartes (1596-1650 AD) is iconic of this advance. This revolutionary new way of thinking of mathematics, unknown to the ancient Greeks, was required for and becomes characteristic of the new physics. The new physics, as we've mentioned, had its first success in Newton's great work *Principia*, the complete title of which is: ***Philosophiæ Naturalis Principia Mathematica*** (The Mathematical Principles of Natural Philosophy).

We give a name to this way of approaching nature, of treating it as quantitative, *especially* as quantitative in this modern symbolized and systemized way. We call it, following the 20th century Princeton philosopher Jacques Maritain (1882-1973 AD), the ***empiriometric*** approach, for it looks at the empirical (empirio) as quantitatively measured (metric). This method of looking at the physical world reflected into the mathematical domain is at the heart of modern physics. We will see that it is powerful in uncovering new understanding of the world and yet, because the world is not simply mathematical, the method, of its nature, leaves things out. Knowing what the method is, how it works and watching it work will enable us to clarify our new understandings and, even when the exact full physical meaning is not yet clear, to put our new understandings into their proper context.

Aristotle was born at Stageira, a colony of Andros on the Macedonian peninsula of Chalcidice in 384 BC. He was taught by Plato (427-347 BC), who in turn was taught by Socrates (470-399 BC). He was the first to clearly articulate that all of our knowledge comes through our senses and to talk about scientific tools to understand more deeply. He recognized the nine categories of properties of a substance and pointed out the simple, but profound principles that all disciplines start with. He is also known as the father of logic, having laid down those principles very clearly. Of the physical sciences, he was most successful in Biology, being praised by Charles Darwin who said, *"Linnaeus and Cuvier have been my two great gods ... but they were mere school boys to old Aristotle."*[*] Through one of Aristotle's students arose the great Museum (a scholarly institution) of Alexandria (see history chart). While Aristotle's foundational insights about the physical world were profound, he faltered in his later understanding of motion, which helped set the stage for the sparse activity in dynamics (activity was in statics) until the advent of the High Middle Ages, where his ideas on dynamics and other areas were, for the first time, challenged without abandoning his important foundational insights. Indeed, the challenge was mounted using those foundational principles. He died in 322 BC.

[*] Quoted from *The Relevance of Physics* pg 30.

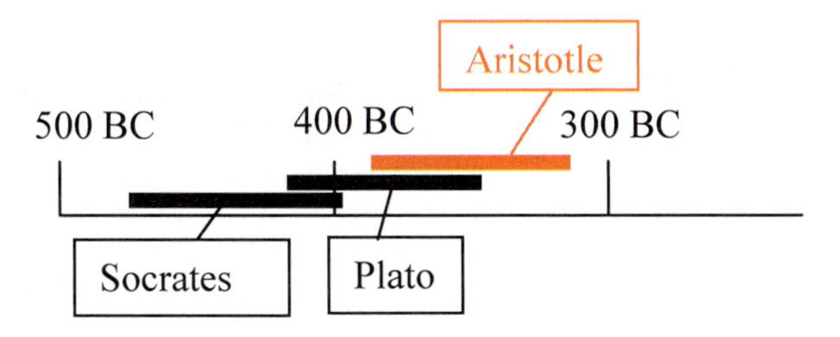

History Note

A healthy seed was planted in Greece; with Aristotle's principles in place, advances are made but eventually cultural forces result in the death of that seed. The chart on the next page shows the history of science in outline. Our study of mechanics will show how science (including the crucial empiriometric method) matured, being conceived robustly in the middle ages, going through labor pains with Galileo and finally born with Isaac Newton's mechanics that is largely the context of our study in this course.

HISTORY OF SCIENCE

Aristotle (384-322 BC):
Conception of science seed planted
Founds Lyceum Research Center

Pupil

Demetrius of Phaleron (c 350-280 BC) (student of Aristotle) founds the Museum Research Center in Alexandria for **HELLENISTIC SCIENCE** (297 BC- 500 AD):
Aristarchus of Samos (310-230 BC)
Euclid (325-265 BC)
Archimedes (287-212 BC)
Strato of Alexandria (268 BC)
Eratosthenes (276-194 BC)
Hipparchus (190-120 BC)
Hero of Alexandria (c 62 AD)
Claudius Ptolemy (85-165 AD)
Seed dies

Sporadic Christian Activity
John Philoponus (~490-570 AD)
Some reactivation of seed

Muslim Period (700-1250 AD)
Take over Christian learning and assimilate Greeks
Seed dies
when Islamic theology finally gains ascendancy in the heart of the culture with its disbelief in secondary causality.

1100-1250 AD *First robust conception of science*
Christianity, for the first time, is able to form its own distinctive culture from the ground up.
First universities founded.
University of Bologna (before 1080), Paris (1150), Oxford (1167), Cambridge (1209), Padua (1222), Naples, Siena, Rome, Prague, Vienna, Florence …
Fibonacci, Grosseteste, Albert the Great, Jordanus, Roger Bacon
c.1250 AD *Thomas Aquinas clarifies and fills out unstated conclusions of Aristotle's understanding of science based in the physical world.*

1250-1600 AD *Growth of science*
Including embryonic development of modern empiriometric science and Copernicus's powerful summary of arguments for heliocentrism
Thomas of Bradwardine, de Marchia , Heytesbury, Dumbeleton, Swineshead, Witelo, Theodoric of Freiberg, Jean Buridan, Nicole d'Oresme, Albert Saxony, George Peurbach, Regiomontanus, Copernicus, Toscanelli, Domingo de Soto, Tycho Brahe, Kepler, Gilbert

Black Plague
1350-ended 1690?
(killed 1/3 to 1/2 of Europe)

1600-1800 AD ***The Scientific Revolution***
Birth of modern science (with full empiriometric method and tools)
Era of Newton
Galileo, Descartes, Newton, Hooke, Boyle

1800-Present ***The Scientific Revolution advances***

Maxwell, Darwin, Einstein, Bohr, and others give Electricity and Magnetism, Thermodynamics, Relativity, Quantum Mechanics, Electroweak Theory and much more, including String Theory.

Measurement

As the empiriometric science looks at the physical world as measured quantitatively, we will need to introduce units of measurements to say how long something is, how heavy it is, how long it takes to change and the like. It is through measurement (done in experiments) that modern physicists bring out the quantitative element in nature. Measurements are *relations.* We first consider measurements which are relations between two quantitative properties or two intensities of qualitative properties of two objects. For example, an average man's shoe fits (the long way) two or three times across a door opening, and 300 times across the length of a football field. As an example of the intensity type of measurement, we can use the temperature of a body. The temperature of an adult man is about 100 times that of the average temperature of the cosmic background radiation and about 1/3 that of the melting point of a copper dime. Here the length of the shoe and the temperature of a man provide an example of an analogical use of units; that is, we use the unit in two ways that are similar but not exactly the same. The length of something is a *quantity,* while the temperature is the intensity of a *quality.* There is a similarity between the *measurement* of length and the *measurement* of intensity, because the comparison yields a number in each case. The analogically general meaning of number includes both types of numbers. Of course, this similarity between the comparisons (measurements) does not make length and intensity the same. Yet, they are related, for quantity is the first property of any material thing--in this way, it is a sort of foundation for the rest of the categories starting with quality;[12] in particular, the length of a thing is dependent on its qualitative state.

By comparing to a standard or standards, i.e. thinking in terms of measurement, we can assign a number (i.e. a quantity in an analogical sense) to any aspect of a thing. Thus, in this limited sense, we can speak of any aspect as a quantity. This is the standard mode of speaking in empiriometric science. For example, in this context, we can and will speak of temperature, force, height, pressure, and light intensity as quantities.

[12] All the eight other categories of properties are somehow conditioned by and reflected in the quantitative base (extension, parts outside of each other, one part here and another there) of a thing.

Number and Units

We use the concept of number to talk about all types of measurements. Properly speaking a number is a whole that is measured by a unit, a unit is by definition that which is not divided. In this way, a unit, that is "one," is not properly speaking a number but is the principle of number.[13] However, as you know, this is not the end of it. We can *by analogy* extend the concept of number beyond this. We can cut the shoe in half, though it's no longer properly a shoe, we can now talk about the number of half-shoes or demi-shoes. We now have a smaller unit. We must, of course, introduce even smaller units if we wish to measure things much smaller than a shoe. Indeed, in an ideal world we'd like to know what the smallest possible actual part is and that would be our unit of length in the most primary sense because we'd never, in principle, have to go back and reconsider our unit.

However, even in such a world, it would still not be practical to just use this one smallest unit. We would, for example not want to say, "I'm 10000000000000000 fundamental units tall." We'd want a unit more commensurate with what we are measuring. We could and sometimes do vary our definition of unit, keeping each unit (while it's used) undivided; even though later we leave it aside as a unit and choose something that is some fraction of its length or a something larger than the original unit. But, in practice, one mixes all the different definition of units to make it easy to express the length. For instance, in the English system of units, we might use an inch as a unit of measure (comparison) when we measure the length of a pencil, but we will use 1/16th's of an inch or ½ inches when we measure the pencil's diameter. In the case of the pencil, we would say it's about 3 ½ inches long, mixing ½ inches (call it a demi-inch) as a unit. In decimal, we may say the pencil is approximately 3.456 inches, in which case we are mixing 1 inch with 1 mil (1/1000 inch) units.[14]

---

[13] The most precise definition of number is a whole whose parts share no boundary, and is completed by a "last" unit. We first get the concept of number in a vague way by counting things. These things do not constitute a true whole, i.e. a substance such as a man, but are, for example, a bunch of substances that are in our field of view (for instance, marbles in a bowl). Indeed, we can easily identify (by analogy) a group of things that are in some kind of relation to each other as in some sense a whole, having some kind of wholeness, and in this way number them. For example, two apples in the same area on the table have a certain definite relation to the space on the table that they both occupy, so that we easily note that there are two apples *there*. More abstractly, all the apples in the world, though physically separated are one, are a whole, in the sense that all of them are apples and all part of the extension of the universe. Of course, I can just keep generalizing (by making use of analogy) and not specify what it is I'm numbering and just say it is three or four whatever and finally even drop the "whatever" and just consider three or four as if they could exist on their own. This is the way of mathematicians that is so powerful and fascinating because of its generality.

To get the most fundamental concept of number, i.e. the most basic sense of "a whole whose parts share no boundary and is completed by a "last" unit," we start with a *single* physical substance, not a group of substances. Namely, the "whole" in the basic definition is obtained by considering a single physical substance, such as a man, absent all categories but quantity and the minimal quality of wholeness; leaving just a (discrete) quantitative unity. (For more on number see my article: *What is Math Really?*)

[14] The English system of units has the advantage of using standards more commensurate with ordinary experience, but the resulting disadvantage of being far from systematic. For example, the English system of fractions of inches on base two for inches 1/2, 1/4, 1/8, 1/16, 1/32 is hard to add and subtract with and is not used for larger units, but is handy for carpentry where one often has to know the center of a board.

The extra bit represented by the numbers to the right of the decimal, in this last case: 456/1000, comes up naturally in real measurement. One first measures something with large units, so many inches or so many miles (usually choosing the units appropriate to its general size). Then, one deals with the leftover, so many fractions of an inch or mile. The length of the remainder requires fractions of the original unit, which you can create by continually dividing as the need arises, limited in the practical realm by the ability or need to continue further division.

This "remainder portion" can be confronted immediately by approaching the problem from a different angle. That is, we can start with a unit that is bigger, say longer, than the thing we are trying to measure. Say we are asked, "How many inches thick is this piece of paper?" A natural response, one rooted in the most fundamental meaning of unit, would be, "None, an inch is too big." Of course, what we mean by measuring in inches extends beyond the primitive definition of unit that we begin with. Namely, the unit is not a simple unit, but one that we divide in whatever way necessary to fit into the thing we wish to measure. For the paper, we note that, say, 2 thousandths of an inch will span the thickness. We then usually find it convenient to define a new unit more appropriate for this scale; in our example, this unit is called the mil, and is one thousandth of an inch; we say the paper is 2 mils thick.

We can summarize the above analysis by saying:

1) The concept of number is crucial to measurement.
2) There are measurements of properties in the category of quality and quantity (and, as we only touched on, mixtures of the two are inevitable). We will see that we can, and do, define measurements touching properties in all the categories.
3) We need a system of units for any given type of measurement, rather than a single unit for each one.

Units also need to be standard, that is available to all scientists wherever they might be, and they need to be universally agreed to so all will understand what they mean. Furthermore, a given base unit should be closely related to something fundamental in the physical world, for this is what we are trying to understand. Of course, tying a measure to something fundamental in nature, such as intrinsic properties of a given type of atom, will make it more independent of time and place, giving scientists the ability to make universal statements true across time and place.

What System of Units?

The units we most need in this text are also the most fundamental units: length, mass and time.

In the US, the National Institute of Standards and Technology (NIST) decides such questions. In this book, we will use the standards, called SI (*Le **S**ystème **I**nternational d'Unités),* most common in engineering[15] for their ease of use in practical situations. In physics, we will often find them and others useful, the choice depending on what is being investigated. SI units are also called mks units, named after the base units:

Length: ***m**eter*
Mass: ***k**ilogram*
Time: ***s**econd*

---

[15] However, English units are also common in certain branches of engineering in the US.

Another very common standard is cgs (centimeter gram second). Of course, one can convert cgs to mks and vice versa; such unit conversions amount to using one standard unit to measure the other; for instance we measure, in the fashion previously described, an inch by the centimeter, saying there are, by definition, 2.54 *cm* in one inch.

Length

The meter, which is equal to about one yard ($1m \sim 1.0094 yd$), was famously defined by the length of a bar made of platinum and Iridium kept in Paris in very stable well-defined conditions. Next, the meter was so many wavelengths of ^{86}Kr. Now **a *meter* is how far light travels in a vacuum during a time interval of 1/299,792,458 seconds**. In so defining our unit of length, we are obviously mixing the categories of *time* and *quantity* (length). Only with the study of relativity will the need for this mixing become clear. At this point, we must keep in mind that the usefulness and efficacy of mixing them in this way doesn't mean they are the same or even that they are indistinguishable, it is only indicative of the deep unbreakable interrelation between them. For our purposes in this text, we can continue to think of the meter as the bar in Paris. The meter is broken up into other units by multiplication and division by factors of 10 (as are the second and gram). We use the prefixes in Table I below to name these smaller and larger units.

Table I

Factor	Prefix	Abbreviation	Factor	Prefix	Abbreviation
10^{24}	yotta	Y	10^{-1}	deci	d
10^{21}	zetta	Z	10^{-2}	centi	c
10^{18}	exa	E	10^{-3}	milli	m
10^{15}	peta	P	10^{-6}	micro	μ
10^{12}	tera	T	10^{-9}	nano	n
10^{9}	giga	G	10^{-12}	pico	p
10^{6}	mega	M	10^{-15}	femto	f
10^{3}	kilo	k	10^{-18}	atto	a
10^{2}	hecto	h	10^{-21}	zepto	z
10^{1}	deka	da	10^{-24}	yocto	y

Using these abbreviations, we can write a new unit by pre-pending the needed prefix to the appropriate standard unit. For instance, the meter can be made into a unit more appropriate for measuring protozoa by pre-pending "micro" giving the unit of *micro*meter, one millionth of a meter, also called a micron for short.

Time

Time, the ninth category of property, refers to the fact that physical things exist for one "now" followed by another. When we say that "time flies," we mean a thing is one way now and only a moment later is very different. When I say, "how much time did it take?" I treat time as measure. Treated as a measure, *time is one motion measured by another motion*. For example, the age of a child is normally measured by how many times the Earth has moved around the sun during the much longer action (motion) of the child growing taller, more dexterous and generally maturing. Or again, the speed of a runner is measured by how many times a watch hand rotates while he completes the motion from the starting line to the finish line. Thus we need a standard unit of motion by which to measure other motions. The base unit is the second.

The *second* is defined as the duration of 9,192,631,770 periods of the radiation corresponding to the transition between two hyperfine levels of the ground state of the cesium 133 atom at 0 *K*.[16] There is much interesting advanced physics implicit in this definition; it involves applying quantum mechanics in the field of atomic and condensed matter physics. For our purposes, we only have to note that it is a very reproducible and very stable unit of motion by which we will measure other motions.

Mass

Lastly, mass needs a standard system of units. What is mass? This topic properly belongs in Chapter 3, so we will leave the definition till then. Mass is related to but not the same as weight. Mass is an intrinsic property of bodies, while weight results from action of the force of gravity on a massive body. We often measure mass by using a balance scale such as shown in the Figure 1-3; such a scale implicitly invokes the approximate proportionality between mass and weight on Earth that we will learn about when we discuss gravity later. On one side of the scale, we put our unknown mass, on the other, enough standard units of mass to balance the unknown.

Figure 1-3: A balance scale shown with standard weights in drawer.

The base unit for mass is the kilogram; it is equal to the mass of the international prototype of the kilogram (see Figure 1-4 showing prototype), which is also made of platinum and iridium and kept at the *Bureau International des Poids et Mesures* near Paris.

---

[16] According to BIPM, the temperature specification means that the cesium atom must be "unperturbed by black body radiation, that is, in an environment whose thermodynamic temperature is 0 *K*."

Figure 1-4: Picture of international prototype of the kilogram kept at the *Bureau International des Poids et Mesures* (*Photo Courtesy of BIPM*)

Table II shown below compares the mass of many well known objects to help you begin to get a feel for associating actual things with mass numbers in kilograms and pounds. As a quick reference, a few granules of salt have a mass of about one milligram, a paper clip's mass is about one gram, and a baseball bat's mass is about 1 *kg*.

Table II

Object	Pounds	Kilograms
Carbon atom	4.4×10^{-26}	2.0×10^{-26}
Virus	1.460×10^{-23}	6.64×10^{-24}
Piece of notebook paper	0.014	0.009
Pencil	0.02	0.01
Marble	0.06	0.03
Apple	0.40	0.18
Baseball	0.31	0.14
Football	0.94	0.43
Baseball bat	2.20	1.00
PFR Textbook	4.00	1.82
Gallon of gas	6.50	2.95
Gallon of milk	8.50	3.86
Lawn mower	100	45.5
Gallon of uranium	150	68
Mini van	4,500	2,045
Concrete truck	60,500	27,500

F-14 jet fighter	65,000	29,545
Space shuttle	4,500,000	2,045,454
Aircraft carrier	214,000,000	97,272,727
Asteroid	5×10^{20}	2×10^{20}
Moon	1.617×10^{23}	7.35×10^{22}
Earth	1.313×10^{25}	5.97×10^{24}

F-14 Tomcat has a mass of about 65,000 pounds.

Shown is the Space Shuttle (STS-112 liftoff).
The shuttle has a mass of ~ 4.5 million

USS HARRY S. TRUMAN (CVN 75)
A typical aircraft carrier has a mass of ~ 214 million pounds.

Dimensional Analysis and Units

In the equations we will make to describe the relationship between various quantities, i.e. measured aspects of physical things, the units need to cancel so as to be finally irrelevant. Units are chosen by us for convenience, and thus are not directly reflective of the physical situation we are trying to describe in a given equation. In short, the laws of physics and mathematics do not depend on such choice of standard. For instance, in the equation for the area, *A*, of a square with sides of length *b*, we write: $A = b^2$. Such a formula does not depend on the particular length units chosen. We can quickly see this by rewriting the equation so as to show the cancellation of the units: $A / b^2 = 1$. Because of this cancellation, the right hand side has no units.

Since all physical quantities (measurements) have units (either in terms of counting the number of a given type of unit or comparison between two things), it is often helpful to check one's answer by making sure the left hand side of the equation has the same units as the right hand side. In the same way, given that a problem requires certain units as input and the answer has certain other units, we can, by looking at the various mathematical combinations, understand something of the possible ways to calculate the answer. For example, given a square of length 8 *cm*, knowing that area is measure in cm^2, one could guess the area was obtained as $8cm \times 8cm = 64cm^2$. Of course, this does not prove the formula, but it can spark insight on how to proceed. This activity of looking at the mathematical combinations of units involved in solving a problem is called *dimensional analysis*.

Statics: Dimensions and Stationary Bodies

The next natural step in our introduction to physics is to discuss some details of objects at rest. After all, motion always complicates a situation—though since it is a fundamental fact about all physical things, it also is, at bottom, the most revelatory about a thing. The Greeks after Aristotle made their contribution in statics, the study of non-moving bodies, not in dynamics, the study of moving bodies; the latter had to wait till the Middle Ages for its lineage to begin in earnest. The simplest things that can be said in statics relate to dimensions. We can already start to do basic statics with this abstraction.

We have already seen how extension is abstracted from the sensorial experience. It is fairly easy to see how from that same sort of experience of, for instance, moving one's hand over a table, one may abstract the three distinct dimensions of height, length and width. Notice how we learn of dimension from motion, but dimension itself has no motion; it leaves motion behind. We learn about the actually existing extension in a given direction, which is here and there simultaneously, not here then there. Simple dimensional considerations coupled with simple physical insight can already yield important understanding.

There is a Difference between Big Things and Small Things

Consider the problem of size or scale of various things. An ant can pick up over 20 times its own weight, but the average man can only lift his own weight. Galileo talks about the problem of scale in his "Two New Sciences."

We further notice that a baby can fall a greater distance than an adult without getting hurt, but a grasshopper or an ant can fall still farther, even neglecting air resistance.

Why is this so? It partly has to do with size. We can begin to explain it by noting the following. As we look at larger and larger amounts of a given material, obviously that material takes more and more space, i.e., is more and more extended. As it gets thicker, we expect it to get stronger. Furthermore, the weight of the body also increases, but the relation between the increase in its extension and its increase in weight and strength is more complex than one tends to first think.

To see why an ant has more strength (when adjusted for weight) than a man, we need to analyze their support structures. For example, your leg or that of an ant is a column-like structure, a cylinder (see Figure 1-5). Then, the question becomes: if we make a cylinder two times bigger in every dimension what happens? In other words, what happens as we increase the leg size from that of an ant to that of a man. To answer, we will need the concept of density and thus also volume, which we now discuss briefly before finishing our ant versus man discussion.

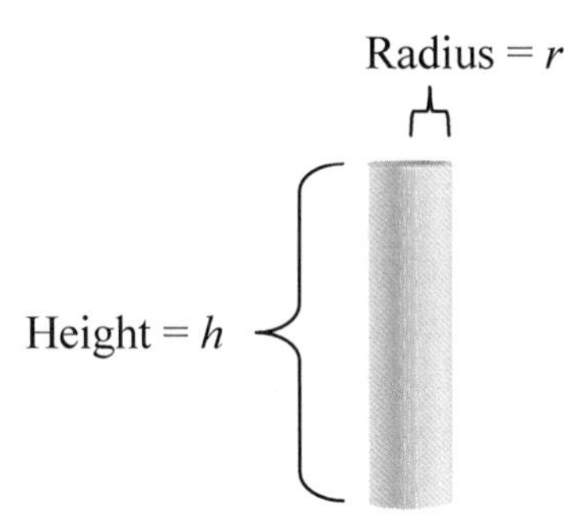

Figure 1-5: We approximate the shape of a leg by a cylinder. We need density and strength to compare abilities of an ant and a man. The volume of a cylinder is: $V_{cylinder} = \pi r^2 h$.

Volume and Density

Density and volume are intrinsic actual properties, not changing properties, so we have no need to consider moving bodies to discuss them as such; i.e., in this sense, they belong to the study of stationary bodies.

We measure density by the number of given units of mass in a given unit of volume. We also say, quantizing the quality of mass (and thus density), that density is a measure of the quantity or amount of mass in a given volume. In mks units, the volume of a region is defined as the number of unit cubes (the unit of measure, i.e., that by which volume is measured, is a cube with sides of dimension 1 meter) that will fit into that region. A unit of density is: 1 unit of mass per unit of volume; this gives $1 kg/m^3$. The density of an object is the number of primary units of density that describe the object. In terms of the primary units of cgs, the density of water is the unit of density because it is $1 g/cm^3$; ceramic is about $4 g/cm^3$ or 4 such primary units of density.

Notice the complication we discussed earlier with respect to length now returns with a new twist. We now have the possibility that we have neither a full 1 m^3 (unit of volume) nor a full 1 *kg* (unit of mass). Suppose we wish to know the density of a thimble of air. If needed, we can still express the volume and mass in these units by using fractions, most commonly expressed as decimals. Also, as previously, we can break up the density units using the prefixes in the table to make smaller units as convenient, speaking, for example, in terms of milli-units of the density of water, or we can convert to mass and volume units more amenable to the situation using those same prefixes (or simply scientific notation).

But, notice that, because of the division of mass by volume in the definition of density, the size of a body is irrelevant in the case of a uniform density body, i.e. a body with a (approximate or average) density that is the same throughout. Thus, even if we had a cube of material with 1 *m* sides, its density may still need decimal representation; it will only depend on the material. In the case of air high in the mountains, the 1 *m* on a side cube may weight .8 *kg*, giving a density of .8 kg/m^3. Yet, the same result would obtain if we only had a thimble of air, for though the volume is much smaller so is the mass.

With the concept of density, notice a further mixing of the categories and of the concept of unit. Mass, which we have yet to precisely define, is a measure of something qualitative (which we will see is a receptivity), while volume is quantitative. Units that are made from other units are called derived units. We can now continue to answer our problem of what explains the difference in strength between ants and men.

Is an Ant Stronger than a Man?

To get a concrete idea of the difference in strength between the ant and the man, we need to write the mass and weight of a cylinder that supports the lifting.

For a body of given volume and approximately uniform density, the mass can be calculated, given $mass = m$, $density = \rho$ and $volume = V$, as follows:

$$m = \rho V$$

Now, the more massive a body is the heavier it is; in fact it is directly proportional, meaning twice the mass gives twice the weight. Thus mathematically, the weight, measured to be *W*, is written:

$W = k\,m$, where *k* is some constant.

This gives, for our cylinder shaped leg:

$$W = k\rho\ V = k\rho\left(\pi r^2 h\right)$$

Now, the weight acts to pull a thing down, trying to crush our cylinder lengthwise, whereas the structure of the leg tends to hold it together, resisting the crushing. Hence, the ratio, *R*, of the strength, measured to be *S*, of a member (one of the legs) to the weight, is indicative of how much further force the member can withstand before breaking. The higher the ratio, $R = S/W$, of a member the stronger it is, for high *R* means *S* is high relative to *W* and so can stand more additional weight before breaking.

We can call this ratio, $R = S/W$, the adjusted (or reduced) strength because it takes into account the effect of weight. Gravity pulls on you and your body pushes back against it. Again, your body has limits to the force which it can resist; using some of the force to counter gravity leaves less to counter other forces, such as those to do lifting.

Now, the strength of a member is, in some range, proportional ($\propto$) to the cross-sectional area $\left(\pi r^2\right)$. That is, we write:

$$S \propto r^2$$

Thus, we can write, substituting for *W* using the previous equation, the ratio as:

$$R = \frac{S}{W} \propto \frac{1}{h}$$

Now, we are ready to answer our question. If we shrink a man (see Figure 1-6) to the size of an ant, say from 2 *m* to 2 *mm*, and keep his same proportions, this means we decrease each dimension by a factor of 1000. Well, the adjusted strength, the ratio of strength to weight, is proportional to the inverse of one of the dimensions; hence, the adjusted strength goes up by a factor of 1000. This explains why an ant can easily lift many times its own weight. Of course, our answer only implies that the strength is many times the ant's own weight and no more, because we have not factored in the differences other than size between a man and an ant.

Figure 1-6: Assuming no shift in composition, if a man shrinks to $1/X$ his original size in each dimension, his weight is reduced by a factor of $1/X^3$ whereas the strength is reduced by a factor of $1/X^2$.

Still, even at this level, we see math gives us the power to say something profound and interesting about our everyday world. It will only get better, more profound and more interesting. It is no surprise that we will need to develop our mathematics abilities in order to climb to these new heights.

Helpful Hints

It's a good idea to memorize and practice writing the Greek alphabet. Greek letters will be used as symbols throughout physics. For instance, the Greek letter ρ (called "rho") is universally used to represent density as we did at the end of the chapter. Facility with making these letters will be helpful.

There are 24 letters in the Greek alphabet; they are:

Lower case	Upper	Name	English
α	A	Alpha	a
β	B	Beta	b
γ	Γ	Gamma	g
δ	Δ	Delta	d
ε	E	Epsilon	e
ζ	Z	Zeta	z
η	H	Eta	h
θ	Θ	Theta	th
ι	I	Iota	i
κ	K	Kappa	k
λ	Λ	Lambda	l
μ	M	Mu	m

Lower case	Upper	Name	English
ν	N	Nu	n
ξ	Ξ	Xi	x
ο	O	Omicron	o
π	Π	Pi	p
ρ	P	Rho	r
σ	Σ	Sigma	s
τ	T	Tau	t
υ	Υ	Upsilon	u
ϕ	Φ	Phi	ph
χ	X	Chi	ch
ψ	Ψ	Psi	ps
ω	Ω	Omega	o

Summary

Physics is the rigorous study of the physical world, the key characteristic of which is its ability to change.

We have two powers of knowing, *sensorial* and *intellectual.* By the first, we know particulars of things, by the second we know their general nature. All we know comes ultimately through the senses. This is why physics in the general sense is the base science.

Some things, called *substances*, exist of themselves; others, called *properties*, can only exist in substances. One can consider a substance, various parts of substances or groups of substances as a system. One can then analogically speak of such systems as having properties, but this meaning of "property" depends on, does not negate, the first meaning.

Physical substances have *nine types or categories of properties*, the base of all of them is *quantity* or extension. The study of this first property, quantity, is called mathematics. This is why mathematics is fundamental to physics.

By *measurement*, we can think of all aspects of physical things as numbers and more deeply incorporate, though by analogy, mathematics into our understanding of the physical world. In this way, all variety of properties and combination of properties and even motion can be treated as quantity in an analogical way. For example, in this way, distance, pressure, speed, temperature, intensity of light can all be called quantities. We must remember this does not reduce these things to quantity, but only makes use of an aspect that is analogical to quantity, ultimately because of the quantitative base of all physical substances.

The major tool of physics is *mathematics*. The *empiriometric* approach, looks at the empirical (empirio) as quantitatively measured (metric), especially using the apparatus of modern symbolized and axiomatized mathematics.

Number, discrete quantity, is a whole that is measured by a unit.[17] Discrete quantity is the most abstract type of quantity that we see directly. It is one step more general than the continuous quantity, which is limited by shapes, and which we study in elementary (Euclidean) geometry.[18]

To do measurements, we need standards or *units* for the key properties of length (quantity), time and mass (a quality, which we will see is important because it relates to motion, especially change of place by which things come into contact to interact).

Dimensions and shapes already give insight into the physical things around us, for example, the insight of Galileo in his "two new sciences" that strength (adjusted for weight) goes as the inverse of a linear dimension.

An analysis of the units needed in a problem, called *dimensional analysis*, reveal something about its solution (see calculational techniques and problems below).

Aristotle was the first to establish the generic foundational principles of the science of the physical world called physics. Greek science prospered in mathematics and statics, but science had to wait till Medieval Christian times to get the principles it needed for dynamics, the study of motion. The empiriometric method in its modern form did not come into its own until the systemization and symbolization characteristic of it were established in the 1600's. The first modern empiriometric theory, i.e. modern scientific theory, of physics was given in Newton's *Principia*. The mechanics that he first introduced there is the topic of this textbook.

---

[17] And, that is completed by the last unit. In axiomatization, this translates to a concept of order.

[18] That is, it is one step more general than continuous quantity, i.e. the simplified three dimensional (Euclidean) geometry we see in the physical world when we leave out all other properties but extension (cf. chapter 2). First, we see the three dimensional continuum of height, width, and length, one part outside the next, but sharing a boundary with the next, for instance, inside of a ball, then we count the simplest parts inside, the units, to get the concept of number. For example, if the ball was made of three simple (partless) parts, we get the number three. In thinking of shapes in geometry, we often think of them as having no parts (in order to focus on their generic shape), but this is not what we actually see; we see extension consisting of parts one outside the next, which is limited by the given shape.

Terms

Ways of Knowing
- *Sense*
- *Intellect*
  - *Being of Reason (Mental Construct)*

Infra-science

Sciences
- ***Pure Science***
  - ***Physics***
    - Primary tool: ***Empiriometric Method***
    - Subfields: Mechanics: *Statics, Dynamics*
  - **Mathematics**
- ***Applied Science***

Key Principles
- *Principle of Contradiction*
- *Principle of Causality*

Substance
- *Property*
  - *Intrinsic:*
    - *Quantity*
    - *Quality*
  - *Relational:*
    - *Relation*
    - *Action*
    - *Reception*
    - *Place*
    - *Orientation*
    - *Environment*
    - *Time*

Measurement
- Number
- Units and Standards: length, time and mass
  - Meter
  - Second
  - Kilogram
- Dimensional Analysis

Statics

Dynamics

Calculational Techniques

***Dimensional Analysis*:** by knowing what units the answer to a problem requires and what units are given in a problem, one can check or even guess at an answer to a problem. As a simple example, if a problem asks for *m/s*, one knows that an equation that yields meters only is incorrect.

Problems by Subtopic

First Principles and Categories of Properties

1. How are sense knowledge and intellectual knowledge different but interrelated?

2. Give examples of each of the 9 categories of properties of substances (at least 3 for each).

3. Some categories have subcategories within them. For example, give examples of three different types (subcategories) of the category of quality. *Hint*: A colored object of a certain shape with the power to move has properties in all three of these subcategories.

4. How would the game of 20 questions be modified by your new understanding of the categories of properties? Come up with a list of things to be guessed. Use this example as a guide.

Is it an animal, a vegetable or a mineral?
None of these.
1) Q: Is it a substance?
A: no
Is it primarily a property of an animal, plant or inanimate object?
inanimate
2) Q: Is the property intrinsic?
A: yes
3) Q: Is it a quantity?
A: no
4) Q: Can you see the quality?
A: no
5) Q: Can the quality be heard with my ears?
A: no
6) Q: Can the quality be felt with my hands?
A: yes

If the answer was no, then we'd ask:
Q: Can you smell the quality?
A: no
Q: Can you taste the quality?

7) Is it the hotness or coldness of a body?
A: no

If the answer was temperature, we would say yes, but you are not done. They would then ask: "Is it the temperature, that is, the degree of hotness of an object?

8) Q: Is it a complex feeling like smoothness or roughness?
A: no

9) Q: Is the quality pressure?
A: yes

Now try to write questions for one of the below.
a) a decade (i.e. ten years)
b) yellow
c) northern direction
d) Rectangle
e) 42
f) Mars

5. Categorize the following as substance, part of substance, group of substances or a simple property and explain why you do so.

Hint: Recall that the first property of all physical substances that we see directly is that they have extension, and thus can be taken apart; therefore, a substance's ability to be taken apart does not mean that it was not a substance before it was taken apart. Because something can be destroyed doesn't mean it never existed.

A substance is distinguished from a part of a substance by its relative independence from things external to it.

For instance, in substances composed of parts that are different from each other (*heterogeneous* substances), crucial top level parts are radically changed (or even destroyed) when removed from the substance. Also, note that continuation of activity proper to the whole is dependent on its parts. When thinking of parts, don't *just* think of atoms; think of the parts most characteristic of the object under consideration. For example, if you pull a leaf off of a tree, the leaf dies. The leaf emanates from the whole.

In the case of *homogeneous* substances, the elemental parts which are repeated are all of one type. For example, a solid copper rod consists of metallically bonded copper atoms.

In the case of a group of interacting substances, when a part is removed, that part is not radically changed. For example, if I remove the engine from a car, the engine is not radically changed.

a) a frog
b) an iron screw
c) a computer
d) your toes and hands
e) the carbon atoms in your body

f) a telephone
g) big
h) clock
i) yourself

Note: in this problem, we have not exhausted what it means to be a substance; we have only given the next generic level of understanding. In general, to decide the correct categorization will: 1) require one to further refine one's understanding and often will also 2) require a detailed knowledge of physics, chemistry and/or biology.

Quantity vs. Quality

6. Give another example of the difference between quality and quantity.

7. Discuss how the shape of a triangle conditions the parts inside.

8. Give an example that shows that mass as such, not as measured, is not a property in the category of quantity. In particular, pick an example that shows that two substances can be the same size (amount of extension) in all dimensions and still have very different masses.

9. What can you think of that is not describable in terms of the categories or analogies to them.

Causality

10. How is the principle of causality related to the principle of non-contradiction? *Hint*: If something were to change itself, what would this mean?

Can yellow change itself to orange?
Can a boat's length change itself to a different length?
Can a granite rock change itself into water?

Measurement, Units and Dimensional Analysis

11. The Mars Observer spacecraft failed because of a simple conversion from English to metric or vice versa that was not done properly. To avoid such a failure on a manned trip to Mars, calculate the equation for converting miles to kilometers and the equation for converting from the unit of force, $1\,pound = 1\,slug\,ft/s^2$ to $1\,N = 1\,kg\,m/s^2$ (note you will learn more about these units in the later chapters). Given: $5280\,ft = 1\,mile$, $1\,ft = .3048\,m$, $1\,kg = .0685\,slug$.

12. If I measure a boat to be 20 feet long and I measure a fish to be 2 feet long how long is the boat in units of fish? Explain why such a unit (comparison) might be helpful?

13. What type of unit might be convenient for measuring planetary distances, like the distance to Mars? What is an appropriate unit of time for the motion of Mars around the sun? In light of this reasoning, why are, respectively the AU, astronomical unit, the distance from the Earth to the sun, and the year, good units? Would AU also be a good unit for distances between galaxies? Why or why not? Look up the distance to the nearest galaxy, Andromeda.

14. A *parsec* is the distance to a star (or other object) that *appears* to move by one second of arc when viewed from the Earth first on one side of the sun then later after a half year on the opposite side of the sun as shown below. It corresponds to about 3.3 light years. The first such apparent motion was detected for 61 Cygni by Friedrich Bessel. a) Given that he measured a value of .314 seconds of arc, how many parsecs away is 61 Cyngi? b) how many light years away is it?

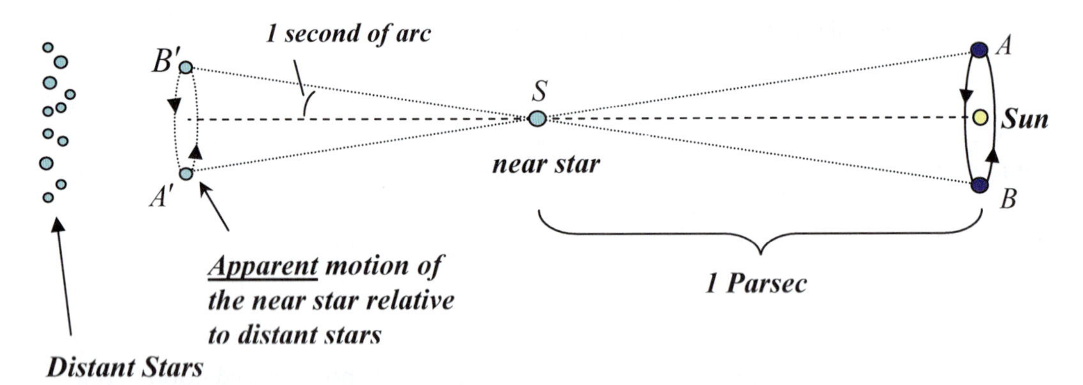

The near star, *S*, *appears* to be at position A' relative to the background stars when viewed from *A*, and it *appears* to be at position B', when viewed from position *B*

15. Given the size of a nucleus is $10^{-14}\,m$, explain why the unit of area of $barns = 10^{-28}\,m^2$ is used in particle physics, the study of subatomic particles. (By the way, the unit is called a *barn* as a take off on the phrase “he could not hit the broad side of a barn”)

16. If the Space Shuttle travels at a speed of 5 miles/second, and a bullet travels at about $2000\,ft/s$, which travels faster?

17. Estimate how many cubic yards of cement is needed to pour a 1 foot deep foundation for a $2000\,ft^2$ house.

18. a) Give an example of a change in property that is in the category of i) quantity (extension), ii) quality and iii) place:
b) How would one measure each of these changes?
c) Explain how a measurement of a uniform speed is a *constant* (unchanging) number although speed is the change of place of an object.

19. Explain how some measurements imply something about the relationship between them and other measurements. For example, consider volume and its relation to length. Give another example in which this is true. Note that this is the basis of dimensional analysis.

20. Using the symbol I for intensity and L for length for the measurements asked for below, express the following measurements in terms of mathematical statements using the proper units:
a) A desk light is 3 times brighter than a given standard candle (note: the unit is approximately equal to the modern unit called the candela).
b) The man is twice as long as the meter stick.
c) The first one is a measurement of a property in which category? What about the second?

21. If given R and S in meters in a figure such as shown to the right, in what way are there four objects being considered in the following equation, $\theta = S / R$?

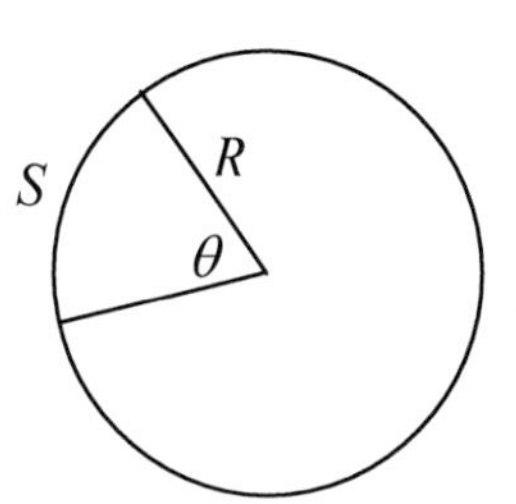

Do the following problems using dimensional arguments:
(Such problems take us from measurements (which are comparisons between objects and standards) to the symbols which represent them, such as x meters, where x represents the length of the object and "meter" the unit or standard. In solving such problems using only the general understanding implicit in the symbols, without the particular meaning of the measurements, we see the problem solving power carried by the measurements and get practice in it.)

Hint: In all of the problems below, take dimensionless constants of proportionality to be equal to one.

22. Construct a relation linking the distance d a person stands from an object, the height d of the object, and the angle, θ, formed between the eye and the object with the eye as the vertex. (See picture to right). Consider the case in which d is a lot bigger than D. *Hint:* Use your common sense knowledge of how the angle depends on d and D.

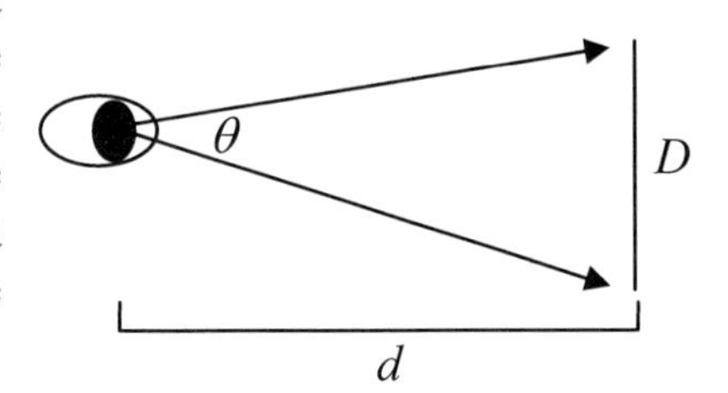

(Problems continued on next page)

23. The number of pixels in a TV determines the clarity of the picture created.
a) Using the result from the above problem, explain why there is a minimum distance of viewing from the TV that gives the best possible view of the detail of the picture (the minimum resolution of the human eye is: $1' \sim 2.9 \times 10^{-4} \, radian$).
b) Assuming a (low definition) TV has 704×480 pixels $\sim 3.4 \times 10^{5}$ pixels and a size of $47cm \times 43cm$ which gives a pixel width, w, of about $7.7 \times 10^{-2} \, cm$, what is this minimum distance, *d*?

c) Assuming brightness is no problem and the contrast is sharp (say white on black to make the problem simple), what is the maximum distance beyond which one gets no discernable view of the *overall* picture (*hint*: assume resolving $1/40^{th}$ of the screen is the minimum that gives any view of the overall scene?)

d) What is the optimum distance for watching *TV*?

This is the type calculation you would do to find out if you could see the American flag on the moon with your home telescope.

24. In a pinhole camera, a small hole in a sheet of paper or other such material is used as a lens to create a picture on another sheet of paper under it. Using dimensional analysis, construct a relation linking the diameter, *d*, of the pinhole, the distance, *D*, between the two sheets of paper, and the wavelength, λ, of the light source. *Hint:* find an equation that looks like:

$$d = some\ function\ of\ \lambda\ and\ D$$

Then calculate the diameter of the pinhole, *d*, that would produce the best quality picture, given the distance between the two sheets of paper is $2m$ and the wavelength of the light is $0.55nm = .55 \times 10^{-9} \, m$ (See picture to right)

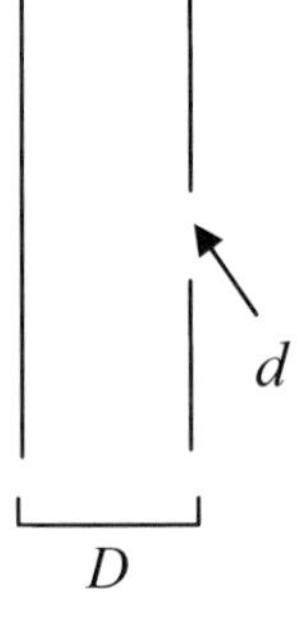

25. Given the speed of light ($c = 3.00 \times 10^{8} \frac{m}{s}$), Planck's constant ($\hbar = 6.626 \times 10^{-34} \frac{kgm^2}{s}$), and the gravitational constant ($G = 6.637 \times 10^{-11} \frac{m^3}{kgs^2}$), using dimensional arguments, calculate the Planck distance (the theoretical smallest length).

26. A light bulb is placed in an empty room.
a) Using dimensional analysis, construct a relation linking the measure of the bulb's intensity, *I*, in units of W / cm^2 with the observer's distance, *d*, from the bulb. (See picture to right)
b) How would you get the value of the constant in the relation above by measurement?
c) Given a bulb of $radius = 3\ cm$ has a measured intensity of $1\ W / m^2$ a distance of $4m$, what is the Wattage of the bulb?
d) What is the intensity generated by the above bulb at $9m$?

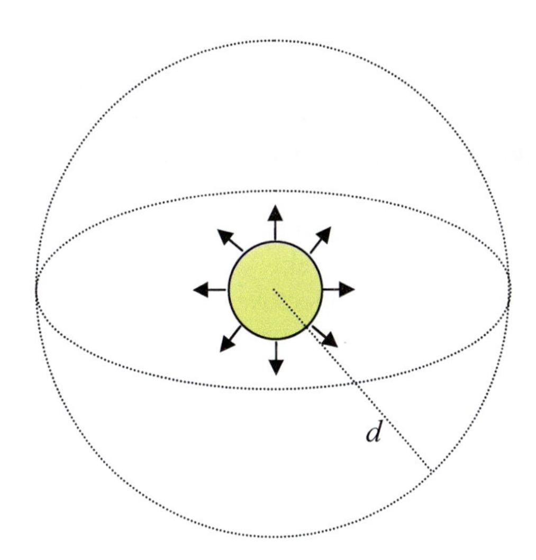

27. An object is released at rest and allowed to fall for a time t_1. a) Assuming the simplest possible equation, write the relation specifying the object's speed, v_1, at time, t_1. b) Relate its speed at that time with its speed, v_2, at a different time, t_2. c) Use the relation obtained in b) to calculate the objects speed after 7 seconds, given its speed at after 1 second is $1\,m / s$.

28. A car is accelerating at a constant rate down the road. a) Give the relation for the speed *v*, in m / s to the acceleration, *a*, in m / s^2 of the car with the time, *t*, in seconds it has spent in at the constant rate of acceleration. b) If a second car is moving in the same direction at a different constant acceleration a_2, what is the relative speed between the two cars? c) A Jaguar (XK8 (2001)) can do 0-60 *mi/h* in 7 seconds, what is its average acceleration in m/s^2? d) A Formula One car can do 0-60 in 2 seconds; plot how the relative speed between these two cars changes if each is constantly accelerating at its maximum rate.
For more car accelerations see:
http://hypertextbook.com/facts/2001/MeredithBarricella.shtml

29. A spacecraft is injected into orbit toward Mars at a speed *v*. a) Write a relation for the momentum *p* in units of *kg m/s* in terms of its speed *v* in *m/s*, and its mass *m* in *kg*. b) Given a craft of different mass injected at a the same speed what is the ratio, p_2/p_1, between the momenta of the two crafts.

(Problems continued on next page)

30. a) Write an equation that *relates* the weight of a body F_{earth}, in Newtons (that is, the measure of the magnitude of the force on a body that results from the action of the gravitational field of the Earth near the surface) *to* the magnitude of the gravitational field at the Earth's surface, G_{earth}, in units of Newtons/kilogram *and to* the mass, *m,* in kilograms of the body being pulled. Do the same for the weight of the body on the surface of the moon which we label: F_{moon} . Use G_{moon} for the magnitude of the field of the moon. b) Given that a $1kg$ object weighs about $10\ N$ on the Earth, find the weight of a gallon of milk (using masses given in table II above) on the Earth. c) Given that the gravity on the moon is about $^1/_6$ of the gravity on Earth, find the weight of a gallon of milk on the moon. d) Estimate how much a drop of milk weighs. Consider eating on the moon. What irritating problem will be exacerbated on the moon because of this difference in weight of milk. e) Convert all these metric forces to pounds weight (*lbf*) (note: the *lbf* stands for pounds force as opposed to *lbm* for pounds mass). f) Convert this answer for the case of Mars, which has 3/7 of the Earth's gravity.

31. A skydiver jumps out of a plane on Mars and is accelerated by gravity at a constant rate g_{mars} in m/s^2. a) Construct an equation for the speed v in meters per second as a function of the time t measured in seconds b) Construct a relation for the momentum, p in *kg m/s* in terms of the skydiver's speed, v, in *m/s* and mass, m in *kg*. c) Construct a relation linking his momentum, p, in *kg m/s* with a given time, t in seconds.

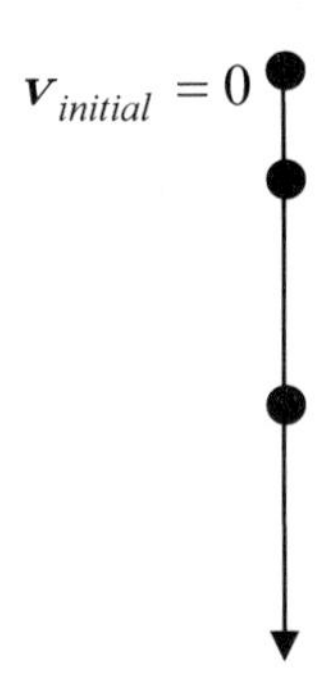

32. An industrial strength fan is blowing air into a hot gym. a) Relate the amount of time, t, in seconds that the fan is running to the amount of energy, E, in Joules consumed by the fan assuming a constant power usage (*Joules/sec*). b) Assuming the fan is the only appliance running in the building and the electrical usage meter shows 3.6 kilowatt-hours used in ten hours, how many Joules of energy were consumed ($1\ Watt = 1\ Joule/s$), and what is the wattage of the fan?

33. Relate the velocity, v, in *m/s* of a cannon ball of mass, m, in kilograms with the amount of kinetic energy, KE, $J = Nm = kg\ m^2/s^2$ the ball possesses. b) Ignoring the dimensionless constant of proportionality (in fact, we learn in Chapter 5, that this constant is ½) how much kinetic energy does a 100*kg* man traveling at 3 *mph* (a brisk walk) have? c) Given that there are 4180 *Joules/*Cal_{food}, how many food calories does this correspond to? We will see in Chapter 5 that this does ***not*** mean walking is not a good way to lose weight.
(Note: The calories that are shown on food product boxes are given in terms of "food calories" are a bigger unit than physicists talk about when we refer to a calorie. Namely, $1\ Cal_{food} = 1000\ Cal$.)

34. a) Write the relation for the energy, PE in $kg\ m^2/s^2$ that would result from an apple of mass, m, in kg, falling to the ground a distance, h, in meters; assume it falls due to gravity at a constant acceleration.. b) Given two different apples of different masses, m_1 and m_2 falling at the same constant acceleration but at two different heights h_1 and h_2, what is the ratio between the energy they have at the end of the fall in terms of the mass and fall distances.

Statics: Dimensions and Stationary Bodies

35. It is often important to note what length scale is relevant to a given physical situation to properly understand it and solve a given problem. Such length scales also turn out to be very useful in giving approximate answers to problems, so called *order of magnitude* answers. For example, the typical length scale for an adult human being is a meter that can be used in calculation; this can be used to get a very rough idea of the volume of a man as about $1m^3$. In this approximate way, calculate the following:

a) Given that at standard temperature and pressure (STP) there are one mole of molecules of air in 22.4 liters, what is the length scale for interaction of air molecules?

b) Given that the density of water $= 1\ g/cm^3$ and that that one Avogadro's number of water molecules has a mass of 18 grams, what is the length scale for liquid interaction?

c) An article in May 2006 *Physics Today* points to a "confluence of energy scales" at around 1 nanometer, where all the forces have about the same energy as the thermal motion of the atoms which is about $k\ T$, where k, the Boltzmann's constant is $k = 1.38 \times 10^{-23}\ J/K\ and\ T \sim 300K$. What force does this imply is acting on this scale?

d) Using the density of iron is about $8g/cm^3$ and its atomic mass as about $56g/mole$ estimate the scale appropriate for solids.

e) Calculate the same for lead (density $11g/cm^3$, atomic mass $\sim 207g/mole$) and compare the numbers with that for iron.

f) Give approximate range of lengths involved in the following:

i) Planck scale
ii) Nucleus
iii) Atoms
iv) molecules
v) known living things
vi) known animals that can see
vii) Asteroids and planetoids
viii) Planets
ix) Stellar Scale
x) Solar Systems
xi) Interstellar Scale
xii) Galactic Scale
xiii) Intergalactic Scale
xiv) Super cluster scale
xv) Scale of Universe

36. Tell how you would go about calculating the density of water, (kg/m^3) using the following hypothetical experiments:
a) Given one drop of water. b) A cube of water with $1m$ sides.

37. a) Given that a black hole has a radius of $2GM/c^2$, where M is the mass of the black hole, $G = 6.67 \times 10^{-11} \dfrac{m^3}{kg\, s^2}$ is Newton's Gravitational constant, and $c = 3 \times 10^8 m/s$, assuming the mass occupies a spherical volume (using standard Euclidean geometry that you learn in high school), what is the density of a black hole as a function of its mass.
b) What is the density of a black hole the mass of the Sun?

38. Discuss how density leaves behind the actual size of the body under consideration. In what sense is size irrelevant to the nature of a body, and what sense is it not. Consider an iron bar and an animal.

39. Explain why can a small child survive a higher fall than an adult?

40. Estimate the tallest structure supporting its own weight that one could make on Earth out of steel given its elastic constant, $E_{steel} \sim 2 \times 10^{11} kg/m/s^2$, its density of about $8\, g/cm^3$, and assuming its height is about as big as its other dimensions. Compare the answer to radius of the Earth and note how these considerations are thus not sufficient to get an accurate idea of what the tallest structure would be.

41. Suppose two robots are made in the shape of men, one of steel, density $8g/cm^3$, and the other of bone, density $2g/cm^3$. Using the equation below, give a very rough estimate of the maximum height of a robot made of steel and one made of bone.

The equation gives the critical height beyond which a structure of circular cross section of radius, r, density, ρ, and elastic constant E, will buckle under its own weight.

$$h = \left(\frac{2.5 E r^2}{\rho g} \right)^{1/3} \quad \text{where: } g = 9.8 m/s^2$$

Take: $E_{steel} \sim 2 \times 10^{11} kg/m/s^2$ $E_{bone} \sim 2 \times 10^9 kg/m/s$ and take the *total* cross sectional area (πr^2, for an effective radius r) of the robot's structural support (i.e. "bones") in its legs to be: $\sim 3.14\, in^2 \sim 2 \times 10^{-3} m^2$

42. Given the ratio of breaking tensile strength of steel to bone of about 3.1, proportionally how much bigger could a steel robot be and still keep the same relative strength according to the definition given in the last section of the chapter.

43. What advantages are there in using Earth motions to measure time? What advantage is there in using the radiation coming from a Cesium atom in the way described in the text?

Chapter II

Vectors and Calculus

Introduction

The queen of the sciences, mathematics, gives us the concept of vector, derivative and integral which are essential for understanding Newtonian mechanics.

As we saw in the previous chapter, quantity, the subject of mathematics, is abstracted from sense experience. Hence, as abstract as they seem, vectors, derivatives and integrals and the like, which are all ultimately quantitative entities, have their root in sense experience.

What are Vectors and Where do They Come From?

We start with the origin of the concept of the vector, because it will help us more fully understand it, as well as prevent us from thinking such entities are solely mental; in fact, they ultimately come from nature. Those who have not realized that such concepts ultimately originate in sense knowledge can, and often do, come to think that mathematics is not about the real world, but about an isolated mental game. Thus, it appears to them that the empiriometric method (looking at the world mathematically, as we do in modern physics) only works by some freak accident.

So, starting with things we can see, we now begin to build up our understanding of vectors. Follow one of the walls of the room you are in until it meets another wall. Draw your attention to where two vertical walls meet. We can see that this is one surface meeting another surface in a corner. By a simple leaving behind (abstraction) of the color and all qualities including any roughness, one sees that the walls are like planes and their intersection forms a line.

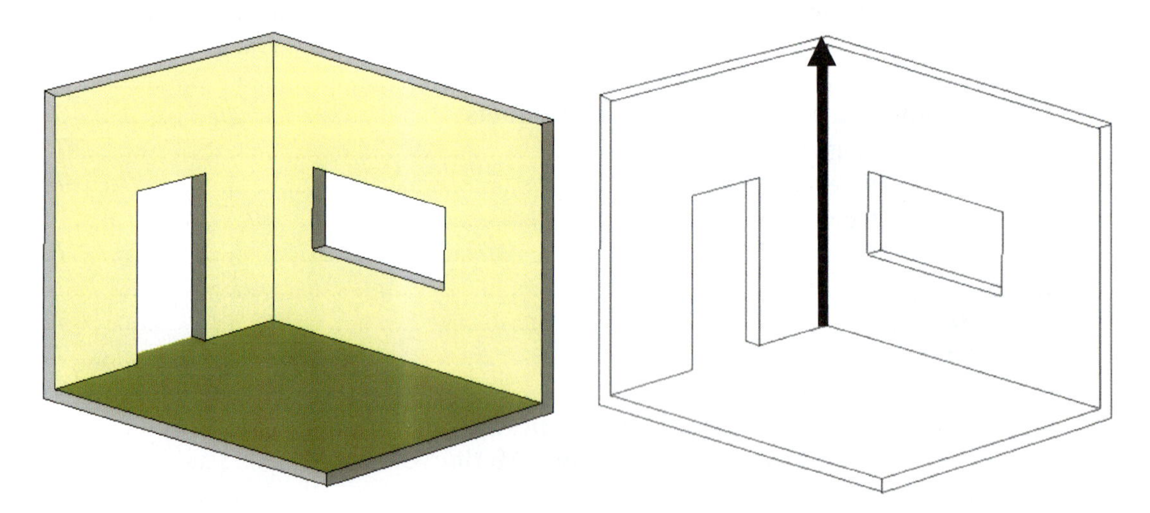

Figure 2-1: A real room has walls with color, texture, certain smells, etc. (left). To proceed to vectors, we leave out all this, keeping only extension and shape (right). Of course, the picture is still white and gray, for we can only suggest such a leaving out (i.e. abstraction) with a picture; it can only finally be done in the mind.

This line of intersection has a vertical orientation with an up and a down end in the actual room. One can thus think of the line as indicating a direction. To manifest this orientation on the line, that is, to establish an "up" in our imagination in which we have deliberately left out gravity etc., extend the line as shown and put an arrow on it as shown to indicate "this end up." Keeping only the ray thus constructed, we effectively reflect the categories of place and orientation into the category of quantity—see Figure 2-1. Of course, up and down are *not* established by rays but by gravity, which, in turn, is manifested by the downward motion of dropped objects. Our considering of just the ray allows us to leave gravity and all other qualities out, but still maintain something of the up/down orientation of the line. Instead of the actual up/down distinction in orientation, we substitute the shape of an arrow to distinguish top from bottom.

Here we have the most basic concept of a vector. It is a line of certain length with a direction. This, of course, is a specific vector; it has a specific length and direction. We can imagine many others. For example, consider the variety shown in Figure 2-2.

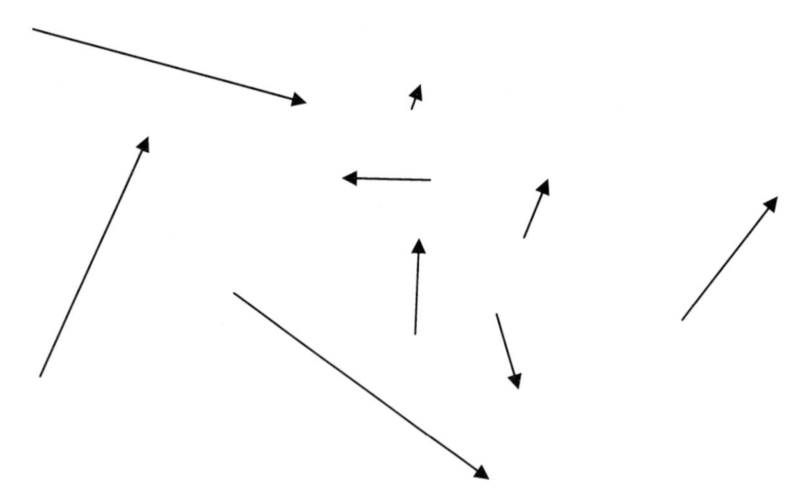

Figure 2-2: Vectors of many different directions and lengths.

Hence, to specify a vector we must give its length and direction; it is a ray of certain length and direction.

Notice "length" has two meanings: it can mean the extension of a line or the *measure* of that extension.[1] We use the first meaning when, for instance, we say a rectangle has length and width, but no height; we focus only on the intrinsic extension, not its comparison to another extension (i.e. measure). In the second usage, as we saw in Chapter 1, length is measured by comparison with a standard unit (such as meters), so that we say the line is so many units long. Again, pay attention to the fact that when we give a measure of the length of a line, we assign it a number, which is what we need in our equations. We will use both meanings of "length," allowing the context to distinguish the appropriate meaning.

Yet, we still do not have the full idea of a vector. Two important facts need to be noticed. *Length* is clearly a quantity and *direction* as given is a relation among the parts of a quantity of a given shape within a larger quantity ("space"), i.e. again it's what's left of orientation in the category of quantity, keeping only shape. More explicitly, length is the extension along a given direction, and direction is defined by the relation between various parts that make up the extension in, for example, a plane such as in Figure 2-2. One can see such vectors have a kind of reference to the concept of place---that is, what is left of it when we take away all categories of properties, including qualities that can cause changes, except extension and shape. Indeed, we say that this vector is here, and the other one there.

[1] In order to have a *real* relation, one must have at least two real things to relate, in this case two lengths. Though this seems obvious, it is often forgotten, since, in our mind, we conceive of logical relations such as "or" and "and" which don't specify real relations within things but only in the mind. Furthermore, they can be thought of without specifying real subjects of their relations but only signs, as, for instance, when we say "*A* and *B*." Such relations are not physically real because they cannot exist outside the mind.

This then is the quantitative idea of a vector, that which has length and direction, which we represent quite precisely by an arrow. But, this purely geometric meaning is not all a physicist means by a vector.

We can talk about how intense a quality, such as a color of something, is and notice that it can be quantified by considering a greater intensity to be a greater length and a smaller intensity to be a smaller length. For instance, in saying this light is twice as bright as that one, we measure the first by the second. We make the first the standard unit, in analogy to the way that six pieces of a pie make up one whole pie. By this analogy, we can treat the intensity of a quality as a length.[2] The first to make this analogy were the already mentioned "Oxford (University) Calculators"[3] of the Middle Ages, who are also called the Mertonians (see box). We need to emphasize that such a treatment does *not* make quality reducible to quantity. It only means there is an analogy between the two, which is not surprising, because they are related; namely, quantity is the first property of all physical things.

We can also give direction to a changing intensity. We can say this is so hot, (which is analogous to length) and getting hotter (analogous to quantitative direction). We can even combine the intensity of quality with direction in space (quantity). We can say this is so hot and increases in temperature along this path.

---

[2] The analogy of the intensity of a quality with the length of a line is based in the fact that both are continuous, having any number of possible states between any two given states. They are not continuous in the same way, but only in an analogous way, for quantity has parts outside of each other, whereas a quality, as such, does not. In what sense then is it analogous? A quality can be more or less intense. This statement contains an implicit reference to change. It's the nature of material things to have a certain quality and then be able to be changed; for example, something can increase the intensity of that quality. The analogical concept of continuity comes from comparing locomotion and intensification of a quality. Imagine yourself walking a mile with a chameleon which is gradually deepening the greenness of its skin to match your sleeve. At any point along the road, the chameleon is a certain intensity of green. The moving from one part of the road to another and the change from light green to deep green have a certain similarity.

[3] Grant, *Physical Sciences in the Middle Ages,* p 55, Cambridge University Press, 1993

The Oxford Calculators were almost all associated with Merton College of Oxford University. Oxford was founded before the beginning of the 1100's. The most important of this group of thinkers were: Bishop Thomas Bradwardine (c. 1290-1349 AD), William Heytesbury, Richard Swineshead (c. 1340 – 1354) and John Dumbleton. Leibniz (1646-1716), co-inventor of calculus, so much admired one of their surviving works that he wanted it republished.

Thomas Bradwardine may have been the first to use functions (before modern symbolic notation) to describe motion. All later scientists that will study the pivotal concept of impetus (inertia) show direct influence from Bradwardine.* In *Tractatus de proportionibus* (1328), Bradwardine extended the theory of proportions, anticipating the concept of exponential growth. The Calculators were the first to formulate the mean speed theorem: namely, for a uniformly accelerated body which begins from rest, the constant speed (the mean speed) that one would have to travel in order to traverse the same distance in a given time as the uniformly accelerated body is given by: $\boldsymbol{v}_{mean} = \frac{\boldsymbol{v}_f}{2}$, where $\boldsymbol{v}_f$ is the final velocity of the accelerated body. They proved this important theorem (in modern terms, a particular case of the mean value theorem). They also distinguished kinematics from dynamics, emphasizing kinematics, and investigated instantaneous velocity. Such are the seeds of the calculus applied to motion that Newton was to use. Physicist Clifford Truesdell says that, due to these Calculators, "the qualities of Greek physics were replaced, at least for motions, by the numerical quantities that have ruled Western science ever since. The work was quickly diffused into France, Italy, and other parts of Europe. Almost immediately, Antonio da Casale and Nicole Oresme (see box below) found how to represent the results by geometrical graphs, introducing the connection between geometry and the physical world that became a second characteristic habit of Western thought ..." Again, this is not a warrant to deny or even forget qualities, just to note the importance of quantity.

Bishop Grosseteste (cf. chapter 1) while not a "calculator," was educated at Oxford.

* Crombie, *Medieval and Modern Science II*, Doubleday 1959, pg 90

So, with the key concepts abstracted from sense experience, and the needed distinctions understood, we can now define a vector. ***A vector*** *is a mathematical entity of a given magnitude and direction, where magnitude and direction are taken in many varied ways so that the concept is applicable to many qualitatively different situations.* We have discussed two basic categories of vectors.

1. The first category is what we first mean by a vector, the one on which the others will, by analogy, be based. It is **purely quantitative**; it is something that literally has length and direction, such as the meeting of two walls or the imaginary line approximating the trunk of a straight tree (see Figure 2-3). A more abstract example is a highway sign that points the way and gives the distance to a city; it refers to the arrow that lies between the sign and the city and points toward the city.

Figure 2-3: Some objects that one can associate with a vector in the first sense.

2. The second category is an analogous use of the first. Here, the vector means the magnitude and direction of **a quality**.

Of course, all kinds of combinations of properties in the category of quantity and quality can occur. Furthermore, other categories can be involved with complex intertwining of the categories possible in the use of a given vector.

For a simple example of the combining properties of different categories, consider a lake that shimmers brightly in the middle, but which falls off in intensity towards the shore. We can assess the brightness of a given patch of a certain area, say by holding a grid at arms length and assigning a number from 0 to 10, where 10 is the brightest patch, to each grid patch (see Figure 2-4). We can then construct a vector that has a direction for a given grid location defined as the nearest neighbor (the ones that share a boundary) with the largest increase in brightness. The magnitude of the vector would be specified as the difference in the apparent brightness (e.g. 0-10) per unit of area and so, like the direction, is a combination of quality and quantity.[4] Of course, if the brightness did not change from

[4] Note, the area of a square can itself be thought of as composed of squares of a given length so that area $\sim$ (length)^2; in other words, area is, in an analogous way, measured by length.

one grid location to any of the next, the vector would be zero magnitude. Further, if, for a given cell of the grid, the brightness increased equally in any two directions, then the direction--and hence the vector--could not be *uniquely* defined for that cell. Such are the complications that results from combining various types of properties into the concept of a vector. Yet, these complications are more than compensated by the ease we buy in ability to manipulate the complicated network of information that we packed into the vector.

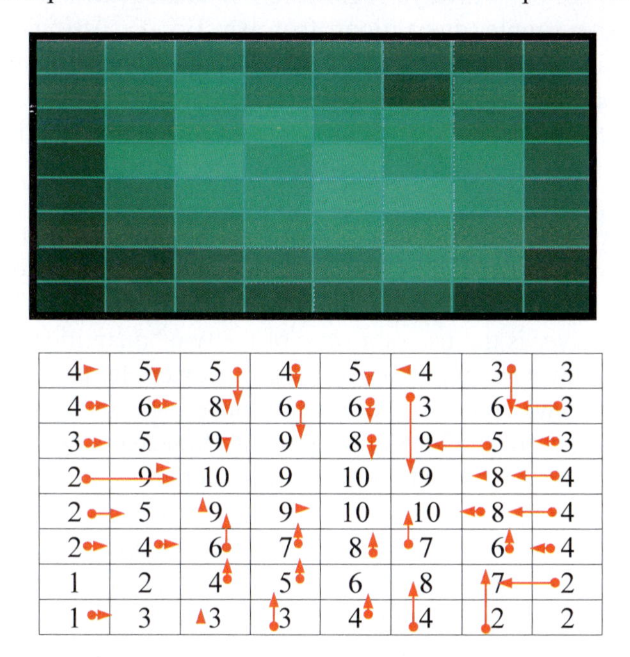

4	5	5	4	5	4	3	3
4	6	8	6	6	3	6	3
3	5	9	9	8	9	5	3
2	9	10	9	10	9	8	4
2	5	9	9	10	10	8	4
2	4	6	7	8	7	6	4
1	2	4	5	6	8	7	2
1	3	3	3	4	4	2	2

Figure 2-4: *Brightness pattern on a lake.* Each number in the grid represents the brightness of the associated rectangular area on the surface of a lake (top). Follow the red arrows to *maximum* increase in brightness. The arrow direction specifies the most direct ascent, via nearest neighbors, up the "mountain" of brightness. The arrow length specifies how steep the ascent in brightness is to that next cell. Some arrows are so small that only the heads are visible.

Since brightness is a quality, the vectors specifying the magnitude and direction of brightness differences can only analogically be thought of as actual length and direction (*the arrows*), the primary analogate for a vector. Again, though the brightness is *not* identical with the length (extension) of an arrow, we *can and do* make the arrow's *measured* length *proportional* to the *measured* brightness difference.

Note that in defining the direction of greatest increase, we only consider nearest neighbors, not diagonal neighbors. Thus, each brightness vector we define has the direction of a given red arrow and a magnitude given by the measured length which is proportional to the *increase* in brightness to the next cell in that direction. Vectors are not defined for cells with nearest neighbors of the same maximal brightness; so, no arrow is drawn in that case. Neither is one drawn when a vector length is zero or negative.

Figure 2-4 above illustrates this analogical use of the primary concept of a vector (i.e., length and direction) and the problems that come in trying to *picture* it. The red arrows have an actual length and direction, but the brightness does not. However, the length is analogically like the brightness, so we can take the measured length of the vector to be proportional to the measured difference in brightness. The diagram exemplifies how the primary definition of vector is always regulating our analogical usage. If we understand and keep in mind the simple distinctions that we are discussing, we can avoid the confusions that students often have in understanding and applying vectors.

Another example is a *velocity vector*, which represents say the speed and direction of a runner at a given moment; it is a combination of length and time, allowing us to "quantify" the amount and direction of *locomotion.* Here we introduce still another level of analogical use of our primary definition vector. The first layer of analogy is the use of time as if it were a quantity, which shows up in the simplest statement about the velocity vector. Namely, the vector, in its definition, is a magnitude (say 2 meters/second) and a direction (say south), both of which do not change; whereas in fact, the runner which it represents, is changing. The second layer is the use of a *ratio* of the two "quantities." We can, of course, make ratios like this with *pure* quantities as well; for example, on the slope of a hill, we might specify how far up we go per unit of total distance traversed.

A further expansion of the idea of a vector is suggested when we think more about motion. All physical things change with time, so why not allow any given vector to change with time? Of course, we cannot make arrows change on the page as we work with them, and, generally, we would not want to. However, we can setup a correspondence[5] (called a mapping) between states of the body and the static world of mathematics. That is, we can make vectors functions of time, making magnitude and direction change as time marches on. We can, to take the simplest example, shift from considering one vector of given magnitude and length at 5:00PM to another at 7:00PM. For example, looking at water near your favorite fishing spot, you may notice that as the sun sets, a wind picks up, moving the water there faster and in a different direction than it had been.

The crucial point to remember is that although we can represent vectors by arrows, they are *not always* identical with arrows, indeed usually are not. The arrow just represents the magnitude and direction under consideration; the types of magnitudes and directions one might consider are very different. Remembering this can save you from trying to picture a ray *actually existing* in the water flow or on the road where the runner is moving or on the lake between cells of different brightness. Such images are then free to be what they are: aids in thinking, helping to bridge the connection between the actual situation and the mathematics describing it.

Another important aspect to note is that, despite our examples which localize a vector to a given place, the abstract concept of a vector, which is something with magnitude and direction, does not, of itself, specify a location.

[5] Such a correspondence between two sets of objects is called a *map*.

Representing and Using Vectors Mathematically

Symbols

A vector is written as a symbol with a ray over it. The symbol is used in the same way as a variable in algebra, i.e., to name the particular vector, while the arrow indicates it is modeled after a ray, having magnitude and direction. For example, a vector for a distance to a nearby town might be represented by $\vec{d}$, whereas a general vector might be written $\vec{V}$.

Some common types of vectors, with their usual symbols, include:

<u>Quantity</u>:

Distance, $\vec{x}$ (primary usage from which others are made by analogy)

Combined:

Change of a group of distance vectors, $\Delta \vec{y} / \Delta x$

<u>Quality</u>:

Gravitational force or any force, $\vec{F}$

<u>Combinations</u>:

Change (gradient) of intensity over a distance, $\vec{\Delta} I / \Delta x$

Motion:

Velocity, $\vec{v}$, acceleration, $\vec{a}$

(E.g., of wind or an individual body)

Mass or charge flow across a surface, $\vec{J}$

To talk precisely about the abstract concept of vectors, we need the concept of a space.

Space

Three-dimensional space is the extension in length, height and width that we learn about through what we see, hear and feel around us. As we've discussed, extension is what one gets when one leaves behind all the other categories of properties; usually this means we also keep the quality of shape, which is a limit on the extension. Typically, one could imagine then a solid figure composed of actual parts, with no color, hardness, vibration, smell or any other such quality. Of course, we always retain some such qualities in our imagination, but it's convenient to minimize them, because they are not what we are interested in, and thus should only facilitate our thinking about the quantitative.

So, can we say space in one dimension is a line composed of so many actual parts? No, not quite. Though everything we see has a certain number of parts, e.g. atoms, space in its modern mathematical sense has no *actual* parts. In this notion of space, we leave out of consideration the fact that extension involves a certain number of parts. Taken as written, this would be contradictory, because having one part here and another part over there is what we mean by extension (at least it's the primary meaning). Instead, we mean here that any given line is *potentially* divided in any way we want. We concentrate on the fact that a given line can have any number of parts, in any size we please. In other words, whatever parts the line has now--picture one in your imagination--we just don't care, because we're going to make it what we want—picture dividing it into parts of your choice. Indeed, we carry this thinking to a limit and think of the line as having no actual parts but only having potential for them. In this shorthand limit, we are back to a contradiction. When we think of it in this shorthand way, the concept of "line" is purely a mental construct, though an invaluable one, since it can only exist in the mind. Purely mental constructs are also called beings of reason, entities existing (being) only in thought (reason). Such constructs are part of our everyday thinking as well. For instance, the concept of "nothing" is such a construct; if taken literally it means something which is not anything,[6] which is clearly a contradiction. Yet, this is a handy construct that has its place, which when unwrapped means some particular thing that we have in mind is not there. Our mathematical concept of space, as we have seen, can be similarly unwrapped. Concretely, we may, for example, label the line in a particular way, which is a way of specifying parts as fine as we want. Of course, we need three such dimensions to discuss all of space.

Also, we would like *space* to have no bounds.[7] Again, this means, in our imaginations, we can potentially stretch the length of our lines to any size we like. In this way, we can setup a coordinate system by dividing space into parts (say cube shaped parts for convenience) that are as small as we want *and* extend this slicing as far as we want. Via shorthand wording, we will say each coordinate line is infinitely long, and the space is made up of points that we label, but keep in mind what is really meant. It should be clear that it would be very easy, if one is not careful here, to lose the *crucial* distinction between actuality and potentiality.

For example, something actually divided an infinite number of times is a contradiction. To see this, suppose there were such a thing. If we divided it again, it would still be divided an infinite number of times. Now, is that last division accomplished or not? It cannot be both. Yet, it is identically the same infinity whether we do this extra division or not; indeed we run up against the very definition of infinity of divisions which is that we can always do more divisions than we've done. If we try to specify the notion of infinity further, we lose the notion of infinity. If we leave it unspecified, we lose reality in which a thing is either there or not, *not both.*

Figure 2-5: There is an implicit contradiction involved in the notion of a completed infinity.

---

[6] Webster's New World Dictionary's first definition is "not anything."

[7] Hence, "mathematical space," in its modern sense, is thought of as not having any actual parts and not having any bounds; part of the power and convenience of this being of reason is that it allows us to avoid the last quality that remains when one considers real quantity, namely, the complementary quality of shape. In particular, the boundary of an extended region is a shape and the *actual* parts that make up that region need, in the physical world, a shape or limit as well.

Vector as Three Numbers: Rectangular Coordinates

We now move to represent vectors in space. Start with the primary definition of vector from which all our others are derived, a ray of given direction and length. We can represent this in *rectangular* coordinates as shown in Figure 2-6.

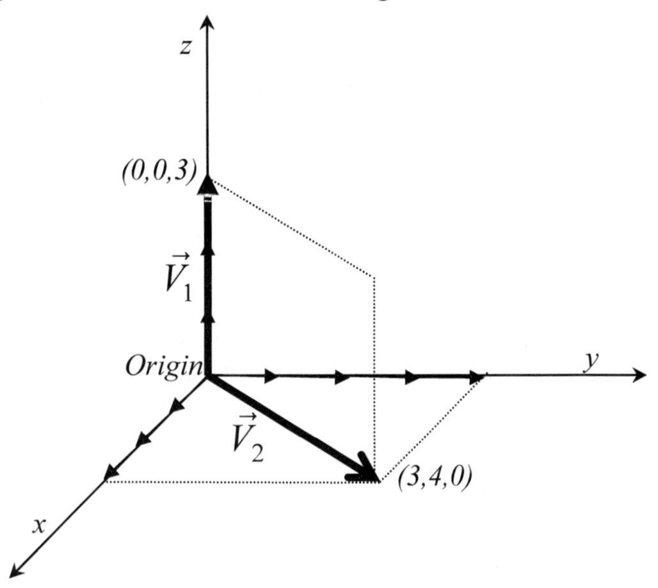

Figure 2-6: Examples of vectors in three dimensional space; three numbers identify them in rectangular (Cartesian) coordinates, notice the rectangles involved.

In the figure, we pick the tail of the first ray, $\vec{V}_1$, to begin at the origin and end at the point that can be reached by "walking" three units along the *z*-axis. This is obviously a vector of length three and direction "up." Hence, we represent it by (0,0,3). The second vector shown, $\vec{V}_2$, starts at the origin and ends at a point that is three units in the *x*-direction and four units along the *y*-axis (and *no* units along the *z*-axis). This vector, written as (3,4,0), has a length of 5, which is easily determined by application of the Pythagorean theorem. Because of the way we've defined our space, we obviously can write any vector of any length and direction in this way.

Given the nature of undifferentiated space as composed of three independent directions, three simple directions are picked out by our rectangular coordinate system. To discuss these directions, indeed any direction, we do not need to emphasize the length. So, we introduce the concept of a *unit vector*, a vector of unit length, the unit being the simplest magnitude we can give to a vector. Thus, we have three primary unit vectors in rectangular coordinates written as: $\hat{x}, \hat{y}, \hat{z}$; think of the carrot marking each letter as a kind of modified arrow.

This makes evident that the coordinate axes we define, excluding any vectors we add later, are the only features that distinguish one part of space from another. In particular, clearly, the intersection of the axes distinguishes the origin from the other points in the space. Since the general idea of a vector doesn't specify where it is, we can put a coordinate origin anywhere (and in our space everywhere is the same) and put our vector ray at the origin or anywhere else, though usually the origin is convenient. Again, the vector is only magnitude and direction, not location, so we only need to specify three numbers; they are, in rectangular coordinates, also called Cartesian coordinates: *(x,y,z)*, implicitly referenced to some arbitrary origin.

Although credit for rectangular coordinates is commonly given to Descartes through the name "Cartesian" for such coordinates, Nicole Oresme (see box below) was probably the first to use rectangular coordinates.

Nicole Oresme (o'rem) was born c.1320 AD in Normandy, France and died in 1382. He was, among other things, a mathematician, a physicist and a priest (made bishop of Liseaux, France in 1377). His major mathematical work is "Tractatus de Figuratione Potentiarum et Mensurarum Difformitatum." Among his accomplishments are the use of rectangular coordinates and graphing of the *intensity of a quality* then called *latitudo* (say, temperature of a rod), against a length then called *longitudo* (e.g., distance along the same rod), on such a rectangular coordinate system. He also discovered that speed versus time graphs can be constructed. Further, he investigated, in his own notation,* fractional powers, saying $4^{3/2} = 8$.

In dynamics, he shows, following Jean Buridan (see history box in chapter 3), that the movement of the Earth is consistent with immediate experience, though it does not seem so at first. And, he points out that the movement of the Earth, not the geocentric hypothesis, is, indeed, the simpler one. These arguments will be employed by his successors, including Copernicus and Galileo.

He also made an argument for an international dateline. His argument fundamentally rests on noticing that only the relative motion of the Earth and sun defines a conventional day (i.e., based on the rising and setting of the sun for a given observer). The principle can be understood by noting that if one moved fast enough to stay "under" the sun, one would never experience a single conventional day. The roundness of the Earth, which is also needed in the argument, was commonly assumed in the Middle Ages.

He proved, using graphical methods, the mean speed theorem (see Figure 2-31) which Galileo imported into his work without reference. He determined the *distance = ½ acceleration time²* law for uniformly accelerated motion that was later applied by Fr. Dominic De Soto (1494-1560…see history box in chapter 4) to free falling objects.

* In his notation, he wrote this fractional power as: $\boxed{1^{\text{P}}\tfrac{1}{2}}\,4$. See for instance, A history of Mathematical Notations by F. Cajori, 2007 Cosimo,.

Vectors in Different Maps or Coordinates

You may have noticed that picking a rectangular grid is not at all necessary. It is only convenient because it works well in describing our homogeneous space, the simplest space where all divisions are only potential, and, thus, we can divide it anyway we want. Dividing space into cubes that respect the natural independence of the three dimensions is best if this is really all we have to consider. However, in an actual place, there would be something that would break the sameness of the space. We call this "breaking" the symmetry.

For instance, the presence of a pie breaks the sameness of space. Yet, it leaves some symmetry; slicing a pie in half through a diameter gives two pieces that are the same because the pie is "symmetrical" around that line.[8] "Space" can be sliced in half in anyway, and the two pieces would be the same; this is because "space" is homogeneous, i.e., the same throughout.

Real objects, such as pies, would not only break this symmetry by their boundaries but, generally, also would not be exactly the same throughout their interior either. Apple pies, for instance, would have different sized chucks of apple in one part than another (see Figure 2-7). In fact, any real thing must have parts in the interior, so that it is not just the boundary that matters. Because, as we showed earlier, there cannot be a completed infinity, there must be, at any given moment, a certain number of actual parts. These parts can be clustered in an infinite number of ways. Recall we've left the qualities, per se, out of consideration here, but they will determine what arrangements actually occur, and so they will be reflected in the parts that we analyze.

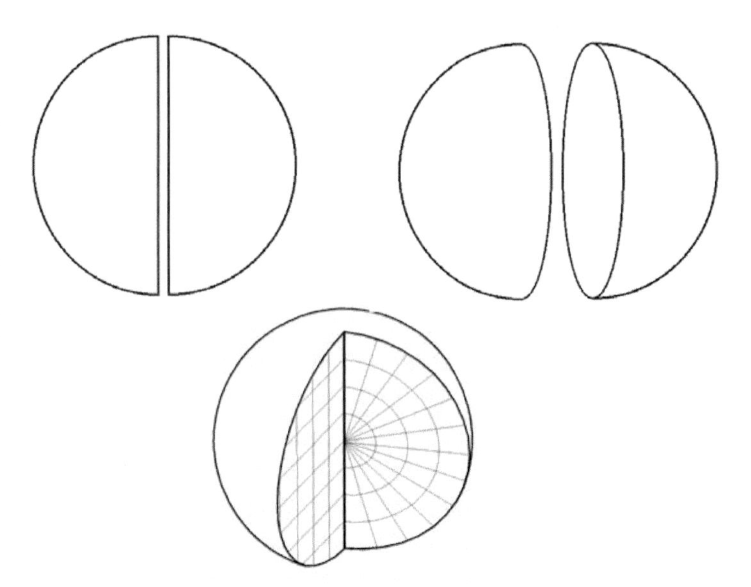

Figure 2-7: *Top*: Slicing a pie, or a grapefruit in half shows the circle and the sphere have symmetry about diameter slices. *Bottom*: Slice of interior of a badly formed fruit highlights the lack of symmetry in real physical objects because of their non-uniform interior.

[8] Symmetry will be discussed in more detail in chapter seven. The word symmetry comes from the Greek *symmetros,* combining syn (together) + metron (measure, from which word metric is derived); so, we have "measure together." The two halves of the pie are the same, so each serves as the same measure of the whole pie; together they make the whole pie.

Rectangular Coordinates Revisited

In the case of a cubical body, we can assume the simplest possible quality forms the body such that it is divided into cubical parts of a particular size which can be as small as needed. These cubical parts can, as already described, become addresses of various locations to arbitrary precision, serving as rectangular coordinates.

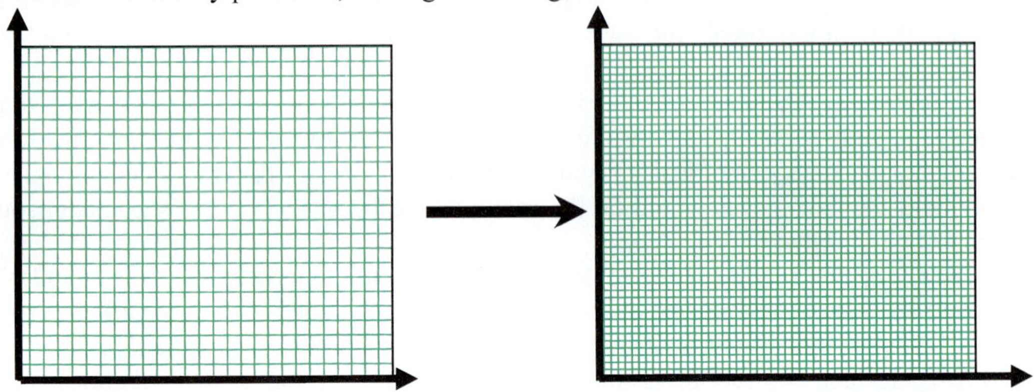

Figure 2-8: *Creating rectangular coordinates in two dimensions.* By continually dividing up space into smaller and smaller squares, one can define position more precisely.

Creating Spherical Coordinates

Similarly, we can assume the simplest possible quality forms a whole spherical body into spherically symmetric parts throughout its interior. For concreteness, we could consider this as an *idealization* of the Earth in which gravity pulls everything together homogeneously. Again, because of the simplicity of such an object, we can make a fairly easily understood coordinate system based on the location of these parts. We assume the parts are as small as we like and, by a purely mental construct, take the parts to be points as we did with the rectangular system. We should always remember that this limit never really happens; one way we remember is by using small quantities. For instance: in dividing up a line, we would consider small Δx pieces.[9] In our present case, we are discussing three dimensional parts, so we need to consider three such small quantities or, said another way, small pieces of the 3-D body. We can construct these spherically symmetric parts in the following way.

Consider the spherical body shown in Figure 2-9. The figure also depicts a concentric sphere of radius smaller than the body. Consider four such concentric spheres with radii spaced so as to divide the sphere into shells of equal thickness. Now, consider dividing the sphere further into an arbitrarily high number N of such shells. In this way, we create an onion effect with each shell of thickness $\Delta r = (Body\ Radius)/N$. By letting N approach infinity, we effectively chop up the *radius* into small pieces of equal length. This then defines one of the needed dimensions; we still need to define the two other dimensions of the little cells that will make up our sphere.

[9] We will later call them differential, in this case dx; however, we will still use small sized, not zero sized things in picturing them, as we must.

We can define a second cell dimension by considering the angle θ shown in Figure 2-9. Each such angle can be imagined existing in a plane chopped through the sphere.

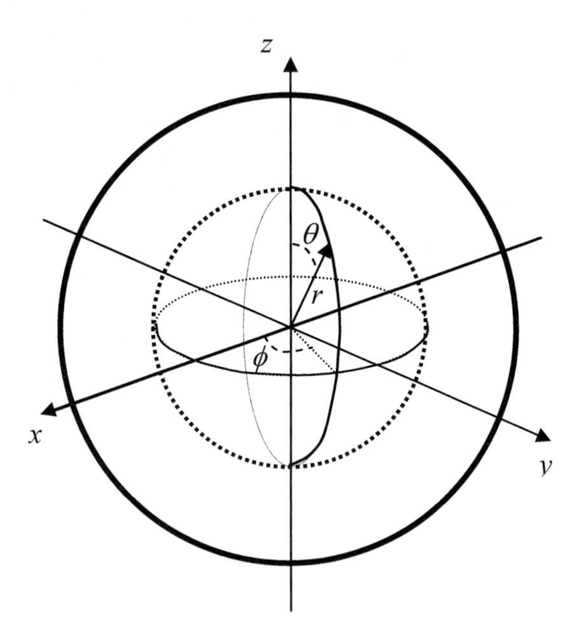

Figure 2-9: *Spherical coordinates in a spherical object.* Note the spherical object with surface of constant radius embedded. A point on the surface of this interior sphere is identified by two angles. The angle measured from the vertical or z-axis is called the co-latitude and is designated by θ, while the angle measured from the x-axis is called the longitude and is designated by ϕ.

Namely, if, after imagining our sphere as divided internally like an onion with more layers than you could ever imagine, we cut the "onion" through the diameter, i.e. through its "north pole," we would see the following pattern.

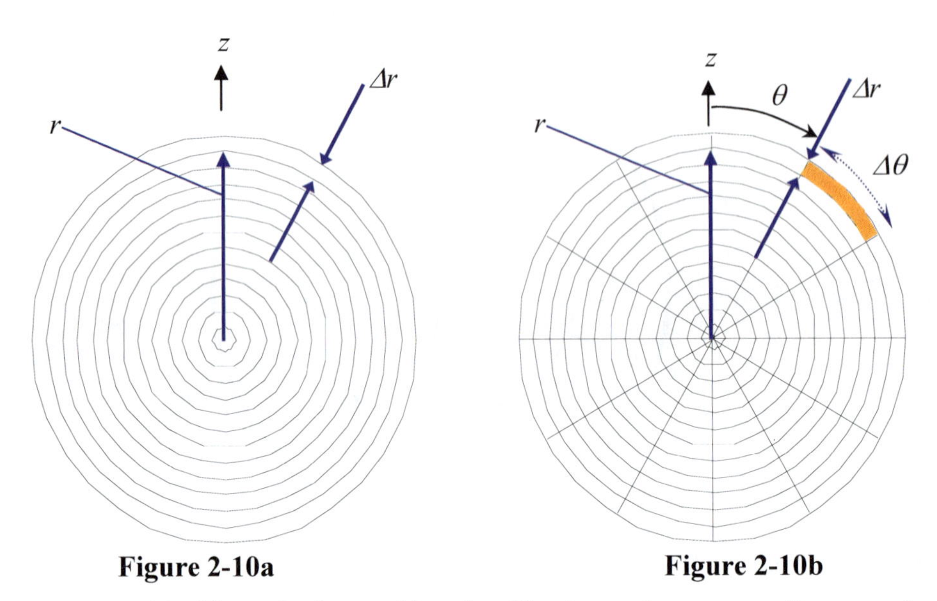

Figure 2-10: Slice of sphere with onion-like-layered structure of concentric spheres inside. The figures show the cells resulting from two sets of divisions. *Left*: *2-10a* Shows the layering corresponding to a subdivision in the radial direction, *r*. *Right*: *2-10b* We divide each such flat ring into equal angle cells so that each ring is composed of two dimensional cells of dimension $r\Delta\theta$ and Δr as shown for the colored cell. The angle around the disk measured clockwise from 12 o'clock is called θ. Notice that, as we increase the number of radial and angular divisions, the cells approach a rectangular structure. In the limit, we specify a point and thus have created our polar coordinate system.

We take the vertical line as the reference ray for the altitude angle θ (also called co-latitude[10]). We see that the length of a cell, such as the colored one shown in Figure 2-10b, is $r\Delta\theta$ and its width is Δr. This creates the spider web-like pattern shown in that figure. As we increase the number of radial layers, i.e. decrease the thickness, Δr, of each layer, and increase the number of angular layers, i.e. decrease $\Delta\theta$, we obtain more and more refinement in our ability to give "addresses" to points on the disk.

Polar Coordinate Interlude

If we, by a purely mental construct, complete this limit, we can label each point by two numbers: r and θ. This is a special coordinate system, a way of labeling every point on a circularly symmetric disk, and is called polar coordinates. Instead of x and y directions, we now have r and θ directions. The x and y-directions are usually called $\hat{x}$

[10] Geographers measure latitude on the Earth from the equator, thus making θ = co-latitude = 90° - latitude .

and $\hat{y}$ respectively. For the angular and radial directions, one uses: $\hat{\theta}$, $\hat{r}$, respectively. As we've mentioned, we want to consider these as vectors, so we need to assign them lengths. As previously explained, since we are mainly interested in specifying the direction, we assign them unit length and call them unit vectors. The relation between the unit vectors for these two coordinate systems, i.e., rectangular and polar, can be written:

$$\begin{aligned} \hat{r} &= \hat{x}\cos\theta + \hat{y}\sin\theta \\ \hat{\theta} &= -\hat{x}\sin\theta + \hat{y}\cos\theta \end{aligned} \tag{2.1}$$

where one can define $\vec{r} = r\hat{r}$ as a vector defined at the origin. (Why these relations obtain is investigated in a problem at the end of the chapter)

Spherical Coordinates Completed

Now, to finish developing our coordinate system for the sphere, we return to the full sphere (i.e., before it was cut in half and looked at in two dimensions) with its small parts (i.e., before they were reduced to points). We, thus, have a sphere made up of rings that come from chopping up the sphere in the θ and radial directions. Namely, at a given average radius, r, and angle, θ, we can vary ϕ and follow along a yet un-partitioned ring. We can picture these rings by mentally creating a solid of revolution by rotating Figure 2-10b about its vertical diameter. Think, for example, of the colored cell in that figure; a given point in that cell will generate a circle as the figure is rotated, and the whole cell will generate a ring. Such a ring is illustrated in Figure 2-11.

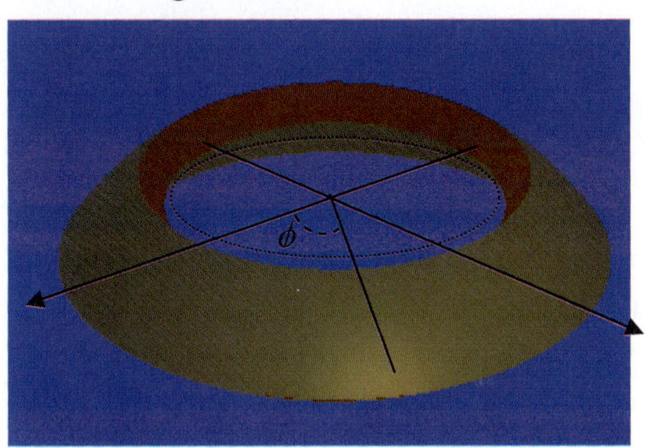

Figure 2-11: A section of a sphere created by chopping a sphere at the boundaries of a fixed radius r and $r+\Delta r$ and at the boundaries of fixed θ and $\theta+\Delta\theta$. Such a cell is a volume of revolution created by rotating a single cell, such as the one colored in Figure 2-10b, about the z-axis. In order to show the ϕ angle, we draw a set of rays parallel to the x and y axes, which intersect at the center of this particular slice.

Each such ring needs to be divided so that we can consider the small cells making up the sphere. Like previous cases, the sphere cannot have different sized cells at different ϕ without violating our assumption that the object is spherically symmetric; hence, we divide each ring into equal parts around the ϕ direction. Thus we have our sphere divided into small cells quantified by three small numbers, Δr, $\Delta\theta$, $\Delta\phi$. Each such cell has flat faces except the r-sliced faces.

By making the cells as small as we please, we can label each region of the interior of the sphere by a given r, θ, ϕ combination. As before we, by a mental construct, take the limit to zero size, and act as if we have actually divided the sphere into points. In this way, we can label each point by the three numbers. This system of labeling or mapping the space is called a *spherical coordinate system*. The three numbers (r, θ, ϕ) tell *how far* out to go and what *direction*, hence they are the "components" of a vector.

If we consider the sphere as an idealized Earth, we can think of the θ as the latitude, ϕ as the longitude and the radius as how far from the center of the Earth we are. Notice, in this way, we can specify the location of any point in or on the Earth. The details of the shape of the cells will become important when we calculate areas and volumes using those cells.

General Facts about Coordinate Systems

Hence, we see in these two examples how we pick coordinate systems. Namely, in the same way that we draw a street map according to the way the houses and streets are actually laid out, we pick our coordinate system by the way the space under consideration is actually laid out. If, for instance, we are in Denver, it is convenient to use a rectangular grid, while if you are in an old city built around a center from which roads radiate outward, a spherical system might be easier.

This street analogy brings out an important point. The point, like the house, is wherever it is, but we can map it in many different ways. In other words, whatever coordinate system we use, the vector[11] (with tail at origin and arrow at the chosen point) is the same, although the numbers used to indicate it will be different. For example, a five unit trek in the z-direction is given in spherical coordinates by specifying that $\theta = 0$ and $r = 5$ (ϕ can remain unspecified), which is radically different from its rectangular representation (0,0,5).

Vector Fields

Now that we have the precise idea of space and how to label it with a coordinate map, we can define a vector field. A *vector field* assigns a vector to every point of space. By this, we again mean that we can always choose a finer and finer partition that will in the limit define a point for us to "place" a vector.

A field of corn is a simple example of a vector field. Each corn stalk points a certain direction toward the sky and is a certain length. Each corn stalk can be replaced by a vector representing the length and direction of the corn growing out of place on the ground. In this way, one can imagine a plane with arrows (vectors) at many points representing the corn. Such a construct is illustrated in one dimension in Figure 2-12. One could imagine making finer and finer corn stalks until one filled more and more of the field, and, in the limit, one has literally a field of vectors. All parts of our two dimensional field thus have vectors.

---

[11] Here we specify the vector to reside in a particular place, but this is not necessary. The magnitude and length of the vector can simply be defined by the vector in that place and then simply thought to be anywhere one wants. Indeed, this can be accomplished by just moving the origin wherever one wants.

Figure 2-12: Top: *Corn field illustrates vector field.* Each region in a plane (i.e. the surface of the Earth) is associated with a vector (i.e. a stalk of corn). Bottom: Schematic illustration of stalks in a row, illustrating vectors assigned to regions of a line. The extension of the stalk can be considered to be the body of the arrow and its tassel can be thought of as the arrow's head. In the limit of closer and closer spacing of corn stalks, one can assign a vector (stalk) to every point on the field.

Note that though, unlike our generic definition of a vector, a vector field associates a vector with a place, this still does not *generally* mean that the vector resides in the space as an arrow; indeed, it still simply represents the fact that something of given magnitude and direction is associated with that point. Our corn example above had the arrows really existing, i.e., as the actual extension of a corn stalk, and the tassel of the corn being the arrow's head. Let's consider another example that is not so literally associated with arrows in the space.

Consider players converging around a soccer goal. We can draw an arrow from each player to the goal. The vector (arrow) indicates the direction the player must kick to make a goal, and the length of the vector is how far the ball must travel. Since players can kick from any point on the field, this method "creates" a vector for every point on the field; that is, it defines a vector field. Notice that, unlike above, the arrows we might draw to represent the vectors cannot be real bodies in the space, because all of them would have to exist in the same place of at least part of other arrows. Thus, remembering that a vector is an analogically general concept, which is modeled after an arrow but is not simply the same as an arrow, will help us use such pictures without confusion.

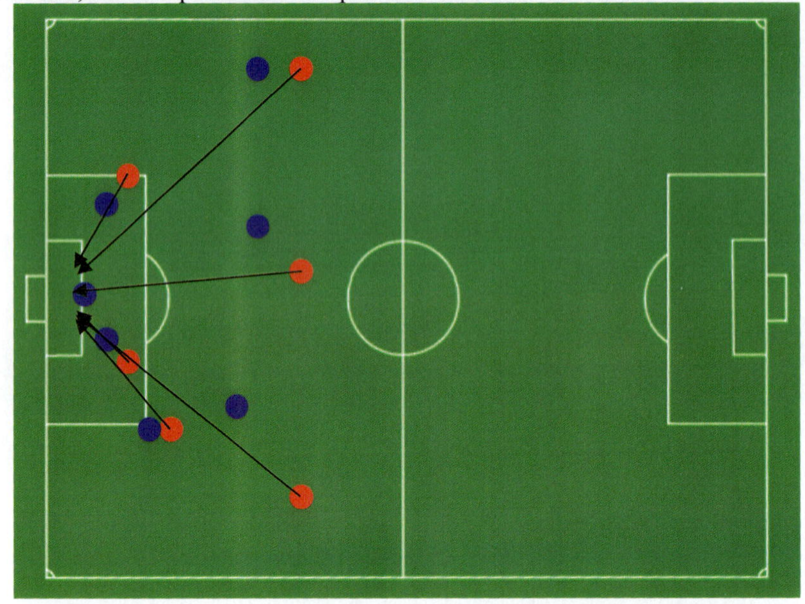

Figure 2-13: *A soccer vector field.* A vector field can be defined on a soccer field by defining the vector that is the distance and length from any given point to the goal. The red players are shown here with arrows indicating how they must kick the ball to make a goal from their given location.

Another more abstract example of a vector field comes from studying a steadily flowing river with leaves floating on it. We note that the speed and motion of the leaves vary. The leaves up river are moving faster and towards the banks, while those down river are moving slower and in a straighter path down river. Meanwhile, those on the near side are faster than those on the far side. Each leaf has a speed and a direction. Thus, by noting the speed and direction of a leaf as it passes a given point, one can assign a velocity vector to each point along the path of the leaf. One can easily imagine extending this process to every point of the river using smaller and smaller leaves to get finer measures of the river's velocity.

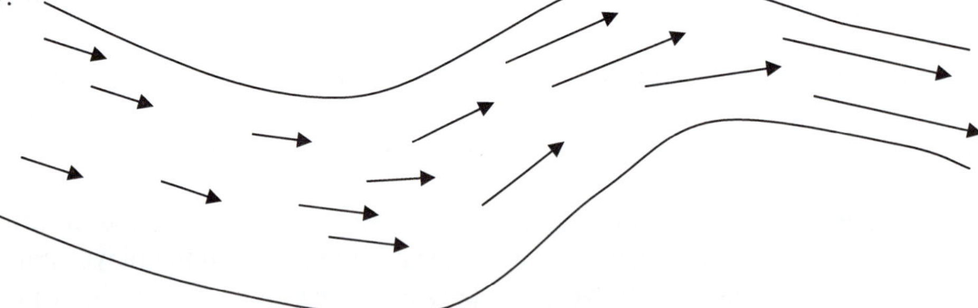

Figure 2-14: *Velocity field of a river.* The velocity associated with leaves floating on a river can be used to define a vector field on the surface of a river. Such a field is called the velocity field of the fluid flow. We define the flow, but we could also choose to follow a given leaf and define the flow in that way.

Labeling and Manipulating Vector Fields

As we've added more structure to our understanding of vectors, a brief word about vector fields and coordinates is needed. Vector fields require, in addition to the three numbers required by a vector to label the direction and magnitude, three numbers to label the location of the particular vector. This can be written $\vec{V}(x,y,z)$ in Cartesian coordinates.

Whether a vector is assigned to a point or not, we've seen its representation changes with changes in the coordinate system. We illustrate this principle in Figure 2-15 below. This time, however, we do so not by moving between spherical and rectangular coordinates, but by rotating one rectangular coordinate system relative to another.

Notice that, by contrast, if we simply used a rectangular coordinate system that was shifted from the first, this would not affect the representation, just the location that the vector is assigned to *in the coordinate system.* And again, in all such *passive* translations, note that we do not move the vector, but only our labeling system, the coordinates.

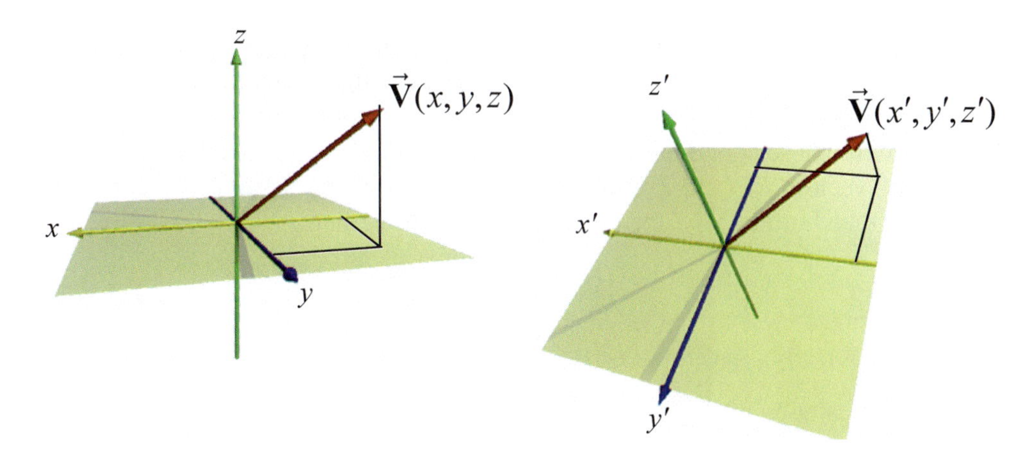

Figure 2-15: *Rotation of coordinate system.* The same vector (shown in red) at the same point in space is shown in two different rectangular coordinate systems. Note that the vector has different components in the two systems. The coordinate system on the *right* (labeled: x', y', z') is rotated relative to the one on the *left* (labeled: x, y, z). In each case, the *x-y* plane is highlighted in yellow to facilitate the illustration of the rotation and of the components.

In general, one could have a rotation *and* a shift. In such a case, to get to the second coordinate system from the first, one would rotate the axes of the first and then shift the origin by a certain amount.

Vector Addition and Subtraction

Having assigned symbols and numbers to vectors in an algebra-like fashion, you may recall that vectors can be added and multiplied like variables in algebra. To consider the most basic operation, addition, and its opposite, subtraction,[12] we first consider an example close to the primary definition of vector, a ray, because the analogous usages will follow it.

Specifically, consider the problem of finding the *net distance* and *direction* a man travels if he makes a journey in two simple parts. Assume he first travels so many feet, L_1, in one direction and then so many feet, L_2, in another.[13] The problem is simple if he confines himself to motion along a line. Assigning forward motion positive units and backward motions negative units, we can say the distance he covered is:

2.2
$$L_{Total} = L_1 + L_2$$

So, for instance, if our traveler walked two steps forward and one step back, we have: $L_{Total} = 2 - 1 = 1$ or a net distance traveled of 1 step forward.

However, if the traveler is allowed to move in all three dimensions (for example, he may climb mountains or fly in a plane as well as move along the ground), we need three equations like equation 2.2. To find the net travel, we must consider each of the three perpendicular directions separately. In saying this, we bring out a fundamental property of three dimensional Euclidean spaces; namely, there are three linear dimensions that cannot extend into each other. Each such dimension behaves just like any one of the others, so we can break down the motion into the different independent directions to get the equations needed.

We can write these equations succinctly, and see how they come about, by recalling the definition of vectors L_1 (and L_2) in rectangular coordinates. Listing the components vertically, rather than horizontally, the components can be written:

2.3
$$\vec{L}_1 = \begin{pmatrix} L_{1x} \\ L_{1y} \\ L_{1z} \end{pmatrix} \qquad \vec{L}_2 = \begin{pmatrix} L_{2x} \\ L_{2y} \\ L_{2z} \end{pmatrix}$$

In thinking about the meaning of these components, remember that vectors *of themselves* are not assigned a place; hence, though we define them in the representation above with the tail at the origin and the arrow at the point given in parenthesis for convenience, they can be moved anywhere.

Now, treating each component just as the one dimensional case gives:

2.4
$$\begin{aligned} L_{Total\,x} &= L_{1x} + L_{2x} \\ L_{Total\,y} &= L_{1y} + L_{2y} \\ L_{Total\,z} &= L_{1z} + L_{2z} \end{aligned}$$

---

[12] In fact, in principle, division is the most basic operation, but it remains true that we first learn how to count, before we see the meaning of counting. By contrast, in terms of meaning (principles), first comes unity or one, then we see we can divide the one object into two parts, so that we have a whole with two parts; we then see, by iteration, that we can repeat this process each time getting a new whole that has an additional part, first three then four and on and on. Thus, from this dividing we understand addition, from there, we see subtraction, the reverse process. We next see multiplication, which is simply a special type of addition; two times four means add two four times.

[13] For simplicity, assume the curvature of the Earth can be ignored; e.g., you are told he doesn't go far enough for the curvature of the Earth to be important.

Writing this in vector notation gives the following simple generalization of equation 2.2 above:

2.5 $$\vec{L}_{Total} = \vec{L}_1 + \vec{L}_2$$

Figure 2-16 below elucidates the meaning of this addition.

Figure 2-16: ***Left***: *Result of a man walking forward and then backward in a line.* Note these vectors are shown as simultaneously present arrows, whereas the man actually traversed the distance in time, thus distinguishing the actual situation from the primary analogate, i.e. actual length and direction at a given moment. Though the arrows representing $\vec{L}_1$ and $\vec{L}_2$ in this first picture are separate, one could simply use a single line segment for both arrows, distinguishing them by the location of the arrow heads. To understand the meaning of this in an actual physical situation, consider again two walls meeting at a corner (cf. Figure 2-1); however, this time let the right wall, but not the left wall, have a distinguishing mark drawn at the intersection of the two walls at some height. The intersection of the surfaces of the two walls is a line; the entire boundary of the surface of the left wall, with an associated upward direction, generates $\vec{L}_1$, and the boundary of the right wall from the ceiling *down to the mark* and pointing down generates $\vec{L}_2$. Alternatively, we could simply imagine one arrow present at one moment followed by the second, which is more like the actual case in which the man *first* traverses the path illustrated by L_1, *then* the one given by L_2. ***Right***: *Two dimensional addition of vectors*. The man walks northwest then southwest, giving the net ground covered shown by $\vec{L}_{Total}$. In this case, unlike the first, physical *3-dimensional* arrows can represent the paths.

Thus, each of the two portions of the man's journey naturally defines a vector. To more completely understand, consider a concrete case. Assume the traveler first moves northwest for 8 miles then southwest for 8 miles. Taking, as shown below, *north* as the *y*-direction and *up* as the *z*-direction (not shown) makes the *x*-direction *east*. This means we can, using simple trigonometry, write:

2.6
$$\vec{L}_1 = \begin{pmatrix} -\frac{8}{\sqrt{2}} \\ \frac{8}{\sqrt{2}} \\ 0 \end{pmatrix}, \quad \vec{L}_2 = \begin{pmatrix} -\frac{8}{\sqrt{2}} \\ -\frac{8}{\sqrt{2}} \\ 0 \end{pmatrix}$$

y (North)
x (East)
$\vec{L}_1$
$\vec{L}_2$
$\vec{L}_{Total}$

$$\vec{L}_{Total} = \vec{L}_1 + \vec{L}_2 = \begin{pmatrix} -16/\sqrt{2} \\ 0 \\ 0 \end{pmatrix}$$

These vectors are drawn in the figure above as if they were associated with the origin. The coordinate notation "forces" this interpretation. Yet, since the *general* notion of a vector doesn't include a reference to a place, it doesn't really make any difference where we imagine the vectors to be; *generally speaking*, they may just as well all be at the origin from that point of view. In some particular cases such as our traveling man, however, once a coordinate system is defined, the arrows *are* associated with particular coordinates.

Because the traveling man's vectors actually exist in the physical space (i.e., they are actual paths along the ground), these results were straightforward. And, because we have established these results for length vectors, we can extend them to all analogical vectors, i.e. those vectors built by analogy from the first quantitative (simple extension) meaning of the word vector. Vector notation then gives us a way to treat the similarities of all the various analogous meanings of vectors in a common formalism. Every analogous use is grounded in the primary one, and this is seen in the formalism, especially through use of arrow pictures. More specifically, the notation allows us to treat each vector as a single object to be manipulated by certain rules.

Indeed, each vector is considered a single object when treated as part of the mathematical formalism.[14] For example, we no longer need to think of the components explicitly when doing the operations of addition and subtraction. We can just apply the commutative, associative, transitive and all the other rules (axioms) of addition directly to the vectors without considering the components, and they will all be true. For example, $\vec{L}_2 + \vec{L}_1 = \vec{L}_1 + \vec{L}_2$.

Vector Multiplication: Dot and Cross Products

We can further develop our formalism by working more on the analogy between vector operators and numbers operators. Specifically, we would like to define a vector operation that is like multiplication.

To find an analogy, first note that in multiplying two numbers (also called scalars[15]), we are dealing with a number *line*. We start with two numbers somewhere along the line and the product yields a third. By contrast, when considering the multiplication of two vectors $\vec{A}$ and $\vec{B}$, we think of the plane naturally formed by these two vectors; this

[14] Of course, one cannot have a real thing that simply is magnitude and direction in the generalized sense that we refer to within the vector formalism. Yet, we can, via a mental construct, carry it around in a package as if it were one entity as we did with completed limits and completed infinity earlier.

[15] After defining vectors as entities with magnitude and direction, it is natural to want a name to specify entities that just refer to magnitudes (which are numbers); we refer to them as scalars.

plane can be visualized, for instance, by moving the vectors so that their tails coincide. If the vectors happen to be parallel, we obviously have a special case, called a degenerate case.

In the non-degenerate case, we notice that there are two possibilities for reducing the vectors to a one dimensional arena, which may, at least, allow us to handle the magnitudes. Pick one of the two vectors to be multiplied, say $\vec{A}$, as a baseline; two possibilities then present themselves. We can multiply the magnitude of $\vec{A}$ by: 1) the component of $\vec{B}$ *parallel* to $\vec{A}$ or 2) the component of $\vec{B}$ *perpendicular* to $\vec{A}$. The first yields the *dot product,* the second the *cross product.*

Dot Product

The dot product, thus, could also be called the parallel product. We define this first kind of vector multiplication as:

$$\vec{A} = \begin{pmatrix} A_x \\ A_y \\ A_z \end{pmatrix} \qquad \vec{B} = \begin{pmatrix} B_x \\ B_y \\ B_z \end{pmatrix}$$

2.7
$$\vec{A} \cdot \vec{B} = A_x B_x + A_y B_y + A_z B_z$$
$$= A\,B\cos(\angle\,\vec{A}\vec{B})$$

where, in the third line, $A \equiv |A| = \sqrt{A_x^{\,2} + A_y^{\,2} + A_z^{\,2}}$, likewise for B

The right hand side shows that the result of the dot product of two vectors is *not a vector*, unlike the case for multiplication of two numbers, which is a number. The definition given in the second line gives: $\vec{A} \cdot \vec{A} = A^2$, which we like because we expect multiplication of A by itself to give A^2. (*Exercise:* show that the magnitude of a three dimensional vector is given by the fourth line of equations 2.7).

The most important insight to be gleaned from equation 2.7 is that the dot product *projects* one vector onto the other. This simply means the equation of the definition considers only the component of one vector that is parallel to the other; that is, it does just what we sought to do. To see this, we can write equation 2.7 in the following suggestive form:

2.8
$$\vec{A} \cdot \vec{B} = A\big(B\cos(\angle\,AB)\big) = A \cdot B_{\|} \quad \textit{where } B_{\|} \textit{ is the component of } \vec{B} \textit{ parallel to } \vec{A}$$

OR

$$\vec{A} \cdot \vec{B} = B\big(A\cos(\angle\,AB)\big) = A_{\|} \cdot B \quad \textit{where } A_{\|} \textit{ is the component of } \vec{A} \textit{ parallel to } \vec{B}$$

To understand such *projection* more concretely, consider a stick poking out of the ground at some angle leaning toward the east. Assuming that the sun is directly overhead, a shadow of the stick appears on the ground. To begin to make the connection with the dot product, lay a second stick on the ground as shown in Figure 2-17. Hence, there appears, as shown in Figure 2-15, a "projection" of the stick's length and pointy end (the direction) onto the second pointy stick (vector) pointing east on the ground.

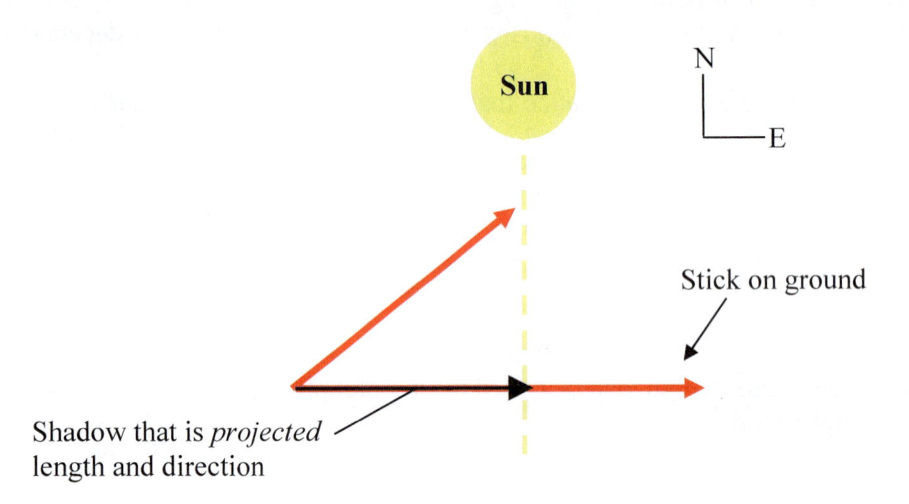

Figure 2-17: Two red sticks illustrate the concept of *projection* used in understanding the dot product. The upper stick casts a shadow on the lower stick.

To complete the analogy consider a simple dot product: $\vec{A}\cdot\hat{x}$. In terms of rectangular coordinates we can write:

2.9
$$\vec{A}=\begin{pmatrix}A_x\\A_y\\A_z\end{pmatrix},\ \hat{x}=\begin{pmatrix}1\\0\\0\end{pmatrix}$$
$$\vec{A}\cdot\hat{x}=A_x$$

The last line is obtained by applying the definition in equation 2.7.

Now, take the tilted stick to be the $\vec{A}$ vector and $\hat{x}$ to be the ground vector, we can see that $\vec{A}\cdot\hat{x}$ is simply how far the stick extends in the *x*-direction. Doing the component multiplication also shows that $\vec{A}\cdot\hat{x}$ just projects out the *x* component of $\vec{A}$, A_x. Of course, since there is nothing special about our choice of one direction as the *x*-direction, such reasoning applies just as well to the *y* and *z* direction (i.e., for $\vec{A}\cdot\hat{y}$ and $\vec{A}\cdot\hat{z}$).

Notice that, in the case we've shown in Figure 2-17, the stick we project onto (which is one unit long) is longer than the shadow projected. However, this makes no difference, for we only project vector $\vec{A}$ onto $\vec{B}$ (or the reverse) to get the two vectors on the same line; once we've done so we can multiply the projected magnitude of the one by the magnitude of the other to get the dot product. Hence, if one vector is a unit vector, the dot multiplication of it and second vector reproduces the length of the second along the direction of the unit vector, as we just saw for the case of $\hat{x}$.

Now, in practical circumstances, one often wants to know what part of one vector lies along another. For example, suppose you go on a trip traveling *northwest* for some distance, it is natural to wonder how far *north* you will go (after all it gets colder as you go north). This can be found by dotting the trip vector, $\vec{T}$, with the northerly directed unit vector (say, $\hat{n}$ *giving* : $\vec{T}\cdot\hat{n}$).

Also, given a force field radiating from the origin, i.e. a force that pulls with a certain magnitude toward the origin at every point, we might want to know what the force directed say in the *x*-direction. Given, for instance, the force vector, in some coordinate system, at a given point, we could obtain this answer by the dot product: $\vec{F}\cdot\hat{x}=F_x$.

Since the unit vectors have special significance (they specify independent directions and the magnitude (i.e. unity) by which all others can be compared), it is helpful to look at the dot product of the all the independent unit vectors in rectangular coordinates.

$$\begin{array}{lll} \hat{x}\cdot\hat{x}=1 & \hat{x}\cdot\hat{y}=0 & \hat{x}\cdot\hat{z}=0 \\ \hat{y}\cdot\hat{x}=0 & \hat{y}\cdot\hat{y}=1 & \hat{y}\cdot\hat{z}=0 \\ \hat{z}\cdot\hat{x}=0 & \hat{z}\cdot\hat{y}=0 & \hat{z}\cdot\hat{z}=1 \end{array} \qquad (2.10)$$

These relations are of primary importance, since any vector we write is made out of linear combinations of these three unit vectors. That is, any vector can be written as $\vec{A}=a\hat{x}+b\hat{y}+c\hat{z}$ for the appropriate choice of variables *a*, *b* and *c*. Hence, equations 2.10 give a complete description of the dot product and it, again, boils down to projecting the parallel component of one onto the other, then multiplying. For example, dotting $\vec{A}=\hat{x}+3\hat{y}$ into $\vec{B}=2\hat{y}$ gives 6, because the *x*-component of $\vec{A}$ gets projected out, leaving $3\times2=6$.

A First Principles Explanation of Dot Product

To see this more clearly let's back up to first principles. In constructing the vector formalism, we are making an analogy with numbers, and since the unit is the principle of number, in our analogy the *unit vector* will be primary. We can see this by noting that if we are given a direction along with a unit of length, we can make any length vector by simply using a certain number of those unit lengths or a certain fraction of those units (including those "fractions" that are limits that we call irrational numbers). Thus, any vector, for instance, along the *x*-direction can be constructed by multiplying the unit vector $\hat{x}$ by the appropriate number; or mathematically: $\vec{V}=V\hat{x}$. As we've seen, since there are three independent directions in space, any vector can be made by thinking of each vector as composed of the extent it spans in each of those three directions. Said another way, any vector can be written as the sum of appropriately weighted three unit vectors: $\hat{x},\hat{y},\hat{z}$; that is, as we've already noted: $\vec{A}=a\hat{x}+b\hat{y}+c\hat{z}$. We begin to see that the dot or parallel product is implicit in the nature of vectors; projection (finding the parallel component) onto the various unit vectors is important in the understanding and representation of vectors.

Now, there is nothing special about the choice of the direction of the *x*, *y* and *z* axes, other than being mutually perpendicular and satisfying the right hand rule. In empty space, where we've left behind all distinctions between the various directions, it's only the definition of those axes themselves that break the bland uniformity of space. Any other directions could be chosen, and we can, in addition, choose any other vector (s), i.e. magnitude and direction. Thus, the *unit vector* of such a new direction, though not one of the rectangular coordinate directions, would still be important because it is by taking real number multiples of it that we can get all vectors in that direction. For instance, a vector $\vec{A}$, can be written as a real number, A, times the unit vector $\hat{A}$ (i.e. the vector that points along $\vec{A}$ with unit magnitude) which can be written symbolically as: $\vec{A} = A\hat{A}$. This notation is suggestive of the very definition of a vector, where A gives the magnitude and $\hat{A}$ indicates its direction. In analogy to the one dimensional absolute value, which picks out the number itself, not which side of zero it's on, we use $\left|\vec{A}\right|$ to represent the magnitude of a vector, leaving out the vector's direction; we write: $\left|\vec{A}\right| = A$.

Given our coordinate definitions, we will want to know how this vector, $\vec{A}$, relates to our coordinate vectors. This is similar to when we wanted to know how far *north* the man traveling *northwest* went. In terms of this example, the components in the coordinate system specify how far we traverse in the x and y directions, when we traverse the path of vector $\vec{A}$, i.e., when we move A units in the $\hat{A}$ direction. For instance, in the x-direction, A_x units of length must be laid along the x-axis to cover the distance spanned by the vector $\vec{A}$ in that direction. Again, for this reason, we say we *project* $\vec{A}$ onto $\hat{x}$ (cf. Figure 2-17), and write: $\vec{A}\cdot\hat{x}$. Figure 2-18 shows how the vectors mentioned just before the beginning of this section, $\vec{A} = \hat{x} + 3\hat{y}$ and $\vec{B} = 2\hat{y}$, are resolved into components and into unit vectors and magnitudes.

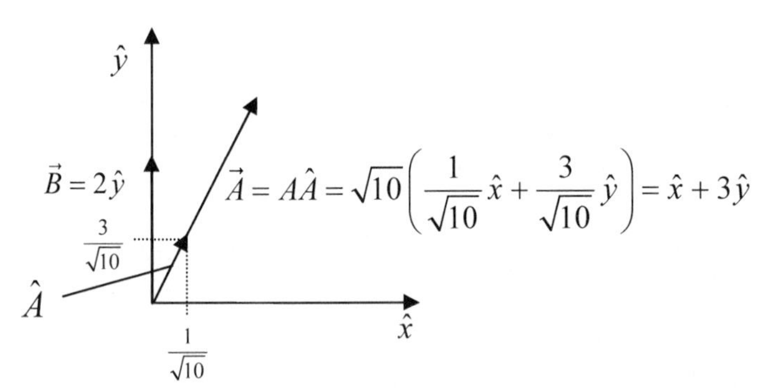

Figure 2-18: The Vector $\vec{A}$ is in the direction $\hat{A}$ and has the length that spans $A = \sqrt{10}$ units. Vectors $\vec{A}$ and $\vec{B}$, like any vector, can also be viewed according to how many units they span in the mutually independent directions: $\hat{x}$ and $\hat{y}$. In such a decomposition, the unit vector in the $\vec{A}$ direction is: $\hat{A} = \frac{1}{\sqrt{10}}\hat{x} + \frac{3}{\sqrt{10}}\hat{y}$.

Suppose instead of projecting $\vec{A}$ onto a unit vector, we projected it on a vector with some non-unit length; say $\vec{B} = B\hat{B}$, then: we write $\vec{A}\cdot\vec{B} = A\hat{A}\cdot B\hat{B} = AB\left(\hat{A}\cdot\hat{B}\right)$. In this case, instead of multiplying A by the unit length, we multiply by the length of $\vec{B}$ (which is, by definition B units). This is the generalization of multiplication we sought. Namely, the parallel multiplication of $\vec{A}$ with $\vec{B}$ is the multiplication of the number of units that $\vec{A}$ spans in the direction parallel to $\vec{B}$ ***times*** the magnitude of $\vec{B}$. Note that the parallel product of the two *unit* vectors, $\hat{A}\cdot\hat{B}$, carries the new information that makes the product $\vec{A}\cdot\vec{B}$ different from standard multiplication along a number line.

Note finally, as we've said, that, when we multiply in this way, the product is not a vector. We have multiplied two vectors, i.e. two directions with associated lengths, and the result is a magnitude with no direction, what we've called a scalar. This is because, in the dot product, it makes no difference which direction we project from or onto (mathematically, we say: $\vec{A}\cdot\vec{B}=\vec{B}\cdot\vec{A}$); hence, there is no reason to keep, nor help in keeping, a mention of the direction in the product.

In this way, we see clearly that the real quantitative content caught in the dot product formalism comes from the analogical multiplication that arises by projecting one vector onto another. Again, because of the definition, the dot product is very like arithmetic multiplication; it is, for instance, commutative, associative, and distributive with respect to addition. However, we have only caught part of the story for we've left out what happens with the perpendicular component, the part not projected.

Cross Product

This is where the second type of vector multiplication, the *cross product* (or we could call it the perpendicular product) needs to be considered. Unlike the dot product, the cross product of two vectors gives a vector. In the case of the cross product, we will see that it matters whether we cross $\vec{A}$ into $\vec{B}$, $\vec{A}\times\vec{B}$, or vice versa, $\vec{B}\times\vec{A}$, and that there are two opposing directions naturally associated with each of these products. The dot product reduces to an operation on *either* one of two lines, whereas the cross product, because it uses the perpendicular component, generally defines a unique plane (or, more properly, if one considers that vectors have no place, a unique series of parallel planes) which, in turn, has an orientation, i.e. a direction associated with it.

Whereas the dot product just helps one capture some aspect of 2-D space, the cross product captures something of the 3-D space that the two vectors "live" in. We say, "*A* cross *B*" and write:

2.11 $$\vec{A}\times\vec{B}=AB\,\sin(\angle AB)\,\hat{n}$$

Here, $\hat{n}$ means the direction perpendicular to the plane formed by $\vec{A}$ into $\vec{B}$ given by the *right hand rule*. The *right hand rule* is defined in the following way. Hold your ***right*** hand as if you were hitchhiking, but only curl your fingers lightly. Then, by closing your grip, but keeping your thumb fixed in place, use your *fingertips* to push $\vec{A}$ into $\vec{B}$ as shown in Figure 2-19; the direction your thumb points is the direction $\hat{n}$ (or $\hat{\perp}$) of $\vec{A}\times\vec{B}$. In this way, $\vec{A}$ is pushed into $\vec{B}$ within the plane that they define. In specifying this plane through the vector perpendicular to it, also called its *normal vector*, we bring in the third dimension.

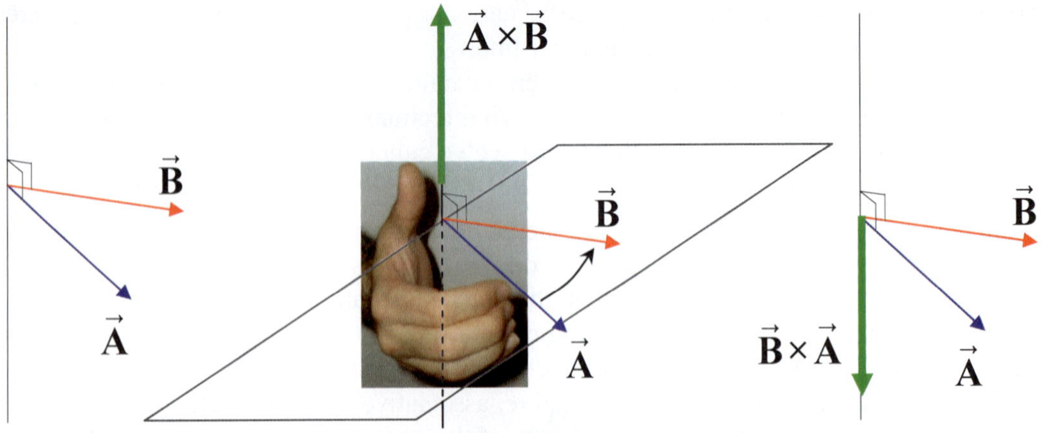

Figure 2-19: *Calculation of cross product direction using right hand rule.* With your right hand as shown, push vector $\vec{A}$ into $\vec{B}$; the direction your thumb points is the direction of the cross product $\vec{A}\times\vec{B}$.

Indeed, from this, we get an insight into the nature of three dimensions; namely, once we try to define a circular motion we, of necessity, single out an axis of rotation. This is a property of 3-D space that we see in our imaginations (which originally came from our senses). The right hand rule itself is one way of capturing this in our formalism. The rule also captures an orientation along the direction of the normal vector, which is a reality in space as long as the vectors have different lengths (which distinguish one from the other); otherwise the orientation is introduced by our choosing. However, the particular direction assigned to the orientation is conventional with respect to the space, since we could have as easily made a left hand rule. It does reflect the incidental fact that most people are right handed,[16] though the fact that most people have a hand/arm that is dominant obviously will not be relevant to the simple physical situations that we will deal with in this course. However, we should note that there is a deep asymmetry with respect to right and left that is a part of physics, organic chemistry and biology, which means, at some point, we will have to deal with this handedness which we have here treated as convention.

To further understand the cross product, we make use of rectangular (Cartesian) coordinates and the definition given in equation 2.11 above (especially, the right hand rule) to obtain the cross product of all the various Cartesian unit vectors among themselves.

2.12a

$$\hat{x}\times\hat{y}=\hat{z}$$
$$\hat{z}\times\hat{x}=\hat{y}$$
$$\hat{y}\times\hat{z}=\hat{x}$$
$$\hat{x}\times\hat{x}=0,\ \ \hat{y}\times\hat{y}=0,\ \ \hat{z}\times\hat{z}=0$$

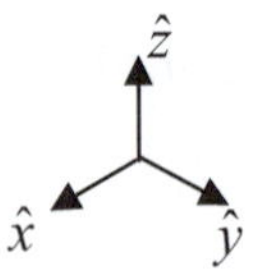

Commuting these gives minus signs; namely:

[16] Hence, the direction we write and other standard conventions. Because of this, left handed people have many extra adaptations. For example, to see what they are writing, they cannot write in the same way as right handers; some, for instance, will write hook-handed to keep their hands out of their line of sight.

2.12b
$$\hat{y}\times\hat{x}=-\hat{z}$$
$$\hat{x}\times\hat{z}=-\hat{y}$$
$$\hat{z}\times\hat{y}=-\hat{x}$$

A close look at the cross product definition explains these results, and it's partially because the definition is the sought after perpendicular product. Rewriting 2.11 we get:

2.13

$$\vec{A}\times\vec{B}=A\left(B\sin(\angle AB)\right)\hat{n}=A\cdot B_{\perp}\hat{n}\quad \textit{where } B_{\perp} \textit{ is the component of } \vec{B} \textit{ perpendicular to } \vec{A}$$

OR

$$\vec{A}\times\vec{B}=B\left(A\sin(\angle AB)\right)\hat{n}=A_{\perp}\cdot B\,\hat{n}\quad \textit{where } A_{\perp} \textit{ is the component of } \vec{A} \textit{ perpendicular to } \vec{B}$$

Figure 2-20 shows the perpendicular component, and simple trigonometry shows us the origin of the sine (and for that matter the cosine of the dot product).

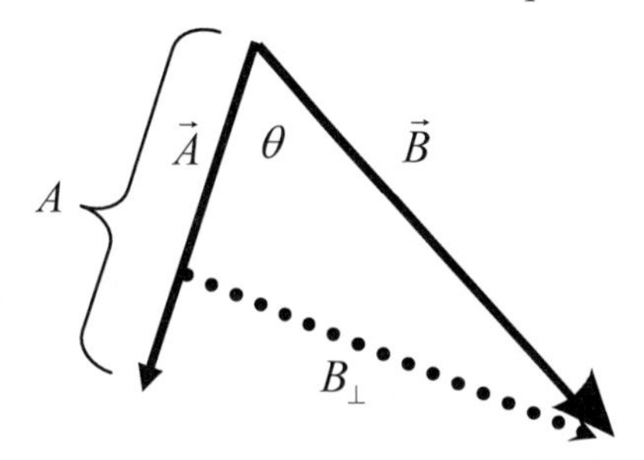

Figure 2-20: Two vectors, $\vec{A}$ *and* $\vec{B}$, are shown making an angle θ, i.e. $\theta\equiv\angle AB$. The component of $\vec{B}$ perpendicular to $\vec{A}$, written $B_{\perp}$, is used in forming a magnitude for the cross product: $\left|\vec{A}\times\vec{B}\right|=A\,B_{\perp}$.

Given this, it is clear that two vectors that have the same direction will yield zero cross product because there is no perpendicular component. The other relations follow from the fact that the vectors are unit vectors and from the right hand rule.

Using the relations 2.12, we can look at a couple of particulars from a different perspective. If, for example, we cross $3\hat{x}$ into $\hat{y}$, we get $3\hat{z}$ which is simply like pushing $\hat{x}$ into $\hat{y}$ three times, analogous to regular multiplication. If we take $\vec{A}=\hat{x}+42\hat{y}$ and calculate $\vec{A}\times\hat{y}$, we get $\hat{z}$. This is because the part that is parallel to $\hat{y}$ contributes nothing (or seen another way, no circle can be defined by only one vector ($\hat{V}\times\hat{V}=0$)); thus only $\hat{x}\times\hat{y}$ survives in the product.

Focusing on our definition of the cross product in the form given in 2.13, we notice that the *magnitude of the cross product is the area of the parallelogram* formed by the vectors $\vec{A}$ and $\vec{B}$ (see Figure 2-21). Recall that vectors (of themselves) are not associated with a place, so we can, without contradiction, think of the vectors as moved and reproduced in the needed way to create the parallelograms in Figure 2-21.

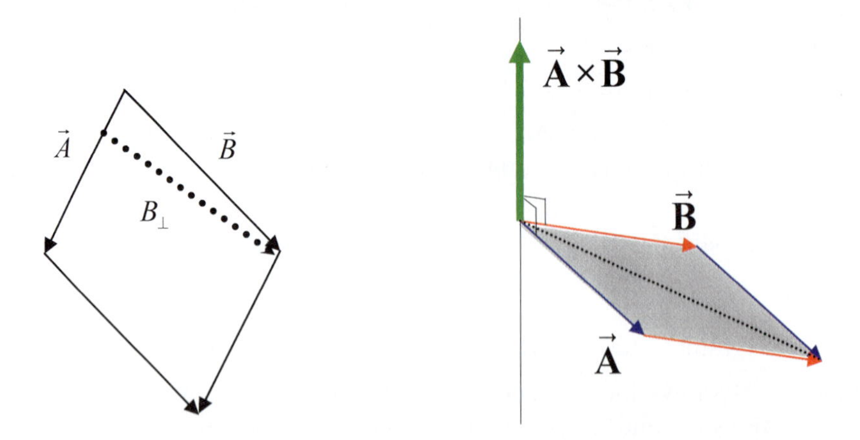

Figure 2-21: The magnitude of cross product of two vectors is equal to the area of the parallelogram formed by the two vectors. *Left*: The area of a parallelogram is the length of an edge times the height (i.e. length of perpendicular line to top of the parallelogram) measured from the edge. *Right*: Cross product of vectors $\vec{A}$ *and* $\vec{B}$ is given by the green arrow of magnitude equal to the area of the parallelogram shaded in the figure.

Measuring Volume using Dot and Cross Products

The area given by the cross product can, when coupled with the dot product, give the volume associated with a solid figure naturally formed by three vectors. Using the rules of pushing one vector into the other given in equations 2.12 and simple algebraic rules, one can show (see chapter-end problem) that the product $\vec{A}\times\vec{B}$ can be written in determinant form as:

2.14 $$\vec{A}\times\vec{B} = (A_yB_z\text{-}A_zB_y)\hat{x} + (A_zB_x\text{-}A_xB_z)\hat{y} + (A_xB_y\text{-}A_yB_x)\hat{z} = \begin{vmatrix} \hat{x} & \hat{y} & \hat{z} \\ A_x & A_y & A_z \\ B_x & B_y & B_z \end{vmatrix}$$

With this, we can show that $\vec{C}\cdot\vec{A}\times\vec{B}$ is the volume of a solid composed of parallelogram sides made from $\vec{A}$, $\vec{B}$ *and* $\vec{C}$ shown in Figure 2-22 below.

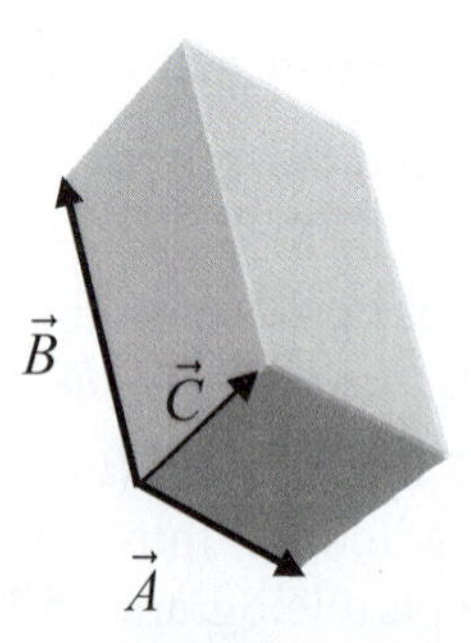

Figure 2-22: The volume of the figure formed by $\vec{A}, \vec{B}$ *and* $\vec{C}$ above is given by $\vec{C}\cdot\vec{A}\times\vec{B}$.

The area of the base parallelogram is the magnitude of $\vec{A} \times \vec{B}$ and the direction is the perpendicular, while $\vec{C}$ is a vector that we can move to the joining point of $\vec{A}$ and $\vec{B}$. Doting $\vec{C}$ into the perpendicular at that point gives the height of the solid, so that the result is height times the area of the base, which is the volume of the solid figure shown in Figure 2-22. We can write this as:

2.15

$$Volume = (\vec{A} \times \vec{B}) \cdot C = \begin{vmatrix} A_x & A_y & A_z \\ B_x & B_y & B_z \\ C_x & C_y & C_z \end{vmatrix} = (\mathrm{B_y C_z \text{-} B_z C_y}) A_x + (\mathrm{B_z C_x \text{-} B_x C_z}) A_y + (\mathrm{B_x C_y \text{-} B_y C_x}) A_z$$

Note the volume is, as it must be, a number or scalar, which is simply another word for a number.

Summary of Dot and Cross Products

Thus, we see that the dot product (parallel product) gives us a way of projecting one vector into another so as to effect multiplication and thus loses all directionality, but tells us something about the "sameness" or overlap of the two vectors. The cross product (perpendicular product) measures the area or the non-overlap of the two vectors and specifies a direction outside of both (see Figure 2-23).

We can thus see heuristically, from this new perspective, why the cross product of two vectors, which gives the non-overlap, i.e. what's outside the two vectors (the area between them), dotted with a third vector, which has some component in perpendicular direction gives the volume. The dot product needs to pick out the overlap if we are to get volume, because the area can only contribute to the volume to the degree that it is not parallel to the area; namely, it must "come out" of the area to contribute.

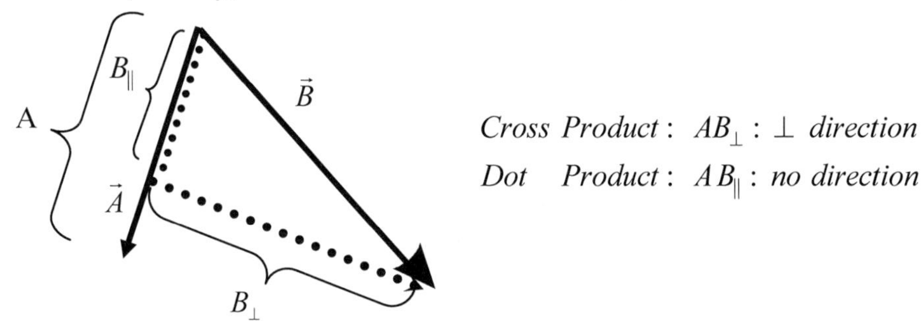

Figure 2-23: *Dot and cross product.* The *dot product* reduces $\vec{B}$ to the component parallel to $\vec{A}$, $B_\parallel$, and proceeds with scalar multiplication of $B_\parallel$ by *A*, yielding only a scalar (number, with no associated direction). The *cross product* reduces $\vec{B}$ to the component perpendicular to $\vec{A}$, $B_\perp$, and proceeds with scalar multiplication of $B_\perp$ by *A*, but assigns a direction to the resulting number by the right hand rule. It is natural to associate a direction with this product, because the two numbers multiplied are, in a real way, from independent dimensions. The direction perpendicular to both is the only one that singles out the plane of the area thereby calculated.

Extend by Analogy to *n* Dimensions

We can generalize vectors further by considering vectors not just in three dimensions, but in n-dimensions. The basic rules of vector addition and multiplication (you will learn more about this in linear algebra) are just extended by analogy to apply to as many dimensions as you may choose. For example, in four dimensions, the vector represented in a Cartesian system would have four components, one for each coordinate axis. One *cannot picture* four spatial dimensions, but one *can think* about it in analogy to three dimensions and the same is true with n-dimensions.

Adding dimensions is very common in physics, because we usually want to reflect our thinking into the quantitative domain,[17] and each different aspect that we want to reflect will have its own kind of independence, an independence that in its own way is similar to that of spatial dimensions.

As a simple example, consider the location of a ball and its temperature, which may change, for instance, as it moves in and out of shade. It might be handy to have the regular three dimensions for the ball's location and one extra for its temperature, thus making four "dimensions."

As another example, neglect the temperature and just consider the ball's movement. Two aspects of a ball in motion that are commonly needed are its location and its speed; this would require six dimensions, three for location in space (x, y, z) and three for the speed in each of the three spatial directions (v_x, v_y, v_z). This analogical use of space is so common it has a name. It is called "phase space."

Position and Difference Vectors

Since locomotion is very important in physics and locomotion is a change of position (place), the position vector is one of the most useful vectors in physics. Typically, we are viewing an object from a particular place and we want to track its motion relative to us. Here, following the powerful procedure of modern physics, we consider what's left of the category of place once it is cast into the quantitative. Specifically, we locate our object in space as defined above. A vector $\vec{X}$ with tail at the origin and head at the point of interest will do this for us. Figure 2-24 below shows the position of two objects of interest, labeled 1 and 2. $\vec{X}_1$ encodes, for instance, how far object 1 is away from the origin and in what direction.

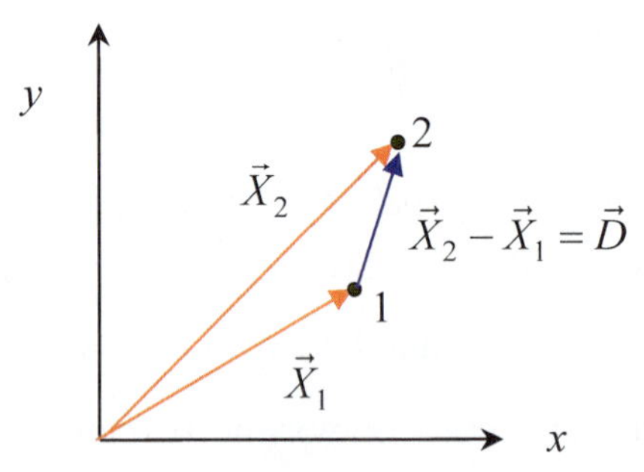

Figure 2-24: Position vectors illustrating location of two objects labeled "1" and "2." The displacement vector, representing the motion from "1" to "2" is shown in blue.

---

[17] We do this in many analogous ways.

The difference between the positions of the two objects is given by $\vec{X}_2 - \vec{X}_1$. It tells us how far and what direction, one must go to get from position 1 to position 2. Unlike, vectors generally, which do not have a location, the position vector comes with a location. We see that $\vec{X}_1$ would tell us nothing about the position if we did not also know where to start. The same is true of the difference vector only so far as it locates object 1 with respect to object 2. Indeed, notice that in taking the difference vector, we have lost some information. Position vectors all start from the origin; this is part of our definition. However, once we've taken the difference, they no longer start from the origin, so we lose this implicit reference. Only by again specifying one of the two position vectors can we locate the objects in the broader space.

Considering the difference vector itself without any association to objects one and two, i.e. considering $\vec{D}$ only insofar as it is a vector, we see that two radically different position vectors, for example, see Figure 2-25 below, can give the same difference vector.

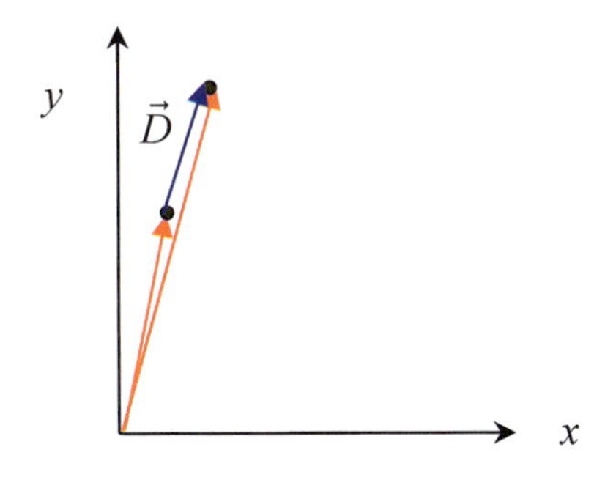

Figure 2-25: Difference vector $\vec{D}$, as a vector (i.e., without position information), can potentially be formed out of an infinite number of different position vectors; it can be anywhere in space. Note how different these two *position vectors* are from those given in Figure 2-24. However, once object 1 or 2 is specified along with the difference vector, the vector is now pinned to a point. Depending on what we are interested in, sometimes we will use the vector alone, sometimes with its location.

Vectors that Depend on Time

Up till this point, vectors sometimes represented motion, but they themselves were not considered to change. We now proceed to changing vectors. How can we make vectors change when mathematics is fixed? We do so analogically, as we've done with so much else. We introduce a new coordinate, the time, to mark each successive moment on a line. We then write *V(x,y,z)(t)*. Time is then treated like a kind of fourth spatial dimension. We will discuss the issue of time in mathematics further when we reach calculus of motion, where we will want to take derivatives of vectors with respect to time as well as space. First, we discuss spatial calculus.

Calculus: Differentiation

Definition of Calculus

Calculus is often said to be the mathematics of motion and change. Since, as we've seen, mathematics itself leaves change behind, this must be an analogical meaning, not the first meaning. The mathematics of calculus, *per se*, properly deals with geometrical objects which are fixed; parabolas, circles, squares, spheres, cubes, etc. as such have no tendency to become something else. Calculus is the mathematics that studies curves and shapes by importing into geometry powerful generalizations from algebra, which are, in turn, deeply rooted in the concept of number. We should, thus, instead say calculus has a powerful analogical meaning, namely, to describe change.

To make it clear that mathematics is about geometry first, we will introduce calculus using fully mathematical objects *only* and then *afterward* introduce its use in describing change. Hence, we start with algebra applied to geometry.

Derivatives

To begin, take as an example, finding the slope of the tangent to a curve at a given point, say the curve given by:

2.16 $$y(x) = h\left(-\left(\frac{x}{d}\right)^2 + 1\right)$$

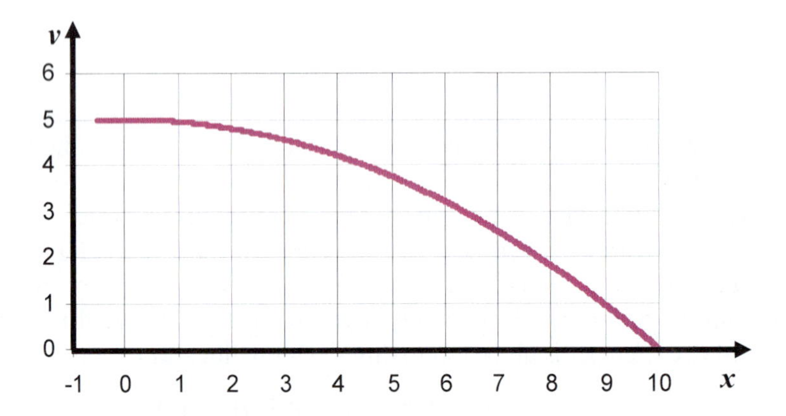

Figure 2-26: Graph of curve represented by equation 2.16, given: $h = 5,\ d = 10$.

For concreteness, consider: $h = 5,\ d = 10$. The curve with these parameters is graphed in Figure 2-26. We seek the tangent to this curve at, say $x = d/2 = 5$. By tangent, we simply mean the line that hits the curve at a point in a glancing fashion, without crossing over it; as shown in Figure 2-27 below. We can see from Figure 2-26, for instance, that the slope of the tangent at $x = 0$ is zero. However, to establish a connection with the equation, and thereby incorporate the power of algebra (in a way) into the geometric analysis, we need more. Though it may seem paradoxical, to do this we need more geometry as well as some algebra. We discuss the further geometry first and then the algebra.

Geometrically, we see that we can approximate the tangent by breaking up the x-axis into little equal (for convenience) parts. Each of these can be used to set up vertical slices that chop up our curve. Each of these slices, in turn, slices a corresponding part of the curve. That part of the curve, see Figure 2-27, is much closer to a line than was the original curve; in fact, it is an approximation to the tangent line in that region. If we want to establish the tangent for a more narrowly defined region, we will have to chop the curve into finer parts. The finer we chop, the more the part of the curve in the chopped region approximates a line, the better the approximation to the tangent we get, *and* the more narrowly defined is the region for which the tangent is defined. Suppose we chop the curve into ever finer slices, setting up a limit toward zero-sized pieces. By acting as if we have completed this limit, we can define the tangent at any given point on the x-axis. For in the completed limit, the piece size is shrunk to a point, and the curve within that piece is a line that is the tangent at that point. From this, it appears we are saying a point can be a line; we are not, because though our mental construct, for convenience, treats the limit via a symbol as if it were completed, in reality it cannot be completed. For, again, this completed limit is a being of reason, a purely mental construct, for objects are not actually divided into points, but *only* potentially divisible into smaller and smaller parts as far as one wishes to proceed. This latter is what we finally really mean.

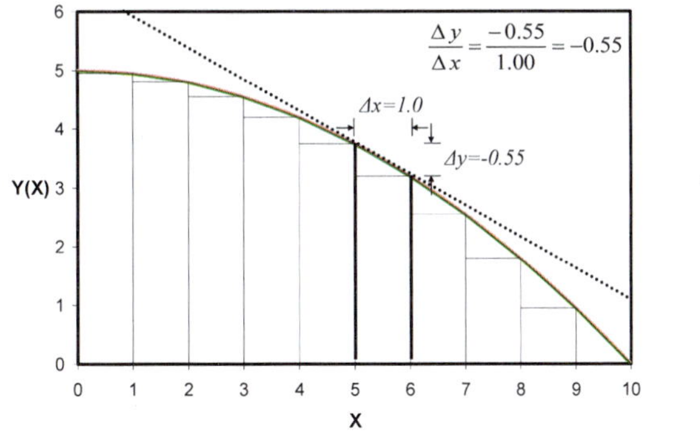

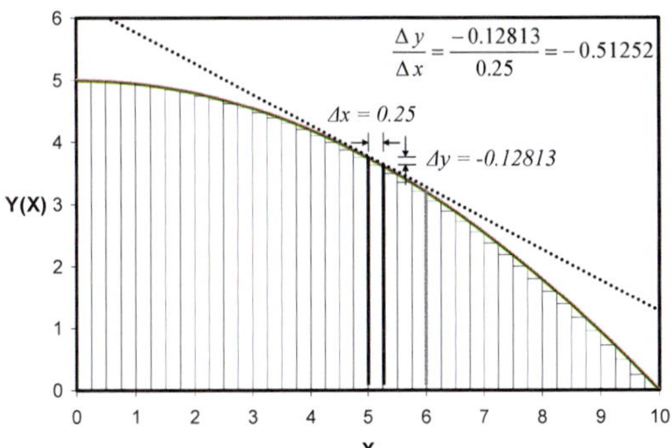

Figure 2-27: Plots showing calculation of tangent to curve given by: $y(x) = 5(1-(x/10)^2)$. The calculation for the tangent at $x = 5$ is shown. The calculation proceeds by approaching limit of zero width rectangles; the left figure shows $width = \Delta x = 1$, right shows $\Delta x = .2$

To begin to calculate the slope of the tangent of the curve given by equation 2.16, i.e., algebraically, we label the graph as shown in Figure 2-27. The length of each of the little parts into which we chop the x-axis is Δx; later, we will let the number of parts go to infinity thereby letting the length of each part become arbitrarily small as above, i.e., $\Delta x \to 0$. The difference in the y coordinates that results when we move the distance Δx across a given part will be called Δy. Recall the slope is defined as:

$Slope = \Delta y / \Delta x$.

Since we plan on making Δx -- and, thus, for "well behaved"[18] functions, Δy -- as small as we like, we introduce a special notation for the slope of the tangent of a curve at a given point. This notation can be thought of as coming about by sending $\Delta x \to dx$ and $\Delta y \to dy$, as $\Delta x \to 0$. Of course, if taken literally, this means $dx = 0$ and the notation seems pointless. However, remember, once again, the limit is not really completed, but as Δx gets smaller and smaller, the slope, $\Delta y/\Delta x$, approaches some limit, some point on the number line, which we then identify with the completed limit as if we'd actually done the infinite subdividing. Again, the completed infinity cannot happen, but we can see where the series is converging to. Hence, in our notation *dx*, (called an infinitesimal in *x*), we really don't mean the limit is completed, but only that we can make it as small as we please, someday completing it in the much different sense just described. All this means that the sought after slope of the tangent, called the derivative, can be written as *dy/dx*.

That is, the slope of the tangent, at a point, say *x*, is:

2.17
$$\frac{dy}{dx} \equiv \lim_{\Delta x \to 0} \frac{\Delta y}{\Delta x}$$

Remembering *y* is a function of *x*, which we write as *y(x)*, and using the definition of Δy, equation 2.17 becomes:

2.18
$$y'(x) = \frac{dy}{dx} \equiv \lim_{\Delta x \to 0} \frac{y(x+\Delta x) - y(x)}{\Delta x}$$

where $y'(x)$ means derivative of the curve at *x*.

In our particular case, i.e. using equation 2.16, this becomes:

2.19
$$\begin{aligned} \frac{dy}{dx} &= \lim_{\Delta x \to 0} \frac{\left(-\frac{h(x+\Delta x)^2}{d^2} + h\right) - \left(-\frac{hx^2}{d^2} + h\right)}{\Delta x} \\ &= \lim_{\Delta x \to 0} \frac{-h(2x\Delta x + (\Delta x)^2)}{\Delta x\, d^2} \\ &= \lim_{\Delta x \to 0} \frac{-h}{d^2}(2x + \Delta x) \\ &= -\frac{2h}{d^2}x \end{aligned}$$

Thus, the answer to our particular question is that the slope of the tangent at $x = d/2$ is $-h/d$, which gives: $-1/2$, for $h = 5, d = 10$.

[18] Such conditions can be rigorously defined. After understanding the points introduced in this section, the student is encouraged to investigate such details further by thinking about what conditions are necessary to have a meaningful definition of the derivative.

It is left as an exercise to use the above definition to obtain the general formula for finding the derivative (i.e. finding the tangent of a curve) of an arbitrary polynomial. The resulting simple formula for an individual term ($y(x) = a x^n$) of such a polynomial is:

2.20 $y'(x) = n a x^{n-1}$

It should be memorized, for it will facilitate calculation in many basic physics problems.

The operation of finding the slope of the tangent, which involves taking the difference in the numerator, is called *differentiation*, while $\frac{dy}{dx}$ is called the *derivative* of *y(x)*, as it is derived from the curve; it defines the closest approximation to the curve, *y(x)*, that one can get using only a line (it's also called a linear approximation). The derivative is defined at every point on a curve; so, in addition to graphing the curve itself, one can graph the value of the slope at each point on the curve: namely, $y'(x)$ versus x.

Obviously, in such a case, the reverse operation can be done. That is, one can move from this later function, $y'(x)$ back to $y(x)$. Whereas the derivative says something about an aspect of the curve at a point, this complementary operation, called integration, says something about the whole curve between two points. It finds the area under the curve. We now turn to this anti-derivative or integral.

Calculus: Integration

To see how we can recover the original function from its derivative, consider an arbitrary "well behaved" curve, $y'(x)$, that is the derivative of another curve $y(x)$. That is, each point of the curve represents the slope of the tangent of another curve, *y(x)*, as shown in Figure 2-28.

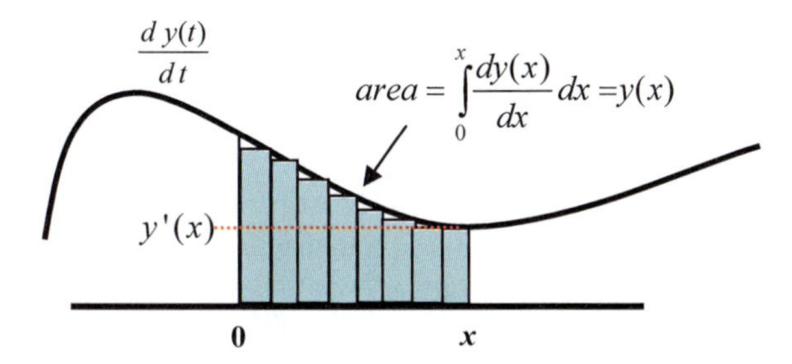

Figure 2-28: *Integration under a curve that represents the derivative of y(x).* Divide region of interest into *N* equal width (*x/N*) rectangles. The area under the curve is *y(x)*.

We begin by dividing the region from $x = 0$ *to* x into *N* equal width regions also as shown in the figure. We choose *N* to be large so our regions are small. We then approximate the area under the curve using our knowledge of the area of these small rectangles, i.e. $y'(x)\Delta x$, and summing all of them up as follows gives the total area under the curve within the prescribed limits:

2.21
$$Area \sim y'(\Delta x)\Delta x + y'(2\Delta x)\Delta x + y'(3\Delta x)\Delta x + \ldots = \sum_{n=1}^{N} \frac{dy}{dx}(n\Delta x) \cdot \Delta x$$

$$where\ \Delta x = \frac{x}{N}$$

These parentheses *do not* indicate multiplication but that the function is evaluated at $n\Delta x$.

This is only an approximation to the area. To get a better approximation, we increase the value of *N*. The larger *N*, the more the slices approach rectangles, and, in the limit, the area of the rectangle gives exactly the slice area. In a similar manner to what we did with derivatives above, we take the infinite limit to be completed and identify it with the number that it is approaching. Thus, the integral, which is the area under the curve, within the given limits, is defined as:

2.22
$$\int_0^x \frac{dy}{dx} dx \equiv \underset{N\to\infty}{Lim} \sum_{n=1}^{N} \frac{dy}{dx}(n\Delta x) * \Delta x = Area$$

Noting the fact that each component of the sum, that is each rectangle, has an area such that: $y'(x)\Delta x = \Delta y$, we see that as we move over Δx in the *x*-direction, the function *y(x)* changes by Δy in height. In other words, starting at $x = 0$, we see that, in addition to giving the area, each term in the summand in equation 2.22 uses the slope, $y'(x)$, to determine what the new $y(x + \Delta x)$ is. Generally: $y(x + \Delta x) \sim y(x) + y'(x)\Delta x$. In the limit, the relation will become precise giving *y(x)* within a constant, $y(x_0)$, thus, recovering the original function of which it was the derivative. Note that the loss of the constant is indicative of the fact that the derivative does not contain all the information of the original function, only its slope at each point. We can see more clearly how the function, *y(x)*, is recovered through adding each little increment as we move to the right in *x* by rewriting equation 2.22 as:

2.23
$$\int_0^x \frac{dy}{dx} dx \equiv \underset{N\to\infty}{Lim} \sum_{n=1}^{N} \frac{dy}{dx}(n\Delta x) * \Delta x$$

These parentheses *do not* indicate multiplication but that the function is evaluated at $n\Delta x$.

$$= \underset{N\to\infty}{Lim} \sum_{n=1}^{N} \Delta y(n\Delta x) = y(x) - y(0)$$

We've just shown that the derivative and integral are inverses; this fact is called *the Fundamental Theorem of Calculus.* Symbolically, it is written:

2.24
$$y(x) - y(0) = \int_0^x \frac{dy}{dx} dx$$

As a simple example of the use of 2.22, consider $y'(x) = x$. We thus get:

$$\int_0^x x dx \equiv \underset{N\to\infty}{Lim} \sum_{n=1}^{N} n\Delta x * \Delta x = \left(\frac{x}{N}\right)^2 \underset{N\to\infty}{Lim} \sum_{n=1}^{N} n = \frac{x^2}{2}$$

In the last equality, we used the equation for the partial sum: $\sum_{n=1}^{N} n = \frac{N^2 + N}{2}$; convince yourself of this by checking it for a few small values of N. Thus, this means, generally, that: $y(x) = \frac{x^2}{2} + y(0)$.

Calculus Applied to Motion

Representing Motion in Static Mathematics

Now, we see both the integral and the derivative of a function are static; they do not change. The derivative of a curve is another curve; for example, in the case of equation 2.16, we get the curve for *y(x)* shown in Figure 2-26 and the curve for *dy/dx* shown below in Figure 2-29. A curve doesn't change, as an equation once written doesn't change.

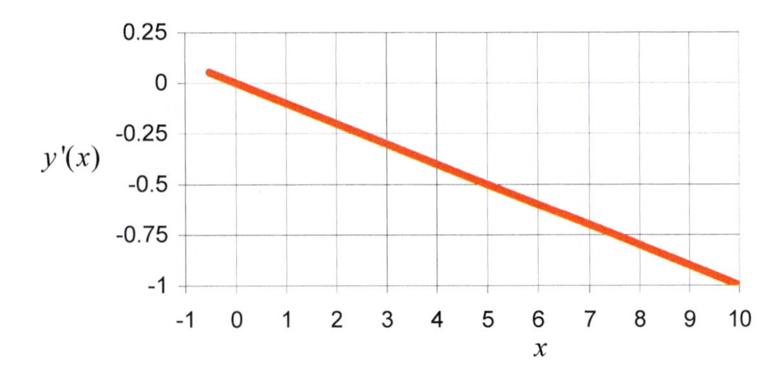

Figure 2-29: *Graph of derivative of equation 2.16.* equation 2.16 is graphed in Figure 2-26. In other words, this figure shows the value of the slope of the tangent at a given x in Figure 2-26, as a value on the vertical axis of this graph. Hence, for instance, since the slope of Figure 2-26 is zero at $x = 0$, the above curve goes through zero at $x = 0$. Or, similarly, the slope at $x = d/2 = 5$, is $-.5$, as previously calculated.

So, how can we use calculus or any mathematics for motion? Remember time is the measure of motion; it is one motion measured by another, especially by some regular simple motion.

Thus, we can graph the *number* of motions that happen, say ticks of a stopwatch, on the x-axis against some measure of the progress of the other motion, say how far a runner goes during the given number of motions, on the y-axis. By simultaneously correlating a number with each of the two motions, we can graph all the pairs of numbers as if they existed all the time using the number lines of, for example, a rectangular coordinate grid.

Said another way, once we treat time as a number, we can graph it as if it were a spatial coordinate and then the resulting curve, say of distance traveled versus time, can be treated formally in just the same way as spatial curves, using the full battery of mathematical tools developed for spatial curves. We only have to remember the meaning of the curve.

Concretely, we might watch a runner and note how many times the second hand ticks (e.g., $\Delta t = 1s$) during each of 10 yards ($\Delta x = 10yds$) of the 100 yard dash. The graph of these points would be very rough; for instance, the tangent may not be as smooth as the motion. The tangent in this case is the speed $= \Delta x / \Delta t$. Hence, we only get a very crude estimate of the speed. To determine the instantaneous speed more accurately, we need a faster time standard and more frequent measurements; we, for example, could count how many ticks of 1/100 second go by in .1yds. We can get increasingly more accurate making the intervals smaller and smaller, approaching the limit of zero time and zero distance, but never actually getting there. However, we can consider the value that the series converges towards. Now, this is nothing but the definition of the derivative applied to motion by numbering it. We thus have a way of talking about motion using geometry and the calculus we developed to analyze that geometry in the last couple of sections.

Consider another example: the flight of a cannon ball as it flies, say vertically, through the air. Its height at a given moment can be graphed against the *number* of clock ticks (see Figure 2-30) that occurred since launch. This plot yields a geometric curve. Indeed, that curve is identical to that of Figure 2-26; the same equation (eq.2.16) describes both a fixed curve and the cannon ball's flight. We only need to recall the difference in meaning, which will lead us to change the variable names in the following way in the x-direction: $x \to t,\ d \to \tau$ which implies $\frac{h}{d^2} \to \frac{h}{\tau^2} \equiv \frac{1}{2}a$ changing equation 2.16 into:

2.25 $$y(t) = -\frac{1}{2}at^2 + h$$

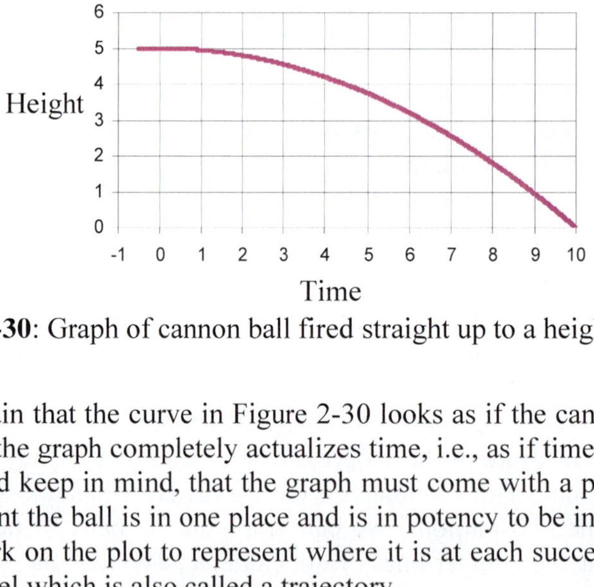

Height

6
5
4
3
2
1
0

-1 0 1 2 3 4 5 6 7 8 9 10

Time

Figure 2-30: Graph of cannon ball fired straight up to a height $h = 5$.

Note again that the curve in Figure 2-30 looks as if the cannon ball is at all points at once. That is, the graph completely actualizes time, i.e., as if time existed all at once. We know, and should keep in mind, that the graph must come with a proper interpretation. At any given moment the ball is in one place and is in potency to be in the next place. We put a *permanent* mark on the plot to represent where it is at each successive moment; thus we get a path of travel which is also called a trajectory.

As we saw earlier, the tangent to the trajectory is the speed. Thus, we can define the speed in each of the independent directions in a rectangular coordinate system as:

2.26 $$v_x = \frac{dx}{dt},\ v_y = \frac{dy}{dt},\ v_z = \frac{dz}{dt}$$

We also are interested in the rate of change of the speed, the acceleration. This is defined for each axis as:

2.27 $$a_x \equiv \frac{d\mathbf{v}_x}{dt} = \frac{d^2x}{dt^2}\,,\ a_y \equiv \frac{d\mathbf{v}_y}{dt} = \frac{d^2y}{dt^2}\,,\ a_z \equiv \frac{d\mathbf{v}_z}{dt} = \frac{d^2z}{dt^2}$$

Uniform Acceleration

In the case of our cannon ball, the speed is, applying equation 2.24 to equation 2.25: $\mathbf{v} = -at$, and the acceleration is: $-a$. So, the curve describes the motion of an object in uniform acceleration.[19] That uniform acceleration results in a parabolic curve such as given by equation 2.25 is often thought to have been a discovery of Galileo. In fact, the proof of Galileo was reproduced by him from the work of Nicole Oresme (see box earlier in chapter) who lived in the middle ages. Figure 2-31 below shows this geometric proof.

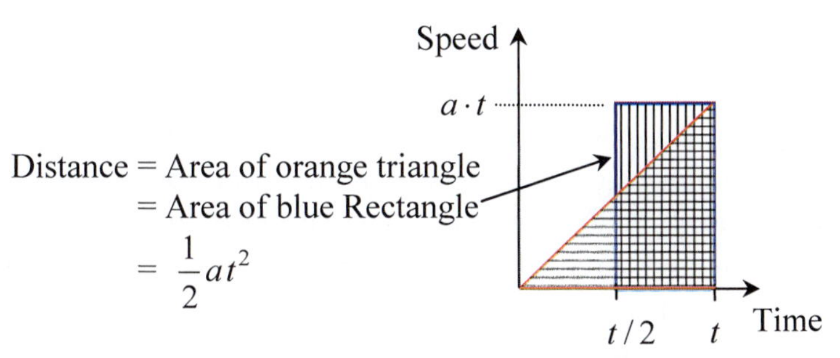

Figure 2-31: The triangle with superimposed rectangle was first used by Oresme in *1300's* to illustrate the $d = \frac{1}{2}at^2$ law of uniform acceleration. Later, Galileo used his proof. Dominic De Soto was the first to apply the relation to falling objects.

Use of Vectors for Describing Motion

Now that we've made the connection between geometry and motion, we can make use of all the axioms and theorems of vector algebra discussed in earlier sections. As discussed then, those axioms incorporate key facts about three dimensional space in a compact notational system, which thus facilitates calculation. For instance, because these axioms succinctly contain the meaning we need, velocity can be written simply as:

2.28 $$\vec{V} = \frac{d\vec{X}}{dt}\text{, where } \vec{X} \text{ is the position vector } \begin{pmatrix} x \\ y \\ z \end{pmatrix} = x\,\hat{x} + y\,\hat{y} + z\,\hat{z}$$

[19] Uniform acceleration is indeed a good approximation for falling bodies near the Earth that are solely under the influence of gravity.

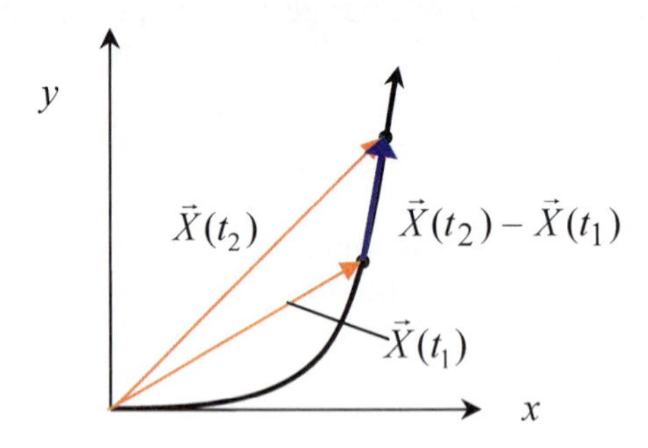

Figure 2-32: Black curve shows path of a particle moving on a table surface. Direction of travel in time is shown with black arrow. Blue arrow shows the displacement vector between the two locations, 1 and 2 at times t_1 and t_2.

Figure 2-32 shows the use of vectors to describe motion. It shows the path an object makes as it moves through a two dimensional *space*, say on the surface of a table. Imagine that it leaves an ink mark as it moves so that it paints the curve shown in black in the figure. The vector $\vec{X}(t)$ represents the position of the object. $\vec{X}(t_1)$ is the position vector at time t_1. As t_1 approaches t_2, we get a smaller and smaller difference vector. In the limit, the difference vector obviously disappears; however, just as we've seen before with the speed, if we divide by the corresponding time interval, as we take the limit, we approach the *velocity* of the object, which is a finite vector including both components of speed of motion (v_x and v_y) at t_1. In this case, the time, t, is called a parameter and the curve is said to be *parameterized* by t.

Note that when the components of the speed are written together, indeed whenever one is talking about the motion of an actual object in its full freedom of motion—up/down, left/right and back/forth---we speak of the *velocity* rather than the speed. Mathematically, we speak of the *velocity vector* and *speed* then is reserved only for the magnitude of the rate of change of position (place).

Acceleration, the rate of change of velocity, can be written compactly in vector notation because it is a vector, i.e. has magnitude and direction. We write it as:

2.29
$$\vec{A}=\frac{d\vec{V}}{dt}=\frac{d^2\vec{X}}{dt^2}$$

Time as a Third Dimension

We can also take *time* to be represented by a third perpendicular axis, in analogy to the z-spatial axis and recover a geometrical visualization. Curves plotted in such a graph that incorporate space and time are called *space-time curves*. Figure 2-33 shows an object departing the origin at $t=0$ at uniform speed, v_y, in the y-direction. In the figure, the speed in the y-direction is just the slope of the "trajectory" line in the t-y plane; this plane is shown shaded to highlight this point. Obviously, at any moment, the velocity can be described by a vector in a plane defined by the direction of the *velocity vector* and a vector parallel to the t-axis. For instance, the object in Figure 2-33, abruptly changes its motion when it reaches y_0; at that point, it ceases motion in the y-direction and takes up uniform motion in the negative x-direction. The figure shows the new plane defined by the direction of the velocity and the time axis. Notice that if the velocity is constant, then this plane is fixed and does not change. However, if the velocity direction changes (other than in simple back and forth motion), then so does the plane. If the object described in Figure 2-33, had accelerated slowly, rather than instantaneously, from motion in the y-direction to motion in the x-direction, then the plane of motion would rotate continuously from the first plane to the second plane rather than "jumping."

In mixed dimensional motion, such as shown in the figure, the usefulness of time derivatives of vectors is clear, for one cannot follow the motion of the particle with a single variable, but needs both the x and y and often must include z as well.

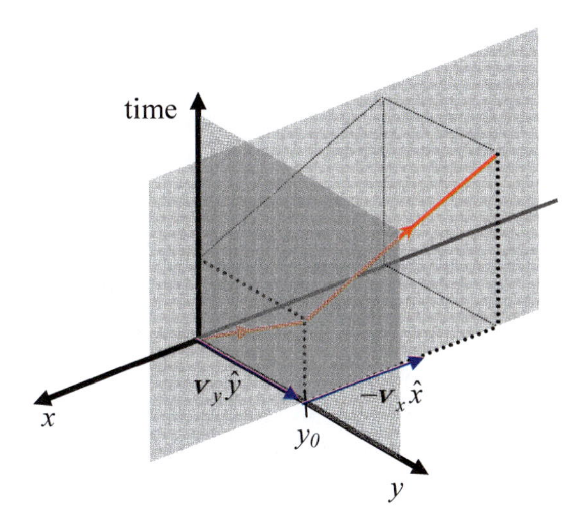

Figure 2-33: *Time represented as the z-axis of a 3-dimensional rectangular coordinate system.* The red arrow shows a particle leaving the origin at time zero, traveling at uniform speed. When it reaches y_0, it ceases its motion in the y direction and begins uniform motion in the $-x$ direction. The two shaded planes show that while the motion is uniform it can be confined to a plane, one dimension for its progress in the given spatial direction and one for counting the number of standard units of motion, say seconds, that occur.

Use of Vectors and Calculus in describing Motion

Position and Velocity with Time dimension

Vector mathematics and calculus are helpful in understanding even simple physical things. Take for example, the case of an object sitting on a merry-go round. It can be described by a position vector:

2.30 $$\vec{X}_{merry} = \begin{pmatrix} x \\ y \end{pmatrix} = \mathrm{x}\,\hat{x} + \mathrm{y}\,\hat{y} = r_o \cos(\omega t)\,\hat{x} + r_o \sin(\omega t)\,\hat{y}$$

Here r_0 is the radius of the merry-go-round, and ω is its rotational speed.

As discussed in the last section, the motion can also be described by a 3-dimensional vector with time as the third axis. We need to distinguish in our notation the new level of analogy that we have introduced by making time a third dimension; again, if we don't recall and understand the distinction, we will get confused about the meaning. As a further example of the different meanings consider that at any one time a particle can only be at one place, though the particle can be at any one place many times; whereas, there is no such distinction between the axes as merely spatial. To mark the fact that we have introduced time as a dimension, we will put a superscript Greek letter, called ***an index***, on the right of the variable and drop the arrow. Vectors that include time are called space-time vectors. If we include the possibility of motion in all three dimensions of space, then we would have four components, t (time) and x, y and z directions. This is the notation for the four dimensional vectors, called four-vectors, in special relativity.[20] The index marks, in our merry-go-round case, the x, y or t component and we get:

2.31 $$X^{\mu}{}_{merry} = \begin{pmatrix} X_x \\ X_y \\ X_t \end{pmatrix} = X_x\,\hat{x} + X_y\,\hat{y} + X_t\hat{t} = r_o\cos(\omega t)\,\hat{x} + r_o\sin(\omega t)\,\hat{y} + t\hat{t}$$

So that: $X^x = x(t) = r_0 \cos\omega t,\ X^y = y(t) = r_0 \sin\omega t,\ X^t = t$

Note that $X^{\mu}{}_{merry} = \vec{X} + X^t\hat{t}$, so that X^{μ} includes the ordinary position vector within it.

Figure 2-34 shows the space-time curve for the child, and a space-time position vector $X_{merry}{}^{\mu}$ at one point, as well as, its derivative, the generalization of the velocity vector, the space-time velocity vector, V^{μ}. Excepting the last chapter, we will be only working with the plain ordinary position and velocity vector that are defined in the spatial plane and are labeled by $\vec{X}$ and $\vec{V}$ in Figure 2-34.

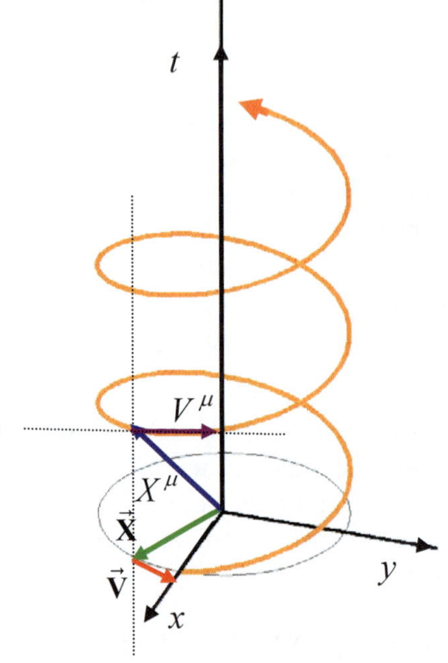

Figure 2-34: The *space-time path* of a child sitting on a horse on a moving merry-go-round is shown in orange. Time is shown on the axis pointing up.

[20] In Newtonian physics, time is treated like a completely independent variable; because of this much of the mathematics of vectors is unused here; only the simplest sorts of operations are used. In Relativity, vectors are further generalized using appropriate analogies.

The child moves around in space in a circle in a plane defined by the x and y axes shown on the plot. The blue vector, labeled X^{μ}, is the space-time 3-vector of equation 2.31, while the green vector, $\vec{X}$, is the ordinary position vector. The purple vector, V^{μ}, tangent to the space-time path, is the instantaneous space-time velocity. The red vector lying in the x-y plane, $\vec{V}$, is the *ordinary velocity* vector. For this course, the ordinary position and velocity vectors, $\vec{X}$ and $\vec{V}$, are the most important.

Notice the trajectory traced on the x-y plane is a circle, so that the direction of the velocity vector is changing; i.e., there is acceleration. We see that if we limit our view to graphing time versus $x(y)$ motion, we get a cosine (sine) wave. Note that the space-time velocity vector, V^{μ}, contains the velocity vector in it.:

2.32 $$V^{\mu} = \frac{dx}{dt}\hat{x} + \frac{dy}{dt}\hat{y} + \hat{t} = -r_o\,\omega\,\sin(\omega t)\,\hat{x} + r_o\,\omega\,\cos(\omega t)\,\hat{y} + \hat{t}$$

It is the first two terms, which can be written as:

2.33 $$\vec{V} = \frac{dx}{dt}\hat{x} + \frac{dy}{dt}\hat{y} = -r_0\sin\theta\,\hat{x} + r_0\cos\theta\,\hat{y} \quad when\ \omega = 1\ and\ \theta = \omega t$$

We illustrate this vector on the two dimensional plot shown in Figure 2-35 below.

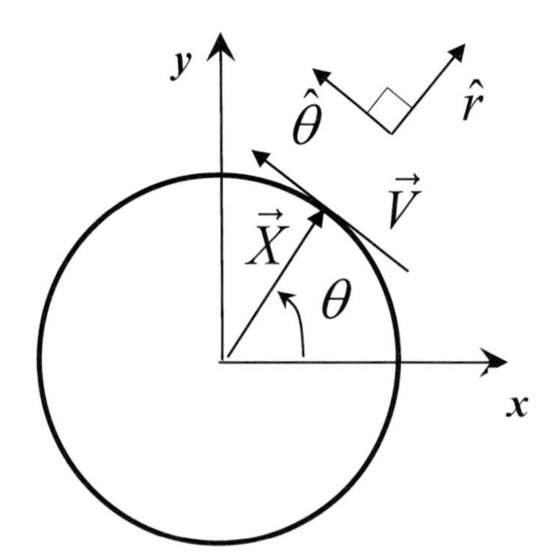

Figure 2-35: *Polar plot of trajectory of child on a merry-go-round*

Polar Coordinates and the Merry-go-round

From Figure 2-35, we see that polar coordinates, without the time coordinate, are the most convenient for this problem. Polar coordinates respect the circular symmetry already present in the physical situation under consideration. And, the time variable can be succinctly included without an explicit dimensional representation. Hence, at this point, we include the time coordinate *only* in so far as the x and y position vectors change in time; we cease to include time explicitly as a 4th dimension until we deal with special relativity. We thus proceed to analyze the merry-go-round in simple polar coordinates.

In polar coordinates, the position vector, usually written $\vec{r}$, for the child on the end of the merry-go-round is:

2.34 $$\vec{X}_{merry} \equiv \vec{r}_{merry} = r_0\,\hat{r}$$

As an aside, we can see that this follows from Cartesian coordinates, using basic trigonometry, in the following way:[21]

2.35 $$\vec{X}_{merry} \equiv \vec{r}_{merry} = x\,\hat{x} + y\hat{y} = r_0 \cos\theta\,\hat{x} + r_0 \sin\theta\,\hat{y} = r_0\,\hat{r}$$

To determine the velocity of the child, we need the time derivative of $\vec{r}_{merry}$, which is shown pictorially in Figure 2-36 below.

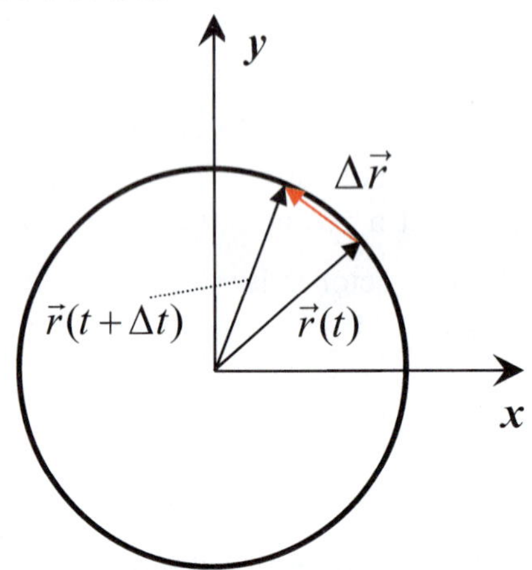

Figure 2-36: In a short time interval Δt, the child moves on the merry-go-round, and, thus, his position vector moves through an angle $\Delta\theta$ making the difference vector $\Delta\vec{r}$ shown. As usual, the velocity is obtained in the limit of smaller and smaller Δt: $\frac{d\vec{r}}{dt} = \underset{\Delta t \to 0}{Lim}\frac{\Delta\vec{r}}{\Delta t}$.

Remaining in polar coordinates, using the multiplication rule, gives the velocity as:

2.36 $$\vec{V}_{merry} = \frac{d\vec{r}_{merry}}{dt} = \frac{d(r_0\,\hat{r})}{dt} = \frac{d(r_0)}{dt}\hat{r} + r_0\frac{d(\hat{r})}{dt}$$

The first term is zero, because r_0 is constant. However, notice that $\hat{r}$ changes direction with time--though the length doesn't, for it always has, by definition, unit length—therefore, the second term is *not* zero. Using the equation for $\hat{r}$ in rectangular coordinates (since $\hat{x}$ and $\hat{y}$ don't change with time): $\hat{r} = \cos\theta\,\hat{x} + \sin\theta\,\hat{y}$, one gets:

2.37 $$\frac{d(\hat{r})}{dt} = \frac{d\theta}{dt}(-\sin\theta\,\hat{x} + \cos\theta\,\hat{y}) = \frac{d\theta}{dt}\hat{\theta}$$

giving:

2.38 $$\vec{V}_{merry} = r_0\frac{d\theta}{dt}\hat{\theta} = \omega r_0\,\hat{\theta}$$

This agrees with the result given in equation 2.33. And, it indicates the child's direction of motion is just what we began with: i.e., around the circle in the angular direction. Further, the magnitude of the velocity is correct, for it follows from the circumference of the circle and the definition of ω.

[21] From which, we note that: $\theta = \omega t$ yields the position vector for our child on the merry-go-round given in equation 2.30.

We can derive similar but more general results and without any direct reference to time. Such a calculation is useful in understanding the nature of polar coordinates (which we will use frequently) and their relation to Cartesian coordinates, so we carry it out briefly.

More on Polar Coordinates

Both $\hat{r}$ and $\hat{\theta}$ are perpendicular like their counterparts in rectangular coordinates, $\hat{x}$ and $\hat{y}$. However, unlike rectangular coordinates, $\hat{r}$ and $\hat{\theta}$ depend on where one is on the polar grid. Such dependence means the derivatives of $\hat{r}$ and $\hat{\theta}$ with respect to r and θ are not zero.

Recalling again the position vector in polar coordinates:

2.39 $$\vec{r} = r\,\hat{r}$$

The derivative of $\vec{r}$ with respect to θ will yield a tangent to the curve that $\vec{r}$ traces out as θ is varied. It is an analogically a velocity, $\vec{V}_{angular}$, that is, a rate of change of $\vec{r}$, as one changes one's position on the circumference. Differentiating $\vec{r}$ in this way gives:

2.40**a** $$\vec{V}_{angular} = \frac{d\vec{r}}{d\theta} = \frac{d(r\hat{r})}{d\theta} = \frac{dr}{d\theta}\hat{r} + r\frac{d(\hat{r})}{d\theta}$$

Again, using the equation for $\hat{r}$ in Cartesian coordinates, making use of $\hat{x}$ and $\hat{y}$ being independent of location, we get the tangent vector:

2.40b $$\frac{d\hat{r}}{d\theta} = \hat{\theta} = -\ \sin\theta\,\hat{x} + \cos\theta\,\hat{y}$$

From which we can quickly show, as an aside, that:

2.40c $$\frac{d\hat{\theta}}{dr} = -\hat{r} = -\cos\theta\,\hat{x} - \sin\theta\,\hat{y}$$

But, our main result is:

2.40d $$\vec{V}_{angular} = r'(\theta)\,\hat{r} + r\,\hat{\theta}$$

By making θ a function of time (referred to as parameterization by the parameter t), one can again regain a time dependent vector. One can imagine the vector as following a circular path (or, in general, an arbitrary path) as time marches on. In this way, we can write any motion in terms of polar coordinates. An important exercise is to calculate at least one further derivative of circular motion in polar coordinates.

Polar coordinates are useful in analyzing many types of motion. They and their relation to rectangular coordinates are important as a starting point for understanding more complex coordinate systems as well as more generalized coordinates that will be central in understand later physics, in particular, in general relativity.

Approximation using Calculus

Calculations in physics can get very complicated, even in Newtonian physics where we simplify things very much. To handle this, we need approximation techniques that will allow us to solve problems in a way that gives approximate numbers rather than exact numbers. This is not as big a drawback as one might first think, because, in actual practice, we cannot get exact measurements for many reasons, not the least of which because nature is so complicated and is constantly changing.

The most used method of approximation is the Taylor series, which is written:

2.41 $$f(x)=\sum_{n=1}^{n=\infty}\frac{df}{dx}(x_0)\,\frac{(x-x_0)^n}{n!}$$

The particular case in which $x_0=0$ is called the Maclaurin series, written as:

2.42 $$f(x)=\sum_{n=1}^{n=\infty}\frac{df}{dx}(0)\,\frac{x^n}{n!}=f(0)+f'(0)x+f''(0)\frac{x^2}{2!}+f'''(0)\frac{x^3}{3!}+....+f^n(0)\frac{x^n}{n!}+...$$

We can somewhat intuitively understand this formula by realizing that each new derivative gives a better approximation to the curve, $f(x)$. The first one is the line that best matches it. Adding the second allows us to get the parabola that best matches it. Or from another view, the second one gives the best line that matches the curve $f'(x)$. As we add more derivatives of derivatives, we incorporate more information about the curve, giving a better approximation.

Now, what about the factors that multiply the various derivatives? Once we've set on expanding in x^n, we see that the factor $1/n!$ is necessary to make the various derivatives come out as they must according to our formula for derivative of a polynomial implied by equation 2.20. Rigorous proof of the convergence of the Taylor expansion is beyond the scope of this text, but the interested student should research it and will find it enlightening.

Taylor Expansion of Exponential

As example, consider the function $y(x)=e^x$. The derivatives of this function are all the same: $y^n(x)=e^x$. Equation 2.42 gives;

2.43 $$e^x=1+x+\frac{x^2}{2}+...\frac{x^n}{n!}+...=\sum_{n=0}^{n=\infty}\frac{x^n}{n!}$$

Taylor Expansion Leading to Sum of Geometric Series

Consider the function $g(x)=\frac{1}{1-x}=\left(1-x\right)^{-1}$. Taylor expansion gives:

2.44 $$\left(1-x\right)^{-1}=\sum_{n=0}^{n=\infty}x^n\text{ , for }|x|<1$$

Derivative Notation Note

Up till now, we have been using Leibniz's notation for derivatives: $\frac{d}{dx}$. However, for convenience we will use Newton's notation for time derivatives, i.e. we will say: $\dot{x}\equiv\frac{dx}{dt}$. We will also use a prime for derivative with respect to the argument, for example, $f'(x)\equiv\frac{df}{dx}$.

We are now ready to proceed to the arena which was the starting point for modern physics, and the main subject of our course: motion.

Summary

Vectors and Calculus are important mathematical tools needed for Newtonian Mechanics.

Mathematics is a study of a real aspect of things. In particular, it is the study of physical things absent all 9 categories of properties, *except* extension. We can walk up to this abstraction by considering only a single substance, which means we have ceased to consider the *environment*. And, since something can only be in *place* in and have an *orientation* with respect to an environment, we thus have left these two categories behind as well. Indeed, there can be no relations between substances, since we are only considering one substance, so we leave behind the category of *relation*. Next, we consider the world "frozen in *time*," so that there is no motion, thus we cease to consider the categories of *action* and *reception*. Lastly, we consider the object with no particular color, smell, or power of any kind; that is, we leave out all *qualities* except shape. This leaves us with pure extension, parts one next to another, limited by some shape, which we study in geometry.

If one starts with two walls in the corner of a room and leaves behind, as above, all the properties but extension, the intersection of the surface of the walls gives a line segment, which, when the distinction between the ceiling on the top and the floor on the bottom is considered, becomes the concept of a ray, that is a directed line segment. In short, in such real things, we see the concept of a directed line segment (an arrow) which is our first meaning for *vector*.

We extend the meaning of vector to the category of quality in the following way. First, note the similarity between the magnitude of a quality's intensity, such as how hot or how cold something is, and the extension of a body in a given direction. Second, we note, for example, how the magnitude of a quality may be greater in one region relative to the region next to it, thus establishing a direction in space. Therefore, we can relate qualities to both how *long* and *in what direction* an arrow points. For instance, an arrow could point in the direction in which the temperature increases from one region to the next, and the arrow's length could be the amount of the increase in that direction. Similar analogies can be made with combinations of the categories and even with motion, giving us an analogically general meaning of *vector*: *that which has magnitude and direction.*

We use mathematical formalisms to capture various aspects of reality in a system of symbols and rules that gives us great power to access those realities if we use them thoughtfully. *Vectors are represented by* a letter with an arrow over it $\vec{V}$, making it pregnant for algebraic manipulation.

In such formalisms, vectors exist in a mathematical space, which is defined as extension without limit, that is not actually divided, but is potentially divided in any way one wishes. We think of *space* as composed of points and of infinite extent, thus, by a mental construct, leaving shape out. In other words, we leave our concept of space "waiting" for a particular physical problem that we wish to solve to specify it. Mathematical formalisms use symbols and concepts such as completed infinities that can only exist in the mind. Such constructs are useful and are called *beings of reason or mind dependent beings.* These constructs are powerful because they make reasoning about the complex realities they encase subject to rigorous logical analysis at a rote level—so rote a computer can often carry the calculations out—and thus making us less prone to mistakes in our reasoning.

We represent vectors by three numbers: (x,y,z) in *rectangular (Cartesian) coordinate systems*. In such representation, the vector is defined as the ray (arrow) which has the same length and direction as the one with its tail at the origin $(0,0,0)$ and its head at the point labeled by x units out, y units over and z units up. A *unit vector,* a vector of a certain direction with *unit* length, is written, for the x-direction, as: $\hat{x}$. The "hat" means one has a vector of unit length. Thus, a general position vector in these coordinates is written: $\vec{x} = x\hat{x} + y\hat{y} + z\hat{z}$.

Like a city which can be represented with different maps, a vector can be represented by different coordinate systems. The symmetry of the situation breaks the symmetry of our abstract "space." *Spherical coordinates* are best for physical situations that have a spherically symmetric aspect to be studied. Spherical coordinates label a vector by its radial distance, r, from the origin, its angular distance, θ, from the z-axis, and its azmuthal angle, ϕ, from the x-axis, (r,θ,ϕ). A position vector in these coordinates is written: $\vec{r} = r\hat{r}$, where, $\hat{r}$ is the unit vector in the radial direction. The relationship between rectangular and spherical coordinates can be summarized as follows:

2.45
$$x = r\sin\theta\cos\phi \qquad r = \sqrt{x^2 + y^2 + z^2}$$
$$y = r\sin\theta\sin\phi \qquad \phi = \tan^{-1}\frac{y}{x}$$
$$z = r\cos\theta \qquad \theta = \tan^{-1}\frac{\rho}{z} = \tan^{-1}\frac{\sqrt{x^2 + y^2}}{z}$$

Polar coordinates are suited for situations which have circular symmetry in a plane. Points and vectors are written: (r,θ). The relationship between two dimensional rectangular and polar coordinates is:

2.46
$$x = r\cos\theta \qquad r = \sqrt{x^2 + y^2}$$
$$y = r\sin\theta \qquad \theta = \tan^{-1}\frac{y}{x}$$

Similar to spherical coordinates, a position vector in polar coordinates is written: $\vec{r} = r\hat{r}$. Also, we get:

2.47
$$\hat{r} = \hat{x}\cos\theta + \hat{y}\sin\theta$$
$$\hat{\theta} = -\hat{x}\sin\theta + \hat{y}\cos\theta$$

A *vector field* is a space in which each point has a vector assigned. In the formalism in rectangular coordinates such fields are written as: $\vec{V}(x,y,z)=V_x\,\hat{x}+V_y\,\hat{y}+V_z\,\hat{z}$.

In order to extend the use of vectors to algebra-like formalism, i.e. to take advantage of the abstractness of the concept of number, we need an analogical definition of *addition and multiplication for vectors*: we need to define the binary operations: $\vec{A}$ *plus* $\vec{B}$ and $\vec{A}$ *times* $\vec{B}$. To do so, we return to our first meaning of vector, that which has length and direction *Vector addition* is simply defined by adding components in rectangular coordinates, so that we add each component as numbers, giving $\vec{A}+\vec{B}=(A_x+B_x)\hat{x}+(A_y+B_y)\hat{y}+(A_z+B_z)\hat{z}$. Vector multiplication is more complicated; there are two definitions, because of all the information contained in a vector.

The first of the two types of multiplication is defined by multiplying only those components of the given rays (arrow) that point in the same direction, and, the second by multiplying only those components which are perpendicular. The parallel product is called the *dot product*. It allows a first vector to have its length, A, reproduced the same number of times as is the second vector units long in the direction of the first. Symbolically:

2.48
$$\vec{A}=\begin{pmatrix}A_x\\A_y\\A_z\end{pmatrix}\qquad \vec{B}=\begin{pmatrix}B_x\\B_y\\B_z\end{pmatrix}$$
$$\vec{A}\cdot\vec{B}=A_xB_x+A_yB_y+A_zB_z$$
$$=AB\cos(\angle AB)$$
$$where\ A\equiv|A|=\sqrt{A_x^{\ 2}+A_y^{\ 2}+A_z^{\ 2}}$$

In rectangular coordinates we get:

2.49
$$\begin{matrix}\hat{x}\cdot\hat{x}=1 & \hat{x}\cdot\hat{y}=0 & \hat{x}\cdot\hat{z}=0\\ \hat{y}\cdot\hat{x}=0 & \hat{y}\cdot\hat{y}=1 & \hat{y}\cdot\hat{z}=0\\ \hat{z}\cdot\hat{x}=0 & \hat{z}\cdot\hat{y}=0 & \hat{z}\cdot\hat{z}=1\end{matrix}$$

Notice that the dot product leaves behind the direction completely; one is left with a number, also called a *scalar*, only.

The perpendicular product carries the direction and other information about the reality implicit in the two vectors. The perpendicular vector product, which multiplies the components of the two vectors that are perpendicular, is called the *cross product*. It is written: $\vec{A}\times\vec{B}=AB\,\sin(\angle AB)\,\hat{n}$

2.50
$$\vec{A}\times\vec{B}=(\mathrm{A_yB_z\text{-}A_zB_y})\hat{x}+(\mathrm{A_zB_x\text{-}A_xB_z})\hat{y}+(\mathrm{A_xB_y\text{-}A_yB_x})\hat{z}=\begin{vmatrix}\hat{x}&\hat{y}&\hat{z}\\A_x&A_y&A_z\\B_x&B_y&B_z\end{vmatrix}$$

The direction, $\hat{n}$, is defined by the right hand rule as shown below:
These rules applied to the rectangular unit vectors give:

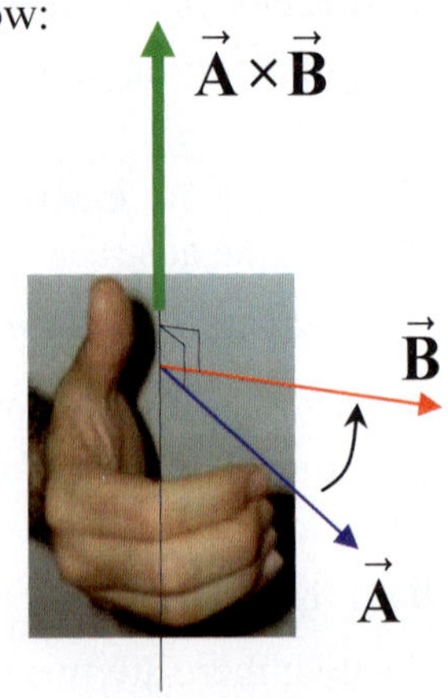

2.51
$$\begin{aligned} &\hat{x}\times\hat{y}=\hat{z} && \hat{y}\times\hat{x}=-\hat{z} \\ &\hat{z}\times\hat{x}=\hat{y} && \hat{x}\times\hat{z}=-\hat{y} \\ &\hat{y}\times\hat{z}=\hat{x} && \hat{z}\times\hat{y}=-\hat{x} \\ &\hat{x}\times\hat{x}=0,\ \hat{y}\times\hat{y}=0,\ \hat{z}\times\hat{z}=0 \end{aligned}$$

The cross product gives the area of the parallelogram formed by the two vector multiplicands. Because of the quantitative realities captured by the perpendicular and parallel product, the volume of solid with parallel sides and edges (parallelepiped) formed by three vectors is given by: $\left(\vec{A}\times\vec{B}\right)\cdot\vec{C}$.

Vectors can be extended to any number of dimensions. These extra dimensions are useful as one includes, by analogy, for instance, measured qualities and motions. Vectors can be thought of as depending on time, that is changing from one moment to the next. So, for example, the position vector of a marble moving across a surface would have one distance and direction from a given origin at one time and another a moment later. Such vectors are written: $\vec{X}(t)$, to make explicit this time dependence.

Calculus is a mathematical formalism applying the power of algebraic formalism (which itself generalizes the abstract concept of number) to continuous quantity, that is, to extension, one part next to another, as described above. Only by analogy is calculus then applied to motion.

Derivatives are defined as the tangent of a curve (for example, graphed on an *x-y* plot) at a point. The tangent or slope is defined for a line as: $\frac{\Delta y}{\Delta x}$. Since a curve is not a line, an approximation has to be made. This is done by breaking the curve up into many, Δx pieces as shown in Figure 2-27, which makes each such piece of the curve look more an more like a line. By making Δx as small as required, one can get as good an approximation as needed to the tangent line near any point. By a mental construct (being of reason), that can only exist in the mind (as there are no completed infinities outside the mind), one can think of completing this limit. Mathematically, i.e. using our symbolized formalism, one writes the derivative as:

2.52
$$y'(x)=\frac{dy}{dx}\equiv\lim_{\Delta x\to 0}\frac{y(x+\Delta x)-y(x)}{\Delta x}$$

The differentiation rule for polynomials is: $\frac{d(x^n)}{dx} = nx^{n-1}$. The product rule for derivatives is: $\frac{d(u\mathbf{v})}{dx} = \frac{du}{dx}\mathbf{v} + u\frac{d\mathbf{v}}{dx}$. The chain rule for differentiation is:
$\frac{d(g(f(x)))}{dx} = \frac{dg}{df}\frac{df}{dx} = \frac{dg}{df}f'(x)$

Integrals or anti-derivatives give the area under a curve by dividing the curve into small Δx segments and approximating the area by many small rectangles of width Δx and height *y(x)*. One can get as close as desired to the area by making the Δx as small as needed. By a mental construct, one can then write the following limit to get the area:

2.53 $$\int_{x_0}^{x_1} y(x)\,dx \equiv \underset{N\to\infty}{Lim}\sum_{n=1}^{N} y(n\Delta x) * \Delta x$$

The fundamental theorem of calculus proves that the integral and the derivatives are inverses.

Since c*alculus can be applied to motion* in a precise way, it is practically indispensable to the physicist. It is applied to motion by treating time quantitatively as if it were a line. Of course, time is not actual all at once like a line, the past is over and the future is not yet. However, time has aspects like a line, for example, its succession of moments, because finally all the categories of properties are related in reality. Treating time like a spatial variable, say *x* and letting $x \to t$, the tangent of a curve plotting position versus time of a uniformly moving object is the speed of the object: $v = \frac{\Delta y}{\Delta t}$. And, thus, the first derivative becomes the instantaneous speed; i.e. $v = \frac{dx}{dt}$. Similarly, the acceleration, the rate of change of the speed, is: $a = \frac{d^2x}{dt^2}$. Including all three dimensions of possible motion, one can then write the vector forms of these quantities, i.e. these quantitative descriptions of motion. That is, given the position vector $\vec{X}$ for a particle moving in space, the velocity and the acceleration are written: $\vec{V} = \frac{d\vec{x}}{dt}$ $\vec{A} = \frac{d^2\vec{x}}{dt^2}$.

The analogy that makes time a line can also be used to plot time as a third dimension for motion on a two dimensional surface; for example, taking *x* and *y* in the horizontal plane and time as the vertical or *z*-axis. For full three dimensional space, one can consider plots of *time as a fourth dimension.*

One can represent motion in any coordinate system. In rectangular coordinates, the unit vectors do not change, but in *polar coordinates* they do. In particular, they change with the angle giving the following derivatives:

2.54 $$\frac{d\hat{r}}{d\theta} = \hat{\theta} = - \sin\theta\,\hat{x} + \cos\theta\,\hat{y}$$

2.55 $$\frac{d\hat{\theta}}{dr} = -\hat{r} = -\cos\theta\,\hat{x} - \sin\theta\,\hat{y}$$

When we apply these to motion we get:

2.56a,b,c,d $\frac{d\hat{r}}{dt} = \dot{\theta}\,\hat{\theta}$ *and* $\frac{d\hat{\theta}}{dt} = -\dot{\theta}\,\hat{r}$ $\quad \vec{v} = \dot{r}\,\hat{r} + r\dot{\theta}\,\hat{\theta} \quad \vec{a} = \left(\ddot{r} - r\dot{\theta}^2\right)\hat{r} + \left(r\,\ddot{\theta} + 2\dot{r}\dot{\theta}\right)\hat{\theta}$

Calculus gives us ways to approximate functions. For example, the *Taylor series* is often useful in this regard:

2.57 $$f(x) = \sum_{n=1}^{n=\infty} \frac{df}{dx}(x_0)\frac{(x-x_0)^n}{n!}$$

Terms

General Math and Vectors

Space

- Symmetry
- Infinity
- Length and direction
- Coordinates
  - Rectangular Coordinates
  - Polar Coordinates
  - Spherical Coordinates
  - Mapping
- Curves in space

Scalars

Measurement of length

Vectors

- Arrows
- Normal Vector
- Unit Vector
- Components
- Addition, subtraction
- Multiplication
  - Dot (parallel) Product
  - Cross (perpendicular) Product
    - Right Hand Rule
  - Degenerate Cases
- Vector Fields

Analogical generalization of Vectors

- Intensity Vectors
- Math applied to Change
  - Displacement Vector
  - Velocity Vector
  - Space-Time Vectors
- General definition of Vector

Calculus

Study of curves and areas.

Limits and Infinity

Differentiation

Tangent Vector

Derivative of functions

Integration

Areas

Integrals of functions

The Fundamental Theorem of Calculus: integral as an anti-derivative

Calculus applied to motion:

Velocity

Speed

Acceleration

Space-Time Curves, Indices to label such vectors.

Approximation techniques:

Taylor Series

Maclaurin Series

Calculational Techniques

The following vector formulas can be useful in calculations:

$$\vec{a}\cdot\left(\vec{b}\times\vec{c}\right)=\vec{b}\cdot\left(\vec{c}\times\vec{a}\right)=\vec{c}\cdot\left(\vec{a}\times\vec{b}\right)$$

$$\vec{a}\times\left(\vec{b}\times\vec{c}\right)=\left(\vec{a}\cdot\vec{c}\right)\cdot\vec{b}-\left(\vec{a}\cdot\vec{b}\right)\vec{c}$$

$$\left(\vec{a}\times\vec{b}\right)\cdot\left(\vec{c}\times\vec{d}\right)=\left(\vec{a}\cdot\vec{c}\right)\left(\vec{b}\cdot\vec{d}\right)-\left(\vec{a}\cdot\vec{d}\right)\left(\vec{b}\cdot\vec{c}\right)$$

In solving problems, remember to:

- Choose a coordinate system that fits the symmetry of the system that is under consideration.
- Draw a diagram of the system and label coordinates to get the problem organized in your thinking.
- Make approximations using Taylor series, or the like, when possible to simplify the mathematics.
- Memorize the basic derivative and integral rules as well as derivatives and integrals of common functions. Develop skills in deriving derivatives and integrals from memorized information.

Problems by Subtopic

General Mathematics and Vectors

1. Where does the idea of a plane come from? Describe how you might help someone see it.

2. The idea of number is one level higher in abstraction than continuous quantity, which we get by leaving out all categories-- to the degree they can be left out-- but extension (quantity). Continuous quantity is what we study in geometry. If we count the number of parts (units) in a given continuous quantity (say bounded by a shape), we get a number. Explain how the use of addition in algebra is a generalization of the concept of number.

3. For the following, list what category(ies) and/or motions are involved and give an example of how a vector might be used.
a) Primary colors of light: blue, green and red.
b) Growth of a plant.
c) Number of marbles in a test tube
d) Earth in spring versus Earth in fall

4. How is it that we can define a location in a space if we have left out the concept of place? *Hint*: In the process of leaving all behind but quantity, consider what remains.

5. Draw the two dimensional cross section for the following vector fields defined on the surface of Mars:
a) the vector field representing the direction and magnitude of the pull of gravity of a region along the equator, assuming that there is a uniform downward pull everywhere.
b) the vector field near a mountain on Mars (such as Olympus Mons) representing the tangent to the surface at every point. (note: no need to draw Olympus Mons, just free hand draw a mountain cross section of any shape)

6. Space is thought of as "waiting" to be divided in some particular way. In this way, one thinks of space as being uniform and undistinguished from one place to the next. For the following, identify what coordinate system would best describe how the uniformity is broken and thus what would be the best coordinate system to use.
a) a bowling ball
b) the trailer on a semi truck
c) a saucer
d) a man standing
e) a surface of a piece of paper

7. Evangelista Torricelli (1608-1647) was a student of Galileo. Torricelli investigated the interesting surface illustrated below, which for obvious reasons is called Gabriel's Horn. He was amazed to discover the infinitely long solid with a surface that he calculated to have an infinite area, but a finite volume! This problem investigates this paradox, of which the philosopher Thomas Hobbes said: "To understand this for sense, it is not required that a man should be a geometrician or logician… but that he should be mad."

Gabriel's Horn is made by rotating the function $f(x)=1/x$, which can be thought of as a curve in an *x-y* plot, about the *x*-axis for the interval $x \geq 1$.

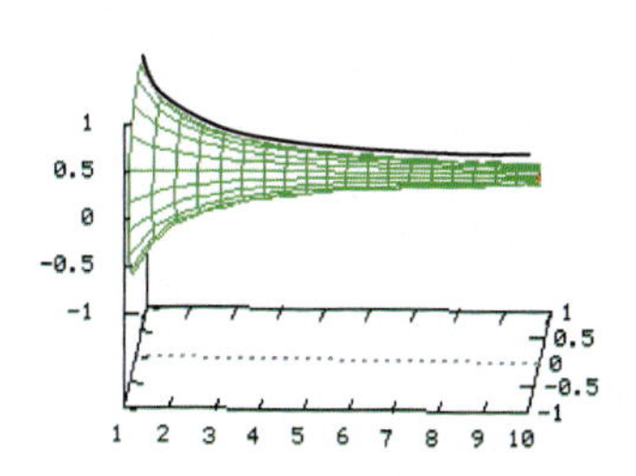

a) Calculate the surface area of the horn in the following way.
Divide the surface up into regions bounded by rings of circumference $2\pi f(x)$ and width, Δs,

where: $\Delta s = \sqrt{1+\left(\frac{d f(x)}{dx}\right)^2}\ \Delta x = \sqrt{1+x^{-4}}\ \Delta x$.

i) Show that the equation for Δs, i.e. the length of a small segment along the $1/x$ curve, follows from the Pythagorean Theorem, treating Δs as approximately the hypotenuse of a triangle of base Δx and height Δy. Integrating over all these regions of approximate area $2\pi f(x)\ \Delta s$ from $x=1$ to infinity gives the following integral: $S = \int_1^\infty \frac{2\pi}{x}\sqrt{1+\frac{1}{x^4}}\, dx$. ii) Show this integral must diverge, i.e. must go to infinity.

b) Calculate the volume of the horn in the following way.
Divide the horn into an infinite number of disks; each disk will have a radius of $f(x)=1/x$, and so will have an area of π/x^2, and each disc will have a thickness of *dx*. Thus, the integral, evaluated from 1 to infinity is:

$Volume = \int_1^\infty \frac{\pi}{x^2}\, dx$, evaluate this integral to get the volume.

c) Explain how it could take an infinite amount of paint to cover the outside surface but a finite amount to fill the horn, though filling the horn implies covering the internal surface, which is the exact same area as the outside surface.

d) Explain how the inside volume can be finite while the surface area is infinite

8. While completed infinities are themselves mental constructs (mind dependent objects), they, like all aspects of mathematical formalism encase realities, (though physicists have found no use for them, some mathematicians have). Gregor Cantor created a system for formally handling completed infinities by using the concept of one-to-one correspondence to define an equality of "size" between various series of numbers. For example, he showed that the real numbers are uncountable, i.e. cannot be put into one-to-one correspondence with the integers. Explain how this is really a restatement of what is implicit in the definition of the real numbers.

9. Explain how in thinking about only the extension of things in the room, we tend to create a mental construct (a being of reason) in relation to the various substances in the room.

10. A line segment cannot be actually divided an infinite number of times, yet to find the area under a curve, we take the limit at infinity. Explain how this can be true using the mathematical definition of a limit.

11. Draw the vector $3\hat{x}+2\hat{y}+\hat{z}$ on a rectangular coordinate grid.
Express the curve $y = m\,x + b$ in polar coordinates
Express the curve $y = x$ in spherical coordinates (r,θ,ϕ):
Express the plane: $y = mx + b$ in spherical coordinates (r,θ,ϕ)

12. The same point can be labeled in any coordinate system:
   a) Express the point (x, y, z) in cylindrical coordinates.
   b) Express the point (r,θ,ϕ) in cylindrical coordinates.
   c) Express the point (x,θ,z) in Cartesian coordinates, then in spherical coordinates.

13. Answer the following using the diagram of Sun, Earth and Mars below where * marks an observer on Earth and given that the east is to the observer's left when he faces toward the page:

a) For the observer on Earth, draw a local coordinate system with unit vectors pointing to the eastern horizon and the zenith (the point directly overhead for the observer). Note that the vector representing the zenith direction is normal to the local surface of the Earth and is called a *normal vector*.
b) Given the configuration shown, where will the Sun appear in the sky for the observer? Is it rising or setting?
c) Where will Mars appear in the sky for the observer?

14. You find a pirate's map that has the following directions:

- Proceed 200 yards north from the center of the island.
- Proceed 50 yards south-east from that point.
- Proceed 20 yards, west and
- dig 5 yards down.

a) Draw the vectors.
b) Write the two dimensional unit vector description of the directions.
c) Add the vectors together graphically
d) Add the vectors using unit vector notion and write the total vector.

15. Crystals are important for many reasons, not the least of which is their importance in semiconductor electronics. Crystals can have many different types of internal grain structure. These are often specified by vectors. a) Draw the most natural unit vectors that would specify the two dimensional structure of atoms shown below. b) Write these vectors in terms of x and y unit vectors.

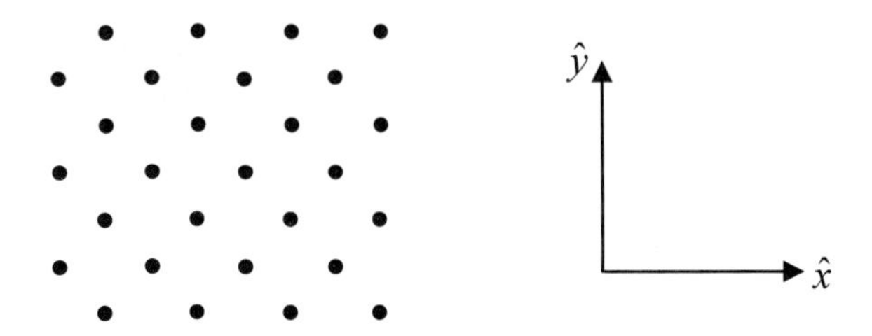

16. To define multiplication by a scalar, α, one simply adds the given vector α times. When we try to generalize this to the multiplication of two unit vectors, we confront the two problems below. Answer the two questions for the dot product and the cross product.

a) How do we interpret adding $\hat{x}$ to itself $\hat{x}$ times ($\hat{x} * \hat{x}$)
b) How do we interpret adding $\hat{x}$ to itself $\hat{y}$ times ($\hat{y} * \hat{x}$).

Where, here, "*" is a general representation of a product.

17. In two dimensions, expand out a) the dot and b) cross product of two vectors and show, using the respective product of unit vectors, how one obtains the standard results. c) Prove the cross product result given in equation 2.14 .

18. Show how the definition of cross product as a perpendicular product (equation 2.13) gives the area of the parallelogram formed by the two vectors.

19. In defining the product of two vectors, we confront the question of how to *analogically* extend our ordinary definition of multiplication of numbers. In particular, we are confronted with the problem of multiplying directions. Consider a flat map of the Earth in answering the following.

a) First, confining oneself to the surface of the map: What is north times east? Give the most natural answer. *Hint:* What, if anything, is common to north and east in the plane of the map?

b) Now, answer the question ***without*** confining oneself to the surface of the map. *Hint:* What is "common", i.e. something both share in fullest sense, in this case, considering what direction we can go without mixing in east *or* north?

c) Explain how the answers to part a) and b) relate to the concept of dot and cross product.

Note: It is such real aspects of quantity (extension in three dimensions) that we have incorporated into our vector formalism.

20. Given a unit vector $\hat{u}$ and another vector $\vec{V}$:

a) Draw a graphic representation of the magnitudes of both the dot product and cross product of these two vectors. b) Which type of product gives the magnitude of $\vec{V}$ along the *u*-axis"? c) Find the projection of $\vec{V}$ onto another axis $\hat{t}$ that is perpendicular to $\hat{u}$.

21. In "crabbing," an aircraft pilot balances the crosswinds at landing by engine thrust. The pilot points the nose of the plane into the wind *as seen from* his moving vantage point. Considering that there is a component of thrust along the runway and another along the direction of the crosswind, draw vectors for each component, and also draw the resultant thrust vector needed at landing. Assume the crosswind is at exactly a right angle to the runway.

22. Graphically show that the area of a parallelogram with sides *A* and *B* is $AB_{\perp}$ by moving the triangle formed by the height line over to make a square (cf. Figure 2-21)

23. a) Write the dot and cross products for all the possible combinations of cylindrical coordinate unit vectors (i.e.,$\hat{r}, \hat{\theta}, \hat{z}$, polar coordinates with the vertical rectangular direction $\hat{z}$ added in the normal manner). b)Write the dot product and cross product for all possible combinations of cylindrical *and* rectangular unit vectors, i.e. each cylindrical unit vector times each rectangular one.

24. Draw a graphic representation of $A \cdot B_{\parallel}$ and $A_{\parallel} \cdot B$ as described in equation 2.7. Are the projections equal in magnitude? Why?

25. Can you determine the magnitude of a vector given the dot and cross products? What about direction (angle) assuming the component vectors are parallel to the axes? Do the same thing for a vector whose tail is not at the origin.

26. Using the cross product to test which side of a line a point is on, determine whether a point is inside or outside the triangle.

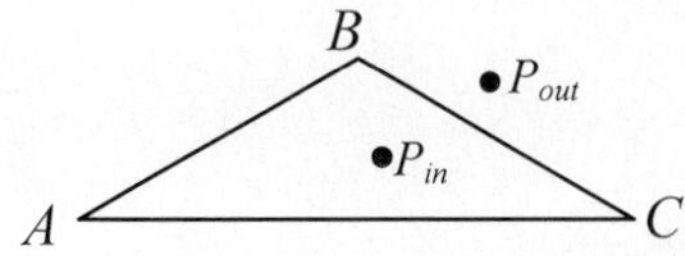

27. Prove that the magnitude of the dot product of a vector with itself is $A \equiv |A| = \sqrt{A_x^2 + A_y^2 + A_z^2}$.

28. Show that the dot product (like the cross product) can be thought of in terms of multiple projections *onto unit vectors*.

29. How would you naturally assign a direction to the dot product, what reality would it capture and what reasons can you give for why this would not be as good of a generic choice as making the dot product a scalar? (Use * to signify this new product)

30. Show that equation 2.14 is true by calculation using the relations 2.12 and simple algebra.

31. Prove the following vector identities:

a) $\vec{a} \cdot \left(\vec{b} \times \vec{c}\right) = \vec{b} \cdot (\vec{c} \times \vec{a}) = \vec{c} \cdot \left(\vec{a} \times \vec{b}\right)$

b) $\vec{a} \times \left(\vec{b} \times \vec{c}\right) = (\vec{a} \cdot \vec{c}) \cdot \vec{b} - \left(\vec{a} \cdot \vec{b}\right)\vec{c}$

c) $\left(\vec{a} \times \vec{b}\right) \cdot \left(\vec{c} \times \vec{d}\right) = (\vec{a} \cdot \vec{c})\left(\vec{b} \cdot \vec{d}\right) - \left(\vec{a} \cdot \vec{d}\right)\left(\vec{b} \cdot \vec{c}\right)$

32. The direction cosine of a vector is defined as the cosine of the angle it makes with one of the three coordinate axes. Thus, a given vector has three direction cosines, one associated with each axis, *x*, *y* and *z*; these are called α, β, γ.
Show: that the sum of the squares of these cosines is unity; i.e. $\alpha^2 + \beta^2 + \gamma^2 = 1$.

33. Given a vector, $\vec{A}$, and a unit vector, $\hat{u}$, show that these vectors satisfy: $\vec{A} = \left(\vec{A} \cdot \hat{u}\right)\hat{u} + \left(\hat{u} \times \vec{A}\right) \times \hat{u}$

34. Demonstrate the law of cosines, $c^2 = a^2 + b^2 - 2ab\cos C$, by using the dot product, the cross product, the Pythagorean Theorem and the right triangle made by adding the *x*-axis and perpendicular shown in blue below :

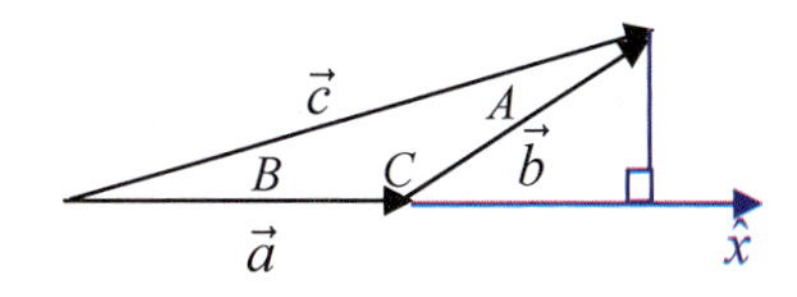

35. Find a unit vector, $\hat{n}$ perpendicular to:
$2\hat{x} + \hat{y} + \hat{z}$ and $\hat{x} - 3\hat{y} + 2\hat{z}$

36. As defined do the dot and cross product have a corresponding division operator?

37. The commutative law of multiplication does not apply to the cross product. We say the cross product anti-commutes, and write $XY + YX = 0$ or $\{X,Y\} = 0$. Where X and Y are operations, that is, actions applied to some other object, so that XB means "do X to B;" if X is a rotation and B is a vector, it means rotate the vector B. The anti-commutation in our cross product, which is part of our vector formalism, arises because of the very nature of rotations in three dimensions. More generally, we can simply say that $XY - YX \neq 0$ *or* $[X,Y] \neq 0$. By analogy, we can find other non-commutative operations. For instance, we can take the product operation $Y * B$, where Y is "put on socks" and B is feet, to mean "put socks on feet."

a) Taking X to be shoes, setup a non-commutative system, i.e. show $XY - YX = 0$, taking zero to mean one has returned to the original state.

b) Give another example, by defining operation and things on which the operation is to be done.

This discussion is important in a branch of mathematics called group theory that is crucial to modern physical theory.

38. Knowing what laws of symbol manipulation apply to vector multiplication (which, in turn, is indicative of some fact about the quantitative realm) is helpful in problem solving.

a) Prove the associative law for cross product

b) Prove the distributive law of multiplication over addition for cross product.

c) Is there an identity element for cross product? What about the dot product? Explain.

39. A scalar field assigns a number to every point (which physically means some average over a particular region) on a surface or volume. The temperature of the surface of Mars was mapped by Mars Global Surveyor. The figure below gives an approximate representation of data collected for temperatures near the southern pole during the southern spring on Mars. Dark blue is the coldest $-190° F$ ($150K$) and dark red is the warmest $26° F$ ($270K$). Assume that the colors in the diagram represent a temperature ranging from 150 to 270 the dark blue end to the dark red end ranging with the whole scale moving through the color hues in this order: blue, green and red. Then, create a table which lists the location of the center of each region in polar coordinates (using coordinates that are centered on the South Pole) along with the region's temperature in Kelvin. Use the notation $\{(r,\theta)$, temperature$\}$. A region is defined as a patch of uniform color separated by white space or color difference. Do not forget to consider the large circular region centered on the pole. In this way, one creates a number representation of the scalar field of the temperature on the surface. A scalar temperature field would mathematically be written: $T(r,\theta)$.

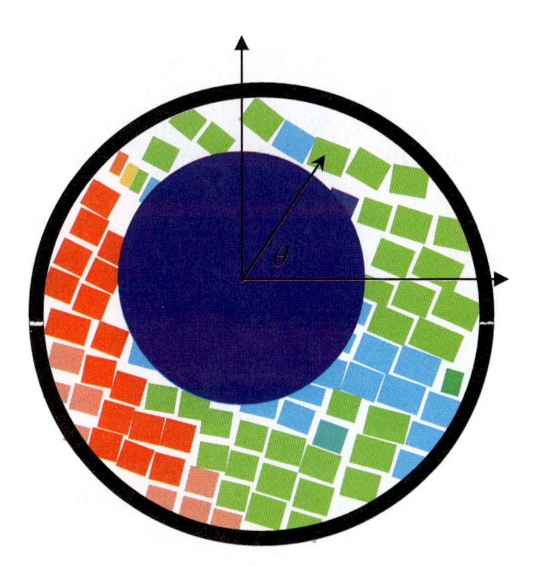

Temperature distribution around Martian southern pole for problem 39.

40. Explain how the vector magnitude $\left|\vec{A}\right|$ can be made the same as a one dimensional absolute value.

41. Using the lake example shown in Figure 2-4, consider the brightness as a vector with a fixed direction, say up for easy visualization, and combine this with the position vector for the cell under consideration. Define the position vector from an origin centered on the lower left corner of the lake. Next, add the brightness vector to the position vector. In this way, create a vector with x and y components that give the position (place) of the cell and a z component that represents the intensity of the light (quality).

a) Discuss in what sense the difference vectors between neighboring cells would correspond with the vectors defined in the figure.

b) For a cell of your choosing in the middle of the lake shown in Figure 2-4, write the rectangular coordinate form of: i) the brightness vector (taking the x and y components to be zero by definition) and ii) the position vector (taking the z component to be zero by definition), and then for iii) the sum of the two.

42. By analogically generalizing space by using it to represent other categories in addition to extension, we can have vectors and spaces of dimension greater than three. Consider the simple case of one dimension of space and one dimension of speed. Such a space is called a phase space. a) In rectangular coordinates, draw a phase space for the motion of a uniformly accelerating ball in one dimension.

b) Watch someone walk for a few seconds, draw a phase space plot for one spatial dimension of their motion.

c) Make a three dimensional velocity vector for the motion of a thrown baseball. Take the y-axis to be north and the z-axis to be vertical.

43. Given the climb up a mountain pass takes you up 100 ft for every 1000 ft:

a) Calculate the angle of the upward slope of the road

b) Calculate the *grade* of a road, the tangent of the slope angle expressed in percentage, down a mountain.
c) Using Cartesian unit vectors, write the vector with magnitude of the grade and direction corresponding to moving upward along the slope.

44. Consider the wind velocity at every point in the air on the surface of the Earth. This can be represented as a vector field. Discuss the limitations of thinking of this as arrows at each point in three dimensional *space*.

45. Consider the vector field representing the corn field. Discuss what happens if we actually were to have a corn stalk at every point so that we could have a continuous vector field on a 2-D plane.

46. In the case of a man walking some distance to the north, discuss in what sense the vector representing his motion exists in the space in which he walked and in what sense it does not. In this way, show that the vector of the man's motion is an analogical use of the word vector based on its primary meaning as actual extension with an associated direction.

47. Derive a rectangular coordinate representation for the vectors for each stage of each of the man's journeys illustrated in Figure 2-16a and b. For each journey, also write the resultant vector in rectangular coordinates.

48. Draw a vector with its tail at the origin and its head at (1,1,1). Perform a translation of the coordinate system by one unit in the (1,1,1) direction. a) write the equation for this passive translation. b) what is the new equation for the tail and head of the vector?

49. Write the following vectors in coordinate form:

a) The position vector for a car at the bottom up a 30 degree incline, given the origin is at the bottom of the incline.

b) The position vector for the car 1 mile up the incline a minute later.

c) The incline ends in another mile, and the car proceeds at the same speed on flat ground for another mile. Write the position vector when the car reaches that point.

d) The displacement vector for each change in place.

e) The velocity vectors that corresponding to each displacement vector.

50. A plane is flying north toward an airport at x miles per hour. The wind is blowing directly west at a constant speed of y miles per hour. At what angle will the pilot have to fly the plane in order to use the least amount of fuel?

51. On a two dimensional circular map of the globe sketch the vectors representing the wind patterns corresponding to the following facts (remember to mark the latitude lines)

- **Polar Easterlies:** From 60-90 degrees latitude.
- **Prevailing Westerlies:** From 30-60 degrees latitude (aka Westerlies).
- **Tropical Easterlies:** From 0-30 degrees latitude (aka Trade Winds)

Kinematics

(The study of the path of locomotion of an object, not the cause of the motion)

52. Draw the following plots of position versus time plots:

  a) uniform speed of $1 m/s$

  b) uniform acceleration, $10 m/s^2$ such as that due to gravity (approximately) near the Earth's surface.

53. Given that a bike, with wheels of radius r, is moving at constant speed V. Further assume the wheels are not slipping. Calculate the equation of motion, i.e. x and y position as function of time, of one of the pump valves in the wheel. Write the answer in vector notation.

Calculus

54. Prove the product rule for derivatives.

55. Prove the product rule for integration. $\int_a^b u dv = \Big|_a^b uv - \int_a^b v du$

56. Find the general formula for the derivative of a curve represented by a polynomial.

57. Show that the derivative of the Maclaurin series gives the correct answer for the nth derivative using the general formula for the derivative of a polynomial.

58. You are attempting to launch a water rocket against the "enemy" who is hunkered far down an incline. Given: the incline makes an angle α with the local horizontal, gravity causes a uniform acceleration g over and above the initial speed, $V_0 \sin\theta$, given in the vertical direction, and the horizontal speed remains unchanged from the initial speed, $V_0 \cos\theta$ given at launch until hitting the ground.

  a) Write the equation of motion for the horizontal direction.

  b) Write the equation of motion for the vertical direction.

  c) At what angle θ must one launch the rocket to maximize the distance it travels in the horizontal direction?

59. Expand the function $(1+x)^N$ for $x<1$, and compare the result to that obtained for the binomial theorem applied to the function.

60. Expand the function $\ln(1+x)$ for $x<1$.

61. Verify the expansion of $\frac{1}{1-x}$ for $x<1$ given in the chapter.

62. Using Taylor series approximate the value of $e^{0.5}$

63. If the yearly interest, r, on a savings of X dollars is compounded every period (say every month), the formula that gives how much money one has in the bank after n months is $P(r) = X(1+r/12)^n$

a) Expand this in a Taylor Series for the first 3 terms and then write the equation for the mth term of the series.

b) Given an interest rate of 5% and $1000 initial amount how much money would you have after 5 years?

c) What does the 3 term Taylor series expansion give and what is the error?

d) If the interest is compounded more frequently say for N times a year the equation becomes $P(r) = X(1+r/N)^{mN}$, where m is how many years the interest is allowed to accumulate. What is the formula if one compounds continuously?

64. Write the equation for the tangent vector to the curve $y = x^2$ valid for each point on the curve.

65. Given that the motion of two particles as a function of time is given by:

i) $y = -\frac{1}{4}t^2 + 1, x = 1$ ii) $y = -\frac{1}{4}gt^2 + 1\ x = \mathbf{v}t,\ where\ \ \mathbf{v} = 1$

a) Draw the x-y plots for the paths traced by each particle.

b) Draw the velocity vectors at several points along these paths for each particle.

c) Given the general equation of motion for a particle $x = f(t),\ \ y = g(t)$, write, in Cartesian coordinates the vector for the tangent at any point and explain why the such velocity vectors are always tangent to the spatial path traversed by the particle.

66. Calculate the acceleration of a particle in terms of polar coordinates (show all steps)

67. Express the curve $y = x$ in polar coordinates

68. Consider an ant walking at uniform speed, v, from the center of a uniformly rotating wheel, in polar coordinates, calculate the position vector and velocity vector for the ant.

69. Show the mean value theorem for uniform acceleration.

70. If a planet's radius were to change by Δr, a small fraction of its radius, r, what would its *approximate* percentage increase in volume be?

71. Graphically prove that, in polar coordinates: $\frac{d\hat{r}}{dt} = \dot{\theta}\hat{\theta}$.

72. A bike with mud on the wheels leaves a zig-zag pattern on the pavement, how can you tell which direction the bike was going. Draw a diagram of the path and use vectors to explain your answer.

73. Given the following equations of motion for a thrown object, write the three dimensional vectors for the velocity and the acceleration of the object.

$x = \mathbf{v}\cos\phi t + x_0 \;\; y = \mathbf{v}\sin\phi t + y_0 \;\; x = -\frac{1}{2}gt^2$

74. Draw a three dimensional plot (with two spatial dimensions and one time) of an object moving at uniform speed along the direction $\frac{1}{\sqrt{2}}(\hat{x}+\hat{y})$.

75. In polar coordinates, if a particle is moving in a path of uniform angular acceleration, is it possible for there to be no radial acceleration? Is there a solution for uniform angular *speed?*

76. To begin to consider artificial gravity for a trip to Mars, think about the following:

a) How fast will a spacecraft that has a diameter of 100 yards have to spin in order to cause an acceleration inward of $10m/s^2$ (which is the same as that due to gravity on Earth)?

b) How big a spacecraft is needed if we want a full $g = 10m/s^2$ (Earth's gravity) but also want to avoid noticing a difference in acceleration between your head and feet? Suppose you cannot sense less than $.2g$.

77. Show that: $\sum_{n=1}^{N} n = \frac{N^2+N}{2}$.

Chapter III

Momentum

Introduction

We are now finally ready to broach the central topic of this text, *motion*. The ability to change is the defining aspect of material things. We are particularly interested in locomotion, motion from place to place. By motion from place to place, objects interact and change; for instance, a chemical may be *moved* into a test tube to cause a reaction. Furthermore, even in the reaction itself, motion is involved. We really cannot say much about physical things if we don't deal with motion. As mentioned earlier, it was not until the Middle Ages that the concepts needed for a *specific* understanding of motion were introduced.

In particular, the key concept in all of modern physics, which extends the understanding motion beyond the foundation given in Chapter 1, is the concept of *inertia*. Aristotle, who had so successfully laid the foundational part of physics, foundered severely when it came to inertia. As we have seen, the issue was first successfully met by Philoponus, an Eastern Christian, and then later by men of the high Middle Ages. The high Middle Ages saw the foundational principles of physics laid by Aristotle not only formulated in a more clear way but assimilated into the culture, indeed thoroughly enculturated. The foundations were truly made part of the thinking of a culture largely by the great thinker St. Thomas Aquinas (1225-1274). Aquinas was also responsible for one of the first formulations about inertial mass, opposing not Aristotle's principles but in subtle ways his mechanics. Aquinas was not to be the one to accomplish the revolution in mechanics, but to lay the firm foundations for science in a culture, whose Christian faith in the goodness and rationality of the physical world was able and ripe to receive it.

Impetus

To understand what inertia is, we start, as we ultimately must, with what we see with our senses directly. We see that if you throw a baseball, the ball keeps moving even after you release it. Aristotle suggested that air might keep it in motion by rushing in behind it, but medieval thinkers debunked this idea. John Buridan, for instance, says:

> *...after leaving the arm of the thrower, the projectile would be moved by an impetus given to it by the thrower and would continue to be moved as long as the impetus remained stronger than the resistance, and would be of infinite duration were it not diminished and corrupted by a contrary force resisting it or by something inclining it to a contrary motion.*[1]

Something must keep it moving. They called it impetus, a quality given to a body that keeps it moving even after you stop pushing it. Buridan even noted that it was proportional to speed and mass. This is what we today call momentum.

Precisely what is impetus? It is a quality of a body that keeps it moving even after all forces cease to act on it. Impetus is not essential to a body, not part of its primary nature, but it must have the capacity to have impetus to be a body; for instance, a baseball can have impetus or not and yet still be a baseball, but it must have the capacity to have impetus, for this is how it moves. Thus, we say impetus is *second nature* as we might say Sam's ability to ride a bicycle is second nature. We might not say so of Joe, though we know both are capable of riding a bike, for that's part of their nature as healthy children. Sam has acquired the skill but Joe has not. Once obtained, it's as if the skill was a part of them, but it could still be lost. Thus, we say:

> ***Impetus*** is the power (category of quality) activated in a body that moves it at a constant speed in a particular linear direction. It is *second* nature to the body, because it can be gained and lost without the body changing its fundamental nature.

[1] XII.9: 73ra of *Subtilissimae Quaestiones super octo Physicorum libros Aristotelis*,(1509) Paris. Rpr. 1964, as *Kommentar zur Aristotelischen Physik*, Minerva, Frankfurt a. M. cf. http://www.science.uva.nl/~seop/archives/sum2002/entries/buridan/#Bib

John Buridan was born c.1300 in France (d. 1358). He taught at the University of Paris in a time when the Catholic Church was the major intellectual force. He recognized, and gave strong arguments in support of what we now call inertia, including stating that the intensity of the impetus is proportional to mass and speed (*mass*speed ~ momentum*). In his physics, he used Aristotelian principles to critique some of Aristotle's conclusions.

He recognized that heavenly bodies are probably similar to earthly bodies. This line of thought, like the idea of inertia, is usually claimed to have originated with Galileo and Newton in the 1600's; in fact, European thought had been digesting these ideas, in particular inertia, for quite some time by then. In addition, before its rediscovery in the *High* Middle Ages, Philoponus had already posited both inertia and the generic nature of bodies in c. 500AD.

Among the first record of inertia reappearing occurs when a monk named De Marchia (1290->1344AD) used impetus as an analogy in explaining the Christian sacraments. Buridan's statement below makes clear his argument against the standard "Aristotelian" explanation and the need for impetus:

> *..., if you should say that the surrounding air moves so great a weight [a mill wheel] circularly after a man ceases to move it, I would object. Because if you should take a rag and wipe the contiguous air away from the wheel, you will not stop the wheel in this way... If the air which I set in motion when I throw a stone can move the stone, why will it be that if I blow the air at you as swiftly as I can without the stone you can hardly feel it?*
>
> *Therefore... the mover impresses on the moved thing not only motion, but along with it a certain impetus or some force or other quality—not the kind of force we usually mean by that name--which impetus has the nature of moving that thing on which it is impressed, just as a magnet impresses on iron a certain force moving the iron to the magnet. And the more swift the motion the more intense the impetus will be. And this impetus in a rock or arrow is continually diminished by the resistance contrary to itself until it is no longer able to move the projectile.*[1]

Buridan's Culture:

Medieval Europe is beginning its first renaissance at the turn of the first millennium, as evidenced by the founding of the very first Universities not in just one city but all over Europe. St. Thomas Aquinas (himself making pregnant statements relative to inertia) and his teacher St. Albert the Great (1200-80), are key figures in enculturating the basics of scientific thinking, especially through introducing, advancing and applying the foundational principles of science as discovered by Aristotle. The medieval culture is able to receive the scientific outlook not only as an achievement of a few individuals of genius but as a part of its own world view. The fertile ground science had in the High Middle Ages is seen in the most quoted bible verse of that age: "Thou hast ordered all things in measure, and number, and weight" (Wis. 11:21).

[1] *Questions on the Heavens and the World* Book II Question 2 by John Buridan in *The Scientific Achievements of the Middle Ages,* Dales, pp. 120-121 from Marshal Clagett's *The Science of Mechanics in the Middle* Ages (1959) pp. 552-556, 570.

The impetus has a direction and an intensity. We call the *measure* of the impetus, the ***momentum.*** There are thus two aspects to momentum, direction and magnitude. We can measure the direction by comparison to a standard coordinate system, say the positive x-direction; this is called the *direction* of the momentum. We can treat the *intensity* of the impetus as a quantity by noticing the *similarity* between an increasing intensity and increasing the separation of two points on a line. In particular, we can measure the magnitude of the intensity by comparing it to some standard unit of intensity thus specifying a number for the intensity. For example, as we've discussed, just as we can specify how many standard units of length (say centimeters on a ruler) make up the length of a given object, we can also specify how many times brighter a bulb in one's house is compared to that of some standard bulb. In this analogous way, we reflect the intensity of the impetus into quantity, the first property of all material things. We've thus enabled a transition to mathematics, the study of quantity, so that everything we've discovered in mathematics can now be applied.

Clearly, as mentioned in Chapter 1, there is a distinction between 1) *the measurement,* a comparison between two things (for instance, how many more times intense one impetus is than another) and 2) those things which are compared, the impeti themselves (qualities). One is a *relation* expressed as a number, the other is a *quality*. However, to avoid cumbersome wording, we will not always make this distinction explicitly, but will allow the measurement (the momentum) which always implies that which is measured (the impetus) and vice-versa to speak silently.

We assign the momentum, the measure of the impetus, the symbol p. This standard use of p comes, as far as I can tell (a very reasonable educated guess), from the p in the word **impetus**, which is Latin and breaks into *im petere*, which means ***in reaching forward***. This phrase describes the impetus (and thus, in a way, its measure, momentum) by noting that a body has acquired a quality ***in*** it and that causes the body to ***reach forward***. Obviously, reaching forward, *petere*, is the crucial element under consideration; hence the symbol p.

Momentum, Speed and Mass

Like all things, impetus is ultimately known via the senses.[2] In this case, this means we will need experiment. In particular, we seek a practical way to measure the intensity of the impetus of a body. Before determining this method, we start simply with ordinary observation and then move to more precise experiments. We need to know the big picture in order to put the details into context.

In everyday experience, everyone knows slowly moving objects, automobiles for example, are less dangerous than fast moving ones. Further, you'd rather be hit by a feather going 10 miles an hour than a bowling ball moving at the same speed. Said another way, a slowly moving bowling ball can cause the same "amount of push" as a fast moving baseball. These are rough observations and rough ways of speaking. Much more can obviously be learned by controlled, well-defined experiments with accurate measurements. We do this in stages. In particular, to understand impetus and determine what p is quantitatively, we do a series of simple experiments which are successively more quantitative.

[2] For instance, even the fact that we know is first known by reflecting on things we know through the senses.

Consider a set of blocks moving on a near frictionless surface—e.g. an air table (see Figure 3-1). Take two blocks of equal mass, m. Set one moving at speed v, the other moving at speed $2v$. To the degree we have eliminated friction, each keeps moving at its initially given speed. If we try to stop them, we confirm that the one moving twice as fast is much harder to stop. Also, we note that the one moving at speed $2v$ is about as hard to stop as one that is moving at speed v but has mass $2m$.

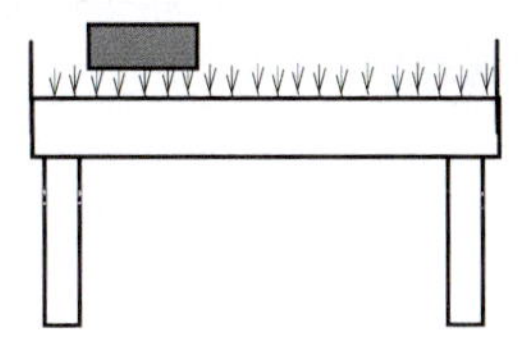

Figure 3-1: An air table has a surface perforated with many small holes laid out in a grid through which air is forced, creating a blanket of air upon which a puck or light block can slide. The blanket of air offers very little resistance to the motion of such objects.

To be more precise, we replace the experimenter stopping the block with his hand with a very massive block, M that is 100 times heavier than the impacting block, which has mass m.[3] Put the M block on one end of the table, and shoot the m block at it at various speeds as shown in Figure 3-2. Assume that they collide elastically; that is, assume they are made of a superball-like (very bouncy) material so their motion is not damped on impact—technically, elastic means no energy is lost to heat (cf. later section in this chapter and Chapter 5). In this case, we notice that the incoming block always bounces off and then returns at about the *same speed* at which it came in. If we vary the speed of the projectile block, we find that, as long as the collision stays elastic (bouncy), the return speed is always pretty much the same (to zeroth order in the ratio m/M) as its ingoing speed.

> *Note:* As mentioned in the preface, Interactive Physics (IP) files and video clips can be found at *www.iapweb.org/interactivephysics*. The IP files require the IP program, while the videos can be played on any computer. To access the directory, you will need your username: *pfr* and password: *m3chanics*. We recommend you spend sometime playing with the given files and then try to make your own. Though this kind of experimentation is *not* a substitute for actual physical experiments, it *is* nonetheless important and can give you insights into the principles of Newtonian mechanics that cannot be as easily attained any other way. In the text, whenever there are such files available for a given discussion, like now, it will be indicated by the orange box shown below.

Interactive Simulation

[3] To do this, we may have to move our experiment to a smooth ice table in order to accommodate the heavy mass. Dealing cleverly with such problems is the key to successful experimental physics. However, we are here primarily interested in the conceptual side, so we will leave such practical details as an exercise (see problems at end of chapter). In the meantime, please refer to simulation *3.1-m elastically collides with M.ip* to help visualize the results we discuss here ; in this simulation, one can adjust the value of both masses. The key is that we are assuming a near frictionless surface.

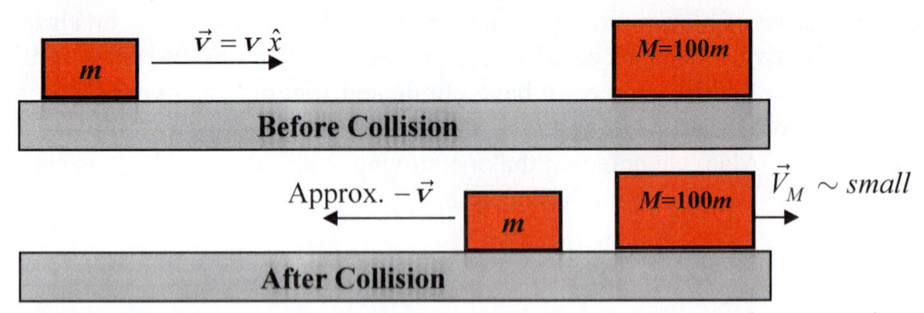

Figure 3-2: Block m on frictionless surface moves at speed v toward M, which begins at rest. After colliding with M in a bouncy (superball-like) way, m moves at about the same speed but in the opposite direction. After the collision, M barely moves, but it moves.[4]

If we increase the mass of the large block, M, keeping the mass and speed of the incoming block the same, we find, after detailed measurement, that the speed of the large block (which is slow) is inversely proportional to its mass. That is: $V_M = Constant/M$. What do we conclude from the above experiments?

Obviously, the impetus, that quality of a body that moves the body uniformly in a given direction, is manifested by the body's speed. Once something activates impetus in an object, it moves at constant speed as long as nothing acts against it. Thus, a given impetus results in a given speed. A faster speed manifests a more powerful impetus, but the speed is also dependent on the mass. In other words, the way a given impetus acts in a given body depends on the mass of the body; the more mass the body has the more resistance it offers to the action of the impetus. Mass is thus a property of a body, a quality that specifies a body's receptivity to being moved by its impetus; mass is property in the category of quality.[5] We will use the word mass, as is often the case in empiriometric physics, for both the quality itself and the measure of that quality. Hence, with an aim towards our mathematical use, we define mass as follows:

Mass is a measure of the receptivity of a body to the action of its impetus, such that the larger the mass, the less speed a given impetus is able to cause.

A body with an intensity p of impetus will move at a speed ($v = p/m$, to be discussed below) because of the way the action of the impetus (category of action) is received by the body (category of reception).

---

[4] The equations for the speeds after collision, which we will study in Chapter 5, are: $v_m = v\left(1 - \frac{2}{1+m/M}\right) \sim -v$, $v_M = \frac{2}{1+M/m}v \sim 2\frac{m}{M}$ to first order in m/M for momentum and energy conservation. To zeroth order in m/M in energy, we have: $v_m \sim -v$ and $V_M \sim 0$.

[5] Mass is also obviously correlative to the bodies' ability to receive impetus for it would be a funny thing to have an object that cannot receive impetus but can resist its action once it has it. If this were the case, why would we ever talk of the latter? Furthermore, mass likely has a relational aspect, e.g. depends on the state of fields around the object. For simplicity, we leave both of these out for now.

To specify the interdependence of mass, momentum, and speed, we must be sure that we impart the same impetus to two blocks of two different masses. We did just this in the above experiment by making the block at rest much more massive than the impacting block. In this way, the interaction is nearly completely determined by the surface of the two colliding blocks *and* the action of the small block. Why so? A collision means two blocks touch and push against each other, which means their surfaces interact. This interaction will, in turn, obviously depend on how the two blocks move into and away from each other. Since the large block's motion is relatively insignificant, at some level, just the motion of the small block needs to be considered.

Thus, if we keep the speed and mass of the incoming block the same and vary the mass of the large block (while always keeping it much larger), each collision will cause approximately the same effect on the large block. Since the key effect on the large block is the activation of impetus in it, it is therefore certainly reasonable to say that, for each such collision, approximately the same impetus is imparted, independent of the size of the large block. Again, this is true as long as the mass of the large block indeed remains much larger than the mass of the impacting body (i.e., $M >> m$), for its much larger mass insures that it cannot move in such a way as to change the situation much.[6]

Again, our experiments above show that for a large block of mass M, a speed V_M results, and for mass $2M$ a speed $V_M / 2$ results; generally, the resulting speed of large mass goes as its mass ($1/M$). Thus, since in both cases the *same* intensity of impetus is activated, i.e. the same magnitude of momentum, the impetus produces a speed proportional to the inverse of the mass. Mathematically, recall p means momentum and $V = p/M$; thus, can rewrite and get:

3.1 $$p = mv$$

where we define the various quantities in such a way that the constant of proportionality is one.

> ***Momentum*** (symbol, p) is the measure of the impetus or more generally (analogically) the sum of all the measures of the impetus of all relevant bodies. It is the mass times the velocity, $p = mv$. In this text, the unit of measure of momentum is the Buridan. We call the unit a "Buridan" because of his important contribution. $1\ Buridan = 1B = 1\ kg\, m/s$.

Now, notice that p (and v) in equation 3.1 can be either negative or positive; these signs correspond to oppositely directed impetus, which thus are qualitatively different impeti. However, since positive and negative numbers are mathematically treated the same, it is helpful to (analogically) extend our definition of intensity to include both signs as if we were measuring a continuous intensity of one type of impetus. Using this extended meaning, we can simply speak of momentum as the measure of the intensity of the impetus. Again, to be precise, we should really say "the intensity of the impetus in a certain

---

[6] The relevant time scales are the comparison between the speed of the small mass block and the large mass block; i.e. mathematically, we want: $\frac{V_M}{v_m} << 1$.

direction," but mathematically, negative numbers and positive numbers are treated the same, so we will sometimes implicitly use the extended meaning of "intensity" that includes a reference to direction in the sign. The usage will become a little less fuzzy when we introduce other dimensions and write the vector equation for the momentum. At that point, we will distinguish the *magnitude* of the momentum, the measure of the intensity of the impetus, from the *direction* of the momentum, the indication of which directional type of impetus (left tending, up tending, right tending, etc.).[7]

At this point, we have not proved that $p = m\mathbf{v}$; we've just shown it is a good hypothesis. But, many, many experiments have verified it, within good approximation, for the motion of all bodies in ordinary experience, including the sun and earth.

Interactive Simulation

Exercise: to familiarize yourself with the nature of impetus use the *Interactive Physics*' software (e.g. *3.1-m elastically collides with M.ip*, *3.2-balls of same impetus hitting wall.ip*) to do some experiments. For example, repeat the above impact experiments. Also, vary the mass of the impacting block and show: $V_M = k'm$, that is, show the speed of the large block is roughly proportional to the mass of the incoming block. By measuring the resulting speed of the mass M after each collision, show that it is proportional to the speed of the incoming block; that is, show $V_M = k\mathbf{v}_m$. Lastly, think about how these results further corroborate equation 3.1.

Even when we leave ordinary experience to consider, for example, quarks and neutrinos, physics still uses, though now by analogy, these basic ideas of inertia. Equation 3.1 is one of the most important equations of physics. But, we have not really seen why yet.

Before we continue developing our understanding of impetus (and momentum as the measure of its intensity and direction), we summarize the idealized phenomenon[8] with which we deal:

> *Every object continues in its state of rest or uniform motion unless acted on by an outside force.*

This is called the law of *inertia* or Newton's first law of motion. In short, with no impetus to move it, an object cannot move, and once impetus is activated in it, the impetus's nature is to move it uniformly, unless something, "a force," which we discuss in the next chapter, intervenes to change its impetus.

---

[7] The impetus of a moving object can be thought of as composed of a mixture, analogous to the mixing of primary colors, of up to three types of impeti (one for each dimension) from the various 6 possible types of impeti, two for each of the three spatial dimensions--one for each direction in the given dimension. Note that while there are only three primary colors, there are six primary species of impeti although no more than three are mixed at once, one from each dimension. Also, unlike color, as far as we know, there is no particular priority of one set of primary species of impeti over another, because one direction is not, as far as we know, unique enough to be given priority over the others. Of course, there are further differences between color and impetus, but most would require a digression into an analysis of color. Furthermore, to say more about the specific nature of impetus, we would need an extended analysis including much beyond the scope of Newtonian mechanics.

[8] We say it's idealized because to conceive "the continuing uniformly in motion," we, for example, have to leave out friction and gravity that, in ordinary experience, usually result in effects which dominate the motion.

Conservation of Momentum

Now, continued experiment would reveal that, in the limit where there are no applied external forces, the total intensity of impetus acting in a given direction is never lost. We say the total momentum *is conserved.* No matter how the intensity of the impetus is split, the total is in some way maintained throughout the universe. It's rather like chopping a worm in many parts of different sizes where each such part immediately becomes a new worm and then adding the strength of each of the new worms to find that the total strength of all the worms together is the same as that of the original worm (of course, worms do live when chopped but the total strength is not necessarily the same). The *net* intensity of the impetus is conserved. To discuss this conservation law, we need to clarify a few important simplifications that we need to make.

Simplifications

We already have discussed how a ball continues in motion even after it is released. In actual fact, it does stop eventually. Why? Air slows it down. Even neglecting air resistance as we have, gravity still pulls it down until it hits the ground which eventually stops it. We'd like to "turn off gravity" to see what happens then.

Remember in understanding physics, we seek to find the most general (universal) properties that we can. This means we have to leave out of consideration as many properties as we can. We thus initially choose to leave out everything except the body's geometric size and shape in space, its mass and its impetus. In other words, pretend we have a bank of switches which control the existence of things and their properties. We shut off all switches except those mentioned. Indeed, initially, we *only* switch on "space" and *one* body. Gravity is "switched off," air and its resistance to motion is off, the ground and its hardness are all off, i.e., not there any more. Hence, under these conditions the ball has momentum and never loses it, because there is nothing to change its impetus. At this level, it is simply the nature of impetus that is responsible for conservation of momentum. Indeed, summing the momentum of any number of such *completely isolated* bodies will always yield a constant—i.e., momentum will be conserved—because each individual momentum does not change.

More than One Body

Now, turn on the switches that make other bodies reappear, but keep all other switches off. The baseball continues with its same momentum until it hits something. Say it is caught by the second baseman, what happens? Just consider the component of momentum in the direction of the ball. The tendency to continue in motion is preserved, so somehow it must be "divided" between the ball and the baseman. Let's say the baseman jumped to catch the ball, so that he is in the air when he caught it. Because the momentum is conserved, the ball and the baseman now, if we add the intensity of each together, have the same net intensity of impetus that the ball had alone. Thinking of the baseman and the ball as a new whole, i.e., as a single body, we say this total momentum impels them toward the outfield. However, now the mass of the new whole (baseman plus ball) is much larger so instead of going the speed of the ball which was say, 70 miles an hour it now goes:

$$V_{ball+baseman} = \frac{p_{ball}}{m+M} \sim \frac{mV_{ball}}{M} = 70mph \times \frac{.1kg}{100kg} = .07mph \tag{3.2}$$

(Here we assume: $m_{ball} \sim .1kg,\ M_{baseman} \sim 100kg$)

This is obviously very slow, given normal walking speeds are about 2*mi/h*. After catching the ball in the air, the baseman falls back to the ground, and the total intensity of impetus remains the same but is now shared by the earth--of course it takes a while for the whole earth to get involved (share the momentum) as indeed it does for the player when he catches the ball.

The momentum can be passed from one object to another but is never lost. Given this, can the action of impetus in a given direction never be stopped? After all, what's there to stop it? In a real sense, the only thing that can stop impetus from moving an object in a given direction is an oppositely directed impetus. Two identical objects hurling at each other at the same speed can come to a complete stop. The intensity of impetus of the one coming from the right is exactly the same as the one coming from the left, and they can neutralize each other. Indeed, only left tending impetus can destroy right tending impetus. This is done by the action of "forces." We will, as mentioned, discuss forces in the next chapter, but notice here we have "switched on" collision forces (interior and exterior). If we had not, the objects would go right through each other, because there would be nothing to stop their forward motion.

In this cancellation of right moving tendency with left moving tendency, we still have *conservation of net momentum*. Indeed, as a mathematical number, momentum can be considered as positive or negative, when it causes motion along a given direction or opposed to the given direction, respectively. Of course, by saying, for instance, that a body has momentum "-1", we don't mean we can have a negative number of something as if there might be "-1" cows along the roadside; we simply mean that there is a right tending and a left tending impetus, which can "cancel" under certain interactions. Mathematically, in the above case, the sum of the intensity of the impetus of a first body and a second body must be zero: $p_{left} + p_{right} = 0$. It is clear that there is no impetus after the collision; before the collision each body has impetus, but the left and right moving tendencies are said to cancel, yielding a net of zero intensity or zero momentum.

In general, for two objects labeled one and two, we can write this fact mathematically as:

$$P_{Total} = p_1 + p_2 = \text{constant}$$

More generally for N objects, labeled by successive integers P_{total} is constant for all time:

3.3 $P_{Total} \equiv \sum_{1}^{N} p_i$, where the "*i*" labels which particle in the system we are considering.

The fact that the total momentum, P_{total}, is constant for any isolated system (that is any system that has no external forces acting on it) is called the ***law of conservation of linear momentum***.

The constancy of the total momentum given by equation 3.3 specifies the conservation law for one dimension. What of the other two dimensions? We can answer this by noting that, at the level of pure extension (continuous quantity) in which no part of space is distinguished from another, there is no distinction between any two linear directions we might pick out.[9] Indeed, we find, by experiment, that P_{total} given in equation

[9] Quantity (in the case of geometry) is simply parts outside of each other that share a boundary. In space as used here, we go further and retain no qualitative distinction, neither by keeping a certain number or shape of parts, nor by limiting the extension by a shape. Thus, each line one conceives in this space (a being of reason)

3.3 is constant for any direction we might pick, when no external forces are acting.[10] As we saw in Chapter 2, we can divide space into Cartesian coordinates and select out three directions $(\hat{x}, \hat{y}, \hat{z})$, which together can capture any direction in space. Thus, using our vector notation, we can write, without loss of generality, the law conservation of momentum as:

3.4 $\vec{P}_{Total} \equiv \sum_{1}^{N} \vec{p}_i$ **is constant always for an isolated system.**

Equation 3.4 is thus the _law of conservation of momentum in its full form._ An example of its use is shown in Figure 3-3.

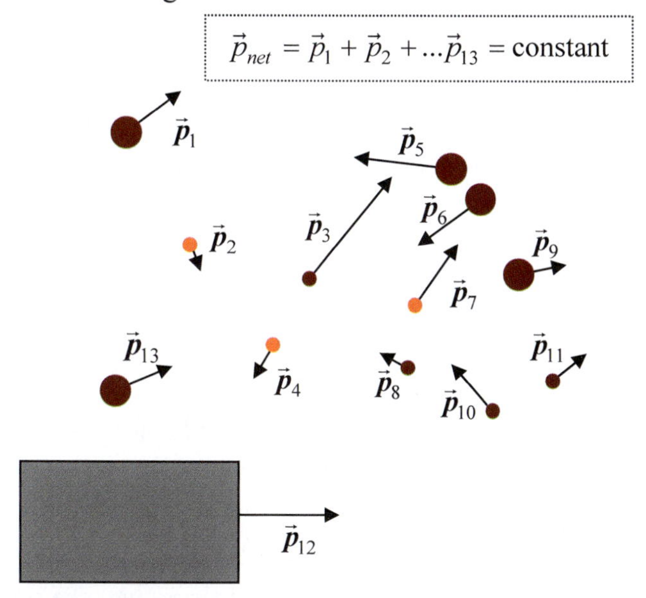

Figure 3-3: Adding the measures of the intensities of the impeti, i.e. the momenta, for all the bodies in a system, accounting separately for each of the three perpendicular directions, results in a "total" or net momentum in each direction, each of which is conserved. In other words, the direction and magnitude of the net momentum is conserved. This is called the law of conservation of momentum.

Equation 3.4 can also be written as three separate equations like equation 3.3, one for the *x*-axis, one for the *y*-direction and one for the *z*. For example, in the *z*-direction we get:

$$P^{z}{}_{Total} \equiv \sum_{1}^{N} p_i^{\,z} = \text{constant}$$

Here as elsewhere in this course, the "*z*" (or *x* or *y*) superscript refers to something, in this case momentum, in the *z*-direction.

---

is identical to any other. Adding even weak qualitative distinctions such as shape can once again distinguish one line from another, but to affect the law of conservation of momentum, we would need something capable of affecting the impetus and/or the way the impetus "propagates."

[10] In such experiments, we see that as we take away frictional forces the object more closely approximates uniform motion. Each new experiment has to be more sensitive to measure the increasingly small changes in motion and has to address smaller external forces.

So, in discussing momentum conservation, fundamentally, we can consider each dimension independently. In this spirit, we investigate the more simple case of one dimension, where the deep principles can be uncovered, and we will then come back armed and ready to learn to attack problems in more dimensions.

Brief Introduction to Energy Conservation

In Chapter 5, we will see that, in addition to the total intensity of impetus in each direction being conserved, the total activity, i.e. the total energy, is conserved. In particular, consider the case in which there is no potential energy, no energy "stored," for example, in a compressed spring. We will see in Chapter 5 that, for an *isolated* system of particles of this type (as viewed from a fixed or uniformly moving state (i.e., frame of reference)), in addition to the momentum given in equation 3.4 being conserved, the total kinetic energy must be conserved. For now, we merely state this experimental fact. Mathematically, we will see ***conservation of kinetic energy*** "demands" that the sum of $1/2mv^2$ (where m is the mass and v the speed of the given particle), for all the particles will remain constant. Mathematically:

3.5 $$KE = \sum_{i=0}^{N} \frac{1}{2} m_i \vec{v}_i^{\,2} = constant$$

Here, m_i and v_i are the mass and speed of the ith particle. From a mathematical point of view, without this equation, the equation of momentum conservation would be underdetermined. That is, given the initial velocities and masses within an isolated system, we could not determine from conservation of momentum alone what the velocities are after collisions occur.

Two Balls in Space

To further our understanding of impetus and mass, we take up another frictionless thought experiment. Consider the simple example of two balls of equal mass with all external forces "switched off." Imagine, for example, these two balls are in outer space far from any other objects so that neither gravity nor any other kind of force is acting.[11] Imagine, as drawn in Figure 3-4, that one ball moves at a constant speed v towards a head-on collision with a second ball which is at rest. What happens?

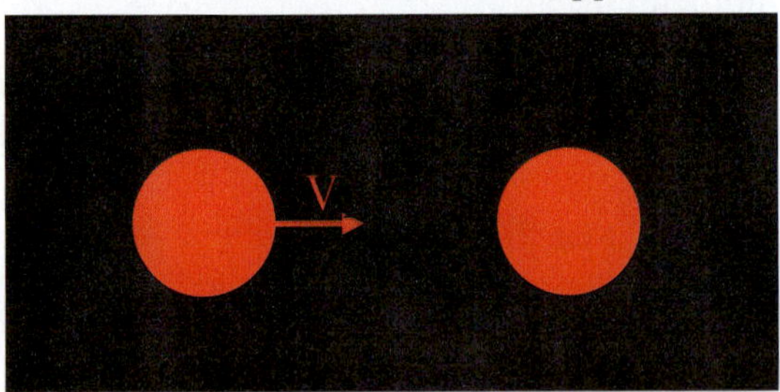

Figure 3-4: Two equal mass balls in outer space far from all other massive bodies. One is at rest; the other moves towards it at speed *V*. Such thought experiments are helpful in understanding the nature of impetus and mass of bodies.

[11] Imagine for example, we are performing an experiment in the vicinity of our spacecraft as it cruises to Mars.

The Moving Ball: How its Parts are Moved

First, notice the moving ball has a certain intensity of impetus, a certain momentum. But, the ball itself has parts. Let's take an extreme case and assume it's a ball filled with marbles each of equal mass. Suppose further that there are *10* identical marbles all jammed together not able to move relative to each other, which means all ten share the speed of the ball itself. If we now consider each marble in the bundle, we see that each has $1/10$ of the intensity of impetus that the whole object has. Each also is $1/10$ the mass of the whole. We could further subdivide each marble into equal mass parts, and once again make the same sort of analysis. Again, the intensity of the impetus, the momentum, must divide itself equally throughout the mass. If it did not divide equally according to mass, we would belie our statement that they move together. If, for instance, one equal mass part (say of mass m) had a larger intensity of impetus than another, it would fly away from the other, because speed is proportional to momentum. Mathematically, using $\mathbf{v} = p/m$, we see that:

$$\textit{If } p_1 > p_2 \textit{ then } \mathbf{v}_1 > \mathbf{v}_2$$

Thus, if the momentum is equally divided between two equal mass parts, there is a relative speed between part one and part two, and they move apart, contrary to our initial assumption that they do not.

We could continue our analysis of the division of the ball into parts. Ultimately, if we kept dividing, we would come to a group of things that are themselves each single substances--that is, each is not itself composed of other substances, and do not have qualities that are *merely* the result of interaction of some deeper substances, but are themselves the base for properties of mass and impetus.[12] Or said a different way, each of these substances will simply have the intensity of the quality, that is, it will not be a compound of qualities of other substances[13] acting together.[14] These substances would also have parts and the above analysis applies to those parts as well. Note that the type of whole involved (that is, whether the body under consideration is a bunch of loosely interacting substances such as a packet of air, a single substance such as a dog, or atoms of a single substance) will affect the way the parts are treated, but only, for our level of consideration, at an insignificant level;[15] thus, we leave such considerations of the types of whole out of account for our purposes.[16] In particular, in order to say the most general things we can

---

[12] This is *not* to say it has no parts.

[13] Recall a quality, unlike a quantity, cannot be made by grouping together various lesser qualities. Grouping together a bunch of green peas doesn't make a more intense green; it makes the same intensity of green, just more of it. However, impetus is an active quality, an ability or power to do something--in this case a power to make something move at constant speed. Hence, it has a quantitative effect, motion from here to there, by which the power can be measured. In the case of forces this is clear. Given 3 men on a job; although no one of them could lift more than 200 pounds, together they could lift 600 pounds; still no individual, i.e. no substance, would have the quality of being able to lift more than 200 pounds (without a mechanical aid). Impetus is similar; two substances of the same mass with the same impetus can be "glued" together and because they have the potential of activating twice the intensity of impetus that is in either body in a third body, we note that they add, in the sense of causing a net effect.

[14] Though, it may itself have parts where its qualities are concentrated, so we could analogically treat it as a compound of qualities.

[15] The change will be beyond the sensitivity the measurements we make at our chosen level.

[16] You may know, for example, that the forces holding two atoms together in a molecule have much less energy tied up in them than the nuclear forces holding protons and neutrons together in the nucleus of an

about momentum, force and mass, we will leave out consideration whether a given body is a substance or not. Of course, this doesn't make substances non-existent; there still are substances, things that exist, not just as a manifestation of the activity of a group, but things that exist of themselves. Neither does leaving substance out of consideration mean that we are doing something incorrect, as long as we realize that at some point we have to answer that question as well. Again, this sort of "leaving out" or "switching off" is part of the process. This means we must, of course, come back at some point and try to say what we can about those things we've left behind. In any case, we define a ***body* as any physical substance, part of a substance, or a group of substances that we choose to consider as separate from the rest**; typically, "parts" of a system will be considered a body because they are separated in distance from the rest and/or because they stick together in some way.

Back to our ball jammed with marbles. Any real ball would actually be more like a container not quite full of marbles. For instance, the parts of an iron ball are atoms that can jiggle around. You can hear it if you knock two of them together; they will oscillate (jiggle), making a "ting" sound. We could go on in more detail. But, do we need a detailed understanding of the structure and motion within each ball in order to understand more about the collision of the two balls?

No. This is exactly the point: the activity of the interior parts of the body, whatever they may be, must result in the "net" intensity of the impetus, the net momentum being the constant. That is, the momentum of the ball stays the same throughout all internal collisions. Because of the conservation of momentum, each of those collisions results in the same net tendency to move in the given direction, thus not changing the net momentum of the ball.

The Collision of the Two Balls

Finally, we can analyze the two balls of equal mass. In fact, we could not go any further in the above analysis without understanding the two-ball problem. Why? It is only by analogy to what we directly see in a two ball collision that we can talk about what we cannot directly see in the inner workings of each ball.

To solve the problem, we use conservation of momentum, but we need to further specify our problem. What happens to the two balls depends on what the balls are made of, especially at the surface which is where the interesting collision interactions occur.

Inelastic Collision

If we assume that the balls stick when they hit and cannot come apart--say they have really thick Velcro glued on them--the surface details cease to matter. By the arguments in the previous subsection, the momentum divides equally between the two balls, and the net momentum remains whatever it was for the original ball. Since the total mass is twice the original, the speed is half. This situation is shown in Figure 3-5.

---

atom. This energy difference has an effect on the mass, for instance, of two protons forming a hydrogen atom versus two protons in an iron atom.

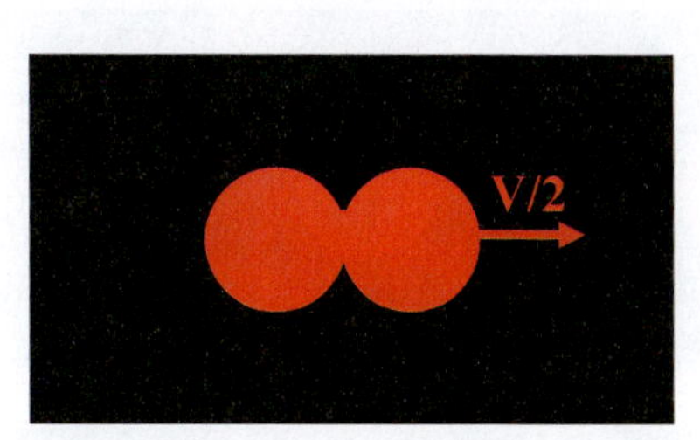

Figure 3-5: *An inelastic collision:* if mass m ball coming from left impacts mass m ball at rest and sticks to it, the two balls will move off together at ½ the original speed of the initially moving ball.

Elastic Collision: Using Moving Point of View

On the other hand, if the balls bounce elastically, like super balls, with no damping, then we need to do more investigation. To find a solution to this *elastic* collision problem, let's try looking at it from a different point of view. Jump into a spaceship and rev-up the speed to $V/2$ along the direction of the moving ball; from this vantage point, the left ball appears to be moving at a speed of $V-V/2=V/2$. The right ball now appears to be moving at: $0-V/2=-V/2$. In short, from the view in our spaceship, it appears as if both balls are moving towards a common center at speed $V/2$ as shown in Figure 3-6. This act of moving from one moving point of view to another is called a change of reference frame. Going to the spaceship frame of reference[17] has given us a much simpler situation; it is symmetric, the same on the left as it is on the right. Furthermore, it seems clear that this symmetry is not changed by the collision. Before the collision, the balls are stuck moving in a left and right line and the head-on collision appears to have nothing in its nature that should introduce anything in the in and out or up and down direction. The problem is one dimensional, and we expect it to stay so.[18] Thus, the question is: After the collision, what type of motion will preserve the left-right symmetry in this line? There is only one way. The balls cannot go through each other, so they must slow down and then each move off with the same speed u but in opposite directions. Conservation of momentum tells us this must be correct.[19] Before the collision, the net momentum is zero, and thus, afterwards, it must also be zero as it will be with the speeds oppositely directed: $\vec{P}_{after}=(mu-mu)\hat{x}=0$. In fact, conservation of momentum is related to this symmetry from one place to the next, called *translational spatial symmetry.*

[17] This particular frame is called the "center of mass" frame, the meaning of which we will discuss later in this chapter.

[18] Indeed, we can say more: The right ball continues for some time with nothing breaking the left-right symmetry at all. One part of "space" is the same as another. The same is true of the left ball until it hits the right ball; obviously the same is true of the right ball when it encounters the left.

[19] u is often a symbol used for speed. The v is used for velocity because the velocity (which is a vector) is often written as: $\vec{v}$ where the "v" is for velocity. "u" is probably used because of its proximity in the alphabet to v.

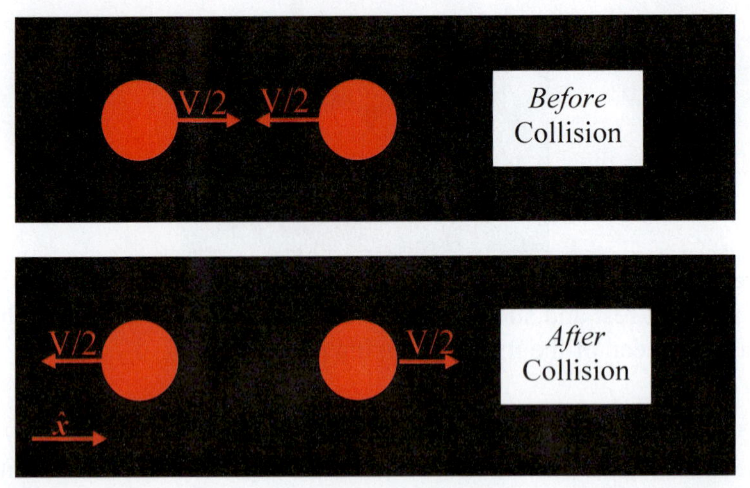

Figure 3-6: a) *(Top)* Two equal mass balls *before colliding* seen from a moving spacecraft, making it appear as if each body is moving at a speed $V/2$ towards the other.
b) *(Bottom)* The same two balls shown *after they collide*; each now, as viewed from the spacecraft, moves in the opposite direction but still at speed $V/2$. Note that the net momentum is zero both before and after collision.

Elastic Collision: Using Energy Conservation

Now, if we *only* use conservation of momentum, we do know that the receding speeds of the two equal mass balls are equal and oppositely directed, but we do not know what that speed is; it could, for example, be zero and still conserve momentum. However, we have assumed an elastic collision, which we've described as "bouncy." Fundamentally, but loosely, as we've mentioned, this means that the activity (energy) in the ball is not lost during the collision. Thus, in this case, because the mass of each ball is the same, and speed is a measure of the activity (energy) in some fashion, we can further say that the speed of each ball, though oppositely directed, must be the same. Hence, the balls come in at a speed $V/2$ and move out at a speed $-V/2$.

From a mathematical point of view, we know that the solution above is the one and only solution, because it can be shown (*exercise*) that the solution satisfies both the equation of conservation of linear momentum (eq. 3.4) and that of kinetic energy (eq. 3.5), and there is only one solution to those equations.

Elastic Collision: Viewed from Rest

Yet, we haven't quite solved our original problem of the colliding balls, for we are still in our spaceship moving at speed $V/2$. Our question was: What does the collision look like from our original vantage point at rest? Specifically, we already know what the situation looks like before the collision (Figure 3-7a), but what does it look like after the collision? We jump out into space (in our spacesuit) and decelerate ourselves from $V/2$ to rest. This effectively adds $V/2$ to all the speeds. Thus, we get the following.
The *left ball's* speed now is zero; mathematically: $-V/2+V/2=0$.
The *right ball* is moving at speed V; mathematically: $V/2+V/2=V$.

Thus, the result of the collision in the original frame is as shown in Figure 3-7b. In summary, in our equal-mass ball case, the left ball moves at speed V and hits the right ball losing all its momentum in the interaction with it. Meanwhile, the right ball ends with the same momentum that the left ball originally had; thus the right ball moves to the right at speed V. The net result is that it appears as if the speed V of the first ball has been drained out of it into the second ball.

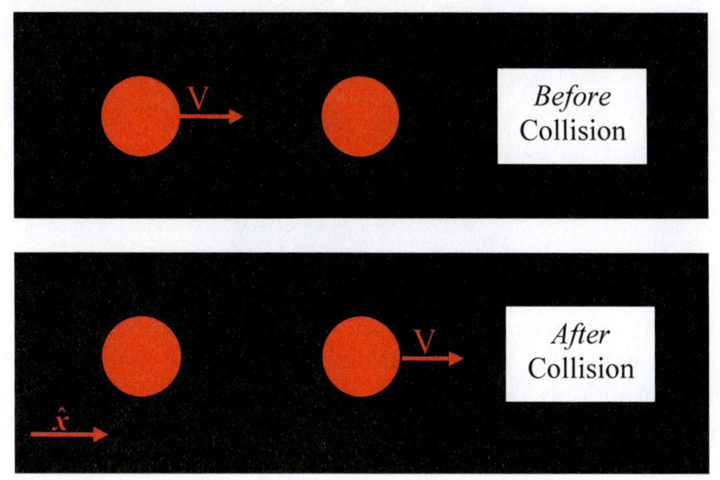

Figure 3-7: a) *(Top)* Two equal mass balls viewed from the original rest frame before they collide. **b)** *(Bottom)* The same two balls shown after they collide. The net effect is that the original velocity, V, of the first ball has been "drained" into the second one. Note that the net momentum in this frame is mV both before and after the collision.

Notice how moving to a different frame, that is observing from a uniformly moving pointing of view, has helped us easily solve this problem. Such frame shifts will often be helpful in solving problems.

Exercise: using *Interactive Physics* file *3.1-m elastically collides with M.ip*, try your own experiment using both sets of initial conditions (one moving at the other and both moving at each other at the same speed) and verify our results; compare this changing of initial conditions to changing the reference frame.

Elastic Collision: Subdividing Each Ball

Continuing our analysis, what would have happened if the left ball and the right ball had instead *each* been *two* balls, making a total of four balls, two moving in from the left towards two at rest on the right. Assume that the each ball has a mass $m/2$ so that the total mass moving on the left remains m, and the total mass on the right remains m. Also, suppose that neither pair of balls is glued together, but are just touching, what happens when the pairs collide?

We know the answer as long as the collision is elastic. Under idealized conditions, it is fundamentally the same as our previous case. The two balls come in from the left and hit the two balls at rest; the moving balls are stopped by the balls at rest and the rest balls are given the speed that the moving ones had.

The key pillar of this reasoning is the left-right symmetry of the *whole* system before, during and after the collision. If we break this symmetry, i.e. make right different than left--for instance, by making one of the balls on the left more massive-- then the result will be different.

Figure 3-8: Newton's Cradle[20] is used to demonstrate the elastic collision of equal-mass balls. Pull one ball to the side, watch it swing down and hit the others, causing only one to swing off the opposite end. Pull two off and watch exactly two swing off the other end. In general, pull off N balls and watch N swing the other direction.

Similar arguments using symmetry apply to more groups of balls, and is aptly illustrated by the office toy called "Newton's Cradle" (see Figure 3-8).

Many results that might take much grinding of symbols in long equations can be found instead by analyzing the physical situation as we have done above; this includes using symmetry and uniqueness considerations in combination with the equations. Such analysis time is well spent, yielding insights into the problem and problems like it, which, in turn, facilitates solving other problems. Oftentimes changing frames of references--as we saw in the cases above-- is helpful in making use of symmetry considerations.

"*Galilean*" Relativity

Notice, we have now done a lot more than solve the equal mass, equal speed two-body collision problem. By introducing reference frames, we have implicitly solved a whole class of problems, each one different depending on what speed one views the situation from. The solutions are related by adding or subtracting whatever our observation speed is. For instance, the two equal-mass ball solution above, which had a left ball moving at speed V toward a stationary right ball, can easily be extended to the case of a ball moving to the right at speed V with the left ball at rest by viewing the situation from a platform moving at speed $2V$. Since such addition and subtraction of speed will occur in any direction, the same analysis carries over when we view the system from reference frames moving at any constant speed in any direction. That is, we can add or subtract any constant *velocity* (vector), and obtain more solutions. What's more, the very concept of reference frame is crucial to understanding Newtonian mechanics and later Einstein's special and general theory of relativity. Indeed, uniformly moving reference frames will be indispensable in the efficient solution of problems in mechanics in this course.

---

[20] See applet at http://www.walter-fendt.de/ph11e/ncradle.htm .

Such preferred reference frames are called *inertial* frames because they are related to inertia, the tendency of a body to continue in motion. Indeed, the preference of the uniformly moving frames arises from the nature of impetus. Because of impetus, bodies all moving at the same speed will tend to keep moving at the same speed. Further, if one is moving at any constant speed relative to a system of bodies, one has a privileged viewpoint in that all the objects appear to behave as if they had the momentum that they are viewed to have in that frame.[21] Buridan in the 1300's noted that: [22]

> "...I make the supposition that local motion can be sensibly perceived only in so far as one may perceive one body to be differently disposed with respect to another.... If a person is in one ship called *a* which is moved ...without pitching or rolling...and this person sees nothing except another ship called *b*, which is moved in every respect in the same manner as *a* in which he is situated, I say that it will seem to this person that neither ship is moving. And if *a* is at rest and *b* is moved it will appear and seem to him that *b* is moved. On the other hand, if *a* is moved and *b* at rest, it will appear to him as before that *a* is at rest and that *b* is moved..."[23]

Obviously, in contradistinction to what is often thought, the relativity of motion was understood, at some real level, in medieval times and *then inherited* by Galileo (see box[24]). Because Galileo brought physics to a new level of maturity by digesting many of the results of the Middle Ages, and by performing detailed first-of-their-kind experiments of his own, his name is associated with this relativity of motion. What is the origin of and what exactly does relativity of motion mean? To answer, we need to further investigate both impetus and locomotion.

---

[21] This results from assuming time is uniform, and mass and space don't change with motion—see Chapter 10: *Introduction to Special Relativity*. Newton assumes space and time are not interconnected; indeed, for him, they exist independent of what's happening. We know this is not true because there is no pure extended quantity (which is what one first means by "space") and motion (of which time is a measure) without physical things that have quantity and undergo change and move in time. (Said another way, quantity, as we've seen, is a property of physical substances, not itself a substance and change happens to substances.) However, we can turn off these interactions in the fashion that we've discussed and see what can be said in this truncated world. Indeed, in the right range of physical situations, we can say a lot; this is the domain of Newtonian physics. In that domain, Newtonian mechanics is true in an approximate sense, as we will make clear as the course unfolds.

[22] Buridan: *Questions on the Heavens and the World* , Book II Question 22 quoted in Dales page 132

[23] Here he is talking about uniform motion (constant velocity). He goes on to talk about accelerated motion in the same manner. It is important to note that such reasoning is true in so far as we stay with the quantitative isolated (abstracted) from qualities except shape. However, when we include other qualities (as we do in the discussion of impetus), the situation is radically different. For instance, because, absent consideration of gravity (where many more distinctions need to be made), there is no quality like impetus that maintains accelerated motion, and one ***can*** sense accelerated motion. From the passages he writes after the quoted section above, it *appears* Buridan did not understand this fact. Still, Buridan's statement as limited to quantity qualified only by shape remains true and, in fact, reappears with the introduction of the equivalence principle, which we discuss briefly in Chapter 8.

[24] Quote in the Galileo history box that begins "gave more financial..." is by Dr. Heilbron professor emeritus of history and vice chancellor emeritus at the University of California at Berkeley.

Galileo Galilei, (born in Pisa, 1564, d. 1642 AD) was responsible for bringing together important parts of the puzzle of the modern scientific method, which consists in the heavy use of empiriometric thinking leaning on quantifiable experimental work, in his own work especially in dynamics (the study of motion). For this and related reasons, he is considered the herald of the scientific revolution. Despite the historical picture that is usually painted, he did not single-handedly bring about that revolution. Complex knowledge, such as modern science, evolves with the work of many and doesn't appear overnight. Galileo was not the first to use mathematics or to do experiments or to do them together. Galileo, a believing Catholic and brilliant scientist, built on the work of medieval thinkers in this and other aspects. In using the information and insights received from his predecessors, he, through his own work, was able to formulate more complete insights into mechanics, including in astronomy. Many of the pieces of the puzzle, not to have its first full features completed until Newton, that Galileo utilized were part and parcel of his cultural milieu as a result of the scientific mode of thinking built up from the middle ages. His experimental work greatly advanced the use of the quantitative method, especially through his experiments with balls rolling down inclined planes verifying the $\frac{1}{2}at^2$ law proposed by predecessors. His use of the telescope (which he did not invent, but improved) in plotting the position of four moons of Jupiter, his study of the sun and the phases of Venus, and his defense of the Earth's rotation around the sun stand out as well, though he had no proof of the latter—that came gradually with Newton's theory of gravity, the measurement of the aberration of starlight in 1728 and the measurement of parallax of the stars in the early 1800's providing proof that convinced nearly all.

Born shortly after the death of Martin Luther and a contemporary of Descartes and Kepler, Galileo lived in a complex time of upheaval and discovery. His work was initially supported by his friend, Pope Urban VII. Indeed, the historian Dr. John Heilbron has said that the Church "gave more financial and social support to the study of astronomy for over six centuries, from the recovery of ancient learning during the late Middle Ages into the Enlightenment, than any other, and probably, all other, institutions."[24] But, his cosmological work famously ran into trouble with Church authorities, being condemned, ostensibly for holding that the earth goes around the sun. Since the condemnation had *not* been formally issued by the Pope, contemporaries knew it and Galileo's censure (after recanting his heliocentric view, he was sentenced to house arrest in his villa) were not to be considered Church doctrine. The condemnation of the heliocentric theory greatly missed the mark in attacking a scientific theory that was still in the making; indeed, the theory had *not* been condemned, despite heavy discussion of it through the high middle ages through the time of Copernicus, *till then*. For Galileo's part, having so acutely seen the profound and essential need for mathematics in physics, he forgot what he knew in his work, that reality was not only mathematical, and said, for instance, in his work *The Assayer*: "Hence, I think tastes, odors, colors and so on are no more than mere names...they reside only in the consciousness."

Galileo was a key figure in the advance of mechanics that culminated in Newton's work, but also inadvertently generated much confusion over the historical and philosophical place of empiriometric science.

Momentum is the measure of the intensity of a real quality (impetus) causing a given object to persist in motion. To better understand this, consider the analogy of a fish swimming through water. The fish pushes against the water, moving it out of the way, and thus moves from one part of the water to another. The fish is moving itself, but, in this case, it requires a continual replenishment of impetus, which we call "a force." The fish is acting (category of action), the water is receiving the action (category of reception).

Or again, consider the following thought experiment in which, as before, gravity and all forces except limited collision forces are "shut off." Furthermore, consider that, in this thought experiment, a monkey sleeping on one side of a large air-filled box *is the whole universe*. By this, we mean that outside the box is not "space," but really nothing, no parts outside of each other, nothing to consider at all because nothing is there. Initially, the air is still, and the monkey is still. In this situation, shown in Figure 3-9, the monkey and the box are truly at rest; they have no impetus. Take the mass of the box to be millions of tons and its length to be many yards. Soon the monkey and the box will move, but *can* the box move?

We have deliberately picked this thought experiment to bring out important issues related to locomotion. In particular, since there is nothing outside the box for the box to move in, the experiment brings in an extended idea of locomotion. Ordinarily, by locomotion we mean that a thing moves from one place in a medium, such as space or air or water, to another within it. When something in a container is moved by moving the container, take, for example, a ball in a box, we say the box moves (say through the air in the room) and, thus, the ball moves along with it. We could also have the box move *without* the ball moving, at least for a time until the ball hits the box wall. However, in our *monkey+air+box* universe pictured Figure 3-9, the box has no place to move to; there is nothing outside of it. Hence, in order to talk about locomotion of the box, we have to assume we can *extend by analogy* the concept of place which we abstract from the ordinary world. Namely, in such a case where there is no "outside" of the box, the place of the box would then mean, for instance, the *box+air* would have a place around the monkey; *so that the box+air would move (category of action) over the monkey (category of reception) but nowhere else.* Again, this is a sort-of inside out place, for normally place is defined by the *external* contacting medium, whether it be space or water or whatever. This concept of moving in this way is not something we can, *a priori*, say is possible (it is odd to talk about moving a box that has nothing to move "in"), but perhaps it is possible. It may simply be that the universe cannot be bounded by a massive object (such as the box) but must be bounded by real space (as distinguished from the being of reason we have discussed earlier), which we will discuss in later chapters. In any case, the analogical extension is consistent with Newtonian physics, and we will assume, for simplicity of analysis, that it is possible for the box to move.

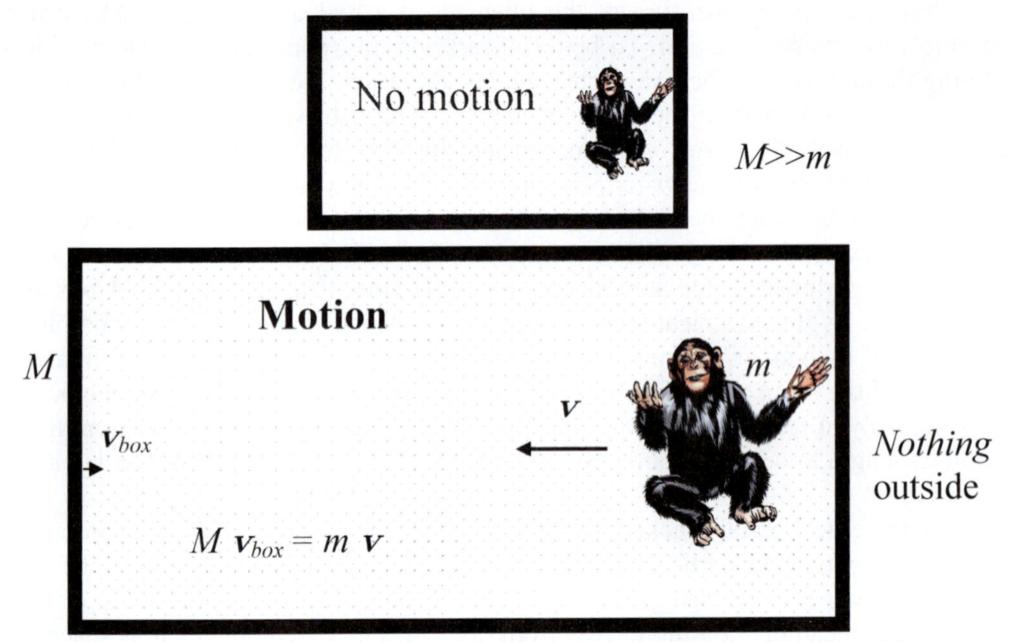

Figure 3-9: Thought experiment on momentum and locomotion considers a monkey and a box filled with air as the entire universe. a) *Top*: Monkey and box are both at rest. b) *Bottom*: Magnified view of the scene after the monkey pushes off the box. The box and the monkey move as shown. The total momentum must be conserved $\vec{p}_{monkey} = -\vec{p}_{box}$, so that the massive box ($M = millions\ of\ tons$) moves only slowly, whereas the 20 pound monkey moves much more quickly. Though they have the same intensity of impetus, the impetus of the box encounters much more resistance because of its greater mass.

Now, the monkey wakes up and pushes against the side of the box as he stretches; he begins to move through the air. Even though, from conservation of momentum, the box has the same momentum as the monkey but in the opposite direction, the box barely responds, since the mass of the monkey is many times smaller than that of the box. Thus, the speed of the box is obviously slow compared to that of the monkey.[25]

The monkey now has an impetus. The air contained in the box eventually is forced by the box to move along at the box's slow speed, thus giving the light air only a very little of its momentum (*exercise:* how much momentum does the air have?). Hence, the air has practically no impetus compared to the monkey. Of course, even after the monkey's push from the wall, the monkey continues to move (category of action)--*by the power of the impetus* generated by his stretch or jump-- from one part of the air to another part of the air (category of reception). However, notice that the monkey cannot pass through the air without the air also passing by the monkey. Now, it's only what impetus and mass the

[25] We can define this quantitatively by working always in the center of mass frame. Both the monkey (mass *m*), and the box (mass *M*) have momentum *p*, thus the monkey moves at speed, $v_{monkey} = p/m >> v_{box} = p/M$, the speed of the box.

monkey and air *actually* have that determine what's moving. After all, the only cause of motion we are considering is impetus. If a thing has no impetus, it doesn't move.[26]

Indeed, suppose we keep only those aspects of the categories that can be expressed quantitatively, which means we leave out key aspects of the other categories –in particular, in this case, quality, action, and reception. We are then just left with a relative notion of motion. That is, limiting ourselves largely to the quantitative (with the assumption that one can analogically extend the definition of locomotion as previously described), we can just as well say the air is moving as say the monkey is moving. Recall the category of quantity is about extension, one part outside of another. So, by a mental construct (being of reason) in which we limit the situation solely to this aspect, we can say one part of the air passing one part of the monkey is *the same as* one part of the monkey passing one part of the air; they are symmetrical with no properties of action and reception to make them different.[27]

Furthermore, the monkey cannot sense the presence of the impetus because, as we showed earlier, it distributes itself throughout the mass of a given body. For example, if you are flying through space at a constant uniform speed, all parts of your body have impetus distributed equally according to mass. Thus, no part is exerting force on any adjoining part, because each part shares the same speed due to its impetus.

So, although impetus is indeed a real quality, when we project it into the mathematical (i.e. quantitative) realm and take advantage of the fact that we cannot directly sense impetus, we can choose to no longer consider what actually has momentum (impetus) and what does not. In this way, we focus only on the relative motion; all that matters in this context is the relative motion; this is what we mean by Galilean relativity. Thus, we will choose our speed in such a way as to make the problem appear simple. On the earth, for example, it's often convenient to choose to match the speed of the earth. It's much easier to describe a baseball game from the earth-fixed seat in the stands than it is from a passing jet. In general, we pick the frame that makes the problem simplest and just deal with the various momenta as they *appear* in that frame. If we have a different problem within the same physical situation, a different frame may be helpful. Say, for instance, that instead of following the baseball game, we now want to see how the homerun baseball is spinning when it hits the wall. In such a case, we switch to the new frame (that of the moving ball) and only deal with *apparent* momenta in that frame, not giving any concern to the momenta as they appeared in the old frame, except possibly to note how they transform one into the other when we change frames. We never ask: "what is the actual momentum (intensity and direction of the impetus)," just: "what momentum does it appear to have in a given frame."[28] Again, this doesn't belie the reality of an object's actual

---

[26]As a concrete example, take the case of a man who is arrested by the police for punching another man in the nose. When the case comes to trial, the defense attorney argues that no one can say for sure that the defendant hit the man's nose, because if you move at the speed of the defendant's fist you will see that his fist is at rest and the man's nose is responsible by hitting the defendant's fist. The defense thus argues that the issue is purely dependent on one's point of view. As long as he is successful in getting the jury not to consider the cause of the motion, but only the symmetrical aspect of motion, he will win the argument.

[27] Again, if the analogically extended definition of place, a sort of inside out definition of place, is possible, we can say impetus of the box/air changes its place over the monkey and impetus of the monkey can change the monkey's place in the box/air. Note that we take the box and air to move as one for simplicity of analysis; there is no loss of generality in doing this as the core principle is still the same in the more complex analysis.

[28] Thus, in this way of thinking, it becomes what it is quantitatively: a totally symmetrical relation: this is to that as that is to this. And, in this way, the theory then incorporates the group or set of all possible frames from which one can view the physical situation.

impetus, any more than looking at something from various different perspectives belies its actual shape. Obviously, we cannot make something disappear by ceasing to ask about it. If the ball is at second base and then at third, something must really *move* (something else must receive the motion) so that the relation between the ball and the bases change; typically it's the ball. We must return to consider the full physical situation at some point; just as, at some point, we must "turn on" the other things that we have "switched off," such as gravity. Ultimately, it's physical reality we are after, not some dissected version of it. Now, it is clear that because of the nature of impetus, it will be hard to determine what actually has impetus and what doesn't; as we've mentioned above we don't directly sense it. In any case, in problems in which we want to discuss the full reality, we can always specify, as part of the problem definition, what has impetus and what doesn't.[29] We will typically not do this in order to focus on that which we do know.

Center of Mass Frame

One frame of reference stands out from the others. It is called the center of mass frame. To define it, we must first define ***the center of mass*** of a given system of bodies. Mathematically it is:

3.6 $$X_{cm} \equiv \frac{\sum_{i=1}^{N} x_i m_i}{\sum_{i=1}^{N} m_i} = \frac{\sum_{i=1}^{N} x_i m_i}{M}$$, where M is the total mass of the system.

For later use, we note the vector equivalent of this is (see Figure 3-10):

3.7 $$\vec{X}_{cm} \equiv \frac{\sum_{i=1}^{N} \vec{x}_i m_i}{\sum_{i=1}^{N} m_i} = \frac{\sum_{i=1}^{N} \vec{x}_i m_i}{M}$$

[29] To really tackle the issue of what actually has impetus, one needs to look at the universe as a whole, the sum total of all physical matter. The universe as a whole cannot move from one place to another because it is not and cannot be in a place for, by definition, there is nothing outside of it. Nor is there an "inside", for if we consider a given part of the universe as inside, then we have split the universe into that part and the rest, and are no longer considering the motion of the whole. Thus, we cannot use the extended definition of locomotion which we discussed earlier.

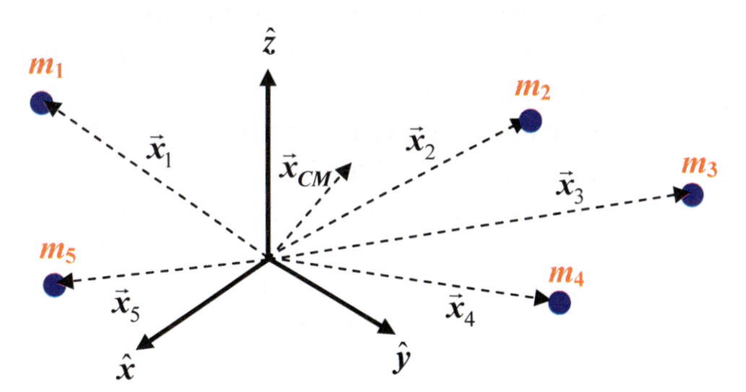

Figure 3-10: Center of Mass is found making the following so called weighted average. Find the distance (from a chosen origin) and direction, i.e. the *displacement vector*, to each object and the mass of each object. Multiply the displacement for each object by that object's mass. Calculate the sum of those "weighted numbers," and divide by the total mass. The result yields a vector, $\vec{x}_{cm}$, with head that marks the "center" of all the mass, the so called *center of mass*.[30]

Remember one can just look at this as three separate equations, one for each of the three Cartesian coordinates.

Take the simple example of our two equal mass balls at any one moment. That is, freeze time--"turn off the time switch"-- so that the distance between the balls is fixed. What is the center of mass at this moment? Define the origin of our coordinate system at the center as shown in the Figure 3-11.

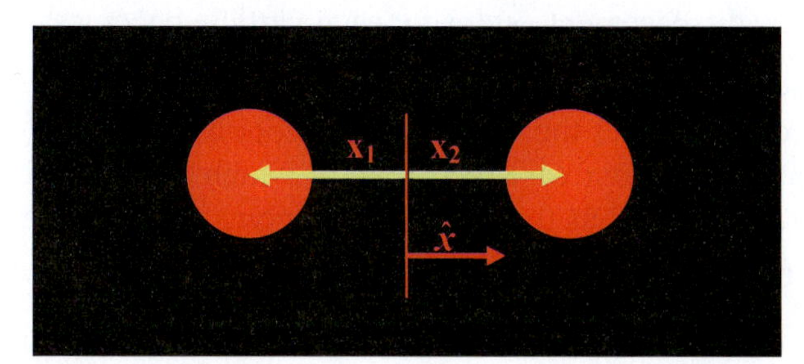

Figure 3-11: Center of Mass of two balls measured from origin midway between them.

Thus, we can write: $X_{cm} = m(-x_1) + m x_1 = 0$ So the center of mass is at the origin. See how it helps to choose your origin in a convenient place! You will learn from experience how to recognize such symmetries and thus judiciously choose coordinate origins among other things.

Time is off. What if we turn it on, what happens to our center of mass? We need to use conservation of momentum to answer this completely. Before we answer specifically,

[30] Notice we make these particles small so we do not yet deal with the question of where in the particle we measures the distance from. In the end, we will want to add up small chunks of all masses separately and then take the limit to infinitely small sized chunks.

let's use the power of mathematics---remember all quantitative facts about reality that we packed into our mathematical formalism in Chapter 2—to make a general statement that we can then apply to our colliding ball case.

In particular, we need a formalism that quantitatively takes account of the aspects we are interested in now: mass, impetus and motion. We have already accounted for mass using M as the symbol for the total mass of the system. But, how do we talk about the net motion of the system? The center of mass at a given moment identifies a kind of dynamic center for the system, because mass is the resistance to the action of the impetus and thus our best indicator of how motion (which is the result of impetus) will proceed. It is thus reasonable to take the momentum of the center of mass as:

3.8 $$\vec{P}_{cm} = M\vec{V}_{cm} = M\frac{d\vec{X}_{cm}}{dt}$$

Assuming the mass of each body, m_i , doesn't change (that is, $dm_i / dt = 0$) and substituting equation 3.7 into the equation 3.8 above gives:

3.9 $$\vec{P}_{cm} \equiv M\frac{\sum_1^N \frac{d\vec{x}_i}{dt}m_i}{M} = M\vec{V}_{cm} = \sum_1^N \vec{p}_i$$

Hence, $\vec{P}_{cm}$ is constant, by conservation of momentum, which mathematically is written:

3.10 $$\frac{d\vec{P}_{cm}}{dt} = 0$$

Now, suppose we pick the center of mass, $\vec{X}_{cm}$, as the origin of our coordinate system. Then, the new displacement vector for each particle becomes:

3.11 $$\vec{x}\,' = \vec{x} - \vec{X}_{cm}$$

We call these primed coordinates the center of mass coordinates.

Let's examine this for our case of two balls, which is summarized in Figure 3-7. If we start with the origin located at the initial center point and let time advance, we notice that the center of mass moves at speed $V/2$ as the right ball advances. That is, $\vec{X}_{cm}$ changes with time. Thus, when we take $\vec{X}_{cm}$ as our new coordinate origin, we are actually changing frames. Our new coordinate system describes a frame that is moving uniformly at speed $\vec{V}_{cm}$ (defined in equation 3.9 above). It is no surprise then that moving to this frame makes the momentum of the center of mass appear to go to zero. We can see this mathematically by substituting $\vec{x} = \vec{x}\,' + \vec{X}_{cm}$ in equation 3.9:

$$\vec{P}_{cm} \equiv M\frac{\sum_1^N \frac{d(\vec{x}\,'+\vec{X}_{cm})}{dt}m_i}{M} = \sum_1^N \frac{d\vec{x}\,'}{dt}m_i + \sum_1^N \frac{d\vec{X}_{cm}}{dt}m_i$$

3.12 *Solving for* $\vec{P}\,'_{cm}$

$$\vec{P}'_{cm} = \sum_1^N \frac{d\vec{x}\,'}{dt}m_i = \vec{P}_{cm} - \frac{d\vec{X}_{cm}}{dt}\sum_1^N m_i = 0$$

Thus, we see again that the net momentum of the system in the frame fixed to the center of mass is zero. This frame is called the ***center of mass frame***. Obviously, it has the advantage focusing attention on the system itself, not the motion of the system through the external environment, which we decided earlier was not of interest because it did not effect the behavior of the system at some level. The latter is what we meant by saying the system is an *isolated* system; recall momentum is conserved only in an isolated system.

Collision of Balls of Different Mass

Having used center of mass coordinates for two balls of *equal* mass, we now do the same for balls of unequal mass. Consider the elastic case in which the balls lose none of their energy of motion (kinetic energy). In the center of mass frame, by definition,

3.13 $$m x'_1 + M x'_2 = 0 \text{ thus } x'_1 = -\frac{m}{M} x'_2$$

Thus, if we solve for the position of one coordinate, we trivially have the other. (See Figure 3-12).

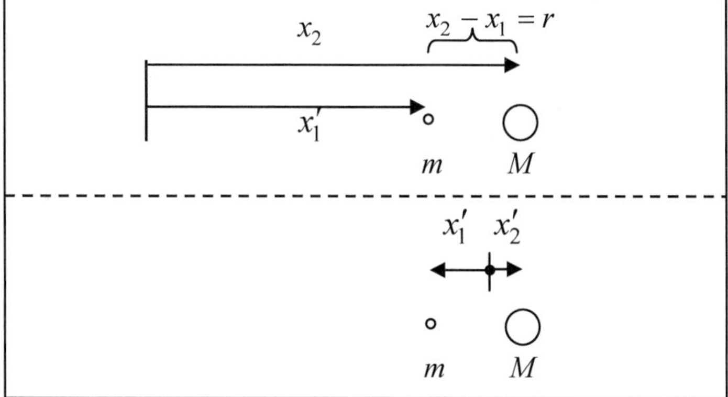

Figure 3-12: a) *Top:* Coordinate system based on arbitrary origin for a one dimensional collision of balls of unequal mass. b) *Bottom:* same collision in center of mass coordinates. Note that though the lesser mass ball is drawn with smaller radius, this convention should not allow one to forget that it's mass that matters here, *not size.*

Using equations 3.7 and 3.11, it can be shown, that

3.14 $$x_1' = \frac{(x_1 - x_2)M}{m+M} = -r\left(\frac{M}{m+M}\right)$$
$$x_2' = \frac{(x_2 - x_1)m}{m+M} = r\left(\frac{m}{m+M}\right)$$

If we assume one ball, say the ball on the right is much heavier ($M >> m$):

$$x_1' \approx -r$$
$$x_2' \approx r\frac{m}{M}$$ where r is the distance between the two balls.

And, if we further assume that, relative to the frame of the first coordinate system, the light ball is moving at speed V towards the heavy ball which is initially at rest, we can write:

Before the collision: $v_1 = V$, $v_2 = 0$, and $dr/dt = d(x_2 - x_1)/dt = -V$ so that [31] in center of mass (CM) coordinates, we get:[32] $v'_1 \approx V$, $v'_2 \approx -V\dfrac{m}{M}$

After the collision, we get: $v'_1 \approx -V$, $v'_2 \approx V\dfrac{m}{M}$

This makes sense because the total momentum must stay zero, and this is a simple way of conserving momentum, which we might have noticed because the symmetry of the solution.

Since we are considering the elastic case, kinetic energy is conserved as well. The fact that the speeds stay the same (though not the directions, which don't show up in the kinetic energy) means that no kinetic energy is lost.[33]

We can clarify the nature of impetus in this problem by using Figure 3-13 below. In it, we use color to illustrate the fact that impetus is a quality. The different colors represent different directional types of impeti: in particular, we pick red and green, since red is contrary to green as forward is to backward. The intensity of the particular color is used as an analogy to the intensity of the impetus (magnitude of the momentum). A more intense color in the picture indicates a more intense impetus. Figure 3-14 shows the relationship between the mathematical representation of the intensity of the impetus, i.e. using vectors, and the qualitative representation using color.

As previously, the first ball is initially moving toward the heavy (second) ball at speed V, impelled by a momentum (measured intensity of impetus) p. This intensity of impetus is shown as half intensity green. The second ball is at rest and much more massive ($M >> m$) than the first. It has no impetus, and thus it is shown black, which is the absence of color.

After the collision, the left ball moves in the *opposite direction* (shown by the opposite color, red), but with about the *same magnitude* of intensity of impetus activated in it. Hence, the left ball is shown with half intensity red. By contrast, the heavy ball (on the right) now moves in the original direction at a very slow speed, but with twice the intensity of impetus (shown by a twice intense green) of the light ball. Still, the *net* momentum is the same as before the collision.

---

[31] Notice that because the heavy mass is at rest and is so much more massive, the center of mass does not move much (compared to the speed, V, of the body of mass m) in the original frame.

[32] Note that keeping terms to order m/M in both we get before the collision: $v'_1 \approx V\left(1-\dfrac{m}{M}\right)$, $v'_2 \approx -V\dfrac{m}{M}$ and after: $v'_1 \approx -V\left(1-\dfrac{m}{M}\right)$, $v'_2 \approx V\dfrac{m}{M}$. We justify the text's keeping of *only* the first nonzero term in expansions by showing that such an approach satisfies both conservation of momentum and kinetic energy to *first* order.

[33] If we do this problem without approximations, the work is not substantially different. We get:
Before collision: $v'_1 = -VM/(M+m)$, $v'_2 = Vm/(M+m)$ implies

$KE = \dfrac{1}{2}mv_1'^2 + \dfrac{1}{2}Mv_2'^2 = \dfrac{1}{2}\mu V^2$ where $\mu = mM/(m+M)$ is called the reduced mass. After collision:

$v'_1 = VM/(M+m)$, $v'_2 = -Vm/(M+m)$, which obviously still gives $\dfrac{1}{2}\mu V^2$ because the magnitude of the velocities are the same as they were before the collision.

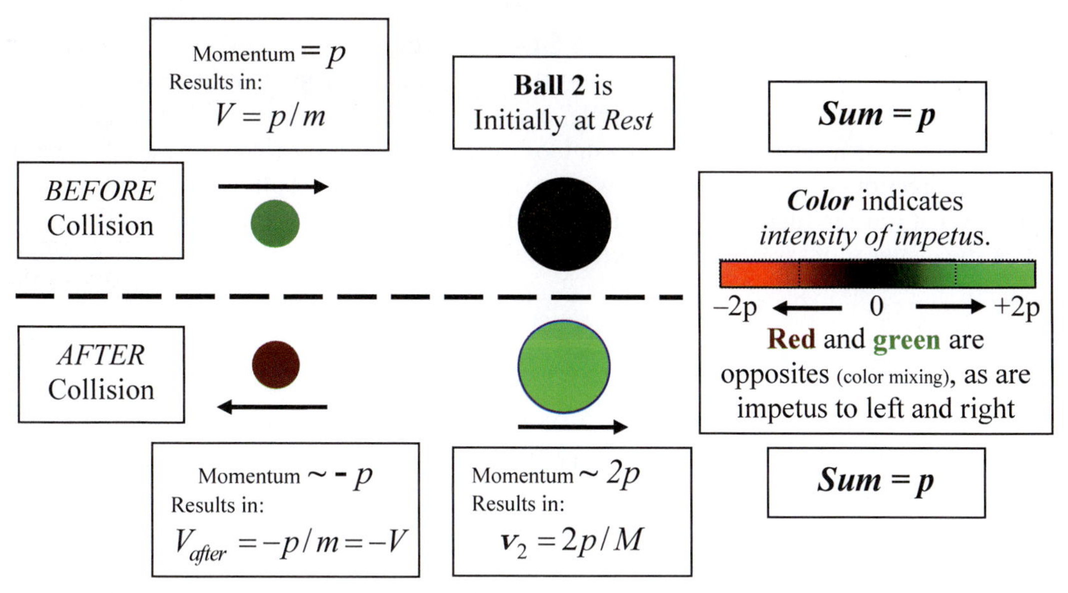

Figure 3-13: A smaller mass ball colliding with a larger mass shown using color to illustrate the intensity of the impetus of each ball. Color is a quality like impetus, so it is helpful to make this analogy to help recall that impetus does not, properly speaking, belong to the category of quantity.

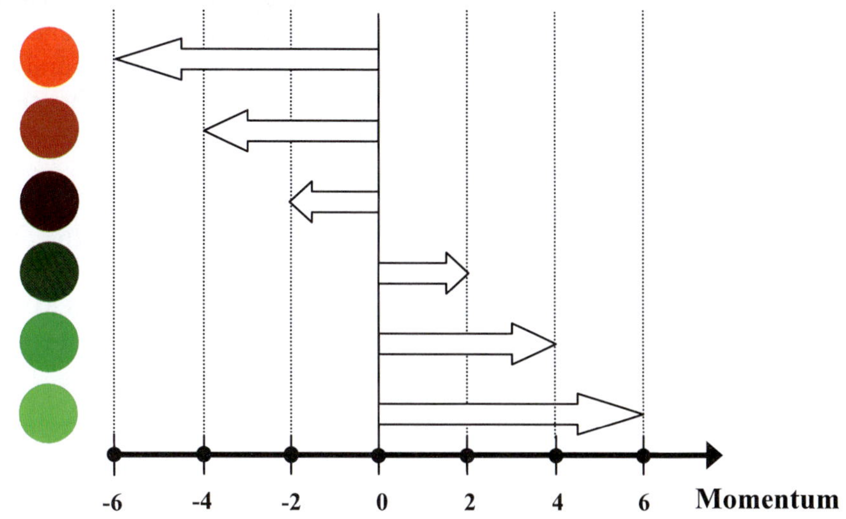

Figure 3-14: Two representations of the impetus. *Left:*: Color represents the impetus of the balls shown. Green indicates a right tending impetus, while red is a left tending impetus. Higher intensity color corresponds, according to the scale in Figure 3-13, to a higher intensity of impetus. *Right*: Corresponding vectors are also used to represent the intensity (and direction) of impetus. Impetus is a quality, like color, but it is helpful to represent it mathematically, especially in comparing one situation to a standard, which we do in measurement. We call the measure of the intensity of the impetus along with its associated direction (which we sometimes simply say, extending the meaning of intensity, as the intensity of the impetus), the momentum.

A Stick on Ice

With concept of center of mass and the conservation of momentum, we can tackle what look like hard problems. For example, consider the situation shown in the Figure 3-15 below. The ball of mass m hits the top edge of the stick, which is oriented perpendicular to the balls direction of motion. Let's assume the stick and the ball weld together upon impact. What motion then results? In the specifics, it is somewhat complicated. But if we look at the center of mass of each object, it is fairly simple

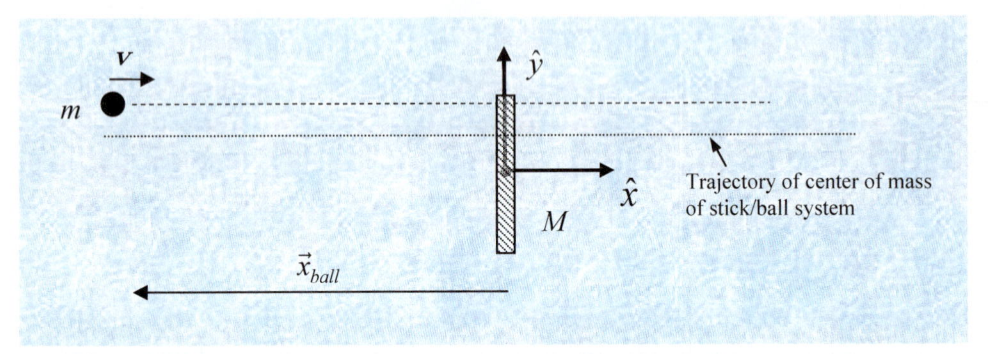

Figure 3-15: A bird's eye view of an ice rink where a ball is heading for impact with a (uniform-density rectangular-shaped) stick at rest. The stick is shown before the collision where its center was chosen to coincide with the origin of the coordinate system fixed to the earth, for example, drawn on the ice. The stick is shown as see-through to illustrate that the coordinate system is fixed to the ground rather than to the stick.

First, we need the center of mass of the stick. To help digest the concept of center of mass and develop habits in its use, we go through the calculation of the stick's center of mass in detail. Using the definition of the center of mass (equation 3.6) and Figure 3-16 below, we obtain the following equation.

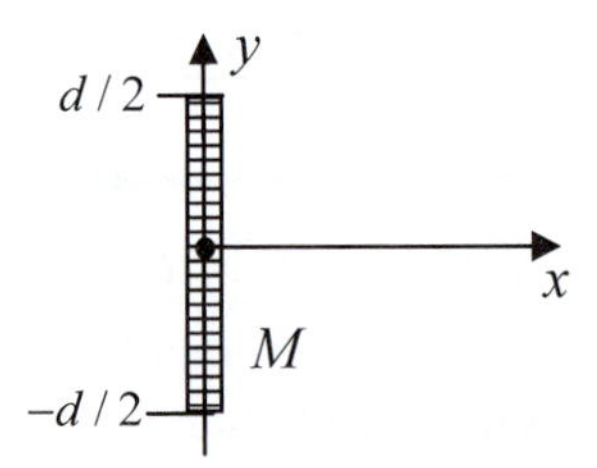

Figure 3-16: We setup a rectangular coordinate system with origin at center of the uniform density stick to calculate the center of mass of the stick. The stick of length d is divided into pieces Δy wide in the y-direction. The pieces must be very small so that the point from which we measure will not matter to a given order and in the limit, according to that mental construct, will matter not at all.

3.15

$$y_{cm} = \frac{\Delta y\,\Delta m + 2\Delta y \Delta m + 3\Delta y \Delta m + ... + N\Delta y \Delta m - (\Delta y\,\Delta m + 2\Delta y \Delta m + 3\Delta y \Delta m + ... + N\Delta y \Delta m)}{M}$$

$$= 0$$

Note, here we have divided the stick into 2*N* equal size pieces (Δy), and we have assumed the stick is uniform density so that each equal size piece is the same mass Δm. Mathematically, $\Delta y = \frac{d}{2N}$ *and* $\Delta m = \lambda \Delta y$, where λ is the linear mass density (mass per unit length) of the stick.

In the limit of $\Delta y \to 0$, equation 3.15 becomes:

3.16 $$\int y\, dm = \int_{-d/2}^{d/2} y(\lambda\, dy) = 0$$

Note the integrand is odd, meaning the area integrated is below the *x*-axis from *0* to $-d/2$ and above it from *0* to $d/2$, but these two areas have the same magnitude so that they cancel. This corresponds to the cancellation shown in equation 3.15. Physically, each side of the stick has equal mass so the balancing act of finding the center of mass yields the geometric center. A similar argument can be made for the width of the stick in the *x*-direction, thus placing the center of mass in the center of the stick.

Now, if we assume $M = m$, we can quickly work the problem. First, calculate the center of mass of the entire system. We have two components to consider at two times. If we take the initial position of the center of mass of the stick as: $\vec{X}_{CM} = 0$; that is, the origin of our coordinate system is taken to be at the center of the stick. We obtain:

<u>Before the collision</u>:

3.17a. Component in the *y*-direction ($\hat{y}$):

$$Y_{system_{CM}} = \frac{m\,Y_{ball} + m\,Y_{stick}}{2m} = \frac{m \cdot \frac{d}{2} + m \cdot 0}{2m} = \frac{d}{4}$$

3.17b. Component in the *x*-direction ($\hat{x}$):

$$X_{system_{CM}} = \frac{m\,X_{ball} + m\,X_{stick}}{2m} = \frac{X_{ball}}{2}$$

<u>After the collision</u> (when the ball sticks to top of stick):

3.18a. Component in the *y*-direction ($\hat{y}$)

$$Y_{system_{CM}} = \frac{m\,Y^{after}{}_{ball} + m\,Y^{after}{}_{stick}}{2m} = \frac{m \cdot \frac{d}{2} + m \cdot 0}{2m} = \frac{d}{4}$$

3.18b. Component in the *x*-direction ($\hat{x}$):

$$X_{system_{CM}} = \frac{m\,X^{after}{}_{ball} + m\,X^{after}{}_{stick}}{2m} = X_{ball+stick}$$

Now, there are no forces on the whole system-- though there obviously are forces within the system. Thus, the center of mass must "remain in its state of rest or uniform motion." Said another way, this means $d\vec{X}_{cm} / dt$ (i.e., V_{xCM} and V_{yCM}) is constant. In this case, there is no motion in the *y*-direction before the collision, so there can be no motion afterward of the single stick-plus-ball body. Again, if it were otherwise, impetus would be activated with no power to activate it, and we would violate conservation of momentum.

Hence, the motion of the center of mass must remain, before and after the collision, along the line perpendicular to the stick that passes $d/4$ from its center (see Figure 3-15).

After the collision we have one body, so the body must spin, if it spins, around this point. It does spin, but calculating how fast requires concepts that we will learn in later chapters. So, for now, all that's left is to determine the motion of the stick-ball body in the *x-direction* after the collision.

The initial speed of the system in the *x*-direction, using equation 3.17b, is: $V_{xCM} = \dfrac{dX_{ball} / dt}{2} = \dfrac{V}{2}$, so it must remain $V/2$ after the collision.

We can also see this by going into the center of mass frame by jumping into a ship going at speed $V/2$. This means before the collision the ball is moving at speed $V/2$ and the stick's center of mass is moving at speed $-V/2$. Thus, after the collision, the new body (formed of the ball plus stick stuck together) must *not* appear to be moving in this frame, i.e. have speed zero. Hence, slowing down our ship back to zero speed (in other words, moving back to our original frame), we again see the ball-stick body moves at speed $V/2$ in the same direction as the ball before the collision. What happens, in this regard, is the same as the behavior of two equal mass balls that collide and stick together at impact. The impetus is not lost but has twice as much mass resisting its action so it goes half as fast.

Notice another important fact. In the ball collisions, we have implicitly been assuming that everything acts at the center of mass of the ball (which we have assumed to be the center of the ball-- *exercise:* show this), and that the center of mass is the proper way to keep track of *the ball as a whole*. To see this, recall, as we already have for the stick, that the ball is composed of many parts with mass, and thus is itself a system of masses. Indeed, to not use the center of mass concept, we have to take the balls to be infinitely small. Of course, as we've seen such a concept is merely a way of talking about a limit. We mean that as we make our ball smaller and smaller compared to other lengths relevant in our system, the point we take as our marker for the ball becomes less and less important. We can as we have discussed in the Chapter 2, think of this limit as being completed for the sake of brevity and convenience. This is fine as long as we realize what we mean and that a really infinitely small material thing is purely a mental being, not able to exist outside the mind for it is a contradiction. Material things, things that can be broken down and become other things, must have quantity, i.e. have parts outside of each other.

Remembering this fact, we can confidently use the helpful fiction that our colliding balls are just points with their mass concentrated at their center of mass. Indeed, we can consider bodies, as we have in integration, as composed of points of mass. This is very useful in solving problems in all areas of physics.

Getting ready for Rockets

We now have the basic tools to solve all kinds of problems. Indeed, we are conceptually ready to talk about how we might power our rocket to Mars. However, we need more understanding of the details of working out such problems before we can tackle it successfully. Let's start with two problems that have aspects that will be helpful in understanding and solving our rocket propulsion problem.

Toboggan on the Ice

First, consider the case of a man on a sled stuck on thin ice in the middle of a glassy smooth lake. The ice is so thin that if he steps out he will fall through.[34] He does, however, have one small rock. What can he do? Put the textbook aside and think a moment before reading the answer that follows; at this point, you can figure it out with a little thought.

He can toss the rock in one direction. This will cause him to move in the other, for this is the only way to conserve momentum. This is just like what a rocket does in ejecting burned fuel. How fast will he go if he throws the rock at a speed v_{rock}? It's a one dimensional problem:

Before collision:

$$P^{Before}{}_{net} = P_{man+sled+rock} = 0$$

After collision, using conservation of momentum:

$$P^{After}{}_{net} = V_{man\&sled} m_{man\&sled} + V_{rock} m_{rock} = P_{net}{}^{Before} = 0$$

Thus, he and the sled will move across the lake at a speed: $V_{man\&sled} = V_{rock} \dfrac{m_{rock}}{m_{man\&sled}}$

The end of chapter problems investigate how effective a mode of transportation across the lake this is, including what happens if multiple rocks are thrown.

Wagon in the Rain

The second problem is in some ways the inverse of a rocket. It shows the importance of careful consideration of all mass in your system. Indeed, it is always important to draw a mental box around all bodies (all masses) that we consider to be in a system, for if we change the system definition in the middle of a calculation, the momentum is obviously no longer necessarily conserved.

Consider the wagon, shown in Figure 3-17, that is initially moving at speed V_0. It is moving in the rain, which is falling at rate of $f\ kg/m^2/s$. We are also told it is falling straight down and that the wagon is traveling slow enough that we can neglect any effect due to the apparent fall angle of the rain (*exercise:* what is this angle?). The wagon has dimensions: w, L, and h and is initially empty and has mass, m_0. What is the motion of the wagon over time?

Obviously, the wagon is fixed in the y-direction by the ground and gravity, so we know y stays constant.[35] To get the motion in the x-direction, we need to calculate the momentum in the x-direction, but first we need to apply the principle of conservation of momentum in that direction.

---

[34] To find out why he might not fall through while in the wagon but would if he steps out, see the problem at end of Chapter 4.

[35] Of course, the third dimension (i.e., into the page, in our figures), is not relevant for this example, because neither impetus nor forces act in that direction.

Figure 3-17: Wagon in rain moving at speed V_0 at a given moment. The wagon fills with rain water at a steady rate and because of this (assuming no friction), gradually slows down.

Our system is the free rain (i.e., the rain not in the wagon) plus the wagon (including whatever rain is in it). Given our assumptions, the rain only enters in terms of the mass it adds to the wagon over time. This is because the free rain has no momentum in the *x*-direction. Thus, since there are no external forces acting on our system, we see that the momentum of the wagon must remain constant. Mathematically, for the *x*-direction, we write:

3.19 $$P_{system} = P_{free\ rain} + P_{Wagan} = P_{Wagon} = \text{constant}\,.$$

To write the *x*-momentum in detail, we need the rate at which the mass of the wagon increases. This rate, with measure, $R = \dfrac{dm}{dt}$, is obviously constant and can be written (*exercise*: justify this result by dimensional analysis): $R = f\,wL$.

Hence, the *x*-momentum at some time: $t = 0$ is:

3.20 $$P_{wagon}(0) = m_0 V_0$$

At some time t later it is:

3.21 $$P_{wagon}(t) = m(t)V(t) = mV$$

Here $V(t)$ is the speed of the wagon and $m(t)$ is the mass of the wagon (again, including whatever water is in it) at the given time *t*.

In the last equality, to decrease clutter, we drop the *explicit* label of the time dependence.

At $t = t + dt$:

3.22 $$\begin{aligned} P(t+dt) &= m(t+dt)\cdot V(t+dt) \\ &= \left(m + R\,dt\right)\cdot\left(V + dV\right) \end{aligned}$$

A possible point of confusion: The functional notation, $f(x)$, which means the value of a function called "*f*" at the point *x* (we say *x* is the *argument* of *f*) can sometimes be confused with $f\,x$. That is, it may look like one is multiplying "*f*" times "*x*", especially when the argument is compound, such as $f(x+y)$. The only way to avoid confusion is to carefully note the context.

Now, applying conservation of momentum in the form of equation 3.19, i.e., $P(t+dt) = P(t)$ gives:

3.23 $$mV + m\,dV + RV\,dt + R\,dt\,dV = mV$$

Notice that in the $dt\,dV$ is a small quantity squared; recall: $dV = V'(t)dt$, so $dt\,dV = V'(t)(dt)^2$. Thus, equation 3.23 in the limit keeping only first order terms in dt is: $m\,dV = -R\,V\,dt$, which, using the integrated version of $\frac{dm}{dt} = R = constant$, gives:

3.24
$$\int_{V_0}^{V(t)} \frac{dV}{V} = -\int_0^t \frac{R}{m(t)}dt = -\int_0^t \frac{R}{m_0 + Rt}dt$$
$$V = V_0 \frac{1}{1 + \frac{R}{m_0}t}$$

We graph this wagon speed versus time in Figure 3-18 below (*exercise:* try an *Interactive Physics* (or other) simulation).

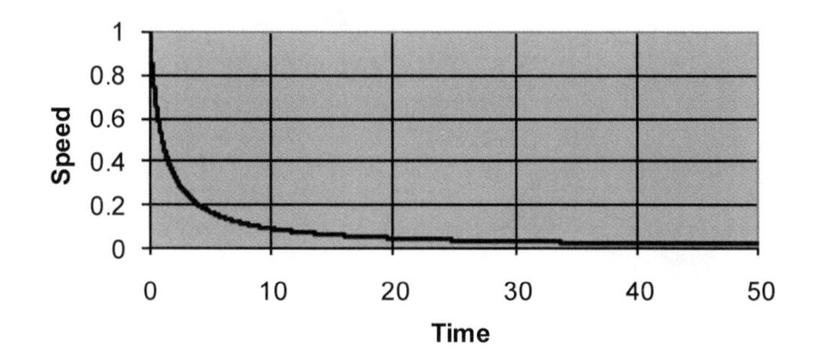

Figure 3-18: Graph showing speed function of wagon pictured in Figure 3-17 as function of time. Time is in arbitrary units and speed is in units of V_0.

Notice the wagon gradually slows down. Physically, the wagon has a certain impetus that causes it to move in the x-direction at a certain speed. As it rains, more massive bodies (water drops) with no impetus fall into the wagon. This mass is stuck in the wagon so ultimately impetus to move in the x-direction is activated in it as well. By conservation of momentum, we have the same net intensity of the impetus for the wagon as we had before, but the wagon is heavier so it moves slower. Alternatively, we can say that the intensity of the impetus per equal mass part is lower (each massive part has less impetus than it did), so the parts of the wagon (hence the wagon) moves slower.

Rocket in Space

We are now ready for our rocket. We will suppose our rocket is somewhere far from all gravity, so that we can avoid the complication of gravity for now. This is very like the part of our trip to Mars in which we are not landing or taking off from Mars or Earth.

We will assume our rocket is powered by expelling water; we will reserve detailed analysis of that water power for the next chapter. Of course, for our trip to Mars we will need a better propellant---for even more extravagant propellants see the end of chapter problems using matter-anti matter and light for fuel.

We will further assume that some kind of mechanism spits the water out the nozzle at a constant speed $\boldsymbol{v}_{\mathrm{p}}$ *relative to the rocket*, which means the propellant moves at a speed of $(\boldsymbol{v}-\boldsymbol{v}_p)$ away from the ground. Note: only when this difference is negative is the propellant's net motion toward the ground.

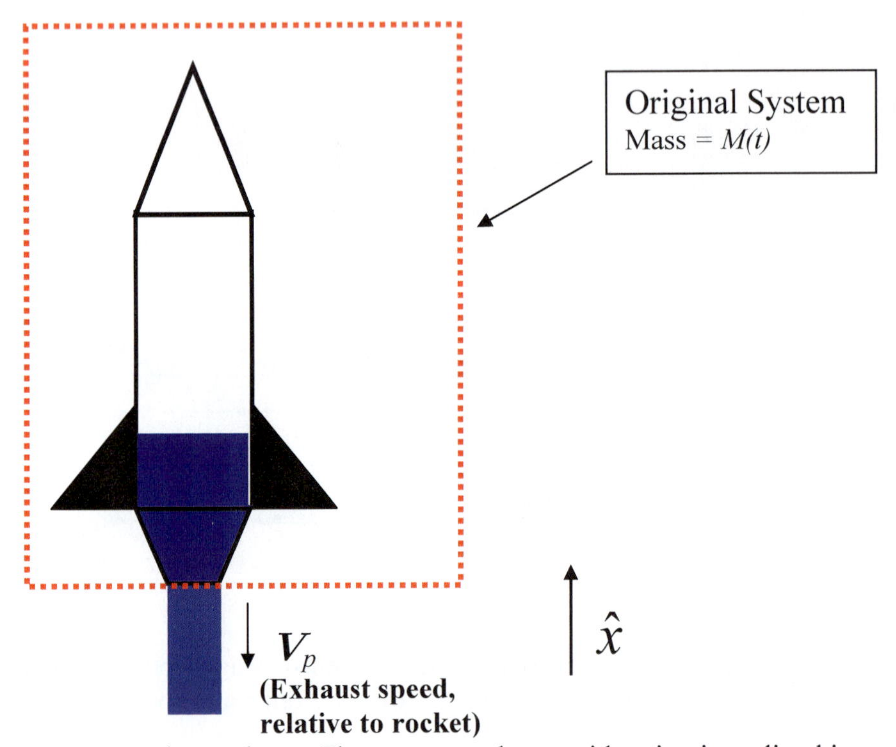

Figure 3-19: Rocket at time *t*. The system under consideration is outlined in red.

Figure 3-19 shows our rocket. The red rectangle defines what massive bodies ("mass") we consider part of our system. We will look at the rocket at a time *t*. All the exhaust that has been expelled before this time can have no further effect on the rocket, so we ignore it.

As with our wagon, after analyzing at time *t*, we analyze our rocket an instant later, $t+\Delta t$. Keep in mind that, eventually, we will want to take the limit as Δt becomes zero, so as to characterize the rocket by the rates of "flow" at one instant.

At time *t*, the rocket mass is given by $M(t)$. The momentum, *P*, at this time can then be written.

3.25 $$P(t)=M(t)\cdot\boldsymbol{v}(t)$$

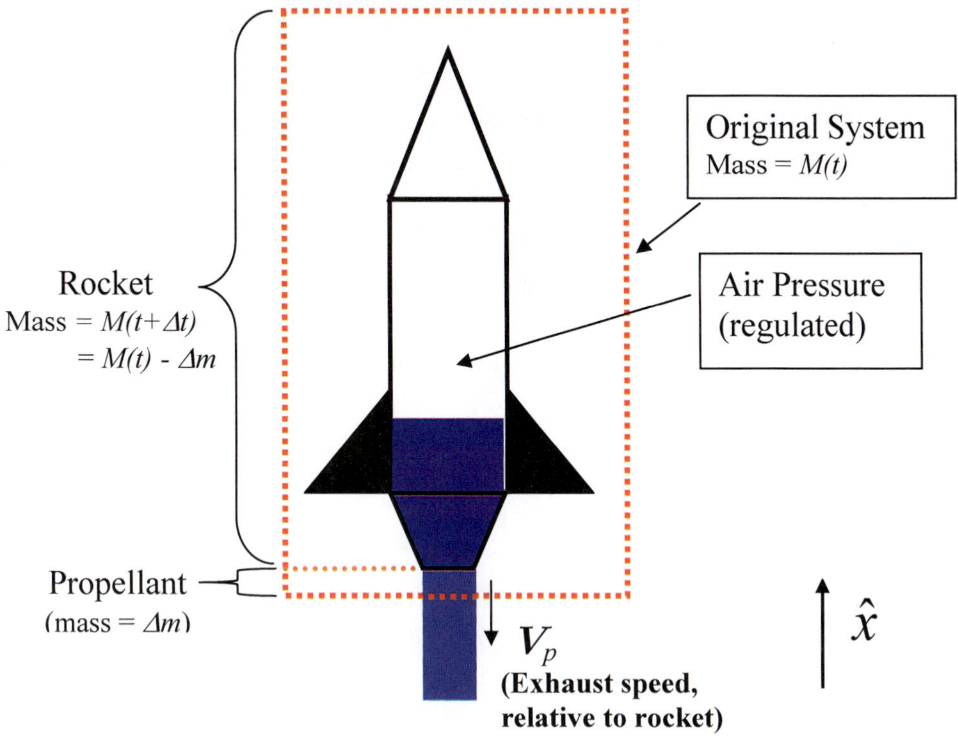

Figure 3-20: Rocket at time: $t+\Delta t$. The system is again shown in red outline, but now part of the original system has left the rocket, having been expelled to propel what's left of the rocket.

We should note the important fact that:

3.26 $$\Delta M = M(t+\Delta t) - M(t) = \left(M(t) - \Delta m\right) - M(t) = -\Delta m$$

This fact that $\Delta M = -\Delta m$ simply states that as the mass (small m) of propellant that has separated from the rocket increases, the mass of the rocket (capital M) decreases.

At time $t+\Delta t$, the system looks as shown in Figure 3-20. At this new time, some of the mass of the rocket has left as propellant. The rocket is thus moving $\boldsymbol{v}+\Delta\boldsymbol{v}$, a little faster in the x-direction. The propellant is thus moving at $\boldsymbol{v}+\Delta\boldsymbol{v}-\boldsymbol{v}_p$ relative to the ground.

Note: This later result follows because the rocket must gain the full increment of velocity $\Delta\boldsymbol{v}$ at the moment that the Δm of propellant is expelled. The rocket could not, for instance, gain any speed *after* the Δm of propellant exited, because after that point, the Δm is no longer able to influence the rocket, and thus no further increase in speed can be had ***due to it.*** Hence, at release, the rocket moves at $\boldsymbol{v}+\Delta\boldsymbol{v}$, and since we've defined the relative speed of propellant at release to be $\boldsymbol{v}_p$, the speed of the propellant is: $\boldsymbol{v}+\Delta\boldsymbol{v}-\boldsymbol{v}_p$.

Dropping the *explicit* functional dependence of M for convenience of writing, we get the momentum at this new time:

3.27 $$P(t+\Delta t)=(M-\Delta m)(\boldsymbol{v}+\Delta\boldsymbol{v})+\Delta m(\boldsymbol{v}+\Delta\boldsymbol{v}-\boldsymbol{v}_p)$$

Now, conservation of momentum demands:

3.28 $$\begin{aligned}P(t+\Delta t)&=P(t)\\&=M\boldsymbol{v}+M\Delta\boldsymbol{v}-\Delta m\boldsymbol{v}-\Delta m\Delta\boldsymbol{v}+\Delta m\,\boldsymbol{v}+\Delta m\Delta\boldsymbol{v}-\Delta m\boldsymbol{v}_p=M\boldsymbol{v}\end{aligned}$$

We now recall that all the Δ's are small quantities, that is eventually we can take: $\Delta t\to 0$; this means, $\Delta\boldsymbol{v}\to 0, \Delta m\to 0$. As in the case of the wagon, we eliminate all the terms that are squares of small quantities and replace finite difference with differentials. We thus get:

3.29 $$M\,d\boldsymbol{v}=\boldsymbol{v}_p\,dm$$

Note we have an equation that has two related masses: *M* and *m*. Using $dm=-dM$ to eliminate *m* gives:

3.30 $$M\,d\boldsymbol{v}=-dM\,\boldsymbol{v}_p$$

By integration, we get:

3.31 $$\boxed{\boldsymbol{v}_{final}=\boldsymbol{v}_i+\boldsymbol{v}_p\ln\left(\frac{M_i}{M_f}\right)}$$

Here M_i and $\boldsymbol{v}_i$ are the initial mass and velocity of the rocket (with whatever fuel it has), and M_f and $\boldsymbol{v}_{final}$ are the final mass and velocity of the rocket.

This calculation shows that the faster the propellant's exit velocity, and the larger the ratio of initial to final mass, the faster the rocket is going by the end of the exhaust sequence.

We've now seen something of the scope of physical situations that we can understand using the simple, but deep principles we've discovered. Link this with the deep aspects of quantity embedded in the mathematics discussed in Chapter 2 and with a developed ability at applying the principles and mathematics in problem solving and many, many more physical situations can be understood. The exercises at the end of the chapter explore some of these.

Still, we are missing important elements. We've seen impetus change. What makes it change? We need to make this explicit, so our study thus turns to *forces* in the next chapter.

Summary

Locomotion, motion from one place to another, is the central topic of Newtonian mechanics. Capacity for motion generally is the defining characteristic of physical things. Locomotion is responsible for bringing substances together to interact. Motion among parts of a substance can (secondarily) also be considered a type of locomotion. Because locomotion is a *change of place*, it requires a cause.

The power (category of quality) of a massive body that keeps it moving at a constant speed and linear direction is called *impetus*. Impetus is activated in a body and is second nature to the body because it can be gained and lost without the body changing its fundamental nature. A *body* is defined as a substance, group of substances or part(s) of a substance or substances, and, thus, we speak analogically of the impetus (and mass) of a body, since properties properly belong to an *individual* substance (for instance, one can say a group of people know math, only because they individually know math). Hence, for example, we may treat the man and the wrench he is holding together as one body.

Mass (inertial mass) is the measure of the receptivity of a body to the action of its impetus. Massive bodies, i.e. bodies with mass, move only by impetus.

Inertia is a term used in two different ways, not often clearly distinguished. It can mean the tendency of a body to move uniformly in a straight line, and, as such, refers to impetus. It also can refer to the resistance of a body to acceleration, which refers indirectly to the *mass*. Mass and impetus are correlative properties; that is, like top and bottom, they cannot be understood separate from one another, but yet are distinct.

In understanding a physical situation it will often be helpful, in order to facilitate focusing on one aspect of reality, to switch "off" certain properties that can be made to have or do have a small effect in some sense proper to the problem. These properties may actually be crucial to our understanding (possibly even essential) at some point, but we leave them out in order to focus attention on the problem at hand. For example, in understanding impetus, we, at first, leave out any resistance to the action of the impetus, by neglecting, for instance, the resistance of the air to a moving baseball or the ice to a sliding hockey puck.

Momentum is the *measure* of the intensity and direction of the impetus and, as such, is a number. As with other measured properties, we will often use the same word to designate the property itself and its measure (comparison to a standard). That is, we will often use the word *momentum* to designate both the impetus and its measure and let the context indicate the meaning. The unit of momentum we use is: $1\ Buridan = 1 kg\ m/s$.

Momentum, symbolically $\vec{p}$ (which can be remembered as the "*p*" in im*p*etus), is a vector; it has magnitude and direction equal to the measure of the intensity and directional tendency of the impetus of a body. Mathematically, i.e. writing the symbolic relation of the measures of the properties (mass and impetus) and motion (change of place) involved, we write: $\vec{p} = m\vec{v}$, where $\vec{v}$ is the velocity of the body. This has been verified by much experiment within the domain of applicability of Newtonian physics and, in analogical ways, beyond.

The generic behavior of massive bodies leads us to ***Newton's first law***: *Every object continues in its state of rest or uniform motion unless acted on by an outside force,* which implies that momentum is conserved for an isolated body in empty space.

Note that impetus cannot be impeded from acting; it can only be transferred or completely deactivated (i.e. destroyed).[36] For example, only left tending impetus can destroy right tending impetus. When there is impetus, there must be motion; if a body has impetus, regardless of how large its mass (the resistance to the action of its impetus) there will always be some motion given by: $\vec{v} = \vec{p}/m$.

Conservation of momentum means that in summing *all* the momentum of bodies in a system not subject to external forces, i.e., a so called isolated system, the *net momentum cannot change.* This conservation law is written mathematically as: $\vec{P}_{Total} \equiv \sum_{1}^{N} \vec{p}_i = constant$, where the index i labels the different bodies in the system, so that p_i is the momentum of the ith body. In this equation, we take the impeti of opposite directional tendency to have opposite sign and, thus, in this sense, allow them to "cancel" each other; and, we get a conservation of tendency to move in a given direction, called conservation of momentum. The momentum of a body can only be changed by the action of an external force. In this way, if the space is translationally symmetric, that is having no force that distinguishes backward from forward along a line then there is no way the momentum can change. Furthermore, conservation of momentum means that two or more bodies isolated from external forces can collide and change each other's impetus, but only in such a way as to keep the total momentum constant. As seen in the vector nature of the equation of conservation of momentum above, this conservation is *separately* true for each of the three linear spatial dimensions.

The nature of massive bodies is also such that the total activity, i.e. energy, of the impetus of an isolated system is conserved. Conservation of Kinetic Energy is written: $KE \equiv \sum_{i=0}^{N} \frac{1}{2} m_i v_i^2 = constant$ and will be studied in detail in Chapter 5 where potential energy will be included in the conservation law. Collisions in which the total kinetic energy of the bodies under consideration is conserved are called *elastic collisions*, those where it is not are called *inelastic collisions*. When the massive body is composed of movable massive parts, those parts can be analyzed to give another viewpoint; for instance, in the inelastic case, some kinetic energy might go into motion of internal parts of the bodies. Taking advantage of our analogical definition of "body," we can also analyze the

[36] To see this, be careful to include all bodies.

internal parts as separate bodies, imposing the conservation of momentum requirement on the center of mass motion of the balls.

The nature of the impetus is such that any mechanical experiment done with objects of given *apparent* momentum in a laboratory that is uniformly moving at speed v will come out the same if repeated with objects of the same *apparent* momentum when the laboratory is moving at *any* other uniform speed or at rest. This principle is called *Galilean relativity* and reveals the inertial frames (uniformly moving frames) as preferred reference frames. Indeed, we cannot sense when we are in uniform motion. Galilean relativity fundamentally arises from the relativity of locomotion at the quantitative level (the level of pure extension that leaves out even some shape (i.e. the quality that limits the extension)) coupled with impetus's nature that, for speeds small compared to that of light, does not affect the body in any way other than moving it uniformly, so that a uniformly moving system, for example, will respond to further activation of impetus as if it had none.

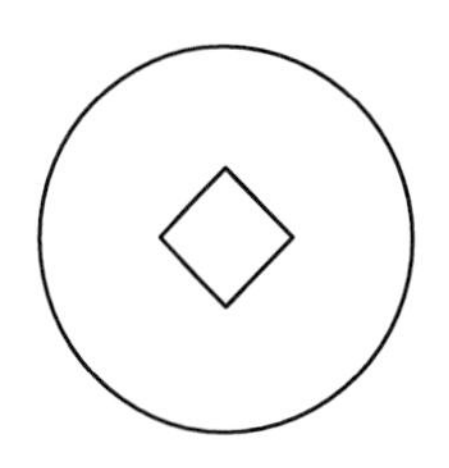

The primary meaning of locomotion, the one we know directly through our senses, is the motion of a body from one place in an external contacting media to another place in that media. For example, the baseball moves (category of action) by its impetus (quality) through the air (category of reception). Of course, whatever contains the object moving might also be moving. For instance, the earth is moving through space. So, to finally understand the complete motion of a body and know how much impetus it actually has, we would have to determine the state of the motion of the ultimate outer part of the universe. In order to focus on the issues at hand and avoid these larger issues, we consider the universe devoid of all but quantity (i.e., to the degree one can, "switch off" all other properties). To further simplify, consider the universe as only having a space of non-regular shape around a massive spherical object with a cubic object inside of it. Ordinarily, the motion of the sphere would be distinguished from motion of the cube, because the cube's motion would require impetus (quality) to move (action) it through the sphere (reception), whereas the sphere's motion would require impetus to move (action) it through space (reception). However, eliminating all qualities (except shape), all action and all reception removes all distinctions but those arising from the cube being inside the sphere in space. These last distinctions can be removed, by leaving out the space exterior to the sphere. Now, in order to give meaning to motion of the sphere with nothing outside of it to move in, we have to analogically generalize the idea of locomotion to include motion of the sphere over the cube with no motion of sphere in anything. Such reasoning allows us to treat any isolated system without considering what, if anything, is outside of it. So, by a mental construct (being of reason) in which we limit our consideration to extension (quantity) with only the shapes of the objects to limit that extension (not any consideration of inside and outside), we can say one part of the sphere passing one part of the cube is *the same as* one part of the cube passing one part of the sphere; they are symmetrical with no properties of action and reception to make them different. In this way, we cease to consider the impetus that a body actually has but only what it *apparently* has. We do this when we "transform" i.e. move to a uniformly moving frame to view a given physical situation.

More generally, when we exclude any consideration of the properties of the media, through which massive bodies of interest are traveling (e.g. air or vacuum), except the distance (the measure of the extension of the media) *between* the bodies, *uniform* locomotion of the bodies becomes relative *if* we exclude the actual impetus and consider only apparent impetus.

The nature of impetus to move things at a uniform speed in a given direction means that we can move at uniform speed in that direction (or the opposing direction) and appear to take away (or give) that amount of speed from each body we are viewing. Further, since impetus does not change other aspects of the body (at our level of approximation), we also change the *apparent* impetus of a system of bodies by the total mass of the system times our speed.

The *center of mass* of a system of bodies is defined as: $\vec{X}_{cm} \equiv \frac{\sum_{i=1}^{N} \vec{x}_i m_i}{\sum_{i=1}^{N} m_i} = \frac{\sum_{i=1}^{N} \vec{x}_i m_i}{M}$.

By taking a time derivative, assuming the mass does not change with time, we get the momentum of the center of mass: $\vec{P}_{cm} \equiv M \frac{\sum_{1}^{N} \frac{d\vec{x}_i}{dt} m_i}{M} = M\vec{V}_{cm} = \sum_{1}^{N} \vec{p}_i$. Where M is the total mass of the system and $\vec{p}_i$ are the momenta of each body of the system.

One transforms positions in an initial frame, labeled by unprimed vectors, to the center of mass frame, labeled by primed vectors, by the following equation: $\vec{x}_i\,' = \vec{x}_i - \vec{X}_{cm}$, for the ith particle in the system.

By use of the *center of mass frame,* that is by viewing the system from a frame moving at the same velocity as the center of mass of the system, we can concentrate on just the internal dynamics of the system, ignoring whatever motion the system might have in the larger world. By transforming back to the rest frame, we can "put back" the impetus we ignored earlier. Separating out the system's internal and external motions in this way is a helpful in reducing the complexity of a problem.

We can use color as an analogy to help us recall that impetus is a property in the category of quality. In particular, we choose the intensity of green to indicate the intensity of right moving impetus and the intensity of its opposing color, red, to indicate the intensity of the left moving impetus. In this analogy, a black (the absence of color) object has no momentum and thus is at rest. Since, for reasons discussed above, we typically only know the relative momenta, these colors will typically represent apparent momenta with respect to a given uniformly moving frame, i.e. inertial frame.

Terms

Motion

- Inertia
- Impetus/Momentum
  - Locomotion
  - Second nature
  - Mass
  - Newton's First Law of Motion

Collisions

- Conservation of Linear Momentum
- Body
- Kinetic Energy and its conservation (introduction only)
- Inelastic collision
- Elastic collision

Galilean relativity

- Change of reference frame
- Translational spatial symmetry
- Reference frames:
  - Inertial frames
    - Moving frame
    - Laboratory frame

Center of Mass Frame

- Center of Mass

Calculational Techniques:

In defining a problem always identify the system under consideration and *keep it fixed* during calculation.

Pick a convenient coordinate system and origin. Pick a reference frame that simplifies the problem. Transform to that frame, solve the problem, and then transform to the original frame to get the answer.

To solve for the motion of bodies, write the expression for the total momentum (and, when necessary, energy) of the system a moment before the interaction and a moment after, then write the statement of conservation of momentum (and energy). In the case of an iterated action, pick a time t and then consider a moment later, $t + \Delta t$ after some amount of activity has occurred.

During momentum transfer problems, remember once a part of a body loses contact with the original body, the part can no longer change the original body's velocity, so the velocity assigned to it must be whatever it is at the moment of release. Also, for ejection of parts of a body, remember it is important to keep track of labels for mass that belong to the body (which might be labeled *M*) versus mass that belong to ejecta of parts (which total might be labeled *m*) of the body. In the case of the rocket, this introduced a minus sign: $dm = -dM$.

In applying conservation of momentum, remember to include the momentum for all interacting bodies; also, contact is a good indication that there is an interaction.

Notation and Vocabulary Notes:

It is important to use convenient and easy to follow notation in solving a problem. In this regard, it is helpful to use standard notational conventions as much as possible. Some notational meanings overlap and must be kept straight. For example, primed coordinates are used *both* a) for properties and motion measured in the center of mass frame (which we call "quantities in the center of mass frame") *and* b) to label quantities as they are *after* an interaction (called, "an event"), such as a collision. Namely, a) the position vector of an object in the laboratory rest frame (the frame in which the lab is at rest) is written: $\vec{x}$, while the position vector in the center of mass frame is written, $\vec{x}'$ and b) the velocity of a body *A* before a collision is written $\vec{v}_A$, while the speed afterward is written $\vec{v}'_A$.

In speaking of frames, problems will often use shorthand wording such as *laboratory frame* and *moving frames*. The laboratory frame refers to observers standing still in the laboratory and the moving frame refers to observers "flying by" at a certain speed and direction. Note the *frame* of an observer specifies his state of motion (generally, orientation and locomotion), while the *coordinate system* specifies the labeling system to be used in identifying object positions in the given frame.

Problems by Subtopic

Momentum

1. We say mass is a measure of the receptivity to the action of the impetus, and add that the more the mass the more the resistance to the impetus. We call it a receptivity to point out that the impetus' movement of the body is in the category of action, while the quality of the body (and media) that receives this action, the mass, is in the category of reception. However, suppose we wanted to emphasize that the mass, as we measure it, is expressed as a resistance. What would be the equation relating the mass to a *direct* measure of the receptivity, symbolized by R_m? *Hint*: The units of this new measure might be a marg (gram spelled backward) in analogy to the way, electrical resistance is measured in ohms and electrical conductance is measured in "mhos."

2. A bird is perched on a boat in the middle of a very large lake. a) Taking (relative) place to be the inner surface of the immediate containing bodies, describe the place of the bird. b) Ignoring the earth's rotation, give one possible scenario of what its impetus might be in a universe that just consisted of our solar system and nothing else.

3. Assume all the massive objects in the universe are identical and *exactly* uniformly distributed and are expanding radially at a rate *exactly* proportional to the distance from a given origin and "space" is also the same throughout the universe. Think, for instance of a spherical loaf of bread rising in the oven. Pick a center from which everything is flying apart.
a) Using radial vectors, write the vector equation for the velocity of any given massive body as seen from the origin.
b) By switching origins by a coordinate transformation, show that one cannot locally (nearby any given point of observation, explicitly *excluding* the edge of the universe) determine what impetus each body has by looking at the motion of the bodies alone. (Recall that the inability to determine something does not make it cease to exist). c) To help understand this result, consider two observers *A* and *B*, one at the center (*A*) and one a distance r_0 (*B*) away. What apparent velocity does a body $r_0/2$ away from the origin have according to *A*? According to *B*? d) Comment on what extent the assumptions above are true and at what length scale for our universe.

4. Explain why uniform locomotion is not a property, but change of a property.

(Problems continued on next page)

5. There are two objects of equal mass and intensity of impetus as shown to the right. If the two objects collide, excluding considerations of conservation of kinetic energy, a) in general, what are the possible outcomes? b) do the pictures below represent possible outcomes of the above scenario? Why or why not? What are some other possible outcomes?

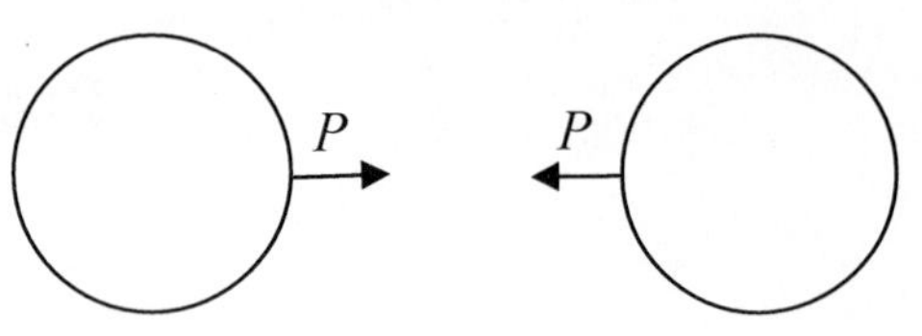

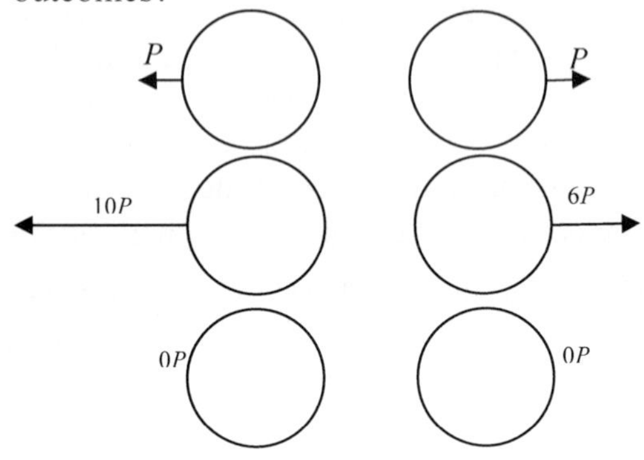

6. What is the only way to cancel an impetus traveling in a specified direction?

7. A small ball of impetus p and mass m strikes a ball at rest with mass $0.5m$ (see figure to right).
a) What will be the result from the collision?

b) Another ball with the same mass and initial impetus strikes a ball at rest with mass m (see figure to right). What will be the result from the collision?

c) A third ball with the same mass and initial impetus strikes a ball with mass $2m$. What will be the result from the collision?

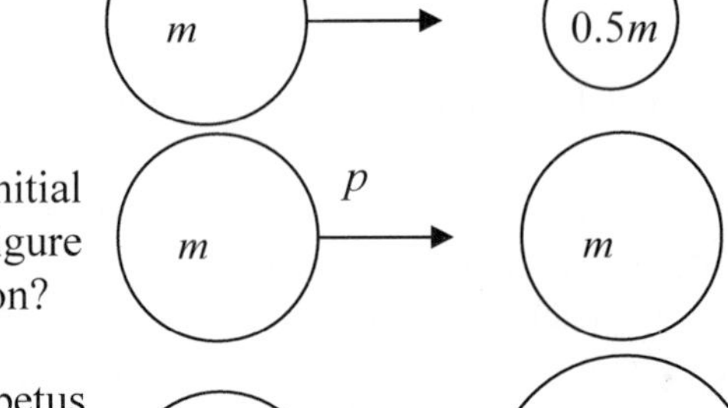

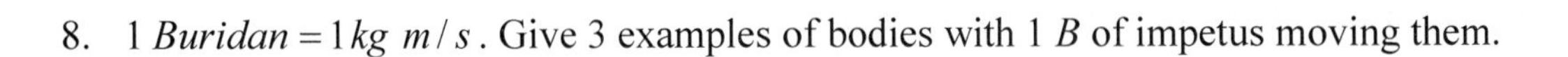

8. $1\ Buridan = 1 kg\ m/s$. Give 3 examples of bodies with 1 B of impetus moving them.

9. Explain the relationship between impetus and momentum.

10. Explain and draw a diagram showing what would happen if a body of mass $2M$ traveling at speed $4V$ to the right collided head on with a body of mass $5M$ traveling at speed $3V$ directly at it. a) Work this problem for the inelastic case in which the balls stick to each other. b) For the elastic case in which kinetic energy is conserved.

11. Suppose conservation of momentum was not true, what would that imply?

12. Explain how the equation for center of mass momentum uses the chain rule by assuming conservation of mass.

13. A boy is walking across a baseball field and holding a ball. If he drops it while he is still walking, describe the motion of the ball. From where does the ball get its forward impetus?

14. A catcher picks up a bunted ball (which is already at rest) and throws it to the first baseman. Describe how the impetus of the ball *and* the players change from before the throw to after the catch.

15. A dump truck of mass M carrying sand of mass m has activated in it a certain intensity of impetus, p, on a flat road. A small hole in the bottom of the truck bed allows the sand to slowly leak out while the truck moves along the road. Assume no friction or air resistance is present.
a) What is the velocity of the dump truck after all of the sand has leaked out?
b) After losing all of the sand, will the truck have a chance of hitting the county store at the end of the road?
c) When only half the sand has been lost from the truck it impacts the county store. What is the speed of the truck under this condition?
d) What is the difference in speed of the truck between the situations described in b) and c)?
e) What, if any difference, in damage to the county store would you expect?

16. A box is moving with speed v. Contained within it is an inner box and inside the inner box is a marble. Both the inner box and the marble are at rest. The box and marble inside are at rest. The inner box and marble will stick to whatever surface they touch, describe the motion assuming the box and marble inside start at the center of the outer box. Assume the mass for each of the three objects is equal to m.

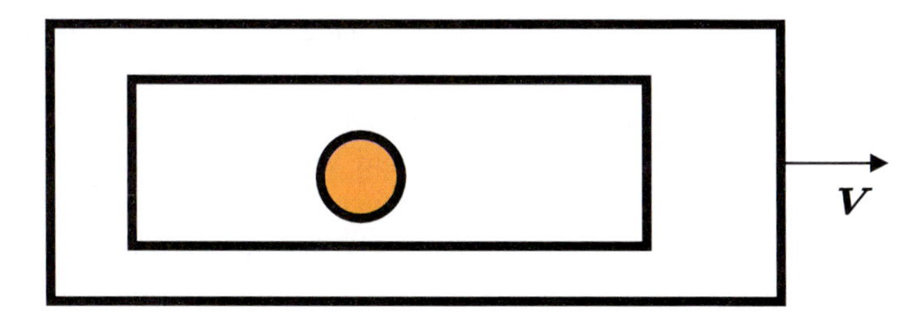

17. Describe in what sense the chopping of a worm to get worms whose total strength adds to the same as the original is an appropriate analogy for the splitting of impetus and in it what sense it is not.

18. Show using conservation of energy and momentum that for the case of two *elastically* colliding, equal-mass balls as viewed in a frame in which the two balls approach each other each at a speed $\mathbf{v}/2$ that after collision they must recede at $\mathbf{v}/2$ each but in the opposite direction.

19. Given a head-on elastic collision, in which a moving ball, *a*, coming in from the left impacts a ball, *b*, at rest on the right. Using both graphical methods and derivatives answer the following: a) What is the condition for maximum momentum transfer from the first ball to the second ball? b) What is the condition for maximum energy transfer?

20. Show that if a ball of mass *m* moving at speed *V* collides in an elastic collision that is *not* head-on with an equal mass ball at rest, the trajectories of the balls after the collision must always make a right angle. (Experiment in *Interactive Physics* with file: 3.3-equal mass balls collide to give right angle.ip)

21. For a head-on elastic collision of two bodies with mass m_a and m_b, moving at speeds $\mathbf{v}_a$ and $\mathbf{v}_b$ at each other, use conservation of *KE* and *P* to calculate the velocity of the bodies after the collision. To simplify the calculations, take $m_a = 1$ and $m_b = m_r$; no generality is lost in this way, since the ratio of the two masses, m_r, is required in the equations.

22. How can a sail boat traveling upwind conserve momentum? *Hint:* consider the keel.

23. Explain at a simple level how "Newton's Cradle" (Figure 3-8 and http://www.walter-fendt.de/ph11e/ncradle.htm) works by answering the following. Consider a group of 5 balls, each of mass *m*, in outer space far from the effects of gravity or any friction. Suppose one of the 5 balls is separated from the others and is initially moving at speed $\mathbf{v}$ toward the others (which are all in contact with each other but not stuck together). After colliding with the other balls, the first ball comes to rest along side balls two, three and four, and the fifth ball moves off at speed $\mathbf{v}$
a) By giving one other such outcome, show that there are other outcomes that conserve momentum and energy and do not result in further ball collisions and do not violate the integrity of the balls by having more than one ball in the same place at the same time. b) Show why this apparently feasible outcome would violate the assumed symmetry of the situation. Note that in Chapter 4, we will discuss the surface forces that change the momenta of the various balls, and, in Chapter 5, we will discuss how potential energy is important in this problem.

24. a) How much impetus does the earth need to move in orbit around the sun; given its mass and speed are: $6 \times 10^{24} kg$ $3 \times 10^{4} m/s$. b) How much impetus (on the average) does a $100kg$ man need to walk at about $1m/s$. c) How much impetus does it take to move your car at 60*mph* down the road? d) a baseball? e) to wave your little finger

25. There is a box of basketballs of total mass *m* hanging from the ceiling of a gym. a) Describe what happens while the balls fall assuming the impetus is activated equally in all the basketballs at once as it falls. b) Describe what would happen if the box was resting on the frictionless ground and someone threw another basketball at it.

26. The LSU and Auburn tigers are getting ready for the national championship game. It begins to rain, soaking the field, the players and the die-hard fans. Besides making the game more interesting, describe how the game would change (in terms of momentum) since it is being played in the mud. a) Which types of players would gain the biggest advantage: the biggest players or the smallest? b) Assume the rain continues uniformly, neglecting wind resistance, what is the equation for the speed of a football of cross sectional radius A that is thrown at $\boldsymbol{v}$ *m/s* move in rain of density ρ, assuming the rain sticks to the ball (not adding significant mass) and then falls off once brought up to speed of the football; discuss the limits of the approximations made in this section.

27. If all the men in China jumped off a table at the exact same time: a) how much would they change the momentum of the earth assuming a rate of acceleration of $10 m/s^2$? b) If they all wore jetpacks and shot themselves directly up into the air about 3 feet and hovered there, how much would they change the momentum of the earth once in the air? c) If they aimed their jetpacks sideways (parallel to the earth) and tied themselves to the earth, how would it change the momentum of the earth? e) If some of the exhaust from their jetpacks was to escape Earth's atmosphere, in general terms, how would this change the momentum of the earth?

28. Assuming an inelastic collision with a nail, which is better for the driving a nail, a heavy hammer or a light hammer? Explain

Galilean Relativity and Reference Frames

29. Solve the problem of a ball of mass *m* colliding head on with a ball mass *M* a) in the lab frame where the mass *M* is at rest b) then, by a simple transformation, solve it *in a frame moving* at a speed $\boldsymbol{v}$ with respect to the laboratory frame. c) Explain why this shows that the results of the following two experiments as viewed, respectively, in the lab frame and the moving frame, would be the same: *i)* A laboratory table on which a mass *m* moves at speed $\boldsymbol{v}_a$ towards a mass *M* at rest *ii)* A platform with a table nailed to it moves at speed $\boldsymbol{v}$ to the right; a ball of mass *m* on the moving table moves to the right at a speed of $\boldsymbol{v}_a$ with respect to the table towards a ball of mass *M* at rest with respect to the table.

30. Consider again the monkey in the box shown in Figure 3-9. Assuming a cubic box with length of 10 meters that weighs 10 million tons, with mostly N_2 air ($\sim 28 g/mole$), a) Calculate what fraction of the momentum moving to the right is from the box and what from the air. b) What is the momentum of the center of mass of the system?

31. Explain how, in the absence of gravity, say on our trip to Mars, one could, in principle, establish a *physical* reference point that would serve to mark the relative rest condition in a given uniformly moving frame.

32. Consider a spacecraft with air, a man, a monkey, and a cat inside. Our definition of body allows us to consider any combination of these things or parts of things as bodies. Though they are parts of different substances, would there ever be an advantage to considering the heart of the man and the heart of the monkey as parts of one body, though they are obviously parts of different substances? If not, why not? If so, what is the most obvious whole of which they are a part, and hypothesize as to why one might focus on the two hearts.

33. A 100 *kg* man is running at 20*mph* on a large thin platform the mass of the earth, how fast does the platform move?

34. Think of outer space as if it had no edge; describe the motion of two bodies moving through it. In what sense can we talk about relative motion alone in this space?

35. Discuss why the universe cannot be *actually* infinite in extension at any one time, but can potentially expand for as long as one likes.

36. Discuss the advantages and limits of treating locomotion simply as the change of the distance between objects; to simplify the discussion, consider the locomotion of two massive objects.

37. Suppose a large asteroid of mass m, which is moving parallel to the direction of the moon's motion, collides with the moon (mass M, speed $\boldsymbol{v}_M$ with respect to the solar system rest frame) at a relative speed $\boldsymbol{v}$, making a huge crater leaving *all* the particles of the collision scattered far and wide on the surface of the moon. You have been called in to investigate what will happen to the moon and us. a) Assuming that it happens instantaneously so that gravity can be ignored during the collision, and given that the energy is not enough to dislodge the moon from its orbit around the earth, what instantaneous speed change will the moon undergo? b) Treating the earth moon as a single "glued together" system and assuming the asteroid motion also happened to line up with the motion of the earth/moon center of mass. What speed change will the earth undergo? c) Assuming $m = 2\times10^{12}\,kg$ (a 1 *km* sized object of density $2g/cm^3$) and $\boldsymbol{v} = 32\,km/s$, give number values for parts *a* and *b*. d) comment on what the real threat of asteroids might be.

Center of Mass

38. Find the center of mass of a triangle, assuming uniform surface density.

39. Two balls with the same mass and velocity are on a collision course. Neglecting gravity, friction, etc, find the center of mass of the two balls.

40. Two cars of equal mass are racing toward each other on a straight, flat road in the middle of nowhere. The one on the right is three miles from a stop sign and the one on the left is seven miles from the same stop sign. a) Assuming the cars are traveling at the same speed, find the center of mass of the two cars. b) How fast is the center of mass moving? c) If the center of mass was at the center of the coordinate system and remained stationary, how would the scenario be different?

41. Consider empty candy boxes of mass M moving down a conveyor belt at speed v_0. As each box moves down the belt it gets a new candy, each of approximately mass m, put in a different section of the box. This happens once every τ seconds. a) How long would it take for a candy box to be moving at half speed if no horizontal momentum was supplied by the conveyor belt? B) What change in momentum per unit time does the conveyer belt need to supply to keep a given candy box moving at uniform speed?

42. Why does a ball elastically colliding with another equal mass ball at rest transfer all its momentum to the second?

Rockets and the Like

43. A boy of mass M_1 sits on a sled of mass M_2 with a collection of N rocks, each with mass m. The boy can throw a rock at a speed $\mathbf{v}$ relative to himself while sitting on the sled.
a) What is the speed of the sled relative to the ground after the boy throws the first rock?
b) What is the speed of the sled relative to the ground after the boy throws the second rock?
c) What is the speed of the sled relative to the ground after the boy throws all N rocks?
d) Show that in the limit as the number of rocks becomes infinite and the total mass of rock remains fixed that the final speed of the sled is described by the rocket equation.
e) Using 10 rocks of mass of about .2 *kg* each estimate how fast a boy and wagon of mass 50*kg* could get going and thus how effective a mode of transportation it is.

44. In a rain falling at rate R on a wagon initially moving at speed V_0, what functional form $R(t)$ does one need to get a wagon speed that decays exponentially as: $V(t) = V_0 e^{-\alpha t}$

45. Consider an interstellar spaceship of mass M_0 that uses collimated light shot out the back engines as propellant. The light has a measured power (energy per unit time) of P_0. Assume that this effectively means that a chunk of mass P_0 / c^2 is ejected at the speed of light c (in the lab frame--not relative to the rocket as done in the text) every second $c \sim 3 \times 10^8 m/s$.
a) What is rocket speed *and* acceleration as a function of time for this non-relativistic ($v << c$) case which we assume above? *Note*: The fully relativistic case is treated in Chapter 10.
b) How long would it take to get going one tenth the speed of light if $P_0 = 1000\, watts$ and $M_0 = 5000\ kg$.
c) How long, under the above conditions, would it take to achieve an acceleration of $g = 9.8 m/s^2$ (i.e. the mean acceleration due to gravity at earth's surface) ?
d) If this light was caused by matter anti-matter annihilation (a positron crashing into an electron and destroying each other), how many electron/positron pairs would it require. Given that an electron/positron pair annihilation causes the same effect as shooting both members of the pair off at speed c. The mass of an electron is: $9.11 \times 10^{-31} kg$ and a positron has the same mass.

46. How is momentum conserved when a bike turns a corner, say to the left? Estimate the momentum change in the turn.

47. A pool ball of mass m is assumed to make an elastic collision with the edge of the table of mass M. At what angle must the ball come off?

Center of Mass and Center of Mass Coordinates

48. If a bullet of mass m traveling at speed $\boldsymbol{v}$ hits a man of mass M wearing a bulletproof vest and the bullet is stopped by the vest, how much momentum is given to the man?

49. Suppose a car of about the same size as your car and moving at speed V hits your car from behind; to avoid whiplash, should you press your brakes or not? Assume that your brakes work very well in keeping your car from moving very fast and that the collision duration is about the same in both cases. (Note that this is a simplification of a complex problem, but it is such beginning analysis that helps bring out the issues that are to be of importance for the complete analysis.)

50. Consider the collision shown in Figure 3-2 in which an object of mass m moving at speed $\boldsymbol{v}$ impacts an object at rest of mass M on a frictionless surface like ice. Show that the following equations for the velocities after the collision will conserve momentum and energy: $\boldsymbol{v}_m = \boldsymbol{v}\left(1 - \frac{2}{1+m/M}\right) \sim -\boldsymbol{v}$, $\boldsymbol{v}_M = \frac{2}{1+M/m}\boldsymbol{v} \sim 2\frac{m}{M}\boldsymbol{v}$. Check *both* the full expression and separately check the first order expansions.

51. Calculate the center of mass speed V_{cm} for the case of two equal-mass colliding balls in the frame where initially the left ball is moving at speed V and the right ball is at rest.

52. As discussed in the text (and in footnote 32) consider two balls of mass m and M which undergo elastic collision. Consider the following expressions (not *necessarily* the correct ones) for the initial and final velocities of the balls to first order in the ratio of the mass $\frac{m}{M}$ in the center of mass frame (indicated by primed symbols):

Initial (before collision): $\boldsymbol{v}'_1 \approx V\left(1 - \frac{m}{M}\right)$, $\boldsymbol{v}'_2 \approx -V\frac{m}{M}$,

Final (after collision): $\boldsymbol{v}'_1 \approx -V\left(1 - \frac{m}{M}\right)$, $\boldsymbol{v}'_2 \approx V\frac{m}{M}$

a) According to these approximations, what is the momentum of the center of mass? How can the center of mass have momentum in the center of mass frame and what are the proper expressions for the initial and final velocities?
b) Show that kinetic energy is also conserved to the requisite order.

c) What expansion would be needed to get the speeds right to the next order in the mass ratios?

Chapter IV

Changing Momentum: *Force*

Introduction

In Chapter 3, we discussed impetus and the fact that the net momentum is conserved in any closed system. Still, we saw that a given body within a closed system can lose or gain momentum in a collision with another body in a given direction. In the process, this other body's momentum (intensity and possibly the direction of the impetus) would change. Thus, it is clear that "massive"[1] bodies have--in addition to mass (receptivity to the ***action*** of the impetus), and sometimes impetus itself--another property. This new property is the ability ("power")[2] to activate impetus in another body.[3] Abilities to act (powers) are in the category of quality. Obviously, along with this property, a massive body must have a receptivity to impetus,[4] for activity and receptivity are correlatives; like top and bottom, one does not make sense without the other. We call the ability to change the intensity[5] of the impetus in some given direction, ***force***.

> ***<u>Force</u>*** is the ability (quality) of one body to change the impetus of another. The measure of the intensity and direction of a force is related to other relevant measures by Newton's second law: $\vec{F} = \frac{d\vec{p}}{dt}$. (See laws below) We call the unit of force a "Newton" because of his important contribution. 1 Newton $= 1\ N = 1\ kg\ m/s^2$.

Note that unlike impetus, the measure of which we called momentum, we will not introduce a different name for the measure of force (or any other qualities). We do not do so, because there is no tendency to forget force is a real power (quality), whereas many forget or never see clearly that impetus is a real quality by conflating it with its measurement. This same confusion can also keep students from ever understanding the radical nature of Newton's First Law.

Newton's Three Laws

A new plateau in the understanding of momentum was achieved by the great physicist (they called them natural philosophers back then) Isaac Newton (1643-1727). Through new discoveries--both theoretical and experimental--modern physicists continue to deepen our understanding of basic concepts such as momentum. However, the three laws of Newton, seen in the proper context, still express the central ideas well. They are:

---

[1] Mass is the most manifest property of this type of body, but the other properties under discussion are intimately related to mass and to each other.

[2] The general word power is appropriate but we should not confuse it with our particular empiriometric use of the term in Chapter 5.

[3] We speak loosely here so as to be more general. Recall that properties exist in a substance and are not simply the manifestation of the properties of the interaction of properties of a bunch of substances. At some point, we must have a substance that really has properties that aren't simply reflective of the interaction of other things that have properties. For example, consider the following analogical case: we sometimes speak of a corporation (which means body) as if it had its own individual existence, when in fact the properties it has spring from the individuals that make it up and have no existence separate from them.

[4] Receptivity to having impetus activated in a body by a force (as opposed to a field, which needs receptivity in a body to act as a force) can be lumped into the definition of the force. More precisely, we at least include it in the definition of the measure of the force.

[5] This is an example of the extended use of the word intensity (that we mentioned and used in the previous chapter), which includes the mental construct of negative intensity.

Newton's Three Laws

I. Every object continues in its state of rest or uniform motion unless acted on by an outside force.

II. $\vec{F} = \dfrac{d\vec{p}}{dt}$ *[= rate of change of the intensity of impetus in the* $\hat{F}$ *direction]* [6]

III. For every action, there is an equal and opposite reaction.

Isaac Newton was born 1643 (d. 1727) in Christian England; his contribution to modern science is extremely hard to overestimate. He is responsible for the first truly modern scientific theory that explained in the light of 1) a mathematical formalism 2) the *motion* of 3) the planets and the sun *as well as* ordinary objects on Earth. In this sense of a universal view of nature that is reflected into mathematics, his was the first modern physical theory.
As if this were not enough, he independently developed Calculus (as did Leibniz around the same time), discovered that white light is composed of colors and invented the reflecting telescope. Because of his great accomplishments, Alexander Pope wrote the hyperbole: "Nature and Nature's laws lay hid in night; God said let Newton be and all was light." More accurately, he was the man of genius to pull together the various threads of his predecessors through a series of profound insights.

His most important work is the *Philosophiae Naturalis Principia Mathematica* (*The Mathematical Principles of Natural Philosophy*) which unveils his theory of gravity and motion. He wrote it from 1665-1666, during the height of the Black Plague, and it was published in 1687.

He said of his work: "If I have seen further than others, it is because I have stood on the shoulders of giants." In fact, this includes medieval giants such as Grossesteste, Bradwardine and Buridan as well as more recent men such as Brahe, Kepler, Copernicus and Galileo.

Newton was a very religious man who considered himself a Christian. Although he entertained heterodox ideas regarding the Trinity and the use of astrology, he was immovably orthodox in his view of nature as orderly and rational, revealing the glory of God. Newton said,
"This most beautiful system of sun, planets, and comets, could only proceed from the counsel and dominion of an intelligent and powerful Being...This Being governs all things, not as the soul of the world, but as Lord over all; and on account of his dominion he is wont to be called Lord God." *Principia, Vol. II page 546*

In an apparent effort to connect the two great scientists, it is often said that Newton was born the year Galileo died. This is not the case if a single calendar is applied. However, the calendar of Pope Gregory XIII (our current calendar) was only adopted in England after the birth of Newton, so that, at Newton's birth, the Julian calendar was used in England, while the Gregorian was used in Galileo's homeland of Italy; in this mixed case, one could say the events occurred in the same year.

---

[6] Comment in square parenthesis is added, not part of the stated law which is empiriometric.

Here, the magnitude of the vector $\vec{F}$ is a measure of the strength (intensity) of the force and its direction is the direction that the impetus it imparts would take an object.

A little thinking about the first two laws shows that Law I describes the effects of impetus; an understanding of impetus through its effects is, in turn, needed to understand Law II. Once we know that a given intensity of impetus (momentum) moves a body at a constant speed, then Law II, in a way, includes Law I. In particular, by Law II, if there are no forces, i.e. $F=0$, then the momentum is constant. This means the "body" either has no impetus and can never get any or has a certain intensity of impetus that never changes. This, in turn, means that the body is respectively at rest (because there is no impetus to move it) or in uniform motion and remains so, thus reproducing Law I. So, Law I is indeed simply telling us the effect of impetus which is then implicitly contained in Law II. The second law is thus the deeper one, but both are rooted in the impetus; they describe its effect and indicate something of how it changes. If there were no impetus, then bodies would need to be pushed all the time by something external to themselves and Law I and Law II would not be true.

To understand Law III consider a man pushing on a wall (see Figure 4-1). When the man pushes on the wall, he exerts a certain force on the wall $F_{man\text{-}to\text{-}wall}$. The wall receives this force and acts back, resisting him with a force $F_{wall\text{-}to\text{-}man}$. The man receives this force and may slide backwards if the floor is slippery. Newton's third law says that whatever force he exerts will be matched by the wall. Mathematically, Law III, in terms of force magnitudes, is:

4.1 $$F_{man-to-wall} = F_{wall-to-man}$$

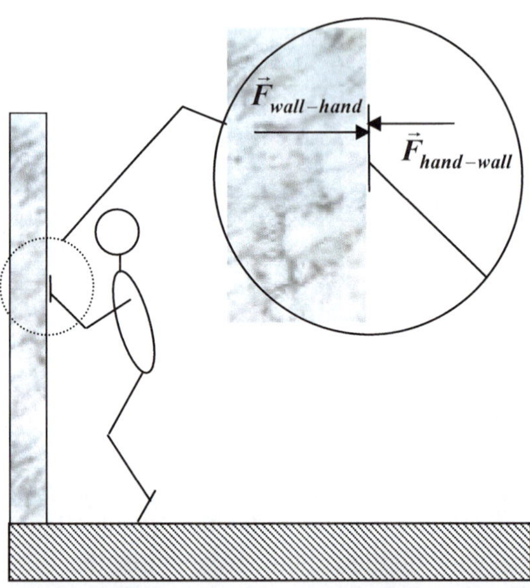

Figure 4-1: Man pushing on wall causes a reaction force of wall pushing on man. These two forces are equal and opposite. In the static case, i.e. when there is no momentum in the system, all forces acting on *each* object cancel so that no impetus is activated. To have such an equilibrium situation, one needs to include the forces between the ground and the wall, and the man and the floor.

Law III, along with permanency of impetus in an isolated substance,[7] insures conservation of momentum for it means whatever impulse is applied to one body is also applied to the one acting on it.

Law II is the one that contains the most information. It is the one that tells us about the force itself, i.e. the ability to change the intensity of the impetus in the direction specified by the force.

4.2 $$\vec{F} = \lim_{\Delta t \to 0} \frac{\Delta \vec{p}}{\Delta t} = \frac{d\vec{p}}{dt} = \dot{\vec{p}}$$

We can write this in component form and thereby emphasize the force changing the intensity of impetus in each of the perpendicular directions of rectangular coordinates:

4.3 $$F_i = \lim_{\Delta t \to 0} \frac{\Delta p_i}{\Delta t} = \frac{dp_i}{dt} = \dot{p}_i \text{ Where } i \in \{x, y, z\}.$$

Note that the last equality here and in equation 4.2, uses Newton's short hand for time derivative; namely, we use a dot for the time derivative: $\dot{\vec{p}} = \frac{d\vec{p}}{dt}$.

We can expand Newton's second law as:

4.4 $$\vec{F} = \frac{d\vec{p}}{dt} = \frac{d(m\mathbf{v})}{dt} = \dot{m}\vec{v} + m\dot{\vec{v}}$$

Thus, when the mass is constant then we get Newton's second law in the familiar form: $\vec{F} = m\frac{d\vec{v}}{dt} = m\vec{a}$.

Since rockets change mass, we might be tempted to use equation 4.4 as a short cut in solving rocket-type problems. However, as we will see shortly, this can lead to confusion because of a lack of a clear and appropriate definition of "*m*." Instead of using the chain rule, it is usually much better to do a careful analysis of the problem using differentials in the way we will describe.

For a glimpse at the significance of the equation 4.4, consider defining m = {*all* mass that is in a system--remember not to redraw system lines once they are drawn; all massive parts that break off are still part of the system}. Then, when there is no force on the moving system, if $\dot{m} \neq 0$, then $\dot{\mathbf{v}} \neq 0$. In other words, if mass disappears into thin air,

---

[7] Note that the interaction of *parts of a substance* is analogically like the interaction of *substances*. If we consider those two analogical types of interactions only to the extent that they can be considered identical with regard to interchange of impetus, then we can show that Newton's third law is required by conservation of the net momentum of the system and thus required also to preserve Newton's first law. If two parts of a body, for example, two parts of a substance could move and share the impetus of the whole in different ways (as they do, for instance, in the vibration of molecules associated with temperature) then if the third law did not hold it would be possible for some momentum of the body to be lost (or gained) solely due to such self interaction of the parts. In particular, consider a part with momentum of p_0, (which is traveling at $V + v_0$ with mass m) which is, at some moment, canceled by the momentum of a distant part of the body which has momentum $-p_0$ (which is traveling at $V - v_0$ with mass m). If the extra momentum of the first particle, p_0, were truly *lost* in an internal collision (not transferred or used to cancel another part's momentum), the body's momentum would become $P - p_0$, i.e. it would decrease, violating momentum conservation. Also, such a decrease would, assuming conservation of mass, mean that it slows down, thus belying Newton's first law that bodies continue in uniform motion if there are no external forces. Note that we could hypothesize that a substance compensates for a violation of Newton's third law in one part by a different violation somewhere else, but this could only be true by violating the analogically general nature of force (i.e., being the same in a substance as in a system) that we would also like to preserve (and, indeed, is, in this domain).

then, in order to preserve momentum (mv), the remaining mass must move faster; that is, it must accelerate, thus violating Newton's first law. Of course, in reality, in the Newtonian domain of slow speeds and low gravity, mass does not disappear. The constancy of a body's receptivity to impetus at some level of approximation seems to be an essential quality of massive bodies. We will learn more about the nature of mass when we discuss special relativity.

The Rocket Problem from the Point of View of Forces

Notice that the changing mass term, $v\frac{dm}{dt}$, of $\frac{dp}{dt}$ would come into play in our rocket (as well as the wagon and toboggan example of the previous chapter) if we wanted to analyze forces. It is important to decide what one is interested in and then pick a method that focuses on that property or that effect. In the rocket example, we first wanted to know the speed. However, at another time, we might want to know the strength of the force that acts on the rocket. Of course, different approaches will not affect the answer, but only how hard it is to get it.

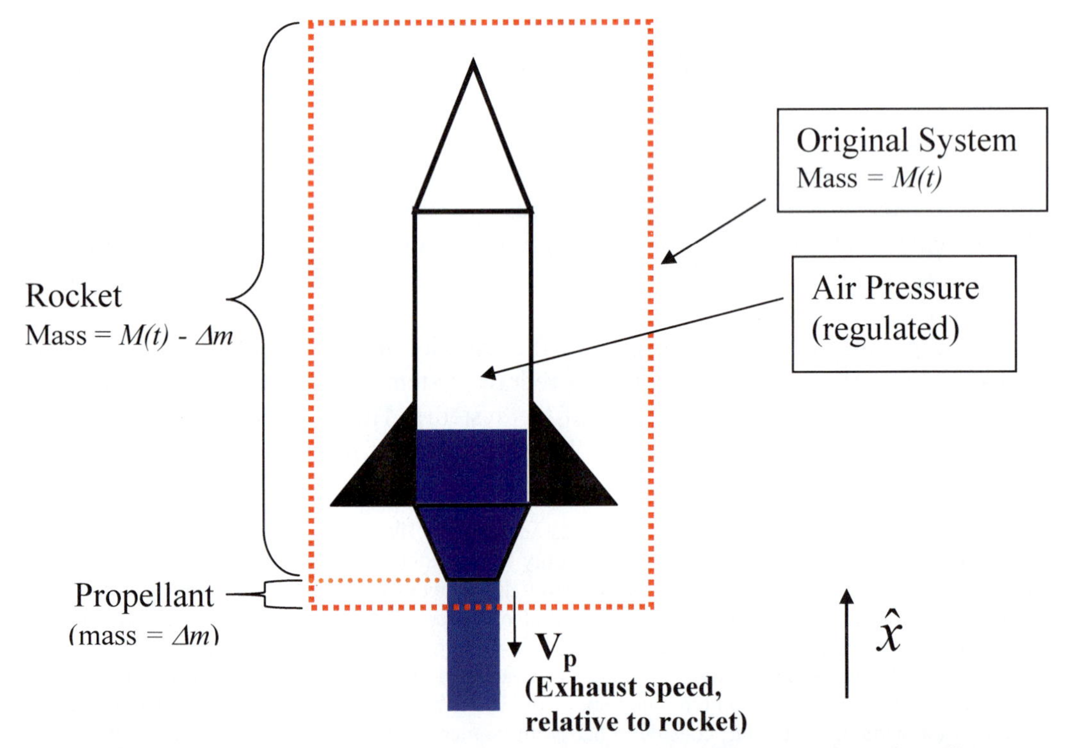

Figure 4-2: Rocket at time $t + \Delta t$. The system is still shown in red outline, but now part of the original system has left the rocket, expelled to propel what's left of the rocket.

Consider now the problem of finding the rocket's speed from the point of view of force (see Figure 4-2). This is particularly simple because the force acting on our whole system is zero in space, i.e. $F_{system} = 0$. Notice the system mass does not vary so that we get, after dropping the vector notation (since this is a one dimensional problem):

4.5 $$F_{system} = \frac{dP_{system}}{dt} = \frac{d(M_{system}V_{system})}{dt} = M_{system}\frac{dV_{system}}{dt} = 0$$

Hence, we have $V_{system} = \text{constant}$, which in turn is a simple statement that the center of mass stays in uniform motion, because there are no outside forces acting. This result tells us nothing interesting about our specific problem. We would like to know V_{rocket}, so we need to look at P_{rocket} rather than P_{system} where:

4.6 $$P_{system} = P_{rocket} + P_{water}$$

Here we have to be careful, for we might be tempted to say, with an eye towards differentiating using the expanded form of Newton's second law given in equation 4.4 above:

4.7 $$P_{rocket} = MV \textit{ and } P_{water} = m\mathbf{v}_p$$

Where M and V are the mass and speed of the rocket (with whatever amount of fuel (water) it may have in it at the given time), and m and $\mathbf{v}_p$ are the mass and speed of the ejected fuel.

What's wrong with this? Well, we have not clearly defined what we mean by "*m*." It cannot mean all of the mass that has been ejected since the thrusters started firing. Why? Each part of those ejecta is moving at a different speed relative to the start frame. For concreteness, let's say we took off from a space station. The first chunk of propellant mass that was ejected did so when the rocket was barely moving, which means that chunk's speed relative to the station is above $\mathbf{v}_p$. The mass ejected later, though it was ejected at the same speed relative to the rocket, is moving much slower relative to the station because the rocket is moving much faster (see Figure 4-3).

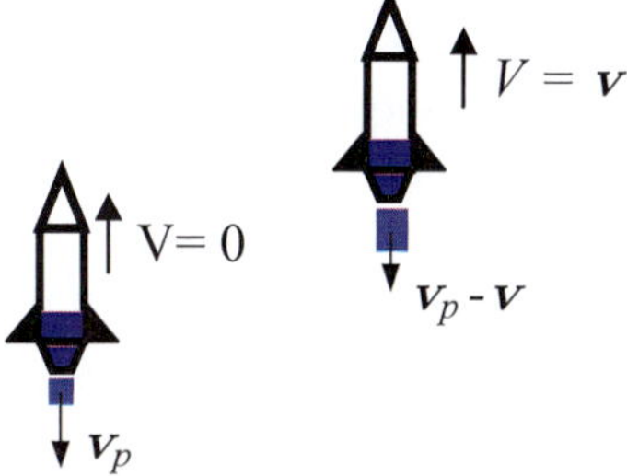

Figure 4-3: Rocket ejecting propellant at speed relative to the rocket $\mathbf{v}_p$ *Left:* The rocket ejects the first piece of propellant when its speed is negligible, so the speed of the propellant seen from the ground is $\mathbf{v}_p$. *Right:* The rocket ejects later propellant when it is moving at a speed $\mathbf{v}$, so that the propellant speed seen from ground is $\mathbf{v}_p - \mathbf{v}$. If the rocket goes faster than the propellant speed the propellant will move upward as well (as can be witnessed in some water rockets)!

This brings out a second problem: We have to be clear about our chosen reference frame. While the rocket is firing, it is constantly changing frames. We need to pick one frame and stay with it. For instance, in the above analysis of ejected fuel, it is convenient to choose our frame to coincide with the frame of the rocket before we turned on our thrusters (i.e., started the water squirting out), which is the frame of the space station. If we don't keep the same point of view things will appear to change, when all that has happened is that we have changed. For example, if I am observing a ball move at constant speed, and I decide to accelerate to some speed, say in the same direction as the motion of the ball, then, while I am accelerating, the ball will appear to be decelerating. Thus, my motion has induced an apparent acceleration. We must pick a reference frame and stick with it for all our calculations. It is okay to switch frames later and analyze everything again from the new frame, but we cannot mix quantities measured (and, if necessary, calculated) in one frame with those from another.

We will see later that we can have non-uniformly moving (accelerated) frames of reference only by introducing the concept of fictitious forces, that is, apparent forces on observed objects that arise from observer acceleration, not from physical interactions on the observed objects.[8]

Let's summarize our two lessons:

1. *Pick a frame of reference and stay with it.* You will introduce apparent forces if you allow your frame to shift, and you will not have accounted for them in any way with what we know up till now.
2. *Define your system carefully throughout the problem.* Keep track of the elements in your system and especially be clear about what you mean by each of the terms in the equations you use and formulate.

We know that things are dynamically changing at each moment with our rocket, so we need to compare one moment to the next moment. We thus look at the change in the system momentum ΔP_{system} (during some short time interval Δt) which we know to be zero:

4.8 $$\Delta P_{system} = \Delta P_{rocket} + \Delta P_{water} = 0$$

Applying our rules:

1) We fix ourselves at rest with respect to the rocket at time t. This means that for all later moments, the rocket will be moving with respect to us. At this point, we are only interested in the next moment. That is, after the short time, Δt, we want to know what the change in P_{system} is in terms of the speed and mass of the rocket and ejected fuel.

Notice our choice of frame will automatically give the Δp's we need. For, in the chosen frame, *initially* the rocket has no momentum *and a moment later* it has momentum:

4.9 $$\Delta P_{rocket} = P_{rocket}\left(t+\Delta t\right) = mass \cdot speed = M\,\Delta V$$

Notice: ΔV is the speed of the rocket after time Δt. Here V is considered positive in the direction of motion of the rocket, $\hat{x}$.

Note also the use of momentum in an empiriometric way here, talking purely in terms of relative momentum, not considering whether it *actually* has impetus or not.

[8] We see, coming back again to Buridan's point mentioned in Chapter 3 (see footnote in Chapter 3), that motion, even accelerated motion, ***in so far as it is merely quantitative*** (extension between objects) is wholly relative. Leaving behind the other categories, for instance, considering only the extension (distance) between the object's center points, the square moving past the line is the same as the line moving past the square. Again, the fully physical distinction (not purely quantitative distinction) comes when I note causes and the properties other than quantity; for example, I move the square body and move it past the line. In such a case, the square body (say a DVD case) is acting by its impetus and the line (say the near edge of a ruler) and the surrounding air and "space" is receiving the action.

2) We choose our system to be all the mass in the rocket at our start time. As the water is part of the rocket, it is not initially moving in our chosen frame. So the momentum of the ejected water at the next moment is ΔP_{water}. For the little amount of water of mass Δm that has been ejected at speed $\boldsymbol{v}_p$, we have:

4.10 $$\Delta P_{water} = P_{water} = -(\Delta m)\boldsymbol{v}_p$$

Notice the minus sign arises because the water's impetus is directed in the $-\hat{x}$ direction. In the end, the reason we picked the frame we did is because of the constantly changing nature of the physical situation and *the specification of the propellant speed* in the rest frame of the rocket.

Now, substituting equation 4.9 and 4.10 into equation 4.8 and taking the appropriate limits, we can write the measure of the strength of the force as:

4.11 $$F_{system} = \frac{dP_{system}}{dt} = \frac{MdV}{dt} - \frac{dm}{dt}\boldsymbol{v}_p = 0$$

Recalling that as the mass of the ejected fuel, m, increases, the mass of the rocket, M, decreases (mathematically: $\frac{dm}{dt} = -\frac{dM}{dt}$), we get:

4.12 $$\frac{dM}{M} = -\frac{dV}{\boldsymbol{v}_p}$$

This is the same equation we obtained in Chapter 3.

Forces on the Rocket and the Fuel

We have found a shorter if more conceptually demanding--it's much easier to make a mistake here--way to solve the rocket problem. However, we have also discovered, maybe without knowing it, something about the forces involved. Note that equation 4.11 has two terms that look, as they should, like forces. Of course, we know the first term is derived from the change in momentum of the *rocket*, and the second comes from change in momentum of the small element of *ejected fuel*. Hence, we have by Newton's second law that the strength of the force the water exerts on the rocket:

4.13 $$F_{rocket} = M\frac{dV}{dt}$$

And that on the fuel element by the rocket is:

4.14 $$F_{water} = \frac{dM}{dt}\boldsymbol{v}_p$$

And, as required by Newton's second law, these two forces are equal in magnitude (strength) but opposite in direction.

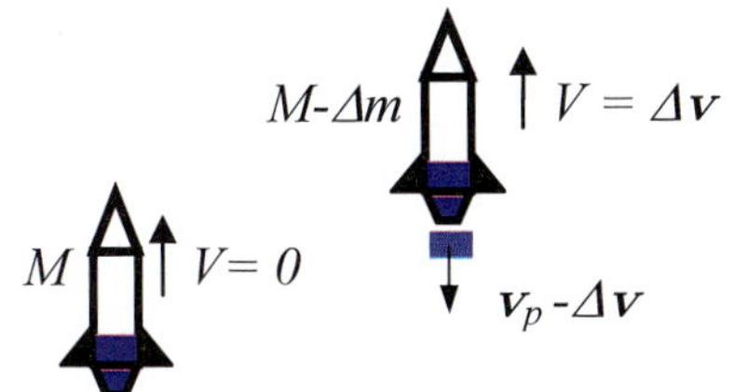

Figure 4-4: Rocket shown in its instantaneous rest frame at time t: *Left* shows rocket at time t and *Right* shows a moment later at $t+\Delta t$.

Keep in mind that these are the forces as _seen in the instantaneous rest frame_ of the rocket. Since the rocket is accelerating, this means we need a new observer for each new speed at each moment as the rocket accelerates. For any ***given*** frame, the rocket will only be at rest for a moment; at the next moment, we have to account for the rocket gaining momentum by ejecting mass. Even if the mass is ejected with no speed relative to the rocket, the rocket momentum is still changed as the ejecta takes with it material and momentum that formerly were part of the rocket. Likewise, the momentum of the ejected water (fuel) is increased by the addition of this momentum as well as the momentum related to the ejection speed. So, in order to get the force in a fixed frame, we need to recalculate at each moment, whereas in the instantaneous rest frame (which is actually a sequence of frames) it always remains as given above. We will indicate the measure of the forces in a fixed frame by the symbols: F'_{rocket} and F'_{water} and calculate the equations for them in the exercises. Those equations obviously reduce to the instantaneous rest frame equations calculated above for the moment at which the rocket is at rest in the chosen frame.

How Forces Propel the Rocket

So, how does the rocket work? In a fixed reference frame, the water pushes with a force F'_{water} against the rocket, activating impetus in the rocket; the rocket pushes back on the water with equal force, activating impetus in it. The intensity of the rocket's impetus is thus increased as each new volume of fuel pushes against it.

To be more detailed in our analysis, we need to know more about action and reaction and the surface forces that allow the rocket to work.

Action and Reaction and Surface Forces

Athletic Contests on the Way to Mars

To understand action and reaction as well as the profound nature of Newton's second law, consider that during our trip to Mars, we have a series of microgravity (gravity off) "Olympic" games. One game involves timing how long it takes an astronaut to "climb" from the end of a rope tied at the opposite end to an interior wall of the spaceship. Another is a "tug of war" in which two astronauts are at the ends of a rope each trying to pull the other across a line drawn in the middle on the "floor" over which the rope is stretched. The men are *not* allowed to push off the floor or any other structure but may only pull on the rope. These games help reveal the nature of force, momentum and mass. Suppose the weakest man challenges the strongest man to each of these events; further assume that the men are about the same mass.

The First Event

The first event is straightforward; its force diagram is shown in Figure 4-5. The strong man is able to exert a maximum force on the rope of magnitude F_S. From Newton's third law, we know that unless the rope is too weak and thus breaks, the rope tension will match that force. That is, as the man pulls on the rope with a force F_S, the rope's tension pulls ***back*** with a force, F_S. Mathematically, we say: $F_S = F_{man-on-rope} = -T_{rope-on-man}$, using the direction convention shown in Figure 4-5.

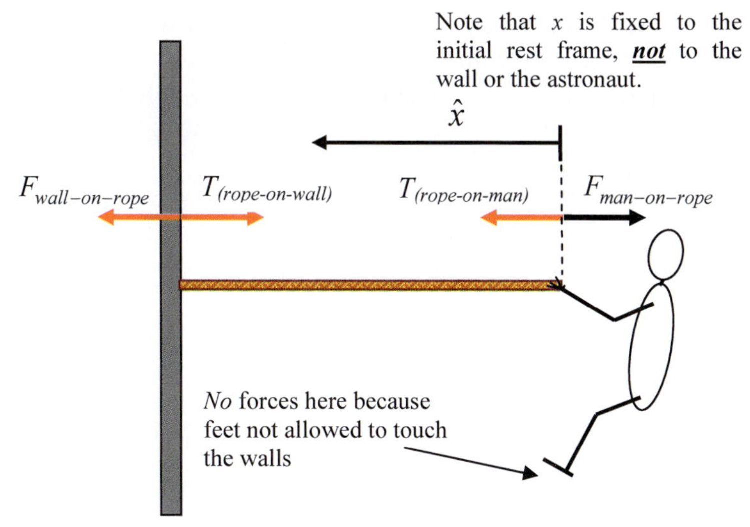

Figure 4-5: Force diagram for astronaut climbing rope from the far end to the wall. Since there is no gravity, the rope will only remain taut while he is pulling on it. To maintain the force as he moves faster and faster along the rope, he will have put one hand over the other at a faster and faster rate; hence, the event also tests one's manual dexterity. Because the *x* coordinate is defined in the initial rest frame, the wall, rope and astronauts start at fixed value of *x*, but after interaction they all have changing *x* values.

The forces on the wall can be analyzed similarly at this same instant[9] when he first pulls on the rope. The rope tension appears at the wall end, and pulls on the wall with a force: $T_{rope-on-wall} = -T_{rope-on-man} = F_S$. Finally, the wall pulls back on the rope with the same force F_S. What is the effect of all this on our event? To answer this, we must look at the forces on each of the bodies of interest.

For example, consider the forces on the wall, which is, effectively, the spacecraft itself. The only force on the wall is the rope pulling with F_S in the $-\hat{x}$ direction. Thus, we have, using Newton's second law: $F = dP_{ship} / dt = M\, dv_{ship} / dt = M\, a_{ship} = F_S$ so that, assuming this state is maintained, using $L = \text{length of rope}$, $M = \text{mass of ship}$, and defining the *x* position in the initial rest frame, we integrate to get:

4.15 $$X_{ship} = L - \frac{1}{2}\frac{F_S}{M}t^2$$

By similar reasoning, the equation for the man's position is:

4.16 $$X_{man} = \frac{1}{2}\frac{F_S}{m}t^2$$

This is the simplest case of force-driven motion: namely, a constant intensity force applied to a body of constant mass. The force constantly intensifies the impetus and thus

[9] This assumes the force propagates instantaneously down the rope. This of course is not true, but we have switched off these very fast speeds and, as we will discuss, they will be brought back later. Empiriometrically, we say we solve the problem under the approximation of infinite speed of light.

constantly increases the speed of the object. Empiriometrically, we say that the momentum and thus the speed linearly increase with time.
When then does the man reach the wall? When $X_{man} = X_{ship}$, which gives:

$$4.17 \quad t_{strongest\ man} = \sqrt{\frac{2L}{F_S}\left(\frac{m}{1+\frac{m}{M}}\right)} \sim \sqrt{\frac{2Lm}{F_S}}, \text{ since } M \gg m$$

By contrast, the time for the weakest man is, who can only exert F_W, is $\sqrt{\frac{2Lm}{F_w}}$. This means the ratio of the times stronger to the weaker is: $\sqrt{\frac{F_w}{F_S}}$, so that the stronger-- no surprise-- is faster and he wins this event.

The Second Event:
When is the Strongest Man Reduced to the Strength of the Weakest?

That's only one event though; will he also win the tug-of-war? To see, we first, draw a force diagram such as Figure 4-6, labeling all forces acting on all bodies. In drawing these diagrams, always remember that *forces act from one body onto another*, even though sometimes (though not in the current case) the body causing a force can be left off the picture; in such cases, just think of a body generating that force in some unseen way--which we will often do with gravity, for example. Using Newton's third law as illustrated in Figure 4-6, we see that the strength of the forces must be equal. Thus, if one man were more massive than the other, the less massive one would accelerate over the line before the other—as long as each kept pulling the rope as he moved forward. If one astronaut pulls harder, the reaction force must increase. But $F_S > F_W$, so what happens when the stronger man pulls with a force larger than F_W? He pulls the rope out of the weaker one's hands, and, thus, no one crosses the line. Hence, by the strict rules of the game, since no one crosses the line, no one wins. The strong man is limited in what he can do by the weaker man's strength.

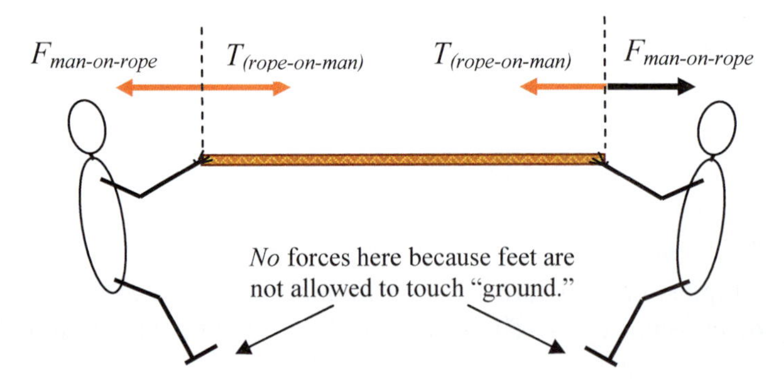

Figure 4-6: Force diagram for two astronauts having tug-of-war. The center of mass of the system composed of the astronauts and the rope cannot move because no external forces act on them. Thus, assuming the astronauts are equal mass, this event must end in a stalemate.

The First Event Again

To understand more deeply the rope-connecting–two-body situations, let's look in more detail at the first event in which the contender climbs the rope to the wall. As the contender pulls on the rope, the force he applies activates impetus in the first part of the rope, which moves it a little causing it to exert a force on the next part, which acts back constraining the motion of the first part. Meanwhile, the second part also moves a little causing it to exert a force on the third part, which acts back constraining the motion of the second part.[10] In this way the force is transmitted down the rope until the last part of the rope acts on the wall of the ship. The wall then transmits the force, part by part throughout the ship until, after some time, the ship has been given impetus that moves it toward the contender. Meanwhile, as the contender pulls on the rope, the rope acts back on him and thus the force of the rope activates an impetus in him as well. In a time Δt, he gets an increase in intensity of his impetus, a larger momentum, which mathematically we write: $\Delta p = F\Delta t$. Such a burst of momentum is called an **impulse**, and results in a speed increase. As long as the force is constant, the contender's speed increases linearly; he gains an equal speed in each equal interval of time. In this way, he moves quickly in the direction of the wall and the wall moves slowly in toward him.

Closer Look at Colliding Balls: Surface Forces

Take a light ball moving towards an elastic (no kinetic energy lost) collision with a very much heavier ball. In contrast to the discussion in Chapter 3, where we were only interested in before and after the collision, we are now interested in the collision itself. To understand the collision, we note that in the collisions it is the surfaces that cause the forces, and they do so by their elasticity. The surfaces deform upon impact and then rebound (as shown in Figure 4-7a) like the surface of a drum but with more powerful reacting forces (see strobe Figure 4-7b of drum head). Indeed, it's similar to two springs being pushed against each other. We can thus think of the interaction of the two surfaces during the collision as somewhat like a single spring being compressed. So, let's model the surface of the ball by a single spring, which we can imagine being attached as shown in Figure 4-8.

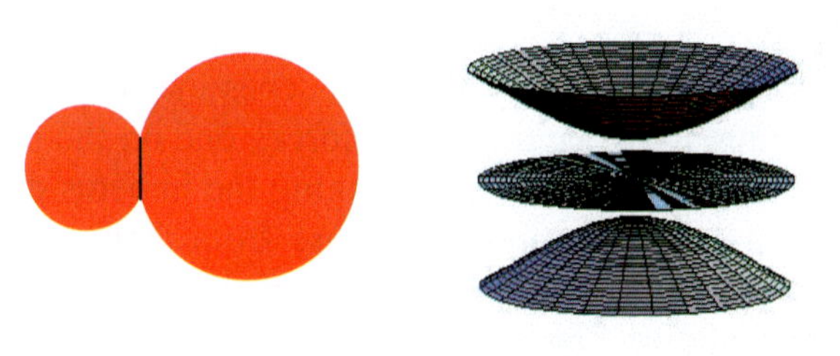

Figure 4-7: a. *(left)* When light ball hits heavy ball, it pushes in the surface of the heavy ball at the region of contact, and the heavy ball, in turn, pushes in the corresponding surface of the light ball, causing the deformed surfaces shown. **b.** *(right)* A drum surface bends then snaps back as shown in a succession of three images. A vibrating drum head is an example of extreme elastic behavior. Elastic deformation is involved in certain collisions in which contact surfaces get pushed in and spring back. Note: pictured is only the lowest frequency mode of drum vibration that causes the sound we hear.[11]

---

[10] Imagine, for instance, a series of springs connected one to the next, so that, when I pull on the first, it stretches the next which stretches the next and so on.

[11] To see higher modes in animation, go to: (see next page)

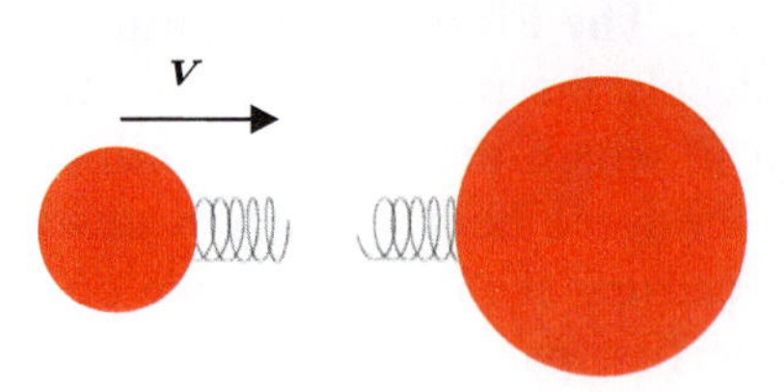

Figure 4-8: Light ball moves towards resting heavy ball. Springs are used as a simple model of the elastic surface of each ball.

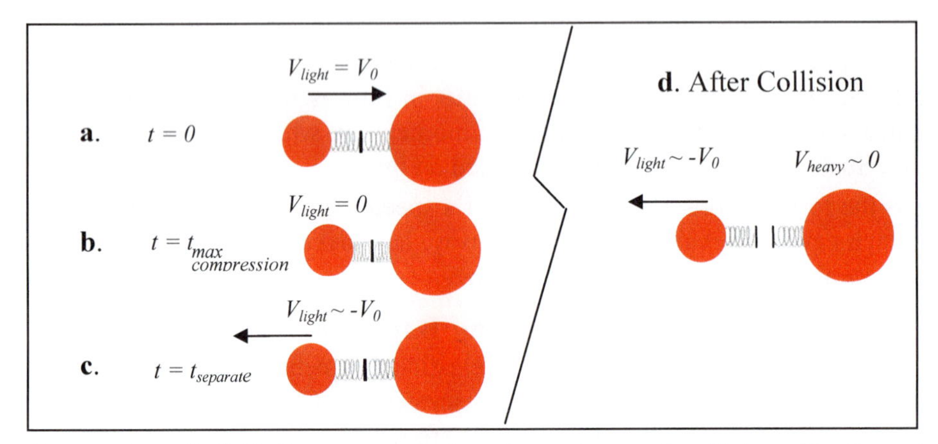

Figure 4-9: *(left)* Balls as seen during three stages of collision. Starting at far left moving to the right: **a.** The springs' ends just meet. **b.** The springs are at maximum compression—note that because of Newton's third law the compression of each (identical) spring must be equal. **c**. The springs again return to equilibrium and lose contact and thus cease to interact. *(right)* **d.** After the collision sequence, the balls continue on with the velocities shown; the springs (the surfaces) may continue to vibrate back and forth for some short time after the interaction ceases.

As shown in

Figure **4-9**, the first ball approaches the second ball until finally the springs touch. Upon touching, they begin to exert forces one on the other. However, it takes time for the forces to change the motion by changing the impetus driving the motion. During this time, the springs are pushed more and more from their equilibrium position, until the relative speed between the balls is zero. Given the much greater mass of the second ball, its speed can be neglected to first order; this means the springs compress approximately until the first ball stops moving. At this point, the spring force can begin to activate impetus to move the light ball in the *opposite* direction. This continues, assuming very low mass springs, until the springs (approximately) again reach their equilibrium point, at which time, the springs would begin to exert forces in the other direction. Hence, at this point, (for springs of negligible mass) the springs cease moving, and lose contact with each other, while the light ball continues on its way moving in the opposite direction from which it first came. Notice that, during the collision, the spring goes through one complete cycle so that

(footnote cont'd from previous page) www.kettering.edu/~drussell/Demos/MembraneCircle/Circle.html. For sound and pictures of various drum modes see: http://www.robertus.staff.shef.ac.uk/ama349/cir_membrane/

(assuming no energy is lost and a very heavy second ball) the speed at which the light ball came in is approximately the speed at which it leaves. This and many other aspects can be studied using the model in the *Interactive Physics* file: *4.1-collision of masses with springs.ip*. In this way, we can more completely understand each of the collision results found in Chapter 3.

Interactive Simulation

Furthermore, the spring model suggests that the cause of the Newtonian third law is the generic nature and the relational symmetry of the interaction forces; namely, the nature of the forces is such that when a force acts from one body on another, the reverse also occurs. We will see this when we talk more about these forces, such as the electromagnetic force, later. However, we are now ready to talk about how our water propelled rocket works.

Pressurized Water Propellant

In particular, we will analyze a water rocket such as the *Monster Rocket* by *Hasbro.* Such a rocket works by pressurized air pushing the water out. We will not get into the thermodynamic details here, but only consider the propulsion in general terms. For instance, we will not discuss how one might regulate the pressure so as to maintain a constant propellant ejection speed $\boldsymbol{v}_p$ relative to the rocket. We will simply take a constant ejection speed as a first cut model of the rocket.

We start by looking at the pressurized air that we pump behind the water and that is maintained in the sealed enclosure *before* the nozzle is opened. We can think of the pressurized air as many pool ball-like substances (molecules) moving around in "space." These "balls" have impetus in random directions, and they activate impetus in the wall. A given molecule does so by pushing in part of the wall's surface which results in the wall responding back in the way described above for the ball collision, bouncing the molecule back into the chamber. A net impetus is thus activated in the wall, and, hence, the rocket. However, on the other side of the chamber, on the average, the impetus of exactly the opposite directional tendency has been activated that exactly cancels motion caused by the first. Remember that these molecules are small and hit frequently. Hence, the rocket shakes back and forth at some small, unnoticeable level, and on the average doesn't move at all.

If we were to consider the fuel as part of the rocket, then we can say it doesn't shake at all, for the center of mass of the *fuel plus rocket system* remains at rest or in uniform motion. In the case in which there is no gravity, this must be so for there is no *net* force acting on the system, only forces between the various parts of the rocket plus fuel system. Hence, there can be no net change in the system's impetus and thus no way for its state of motion to change.

However, at some level, we want to consider the rocket and its fuel *separately*. As we've said, when they are considered separately, we can still say that, *on the average*, the rocket is not vibrating. This is so because, with a sealed tank, there is no place for the fuel to go; thus, the system can only conserve its center of mass motion by shifting the rocket back and forth in such a way that the average position is the same as if it weren't shaking. Empiriometrically, in our equations, we can use this average to say the force due to the air molecules on one side is cancelled by force on the other side.

However, when the nozzle is opened, the water can now move and some of the impetus activated in the water by the air molecules goes into water that leaves the rocket. This means some molecules acting on the top of the chamber activate impetus upward in the rocket that is never countered by impetus of the downward variety acting on the bottom chamber--and thus on the rocket. Thus, through each molecule that is not balanced by a partner on the bottom of the chamber, the rocket is given little increases of intensity of impetus that cause it to move faster in little imperceptible spurts. However, when all the molecules acting along the whole top surface of the chamber are considered acting for some time, it can amount to a massive increase in rocket velocity.

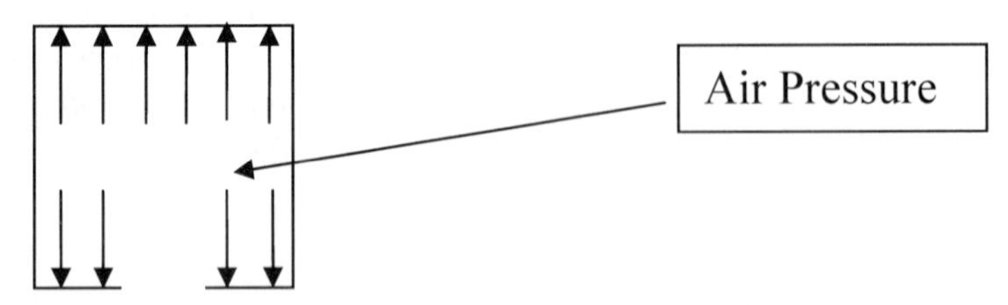

Empiriometrically, we would average the "spurts" out and say that there is a net force acting on the rocket because the force of the air acting on the top wall is not balanced by an equal strength force acting downward on the bottom.

What then is the summary of how the rocket moves? The water provides something for the air to push against and thereby allows the air to maintain contact with the top of the chamber longer so as to activate more forward impetus in the rocket. Without the water, the air would still push against lower parts of itself; however, the air is not very massive so it would move out quickly, not allowing nearly as much time for impetus to be imparted to the rocket. We'd either need to use a lot more air or the air would have to do its work a lot faster-- for example, by starting with a lot higher temperature air.

Empiriometrically, i.e., reflecting into quantities, we'd simply say that the momentum must be conserved. So, if the rocket starts out at rest, and the rocket ends with momentum P, then we must give momentum $-P$ to something else to keep the total $P=0$. We can give the $-P$ to water or air. However, since water is denser than air, to use air, we will either need a lot more air or we will need a lot faster moving air.

Having solved the rocket problem in the absence of gravity, we now would like to include gravity. We thus need to understand something about gravity. We also would like to better understand the key forces operating in the rocket as well as in other physical situations. Gravity and the electric force, which plays an important role in all ordinary interactions, are the two most important players.

Two Types of Forces

Different forces have different mathematical equations for the measure of the strength of the force, F. We will consider only two, uniform gravity and the electric force. The simplest is an approximation to Earth's gravity which is called uniform gravitational field. A uniform gravitational field is one in which the pull of gravity on a body of given mass is the same every place. Wherever you go, be it far or near, up or down or left or right or back and forth, the gravity force acting would be the same. A uniform gravitational field, an example of a constant vector field that we discussed in Chapter 2, is shown in Figure 4-10. Now, the strength of the force of gravity is proportional to the mass of the object being pulled. In this case, we call the constant of proportionality g. Mathematically, we write: $\vec{F}_{gravity} = -m\,g\,\hat{y}$.

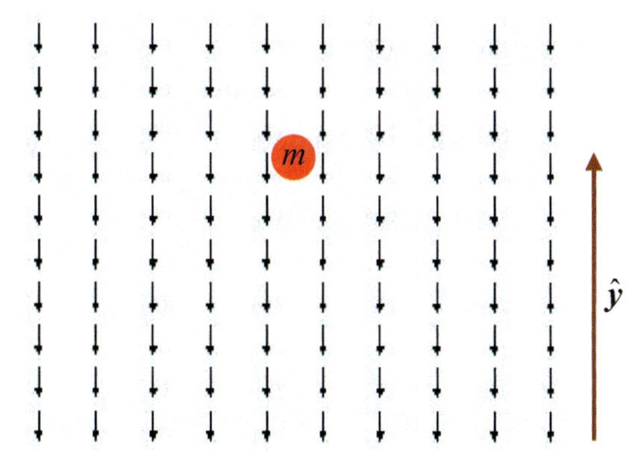

Figure 4-10: A test body of mass m in a uniform gravitational field. The field acts on the test body with the force: $\vec{F}_{gravity} = -m\,g\,\hat{y}$.

Applying Newton's second law, this equation implies that the rate of change of the momentum (measure of the impetus' intensity in the given direction) is proportional to the mass and to g. As mentioned earlier, applying a constant force to a body means that, in any given interval of time it undergoes an equal incremental increase in momentum and, when the mass is constant, a uniformly increasing speed. The Oxford "Calculators," Oresme and others understood the latter fact about uniform acceleration, but it was not until the late 1400's that the Dominican De Soto (see box) applied it to free-falling bodies for the first time. It is of prime importance to note that the magnitude of the gravitational force is proportional to the mass and the acceleration resulting from a constant force is also proportional to the mass.

Dominic De Soto was a Dominican priest born 1494, in Segovia, Spain (d. 1560). He followed Aristotelian thinking, and, not being a slave to a tradition for its own sake, he was the first to apply the law of uniformly accelerating velocity to free falling bodies. Going the next step beyond his distant predecessor Bishop Albert of Saxony (1316-1390, who was a student of Buridan), he took the velocity of a free falling body as increasing with time; i.e., in modern symbolic notion: $v = \text{constant} \cdot time$ and noted that it was uniformly accelerating. He went on to explain that, in contrast to when an object is dropped, when it is fired upward, it undergoes uniform deceleration. He applied the mean speed theorem (see Oresme box in Chapter 2) to both cases. In so doing, he focused attention on the impetus, ignoring its origin. That is, he ignores the fact that a second and different type of force (electromagnetic) operates in firing an object upward that does not operate when one simply drops a body. In this way, he further established the abstractions necessary to formulate general laws of motion such as Newton's. He, along with others, tried unsuccessfully to explain the trajectory of bodies by a combination of impetus and gravity.

De Soto was also a theologian and economic theorist.

Indeed, from Newton's second law, we write:

4.18 $$F = m a_y = -m g \text{ which implies } a_y = g$$

In words, the acceleration is the same independent of what the mass of the object is. That is, if we drop any object, as long as *only* gravity is operating, no matter what it is made of, or what it's mass or shape, it will drop at the same acceleration. This fact is called the *principle of equivalence*. We will discuss it more in Chapter 8, but perhaps you can see already why it will be useful in bringing back a new preference to another class of reference frames.

By careful measurement, g on the surface of the Earth at sea level is known to be $9.80665\ m/s^2$. For most of our problems, $9.8\ m/s^2$ will be accurate enough. Thus, the vector at every point in the uniform gravitational field approximation shown in Figure 4-10 has magnitude of g everywhere and points down everywhere. The path of motion of a body in such a field is easily found:

4.19 $$\frac{d^2 y}{dt^2} = -g, \quad so \quad y = h + \mathbf{v}\, t - \frac{1}{2} g t^2$$

Here h and $\mathbf{v}$ are the initial height above the ground and initial upward speed of the object.

Of course, gravity is not really perfectly uniform, it changes from place to place; in particular, it decreases as one moves away from the Earth. Indeed, if the Earth were a perfect sphere, it would decrease exactly as square of the distance from the Earth, namely, $1/r^2$. The gravitational force law (which we will discuss in more detail in Chapter 8) has the same form as a totally different type of force, the electric force. The electrical force strength between two charges, q_1 and q_2 can be written:

4.20 $$\vec{F}_{electric} = \frac{q_1 q_2}{r^2} \hat{r}$$

Using this equation, we plot the electric field around a central positive charge in Figure 4-11; note the inverse square fall off. In the figure, we assume the charge is anchored to the table so that it cannot move and thus its field doesn't move; we also assume that its charge is much larger than any charge we bring near it, so that its field is not distorted by their presence.

In words, equation 4.20 says that the larger the charges of the given bodies and the closer they are together, the higher will be the rate of change of impetus of the bodies.

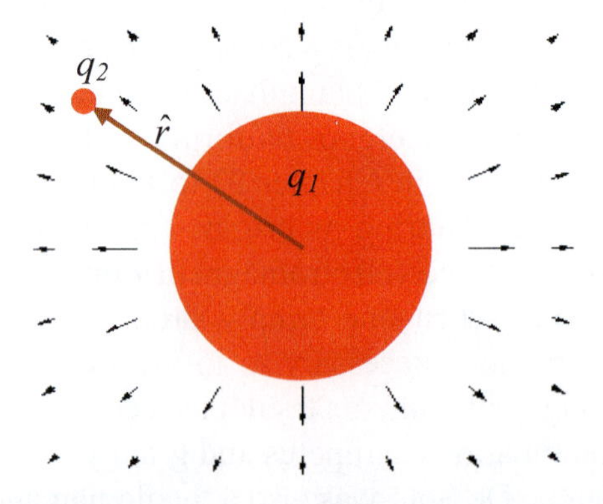

Figure 4-11: A test body with a small positive charge, q_2, in the field of a highly charged particle (charge q_1). The field acts on the test body with a force: $\vec{F}_{electric} = -\frac{q_1 q_2}{r^2} \hat{r}$. Notice how the field, while always pointing in the radial direction, decreases in magnitude as the charge is moved away from q_1.

Some questions immediately come to mind about the force fields involved in both the gravitational and electrical forces. First, it appears these bodies are acted on by fields not bodies; so, how can we say forces are *only* applied by a body? And, thus, secondly, how can we say: for every reaction there is an equal and opposite reaction?

To answer concretely, consider the electrical force case, keeping in mind that identical reasoning applies to gravity. One charged body does indeed act on the other and vice versa, though this action is mediated, that is transmitted via "space." Each charge experiences the same strength force (given by equation 4.20) acting on it. So, in one way, far from denying action and reaction, that equality coming from this fundamental force suggests that the action-reaction law comes from the symmetrical nature of everyday forces as magnetism and electricity (which, at some important level, is what is responsible for the surfaces forces) and gravity. In the case of the electrical force, the symmetry between the force acting on each of two charge particles is manifest in equation 4.20 by the symmetrical relation between the two charges in the equation; namely, one can switch q_1 for q_2 without changing the force.

Still, there remains the real complication of the mediation through "space"--we'll talk about what "space" means in Chapter 8; note though it can't mean pure quantity, which is only one of the ***properties*** of a physical thing.[12] In particular, one expects it to take a certain time for the field generated at a first charge to act on the second. In fact, it does. This would be a real complication of Newtonian physics; however, in the spirit of "turning off" as much as possible, we will turn off this lag and assume that the speed of travel of the field is infinite. We do this for both gravitational and electric forces. We can do it only because the speed is so fast that we can't notice it in ordinary life.

Neglecting the travel time of the fields is a good approximation which also keeps a very simple sense to Newton's third law. That is, with the assumption that fields travel at infinite speed, a charge applies a force F *instantly* on a second charge and the second acts *instantly* back on the first with an equal but opposite force (see Figure 4-12). You may wonder how one knows when to make such approximations. That ability and related ones, such as knowing what to include and what not to include in understanding a given physical situation, are only learned by thinking and solving problems. Like a good athlete, a physicist must learn fundamental skills by doing. And, the skill of judiciously making choices about the system being analyzed and of making reasonable approximations is critical to successfully solving physics problems.

Figure 4-12: Two positively charge particles generating fields in the surrounding region. Each charged particle acts, in our approximation, instantly on the other with a force, $F = \left|\vec{F}\right|_{electric} = \frac{q_1 q_1}{r^2}$.

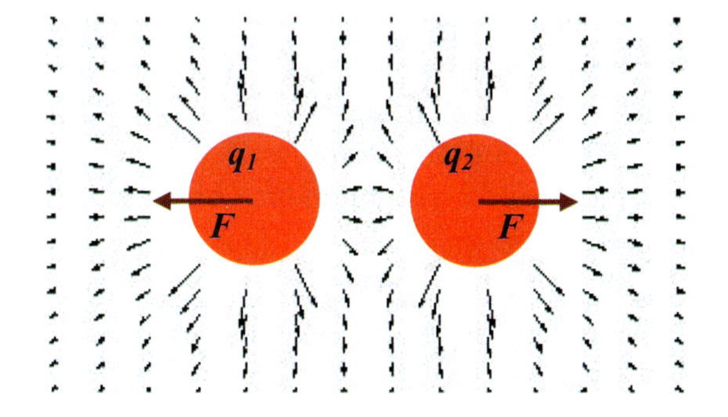

[12] Indeed, in this context, understanding space is the key to formulating a precise statement about what acts directly on what.

Now, in the case of uniform gravitational field shown in Figure 4-10, where is the massive body that we say acts on bodies introduced into the field? Recall the uniform field is an approximation of the Earth's gravity; we left out the Earth as generator of the field to focus on the field and its effect.

In general, even if the Earth is not a focus of the problem, in the vicinity of the Earth's surface (except in thought experiment cases where we consider gravity "switched off"), we must note a gravitational force $F = m\,g$ acting on a body of mass m due to the Earth. Further, momentum will only be conserved if I include the Earth, but Newton's second and third law, in a way, does this for us. For those laws tell us that if momentum is "lost" in one place, it will appear some place else, because wherever and whenever there is a force there is an equal magnitude but oppositely directed force somewhere else.

With this knowledge of force, we can return to our rocket to discuss its launch from Earth (or Mars) where gravity is present.

Launching the Rocket

Note that this can be done very easily by starting with equation 4.11, without setting the left hand side equal to zero. That is, we have:

4.21
$$\frac{dP_{system}}{dt} = \frac{MdV}{dt} - \frac{dm}{dt}\mathbf{v}_p$$

Because of the generality of our formalism, we do not have to redo all the steps we have taken; they still apply. We only need a mathematical expression for the strength of the force, for this would tell us the rate at which the momentum of the system changes. We have just seen that the gravitational force near the surface of the Earth for an object of mass M is approximately $M\,g$ everywhere.[13] Recall M is the mass of the rocket (including whatever fuel it has at the moment). Note that the equation is written relative to an inertial observer to which the rocket appears to be at rest at the given moment. But wait! How can we have an inertial observer (i.e. a uniformly moving force-free observer) when gravity (a force) is always pulling on us? Indeed, in order for a massive body such as you or I to move uniformly, a counteracting force must be applied to halt the momentum drain that would be caused by gravity. Such a frame is not an inertial frame; objects in it would not continue in their state of rest or uniform motion. For example, people in an elevator that was allowed to fall at a constant speed but was *prevented from accelerating* would still feel the pull downward of gravity. Now, noticing this fact puts you in the company of Albert Einstein (1879-1955) who also noticed such things, and those insights put him on the path to his general theory of relativity. We are not ready for general relativity, so let's simply "shut off" our own (the observer's) gravitational mass (which we can do as a thought experiment though not in reality), so that gravity can't affect us; this amounts to simply leaving a more clear understanding of inertial frames till later.[14] Hence, equation 4.21,

---

[13] We can imagine all of this as concentrated at a point, the center of mass, and just track the problem as if it were a point problem.

[14] If this "Newtonian" method, doesn't satisfy you-- as indeed, it shouldn't because it is obviously and intentionally incomplete-- a short explanation of the method used in general relativity might help. In particular, we simply consider a very small mass body (so that it doesn't affect the gravity field of the main body too much) that is in free fall at that speed at that particular moment; then, if we define the frame that falls with it as "rest," and if we consider a small enough region of space and a small enough duration of time, things viewed in that frame will behave approximately like Newtonian mechanics.

which, again, is defined in the uniformly moving frame that matches the speed of the craft, becomes, after using $dm = -dM$:

4.22 $$-Mg = \frac{MdV}{dt} + \frac{dM}{dt}\mathbf{v}_p$$

We can easily solve this to yield:

4.23 $$V_{final} = V_i + \mathbf{v}_p \ln[\frac{M_i}{M_f}] - g t_f$$

Where t_f is the time after the last amount of mass is ejected. The launch and flight of a water rocket is shown in Figure 4-13.

Figure 4-13: Launch and beginning of flight of a seven foot long water rocket.

Statics, Motion, and Parts of a Body

With the above background, there are many other problems that we can now solve. But, before we solve these more advanced problems, we need to fill out a little more of our knowledge of the simple case when objects are not in motion but are in contact with one another. For instance, the books resting on the shelf and the shelf resting on the floor are not moving but are interacting. In each case, an equilibrium state has been reached so that there is no locomotion. In saying this, we, of course, only consider the locomotion of each "body" as a whole; we do not consider any relative motion of the parts of the book, for example thermal motions, [15] that don't contribute to the motion of the book on the shelf. We turn all such effects off; i.e., we simply ignore them for the time being.

Now, the book on the shelf would fall to the ground if it were not prevented from doing so by the shelf. We say the force of gravity is canceled by the force of the surface of the shelf; the forces are in static balance. Note, by the way, this static balance of forces on a single body should not be confused with the equal and opposite forces acting between two bodies described by Newton's third law, which generally results in motion of one or both bodies.

When I put the book on the shelf, gravity acts to activate impetus in the book; however, the surface forces of the shelf act back, and the book actually only is minimally successful in moving the surface (the book only slightly bends it) before the shelf force is sufficient to impede the action of gravity. We say that, at this point, no further work is done on the shelf. The idea of work will be explored in the next chapter, but we touch on it here, because of the important distinctions it brings out. From a common sense point of view it seems wrong to say no work is done in some such cases. You may say, for instance, "If I push on the wall I will not succeed in moving the wall, but I will certainly do some work." In fact, you will do work in the general sense of expending energy for you will at least generate heat, which is associated with the motion of the parts of the body. However, you will not do *work in the technical sense* used in this text (which we define completely in the next chapter) that relates to motion of the wall, not its parts. This distinction focuses our attention on aspects of bodies more directly accessible to our senses; heat, for example, doesn't appear to our senses to be related to motion of bodies. In fact, when we say the book sitting on the shelf is a static situation, we are obviously leaving out, for instance, the thermal motion of atoms making up the book and shelf.

Now, since we are, by definition, not considering intra-body (especially, thermal) effects, but the whole body motion only, the only way for work to be done is by moving a body from rest or to a new state of uniform motion by giving it impetus that it did not already have. Empiriometrically, where we leave out any reference to what actually has impetus and what doesn't and only consider relative motion, we simply say that, relative to any uniformly moving point of view (frame), a force that causes a change in the state of rest or motion of a body does work on that body. This definition of work does differ, in some important cases, from ordinary usage, so we should be aware of the thinner, more precise meaning we are invoking here.

In the problem in the next section, like in most others, we will use the distinction drawn above, and ignore all motion due to thermal effects, concentrating our attention on the motion of *identified* bodies only--this means parts of such bodies are ignored, unless they themselves are identified bodies.

An Atwood Machine

We now further hone our skills and understanding by solving for the motion of bodies in one example of an "Atwood Machine."

Consider a light sturdy cord that is attached to two massive blocks m_1 and m_2 in the manner shown in Figure 4-14. The cord passes over a *stationary* frictionless rod, which is

[15] Thermal motion will be studied in statistical physics and thermodynamics. It doesn't generally contribute to movement of the book across the shelf.

fixed to a wall. This experiment is done in a laboratory frame; such frames are typically taken, as we do here, to be at rest or in uniform motion, i.e. they are taken to be inertial frames.

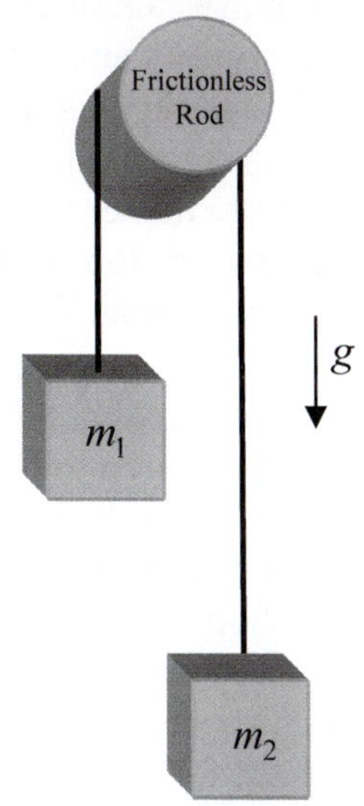

Figure 4-14: A simple Atwood machine in which a *stationary*, *frictionless* rod mounted to the wall holds two masses suspended by a cord.

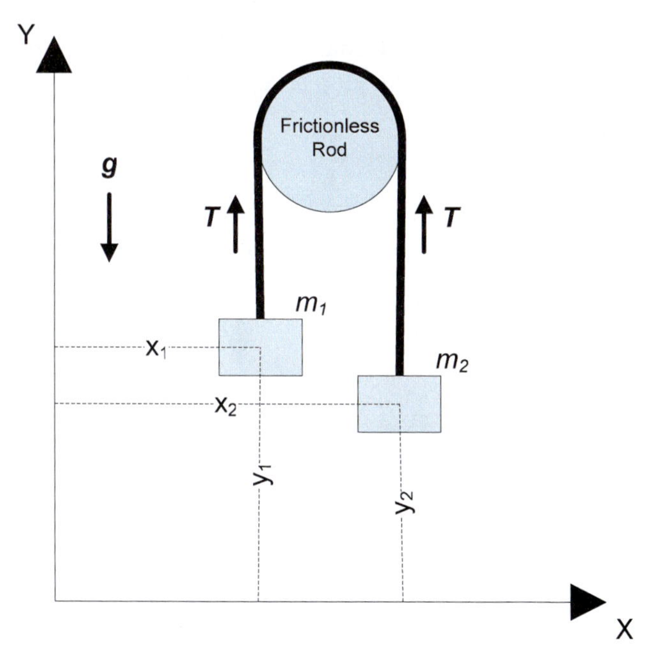

Figure 4-15: Our Atwood machine with forces shown.

As usual, we construct force diagrams, which are shown in Figure 4-15 and Figure 4-16. In the latter, we conceptually isolate the forces applied to each block; such diagrams are, for this reason, also called *free-body diagrams.* For each block, the force of the Earth's gravity "pulls" with a magnitude of g times the mass of the particular body, while the tension in the cord pulls up.

We assume the cord is very much lighter (less massive) than the blocks; in fact, we consider that it is as light as we please and then make it lighter and then do it again. Indeed, we imagine that we can actually accomplish this limit and take the cord to be massless. We are, by now, familiar with this type of purely mental construct (being of reason), which is, if taken simply as given, a contradiction (in this case, a "massless mass"), but, when understood as a convenient mental "carrying case," is extremely useful. To understand this particular mental construct, see the end of chapter problem in which this limit is discussed. Now, in the limit of a massless cord, the tension in the cord is transmitted undiminished through its entire length. Therefore, the tension, T pulling upward on each block is the same, as shown in the figure.

The rod is important in satisfying Newton's third law. Notice that while the cord "transmits" the force from one block to another, the pair of tension forces in Figure 4-15 *is not* a direct action-reaction pair of forces between blocks 1 and 2. The tensions are equal in both magnitude, but are *not* oppositely directed. How then are these parallel tensions related to Newton's third Law? An action-reaction pair of forces exists between the cord and block 1. *Another* pair of action-reaction forces exists between the cord and block 2. The cord itself interacts with the blocks *and* the supporting rod. The cord pushes against the rod and the rod pushes back. The net force of the cord on the rod is downward, while that of the rod on the cord is upward. The rod must be in static force equilibrium, since it does not move (zero net force: $F_{net} = m_{rod}\, a_{rod} = 0$). The tensions pulling down on either side of the rod must be counteracted by a net opposite force acting on the rod at its point of support on the wall as shown in Figure 4-16a. *It is this interaction with the ground via the rod and wall* which allows the cord to develop an upward tension on the blocks.

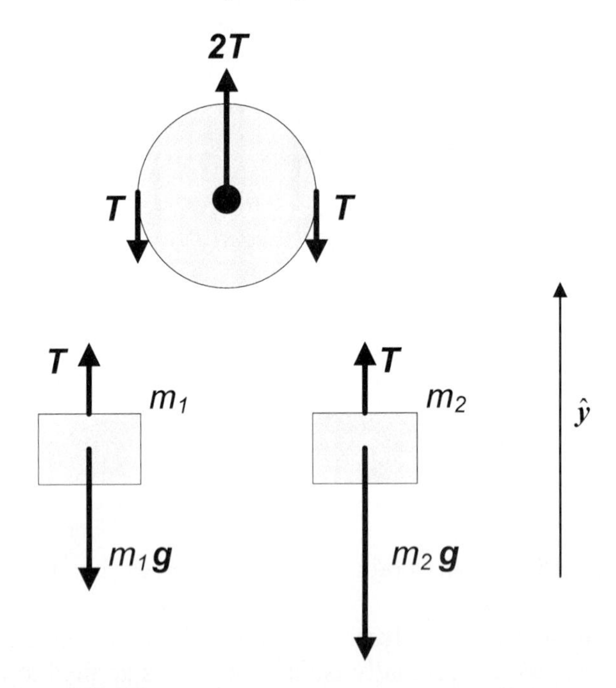

Figure 4-16: a. *(Top)* The forces on the rod. Note the forces cancel since the rod is stationary. **b**. *(Bottom)* The forces acting on each mass.

So, we repeat, the tension acting upward on each block is the same as shown, along with the gravity forces, in Figure 4-16b. It is helpful in making such diagrams to remember 1) forces act on bodies and 2) forces are abilities (qualities) of bodies.

We now set up the equations for Newton's second Law, in the form $F = ma$—recall this is the form for a constant mass body—applied to each block. We only need to consider motion in the vertical direction parallel to the *y*-axis, since no forces are applied horizontally:

4.24
$$a)\ m_1 a_1 = -m_1 g + T$$
$$b)\ m_2 a_2 = -m_2 g + T$$

Where: $a_1 = \frac{d^2 y_1}{dt^2}$ *and* $a_2 = \frac{d^2 y_2}{dt^2}$

Note that the *y* coordinates (y_1 and y_2) in these equations are defined as positive in the upward direction for both blocks.

This system also has a constraint on how the blocks move together. Suppose block 1 moves down a distance of 1 cm. Because the cord is in tension, block 2 will move upward a distance of 1 cm. Every displacement by one block is mirrored by an opposite displacement by the other. This is also true for the velocities and accelerations. Thus, in whatever way the system may move within the bounds of the constraint, we must have $a_1 = -a_2 \equiv a$ yielding:

4.25a,b
$$m_1 a = -m_1 g + T$$
$$-m_2 a = -m_2 g + T$$

We now subtract the bottom equation from the top to obtain:

4.26
$$(m_1 + m_2) a = -(m_1 - m_2) g$$

4.27
$$\boxed{a = \left(\frac{m_2 - m_1}{m_2 + m_1}\right) g}$$

What does this show about the motion of our Atwood machine? First, the blocks will move with constant acceleration. In particular, we notice that if the masses are equal the acceleration is zero – they are in balance. If $m_2 > m_1$, then m_1 will accelerate *upwards* and m_2 will accelerate downwards. If the masses are very different from each other, for example: $m_1 \gg m_2$, the limiting acceleration is $-g$ for mass 1; that is, it will never be able to accelerate faster than the gravitational free-fall of about $9.8\ m/s^2$.

We can also solve for the tension, *T*, in the cord by substituting our expression for the acceleration back into equation 4.24 a or b:

4.28
$$T = m_1 \left(\frac{m_2 - m_1}{m_2 + m_1}\right) g + m_1 g$$
$$= \left(\frac{m_1 m_2 - m_1^2 + m_1 m_2 + m_1^2}{m_2 + m_1}\right) g$$
$$= \left(\frac{2 m_1 m_2}{m_2 + m_1}\right) g$$

Again, it is useful to examine how the tension changes for different values of mass. If the masses are equal ($m_1 = m_2 = m$), the machine is in balance, and we see the tension T acting on each block is equal to $+m\,g$; this is just as it should be to counteract the force of gravity, $-m\,g$. Hence, given that the each force successfully prevents the attempt of the other to activate impetus, there is no activity; we say, as previously mentioned, that the work done is zero. If $m_1 \gg m_2$, the tension T is approximately equal to $2\,m_2\,g$. We then see that while a gravitational force of intensity $m_2\,g$ is available to act on block 2, that force is completely impeded by the tension; indeed, after counteracting the gravitational force, that tension is still able to apply a force of intensity $m_2\,g$ to block 2 (equation 4.25b) This remaining or "net" force intensifies the impetus of block 2 in the *upward* direction resulting in an upward acceleration of *g*. Said another way, this acceleration of block 2 is the free-fall acceleration rate determined (via the cord) by gravity acting on the much more massive block 1.

The Atwood machine is obviously useful for understanding the principles of physics, but it is more than this. Notice from equation 4.27 that, whatever the ratio of the masses, the magnitude of the acceleration of the blocks is proportional to the free-fall acceleration due to gravity. By using accurately known masses, we can measure the acceleration rate of the machine over a longer period of time and thus infer the local value of free-fall acceleration of gravity; that is, we can find the value of *g*.

More elaborate versions of the Atwood machine, including placing the block(s) on an inclined plane as well as incorporating friction and the motion of the wheel, will reveal many interesting phenomena. We will return to solving particular Atwood machines in the end of chapter problems.

For now, by revisiting the center of mass formalism with the concept of force in our quiver, we can learn more not only about the Atwood machine but machines in general.

Center of Mass and Force

Position, Velocity and Acceleration

Consider a system of small particles described by a set of position vectors relative to some origin in an inertial frame. Figure 4-17 shows a set of four masses (not necessarily equal) scattered in the *x-y* plane. However, we can think of this system in 3-dimensions as well and containing as many particles of whatever mass we want.

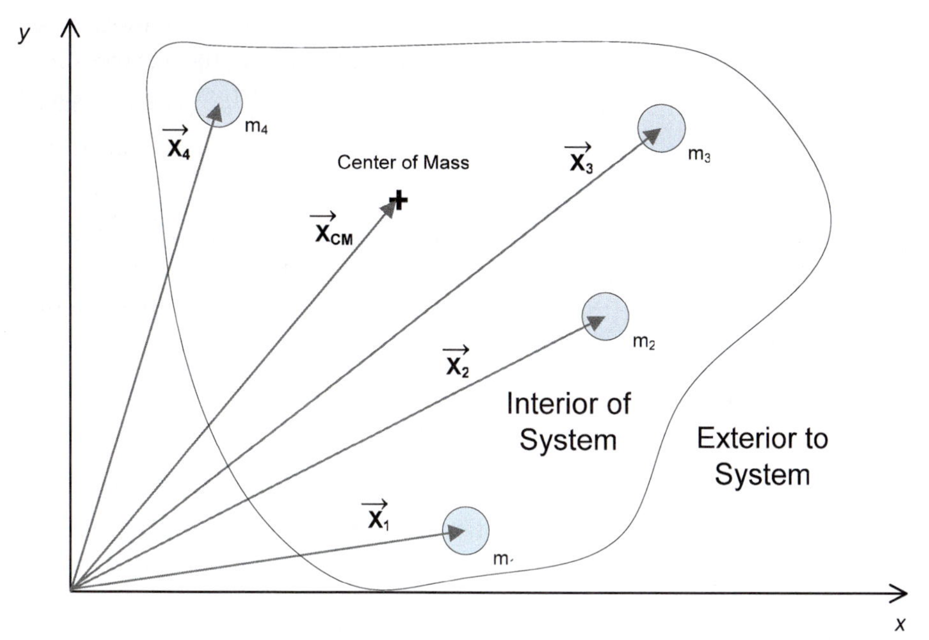

Figure 4-17: Four masses making up a system of particles, with their respective displacement vectors shown. Note that the center of mass of this system is in empty space.

Recall the definition for the center of mass given in the last chapter.

4.29 $$\vec{X}_{CM} = \frac{\sum_{i=1}^{N} m_i \vec{x}_i}{\sum_{i=1}^{N} m_i} = \frac{\sum_{i=1}^{N} m_i \vec{x}_i}{M_{total}}$$

We can take the time derivative of this equation to find the velocity of the center-of-mass in terms of the velocities of the individual particles.

4.30 $$\vec{V}_{CM} = \frac{d}{dt}\vec{x}_{CM} = \frac{\sum_{i=1}^{N} m_i \frac{d}{dt}\vec{x}_i}{\sum_{i=1}^{N} m_i} = \frac{\sum_{i=1}^{N} m_i \vec{v}_i}{M_{total}}$$

Likewise, we can calculate the acceleration of the center-of-mass.

4.31 $$\vec{A}_{CM} = \frac{d}{dt}\vec{V}_{CM} = \frac{d^2}{dt^2}\vec{X}_{CM} = \frac{\sum_{i=1}^{N} m_i \frac{d^2}{dt^2}\vec{x}_i}{\sum_{i=1}^{N} m_i} = \frac{\sum_{i=1}^{N} m_i \vec{a}_i}{M_{total}}$$

Equations of Motion

To determine the motion of the particles, we need to know what forces are acting, which will allow us to fill in the left hand side of Newton's second law. First, consider the forces acting on any one particle. These forces will be of two types: 1) forces originating from other particles *internal* to the system and 2) forces originating *external* to the system. To help visualize, we imagine a boundary separating masses within the system from everything else, as shown in Figure 4-17. If we look at, for instance, particle 3 in the above diagram, its equation of motion, given by Newton's second Law, is:

4.32 $$m_3\vec{a}_3 = \sum_{j=1,2,4} \vec{F}_{3-j} + \vec{F}_{3-ext}$$

The first term is a vector sum of all forces originating from all particles internal to the system acting on particle 3. The second term is a vector sum of all external forces acting on particle 3. We can easily generalize this equation for a system of N particles acting on an arbitrary particle i in the system:

4.33 $$m_i\vec{a}_i = \sum_{\substack{j=1 \\ j\neq i}}^{N} \vec{F}_{i-j} + \vec{F}_{i-ext} \qquad i = 1,\ldots,N$$

Since we want an equation for the center of mass motion, not the individual particle motions, we need to sum the individual particle motions. So, with an eye on center of mass equation 4.31, we thus add all N equations associated with each particle together into a single vector equation.

4.34 $$\sum_{i=1}^{N} m_i\vec{a}_i = \sum_{i=1}^{N}\sum_{\substack{j=1 \\ j\neq i}}^{N} \vec{F}_{i-j} + \sum_{i=1}^{N} \vec{F}_{i-ext}$$

Equation 4.31 then allows us to rewrite the left-hand side to get:

4.35 $$M_{Total}\vec{A}_{CM} = \sum_{i=1}^{N}\sum_{\substack{j=1 \\ j\neq i}}^{N} \vec{F}_{i-j} + \sum_{i=1}^{N} \vec{F}_{i-ext}$$

The first term on the right seems to be the most complicated. However, using Newton's third Law we find it is easy to evaluate. It is a vector sum of all internal forces acting within the system. But we know from Newton's third Law that the force of particle i acting on particle j is equal and opposite to the force of particle j acting on particle i:

4.36 $$\vec{F}_{i-j} = -\vec{F}_{j-i}$$

Thus, their vector sum is zero (i.e., $\vec{F}_{i-j} + \vec{F}_{j-i} = 0$), and this is true for every pair of particles acting *within* the system. Thus, the double summation term is precisely equal to zero. The second term is equal to the net vector sum of all external forces acting on the system, so we get:

4.37 $$\boxed{\vec{F}_{net-ext} = M_{Total}\vec{A}_{CM}}$$

This is a remarkable result. The center-of-mass of the system behaves like a single particle acted on by a force (the net force, i.e. the vector sum of all external forces). We do not need to know any of the details of internal forces within the system.

But are not the internal forces important for understanding how the system behaves? Yes and no. If we want to understand how the parts of the system move individually then yes. If we are only seeking the behavior of the system as a group, then the answer is no. As we pointed out in Chapter 3 and as we see more completely in equation 4.37, the center-of-mass is the center of dynamics. Its motion describes the motion of the parts of the system to the extent that they move as a whole. As we'll prove in a moment, this motion of the system is only changed from its state of rest or uniform motion by external forces.

Because of its importance, we need to explore further the meaning of this center of mass point. First, note that the center of mass doesn't necessarily correspond to a point within one of the masses that we are describing. We've seen this already in our example in Figure 4-17. Pluto and its moon Charon is a real life example (see Figure 4-18). The Pluto/Charon center-of-mass lies between them on the line that joins them. It lies closer to Pluto but outside of it. The center-of-mass is a point that represents the system and acts as if it had a mass equal to the mass of the system, but is not itself a body. Indeed, in so far as the motion of the system as a whole is concerned, we can think of all the mass of the system condensed to the center of mass point and treat the system as a point mass. If we think of the center of mass as an actually existing point mass, which is often convenient, then obviously it is purely mental construct, not able to exist outside the mind (a body cannot exist as a pure geometrical point, and the mass associated with the center-of-mass *does not exist* at that point.). Still, as long as we remember what we mean-- in particular, what we are leaving out (switching off) in conceiving the center of mass in this way-- we can use this mental construct to carry key pieces of information about a system and greatly simplify calculations and thinking in so doing.

Figure 4-18: Pluto and its three known moons (satellites) as seen through the Hubble Space Telescope. Charon orbits Pluto in a circle of radius *~20,000 km*. Pluto is *1200 km* in radius, with a mass of 10^{22} *kg*, while Charon is *600 km* and 10^{21}*kg*. A more accurate description of the motion (kinematics) would be to say Pluto and Charon rotate around their common center of mass, which is in the space *between* them. Pluto is currently classified as a dwarf planet.

We can now make it clear that the internal interactions among the parts (such as the thermal motion discussed earlier) of a system do not affect its center of mass motion. We do this by rewriting equation 4.37 using the definition of velocity and momentum and assuming the mass of the system is constant. We get:

$$M_{Total}\vec{A}_{CM} = M_{Total}\frac{d}{dt}\vec{V}_{CM} = \frac{d}{dt}\left(M_{Total}\vec{V}_{CM}\right) = \frac{d\vec{P}_{CM}}{dt}$$

4.38

$$\Rightarrow \frac{d\vec{P}_{CM}}{dt} = \vec{F}_{net-ext}$$

The rate of change of total center of mass momentum of the system is equal to the vector sum of all external forces; that is, only external agents can cause the change.

Substances and Parts of Substances in CM Formalism

Notice that in using the center of mass formalism, we are able to extend the application of our generalization that ignores whether we are dealing with a substance (such as a man), a part of substance (such as one of his organs) or a group of substances (such as two or three men) to include forces, so that we can *just* speak of ***net forces*** on the system "center." In this way, it is as if each thing that acts can be treated as if it were a substance, existing on its own. Again, this analogical generalization doesn't do away with those distinctions, but only allows us to put our focus on the system of interest; this sort of "flattening" of reality is helpful and good as long as we do not forget this fact.

Revisiting the Atwood Machine

Now, we can further our understanding of the Atwood Machine by verifying equation 4.38 for that case by substituting the appropriate masses into equation 4.31, using the relation between the individual accelerations and substituting for a from equation 4.27:

4.39 $$A_{CM} \equiv \frac{m_1 a_1 + m_2 a_2}{m_1 + m_2} = \frac{(m_1 - m_2)a}{M_{Total}} = \frac{(m_1 - m_2)}{M_{Total}} \times \left(\frac{m_2 - m_1}{m_2 + m_1}\right) g = -\left(\frac{m_2 - m_1}{m_2 + m_1}\right)^2 g$$

This means:

$$M_{Total}A_{CM} = \left(\frac{m_1 m_2 - m_1^2 - m_2^2 + m_1 m_2}{m_2 + m_1}\right) g$$

4.40

$$= \left(\frac{4m_1 m_2 - (m_1 + m_2)^2}{m_2 + m_1}\right) g$$

$$= 2\left(\frac{2m_1 m_2}{m_2 + m_1}\right) g - m_1 g - m_2 g$$

$$= 2T - m_1 g - m_2 g$$

The last equality uses equation 4.28 to introduce the tension, T.

Referring to Figure 4-15, notice that if we take the blocks as our system, all forces are external. Thus, they effectively act on the center of mass of the two blocks, and as long as the net force is negative, there will be a downward acceleration of the center of mass. From this force perspective, we see from Figure 4-16, that $F_{net} = 2T - m_1 g - m_2 g$. Reinserting the tension given in equation 4.28, we see that the inequality in the strength of the gravitational force on blocks of unequal mass (coupled with the equality of the tensions—which yields the $2T$ in the equation for the net force) means there is a net downward force on the center of mass. In particular:

$$F_{net} = 2T - m_1 g - m_2 g = -\frac{(m_1 - m_2)^2}{m_1 + m_2} g$$, which is less than zero whenever $m_1 \neq m_2$.

Inclined Plane

We now analyze one final problem to further our understanding: in particular to illustrate the vectorial character of acting forces and the advantages of choosing a suitable coordinate system for solving problems. Referring to Figure 4-19, we see a block of mass M placed at the top of a ramp (an inclined plane) that is fixed to the ground. The ramp is inclined from the horizontal by an angle θ, which we measure in radians. Gravity is also present and acts directly downward. We examine how Newton's laws of motion apply in the very simple mathematical case where the interaction between the block and the ramp is frictionless.

Coordinate Choice and Solution

We know from ordinary life experience that the block will slide down. Indeed, the block can't move into the ramp, so it must move along it if it does anything. Knowing such physical facts, we choose a coordinate system along the ramp and a reference frame at rest with respect to the ramp. Choosing different coordinates and frames is certainly possible but is artificial for the problem given as we will now show more specifically.

To further understand and make use of our coordinate choice, recall from Chapter 2 that empty space is what one gets when all properties of a thing or group of things are left out; even the shape that normally bounds the area of space is pushed out as far as one wishes so as to not be relevant. Lastly, leaving the space only potentially divided, not actually so, means it is completely simple and symmetric throughout with no direction or place distinguished from another.

Now, in the case of our present experiment, the symmetry of simple space is broken by our switching on gravity and by introducing the hardness of the block, the ramp, and the ground as well as the anchoring of the ramp to the ground. These things limit where and how the block can move. The surface of the ramp, that to which gravity pulls the block, is two-dimensional, with each dimension independent of the other--this is the most basic definition of dimension.[16] Since we are interested in forces insofar as they cause activity (i.e. motion through activated impetus), and these forces can only activate along the surface of the ramp, it behooves us to place the coordinate axis along the surface of the ramp as in Figure 4-19. Again, in this way, we respect the natural divisions induced by the surface forces and the gravitational force. Any two perpendicular directions are independent, so we divide the force in accord with the way the bodies and their powers most simply induce divisions of space.

In particular, the origin of our *x-y* coordinate axes is chosen at the bottom of the ramp with the x-axis parallel to the surface of the ramp, pointing upward along it, while the y-axis is perpendicular to the surface of the ramp. Free-body diagrams of the forces acting on the block are given in Figure 4-20 and Figure 4-21. The setup is shown rotated so that the x and y-axes will be oriented in standard fashion; this of course skews the angle of the floor.

---

[16] In later courses, you will see that the word dimension is also be used in analogous sense for certain system of generalized coordinates.

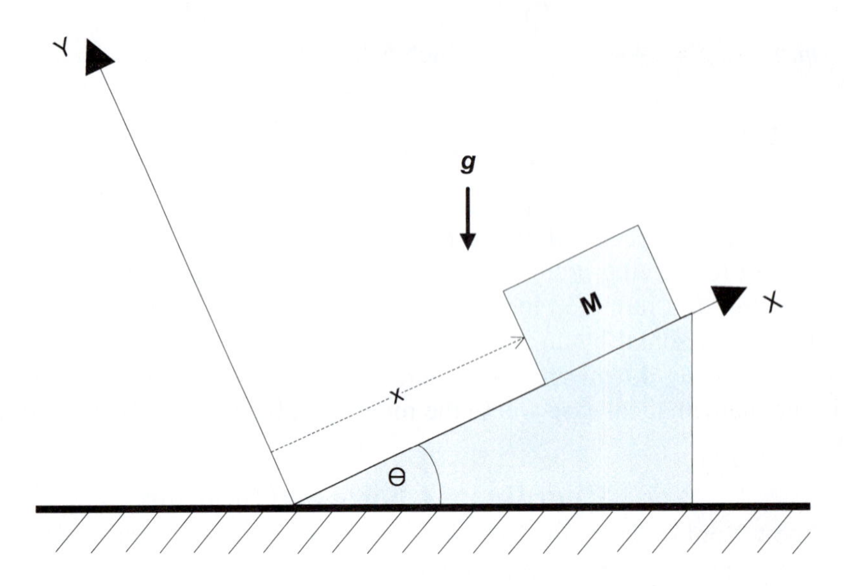

Figure 4-19: A block is placed at the top of a ramp and allowed to slide under the action of gravity. The interaction between the block and the ramp is assumed to be frictionless. The lowercase ***g***, along with the arrow, indicates the magnitude, and direction, of the acceleration due to gravity.

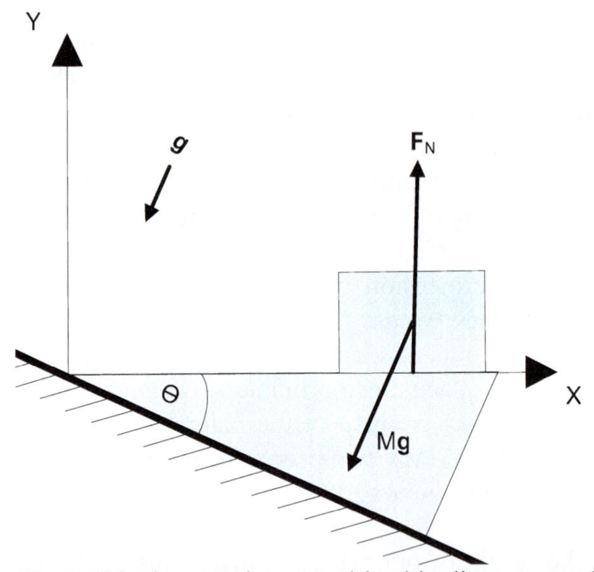

Figure 4-20: Ramp/block setup is rotated in this diagram so that the *x*-axis is horizontal and *y*-axis is vertical. This means the ground appears tilted when it's really we that are tilted in order to view the ramp as a horizontal surface--this is equivalent to standing on the ramp perpendicular to its surface, rather than parallel to gravity.

The ramp applies a normal force of strength, F_N, to the block along the *y*-direction in reaction to gravity pulling the block to the surface of the ramp. Gravity and this normal force, which effectively act at the block's center of mass, mutually impede one another from acting, so there is no motion in the y direction. With no friction, there is no force exerted *by the ramp* parallel to the x-axis. Because of our choice of coordinates, take note of the angle of inclination of gravity; it is inclined from the y-axis by the angle θ.

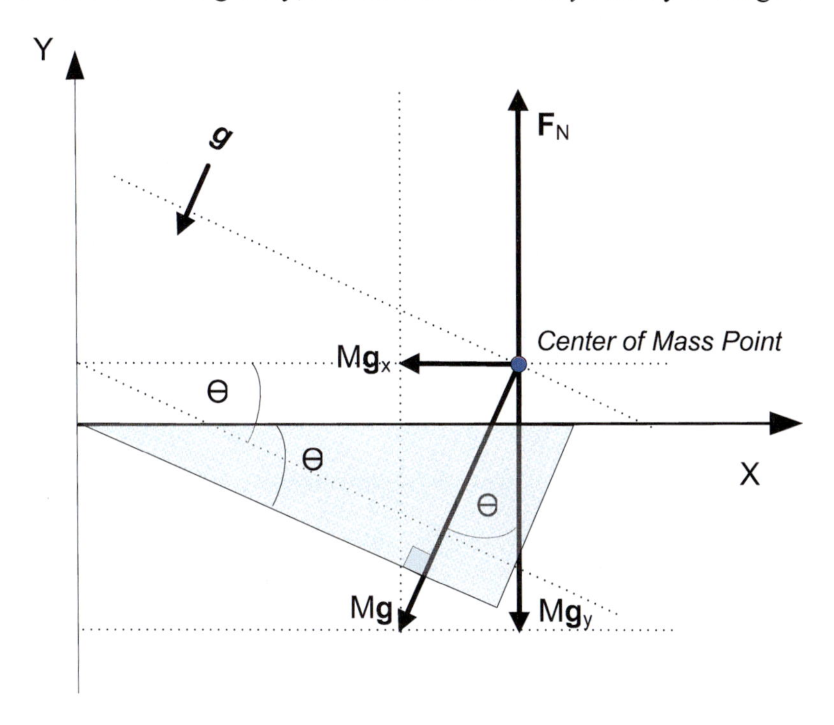

Figure 4-21: A further abstracted version of Figure 4-20. We consider all of the forces acting on the block as though they were acting at its center of mass point. We resolve the force of gravity on the block, $M\vec{g}$, into components along, and perpendicular to, the surface of the ramp.

Geometric considerations illustrated above in Figure 4-21 allow us to resolve the force of gravity into two components, one applied parallel to the x-axis, one to the y-axis. We see by inspection that the magnitude of these components are:

4.41
$$F_x = M\,g_x = M\,g\sin\theta$$
$$F_y = M\,g_y = M\,g\cos\theta$$

Noting carefully the signs of these components relative to the established axes, we write using Newton's second Law:

4.42
$$-M\,g\cdot\sin\theta = M\,a_x$$
$$-M\,g\cdot\cos\theta + F_N = M\,a_y$$

Since the block cannot move in the y direction, the y-coordinate of the block is $y(t)=0$ for all times ***t***. Thus, the block is in equilibrium along the y-axis, and the second equation above gives: $F_N = M\,g\cos\theta$. The first equation in 4.42 shows the block will move with constant acceleration in the negative x-direction ("down" the ramp) at a rate: $a_x = -g\, Sin\theta$. The speed and position are easily found by integration.

It's wise to check answers at the limits where we already know the answer. Take the cases:

4.43 $$\theta = 0 \text{ radians}: \quad \begin{aligned} a_x &= 0 \\ F_N &= m\,g \end{aligned}$$

4.44 $$\theta = \frac{\pi}{2} \text{ radians}: \quad \begin{aligned} a_x &= -g \\ F_N &= 0 \end{aligned}$$

These results are predicted by our result, and we know them to be true by the following reasoning. In the case of no ramp inclination, the block simply rests in equilibrium with the force of gravity fully counteracted by the normal force from the surface of the ramp. When the ramp is vertical, no normal force is present (no interaction with the ramp surface) and the block simply falls at the free-fall acceleration due to gravitation.

Summary

A *Force* is the ability[17] (a property in the category of quality) of a body to change the intensity and/or directional type of the impetus of another body. A body exerts a force on a second body, changing its momentum (impetus) and thus changing its state of uniform motion or rest. Massive bodies thus have the capacity to receive impetus (they do not have to have it) and a certain resistance to the action of the impetus; more massive bodies with same momentum move slower according to $\vec{v} = \frac{\vec{p}}{m}$.

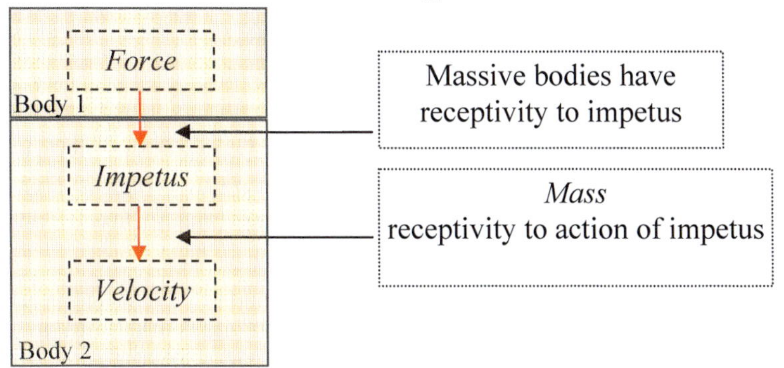

A force can activate impetus in a body that has none. A force that acts along the direction of a uniformly moving body will increase the intensity of the impetus, increase its momentum. A force that acts perpendicular to the direction of motion will activate impetus of the new directional type, i.e., cause momentum in the new direction. A force that acts in a direction that has components parallel and perpendicular to the uniform motion will change the momentum of those components as just described.

Newton's Three Laws of motion describe the interaction of forces, momentum (impetus) and mass, the three central properties considered in Newtonian physics and are used by analogy explicitly or implicitly in all of physics, force being less used, or even ignored, in certain advanced empiriometric areas.

I. Every object continues in its state of rest or uniform motion unless acted on by an outside force. This law reveals the nature of impetus by succinctly stating its manifest effects.

II. The force acting on a body is equal to the time rate of change of its momentum

$$\vec{F}_{net} = \frac{d\vec{p}}{dt}$$

Once impetus is understood, this law encompasses the first law as a special case and further describes how the net strength of forces applied to a body affects its intensity and direction of impetus. Empiriometrically, we don't consider what the actual

[17] As mentioned for the quality of impetus, a force might also be called a "power" if it is not confused with the technical (empiriometric) use of the word power as energy per unit time. As an aside, note the distinction between the force (quality) and the acting of a force (action).

intensity of impetus is, but only consider the relative motion and the relative momentum associated with the motion as it appears in a chosen uniformly moving frame. Expanding the second law using the chain rule introduces a time derivative of the mass which can be ignored in the case of constant mass giving an important form:

$$\vec{F} = m\vec{a}$$

Note that *conservation of mass* of an isolated system is assumed in Newtonian physics.
III. For every action, there is an equal and opposite reaction. For each force of strength F_{AB} applied from body A to body B, there will be a force of equal strength acting against body B on body A, F_{BA}.

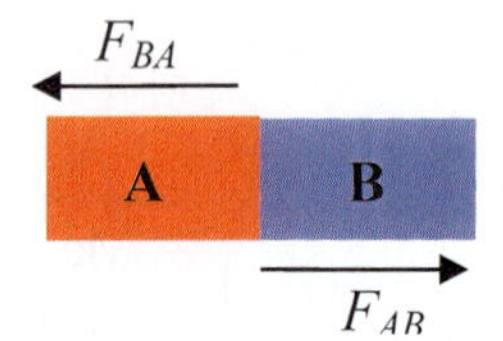

This law is known as the action-reaction principle and, coupled with Law II, is responsible for the conservation of the net momentum of a closed system. Conservation of total kinetic energy, discussed in the next chapter, also follows from it in a similar way.

If we consider the force and impetus as acting the same both in substances and groups of substances, then Newton's third law follows from the second law. In particular, when $F = 0$, the second law indicates that the momentum of a body does not change. And, if by "body" one can mean a group of substances, say two substances, the only way momentum can be conserved is that when the two interact, one gives the other whatever momentum the other gets (which also means one may "cancel" some or all of the momentum of the other). This simply means that the same average force acted from body A on B as did from B on A. (If we allow an arbitrarily small interaction time and any force strength to occur during that time, we can say, in the limit, that the forces are equal and opposite, which is Newton's third law.) Law III actually reflects something of the nature of the fundamental forces involved, in ordinary cases, the so-called electric and gravitational forces.

Newton's laws are vector laws, meaning that all the aspects of geometry captured in the laws of vector analysis are relevant, including that we can treat each of three perpendicular components independently. Newton's Laws must be applied in a particular uniformly moving (or rest) frame of reference.

The net change in momentum in an interaction ($\Delta p = F\Delta t$) is called an *impulse*.

In collisions, surface forces change the momentum of the colliding bodies, but not the momentum of the system. Elastic collisions can be modeled by ideal springs, which are discussed in detail in the next chapter.

Gravity and electric forces[18] are the two most important forces in ordinary interactions. These forces intrinsically have the equal and opposite nature described by Newton's third law, as long as we consider the interacting bodies acting through space so quickly that we can, for now, ignore the activity in space. Indeed, we treat them as if they act *instantaneously* across space, so that, for instance, a positive charge pulls on a negative charge (and vice-versa) the moment the charges appear. The general form of both the gravitational and electric forces is the so called inverse square law. The electric force is written: $\vec{F}_{electric} = \frac{q_1 q_2}{r^2}\hat{r}$, where $r\hat{r}$ is displacement vector between the two charges and q_1 and q_2 are the measures of the charge on each of the two bodies. The gravitational force and the meaning of the various terms of such laws are discussed in detail in Chapter 8.

In connection with electric and gravitational forces, we speak of fields of force throughout some region of "space" that act on bodies in that region. We remember these fields are generated by a body and are properties of "space" which we will discuss in Chapter 8. Near the Earth we approximate the Earth's field as uniform, causing $\vec{F} = -m g \hat{z}$ at each point that a mass m is placed (where z is the vertical direction and $g \approx 9.8 m/s^2$).

In static cases, all forces are impeded from action, no impetus and thus no motion occurs. We say the forces cancel or are balanced. This cancellation of static forces by equal and opposite directed forces on a *single* body should not be confused with the equal and opposite forces of interaction *between* bodies specified by Newton's third law.

We generally ignore internal motions (such as thermal motions) of what we have defined as bodies; however, we can come back and analyze these as well if we keep a consistent definition of the bodies in our new analysis.

We see further the advantage of turning everything off and gradually adding qualities starting with the most generally applicable.

An isolated system of objects has an associated geometric location called the *center of mass*. By a mental construct (being of reason), we can consider all the mass of a given object to be concentrated into this point as if it were a point particle of the same mass; we amalgamate all but internal forces into our point mass, leaving only external ones. This point particle, a stand-in for the center of mass of the system, will behave according to Newton's second law: $\frac{d\vec{P}_{CM}}{dt} = \vec{F}_{net-ext}$.

Note: With the new information learned in this chapter, we can comment further on one use of the term *inertia* that we mentioned in Chapter 3, i.e., the resistance to acceleration. There are two ways of interpreting the word resistance here.

One is the resistance we feel when we accelerate a body; that is, the active resistance of the body that occurs when we accelerate it. This resistance is the *response* of

[18] More generally, we would say electromagnetic forces; however, details of this are outside the scope of this text and belong traditionally to the second half of introductory physics.

the body to an applied force; thus inertia refers, in this context, directly to Newton's third law. However, the third law, in combination with the first and second, describes the realities of force, impetus, and mass. In particular, a body, call it *A*, with no impetus or with impetus of a certain type and intensity needs to have a force applied to it to activate impetus in it or change its already existing impetus to cause it to accelerate. And, when this happens, body *A* responds with its own force against the body applying the original force, call it body *B*; i.e. body *A actively* resists the action of body *B*. And, *A* can only do this because the first little piece of impetus activated in *A* only moved *A* at some *finite* speed because of *A's* non-zero mass, thus keeping *A* around, however briefly, to react back on *B*, which might be you or me.

Instead of focusing on the result of the limited acceleration, the second way of interpreting "the resistance to acceleration" focuses on the reality that limits the acceleration itself, namely, the resistance to the action of the impetus, the mass. When any little piece of impetus (or little increase in intensity of impetus) is activated in a body, the body moves, as just mentioned, only at finite speed. Recall the relevant quantities (measures of the various qualities) are related according to: $\vec{v} = \frac{\vec{p}}{m}$. Thus, for a constant mass, a little change in impetus causes a finite (non-zero) little change in velocity. Now, a constant force causes a certain small change of impetus in every short interval of time, causing a corresponding small change in velocity in that same interval of time. Thus, we have, via the realities expressed in Newton's second law, $\vec{F} = \frac{d\vec{p}}{dt} = m\frac{d\vec{v}}{dt}$; that is, for a given force, the acceleration is proportional to the mass because of the way the mass resists the action of the impetus (mathematically, $\vec{p} = m\vec{v}$). Indeed, as we have just seen, both meanings of resistance ultimately lean on the mass resisting the action of the impetus.

Terms

Force

- Uniform Force
  - Uniform acceleration
- Newton's Three Laws of Motion:
  - Action-Reaction Pair
  - Net Force
  - Free-body diagrams
- Impulse
- Force fields
  - Uniform force fields
- Inverse square law forces
  - Gravity
  - Electric

Reference Frames (originally introduced in Chapter 3)

- Inertial (Chapter 3)
- Center of mass (its relation to force)

Accelerating
Apparent Force

Atwood and other machines

Calculation Techniques

Identify the system under consideration; be clear what you mean by the various elements and label forces clearly only applying forces to bodies, keeping in mind also that they ultimately come from bodies as well. When needed, to help clearly sort out what forces act on each body, draw free-body diagrams for each body.

The solution of many problems is greatly simplified by an appropriate choice of coordinate system (the type of system, rectangular or circular, as well as the orientation and location of the origin) and reference frame (how fast the observer is moving).

Unless the mass is constant, it is not often helpful to use the chain rule expansion of Newton's second law to solve problems; it is better to write the momentum at two successive moments and form differentials from them, as done in this chapter and in Chapter 3.

Problems can be solved with a view toward forces or toward momentum; choose the one which most naturally fits the given problem.

When the time scale of the problem is long compared to the time scale of the collision, we can neglect the latter.

Very near the surface of the Earth, the force of gravity acting on a body of mass m, can be approximated as a uniform force acting vertically downward with a magnitude of $m\,g$. Bodies acted on solely by the force of gravity near the surface of the Earth fall at a constant uniform rate independent of their mass (principle of equivalence).

It is often helpful to assume a rope or other connecting element in a problem is massless, so that the tension or force applied to one side is transmitted undiminished to the other.[19]

Notation and Vocabulary Notes:

$$\dot{m} \equiv \frac{dm}{dt}$$

[19] In the mental construct of a completed limit of a massless body, the forces are transmitted undiminished and the forces on it are exactly balanced. This should be used only when the body in question is between two heavy bodies; remember it's the ratio of such relevant masses that define small and such beings of reason only make sense in light of such limits. The danger of unconsidered use of such beings of reason is evident in the case of applied a finite force to the massless particle; infinite speed acceleration results.

Problems by Subtopic

Force

1. Given that the following pressures in the English units (of pounds per square in (psi)) convert to metric units:
a) bike tire: 40 psi b) football: 13 psi c) car tire: 30 psi

2. What average force, in *B/s*, is needed to accelerate a 4000 pound vehicle to 60mph in 12 seconds?

3. You are involved in a marksmanship contest in which the target at height *h* is rigged to the trigger so that when your gun, which is at the same height, fires, the target drops. At what angle to the horizontal should one aim the gun? Explain.

4. Suppose that, contrary to reality but as the common spontaneous explanation goes, external forces are responsible for the uniform motion of bodies, so that if one threw a baseball, rather than going over the home plate, it would stop as soon as one let go. a) In such a world in which there is no impetus, what would be the analogical equivalent of Newton's second law? Solve for the position as a function of time. Explain what it means. b) Now, consider a world in which, not only is there no impetus, but in which a force is directly responsible for a change of place, so that, for example, when one applies a force, the object gets there instantaneously. The stronger the force applied, the further away the place to which it is beamed. Write the equivalent of Newton's second law in this case. Solve for the position. Explain what it means.

5. Explain why mass is not fundamentally defined by $\vec{F} = m\vec{a}$ but by $\vec{p} = m\vec{v}$

6. Look up the masses for an electron and a proton as well as the typical size for a hydrogen atom. For purposes of this problem we will treat these objects as point particles. a) Compute the electrical force of attraction between the proton and electron. b) Compute the gravitational force of attraction between them. c) What is their ratio? Does it depend on distance? How can gravity ever become important?

7. The force field of an electric dipole of a body of charge *e* is given by: $\vec{F} = \frac{qd}{r^2}\left(3\left(\hat{d}\cdot\hat{r}\right)\hat{r} - \hat{d}\right)$. Where $\hat{d}$ is the unit vector between two charges of opposite sign both with charge *q*. $\hat{d}$
Roughly sketch this force field. *Note*: For magnetic fields, this dipole field is the simplest field one discovers in nature under ordinary situations (and possibly under any situation).

8. Suppose we continue our Olympics on the surface of Mars and they include a marksmanship event using a water ball gun. A water ball gun fires a spherical ball of water. Water sticks to itself, causing a surface tension that makes the water form into a spherical shape. a) Under the force of uniform gravity, g, what are the equations for the vertical and horizontal path of the ball as a function of time if it is launched at an angle θ at speed v from the horizontal? Include in your answer a free-body diagram of the forces on the water ball while it is in the air. b) Assuming the gun is at a height h and assuming one aims correctly, how long does it take for the water ball to hit a target a distance d away, which is at height h_t above the ground? Given those parameters, at what angle should one aim? c) Given all the other parameters and how long it took, how much momentum would it have initially (in the direction of fire) if the mass of each ball was m. If you can get hold of such a water gun do a series of experiments to see how long it does take and calculate the required momentum. Make sure that the gun launches a ball each time, not a smattering of two or more balls. d) Why does it sometimes do this? And hypothesize as to why does it do it more at 0 than 90?)

9. A basketball of mass m is bounced down at a speed v from height, h. Assuming that the collision with the floor is completely elastic and that it occurs in time T, what average force is experienced by the ball during the collision?

10. Water shooting at an angle coming out of a water fountain gets wider as it approaches the top of its path. a) Why is this so? (Note: the purpose of the second smaller nozzle is to provide impetus to check this widening so the stream does not break up and is thus more easily drunk). b) Water dripping out of a sink breaks up as it falls further; why?

11. Consider a balance with an hourglass resting on *each* side. The hour glass on the left has run out, having all the sand on the bottom. The other side has all the sand in the top. What happens to the balance as the sand begins to come out? Suppose the birds were to suddenly ball up and allow themselves to fall, without air resistance, to the ground, what would the weight be then?

12. A truck with mass M when empty is currently full of birds (N of them) and nothing else is about to be weighed. With the noise of the braking truck stopping at the weigh station, the birds, previously at the bottom of the truck asleep, suddenly wake up. Assuming each bird has mass m, what will the truck and birds weigh while they fly in the hermetically sealed truck with proper environment maintained by an automated system?

13. Draw the forces involved in holding a cup of coffee in the air. Explain which forces are equal in magnitude and opposite in direction because of the action-reaction described by Newton's third law and which are equal and opposite maintaining the coffee cup in a static place.

14. On the trip to Mars, when you are in free-fall and gravity can be neglected within the spaceship, you push on a piece of equipment: a) Explain, in terms of forces, what happens if the equipment is the same mass m as you. b) Explain what happens if you try this when the equipment is up against a wall and you use a foothold on the floor to push the equipment. Distinguish clearly between Newton's third law reaction forces and those of static equilibrium.

15. A man slaps his hands together at speed V, holding them together on impact so they cannot separate. In what ways is this collision like a completely inelastic collision of two equal mass putty balls, and in what sense is it different? Consider the nature of the bodies as well as the motion.

16. In the absence of gravity, calculate expressions for forces F'_{water} and F'_{rocket} seen from initial rest frame (i.e. before starting thrusters). How does this help us understand the partial usefulness and potential pitfalls of using: $F = \frac{dp}{dt} = \frac{d(MV)}{dt} = M\frac{dV}{dt} + V\frac{dM}{dt}$ in the rocket problem?

17. A rocket launches ejecting fuel from a first stage. Then after all the first stage fuel is spent, the rocket drops the tank that held the fuel of this first stage and proceeds to burn the second stage fuel and finally to drop that tank. Assume: the speed of fuel ejected relative to the rocket is v_p, the mass of the rocket with none of its stages attached is m_r, the mass of each empty stage is m_s, the mass of the fuel in each stage is m_f; stage one and two burn respectively for T_1 and T_2 seconds. a) Under uniform gravity, give the equation for the speed of the rocket after the last burn is completed. b) Compare this to the formula for the final speed of a rocket with no stages. Assuming the rocket's mass when filled with fuel is $2(m_f + m_s) + m_r$, where $2m_f$ is the mass of the fuel, and assuming the rocket fuel is much larger than the other two (that is, take $m_f >> m_r$ and $m_f >> 2m_s$), which is better, a rocket with two stages or none?

18. In baseball, a balk is called when the pitcher has begun a motion toward the plate and changes it. What defines the change in motion here? Is it primarily impetus or force? Explain in some detail.

19. The air is pulled down by gravity to the Earth. Its net effect is to cause a pressure of about $14.7 lbs/in^2 \sim 10^5\,Pa = 10^5\,N/m^2$ at sea level. How deep must a scuba diver go to experience double this pressure?

20. When a boat sits on a solid surface, the force of its weight is transmitted via that surface to the Earth. However, when it sits on water, the water can move out of the way. If the water is contained in some way, it can only go up the sides of the container, however slightly. The mass of the displaced water, Δm, is pulled down with a force $\Delta m\, g$. This is transmitted through the water to the bottom of the boat to hold it up. Thus, the boat must

displace its weight in water. Consider similar reasoning for a helium balloon. In particular, how much weight can a typical helium birthday balloon lift? How big a balloon is needed to lift a 100kg man?

21. Knowing the pressure in the tires of a car, how can one use the footprint, i.e. the surface area resting on the ground for each tire, to estimate the weight of the car?

22. Consider again the example discussed in Chapter 3 in which a man is stuck on thin ice and must use a toboggan to get safely to shore. The ice is so thin that if he steps out, he will fall through. a) Since the force the man exerts on the ice is actually more in the toboggan, why is it that the man doesn't fall through the ice when in the toboggan but would without it? Explain, given the mass of the man, M, the mass of the toboggan m.

23. Where is the center of mass of the Earth-Moon system?

24. Where is the center of mass of the Sun-Earth system?

Reference Frames

25. Consider a region with two bodies; body A is at rest and has mass m and body B has a mass M moving at speed u to the right. a) Consider an observer moving at a constant speed $\mathbf{v}$ to the left. Describe what impetus each object appear to have from the moving frame. How is it that body B striking body A in the rest frame would behave the same if one did the same experiment in the moving frame with the same relative motions?
b) Describe the impetus that each object appears to have as seen from a uniformly accelerating frame at a rate a to the left that begins accelerating from zero speed.

26. In a place, where gravity and all other forces can be neglected, a machine releases (without giving it any relative momentum) a ball once every T seconds. In its rest frame there are no forces. Explain what one sees in a frame moving to the right at speed $\mathbf{v}$. In particular, in what *extended* sense can we say the machine has a force acting on it. We must be careful with such different uses of the word force. For example, when we say no new apparent forces appear in moving to a new *uniformly* moving frame, but they do in accelerated frames, we obviously exclude such meanings.

27. Consider two *curves* that are identical with each other until a certain point at which they diverge. a) Mathematically, what must be discontinuous at this point? (*Hint*: think of a Taylor series expansion) b) Consider a subclass of such curves; consider one of the two curves as a trajectory of a moving body (motion in category of place) and the other curve as the shape of a hill that constrains, until a certain point, that body to move down the frictionless hill under the force of gravity. (For convenience, assume the body starts on the top of the hill with negligible speed.) What is happening in terms of the motion and causes of motion of the body that is revealed by the mathematical (category of quantity) differences in the curves?

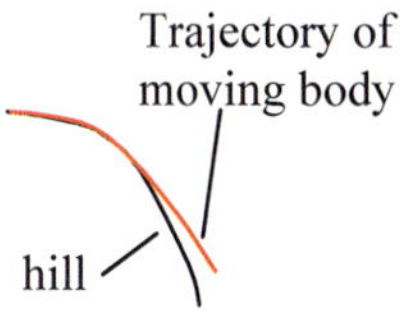

Trajectory of moving body

hill

Note: this problem exposes the analogies we make between the mathematics itself and our use of it to describe motion in the category of place (and indirectly its causes).

28. On Mars (where $g_{mars} = 3.7m/s^2$), an elevator of mass M (when empty) is pulled up by a cable at an acceleration of a, an automated (massless) scale indicates that a small package of mass m resting on the scale on the floor of the elevator has weight W. What tension does the rope have? Draw a diagram labeling forces. Solve this problem both in a) an accelerated frame and b) in the rest frame of the lab. c) Taking the lab to be actually at rest, explain physically, i.e. in terms of causes, what is happening.

Atwood and other Machines

29. A man of 80 kg climbs a massless cable that passes over a massless frictionless wheel and connects to a 200 kg piece of steel. What is the least acceleration the man must have to lift the piece of steel off the ground? What is the tension in the cable?

30. An Atwood machine is in equilibrium with two masses of 1 kg each. How much mass must be moved from one to another so that the downward acceleration of the heavier mass is 100 times slower than the free-fall acceleration on Earth? On Mars? note: $g_{Mars}=3.7m/s$

31. A block of mass M is pushed at constant speed up a ramp inclined by an angle θ by a force directed parallel to the ground. What is the strength of the force?

32. Equipment for exploring Mars needs to be designed to take some wear and tear and dropping. It is decided that certain equipment must be able to withstand a fall down mount Olympus of about 10 meters. Mount Olympus is X feet tall and X feet wide. The equipment is 50 kg and, to get worse case values, we assume that the equipment falls on a place where the hard rock is lightly dusted with Martian soil making a low friction surface. What is the rate of acceleration of the equipment? How fast is it moving at the bottom of the ramp? If it hits a pile of sand that stops it in Y seconds how much force must the equipment be able to take?

33. Two blocks are connected by a massless cord that passes over a frictionless immovable rod. One of the blocks is placed on a fixed frictionless ramp inclined at an angle θ from the horizontal. a) What is the ratio of the masses needed for the system to be in equilibrium? b) Find the expression that describes the acceleration of the system.

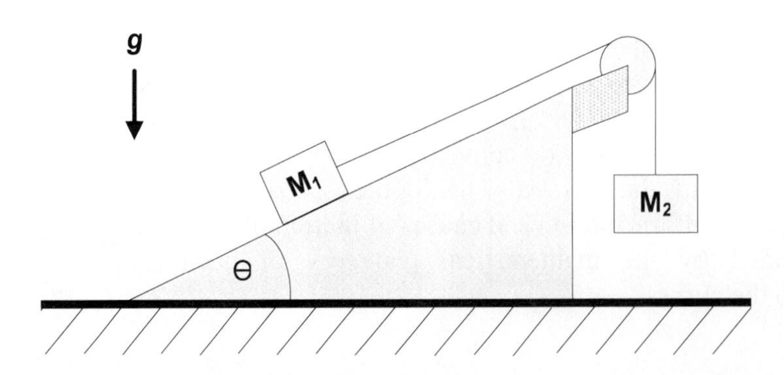

34. A block is pulled by a cord inclined by a fixed angle θ along a frictionless surface. What is the greatest force that can be applied by the cord before the block is lifted off the Earth?

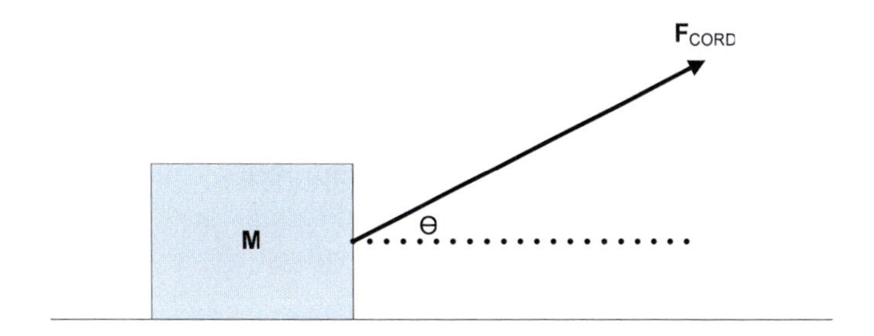

35. a) Idealize a section of rope as a series of four connected equal masses. Imagine the rope is pulled along a frictionless surface by a constant horizontal force. Determine the forces acting at the interface of each section of mass.
b) Determine the forces acting on mass element 3 in the limit that its mass goes to zero and the others remain the same. What are the forces on a body in the limit that its mass goes to zero even while it is moving?
c) Assume the rope has a total mass M, length L, and mass per unit length of μ. Find the force as a position of length along the rope as in part a) in the limit that the number of mass segments in the rope becomes unbounded.

Note: This problem shows the meaning and importance of the idea of a massless string or rope that is used so often in Newtonian physics. When we complete the limit and think we have an actual massless rope, we make a being of reason (a mental construct), a casing, in which to carry our primary thought around. The primary idea is that we can generally conceive a rope even less massive than the one we currently have, making the effect of the rope as insignificant as we like. Again, it is only the actually completed limit that is the being of reason, because massless ropes cannot exist outside the mind; in reality, one can only go so far before one does not have a rope at all.

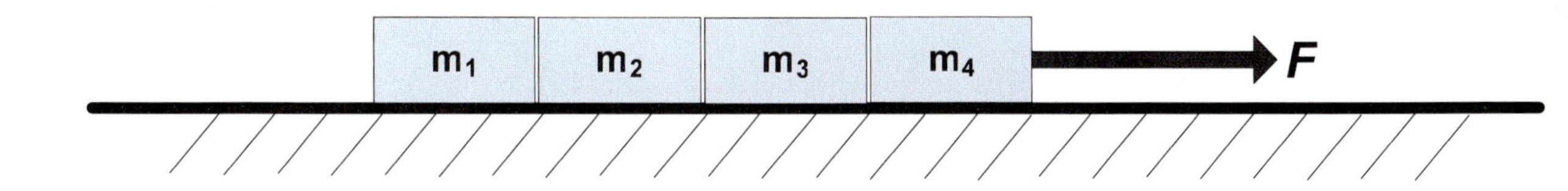

(Problems continued on next page)

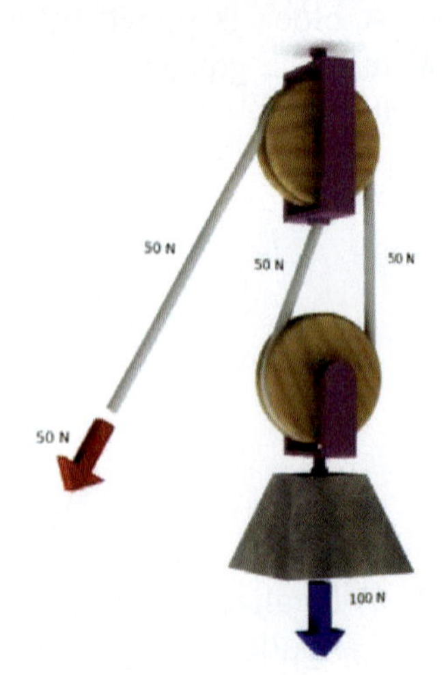

36. Consider the Atwood machine shown Figure 4-14.
a) What mechanical advantage does it offer. Hint: this Atwood machine is, in a way, an upside down pulley. Draw the pulley along with the force diagram that make clears the mechanical advantage. b) Given the pulley shown in the figure[20] to the right, draw a force diagram and explain the mechanical advantage provided by this pulley. c) Draw a four pulley system that would reduce the mechanical force need to raise a weight by a factor of four. Indicate the key forces involved.

37. a) What are the hurdles involved in an experiment using the Atwood machine of the text to measure the *g*? b) Come up with an Atwood-like instrument of your own that could be used for measuring acceleration due to Earth's gravity at the surface. c) Try your idea and the text example in *Interactive Physics*.

Mixed Problems

38. Insert a straw into a glass of water most of the way, leaving a length of the straw h filled with air as illustrated below. Cover the top of straw with your finger tip. At this point, the air in the top of the straw is: $P_{top} = P_{air}$. Now, lift the straw the out of the water. Note that h decreases by Δx. Calculate: Δx.

---

[20] Figure used under GNU free documentation license

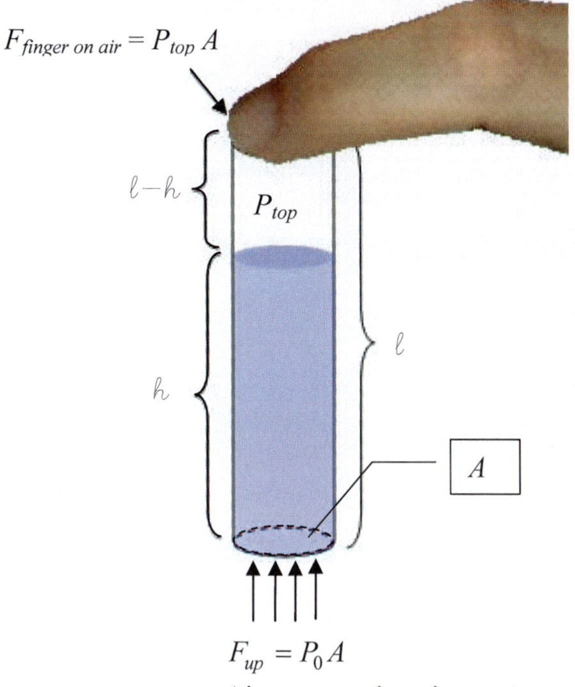

Air pressure keeping water up.

39. On icy roads, why is it best to pump brakes (anti-locking brakes do this automatically as needed) rather than hold the brakes down?

40. Consider an empty railway car of M_0 moving at speed V_0 that is pushed with a force F while sand (which is at rest) is being dumped in vertically at a constant rate of $\frac{dm}{dt} = \dot{m}$. a) What force is needed to keep the car moving at uniform speed V_0? b) Suppose the sand enters at a speed u at an angle θ from the vertical against the direction of the car's motion, what force is needed then? c) Explain the situation b) result from an inertial frame moving at $u \sin \theta$ against the direction of the car.

41. a) If a bullet of mass m traveling at speed $\mathbf{v}$ hits a man of mass M wearing a bulletproof vest and the bullet is stopped by the vest, how much momentum is given to the man? b) Assuming the bullet stops in Δt seconds, how much force did the man experience? c) Assuming the area of the vest is A and that force was evenly distributed over it, how much pressure did he feel? Redo the parts of this problem with the bullet hitting Superman, where the bullet bounces off elastically, except for part b) use the area of the bullet A_b.

42. Suppose that, as in the Chapter 3 problem, a car of about the same size as your car and moving at speed V hits your car from behind; to avoid whiplash, should you press your brakes or not? Explain this time by modeling the collision surface by a single spring that pushes back in standard fashion but gets stuck at its most compressed position, not returning to an uncompressed state.

Chapter V

Energy and Work

Introduction

Newton's three laws of motion are often adequate to explain motion in ordinary experience. However, there is more implicit in those laws than we've unpacked so far. They contain a very important concept called *energy.*

In common daily usage, energy means the power behind some change, especially a motion. We say a man doing heavy physical labor is using a lot of energy, but one sitting on the couch is not. At the end of the day, assuming they are equally fit, one has used much of his reservoir of strength, i.e. his energy; the other has not. We might then think of using the resulting amount of activity to measure the energy. This is just what we will do.

Given what we've seen in our study of momentum and force in the previous chapters, we should not be surprised to find that the study of the simple case of motion of a body will give us a very general empiriometric concept of energy. The measure of the degree of locomotive activity of a body (or bodies), in a sense to be defined, is called its (their) *Kinetic Energy.*

To develop this notion, consider again body A coming into contact with a body B (Figure 5-1). The impetus of A (symbolized by the green intensity) brings it into collision with B whereupon A exerts a force on B, and vice versa. The force from body A on body B intensifies the impetus in B. Yet also, for whatever force body A applies to body B, an equal and opposite force is exerted by B on $A,$ and the increase in impetus given to body B has a corresponding equal and opposite intensification of impetus in body A-- Newton's third law concisely describes this mutual exchange.

Remember that the net acting force applied by a body to another is equal to the time rate of change of the momentum (intensification of impetus) in the body it is acting upon. In the x-direction, we have:

5.1 $$F_x(t) = \frac{dp_x}{dt} \quad \text{(Newton's 2nd Law)}$$

Breaking up the action of the force of one ball on the other into time intervals, Δt, we note that at the end of each such interval, there is a jump or incremental intensification of impetus; in Chapter 4, we called the cause of this jump an "impulse." If we integrate equation 5.1 during the *time* from t to $t+\Delta t$, we obtain the equality between total impulse caused by the action of the *affecting* body and the change in momentum (change in the intensity of impetus) of the *affected* body.

5.2 $$\int_t^{t+\Delta t} F_x(t)\, dt \equiv \text{Impulse}_x \approx F\,\Delta t \text{ (from acting body)}$$
$$= \int_t^{t+\Delta t} \frac{dp_x}{dt} dt = \int_{t_0}^{t_1} dp_x = \Delta p_x \text{ (to body acted on)}$$

Thus, whatever body acts by force F on a second body induces a reciprocal equal but opposite reactive force by that body on the first body causing, as in the case of the collision (Figure 5-1), a decrease in momentum in the first body that is exactly the same in magnitude, Δp_x, as the increase in momentum caused in the second body by the first.

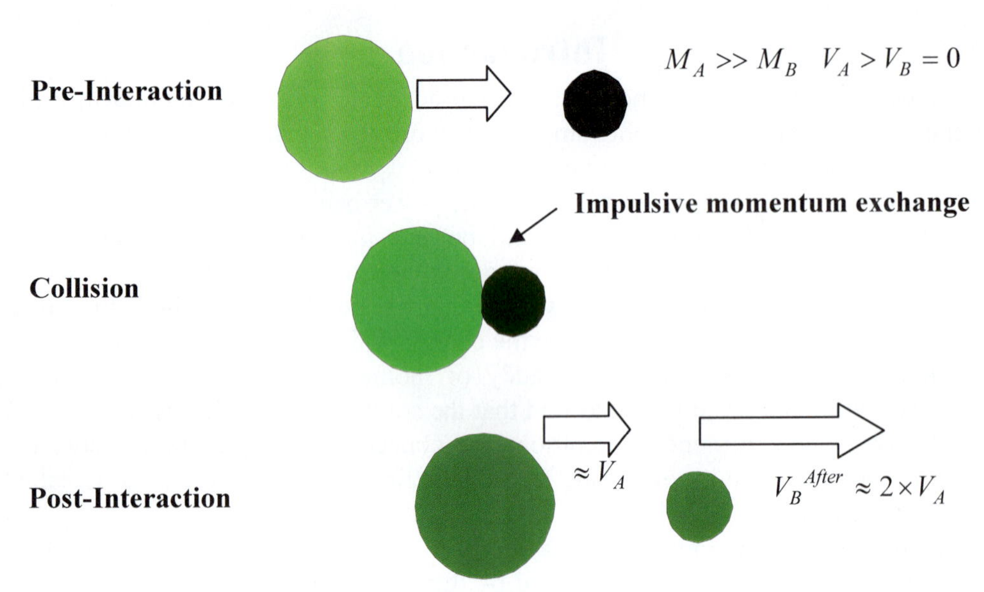

Figure 5-1: Interaction between bodies *A* and *B*. *(Top)* Before the collision the large ball moves at speed V_A , while the smaller ball begins at rest. *(Middle)* Ball *A* applies a force *F* on ball *B*, and ball *B* applies an equal opposite force $-F$ back on *A*. This continues for a period of time so each ball applies the same magnitude of ***impulse*** to the other *(Bottom)* Thus, when the collision is over, the total momentum is unchanged, but the large ball moves slightly slower than, but still very close to, its originally speed V_A , while the small ball moves off to the right at about $2\ V_A$. In terms of momentum, the larger ball loses momentum Δp to the smaller one.[1]

Example 5-1: A 140 *gm* baseball pitched at 100 *km/hr* rebounds from a 560 *gm* bat at 220 *km/hr*. What is the impulse applied by the bat to baseball? What is the impulse from the baseball to the bat?

Answer:

1. Convert the initial and final velocities of the baseball.

$$\boldsymbol{v}_1 = -100\frac{\text{km}}{\text{hr}} \times \frac{1\text{ hr}}{3600\text{ sec}} \times \frac{1000\text{ m}}{1\text{ km}} \sim -28\frac{\text{m}}{\text{sec}}$$

$$\boldsymbol{v}_2 = +220\frac{\text{km}}{\text{hr}} \times \frac{1\text{ hr}}{3600\text{ sec}} \times \frac{1000\text{ m}}{1\text{ km}} \sim +61\frac{\text{m}}{\text{sec}}$$

2. *Impulse applied from bat to baseball* $= I = \Delta p$

$$I = \Delta p = p_2 - p_1 = m\boldsymbol{v}_2 - m\boldsymbol{v}_1$$
$$\approx 0.14\,\text{kg} \cdot \left(61\,\text{m/sec} + 28\,\text{m/sec}\right)$$
$$\approx +12\ \text{kg} \cdot \text{m} \cdot \text{sec}^{-1} = 12B$$

3. From Newton's third law, the impulse applied from the baseball to the bat will be *-12B*.

---

[1] The colors were calculated based on initial values: $M_A = 4, M_B = 1, V_A = 1, V_B = 0$. The approximation is only true to one digit accuracy with these numbers. Of course, the impetus indicated by the color is thus only approximate as well.

Now, the capability of a body in motion to act on other bodies is, in some way, determined by the intensity of its impetus, as well as its receptivity to the action of the impetus (i.e. the mass), its surface forces and those of the other bodies.

Further, we can see, through careful study of collisions, that a body's kinetic (i.e. moving) activity is not given *only* by how fast it goes, or simply the intensity of the body's impetus, or just its mass; rather, it has a capacity for action that is tied up in its various properties.

The kinetic energy, this activity we seek to understand and define, of a body can be measured by its effectiveness in causing activity in other bodies. For concreteness, suppose an object of mass m is at rest on one side of a room and then receives a very small, sharp impulse with a consequent change in momentum $\Delta p = p$. Its speed after the impulse is $v_0 = p/m$ (because there was no initial momentum). Assuming no other forces act, the object moves across the room at this speed. On the far side of the room, it then collides and interacts with some other object(s) and ends up at rest. It imparted *to* these other objects the total momentum (intensity of impetus in the given direction) it possessed while traveling across the room.

Clearly, as a body moves from place to place, the activity it *can* cause in other bodies follows with it. Moreover, a body with a given impetus and mass clearly brings more activity to other places and does so more rapidly than a body of the same mass with less impetus.

In this way, we come to distinguish two types of activity: 1) The rate of change of place of the body itself, which is given by its speed; 2) The rate of change of place of the impetus (category of quality) of a body, which is, in turn, an indication of the body's capacity to cause (or change) impetus in another body some distance away. Because the latter definition incorporates the qualitative cause, it is a more complete definition of activity. It also makes sense, because when we talk about changing place of something, we want to know what it is that is changing place; we know it's a body, but to specify the motion further, we need to know what type of body. Now, a body's type is revealed by its essential properties. Here we are interested in dynamics, and no property is more relevant to dynamics than the impetus of the body.[2]

Note also the iterative nature of the definition which can keep referring to the next possibility of interaction, for instance, down a line. The last body given an impetus can bring that body near enough to the next body for its surface forces to cause impetus in it and so on. This is important because, of course, interactions occur and are an essential part of dynamics, so that a good definition should, at least implicitly, reference the fact that any given body will eventually interact with other bodies or will undergo reactions within itself.

With these insights, we can define and discuss in detail the concept of kinetic energy of a body (or system of bodies—remember most bodies can be conceived of as a system of bodies), which, from the above, is roughly the amount of activity of a body as seen through its ability to affect its environment.

---

[2] Technically, a change of a quality itself is called an alteration, while the change of place of a quality is considered an "accidental" change, because it is not the quality itself that is changing but the substance, with the given quality, that is changing place. However, here we wish to focus on the dynamics alone so, as we have done before, we leave out such considerations.

The Definition of Kinetic Energy

a. General Definition

Kinetic Energy can be defined as a measure of the **rate of transfer of the intensity of impetus** of a body (or bodies) across a distance (i.e. rate of transfer of momentum). That is:

5.3 Kinetic Energy ≡

$$\text{“Rate of transfer of intensity of impetus”} \propto \begin{pmatrix}\text{Speed at which momentum} \\ \text{is communicated } \textbf{\textit{to}} \text{ region} \\ \text{of interaction}\end{pmatrix} \times \begin{pmatrix}\text{momentum imparted} \\ \textbf{\textit{within}} \text{ region of interaction}\end{pmatrix}$$

$$\propto \boldsymbol{v} \cdot \mathrm{p} = \frac{\mathrm{p}}{\mathrm{m}} \cdot p = \frac{p^2}{m}$$

Clearly, a *larger* impulse applied to a body, means a *larger* momentum, *p*, is given to it, which in turn implies the body has greater capacity to act on other bodies at the other side of the room, hence larger energy. Greater intensity of impetus means greater speed and thus its activity is communicated across the room *faster*. Both these facts are taken into account in the equation. An intuitive example that brings these facts to the fore will help in understanding the definition. Compare a strong man pushing a refrigerator across a room at top speed to a toddler doing the same with a toy plastic refrigerator. Though they go at the same speed, you'd much rather be hit by the child's toy than the fridge, for the toy will have less impetus. Although each gets to the other side of the room at the same rate, one has much more ability to do damage than the other. Speed is not all that is relevant in activity; what is moved at that speed is also important.

A different type of example will help us better understand some of the consequences of the above definition. Consider two particles, say particle *A* and particle *B*, both of mass *m* shown in Figure 5-2a. *A* has momentum *p* and *B* has momentum *2p*. Let us define particle *A* to have energy *E*. Cut the *2p* particle (*B*) in half giving two mass $m/2$ particles moving at the same speed as before but now not connected. Because impetus divides uniformly according to mass in a body of constant speed, each sub-particle will have equal momentum *p*, as shown in Figure 5-2b. Since particle *A* and the two sub-particles of *B* now each have the same momentum and since $E \propto p\ \boldsymbol{v}$, the speed alone determines the energy. Now, each of the new sub-particles has twice the speed, and hence twice the energy, of particle *A*. Hence, the particle *B* (composed of the two sub-particles) has a total of four times the energy of the particle *A*. That is, $"Energy\ of\ B" = 2(2E) = 4E$. This approach can be generalized to show, from a different angle, that the activity, the energy, goes as the square of the momentum.

Similarly, given two bodies of equal speed, but one with twice the mass of the other, we know that the one with twice the mass has twice the intensity of impetus and thus requires twice the work, twice the energy, as we'll see below, to move it at that speed.

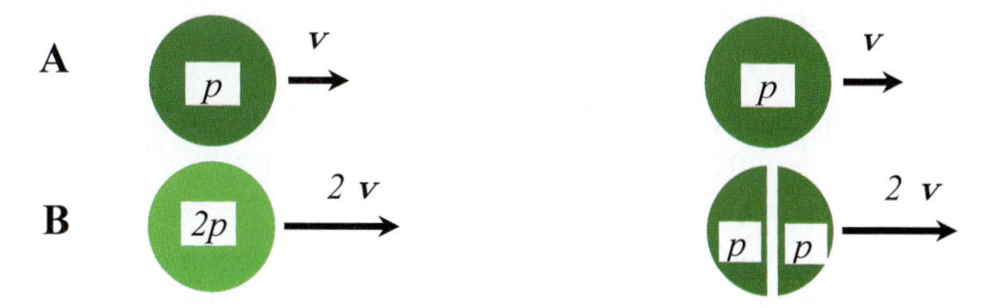

Figure 5-2: *(left)* **a.** Two bodies of equal mass, one with twice the momentum of the other moving to the right *(right)* **b.** We divide the second mass into two equal mass parts. Note that the ***net*** momentum can't change (since there are no external forces), nor can the net speed. Hence, the momentum divides equally between the two halves. Since all three massive bodies that we've defined now have the same momentum, the energy, the rate of transfer of the impetus, is simply determined by the speed. Thus the energy of each can easily be compared to the others. In particular: $E_A \sim pv, E_B = E_{B\,left} + E_{B\,right} \sim p(2v) + p(2v)$. Thus, the relation of the energies of the two original particles is: $E_B = 4E_A$.

b. A Complete Quantitative Definition: A Body which Spends All of Its Impetus

To make the relation 5.3 above an equality rather than a proportionality, we consider "draining" the impetus of a body by successive impulses. We could drain it all at once, but this is not the only way to drain it; there are many ways. Indeed, a body of given mass and impetus has a given amount of activity (energy) that could just as well be expressed as the activity of a group of particles of varied mass and momentum, which indicates that there are many ways that the energy of the original body can be expended. Our definition should incorporate in some generic way the body's capacity for interaction and "spending" of its kinetic energy in various ways.

Now, there is something to be said for looking at the impetus as acting on the smallest mass available for the smaller the mass, the more the impetus acts freely, ultimately reaching its maximal capacity when activated in the smallest mass particle that ever exists. The body cannot give *all* its impetus to a single small particle, for this would clearly create energy by suddenly allowing the given intensity of impetus to travel much more quickly. In fact, in any collision the first little bit of momentum must be drained before the next bit is drained. So, we first consider draining a little momentum at a time, until all the momentum is gone. (Note that in interacting with a succession of individual, very small masses, the body would lose only a little of its momentum at a time. In following this reasoning in detail, we might think of doing this by allowing the mass to hit little tiny balls one at a time (the smaller the balls, the less momentum that is drained)—see end of chapter problem).

In particular, to start, consider again our object of mass m starting from rest and receiving a large initial impulse in some Region 0. After the initial impulse, the body will have a large velocity $\boldsymbol{v}_0$ given by:

5.4
$$\boldsymbol{v}_0 = p_0 / m$$

Next, assume the following happens: following Figure 5-3a, the object crosses through a Region 1 with no interactions and thus at constant speed. The object then enters Region 2 and imparts a fixed portion of its initial impetus, p_0 / N, where N is an integer, to objects in that region. It then crosses Region 3 without interaction. Entering Region 4 it imparts the same fixed portion of impetus as before, and so on. The object loses a degree of its momentum in each region of interaction corresponding to the impulse it imparts to that region. This continues through a set of $2 \times N$ successive regions until the initial impetus is completely transferred to N regions of interaction.

Before entering a given even numbered region *2i* of interaction, the body has an amount of momentum, p_0 / N, which it will lose upon crossing that region. This momentum is communicated to that region at speed $\boldsymbol{v}_{2i+1}$ through the immediately preceding region of non-interaction. Figure 5-3b shows how the initial momentum and energy of the body is imparted to each of the regions of interaction. The horizontal axis is divided into three equal-size parts corresponding to three interaction regions (regions 2, 4, 6-- the *2i* regions), each of which takes away momentum $p_0 / 3$ from the moving body. The vertical axis shows the speed which carried the given $p_0 / 3$ through the immediately preceding non-interaction region (regions 1, 3, 5-- the $2i-1$ regions). Notice that the shaded area of each section of this plot corresponds to the energy lost in the indicated interaction region (i.e. regions 2, 4 or 6).

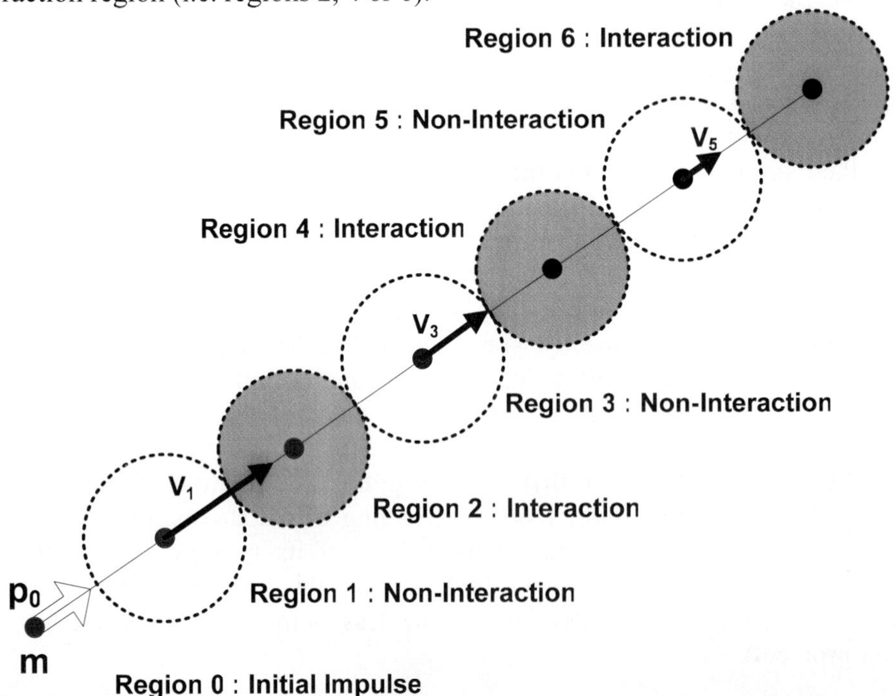

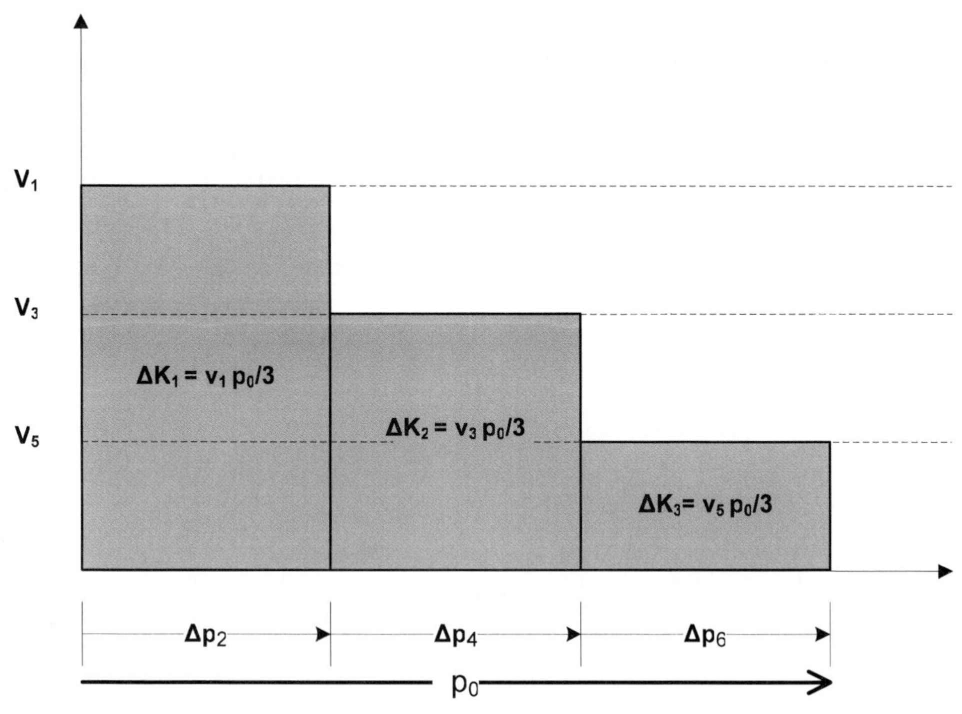

Figure 5-3: *(Top)* **a.** A body starts in region zero with initial momentum p_0 and moves through seven different regions. The body interacts in all *even* number regions and moves unimpeded through all *odd* numbered regions. A fixed portion of the body's momentum is lost in each interaction region. *(Bottom)* **b.** The sequence of lost momentum communicated by the body to the successive even numbered regions is plotted against the speed at which that portion of the lost momentum is communicated. The area of each region on the plot is the kinetic energy of the body lost in the indicated region. Since the body is moving slower when it deposits the last $\Delta p = \dfrac{p_0}{3}$ of momentum, the loss of that last bit of momentum is associated with a lower energy loss.

Thus, a measure of the rate of transfer of the intensity of impetus through a region of non-interaction $2i-1$ to region $2i$ will be

5.5 "activity measure" = speed at which activity is communicated *to* region $2i$
× intensity of impetus imparted *within* region $2i$

$$= \boldsymbol{v}_{2i-1}\ \Delta p_{2i} = \boldsymbol{v}_{2i-1}\left(\frac{p_0}{N}\right)$$

A total "measure of the activity" imparted by the object over all regions of interaction will be:

5.6

$$\text{"total measure of activity"} = \boldsymbol{v}_1 \Delta p_2 + \boldsymbol{v}_3 \Delta p_4 + \boldsymbol{v}_5 \Delta p_6 + \ldots + \boldsymbol{v}_{2N-1} \Delta p_{2N}$$

$$= \sum_{i=1}^{N} \boldsymbol{v}_{2i-1}\, \Delta p_{2i}$$

$$= \sum_{i=1}^{N} \boldsymbol{v}_{2i-1} \left(\frac{p_0}{N}\right) = \sum_{i=1}^{N} \left(\frac{\left(1 - \frac{i-1}{N}\right) p_0}{m}\right) \left(\frac{p_0}{N}\right)$$

$$= \sum_{i=1}^{N} \left(\frac{N-i+1}{N}\right) \left(\frac{p_0}{m}\right) \left(\frac{p_0}{N}\right)$$

$$= \left(\frac{p_0^2}{m}\right) \left(\frac{1}{N^2}\right) \left\{ \sum_{i=1}^{N} N - \sum_{i=1}^{N} i + \sum_{i=1}^{N} 1 \right\}$$

$$= \left(\frac{p_0^2}{m}\right) \left(\frac{1}{N^2}\right) \left\{ N \sum_{i=1}^{N} 1 - \sum_{i=1}^{N} i + \sum_{i=1}^{N} 1 \right\}$$

$$= \left(\frac{p_0^2}{m}\right) \left(\frac{1}{N^2}\right) \left\{ N^2 - \frac{1}{2} N(N+1) + N \right\}$$

$$= \left(\frac{p_0^2}{m}\right) \left(\frac{1}{N^2}\right) \frac{1}{2} N(N+1)$$

$$= \left(\frac{p_0^2}{2m}\right) \left(1 + \frac{1}{N}\right)$$

In writing the third equality, we used the fact that after each region of interaction the momentum is reduced by p_0 / N.

In the limit as the number of regions becomes arbitrarily large $(N \to \infty)$, and thus as the size of the impetus drained off in each interaction (p_0 / N) goes to zero, we have:

5.7 $$\text{"total measure of activity"} = \lim_{N \to \infty} \left(\frac{p_0^2}{2m}\right) \left(1 + \frac{1}{N}\right) = \left(\frac{p_0^2}{2m}\right)$$

This measure of the object's activity, the rate of transfer of imparted impetus, is then called **kinetic energy** of motion.

Thus, in the continuous limit the kinetic energy, K, is:

5.8 $$K \equiv \int_0^p \boldsymbol{v} \cdot d\boldsymbol{p} = \int_0^p \frac{p}{m}\, dp = \frac{p^2}{2m}$$

We can also write the kinetic energy in terms of the speed :

$$K \equiv \int_0^p \boldsymbol{v} \cdot d\boldsymbol{p} = \int_0^v m\boldsymbol{v}\, d\boldsymbol{v} = \frac{1}{2} m\boldsymbol{v}^2$$

To get the second equality, we assume a constant mass. Thus, we get:

5.9
$$K = \frac{p^2}{2m} = \frac{1}{2} m v^2$$

Note that kinetic energy depends on the square of the momentum or speed, so it can *never* be negative. As we will show more clearly below, for any given reference frame, kinetic energy also does not depend on the direction of travel (i.e. it is a scalar).

The above example of a series of regions of interaction gives the kinetic energy in terms of the accomplished activity of a moving body brought to rest. Yet again, kinetic energy represents a measure of the activity the body has and *can* impart due to the impetus it possesses (which also implicitly includes how quickly this *capacity for activity* is brought from one place to another).

At this point, with the experience of the above worked example under our belt, it's helpful to recap. A body has its own proper kinetic activity of which the kinetic energy is a measure. This activity is caused by its quality of impetus, but its action is limited by the receptive quality of the mass,[3] which only allows the body to move so quickly and no more (or less). As we've seen many times by now, the same quality in another body of larger mass would result in less movement. Said another way, speed is an indication of how effective the impetus is in carrying out its activity; impetus is what causes the activity. So, a body's kinetic activity as well as its ability to cause kinetic activity in other bodies is related to both mass and momentum as equation 5.9 clearly shows.

The factor of ½ comes about because of the way the impetus can be given away and from the way the impetus comes to cause the body's motion. Each little piece of impetus, *p*, causes an increase of speed, say v, but the impetus generally doesn't come all at once, but ramps up through time to its total value *p*, accomplishing an increase of speed from zero to v, so that speed that the whole *p* is responsible for is really an average of what happens in that interval of time: $v_{average} = \frac{0+v}{2} = \frac{v}{2}$. In other words, different parts of the p are responsible for different parts of the motion. The impetus of intensity p did the "work' of getting the body *from* 0 *to* v and we don't want to forget that. Hence the product representing the energy is: $\frac{v}{2} p = \frac{1}{2} m v^2$ (see footnote 11). This showed up in the treatment of kinetic energy in this section because we broke the momentum into parts. Notice that we've included in our measure of activity, the energy, activity that is only implicit in the motion of a body already moving uniformly; that is, we always include the potential for the body to be slowed down and the fact that it has already been accelerated at some point.

Thus, the kinetic energy of an object reflects, in a limited way, how the object can affect the impetus of its environment through its locomotion over a *certain* distance. Yet because of the action-reaction principle, the object's activity on its environment associated with its locomotion can only occur when a net force acts on it.

---

[3] Note mass is a relational quality dependent on the media it is in. Typically, when the media is what we call vacuum, or in this book sometimes plana, it is simply treated as an intrinsic property, because the reference to the vacuum is considered standard and implicit. When in other media, we quickly come across the concept of effective mass, which directly incorporates effects in addition to the vacuum or instead of (the full sense of this requires some explaining outside the scope of this text).

In summary, we can say that *kinetic energy is a measure of a body's locomotive activity (rate of transfer of its impetus) including*[4] *how it generally attains its given state of motion and how the body generally* ***can act*** *on its environment through its act of locomotion.*

> In the *MKS* system work and kinetic energy are measured by a unit called the "Joule" with units of $kg \cdot m^2 \cdot s^{-2}$. A quantity of $1\,Joule$ is the energy needed to accelerate a $1\,kg$ mass from rest to a speed of $1\,m/s$. Another common unit used in atomic or subatomic physics is the electron volt: $1\ electron\ volt = 1\,eV = 1.6 \times 10^{-19}\ Joules$.

Example 5-2: A $145\,gram$ baseball is moving at $130\,kph$ toward home plate. What is its kinetic energy?

First we put these quantities in a consistent set of *MKS* units.

$$m = 145g = 145g \times \frac{1kg}{1000g} = 0.145kg$$

$$v = 130kph = 130kph \times \frac{1000m}{1km} \times \frac{1hr}{3600s} = 36.1m/s$$

The kinetic energy of the ball is:

$$K = \frac{1}{2}mv^2 = \frac{1}{2} \cdot 0.145\ kg \cdot \left(36.1\ m/s\right)^2 = 94.5\ Buridan\,m/s = 94.5\ Joules$$

Kinetic Energy, Pressure and Temperature

Though abstract, kinetic energy is important in all everyday physical phenomena. As an example, consider the following simple thought experiment that illustrates, though imperfectly, the connection between kinetic energy and pressure, such as that in a balloon or a basketball.

Example 5-3: Consider a solid cube with edges of length L and perfectly elastic walls (i.e. no kinetic energy is lost to the wall by collisions). A microscopic particle of mass m inside the box moves with momentum p_0 in a direction parallel to one set of edges of the cube and bounces perpendicularly off the immobile walls of the cube.

---

[4] These two consequences are "included," that is implied, in the definition because we take "transfer of impetus" to mean transfer of impetus from point A to point B in so far as the impetus has the ability to interact and cause new activity at point B in other bodies or, in the first consequence, to receive impetus at point A to increase the body's speed so that it may cause activity in the future at point B. The kinetic energy of a body, which is ultimately an expression of the body's impetus, mass, and forces, is always implicitly tied up with past, present or future interaction with other bodies.

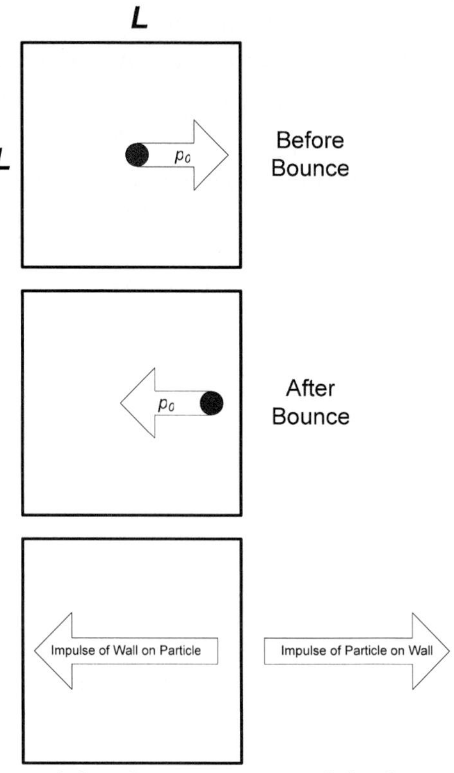

Figure 5-4: A single particle of mass *m*, a model of a molecule, bouncing around in a cubic box causes a force on the sides of the box off of which it bounces.

A two dimensional cross section of the box with a particle moving in one direction is shown in Figure 5-4. At each impact with the wall the particle has a change in momentum equal to $-2p_0$. Thus, an impetus of $+2p_0$ is imparted by the particle to the wall. The particle crosses the cube in a time $t_0 = L / \mathbf{v}_0$. The time between impacts on a wall is *2t.* Thus, the *average* force of the ball acting on the wall over time is given by:

5.10 $$\langle F \rangle = \frac{\Delta p}{\Delta t} = \frac{2p_0}{2t} = p_0 \frac{\mathbf{v}_0}{L} = \frac{{p_0}^2}{mL} = \frac{2K}{L}$$

This equation is illuminated by our understanding of kinetic energy, *K*, as a rate of transfer of momentum, because, given a certain distance *L* to traverse, we can get the rate of transfer of momentum for *that distance, namely the momentum per unit time,* simply by dividing *K* by *L* as shown in the equation; by contrast, to get the numerical factor "2," we have to specialize to the particular case of bouncing off the wall as done above.

Now, suppose you had *N* particles randomly distributed in the box all moving parallel to one side and always bouncing perpendicular to one face. The average force on the impacted side will be $\langle F \rangle = 2 \cdot N \cdot K / L$.

The pressure *P* is defined as the force per unit area acting perpendicular to a surface; you have seen it on your basketball or your car tire on which there is a specification in *pounds per square inch*. In this case, the area of the side of the cube is L^2. So the "pressure" (force per unit area) imposed on a wall will be:[5]

$$P = \frac{\langle F \rangle}{L^2} = 2 \cdot \frac{N}{L^3} \cdot K = 2 \cdot \boldsymbol{K}_{Total} \tag{5.11}$$

$\boldsymbol{K}_{Total}$ (i.e. bold K) is the total kinetic energy per unit volume (kinetic energy density) in the cube. From this we see that the pressure felt by the wall is proportional to the kinetic energy within the box. What we sense on a "macroscopic" level (i.e. pressure) is indicative of the aggregate activity of the microscopic particles.

Furthermore, we also have

$$\begin{gathered} P = \frac{\langle F \rangle}{L^2} = 2 \cdot \frac{N}{L^3} \cdot K \\ \Rightarrow \\ PV = 2NK \end{gathered} \tag{5.12}$$

$$\textit{where } V \textit{ is volume of cube}$$

Now, temperature is a quality associated with the random motion of the parts of a system; in fact, temperature is associated with the average kinetic energy of particles in a volume. That is, the K given is a rough substitute for temperature, which we represent by T, and thus we come to a very important formula in thermodynamics. It is called the ideal gas law: $PV = NRT$, where R is a constant that we have not determined here.

What does the ideal gas law have to do with every day life? It implies, for instance, that if we cool down a car tire, unless the tire contracts proportionally in volume, its pressure will go down. Check your tire pressure in the winter! The concept of air pressure (and its place in the ideal gas law) is just one important example of the essential place of the concept of energy, the measure of the ability of a body or bodies to act.

Conservation of Energy

Historically, the concept of energy and momentum were not clearly distinguished until James Joule and Von Mayer connected it with temperature. Leibniz (see history box) argued that energy was the most important concept. Newton thought, however, that it was a phenomenon only of elastic collisions and thus of minimal importance. The experiment that made clear the importance of energy is shown in Figure 5-5 below. In this experiment, the kinetic energy of a body moving in water was shown to raise the temperature of the water. The kinetic energy was made available as heat rather than in moving the whole body.

---

[5] The example problem leaves out many details important for understanding a real gas. Such an analysis for a so-called "ideal" gas gives $P = \boldsymbol{K}_{Total}/3$. Thermodynamics is that part of physics that studies the relationships between heat, energy, pressure, temperature, etc. Statistical mechanics shows, in a way not unlike here, how the dynamic activity of microscopic particles (such as their kinetic energy) gives rise to the macroscopic phenomena (such as pressure) of a gas.

Gottfried Leibniz (AD 1646-1716) was born July 1 in Leipzig Germany. Independent of Sir Isaac Newton, He invented the differential and integral calculus. He discussed mathematics, physics, logic, history, law, and theology.

He argued against Newtonian's that it was not $m\ v$ that was important but $m\ v^2$ which he called *vis viva* (Latin for: *living force*). Leibniz said, "According to (Newton's) doctrine, God Almighty wants to wind up His watch from time to time: otherwise it would cease to move. He had not, it seems, sufficient foresight to make it a perpetual motion." For Newton, Leibniz's *vis viva* was nothing more than an interesting quantity which remained constant in elastic collisions and therefore had no universal physical import. At the time, it seemed to be an either/or issue, but we know now that *both* momentum and energy are conserved and thus important. It was not until James Joule (1818-89) and Julius von Mayer (1814-78) that the energy concept was really accepted; In fact, the unit of energy is now the Joule in his honor (Mayer unjustly seldom gets any credit).

Joule heated water by spinning a wheel as shown in figure below, showing that the temperature of the water went up by kinetic activity of a rotor.

Leibniz so admired one of the manuscripts of the Oxford calculators of Middle Ages that he wanted it republished in his day. He thought it was by Richard Swineshead, though it was by another Oxford calculator, Ricardus of Ghlymi Eshedi.

In natural theology, Leibniz made famous, once again, the question: "Why is there something rather than nothing?" It is a primal question that makes one realize that all of creation exists, but it did not have to. Unfortunately, Leibniz's philosophy generally tended toward a logicism that mistakes the sign for the thing of which it is a sign.

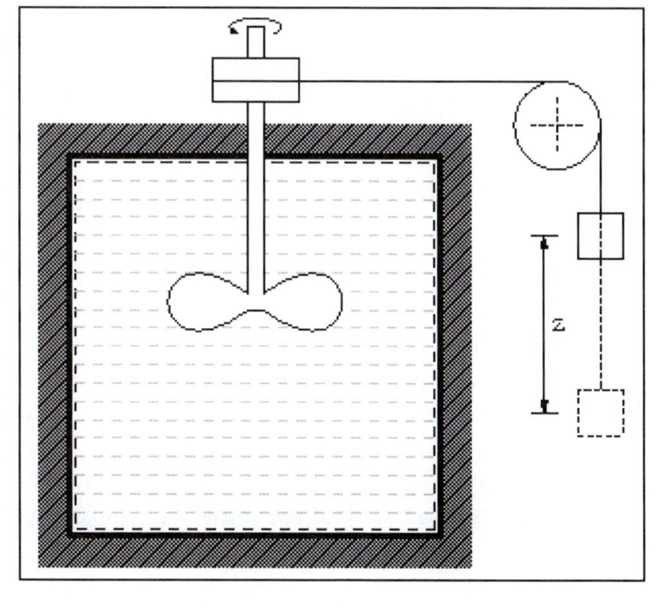

Figure 5-5: An experiment for demonstrating the conversion of kinetic energy to heat. More precisely, the kinetic energy that gravity excites in an object can be turned into heat. The square object hanging from the pulley would normally fall, picking up speed as it did so. However, in this apparatus, this energy is transferred to turning the blade immersed in water in the large square tank. The tank is isolated from other heat sources, yet, after the blade turns, careful measurement of the temperature reveals a temperature increase.

We can here see that, though Leibniz was more right, Newton's point about elastic collisions remains true if we once realize that the bodies of ordinary experience are composed of smaller bodies and Newton's law also applies to these. These bodies can, with the help of forces that hold the parts together-- say forces between molecules that make up the body—have back and forth motions but share the common translational motion. Once known, this appears as a relatively simple answer, yet the fight between the two camps was protracted.

What then, beyond the very nature of the work required to understand such things and the pride and bad will that often plague human activity, caused this to be such a long and hard-fought debate? As is so often the case in attempts to discover general laws in physics, something was left out of consideration that sooner or later should have been brought back but, instead, having been left it out for so long, was forgotten. *Within the appropriate domain of applicability,* we can and should get accustomed to working without whatever it is we leave out, but we must always remember that our leaving it out doesn't make it cease to be.

What had been left out? The same thing we have often left out. In the case of colliding bodies, we considered each colliding body--and will continue to do so when appropriate-- as if it were a single monolithic thing, *not composed of parts*. Yet, we already know the very term body (its first and main usage) implies that it is extended, which means it has parts one outside the next. The discovery of atoms and molecules brought this fact back in a dramatic way. Joule and Mayer had no such help in thinking about the subject because many scientists in Joule's and Mayer's time were skeptical about the existence of atoms. Belief in the existence of atoms did not become mainstream science until the turn of 20th century.

We can quickly understand the importance of Joule and Mayer's discovery through a simple thought experiment. Consider a particle of mass, m and speed v, which enters a volume of water in a thermally insulated container. As it travels through the water, it loses its momentum to water. Now, the idea of conservation of energy is that the activity level is moved around, but never lost. The activity may even be present through the activity related to a different quality, but it is never lost. ***Energy is neither lost nor created.*** Its measure, E, is constant through time. Symbolically, $E = \text{constant}$.

What then happens to the energy in the case of water stopping a moving particle? From the point of view of (efficient) causality, the linear motion of the particle is caused by the impetus and is transferred to the water molecules via the intermolecular forces in the water and the particle. The kinetic energy of the particle $\frac{1}{2}\, mv^2$ is now transferred to motion *within* the body of water. We say heat has been dumped into the water. This is an increase in internal activity or energy rather than external. That is, it contributes to an increase in temperature, not an increase in forward motion of the entire body (see end of chapter problem).

The conservation of energy is a key concept in physics. If energy wasn't conserved, i.e. if energy could be lost or gained, then activity would respectively die out or spiral out of control. But, how does this come about? How does it happen from the point of view of efficient causality (i.e., ordinary cause and effect), that kinetic energy, which is a measure of the transfer of the activity of the impetus, is conserved?

Before we answer this, we should note that energy, as defined above, is only conserved in a limited domain. Later, we will generalize, by analogy, the concept of energy to include potential energy and, in special relativity, to include even the mass. In other physics courses, the concept of energy will be further generalized by analogy to include fields, for example. Still, what we do here is very important because it will be the basis from which all future analogies will grow. As you advance in your understanding of physics, so will your understanding of the importance of the law of conservation of energy. To understand the fantastic realm of color fields, gluons, quarks, black holes and the like you will need to understand conservation of energy. The first step in proving the conservation of energy is to understand work and how it is related to energy. Then, by application of Newton's second and third laws, we can complete the proof of its conservation.

a. Impulse, Work and Energy

The Effect of Force Acting through Time and across Space

Returning to the collision illustrated in Figure 5-1, we see that the bodies are in contact for some *amount of time* and that during that contact time one body pushes in the surface of the other a certain distance, i.e., deforms it. The other body does the same to the first. This mutual action of the bodies on each other continues until the contact ceases. We saw already that, by applying a force for a period of time, one body generates a certain intensity of impetus, momentum in another body; one body provides an *impulse* to the other.

A similar statement can be made about the action of the force *over a distance*. The action of the force over a distance is naturally related to energy. As the force acts, it intensifies the body's impetus over a certain distance; after which, it thus gains a certain amount dp of impetus traveling at the body's speed; this is just what kinetic energy measures. We can make this quantitatively clear by the following.

As force acts on the body through a time dt, the body traverses an increment of distance dx. Similar to the way we derived the equation for the impulse, we construct the following from Newton's second law in the x-direction:[6]

5.13

$$F_x = \frac{dp_x}{dt}$$

The force F_x activates impetus in the body for dt seconds.

$$F_x \cdot dx = \frac{dp_x}{dt} \cdot dx$$

During this time, previously acquired impetus carries the body thru a distance dx at a speed $\mathbf{v}$.

$$F_x \cdot dx = dp_x \cdot \frac{dx}{dt}$$

$$F_x \cdot dx = dp_x \cdot \mathbf{v}$$

The net result is that a momentum of dp_x is transferred at a speed $\mathbf{v}$.

$$F_x \cdot dx = dK_x$$

[6] For the analysis here think of the infinitesimals as finite. Note that, for mathematical exactness, we have to use finite differences and then take limits (not just manipulate the infinitesimals) and restrict our function in some general way to prove these statements. For "reasonable" functions, the statements can be proven rigorously.

Thus, a small change in kinetic energy (measure of activity of the body) is related to the action of the net force applied to it over the interval of distance *dx*. We call the effort expended by applying the force over a distance, ***the work***. The small quantity of work performed by a force F_x acting over a distance *dx* is symbolized as:

5.14 $$dW_x \equiv F_x \cdot dx$$

A force applied over time results in impetus; shifting one's attention to how the force is applied over a distance (because of this impetus), leads one to define energy.

An amount of work dW_x is performed *on* an object when a net force acts on it over a traversed distance *dx*. This work causes a change the body's kinetic energy. That is, by simple substitution we see:[7]

5.15 $$dW_x = dK_x$$

It's appropriate here to pause and answer the important question raised in the previous chapter that is always asked by physics students. Namely, "Are you saying when I push against a wall I'm doing no work because it's not moving?" In fact, you are doing no further work *on the wall* once you have pushed its surface in as far as you are able, but you are doing work in keeping your muscles tight. It is work at the cellular (this is probably small) and/or molecular level; finally it is at the particle level that the motion (and hence the work) is occurring; namely, we are generating heat. The first law of thermodynamics splits out the various components. When split out in this way, we can see that it's merely a matter of keeping track of the various kinds of work.[8] We pointed out in Chapter 4 that the technical definition of work focused our attention on whole bodies, leaving the random motion of the parts for thermodynamics.

Now, we can make a very useful definition that extends our idea of work to include negative as well as ordinary positive work. We say an increment of work dW_x done by a force is positive or negative depending on whether the force acts in a direction parallel or anti-parallel to that of the direction of motion. If the increment of work is positive, the body gains kinetic energy, having received an intensification of impetus. If the increment of work is negative, the body loses kinetic energy, having suffered a loss in intensity of impetus, which results in a gain of kinetic energy for its environment. An example of the latter is a baseball being caught by a first baseman; the ball is slowed down by the force of the glove. The force of the glove acts *against* the ball's motion. We say the glove did negative work on the ball. On the other hand, the ball did positive work on the glove in giving motion in the direction of its force to it. In fact, and this is the key to conservation of energy (and momentum), one body is acting through a certain distance pushing in the surface of a second body, while the first body has an equal and opposite force acting on it. In this way, whatever energy is taken out of the first is exactly matched by what is put into the second. By assigning one change to be positive and the other negative, we can conveniently state that:

---

[7] In other words, the work done on a body is equal to its change in kinetic energy.

[8] We can associate macroscopic changes in the motion of a body with work and the internal ones that involve *many particles in random motion* with "heat." Yet, we could analyze each internal particle as work being done on a microscopic scale as well, thus showing the power of our abstraction, our leaving out all but consideration of locomotion. In particular, there are further qualifications that distinguish *heat* that we know directly through our senses, but, again, we focus on locomotion of bodies ignoring such specifications, since our study here is dynamics. An interesting way to think about the issue of conservation of *p* and *KE* in a 2-body collision is to say that *KE* conservation is what decides how momentum splits (see SBS).

5.16
$$dW_{total} = dW_1 + dW_2 = $$
$$dK_{total} = dK_1 + dK_2 = 0$$

That is, the net work done by both bodies combined is zero[9] and ***energy is conserved***.

TIP: Always keep track of the meaning of each of the symbols you use in an equation. Know, for example, whether the symbol W, depending possibly its subscripts, is work done by one object or is work done by different objects (as is the case above). This is crucial in setting up the problem correctly as well as in interpreting your answer.

Work-Energy Theorem in 1-D

We can now summarize what we've said about work done by a body on another resulting in an equal amount of energy in the second body. As an object moves along a straight line we can add up all changes in kinetic energy due to the force acting on it as it moves from x_1 to x_2.

5.17
$$dW_x = dK_x$$
$$\int_{x_1}^{x_2} dW_x = \int_{x_1}^{x_2} F_x(x) \cdot dx = \int_{K_1}^{K_2} dK_x$$
$$W_x = K_{2x} - K_{1x}$$
$$W_x = \Delta K_x$$

This is the *one-dimensional* version of the **work-energy theorem** which relates changes in kinetic energy of a body moving along the x-direction and the action of forces over space expressed as work. Expressing the kinetic energy in terms of momentum and velocity we have:

5.18
$$\int_{x_1}^{x_2} F_x(x) \cdot dx = \frac{p_{2x}^{\ 2}}{2m} - \frac{p_{1x}^{\ 2}}{2m}$$
$$\int_{x_1}^{x_2} F_x(x) \cdot dx = \frac{1}{2} m v_{2x}^{\ 2} - \frac{1}{2} m v_{1x}^{\ 2}$$

Conservation of Kinetic Energy

The change in the total magnitude of the momentum is related to the work done on a body by:

5.19
$$W_{total} = \frac{p_{2x}^2}{2m} - \frac{p_{1x}^2}{2m}.$$

which can also be written as:

5.20
$$K_{final} = K_{initial} + W_{total}$$

[9] The sum of all the work being zero is generalized to the principle of virtual work in advanced courses in classical mechanics.

An important special case of the work-energy theorem occurs when the net work performed on a body is zero; in that case, the change in kinetic energy is zero. In other words, *if the net work acting on a body (or system of bodies) is zero, its kinetic energy is conserved.* Thus, we have **conservation of kinetic energy,** $K_{final} = K_{initial}$, a particular case of the law of energy conservation that we will develop.

b. Work-Energy and Impulse-Momentum Theorems in 3-Dimensions

The impulse-momentum (Chapter 4) and work-energy theorems applied to a body moving on the *x*-axis appear similar to each other.

5.21
$$I_x = \Delta p_x$$
$$W_x = \Delta K_x$$

Recall the impulse-momentum theorem applies over the *time period* during which a force acts on a body. For the *x, y,* and *z* axes:

5.22
$$I_x = \Delta p_x$$
$$I_y = \Delta p_y$$
$$I_z = \Delta p_z .$$

These equations can be written as a single *vector* equation, giving a 3-dimensional impulse-momentum theorem:

5.23
$$\vec{I} = \int \vec{F}(t)dt = \Delta\vec{p}$$

The net impulse can be positive or negative along a given coordinate axis and the different components combine according to the usual rules of vector addition.

The work-energy theorem applied to the *x*, *y*, and *z* axes is:

5.24
$$W_x = \Delta K_x = \frac{p_{2x}^2}{2m} - \frac{p_{1x}^2}{2m}$$
$$W_y = \Delta K_y = \frac{p_{2y}^2}{2m} - \frac{p_{1y}^2}{2m}$$
$$W_z = \Delta K_z = \frac{p_{2z}^2}{2m} - \frac{p_{1z}^2}{2m}.$$

The total change in kinetic energy is obtained by combining these equations. The result depends solely on total impetus at the start and end of the interaction:

5.25
$$\begin{aligned}\Delta K_{total} &= \Delta K_x + \Delta K_y + \Delta K_z \\ &= \frac{1}{2m}\left(p_{2x}^2 + p_{2y}^2 + p_{2z}^2\right) - \frac{1}{2m}\left(p_{1x}^2 + p_{1y}^2 + p_{1z}^2\right) \\ &= \frac{\vec{p}_2^{\,2}}{2m} - \frac{\vec{p}_1^{\,2}}{2m} \\ &= \frac{1}{2}m\mathbf{v}_2^2 - \frac{1}{2}m\mathbf{v}_1^2 = \frac{1}{2}m\left(\mathbf{v}_{1x}^2 + \mathbf{v}_{1y}^2 + \mathbf{v}_{1z}^2\right) - \frac{1}{2}m\left(\mathbf{v}_{2x}^2 + \mathbf{v}_{2y}^2 + \mathbf{v}_{2z}^2\right)\end{aligned}$$

Where $\vec{p}_i^{\,2} \equiv \vec{p}_i \cdot \vec{p}_i = p_{ix}^2 + p_{iy}^2 + p_{iz}^2$ *and* $\vec{\mathbf{v}}_i^{\,2} = \mathbf{v}_{ix}^2 + \mathbf{v}_{iy}^2 + \mathbf{v}_{iz}^2$

The work terms combine as a *line integral*:

$$5.26 \qquad W_{total} = W_x + W_y + W_z = \int_{Path} F_x^{net}\,dx + F_y^{net}\,dy + F_z^{net}\,dz = \int_{\substack{Path\\from\\1\,to\,2}} \vec{F}_{net} \cdot d\vec{s}$$

In the last line, we use: $d\vec{s} = dx\,\hat{i} + dy\,\hat{j} + dz\,\hat{k}$. Thus, we have the *three-dimensional scalar* **Work-Energy Theorem**:

$$5.27 \qquad \boxed{W_{total} = \int_{\substack{Path\\from\\1\,to\,2}} \vec{F}_{net} \cdot d\vec{s} = \Delta K_{total}}$$

The work-energy theorem applies over the distance over which the force acts on a body.

As a particle moves through space, we can approximate its path as a series of small vector displacements $d\vec{s} = dx\,\hat{i} + dy\,\hat{j} + dz\,\hat{k}$. Figure 5-6 shows the path of a body moving through the *x-y* plane in orange. That same path is shown approximated by a series of small displacements, $d\vec{s}$ of black vectors. A hypothetical force acting on this particle is shown using red vectors. As usual, the length of the force vectors represents the force magnitude and the direction gives the direction in which the impetus it imparts tends.

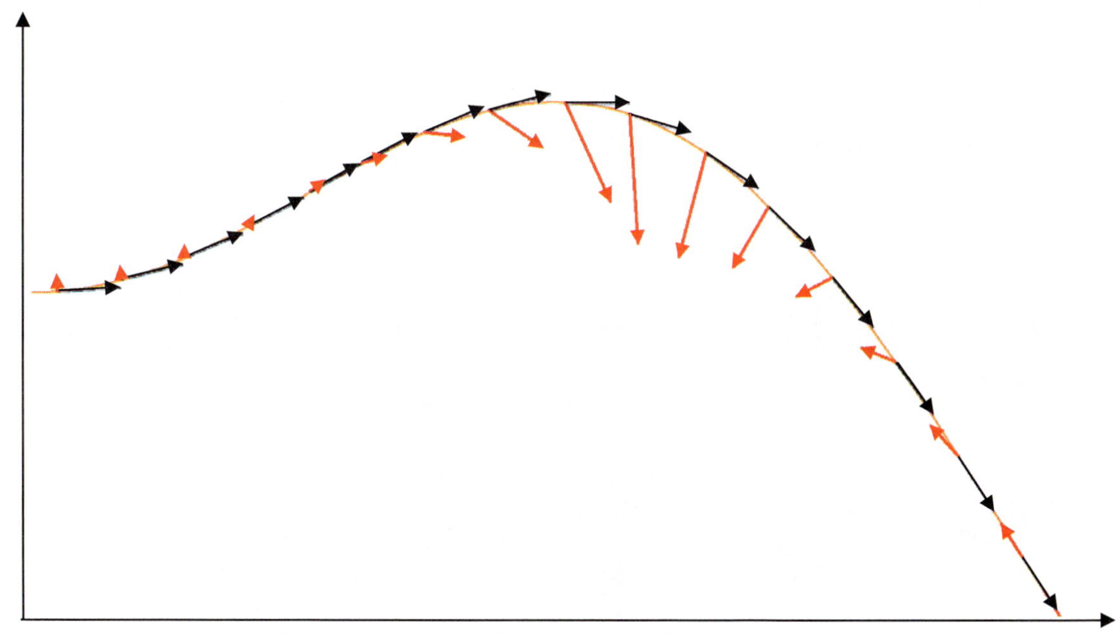

Figure 5-6: The orange curve is the path of the object through a two dimensional **space** (note: this is not a distance versus time plot), say on the surface of a desk. The black arrows indicate the tangent vector to the curve, while the red arrows show the forces acting on the object at the given point in its traverse of the path. During the times when the force (red) is *parallel* to the tangent vector (black), the maximum work is done, while when the force is *perpendicular* to the tangent vector, there is no work done.

The Dot Product in 3-D Work-Energy Theorem

The impulse-momentum theorem shows that whenever an impulse is applied to a body there is always a change in its momentum either in magnitude or direction. However, from the definition of the work-energy theorem, we find that we can have a body with a *net force* acting on it, and yet have *no instantaneous work* done on it? How can this be so?

To begin to answer, first consider, in a little more detail, the dot product of equation 5.27. In particular, consider a net force $\vec{F}$ acting on an object as it traverses a small displacement $d\vec{s}$ along a path. The small amount of work done on the body will be:

5.28 $$\Delta W = \vec{F} \cdot d\vec{s} = F \cos\theta \, ds$$

where $ds = \sqrt{d\vec{s} \cdot d\vec{s}} = \sqrt{dx^2 + dy^2 + dz^2}$ and θ is the angle between $\vec{F}$ and $d\vec{s}$.

Now, take the two cases illustrated in Figure 5-7 below. In case *A,* a component of a force $\vec{F}$ acts *parallel* to the displacement vector $d\vec{s}$ with magnitude $F\cos\theta$. There will thus be a corresponding change in the speed of the object. However, in case *B*, no component of $\vec{F}$ is parallel to the displacement; there is only a perpendicular component. Thus, the instantaneous work done is zero, and, by the work-energy theorem, the kinetic energy does not change. Still, though the object does not change in *speed,* it does change its *direction* of travel.

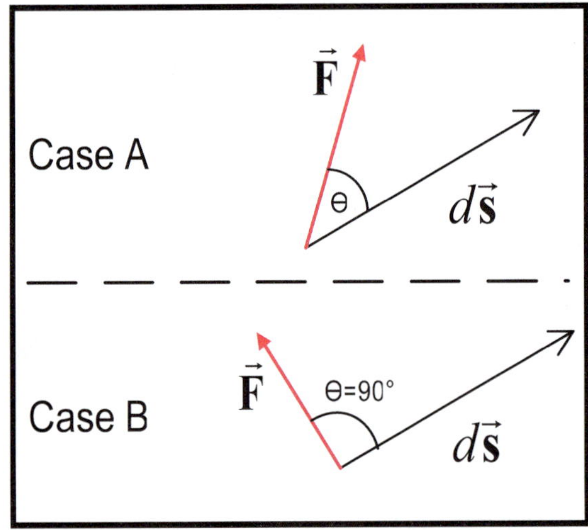

Figure 5-7: Two cases of a force acting on a body. In case *A*, the force has a component along the body's path, which means along the direction of an established impetus. In case *B*, the force has no such component, so its effect is to establish a new impetus. Obviously, such analysis at points is incomplete; to understand a given motion, one must have whole motion.

Leaving details for a later chapter, we can now note that the conundrum of a force acting but exercising no instantaneous work is related to our considering more than one dimension. In any one given dimension, the conundrum does not arise, for the force can obviously only affect that single dimension. In the case of two dimensions, we can define coordinates in such a way that a force moves across the two dimensions, thus in a way mixing the dimensions relative to that force. Such a force continually changes not just the impetus's intensity but its nature by changing the ***line*** along which it tends to push the object.

Of course, our mixing of the dimensions doesn't cause the dimensions to cease to be distinct. As useful and even as revelatory of the qualified (by the force, for instance) condition of the extension as such mixing might be, it cannot change the fact that perpendicular dimensions, while part of a whole, are distinct. Indeed, perpendicular directions or dimensions, such as depth and length, are easily generalized, by analogy, to higher dimension. The "non-overlapping" character of orthogonal dimensions, called *orthogonality*, is the key property used in such analogies that makes them so powerful in the realm of quantum mechanics and many other areas. In short, to answer the conundrum completely, we have to do an analysis keeping in mind their separateness, their *orthogonality*, and thus the distinctness of impetus along perpendicular directions.

When we discuss angular momentum in Chapter 7, we will see that analyzing circular motion, keeping track of impetus along the orthogonal directions, will completely unwind the paradox.

c. Why is Energy a Scalar and Momentum a Vector?

We now have enough understanding of impetus, force, and energy to ask and answer an important question: what does it mean and why is it important that *energy is not a vector* but *momentum is*?

Recall that *momentum* is the measure of the impetus; as a vector, it includes the measure of the intensity and the direction in which the impetus moves a body, for instance in the positive or negative x-direction; this is what makes it a vector. We saw in the previous section that the line along which the impetus moves distinguishes the given impetus from one acting along a different line. Further, we mentioned earlier that the possibility of motion in the positive and negative directions along a given line also distinguishes two different types of impeti. This is why we only considered draining the impetus to zero, not any further, for once the intensity of the impetus becomes negative it is not a lower intensity of the same impetus, but a different impetus because it acts in the opposite direction. In a given rigid body, to get this new impetus, the first impetus must be destroyed, brought to zero intensity. In this way, by focusing on the generic moving power of impetus, the kinetic energy captures something of the activity generally without regard to direction.

Now, notice that, in the case of multiple dimensions, a body can maintain a given *net* magnitude of momentum, preserving the total activity as long as the mass is constant, while the balance between the intensities of the different types of impeti along the different directions shifts. Conservation of momentum prevents such shifting from happening arbitrarily.

Conservation of momentum means that the net motion along a given line is preserved, but because it's a vector conservation, in isolation and given the right initial conditions, it implies that all motion might cease in any given direction and thus in all directions. For instance, two equal mass particles moving at each other under the action of impeti of equal intensity and opposite directional sense would have a *net* momentum of zero and thus could meet and satisfy momentum conservation by stopping, that is, by annihilating all impetus along the given line. Thus, another conservation law is needed to preserve activity (this is called the final (end) causality which simply means it explains the purpose, end or sense of the law in the larger system). Energy conservation does this.

Obviously, this law is not imposed from the outside, like a man-made law imposed by a police department. Rather the law is a manifestation of the way forces, masses and impeti of various bodies interact. Still, indeed, because of this, the various laws make sense together and this is what we are exploring in this section.

Energy conservation, the constancy of the net rate of transfer of impetus, means that, for example, when two balls come together and stop as in the above example, heat[10] must be generated so that the activity continues in some way so that further interaction of some type can still occur. If it weren't for energy conservation, as we've mentioned, things might either spin out of control or stop interacting before anything interesting happened.

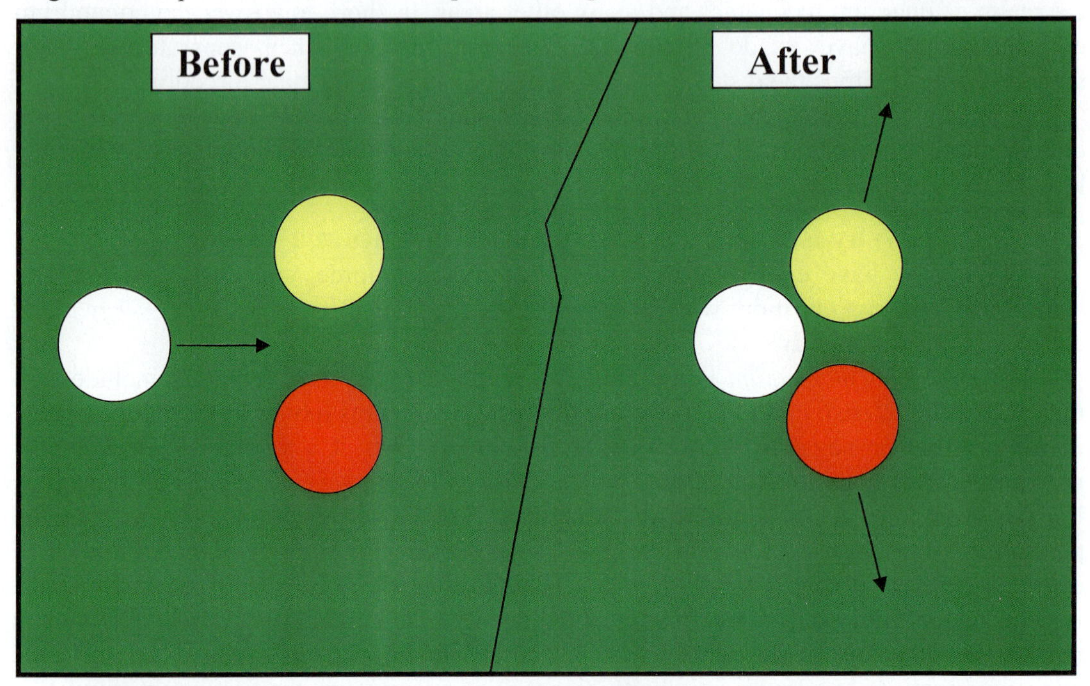

Figure 5-8: Right moving pool ball, with impetus acting only along the positive x-direction, strikes two slightly separated balls spreading the impetus into the y-direction. Energy is conserved by activating impetus in the other direction while draining it in the first in such a way that the energy, the *net rate of transfer of impetus,* is the same. Momentum, the net intensity of the impetus, and its associated direction, is conserved along the y-axis by having the impetus activated in equal amounts in the positive and negative y-direction. It is conserved along the x-direction by sharing the initial positive x directed momentum among the three masses in a way that can be predicted by solving the simultaneous energy and momentum conservation equations.

Energy needs to be a scalar because of the above need to rightly identify and inventory the activity in a line. Its scalar nature (i.e., absent directional information) is thus important in that, when it is conserved, activity is maintained in such a way that the activity can be spread along a line or in multiple directions (this latter is, of course, a real

[10] In addition to, or instead of heat, potential energy (which, while not currently causing a rate of transfer of impetus, can, in principle, do so at some point) could be generated.

possibility though we exclude it in simple examples). Again, this spreading is possible because activity can be maintained when the intensity of a given impetus splits to become impeti acting in different directions, as long as the *net* intensity of them all stays the same. Thus, we can have collisions, such as shown in Figure 5-8 (also see interactive physics model: 5.1-dispersion of momentum into two dimensions.ip), during which the impetus of a particle along the x-axis spreads to other bodies in a way that generates impetus along the y-axis. This spreading of impetus into other dimensions, like so many other issues in mechanics, can be more effectively analyzed in the center of mass coordinates, which we discussed in Chapter 3 and discuss further at the end of this chapter.

Interactive Simulation

In summary, conservation of energy preserves overall activity (roughly by keeping track of the receptivity ($\sim 1/mass$) and overall strength of action ($\sim p^2$), while conservation of momentum, keeps the net intensity of impetus (p) along any given line from gradually building in one direction in a line over another. It also keeps a body from moving arbitrarily, for, speaking metaphorically as we have been, it is conservation of momentum applied to an individual body that keeps it from stopping suddenly and reversing direction.

Power--the Rate of Energy Transferred

In everyday use the word "power" is often heard along with the word "energy." For instance, consider the following statement: "That pick-up truck attempted to pull the car out of the mud. The pick-up did not run out of energy in the attempt, for they had just filled the tank. It just didn't have enough power." In this example, there was plenty of energy but not enough "power."

In this use of "power," we are ultimately really referring to not having a strong enough ***force***. But, we must apply the force across a certain distance to get the car out of the mud. So, the force must be maintained at least over this distance. This means a minimum amount work, a minimum amount of energy, is required along with the minimum force, ($W = \int F dx$). This notion of "power" then is already taken care of by what we've already discussed.

But, we have still another meaning of power. For example, suppose that one truck pulled the car out quickly and another slowly; we would say the quicker truck was more powerful. This notion of power is the starting point for a new measure. The measure of how quickly energy is made available is **power** in this second sense.

> ***Power*** *coming into a body or system of bodies is the measure of the time rate of change of its activity level, i.e. its energy.*

An incremental change in kinetic energy is associated with an intensification of impetus of magnitude, Δp, caused by a force acting across an element of distance $\Delta x = \boldsymbol{v}\Delta t$.[11]

$$\Delta K = \Delta p \cdot \boldsymbol{v} \tag{5.29}$$

[11] This is yet another way to derive the ½ coefficient in the KE formula.

$$W \equiv F\,\Delta x = \frac{\Delta p}{\Delta t}\Delta x \underbrace{=}_{use:\,\Delta x = \boldsymbol{v}_a \Delta t} p_0 \boldsymbol{v}_{avg} = p_0 \frac{p_0}{2m} = \frac{p_0^{\,2}}{2m}$$, taking initial momentum to be zero.

The rate of change of the kinetic energy is obtained by dividing this incremental change by the time, Δt, to cross the element Δx.

5.30
$$\frac{\Delta K}{\Delta t} = \frac{\Delta p}{\Delta t} \cdot \mathbf{v}$$

This rate of change of kinetic energy is the **power**. From Newton's second law, we know the rate of change of the momentum on the right hand side of this equation is equal to the net force acting on the body. Thus, we thus have:

5.31
$$P = \lim_{\Delta t \to 0} \frac{\Delta K}{\Delta t} = \frac{dK}{dt} = \frac{dp}{dt} \cdot \mathbf{v} = F \cdot \mathbf{v}$$

We have a corresponding result from the work-energy theorem regarding work.

5.32
$$P = \lim_{\Delta t \to 0} \frac{\Delta W}{\Delta t} = \frac{dW}{dt} = \frac{F \cdot dx}{dt} = F \cdot \mathbf{v}$$

Thus,

5.33
$$P = \frac{dW}{dt} = \frac{dK}{dt} = F \cdot \mathbf{v}$$

Thus, the rate of transfer of activity to a body, i.e. power, is larger when there is a greater force on it or when the given force acts on it while it is moving faster (i.e. has greater activity).

The Units of Power

In the *MKS* system mechanical power is measured by a unit called the "Watt" with units of $kg \cdot m^2 \cdot \sec^{-3}$ or $J \cdot \sec^{-1}$. A quantity of 1 Watt imparts an energy of 1 Joule in 1 second of time. Equivalently, it corresponds to a rate of energy needed to accelerate a 1 *kg* mass from rest to a speed of 1 *m/sec* in 1 *sec* of time. Other common measures of power are kilowatts (1,000 Watts), megawatts (1,000,000 Watts), or gigawatts (1,000,000,000 Watts). When used with intervals of time, power can also be used to measure energy. For example, electricity usage in a home is often measured in kilowatt-hours. A *kW-hr* is the amount of energy delivered by a 1 *kW* power source acting over 1 *hr* of time = $1000W * 3600 \sec = 3.6 \times 10^6 J$.

A Note on the Ordinary vs. Empiriometric meaning of the term Power

The term "power" as used in this text has a precise empiriometric meaning different from the broader ordinary usage of the term "power" one uses when speaking in other contexts. For example, your ability to read this page is possible because you have the power of sight to see it with your eyes, the power of intellection to understand it, and the power of will to fix your attention on the task of learning modern physics. Or, using what we have learned in the text, we can say a body has the power of impetus. These examples of the term "power" refer to abilities that people and things really possess. We should note the different analogical uses of the term --which is generally one of the devices that makes language so powerful—and let the context decide which is implied.

In modern physics, the empiriometric meaning is the one that is used for good and important reasons, but this does not obviate, in any way, the need for the other more primary meanings. Indeed, it is ultimately from these latter, ordinary uses of the word that the empiriometric uses were, by analogy, derived.

We are now ready, having laid down abstract principles and given mathematical expressions of those principles, to include an important real world effect: friction.

Point to Remember: We always want to establish the most general principles first, then we can go back and specify further. Unlike the first things we know, which we see very directly such as the principle of contradiction or that things have a cause, or simply that things are there, which are instantly and directly seen by our mind as generic principles that must be true, other principles take time and effort to come to. But, once we know those principles understanding just begins; we must specify them for reality is specific not general.

Friction: A Non-Conservative Force

It is usual at this point to discuss two subcategories of forces: *conservative* and *non-conservative*. Usually, conservative forces are discussed in detail first, so that the meaning of the word "conservative" can be exhaustively defined. A conservative force is, roughly speaking, one that is able, at least potentially, to convert all its work (and no more) into kinetic energy (usually, as opposed to heat). However, because *non-conservative* forces are always acting in ordinary life, and because we ultimately need to start with what we directly sense, we will start with these types of forces and *then* make the abstraction, the process of leaving out, that brings us to consider conservative forces. Again, all ordinary physical situations involve non-conservative forces. With these forces acting, kinetic energy is not conserved, e.g. *inelastic* collisions, and work doesn't always produce kinetic energy or even the potency to regain lost kinetic energy. In short, in ordinary life circumstances, non-conservative forces are dissipative (usually heat generating).

For instance, there are many situations in which applying a constant force produces a constant speed, not an increasing forward motion and kinetic energy as might be expected from the impulse-momentum theorem and the work-energy theorem; after all, there is a force acting *for* a time and *through* a distance.

In particular, take, for example, sliding a heavy box across a floor. Pushing the heavy load across the floor with a constant force, especially a carpeted floor, does not always accelerate the box. Indeed, historically, the fact of friction impeded the discovery of the concept of inertia. When you push something, like the box, you must continue to exert a force to keep it moving at a constant speed, and, thus, it appears to contradict Newton's first law. When friction acts, the kinetic energy of forward motion dissipates. Up to this point, we have overlooked such effects.

All friction generates heat and vibrations (generically called sound, but are not necessarily audible) that eventually dissipate as heat. For two surfaces that we attempt to slide, or are sliding, past each other, such as the surface of the heavy box along a floor, there are two idealized types of friction: *static and kinetic*.

Everyone is qualitatively familiar with these types of friction. For instance, take a book and place it on the table. Push it lightly with your finger; the book does not move. A force acts on the book yet it does not move: How does Newton's *second* law apply? Since the book didn't move, there must be another equal and opposite force acting *on the book*; this is the force of static friction (see Figure 5-9a). Notice that this force is distinct from the force reacting back *on your finger* that is required by Newton's *third* law. The static frictional force is a reactive force from the table *acting on the book*, which resists one's attempt to move the book. Of course, as required by Newton's third law, there is *also* an equal an opposite force applied by the block *on the table* in response to the frictional force applied by the table acting on the block.

Now, by gradually increasing the force applied parallel to the table's surface, you will, at some point, overcome the static friction, and the book will start to move. This *maximum force* that the static friction can exert to counter your push is approximately proportional to the ***normal*** *force*, i.e. the force acting perpendicular to the surface. Mathematically, we write:

5.34 $$F_{Static\ Max} = \mu_S F_N = \mu_S N \,,$$

or generally, including the sign: $|F_{static}| \leq |\mu_S N|$

where: $F_N \equiv N$ is the normal force. μ_S is a constant called the coefficient of static friction.

Now that the book is sliding, you can actually decrease the applied force, and the book *will continue to slide* and possibly even continue to accelerate (see Figure 5-9b), though less quickly. When you push with just the right amount of force, the book will slide on the table at constant velocity. At that point, you are pushing with a force equal and opposite to the force of *kinetic* friction. The kinetic friction is also proportional to the normal force, in this case, the weight of the book. Mathematically, we write the kinetic frictional force as:

5.35 $$F_k = \mu_k F_N \qquad \mu_k < \mu_s$$ where: μ_k is a constant called the coefficient of static friction.

If you push with less force than this force of kinetic friction, the book will slow down, with a deceleration proportional to the difference, and eventually stop. Once stopped, the book is again in static equilibrium with your applied force balanced by the static friction. It is interesting to consider piling other books on top of the first; the same general effects would occur, but the details of which are left to an end of chapter problem.

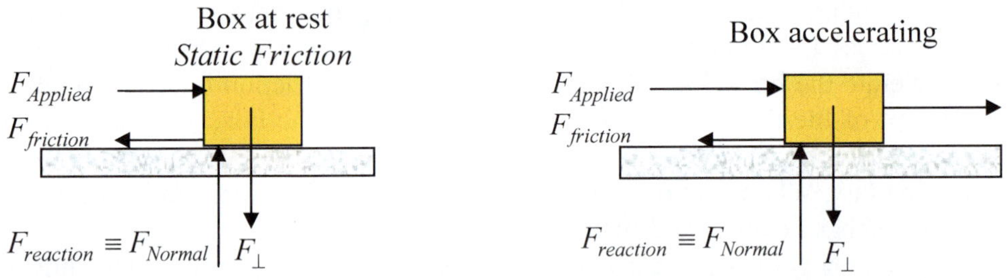

Figure 5-9: *Forces acting on a box (or book) on the floor*: The normal force (gravity) pulls down on the box and is exactly balanced by reactive force of floor, so that the box does not move vertically. **a.** (*Left)* The applied force is balanced by the static friction. **b.**(*Right)* The static friction has been overcome and the box is moving with applied force larger than the reactive frictional force, so the box accelerates in the horizontal direction. Friction encountered while an object is moving is called kinetic friction.

We can summarize the above discussion of friction between two bodies in contact along parallel surfaces as follows:

First, mathematically, we write:

$$F_s \le \mu_s F_N$$
5.36
$$F_k = \mu_k F_N$$
$$\mu_s > \mu_k$$

where: s and k subscript refer to static and kinetic, respectively

Other key related points:

1. Friction resists relative motion.
   a. The force of static friction always acts to oppose attempts to move the body relative to the contacting surface.
   b. The force of kinetic friction, which operates when the surfaces are moving relative to each other, always *acts in the direction opposite the body's velocity relative to the surface.*
2. *Kinetic friction* is always accompanied by the *generation of heat* at the plane of contact of the surfaces. Note that the energy imparted by the body acting against the frictional force of the surface of the table goes into heat and into increasing the momentum of the surface and ultimately the Earth. Of course, this momentum change of the Earth is not noticeable because the Earth is so massive.[12]
3. In our approximation, μ_s and μ_k are constants. Typically "rougher" contact surfaces have a larger μ*'s*.
4. The proportionality of the force of kinetic friction and the maximum force of static friction to the normal force generally means that if the weight is kept constant, increasing the surface area of the sliding body will not change the frictional resistance it presents (cf. end of chapter problem).

To get a better quantitative understanding of static and kinetic friction in various circumstances, we revisit the pushing of a book on a table. Figure 5-10 below shows what happens in the case of the book with a changing applied force. We assume the book has weight of 8.5 *N* (about .85 *kg*), and $\mu_s = .59, \mu_k = .47$.

[12] Notice again that energy goes into various modes of vibration within the object rather than the motion of the whole. If there were no such parts to vibrate all collisions would thus be elastic, because all the energy would have to go into motion of the wholes rather than their parts.

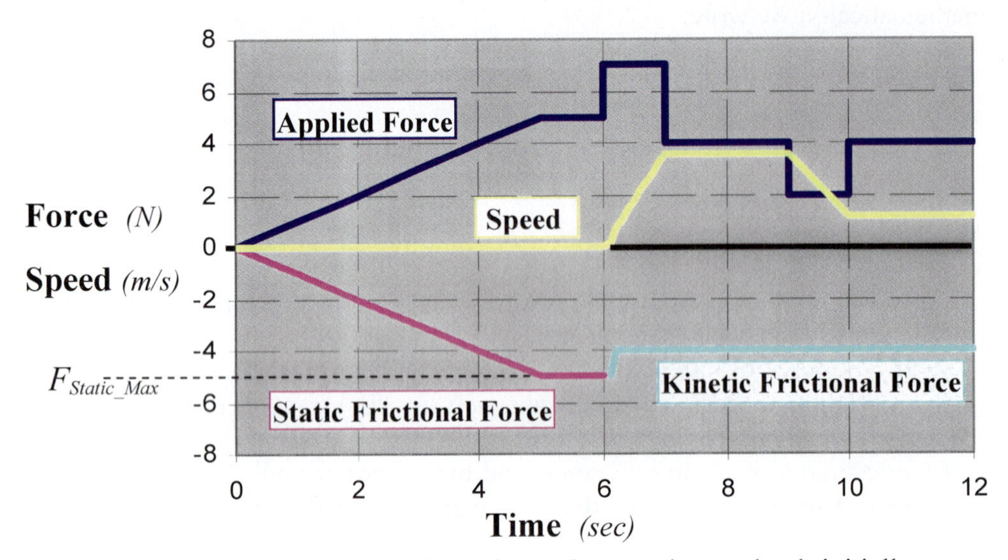

Figure 5-10: Top curve (dark blue) shows force acting on book initially at rest. The middle curve (yellow) shows the resulting speed of the book in response to the action of the applied force. The first portion of the bottom curve (purple) shows static curve that arises in response to the applied force. The static friction successfully resists its action on the block up to a maximum. At that point the frictional force drops, which is now called the kinetic friction (light blue curve), to a lower value that remains constant, in our idealization, though the applied force changes. (Note: it's very fast moving, even dangerously fast, at 7 seconds.)

Are You Directly Responsible for Moving the Things You Move?

Because friction is an integral part of ordinary experience, it can and has led many to miss an important fact about the nature of impetus. In particular, while you are pushing an object moving at a constant speed, you are ***not directly*** responsible for the motion. You gave the object the impetus it has and that, in turn, moves the object; you are not directly moving it; its impetus is. True, you are the cause of its motion (because you caused its impetus) in a deep sense, but not directly.

To vividly illustrate this fact, consider a block sliding as shown in Figure 5-11. In this experiment, we consider an ice cube sitting on top of the block. As shown in Figure 5-11a, the ice has a hole bored in it that is the right size to receive a metal rod inserted by a mechanism in the block. As you push the block up to speed (cf. Figure 5-11a), the mechanism holds the rod inside the ice, thus keeping the ice in place. You are pushing the block with more force than the force of kinetic friction, and this means that part of your applied force is impeded by the action of the friction of the table, but there is a portion of the applied force that is unimpeded. That unimpeded portion of the force activates impetus in the block at a rate given by its magnitude. Because the ice is attached to the block by the rod, the ice also gets whatever impetus is required to keep its speed equal to that of the block.

Once you've reached the desired block speed, reduce your applied force to match the force of kinetic friction so that no new impetus is imparted. Thus the block and ice keep moving, but not because your pushing them but because their impetus moves them. Your role is in countering the frictional force so that their impetus is not drained by it.

To illustrate this, once the block is moving at a constant speed, the mechanism withdraws the metal rod, leaving the ice free to move relative to the block (cf. Figure 5-11b). Of course, it does not do so because its impetus is not changed by this action. However, if you now remove your hand, the frictional force, now unimpeded, operates and slows down the block rather quickly and finally stops it. By contrast, the ice, which moves with very little friction on the surface of the block, will continue to move forward under the action of its impetus (cf. Figure 5-11c). Your hand is nowhere near it, yet the ice keeps moving. This is a vivid reminder that impetus *was* and is moving the ice (and the block), not you.

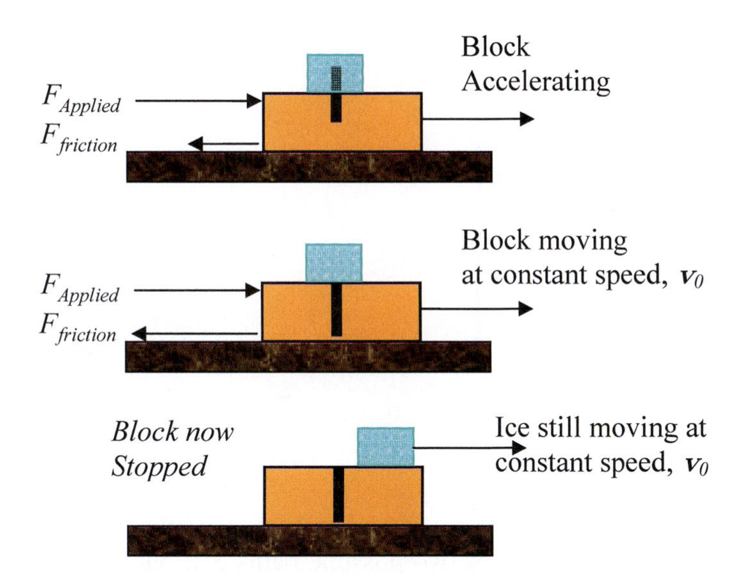

Figure 5-11: a. *(Top)* A cube of ice is held in place on the top of a block by a rod inserted in a bored hole in the ice. The block is then accelerated by the application of the force as shown. **b.** *(middle)* Once the block/ice reaches the desired speed v_0, , we reduce the applied force to match the force of kinetic friction, so that this speed is maintained. We then pull the rod out of the ice. **c.** *(top)* When the applied force is removed, the block quickly comes to rest as shown here. Notice that the ice, assuming it experiences no friction, keeps moving at speed v_0 because its impetus has not been changed.

Path-dependence of Friction

Non-conservative forces such as friction have the property that the work they perform depends on the path taken by the body they act on. For instance, given the job of sliding a piece of furniture to the other side of a room, would you rather go straight across or via a curved course? The answer is obvious. Let's see quantitatively how much different the work required is. This will help us better understand the mathematics of line integrals used for calculating work.

For simplicity, suppose you push a block of mass *m* across the floor surface at constant speed. The normal force of the block acting on the surface is its weight, *mg*. The magnitude of force of kinetic friction is $F_K = \mu_k m g$ and acts in the direction *opposite* the velocity of the block. We apply enough force to exactly oppose, we say cancel, the action of the force of kinetic friction along two different paths bounding a square. See Figure 5-12 for the path. We first move the block along edges marked *A-B*, *B-C*, and *C-D*, the hard way. We next move the block along segment *A-D only*, the easy way. What is the work done by the force of kinetic friction in each of these cases?

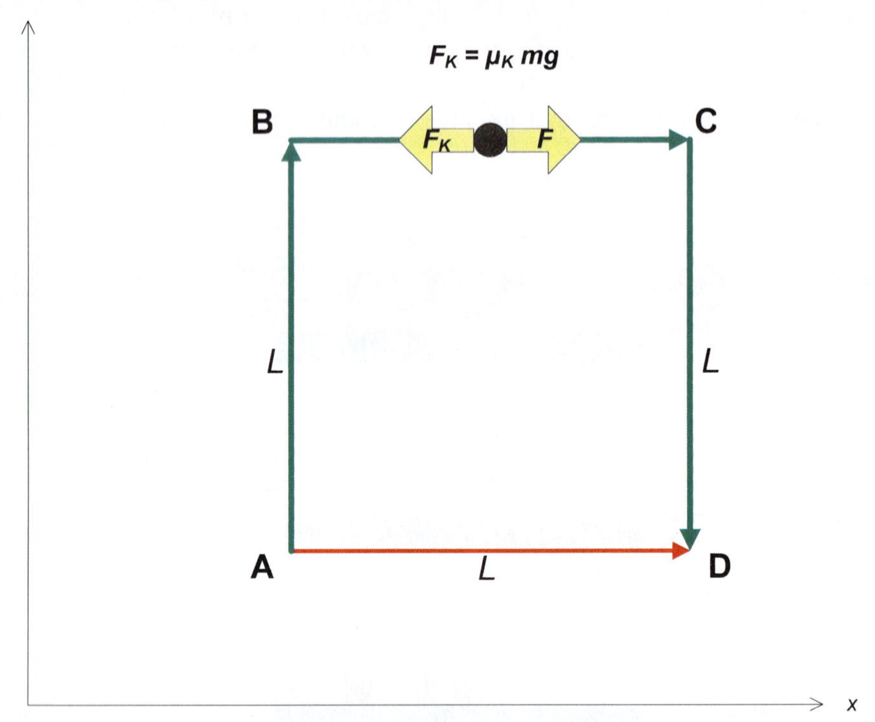

Figure 5-12: Top view of two possible paths that one could push an object across a floor to get it from point *A* to point *D*. Path one is shown in green; path two in red. For non-conservative force such as friction it takes more energy to take the longer path. However, for conservative forces that we study next, it makes no difference. This brings out the abstraction from ordinary circumstances involved in conservative forces.

Along path *A-B-C-D*, the work performed by the force of kinetic friction is:

$$\begin{aligned}
W_{A-B-C-D} &= \int_{Path\ A-B} \vec{F}_{Friction} \cdot d\vec{s} + \int_{Path\ B-C} \vec{F}_{Friction} \cdot d\vec{s} + \int_{Path\ C-D} \vec{F}_{Friction} \cdot d\vec{s} \\
&= \int_A^B \left(-\mu_K mg\,\hat{j}\right)\cdot d\vec{s} + \int_B^C \left(-\mu_K mg\,\hat{i}\right)\cdot d\vec{s} + \int_C^D \left(+\mu_K mg\,\hat{j}\right)\cdot d\vec{s} \\
&= \int_A^B \left(-\mu_K mg\,\hat{j}\right)\cdot\left(dy\,\hat{j}\right) + \int_B^C \left(-\mu_K mg\,\hat{i}\right)\cdot\left(dx\,\hat{i}\right) + \int_C^D \left(+\mu_K mg\,\hat{j}\right)\cdot\left(-dy\,\hat{j}\right) \\
&= \left(-\mu_K mg\right)\int_A^B dy + \left(-\mu_K mg\right)\int_B^C dx + \left(+\mu_K mg\right)\int_C^D dy \qquad (5.37) \\
&= \left(-\mu_K mg\right)\left(y_B - y_A\right) + \left(-\mu_K mg\right)\left(x_C - x_B\right) + \left(\mu_K mg\right)\left(-y_D + y_C\right) \\
&= \left(-\mu_K mg\right)(L) + \left(-\mu_K mg\right)(L) + \left(+\mu_K mg\right)(-L) \\
&= -3\mu_K\, mg\, L
\end{aligned}$$

Along the second path, *A-D*, the work is:

$$
\begin{aligned}
W_{A-D} &= \int_A^D \left(-\mu_K mg\,\hat{i}\right)\bullet d\vec{s} \\
&= \int_A^D \left(-\mu_K mg\,\hat{i}\right)\bullet\left(dx\,\hat{i}\right) \\
&= \left(-\mu_K mg\right)\int_A^D dx \qquad (5.38)\\
&= \left(-\mu_K mg\right)\left(x_D - x_A\right) \\
&= \left(-\mu_K mg\right)\left(L\right) \\
&= -\mu_K\, mg\, L
\end{aligned}
$$

Again, the block begins at A and ends at D in both cases, but *the work performed by the force of friction is not the same along the two paths*:

5.39 $$W_{A-B-C-D} \neq W_{A-D}$$

Again, the work done by the force of kinetic friction on the block is **path-dependent.**

Making a rough estimate of the coefficient of friction of your hands, you will get a chance to calculate the amount of heat you can generate in your hand in chapter end problem.

Note that the negative sign in the results of the work calculations above (equations 5.37 and 5.38) means that the friction does negative work, work against the motion of the particle. That is, the frictional force always results in slowing down of the ball in the direction of the friction's action, never in speeding it up. With the frictional force defined as the positive direction, we have: $dx < 0$. This is a result of the force of kinetic friction always being opposite the velocity of the object. During an interval of time Δt, the object moves: $\Delta x = -\mathbf{v}\Delta t$. Meanwhile during that same time, the force removes momentum Δp (which is >0 because of our chosen directional convention) from the object, thus the work done is: $\Delta W = F\Delta x = \dfrac{\Delta p}{\Delta t}\Delta x = -\Delta p\,\mathbf{v} < 0$.

Conservative Forces and Potential Energy

In the previous section, it probably seemed too obvious to mention that it makes a difference which path you take in doing a job. By leaving out certain complications[13] which only slightly alter the outcome (our favorite trick), we will now examine such forces. For *conservative forces, the work performed is* ***path independent***, specified amazingly enough only by the endpoints of the path, where you start and where you end.

For example, a spring acts on a block shown below in Figure 5-13. The block has mass m and slides without friction on the smooth flat horizontal surface, say an air table or ice. A spring acts on the block with a force proportional to the distance displaced from equilibrium. The proportionality is given by the spring constant k. The force is given by: $F = -k \times (displacement\ from\ equilibrium)$. When the spring is relaxed, the edge of the block is at $x = 0$. What is the amount of work done by the spring as the block moves between two locations measured by x_1 and x_2?

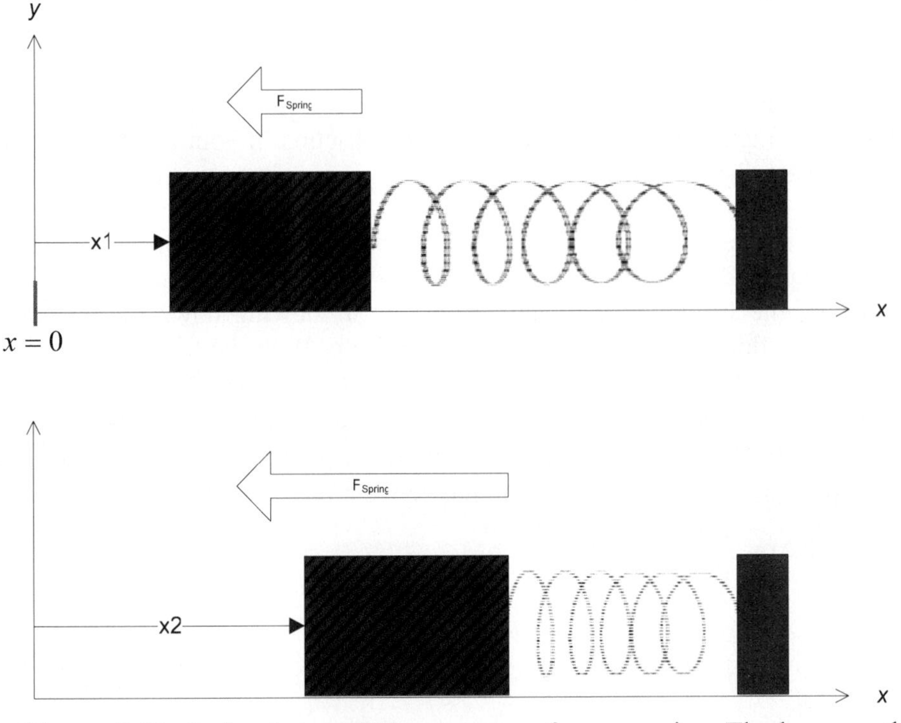

Figure 5-13: Spring in two different states of compression. The larger x, the more force (arrow over block) the spring exerts on the wooden block back towards its $x = 0$ (y-axis) equilibrium point, where if the block had no momentum it would stay.

The force exerted by the spring on the block will be given by:

$$F_{spring} = -k \cdot x \qquad \textit{Hooke's Law} \tag{5.40}$$

[13] Of course, it is actually not a trick but a skill in recognizing what is important in a particular situation and what is less important.

Note that the negative sign indicates a restoring force. The work done by the spring on the block as the block moves from x_1 to x_2 is given by the work integral:

5.41
$$\Delta W_{spring} = \int_{x_1}^{x_2} (-k \cdot x) \cdot dx = -\frac{1}{2} k (x_2^2 - x_1^2)$$

By the work-energy theorem, the change in kinetic energy of the block as it moves from position x_1 to x_2 will be:

5.42
$$\Delta K = K_2 - K_1 = \Delta W_{spring} = -\frac{1}{2} k (x_2^2 - x_1^2)$$

Clearly, this result only depends on the end points, not, for example, on when or how fast it got there.

To bring in an extra dimension, we put the spring/block system on a well oiled rail that allows it to move freely in and out of the page without affecting the coil of the spring. In this way, we can establish a second path independent force. Let's say we first move the block (and spring out) of the page 1 meter, then we compress the spring, then we move the block (and spring) back level with the page. The question is: How much work (in the technical sense in which we neglect internal energy loss such as heat generation) did we add to our task in moving the block out and back?

We had to bring the block up to speed to come out of the page with work: $W_1 = \int Fdx$, where F is the force applied to the system, and we had to stop this motion to come back with work: $W_2 = -\int Fdx = -W_1$, resulting in a net required work of zero to get the block out of the page 1 meter. To get the block back to its original position requires the exact same steps in reverse, so that whole process adds exactly zero to the net work, leaving *no change* in the work required. Given the assumptions we made, we thus have a path independent force.

a. Potential Energy: Conservation of *E=K+V*

Path-independent forces such as this allow us to talk of **potential energy** and thus of conservation of energy in a further extended sense.[14] Recalling the work-energy theorem, we see: $dW = dK$, we note that we can rewrite this as: $dK - dW = 0$ and by defining a new function:[15] $dV = -dW$

[14] In more advanced treatments, one starts with differential work done by forces and tries to derive the potential energy. If the forces (called monogenic forces) are derivable from a single scalar function (it's not always possible), called the work function, then many special mathematical treatments become possible. When the work function is time-independent, one has a conservative force. When it is time *and* velocity independent, one can construct the potential energy as the negative of the work function and the force as the gradient of the potential (see later in this chapter).

[15] Again, the integration cannot always be accomplished; when dV can be integrated to give a function V, it is called a perfect differential.

$$dK + dV = 0$$
$$d(K + V) = 0$$
$$dE = 0$$

In the last line, we see that by defining: $E = K + V$, we see E, called mechanical energy[16] is always the same, doesn't change as time passes; i.e. it is conserved. Furthermore, integration gives:

5.43 $$V(x_{final}) + \frac{1}{2} m v_{final}^{\ 2} = V(x_{initial}) + \frac{1}{2} m v_{initial}^{\ 2}$$

But, what do we mean by the potential energy, V, we've introduced?

To understand, return to our spring. Compress the spring by a certain amount x_0 and hold it there. When you release it, the block will move, and when it is at $x = 0$, it will have a kinetic energy given by substitution into equation 5.42 above. Assuming it starts from zero speed, using $dK = dW$ with careful attention to signs, we then get:

5.44 $$\frac{1}{2} m v_f^{\ 2} = \frac{1}{2} k x_0^{\ 2}$$

This mathematical result comes from the physical fact that the compressed spring pushes on the block. However, like all physical forces, the spring force acts over time; with no time, there is no effect. Said another way, an object's capacity to do work can only be seen when sufficient time passes. However, initially the spring was completely impeded from acting, from activating impetus in the block because your hand supplied an equal and opposite force. Yet, the spring has the power (fully generic sense), the potency, to act. It's just not acting because impeded. This is why in this form we call it *potential* energy. Potential energy is energy or power to act that is *not* actually able to cause its proper activity, at least at the moment. If nothing impedes the force that is responsible for the potency under consideration, it will activate impetus in a body and thus become "active" energy measured by kinetic energy. To obtain our mathematical result, equation 5.44, using potential energy, V, we need to define it for the spring; from which, we could verify equation 5.44 using equation 5.43.

We define the "potential energy function" as:

5.45 $$\begin{aligned} V(x) &\equiv \int dV = -\int dW \\ &= -\int_{path} F \cdot dx + \text{constant} \end{aligned}$$

So, in the case of the spring, we have:

5.46 $$\begin{aligned} V(x) &= \frac{1}{2} \cdot k \cdot x^2 + \text{constant} \\ &= \frac{1}{2} \cdot k \cdot x^2 \end{aligned}$$

Potential energy typically has a zero reference point. In this case, we chose the equilibrium point of the spring (i.e., where it exerts no external force) as the zero reference point. We could, as a matter of convenience also choose another zero reference point (taking constant $\neq 0$), but it is clear that the one we chose is natural to the spring and the given circumstances and, thus, makes the mathematics simpler.

---

[16] We will see in Chapter 7 that K should include both translational *and* rotational kinetic energy.

If we let the block continue to move past the $x = 0$ point until it stops on its own, we will see the block stop at $-x_0$. (The student is encouraged to check this by plugging into $E = K + V = \text{constant}$). It then swings back to the $x = 0$ point at which point the kinetic energy is again maximum.

We can say that the kinetic energy resulting from the action of the spring has been "stored" in the spring as potential energy which is available to the spring to perform work on the load attached to it. This potential energy obviously depends on the spring constant (higher values of k are stiffer, that is, harder to compress) as well as the amount of compression. The potential energy function of the spring is shown in Figure 5-14.

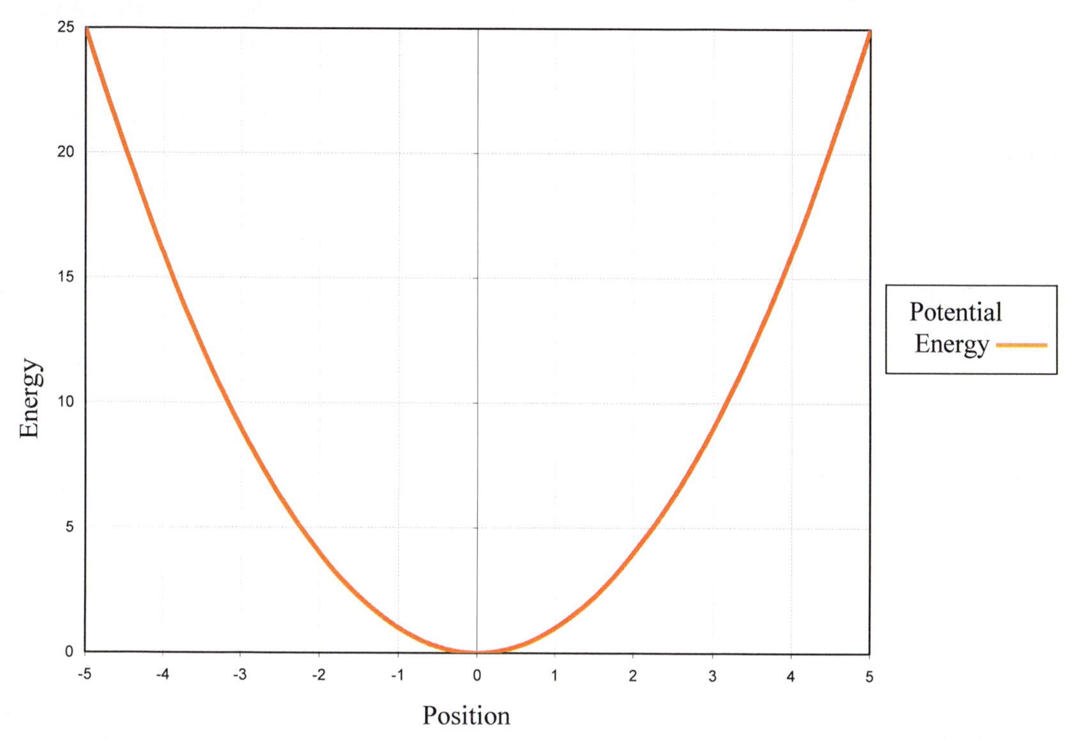

Figure 5-14: Plot of the potential energy of the spring shown in Figure 5-13 against its displacement from equilibrium. Notice the symmetry with respect to the equilibrium position.

The potential energy function, as we can see by looking at its definition, can easily give us the force. Indeed, for conservative forces that are independent of velocity, the relation of the potential energy function to the acting force is:

5.47
$$\boxed{\begin{aligned} F &= -\frac{dV}{dx} \\ \vec{F} &= -\vec{\nabla} V \end{aligned}}$$

The second equation generalizes the first to three dimensions. The force is the negative of the gradient of the potential energy function. In other words, using the example of Figure 5-14, it is the negative of the slope of the potential energy curve. Moving to the right of the Figure 5-14, the slope of the potential energy function increases linearly with increasing displacement x, and this reflects the increasing magnitude of the restoring force of the spring that tends to move the block back toward the position of equilibrium.

Notice again that the sum of the kinetic and potential energies is a constant throughout the motion. We call this quantity the total mechanical energy $E = K + V = constant$. We see that under these conditions where a conservative force is acting on the body (i.e. the work performed does not depend on the path taken but only on the starting and ending coordinates) a potential energy function can be defined and the total mechanical energy will be a constant of the motion. That is, we will have the **conservation of mechanical energy**.

The behavior of such systems can be understood in more detail by looking more closely at the potential energy curve functions. We can draw the total energy as a constant horizontal line in the diagram (Figure 5-15). At each position, x, this reflects the constant energy of motion. The contribution of the potential energy to the total energy is given by the potential energy curve. Thus, the difference between the total energy and potential energy is the kinetic energy of the object at that position. Two locations of interest appear in the diagram where the total energy curve crosses the potential energy curve. These are the stopping points of the motion, or turning points, for the object, i.e. where it comes to momentary rest and then reverses its motion. Likewise the equilibrium position at $x = 0$ is where the potential energy reaches a minimum. When the block is at this location, we say all of the available energy is in the form of kinetic energy; the object is moving at its greatest speed.

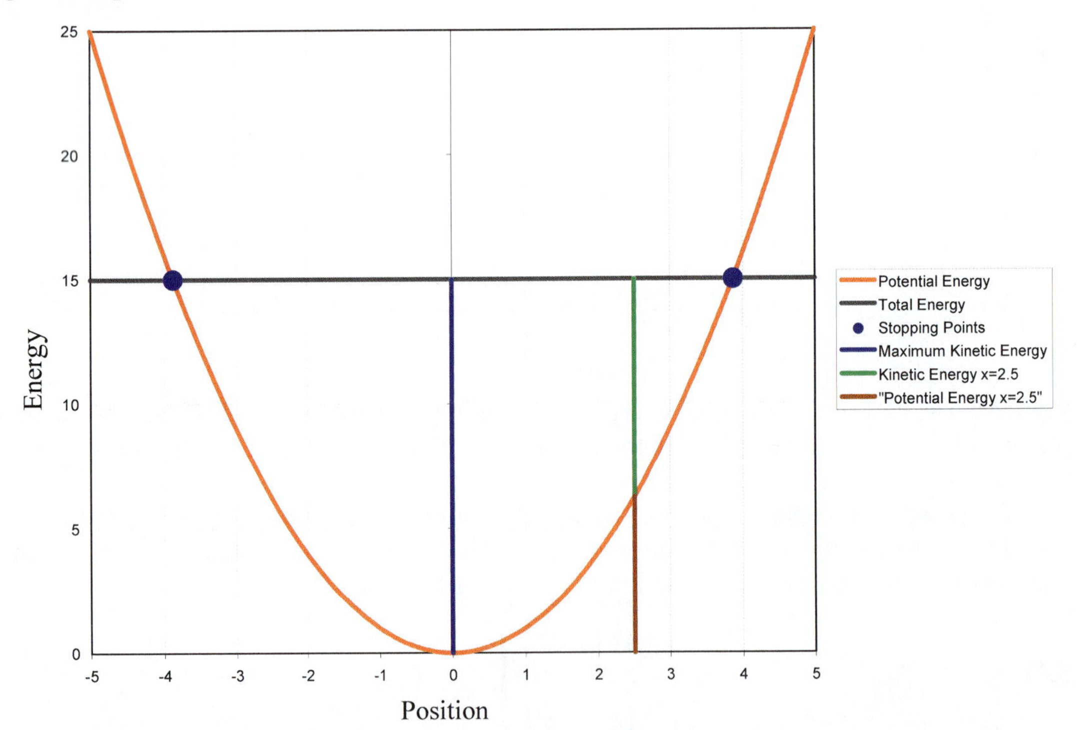

Figure 5-15: Plot of the potential energy (*orange* line) and total energy (*black* line) of the spring shown in Figure 5-13 against its displacement from equilibrium. *Blue* vertical line shows maximum kinetic energy. *Green* and *brown* vertical lines show the kinetic energy and potential energy at $x = 2.5$. Large blue dots show the turning points where the block is drained of impetus due to the restoring force of the spring; at that point, the spring begins activating impetus for movement in the opposite direction.

b. Mathematics of Conservative Forces

If we have a force, how do we know if it is conservative? We can graph it and look to see if a closed path integral is always zero. But, we can also look at the curl $\vec{\nabla}\times\vec{F}$ of the force field, $\vec{F}$, which can be written in the following way:

5.48 $$\vec{\nabla}\times\vec{F}=\left(\hat{x}\frac{\partial}{\partial x}+\hat{y}\frac{\partial}{\partial y}+\hat{z}\frac{\partial}{\partial y}\right)\times\left(F_x\hat{x}+F_y\hat{y}+F_z\hat{z}\right)$$

The curl tells us how much the force "curls" around a given point arbitrarily close to that point. We may have guessed this by recalling that $\vec{A}\times\vec{B}$ introduces the notion of rotation that implicitly brings in 3-D through the concept of cross product (see Chapter 2). We can show explicitly how $\vec{\nabla}\times\vec{F}$ yields amount of curling.

Take the loop integral shown in Figure 5-12 to be very small. We've reproduced the loop in Figure 5-16 below with some new notation to facilitate our present task.

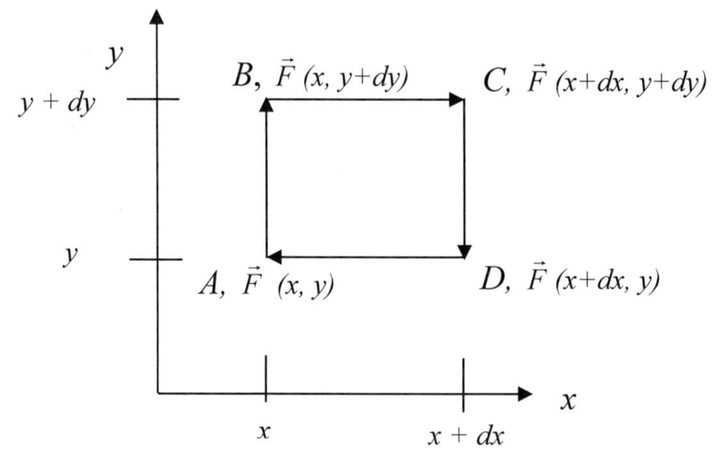

Figure 5-16: Small rectangular path of side length *dx* and *dy* around which we calculate the work.

Note lines *AB* and *CD* each extend *dy* while lines *BC* and *DA* extend *dx*,
Consider first the *AB* and *CD* portion of the line integral: $\oint\vec{F}\cdot d\vec{r}$. Using F_y to mean the *y* component of the Force, $\vec{F}=F_x\hat{x}+F_y\hat{y}+F_z\hat{z}$, and using the average of the forces at the two end points to calculate the integral along each segment, we get:

AB: $\left(\frac{F_y(x,y)+F_y(x,y+dy)}{2}\right)dy\approx F_y(x,y)dy$ because the second term differs from the first only in second order in *dy*.

CD: $-\left(\frac{F_y(x+dx,y+dy)+F_y(x+dx,y)}{2}\right)dy\approx -F_y(x+dx,y)dy$ because the first term differs from the second only in second order.

So, the $AB+CD$ term of line integral gives: $\dfrac{\left(F_y(x,y)-F_y(x+dx)\right)}{dx}dx\,dy \to \dfrac{\partial F_y}{\partial x}dx\,dy$

Similarly for the two horizontal components, *BC* and *DA* of the line integral

$$BC\text{: } \left(\frac{F_x(x,y+dy)+F_x(x+dx,y+dy)}{2}\right)dx \approx F_x(x,y+dy)dx$$

$$DA\text{: } -\left(\frac{F_x(x+dx,y)+F_x(x,y)}{2}\right)dx \approx -F_x(x,y)dx$$

So, $BC+DA$ term of line integral gives: $\left(-F_x(x,y)+F_x(x,y+dy)\right)dx \to -\dfrac{\partial F_x}{\partial y}dy\,dx$

Giving:

5.49 $$\oint_{ABCD} \vec{F}\cdot d\vec{r} = \frac{\partial F_y}{\partial x}dy\,dx - \frac{\partial F_x}{\partial y}dy\,dx = (\vec{\nabla}\times\vec{F})_z\,dx\,dy\,.$$

Recall this line integral, $\oint_{ABCD} \vec{F}\cdot d\vec{r}$, is a sort of differential line integral that needs to be summed with other line integrals to make a finite sized path.

Exactly the same argument applies for the *x-z* plane and the *y-z* plane showing that the line integrals in those planes correspond to the respective components of the curl of *F*. This, in turn, shows that the curl does give an indication of the "curliness" of the field, because, near a point, this is precisely what one means by curliness, i.e. that the line integral is non-zero. So, if curl is zero, the line integral around the region arbitrarily close to the point is zero.

To move to a finite loop integral, we complete the previously mentioned sum. The path of a finite line integral is shown in blue in Figure 5-17; the smaller black loops approximate the larger loop better and better as one approaches infinitesimally small loops. In short, in the limit, these small line integrals sum up to the line integral around the area because where the regions touch, i.e. in the middle, leads to a net integral of zero because each integral path in the interior is paired with one of the opposite sign. In the diagram, only those with only one arrow on a line segment contribute to the sum of all the differential line integrals.

It is intuitively clear that this argument will work for any Euclidean surface. Combining this with *equation* 5.49 , we, thus, obtain the finite version of that equation, which is also called ***Stokes Theorem:***

5.50 $$\oint_{Path} \vec{F}\cdot d\vec{x} = \oiint_{\text{Surface defined by Path}}\left(\vec{\nabla}\times\vec{F}\right)\cdot\hat{n}\,dA$$

Exercise: re-derive this in detail without referencing the textbook

Stokes theorem, in turn, implies that when the curl of a force field is zero, its line integral around any path is zero and, thus, that force field is path independent and thus conservative.

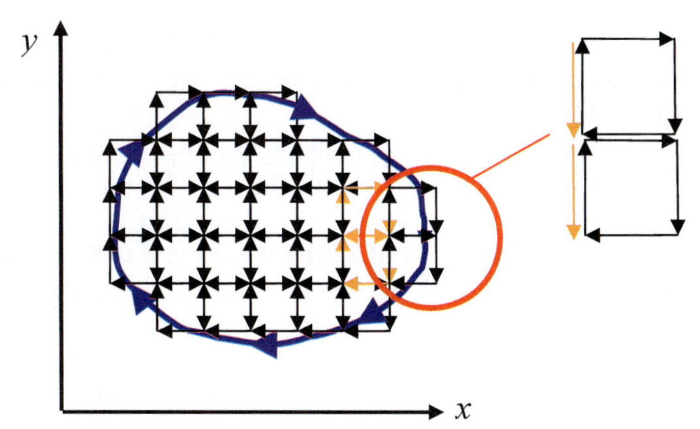

Figure 5-17: We add up the line integrals of all the small square loops that approximately cover the *interior surface bounded by the blue line*. In the limit of infinitesimally small loops, this sum equals the integral along the blue curve because the integral along *each* little line segment in the interior has an opposite sign pair that cancels its contribution to the sum. This cancellation is represented in the figure by arrows of opposing directions on a given segment. The upper right shows a close up of two such small path integrals with the cancellation at the shared boundaries shown. The two orange arrows in this magnified view represent contributions from the neighboring loops not visible in the magnification.

Mathematical points of relevance: *(on Divergence, Gradient and Curl)*

A vector field $\vec{A}$ that has $\vec{\nabla}\times\vec{A}=0$ means there is a scalar ϕ phi such that $\vec{A}=\vec{\nabla}\phi$

A vector field $\vec{B}$ that has $\vec{\nabla}\cdot\vec{B}=0$ means there exists $\vec{A}$ such that: $\vec{B}=\vec{\nabla}\times\vec{A}$.

A vector field $\vec{E}$ with $\vec{\nabla}\cdot\vec{E}\neq 0$ means that there is a "source" of $\vec{E}$ somewhere.

As an example of the application of these ideas, consider the following example problem.

Example 5-4: Is the force field $\vec{F}=F_0\hat{\theta}$ conservative? By drawing the field as below we make clear the nature of this field.

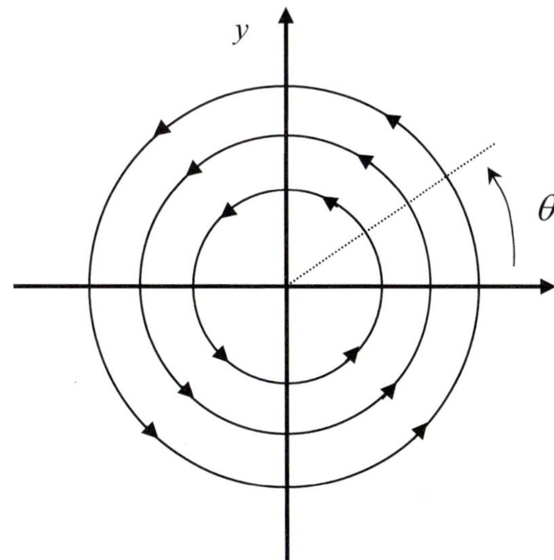

Figure 5-18: Representation of the force field: $\vec{F}=F_0\hat{\theta}$. Every point in space has a force that acts with strength F_0 and in the direction of increasing θ. Shown are a few circles that manifest the angular nature of this force field. The vectors, which we represent as straight arrows, themselves are obviously not curved, rather they are tangent to each point of the circles shown.

The curl of this field is obviously not zero. Further, because of this, we immediately see that the field is not path independent; in particular, the line integral around a circle centered on the center of the field gives:

5.51 $$\int_0^{\theta_0}\left(F_0\hat{\theta}\right)\cdot\left(r\,d\theta\,\hat{\theta}\right)=\int_0^{\theta_0}F_0 r\,d\theta=F_0\cdot\left(r\,\theta_0\right)\neq 0\,.$$

c. Examples of Conservative Forces

Many important forces, such as gravity, electric force and magnetism, taken as defined simply in beginning level physics, are conservative. Let's look briefly at gravity and electric force.

Gravity: Near the Surface of Earth

This first case of gravity is very simple and approximates what we see near the surface of the Earth in everyday life. To test the force near the ground, we make what we call a test mass of mass *m*, say $1\,kg$. Since, as we will see in detail in the next chapter, every mass generates its own gravitational field, we choose the mass small enough (roughly, $m_{test} << m_{Earth}$)[17] so that its field does not interfere with the gravity field of the Earth that we are trying to measure. (That is, it does not interfere at our level of interest, also referred to as the order of magnitude of our approximation, or, experimentally, we say we work within the error bars of our measurement).

We also make the test mass very small in dimensional size so that we can very finely sample throughout space how strong the gravity is at each point. We measure how strong the gravity is at each point. When we graph this strength as a vector field, as shown in Figure 5-19, we see that it is simply a uniform field; the direction is downward at $9.8\,N$ for our $1\,kg$ test mass. We can easily calculate the potential energy by integrating along a vertical line to get the work; we leave this as an exercise for the student, who should take a moment to do this calculation. The result is: $V(x)=m\,g\,x$; we can check this by differentiating to see the force is thus $\vec{F}(x)=-m\,g\,\hat{x}$. This makes problems that seem hard easy to solve.

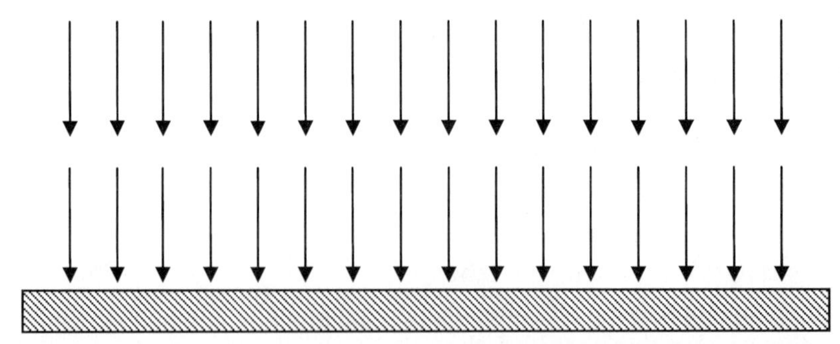

Figure 5-19: The gravitational force field near the surface of a small region of the Earth is everywhere approximately vertical and of magnitude $9.8\,N$ for a 1 *kg* test mass.

---

[17] More precisely, $\left|\vec{F}_{test\ on\ body}\right| << \left|\vec{F}_{earth\ on\ body}\right| \Rightarrow \dfrac{G\,m_{test}}{(\text{distance to test mass})^2} << \dfrac{G\,m_{earth}}{(\text{distance to earth})^2}$. See end of chapter problem, for converting this to an inequality of the masses. Note the implicit reference to third body.

For example, how high will a ball go that is tossed from the ground at a speed v (see Figure 5-20)? We know that initially the ball has: $E = \frac{1}{2} m v^2$, because the potential energy (*PE*) is zero at the ground. At the point of maximum height, say *h*, the ball has no kinetic energy, so $E = m g h$. Since E is the same forever, we can write $E = E_0 = \frac{1}{2} m v^2 = m g h$, which gives $h = \frac{v^2}{2g}$ *or* $v = \sqrt{2gh}$. The relation between the total energy E_0 , the potential energy and the kinetic energy is illustrated in Figure 5-21. Plugging in the numbers: if you want to throw the ball 10 *meters* ($\sim 33\, ft$) high, we must throw it roughly $14\, m/s$ or about $30\, miles\ per\ hour$. If you throw it $100\, mph$ ($\sim 45 m/s$), it will go $300\, ft$ ($\sim 100 m$).

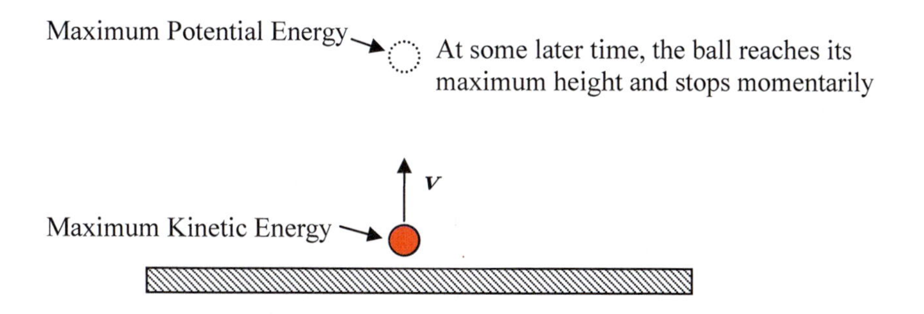

Figure 5-20: Ball starts at speed v directed vertically upward.

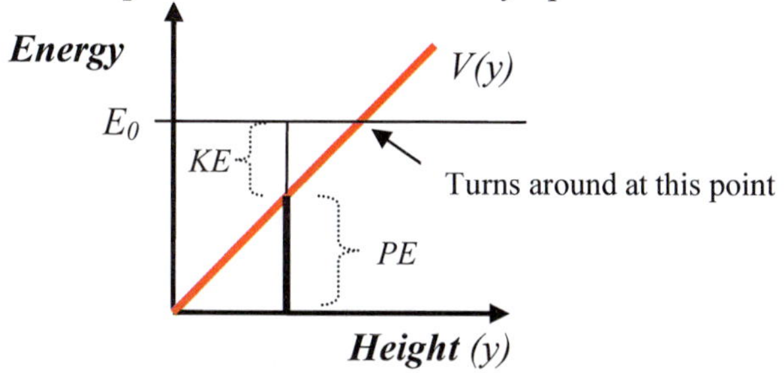

Figure 5-21: Energy versus height diagram for illustrating the motion of a ball under the influence of the gravitational field near the Earth's surface. The red line shows the graph of the ball's potential energy versus its height. E_0 is the total energy initially given the ball. The ball reaches its peak height where the red line crosses the $E = E_0$. Shown are the kinetic energy (*KE*) and the potential energy (*PE*) at an arbitrarily chosen point in the ascent (or descent) of the ball.

The next level of complication is to consider this gravity field as it acts on an object constrained by another force such as the surface of the ground. In particular, consider a marble *sliding* up and down the sides of a smooth (frictionless) round bottom bowl or a bicycle going down into a valley and back up the other side. The mechanics of this situation is similar, in some ways, to the motion of the spring as we will now see.

Consider motion in the x-direction. Because the object is confined to the valley surface, once we know x, the horizontal position, we also know y, the vertical position. In particular, take the mass of the bike (or marble) to be m and the floor of the valley (or bowl) to have equation $y(x) = \frac{1}{2}\frac{k}{gm}x^2$ (Figure 5-22 shows a graph of such a parabolic function). This means the potential energy (PE) for the ball is: $V(x) = mgy = \frac{1}{2}kx^2$, exactly like the spring. Similarly, the total energy, which is unchanging, is: $E = T + V = \frac{1}{2}m\mathbf{v}^2 + \frac{1}{2}kx^2$.

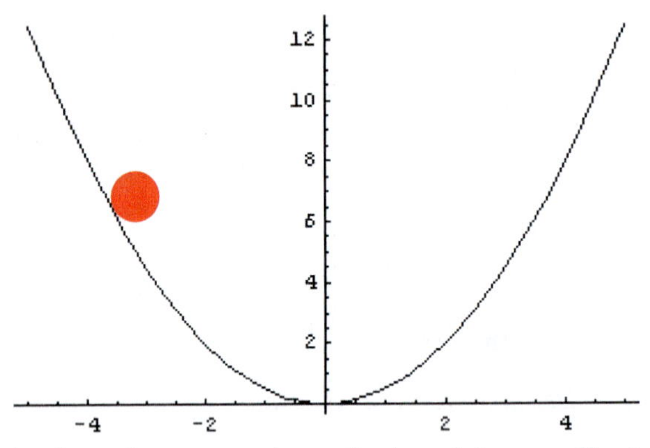

Figure 5-22: A plot of cross-section of a bowl (or a valley) with a marble (or a bike) sliding on the curved surface under the influence of gravity. Arbitrary units are used on the axes.

In advanced treatments, one will re-write this as: $E = \frac{p^2}{2m} + \frac{1}{2}kx^2$, eliminating the time derivative of position, that is the speed, in favor of the momentum. [18] At this point, one only need notice the equivalence of the two representations of the energy and that there is something more fundamental from a point of view of principle about the second (advanced) formulation with its inclusion of the measure of the impetus intensity (p) rather than its effect, the speed ($\mathbf{v}$). *Exercise:* Calculate the force in the x-direction from the PE.

Gravity: Far from the Nearly Spherical Earth

For gravity in Newtonian physics for point objects and nearly spherical objects like the Earth, we have: $\vec{F}(r) = \frac{-GmM}{r^2}\hat{r}$, where r is distance from the center of the Earth[19], where $G \sim 6.67 \times 10^{-11}\, Nm^2/kg^2$ is a universal constant, and the masses, m, M are given in kilograms. We graph this function in Figure 5-23a. Note that the field starts at Earth's surface, r_0. If we think of the Earth as a round ball, then the force pulls toward the center of this ball, leaving field lines that look like a balled up porcupine with its needles straight

[18] In advanced classical mechanics, this functional form of the energy will be called the Hamiltonian and be symbolized by H when treated as a function.

[19] This only applies beyond the radius of the spherical object, in this case above the Earth's surface, so that $\vec{F}$ for $r < r_{surface}$ cannot be obtained from the formula.

out. Even without knowing the Earth was the source of such a field, measuring the field would indicate that the field is emanating from an object (source) at the center. Indeed, the divergence of the field is not zero and the source is the mass of the Earth (see Chapter 8). However, the curl is zero and thus we can define the potential energy. It is:

5.52 $$V(r) = -\frac{GmM}{r}$$, which is illustrated in Figure 5-23b.

In this expression for the gravitational potential energy, we have implicitly defined the potential to be zero very far from the Earth. Indeed, through the use of the being of reason of an actualized infinity (which, as we have discussed in Chapter 2, in the final analysis, refers to the limit), we say that the potential energy is zero at infinity.

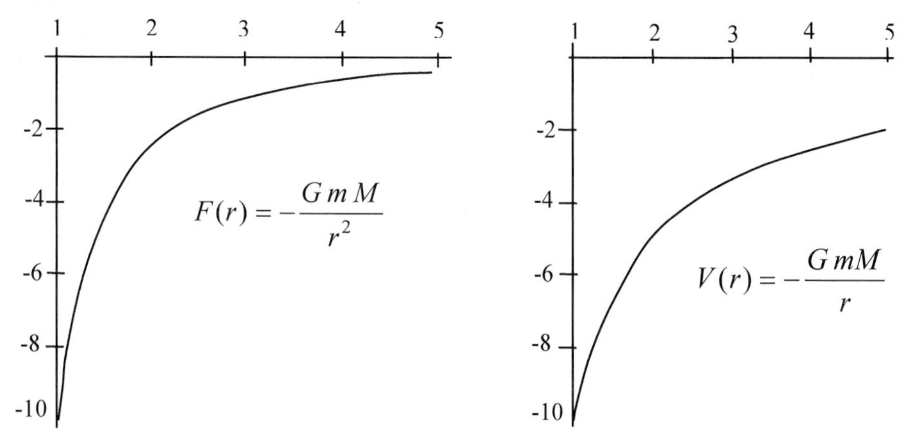

Figure 5-23: a. *(left)* Gravitational force as a function of distance from the center of a spherical body (such as the Earth) generating the force field. **b.** *(right)* Gravitational potential energy as function of distance. The horizontal axis is given in units of r_0, the radius of the spherical body. Vertical units are chosen for easy understanding of the plot.

Why do we define it in this way? After all, mathematically we could define it by adding any constant. In general, the reason we pick the constant to be a certain value is to capture what's essential in the given set of substances under consideration in the given situation (leaving out the many, many details in order to capture what we can at a very general level[20]). The reason for the choice in this case is that given only the Earth and no other massive bodies, there are only two preferred radii, $r = \text{infinity}$ and $r = r_{earth}$. We could pick: $V = -k/r + k/r_{earth}$, making $V = 0$ at r_{earth}; however, this involves carrying an extra term around that really does not serve any purpose. Indeed, it makes sense when one is "at infinity" to not have a potential at all since the force is nil there. Adopting the zero-at-infinity convention, thus, is natural to the physical situation with which we are confronted. However, if the situation were otherwise, we would then choose what is best for understanding and solving that particular problem--the two go together.

With the above definition of potential energy, we see that an object that starts at rest from far away has little potential energy and no kinetic energy; in fact, "at infinity," it has $E = 0$. Gradually, its kinetic energy builds as the gravitational field acts on the massive

[20] Again we try to leave out those whose influence is much less significant under certain conditions.

body under consideration, giving it impetus at increasing rate, pulling it toward the Earth. In the process, the potential energy becomes more negative (cf. Figure 5-23b), "canceling" the kinetic energy, thus keeping the total energy zero. Note that, except at infinity which is a mental construct, the potential energy is always negative, becoming more negative, that is lower, as the object falls inward. In the limit of a body falling in from rest at infinity, the final speed is:

$$v = \sqrt{\frac{2GM}{r}} \tag{5.53}$$

In accelerating the body to this speed, we say that potential energy has been converted to kinetic energy; potential energy has been allowed to activate impetus in the falling body. The potency of the gravity field has been allowed to act over the distance to the Earth, but further action was blocked by the surface of the Earth. Again, notice how the potential energy is defined relative to the given physical situation. In particular, notice that what fields are where and how far the fields can act is determined by what actual physical bodies are in what places.

Now, suppose *instead* of letting the body *fall* from infinitely far away, we retard its motion as it comes in, *not* allowing it to pick up any appreciable speed. Imagine further that we take the limit of slower and slower speeds. Then, the work we do in inhibiting its motion will be precisely given by: $V(r_0)$, (cf. equation 5.52) for this is the definition of the potential energy. This work is negative, so, $|V(r_0)|$ is the amount of energy given *to* us that we must "absorb" to bring the body from infinity to $r = r_0$. On the other hand, $-V(r_0)$ is the (positive) work exerted by the body on us as we inhibited its inward motion. The body exerts positive work on us, giving us energy (i.e. ultimately activating impetus of a certain amount) that we can use to, say, turn a flywheel (see end of chapter problem in Chapter 7).

If we have a system of bodies in a given arrangement, the potential energy is the work required to assemble the system. We will discuss gravity in depth in Chapter 8.

Electric Fields

What we have said about gravity fields easily extends to the case of electric fields, where the electrical force is written: $\vec{F}(r) = \frac{k\, q_1 q_2}{r^2}\hat{r}$. The difference is in the sign of the force and the fact that the charges q_1 and q_2, unlike the masses, can be negative.

Conservative forces are obviously very important because of the important forces, such as electric and gravitational, that are conservative and because of the simple mathematical properties of conservative forces.

d. Elastic Collision Re-examined

With our understanding of energy and conserved forces, in particular springs, we can advance our understanding of elastic collisions. Namely, using a spring to model the surface forces, as we did in Chapter 4, we can investigate how various aspects of the collision depend on the initial energy and mass. In Figure 5-24, we reproduce the diagram of Chapter 4 that shows the interaction during the collision of two balls using this spring model.

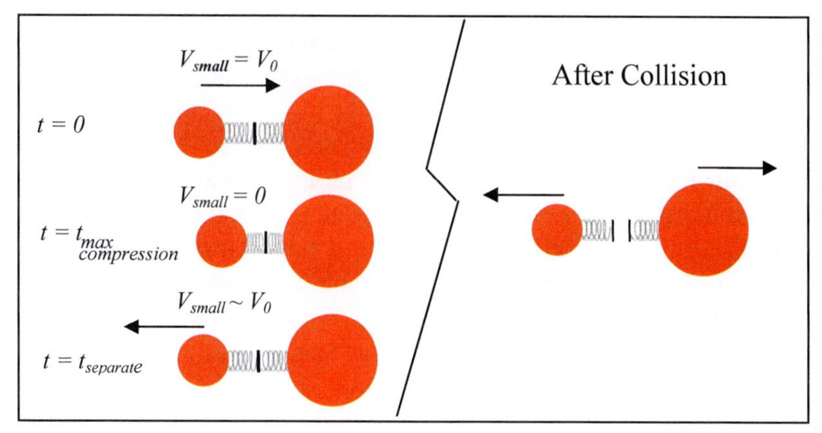

Figure 5-24: *(left)* Balls as seen during three stages of collision; the ball masses in our experiments are varied. Starting at far left moving to the right: **a.** The springs' ends just meet. **b.** The springs are at maximum compression—note that because of Newton's third law the compression of each (identical) spring must be equal.[21] **c.** The springs again return to equilibrium and lose contact and thus cease to interact. *(right)* **d.** After the collision sequence, the balls continue on with the velocities shown; the springs (the surfaces) may continue to vibrate back and forth for some short time after the interaction ceases.

Figure 5-25 shows the results of colliding experiments run using the interactive physics file *4.1-collision of masses with springs.ip*. We take the spring constant, *k*, to be $10\,N/m$.[22]

The *first* set of experiments consists in colliding two balls with masses that sum to $100\,kg$. In each case, the impacting ball is moving at $1\,m/s$ from the left, and the ball into which it collides is at rest at the right. First, we collide a lighter mass against a heavier one, then, in each successive experiment, we repeat the experiment with some of the mass shifted from the rest mass to the moving mass, until the last experiment sends a $99.1\,kg$ body moving at $1\,m/s$ towards a $0.1\,kg$ body at rest.

The *second* set of experiments maintains a large impacting kinetic energy and mass and varies the mass of the second body.

For both sets of experiments, we record the kinetic energy of each ball after the collision, the time of interaction and the maximum spring compression, which is directly related to the maximum force applied.

Notice that the *interaction time* is greater when the two balls are close to the same mass. In fact, it can be shown that, when keeping the sum of the masses the same, the

[21] Note that by making the spring massless (i.e. as close to zero mass as we like), we can avoid consideration of the dynamics peculiar to the spring.

[22] For completeness, note that the equilibrium length of the spring is $1.4\,meters$.

greatest interaction time occurs when the balls have equal mass.[23] To understand this, recall that during the collision the forces applied by one body on the other, which are the only forces, must be equal by Newton's third law. This means that each ball's impetus changes by the *same* amount in a given interval of time. However, if one ball is of significantly lower mass than the other, the change in the intensity of the impetus causes a bigger change in its speed. The result of this is that this smaller mass ball gets an impetus to move away faster than the other. By contrast, if both balls equally share the mass, they spend more time near each other, one not being able to escape the interaction more quickly than the other. Of course, for this same reason, two equal mass balls of larger mass will spend even more time near each other than two of less mass.

This fact also explains why a greater force is seen during collisions of *equal-mass* bodies. Namely, since the impetus of the impacting ball carries it far, but not so far that it pushes its partner out of the way as quick as it would if it were much more massive. Nor is it turned around as quickly as it would be if it were much less massive. In this way, the *equal- mass* balls keep further compressing the spring for longer, needing more impetus to change each of their motions in such a way to push them out of contact with each other. Of course, heavier *equal-mass* balls will need even more force and, as mentioned, have even longer interaction times as is verified in the data in Figure 5-25. The student is encouraged to further analyze this data and to try some of his own experiments.

Obviously, during an elastic collision such as illustrated in Figure 5-24, energy and momentum are transferred in some way from the impacting body (left ball) to the target body (right ball). Experiments such as the above can tell us what mass ratio gives the maximum energy transfer. By such continued experimentation and/or by the calculation one does in a problem at the end of Chapter 3, we can generate the graph, shown in Figure 5-26, of the energy transfer ratio, $\frac{E_{right}^{After}}{E_{left}^{Before}}$, the ratio of the energy of the right ball after the collision to the initial energy of the left ball, versus the mass ratio $\frac{m_{right}}{m_{left}}$. At the mass ratio of one, the energy transfer is maximum. This is the point at which all the momentum of the first ball is transferred to the second ball, and since the masses are the same, this means the energy of the first ball is transferred to the second ball. This is obviously the maximal case, for, to transfer more energy it would have to be created, thus violating conservation of energy.

[23] Interaction time $\sim \sqrt{\mu / k} = \sqrt{m_1 m_2 / (k(m_1 + m_2))}$. Since the square of the interaction time t_i will have its maximum in the same place as t_i itself, we maximize $t_i^{\,2}$ to simplify the math. Taking $c = m_1 + m_2$, we need to maximize a function of the form: $f(m) = m(c-m)/c$; $f'(m) = (c-2m)/c = 0 \Rightarrow c = 2m$.

Mass (kg)		KE (J) Before		KE (J) After		Maximum Spring Compression	Interaction time
M_{Left}	M_{Right}	KE_{Left}	KE_{Right}	KE_{Left}	KE_{Right}	ΔX_{min}	$\Delta t_{Interact}$
1	99	.5	0	.47	.02	0.23 m	1.45 s
10	90	5	0	3.2	1.8	0.68 m	4.25 s
30	70	15	0	2.4	12.6	1.01 m	6.45 s
40	60	20	0	0.8	19.2	1.10 m	6.90 s
50	50	25	0	0	25	1.12 m	7.05 s
60	40	30	0	1.2	28.8	1.10 m	6.90 s
90	10	45	0	28.8	16.1	0.68 m	4.25 s
99.1	.1	49.95	0	49.7	.2	0.08 m	.45 s
100	100	50	0	0	50	1.31 m	8.05 s
50	50	50	0	0	25	1.12 m	7.05 s
50	30	25	0	1.6	23.4	0.97 m	6.10 s
50	20	25	0	4.6	20.4	0.85 m	5.35 s
50	10	25	0	11.1	13.9	0.65 m	4.05 s
50	5	25	0	16.7	8.2	0.49 m	3.00 s
50	1	25	0	23.1	1.9	0.22 m	1.40 s

Figure 5-25: The results of a series of collisions of two balls, one on the right the other on the left. We record simply the compression of the left spring (recall that because of Newton's third law they should be equal; however, there will be some vibration so the spring compression at any point is slightly different between the two—we ignore this for our current discussion). Note that $50\,J$ for $100\,kg$ mass corresponds to $1\,m/s$ speed. Also, the spring constant is $k = 10\,N/m$.

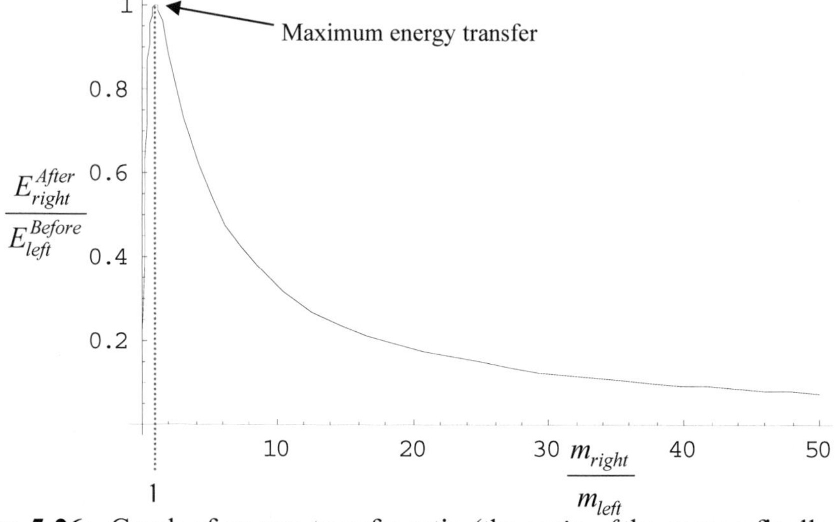

Figure 5-26: Graph of energy transfer ratio (the *ratio of* the energy finally given the right ball after the collision *to* the initial energy of the impacting ball) against the mass ratio (the ratio of the mass of the right ball to that of the left ball).

Kinetic Energy, Collisions and Center of Mass Frame

Many problems can now be solved with the concept of potential energy and kinetic energy in hand. A couple more general conclusions will be helpful in increasing our understanding of these concepts and their applicability.

a. Relativity of Kinetic Energy

First, we should note that kinetic energy, like momentum, will appear different in different frames of reference. For instance, as we noted in early chapters, if we speed up along side of a baseball that is moving at speed v, we can make it look as if it is stopped, making it look as if it had no impetus. If on that baseball another object, say an ant, was moving at speed v_{ant} relative to the baseball (making its total speed $v + v_{ant}$), we would only see it as moving at speed v_{ant}. Since we have assumed the mass doesn't change (an excellent approximation[24] for those things we can directly access with our senses), it is clear that the measure of activity, the kinetic energy, being a function of speed and mass, would also *appear* to be changed. The ball would have no kinetic energy and the ant would change from $\frac{1}{2}m\,(v + v_{ant})^2$ *to* $\frac{1}{2}m\,v_{ant}{}^2$. Again, as we've noted, just because we find it convenient to do such "transformations" and ignore the impetus the object actually has doesn't mean it doesn't have it. Still, such transformations are facts (real symmetries) about the world, and when we are concentrating on them, it is certainly reasonable to do just that, leaving out what is not relevant to the problem. In this way, complex problems can be isolated somewhat from their environment. One would not want to have to take into account the motion of the Earth around the sun, for instance, in finding the speed of a baseball falling from the sky. We say we "transform" to the "frame" in which the Earth is at rest.[25]

We will investigate this "transformation" and "frame" issue more in Chapter 10 on Special Relativity. For now, only be aware that when simplifications can be made by moving to the viewpoint of a convenient object, or even abstract reference system of coordinates (i.e. speed of our x, y, z axis origin), that is moving at constant speed, all calculations of kinetic energy and momentum can be done relative to that frame as if it were at rest.

b. Center of Mass Frame

The center of mass is one abstract frame, or reference system of coordinates.[26] It's often quite helpful to look at collision problems from the center of mass frame. In particle physics, for example, where the search for the Higgs boson is now, in 2006, becoming more pressing, it is useful. In the so-called "collider physics" that might produce Higgs bosons such as lepton colliders (*e+* rammed into *e-*) or hadron colliders (protons and

---

[24] We will see in Chapter 10 that it is not true outside a certain range.

[25] Of course, this only applies to constant linear motions, not, for example, to the rotational motion of the Earth. There are important issues about acceleration caused by gravity that need to be addressed as well, but they, as we will see, are different.

[26] Note that the difference between a frame and a coordinate system is subtle. We can change our frame, move at a constant speed, but we can pick any number of different coordinate systems to label different places and times once we are moving at that speed. We might choose a spherical one if we were on a spherical surface; we would probably choose a Cartesian one if we were in space with no significant forces or other bodies to distinguish one region of our environment from another.

nucleons collided into each other)[27], there are optimal ways to create particles. To explain the use for Higgs boson creation we briefly discuss such "creation."

In special relativity, you will learn that a given amount of activity, energy, can indeed be lost as long as it results in the creation of a body or part of a body which has a mass, given by the $m = E/c^2$. Now, given a certain amount of energy available to accelerate the particles, say of $e+$ and e-, the maximum amount of energy is available for creation of new particles, such as the Higgs, when the particles share the energy in such a way that they move at each other with the same momentum. Said another way, a given amount of energy spent in colliding two particles is best spent in the center of mass frame. This is easy to understand, for any momentum that one particle has over that of the other cannot be put towards creation of massive bodies, but most go towards momentum of the old or newly created particles. In terms of the center of mass frame, the "excess" energy shows up as motion of the center of mass.

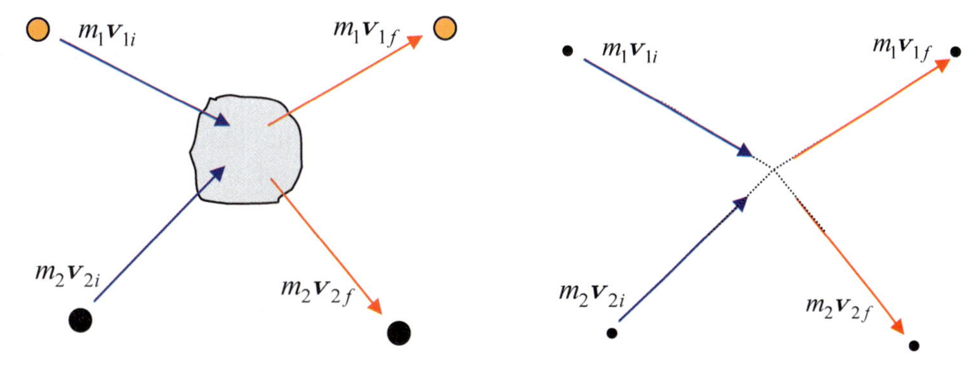

Figure 5-27: **a.** *(left)* Elastic collision of two objects. **b.** *(right)* Same particles considered as colliding point particles.

To further understand the center of mass frame's importance for collisions, consider, first in an arbitrary frame, two bodies of arbitrary mass and velocity that collide elastically as depicted in Figure 5-27a. The general place of the collision is depicted in the figure by the light gray region. Although the specific details of the collision are unknown, we do know some general aspects of it. Most generally, within a certain region, the objects interact with each other in such a way that their respective momenta are changed.

Figure 5-27b shows the same collision leaving out the spatial extension and qualitative details of the collision, concentrating solely on the state of the particles before and after the collision. The bodies are treated as "point particles" resulting in a point "region" for the collision. By introduction of these mental constructs, we focus our mathematics on the momentum of the particles and their mass, speed and direction before and after the collision. Furthermore, both Figure 5-27a and Figure 5-27b depict the bodies as remaining in the same plane after colliding; this simplification is helpful but not necessary.

---

[27] The Large Hadron Collider (LHC) will collide beams of protons given KE of $14\,TeV$. Beams of lead nuclei will be also accelerated, smashing together with a KE of $1150\,TeV$. $1\,TeV = 1.6\times10^{-7}\,J$

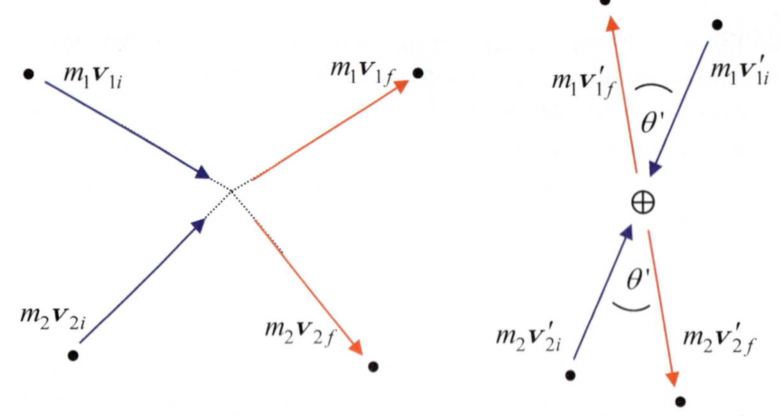

Figure 5-28: a. *(left)* The elastic collision in the laboratory frame and **b.** *(right)* The center of mass frame. Note the use of primes for measurements made in the center of mass frame.

The center of mass frame (cf. Figure 5-28b) can give further insight into the nature of the collision because it leaves out momenta not relevant to the changes caused by the collision. Recall that the center of mass frame is defined as that frame in which the total momentum of the system of bodies is zero. Thus, the center of mass frames moves at velocity of $\vec{v}_{cm}$ relative to the laboratory frame, where $\vec{v}_{cm}$ is the total momentum in the laboratory frame *divided by* the combined mass:[28]

5.54 $$\vec{v}_{cm} = \frac{\vec{p}}{M} = \frac{m_1\vec{v}_{1i} + m_2\vec{v}_{2i}}{m_1 + m_2}.$$

To imagine yourself in the center of mass frame, jump into a craft moving in the direction and speed specified by $\vec{v}_{cm}$.

Since the combined momentum in the center of mass frame is zero, the two bodies must approach each other with momenta that are equal in magnitude but opposite in direction. This must also be true of the retreating particles because of conservation of momentum. Further, the collision is elastic, i.e. sum of kinetic energy is the same before and after ($\frac{1}{2}m_1\vec{v}'^{\,2}_{1i} + \frac{1}{2}m_2\vec{v}'^{\,2}_{2i} = \frac{1}{2}m_1\vec{v}'^{\,2}_{1f} + \frac{1}{2}m_2\vec{v}'^{\,2}_{2f}$),[29] and so each body must conserve the magnitude of its momentum, but not its direction. The result is that the bodies can "scatter" at any angle relative to their incoming velocities, provided that they scatter in opposite directions so as to conserve momentum (see Figure 5-28b and Figure 5-29b).

This simple view of scattering only occurs in the center of mass frame. As shown in the figures, the center of mass frame reduces the *incoming* velocity vectors and the *outgoing* velocity vectors *each* to one dimension. In the *lab frame*, the "head-on collision" results in a simple reversal of direction, i.e. the particles, both before and after, all stay on a single line. By contrast, *all* collisions are "head-on" (in the sense having anti-parallel incoming velocity vectors) in the center of mass frame. However, only in special cases do the particles come in and go out *on the same line*. In general, in the *center of mass frame*,

[28] The center of mass of a system is: $\vec{x}_{cm} \equiv \sum_{i=1}^{N} m_i\vec{x}_i / \sum_{i=1}^{N} m_i$

[29] Note that primes indicate quantities evaluated in center of mass frame.

the collision occurs in a plane defined by the intersection of the lines (and, again, they are always lines not planes) defined by the initial and final velocities. Of course, because of this, we can always define our axes such that the collision occurs on a convenient plane, say, for example, the *x-z* plane (see Figure 5-29b).

By contrast, the collision as viewed in the *laboratory frame* is generally three dimensional, since the incoming velocities define a plane, which does *not have to be the same* plane defined by the outgoing velocities (see Figure 5-29a).

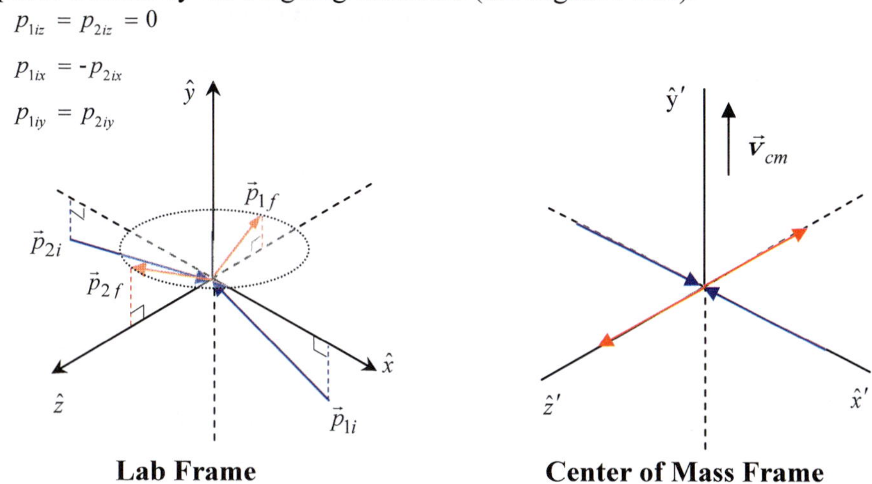

Figure 5-29: a. *(left)* An example of elastic scattering of two bodies colliding in the *lab frame* in the *xy*-plane with equal and opposite momenta in the *x*-direction. The scattering is shown for the particular case when the momentum along the *x*-axis is entirely shifted to the *z*-axis. The dotted circle indicates where the final momentum vectors (shown in red) can fall; they can be rotated, as a pair, anywhere around the *y*-axis. In the lab frame, the final and initial particles have components of momentum in the *y*-direction that are "transformed away" in the center of mass frame. **b.** *(right)* In the *center of mass frame,* the scattering is reduced to a two-dimensional problem.

To illustrate further the use of the center of mass frame, we will solve the collision for the simple case: $\theta' = 0$, i.e. where all the activity is confined to one dimension in the center of mass frame. The incoming velocities for each object in the center of mass frame are (in terms of the incoming velocities in the lab frame):

5.55a,b $$\vec{v}'_{1i} = \vec{v}_{1i} - \vec{v}_{cm} = \frac{m_1 v_{1i} - m_2 v_{2i}}{m_1 + m_2}; \qquad \vec{v}'_{2i} = \vec{v}_{2i} - \vec{v}_{cm} = \frac{m_2 v_{2i} - m_1 v_{1i}}{m_1 + m_2}$$

Similarly: for the final velocities:

5.56 $$\vec{v}'_{1f} = \vec{v}_{1f} - \vec{v}_{cm} \qquad \vec{v}'_{2f} = \vec{v}_{2f} - \vec{v}_{cm}$$

where *the prime indicates the quantity is evaluated in the center of mass frame.* Because of the symmetry of the problem, the equations are identical. We will sometimes make use of this fact, writing only one general equation that can be applied to each particle.

Using the relationship between the velocities after the collision for $\theta' = 0$ we get:

5.57 $$\vec{v}'_{1f} = -\vec{v}'_{1i} = \vec{v}_{cm} - \vec{v}_{1i} \qquad \vec{v}'_{2f} = -\vec{v}'_{2i} = \vec{v}_{cm} - \vec{v}_{2i}$$

From equation (5.56), the final velocities in the laboratory frame are:

5.58 $$\vec{v}_{1f} = 2\vec{v}_{cm} - \vec{v}_{1i} \qquad \vec{v}_{2f} = 2\vec{v}_{cm} - \vec{v}_{2f}$$

Solving for the final velocity in the laboratory frame of the first particle gives:

5.59 $$\vec{v}_{1f} = \left(\frac{2m_1}{m_1+m_2} - 1\right)\vec{v}_{1i} + \frac{2m_2}{m_1+m_2}\vec{v}_{2i}$$

Simplification and repeating the calculation for the second particle gives the following final velocities in terms of the particles' masses and initial velocities:

5.60 $$\vec{v}_{1f} = \frac{m_1 - m_2}{m_1+m_2}\vec{v}_{1i} + \frac{2m_2}{m_1+m_2}\vec{v}_{2i} \qquad \vec{v}_{2f} = \frac{m_2 - m_1}{m_1+m_2}\vec{v}_{2i} + \frac{2m_1}{m_1+m_2}\vec{v}_{1i}$$

For $m_1 = m_2$ (the elastic collision of equal masses), the particles exchange velocities:

5.61 $$\mathbf{v}_{1f} = \mathbf{v}_{2i} \qquad \mathbf{v}_{2f} = \mathbf{v}_{1i}$$

This simple exchange of velocities corresponds to the expected result for a one-dimensional elastic collision of equal masses.

We can obviously write the energy in the center of mass frame as:

5.62 $$E' = \sum_i \frac{1}{2} m_i \mathbf{v}_i'^{\,2}$$

and it is straightforward to show (using equations 5.54 and 5.55) that the energy in the lab frame is:

5.63 $$E_{Lab} = \frac{1}{2} M \vec{v}_{cm}^{\,2} + \frac{1}{2}\sum_{i=1}^{N} m_i \vec{v}'^{\,2}_i \; with\, M = \sum_{i=1}^{N} m_i \qquad for\ N = 2$$

One can also show the above for all N (see end of chapter problem).

To get a concrete appreciation of the importance of the center of mass system, the student is encouraged to calculate the energy required to make a Higgs boson (assume a mass-energy of $200\,GeV$) from colliding $e+$ and $e-$ when 1) one is at rest and 2) if they are coming at each other at equal speed. It will be seen that the energy required is less for case 2 for reasons previously discussed, but not significantly less; it will become significant in special relativity (see end of chapter problem).

The collision can be solved for arbitrary θ' (cf. Figure 5-28b). In this case, equations 5.57 are no longer valid since $\vec{v}'_f = \vec{v}'_f(\theta')$, i.e. each of the two particle final velocities, generically represented by $\mathbf{v}'_f$, is a function of θ'. The solution in the center of mass frame can be found by rotating the head-on solution ($\vec{v}'_f = -\vec{v}'_i$) by the angle θ'

5.64 $$\vec{v}'_f(\theta') = (-\mathbf{v}'_{ix}\cos\theta' + \mathbf{v}'_{iy}\sin\theta')\hat{x} + (-\mathbf{v}'_{ix}\sin\theta' - \mathbf{v}'_{iy}\cos\theta')\hat{y}$$

Note that such an equation applies to each particle. (Note that for $\theta' = 0$, it reduces to equations 5.57)

c. Conservation of Kinetic Energy for a System of Particles.

Another important tool, and a simple extension of what we've seen earlier, is the generalization of the work-energy theorem to many particles.

The work-energy theorem (where we assume no potential energy and no internal motions--so that all collisions are elastic) is, for a single particle that we label *i*,

5.65 $$\int \vec{F}_i \cdot \vec{x}_i = \frac{1}{2} m_i \boldsymbol{v}_i^2 - \frac{1}{2} m_i \boldsymbol{v}_{i0}^2 \, ,$$

where $\boldsymbol{v}_{i0}$ is the speed of particles before action of forces. If we have a collection of *N* bodies, we then let *i* be any number between 1 and *N*, so that if i=1, we deal with body number 1. This means we now have *N* equations, which are the *N* statements of the work-energy theorem; one for each particle. We simply add together these equations; i.e. sum the work of each particle on one side and the energy on the other side to generate to get:

5.66 $$\sum_{i=1}^{N} \int \vec{F}_i \cdot d\vec{x}_i = \sum_{i=1}^{N} \frac{1}{2} m_i \boldsymbol{v}_i^2 - \sum_{i=1}^{N} \frac{1}{2} m_i \boldsymbol{v}_{i0}^2$$

Now, as we said before, if there are no forces acting from outside the system, we know that for every body *i* that has $\vec{F}_i$ there must be an $i+1$ body that has $\vec{F}_{i+1} = -\vec{F}_i$. This means there must be an even number of bodies with forces, which is clear from the physics, for the forces go together, one on the other. Note, also, that these force pairs share a $d\vec{x}_i$; this means that the left hand side must sum to zero.[30] So, that we get the *general statement* of the Conservation of Kinetic Energy.

$$\sum_{i=1}^{N} \frac{1}{2} m_i \boldsymbol{v}_i^2 = \sum_{i=1}^{N} \frac{1}{2} m_i \boldsymbol{v}_{i0}^2$$

$$K_{Total} = \text{constant}$$

The sum of all the kinetic energies of an *isolated* system of bodies never changes.

d. The Rocket Problem

We can now say a couple new things about our rocket trip to Mars. In Chapter 4, we calculated that a rocket under the influence of gravity would reach the following final speed, after time t_f.

5.67 $$\boldsymbol{v}_{final} = \boldsymbol{v}_i + \boldsymbol{v}_p \ln[\frac{M_i}{M_f}] - g\,t_f$$

It is not particularly helpful to re-derive this relationship from an energy perspective, but other information can be extracted.

Escape Velocity of Earth

For instance, using conservation of mechanical energy, we can quickly calculate the answer to a very important question in space travel. How fast must a rocket go to escape the gravitational pull of the planet? More specifically, at a given radius from the center of

---

[30] Notice that these dx_i can be any chosen motion (when we want to indicate explicitly that this is what we mean we use "δ" rather than "d"). In this case, the fact that $\sum_{i=1}^{N} \int \vec{F}_i \cdot \delta\vec{x}_i = 0$ is called the principle of virtual work.

the Earth, how fast must a rocket be going to just barely be moving once it "gets to infinity?" This speed is called the *escape velocity*. If it is traveling any slower than this, it will eventually reach a point where the outgoing impetus will be drained from the spacecraft and gravity will begin to give it impetus to come back to Earth. In other words, an object launched faster than the escape velocity will violate the erroneous, but popularly acclaimed, "whatever goes up must come down" rule. In fact, we've seen the answer to this question in our discussion of potential energy:

5.68 $$v_{escape} = \sqrt{\frac{2GM}{r_0}}$$

In this equation, we use the mass of the rocket for M. We further assume, M is constant for the calculation by assuming all of the rocket's propellant has been expended by the time it reaches a radial distance r_0 from the center of the Earth. If the fuel is used up relatively close to ground then we can approximate: $r_0 \sim r_{earth}$. The escape velocity from the surface of the Earth is:

5.69 $$v_{escape} \sim 1.1x10^4\, m/\sec = 11\, km/\sec \sim 7mi/\sec$$

(using: $M_{earth} \sim 6.0\times10^{24}\, kg$; $r_0 \sim 6400\, km = 6.4\times10^6\, km$)

How High Does a Spring-Loaded Rocket Go?

A review of our analysis of the rocket in Chapter 2 and 3 will remind us that force and momentum are primary in these calculations. A quick attempt to solve the rocket problem according to energy conservation methods appears to indicate that such methods are less potent than momentum and force. Nonetheless, understanding energy does give new insights into the action of the rocket. Trying to solve the rocket problem by energy methods immediately leads to the question: where did the energy come from? There is much more kinetic energy after the rocket fires than before it starts firing. After the rocket engine is emptied, it is also clear that the rocket has much more gravitational potential energy than before, for afterward, it is up in the air. Further, the fuel also gains gravitational potential energy, because it is released at various heights, and most all of the molecules of fuel have kinetic energy not only from being expelled, but from falling (having their potential energy converted to kinetic energy) and from sharing some of the speed of the rocket.[31]

So, where did all the energy come from? It comes from the energy stored in the propellant, i.e. the potential energy of the propellant. In the case of our water rocket, the pressurized air contains the stored energy. We will simplify the problem to expose the root principles and save the problem of the water and chemical propellants for end of chapter problems.

We take the fuel to be many particles that are expelled by one giant spring as illustrated in Figure 5-30 below. We also neglect the $1/r^2$ fall-off of the Earth's gravitational force. This is an important effect that cannot be ignored in interplanetary travel, so it will have to be included in that case, but not here.

[31] Note here all the quantities are defined in terms of the rest frame of the Earth and any motion of the Earth would have to be accounted for later if we were interested in it. Here, as we will do in most cases, we will ignore this aspect and just treat the motion relative to the Earth.

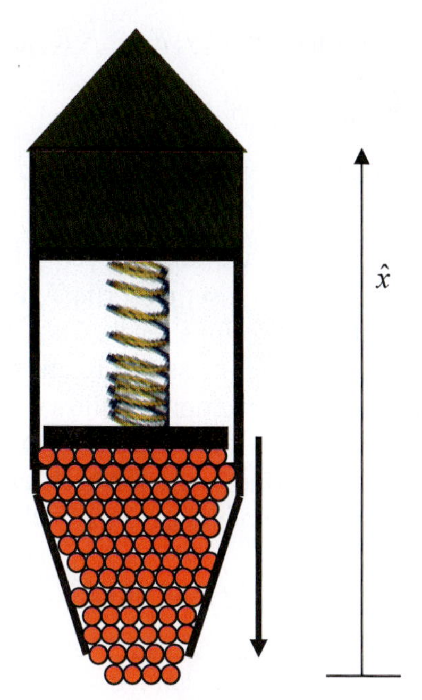

Figure 5-30: Rocket powered by expelling many tiny pellets under the pressure of a compressed spring.

Before the launch, assuming the rocket starts at rest on the ground ($x = 0$) and assuming a spring constant of k and a spring displacement of d_0, the energy is:

5.70 $$E_{before} = KE + PE_{spring} + PE_{gravity} = 0 + \frac{1}{2} k\, d_0^{\,2} + 0 = \frac{1}{2} k\, d_0^{\,2}$$

After all of the fuel is expelled, assuming the spring is left uncompressed after all the propellant is expended and each of the N fuel pellets has mass m, the energy is:[32]

5.71 $$\begin{aligned} E_{after} &= (KE_{fuel} + PE_{fuel}) \quad + (KE_{rocket} + PE_{rocket}) \quad + PE_{spring} \\ &= \sum_{i=1}^{N}\left(\frac{1}{2} m \mathbf{v}_i^{\,2} + m\, g\, x_i\right) + \frac{1}{2} M_{final}\, \mathbf{v}_{final}^{\,2} + M_{final}\, g\, x_{rocket} + 0 \end{aligned}$$

To get a more concrete answer, we have to make some assumptions. Let's say that:

5.72 $E_{rocket} \sim KE_{rocket} + PE_{rocket} >> E_{fuel} = KE_{fuel} + PE_{fuel}$, which is an ideal design, for moving the rocket requires that energy go into the rocket, not the fuel.

We can rewrite the right hand side of equation 5.71 as:

5.73 $$\begin{aligned} E_{after} &= E_{fuel} + E_{rocket} + PE_{spring} = E_{rocket}(1 + \frac{E_{fuel}}{E_{rocket}}) + PE_{spring} \\ &\approx E_{rocket} + 0 = E_{rocket} \end{aligned}$$

This manipulation shows very clearly the level of approximation that we are making; that is, it shows what we mean by saying E_{fuel} is small.

---

32 $\sum_{i=1}^{N} m = N\, m = M_{initial} - M_{final}$, so that $m = \frac{M_i - M_f}{N}$

Then, using conservation of energy, $E_{after} = E_{before}$, we have:

5.74
$$\frac{1}{2} M_{final} \, v_{final}^{\;2} + M_{final} \, g \, x_{rocket} = \frac{1}{2} k \, d_0^{\;2}$$

Now, to solve for x_{rocket}, given the method of propulsion described above, we can *not* use equation 5.67 to substitute for v_{final}; we would need to calculate v_{final} for this case (see end of chapter problem). So, we simplify still further. Assume our propulsion is instead provided by an impulse from a *set* of springs giving each pellet a speed v_p relative to the rocket. In other words, the springs, each with initial compression d_0, act on each pellet of fuel so as to eject a little fuel mass at a relative speed v_p at some constant rate till all the pellets are ejected. This assumption allows us to use the previously derived relationship (equation 5.67) for the final speed, v_f of the rocket.[33] Now, since, before launch there are N compressed springs, one for each element of fuel, we should multiply the PE_{spring} term by N. Conservation of energy then yields at time, t, after launch:

5.75
$$x_{rocket}(t) = \frac{N \, k \, d_0^{\;2} - M_f \left(v_p \ln\left(\frac{M_i}{M_f} \right) - g\,t \right)^2}{2 M_f \, g}$$

Note again that the equation is only valid after all the fuel is expelled.

These are just some of the many uses of the empiriometric concept of energy. In the chapters that follow, we will deepen our understanding of energy as well as the ways that it reveals the quantitative nature of important physical situations. The next chapter, *Harmonic Oscillators*, is particularly important in this regard.

---

[33] To make our $E_{rocket} >> E_{fuel}$ assumption somewhat more specific, we can assume the ejection happens close to the ground zero potential so that the potential energy of the fuel can be neglected.

Summary

The three main qualities involved in Newtonian mechanics are impetus, mass and force. Impetus brings bodies into contact, forces of bodies change the impetus of other bodies, and the masses of the bodies determine how they respond to the impetus.

Mechanical activity due to impetus in one region is transferred to another region by collisions. Collisions are the interaction of forces of two or more bodies. One body acts on another for a period of time applying an impulse, which changes the second one's momentum; mathematically, we write: $\int_t^{t+\Delta t} \vec{F}(t)\,dt = \Delta\vec{p}$. For small times, $\vec{F}(t)\Delta t = \Delta\vec{p}$.

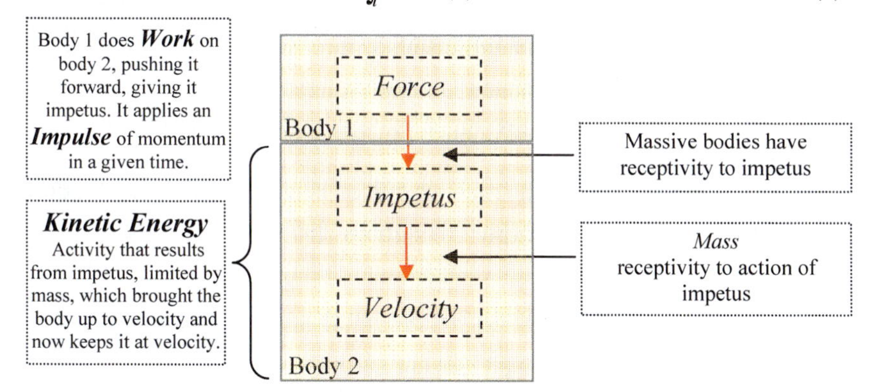

Since the categories of action and reception are correlative, like top and bottom, the active quality impetus only gives half the story of the activity of a body. The receptivity (1/*mass*) to the action of the impetus must also be included to complete the understanding. To specify the activity, we include the magnitude of the impetus as well as how fast that impetus moves, which thereby incorporates the mass via its most important effect. Activity is revealed by the motion it can cause as well as the causes that led to it, and, defined appropriately, *the rate of transfer of the impetus over a distance* includes both.

Thus, the total locomotive activity of a body (and its measure), called its ***kinetic energy***, is defined as the rate of transfer of the body's impetus over a distance, including how the body generally attains its given state of motion and, because a body loses its motion in a symmetrical way, how the body generally can act on its environment through its act of locomotion.

Bodies are usually brought up to speed continuously (in Newtonian mechanics), not born at a speed and cannot be instantaneously stopped but must be slowed down continuously. One should not only think of the impetus as maintaining the body's current speed, but also remember that *the impetus is also responsible for bringing the body up to its current speed*, one incremental piece of intensity of the impetus at a time. The kinetic energy of a body manifests this important aspect of a body's past (or potential loss of) activity by tracking the speed at which each piece of impetus is given to (or taken away); so that, most generally, we write the kinetic energy in differential form: $dK = v\,dp$. Given that the body's mass does not vary with the intensity of impetus activated in it (an assumption that we will see later is not true, except in the sense that the change is small at ordinary speeds--i.e. a lot less than the speed of light), we get the following result when we integrate all the little incremental increases in intensity to a finite intensity.

Mathematically, we write the ***kinetic energy*** as: $K = \frac{p^2}{2m} = \frac{1}{2}m\mathbf{v}^2$. The sum of the activity of all the bodies of a system defines the system's total kinetic energy; mathematically: $K = \sum_{i=1}^{N} \frac{1}{2} m_i \mathbf{v}_i^2$.

A body's change in impetus (and the resulting change in velocity), must come from a force in another body. This force must act for a time, Δt, and over a distance, Δx, for a finite change in the intensity of the impetus (Δp) to be imparted. Mathematically, we write: $F\Delta x = \frac{\Delta p}{\Delta t}\Delta x = \Delta p \frac{\Delta x}{\Delta t} = \mathbf{v}\Delta p = dK$, where in the last equation, we have used the differential form of the kinetic energy which also reveals $dK = \mathbf{v}\,dp$. In words, the force imparts impetus, which causes a certain amount of activity in the body; that amount depends on the distance over which the force acts, which, in turn, depends on the masses and surface forces involved.

The integral of a series of such small impulses is called the work, W, done by the force on the body: $W = \int_{path} \vec{F} \cdot d\vec{s}$ (where, $d\vec{s} = dx\hat{x} + dy\hat{y} + dz\hat{z}$, so $ds = \sqrt{d\vec{s} \cdot d\vec{s}} = \sqrt{dx^2 + dy^2 + dz^2}$) or $W = \int_{path} F_x dx + F_y dy + F_z dz$. We say a force does *positive work* on a body when the force is successful in moving the body through a certain distance in the direction of the action of the force; if not, we say the force does negative work on the body.

By applying the definition of kinetic energy to each independent linear direction for N bodies in a system, we get the ***work-energy theorem***:

$$\sum_{i=1}^{N} \int \vec{F}_i \cdot d\vec{x}_i = \Delta K = \sum_{i=1}^{N} \frac{1}{2} m_i \mathbf{v}_{fi}^{\ 2} - \sum_{i=1}^{N} \frac{1}{2} m_i \mathbf{v}_{0i}^{\ 2}$$

The work-energy theorem only applies when we assume no potential energy and no internal motions (this latter implying that all collisions are elastic). This theorem tells us that, under the specified conditions, when no external forces are applied, the *total kinetic energy is conserved.*

The impulse-momentum theorem analyzes the action of the force during the time, Δt and gives: $\vec{I} = \int \vec{F}(t)dt = \Delta\vec{p}$. Instead of giving activity passed on to the body by the force, this theorem tells of the intensification of impetus, i.e. the increase in the cause that is responsible for whatever activity one then observes. This theorem can also be used to prove conservation of momentum for an isolated system.

The unit of energy is 1 Joule, named after the man who first showed that the temperature of a body of water is raised by movement of a body in water. We know today that locomotion of bodies in a fluid can cause motion of internal microscopic parts that is associated with temperature.

Kinetic energy of the molecular motion of an ideal gas is related to the quality of temperature. The relationship between measures of temperature and pressure is expressed in the equation $PV = NRT$, P is pressure in N/m^2, N is the number of particles in moles, T is temperature in K, V is volume in m^3 and $R \approx 8.3\,J/K/mole$. Since kinetic energy is a measure of activity, which is ultimately caused by impetus, it should be no surprise that force per unit area, pressure, is related to it.

Power, not to be confused with some everyday usages of the word, such as when we speak of the "power" to move or think, is the rate of change of activity of a body or system of bodies. Mathematically, ***power*** is $P = \frac{dE}{dt}$; for cases where there are no internal motions and no potential energy, we have $P = \frac{dW}{dt} = \vec{F} \cdot \vec{v}$.

Up till this chapter, we have left behind the omnipresent real-life force of friction; friction obscures the existence of impetus because it is constantly acting to drain impetus to heavy objects such as the Earth. When you push a body along a surface at a uniform velocity, the body is ***not maintained*** in motion by you, but by the body's impetus (which was activated by us when we started it moving); the force you exert serves only to prevent frictional force of the surface from deactivating that impetus, which happens quickly when you stop your pushing.

Friction between surfaces is characterized by the coefficients of static, μ_s, and kinetic, μ_k, friction: $|F_s| \le \mu_s |F|_N \quad F_k = \mu_k F_N \quad \mu_s > \mu_k$.

We call forces such as friction whose work done on an object is path dependent non-conservative. If the integral of the force around a closed path is zero, we say the force is conservative: $\oint_{path} \vec{F} \cdot d\vec{x} = 0$.

A spring is a good example of an object that can store energy. The force of an ideal spring is $\vec{F} = -k\Delta x\,\hat{x}$, where Δx is the stretch (or compression) of the spring from its equilibrium position. The total work done by a force that compresses a spring from position x_2 to x_1 is: $\frac{1}{2}k\left(x_2^{\,2} - x_1^{\,2}\right)$.

The force exerted by an ideal spring is an example of a conservative force. In conservative forces, unlike frictional type forces, no energy is lost to internal motions, though some may be temporarily stored (i.e. inhibited from acting). For conservative forces, one can write the work done that is not converted immediately to kinetic energy as a single function. Namely, one can use the work-energy theorem: $dK = dW = -dV$, where $V(\vec{r})$ is a function, called ***the potential energ***y, such that one writes the total mechanical energy, E, as: $E = K + V$, where $\Delta E = 0$. i.e. *total mechanical energy is conserved.* $V(\vec{r})$ is called the potential energy associated with the force $F = -\frac{dV}{dx}$ or in three dimensions: $\vec{F} = -\vec{\nabla} V$.

The uniform force field of gravity near the surface of Earth has a potential energy, $V(y) = mgy$, where the zero potential is chosen at the surface of the Earth, and the y direction is defined to be positive upward. A spring has: $V(x) = \frac{1}{2}kx^2$.

The curl of a force field tells how much the field "curls" around a given point, near that point. If the curl is zero the force field is conservative. Stokes theorem conveys this information: $\oint_{Path} \vec{F} \cdot d\vec{x} = \oiint_{Surface\ defined\ by\ Path} \left(\vec{\nabla} \times \vec{F}\right) \cdot \hat{n}\, dA$

Conservation of momentum is a vector conservation law, dealing with impetus *alone*, though with both its intensity and directional type. This law says that the net impetus can never be lost. An impetus of a certain intensity in a given direction can only be destroyed (via forces) by an impetus of equal intensity that is oppositely directed. Thus, via the nature of forces and masses, the tendency to move in a given linear direction is in that sense preserved. By contrast, conservation of energy is called a scalar law, not depending on direction. It deals with activity generally, considering the object's rate of transfer of its impetus over a distance, including how it got it (and how it might lose it). It thus considers the impetus without regard to what direction it has, but just its intensity and how effectively that impetus is in acting. Thus, via the nature of forces, masses and impetus of bodies, activity is neither gained nor lost, but spread around.

A problem can be solved in any uniformly moving frame and, because of the nature of impetus, forces and mass, conservation of momentum and kinetic energy will still apply. One must apply calculations consistently in one frame and then transform the results to other frames, not mixing results between frames.

To defocus attention on the collision itself, we consider the colliding particles to be points, thereby ignoring the details of the interaction and only considering ingoing and outgoing angles as if the particles met at a single point.

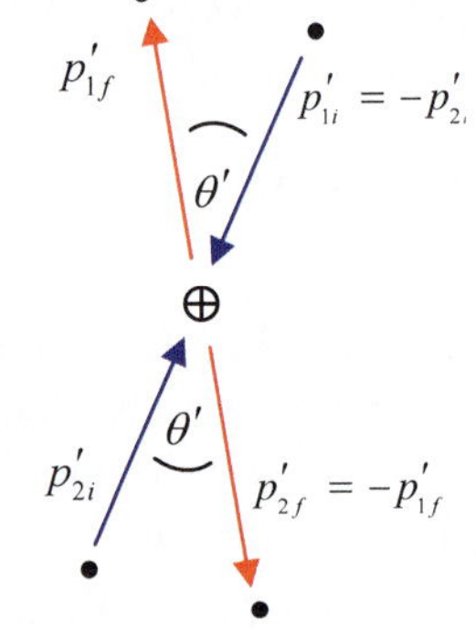

Collisions in the center of mass frame (right) appear simpler because in that frame, we only focus on the impetus relevant to the interactions. By definition of the center of mass frame, the momenta of the incoming (and outgoing) bodies must be equal and opposite. In the center of mass frame, incoming particles always come in on a single line and leave on a single line, though in general the incoming and outgoing lines are not the same. This means, in the center of mass frame, *all* two-body collisions happen in a plane, which is not true in the lab frame.

Furthermore, in the center of mass frame, when we demand conservation of kinetic energy (elastic collisions), the magnitude of the momentum of each body must be conserved.

Terms

Energy

Kinetic energy
Law of Conservation of Kinetic energy
Work-energy Theorem
Work
Power

Geometry terms:

Anti-parallel
Orthogonality

Non-conservative forces

Static Friction
Kinetic Friction

Conservative Forces

Potential Energy
Mechanical energy
Law of Conservation of Mechanical Energy ($K+V$)
Gradient
Curl

Calculation Techniques

In using the concepts of work and potential energy, but also more generally (for instance in defining equations for Newton's laws), it is important to keep track of the definition of the signs of all relevant quantities.

It is always important to take note of particulars of a problem that can simplify the equations to be solved. For instance, define a reference for potential energy that is convenient for the problem at hand.

Often times, problems are practically insoluble until one identifies what the most important elements are. Too much information can be a real deficit if it is an indication that you have not sifted out what is most relevant.

Make approximations as needed by noting parameters that can be considered small. By small, we generally mean the *ratio* of the given parameter a to a key parameter b is much less than one, i.e. $a/b \ll 1$. Sometimes it is helpful to solve a problem at the most approximate level getting an order of magnitude answer. This can give insight into how to proceed to a more accurate approximation and also give a check to your final answer.

In solving some problems, it may be helpful to introduce a test body. It should be taken to be arbitrarily small in size, mass, charge etc. in order to accurately sample the field in a small region and to avoid substantially changing the physical situation by altering the gravitational field or the like.

Notation and Vocabulary Notes:

In labeling components of forces or momentum or other vector quantities, use subscripts that mark the particular component, for example, x, y and z for the components of rectangular coordinates. A second subscript can label which object is under consideration. Other markings can also be added to indicate the frame or whether it is before or after a collision or other important aspects. For example, p'_{1x} means the momentum of the first particle in the x-direction in the center of mass frame, but, for us, can also mean after a collision or other event. Note that "prime" can also mean derivative with respect to position or argument explicitly shown; this meaning is made explicit by showing the argument, for example as $y'(x)$.

Generally, as we have mentioned, just as words that have various meanings have their meaning clarified by the context, so also will the meaning of symbols be set by context.

Mixed Problems

1. Calculate how much kinetic energy the following objects have in *B m/s*:
a) a 100*mph* baseball, b)a 100kg man walking at 3*mph*, and c) the Earth in its orbit around the sun.

2. The space shuttle (4.5 million pounds) enters the atmosphere at a speed of 18,000*mph* by the time it lands it's down to about 200*mph.* a) Calculate the difference in momentum and kinetic energy between entry and landing. b) What happens to the momentum? c) What happened to the energy?

3. Your car gets about 25 miles per gallon of gas. Gas provides about 140M*J/gal*. a) How much energy does one use in traveling 3000 miles across the country? b) Suppose instead of gas, one used electricity which costs 2 cents per 1000 W-hours of energy, how much would it cost to travel across the country? c) Compare this to the cost based on 2007 cost of gas of $3/gallon.

4. A ball is dropped from a height h on a spring of constant k pushing down the spring. How far does the spring compress?

5. What effect do the following have on the kinetic energy of a body: a) doubling the speed of a body, b) doubling the body's mass, and c) doubling the intensity of impetus of a body?

6. Consider two bells of mass m one moving from left to right at speed V, the other coming from the other direction with the same speed. a) Assuming the bells collide completely inelastically, dumping all the energy into the ringing of the structure of the bell, and that they are separated immediately so that they can ring, how much energy is available as sound energy? Suppose $m = 1kg,\ V = .5\,m/s$, b) how much energy in Joules is available?

7. Given a body with 100 calories of kinetic energy, on average, how many Buridans of momentum traverse 1 meter in a second? Notice the factor of two that shows up in this calculation because of our definition of kinetic energy which includes the fact that the body gets up to speed.

8. Suppose one has set up a very large (miles in diameter) scale which has been zeroed out. The air is completely still when a plane of mass M flies over at an altitude of about 100 *ft*, what if anything does the scale register?

9. A plane flies by forcing air downward, via the tilt and shape of the wing; the air, in turn, acts back with equal force supporting the flight of the plane. Given a plane that weighs 5000 *lbs* flying at 150*mph* with the downward deflected air moving at an effective speed of 10% of that velocity, what air mass flow rate in *kg/sec* (and in *tons/s*) is needed to create the requisite upward force?

10. In advanced mechanics, you will study the Hamiltonian function. There you will learn.

$$\begin{aligned}\frac{\partial H}{\partial q} &= -\frac{dp}{dt}\\ \frac{\partial H}{\partial p} &= \frac{dq}{dt}\end{aligned}$$

. H can be thought of as the energy[34], while, for our purposes, q, the so called generalized coordinate, can be thought of as the x-coordinate. Given this, in terms of the concepts of this chapter, describe what these equations mean.

11. Explain the meaning of the equation that connects the energy in the lab frame and that in the center of mass frame.

12. Describe in what sense energy is not a simple property of a body.

13. Take the case of two equal-mass hard bodies that collide with equal and opposite velocity. By considering the surfaces of the bodies during the collision, explain why the nature of impetus leads to the conclusion that bodies must be elastic, i.e., cannot be perfectly stiff and unbending.

14. Do we normally measure kinetic energy directly? If so explain, if not, how do we normally get it, and could we measure kinetic energy directly?

---

[34] When the transformations defining the generalized coordinates do not explicitly depend on time and the potential function is not a function of velocity, then the Hamiltonian is the total energy.

15. In generalizing the definition of energy, we consider new situations and new discoveries. Suppose that we find an entity that can heat water and can move objects. From the latter, we know it has the capacity to activate impetus in massive bodies. However, further suppose that it can only completely interact and destroy itself or not interact at all, thus never losing any of its speed. That is, it can never, while remaining itself (at least to appearances), activate a little impetus in this object and then a little in the next and so on, continually depleting itself and its capacity to impart impetus till it and that capacity disappears; that is, it gives all or none. For such an "object," what would the relation between the intensity of its impetus (or better said, the analogical aspect that we still might call impetus in view of its similarity to actual impetus) and its analogically generalized kinetic energy be?

Note this entity is what we call (again by analogy to our pool ball-like particles[35]) a "photon" and its speed is c.

16. a) Calculate the time it takes for a girl on a swing to get from the high point of her swing through one cycle back to that high point. Assume that the swing has length l and that she is swinging very gently. b) What is the result for $l = 2m$. c) What is the ratio of the periods for a swing on Mars and one on Earth?

17. Suppose you are traveling down the highway at $60mph$. Consider whether it would be worse to crash head-on into a second car moving the other direction at $60mph$ or to hit an immovable brick wall. a) From the point of view of change of momentum, which is worse? b) From the point of view of energy, which is worse? Explain your answers completely.

18. Consider rubbing your hands together quickly three times back and forth across their full length. a) Assuming one exerts a force of 10 pounds and that the frictional coefficient of your hands is $\mu = .6$, very roughly how much heat is generated? Approximate the heat capacity of your skin/hand as about the same as that of water, 1 gram of which raises one degree C in temperature for every 4 J of heat added to it. Assume that the skin is all that heats up and take its thickness to be about 1 mil=$25.4\mu m$.

19. An electrically charged body (of charge $N\ e$) that is uniformly accelerating at rate a radiates electromagnetic energy according to the following formula:

$P = \frac{dE}{dt} = \frac{1}{6\pi\varepsilon_0}\frac{N^2e^2}{c^3}a^2$, in mks units, where N is the number of elementary charges in the body and e is the charge on each; namely, $e = 1.6\times10^{-19}C$; also $c = 3\times10^8 m/s$, and $\varepsilon_0 = 8.85\times10^{-12}F/m$. How much energy does a $100kg$ man falling under the uniform acceleration g of gravity who is charged with 1 mole of excess electrons radiate per second?

[35] Notice that our use of the term particle for electrons and protons (for instance) is itself an analogous use.

20. Gravitational radiation is much different than electromagnetic, but it does radiate energy according to the formula $P_{GW} = \frac{G}{5c^5}\left(\dddot{I}\right)^2$, where I is the so called reduced moment of inertia. Roughly, one can say that the power radiated is:
$P_{GW} \approx P_{source}\frac{P_{source}}{c^5/G} \approx P_{source}\frac{P_{source}}{4\times10^{52}W}$. How much gravitational energy is radiated per second while a baseball is being thrown at 100*mph*?

21. Recent analysis of photosynthesis, water pressure and other factors have led biologists to conclude that redwoods cannot grow higher than 426 *ft*. a) Assuming a cylindrical capillary 100 microns in diameter that runs from the bottom to the top of the tallest possible redwood, how much work does it take for the tree to pump the water into that capillary. b) Given a power rate of light coming from the sun of $1300 Watt/m^2$ and the area of a single leaf as $100cm^2$, and using this energy directly, how much water could be pumped up such a capillary in a second?

22. A particle of mass *m* moving under an impetus of intensity *p* plunges into the surface of a lake and all of its energy is lost to the water. Explain in some detail what the impetus causes to happen in the water. Does all the forward tendency of motion of the particle get lost? *Hint*: Start with conservation of momentum and then consider how the momentum of the particle directed in a single linear direction can be converted to randomly directed motion of the parts of the water.

23. a) How much work does it take to move a body of mass *m* a distance *d* along the surface of a ramp that makes an angle θ with the horizontal, assuming no friction? b) Assuming a coefficient of kinetic friction of μ_k? c) Given a fixed height *h* to lift the box and assuming $\mu_k = .2$, and a max available force of .5 *mg*, what is the (approximate) optimum angle of the ram?

24. Consider the system shown below. Assume the bodies each start from rest. Find the equation for the speed of body *A* as a function of the distance *x* from the initial rest position.

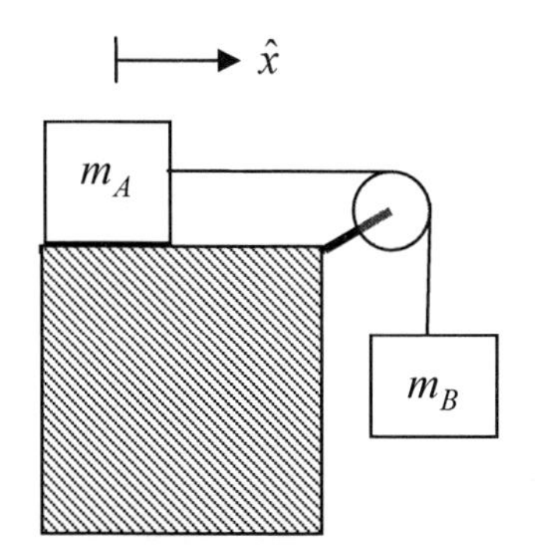

25. A spring of constant $k = 100N/m$ is initially compressed by $d_1 = 10cm$ and pushes a block of M=$1kg$, which moves $d_2 = 30cm$ further before stopping. What is the coefficient of friction of the surface?

26. Calculate the work done in the radial direction by a force that holds a body in circular motion.

27. a) What force is needed and how much work is done in lifting a box of mass M a height h. b) How much force and work are needed to use the pulley shown below to lift the body and what is the difference or similarity in the way the work is done between the two methods of lifting the body.

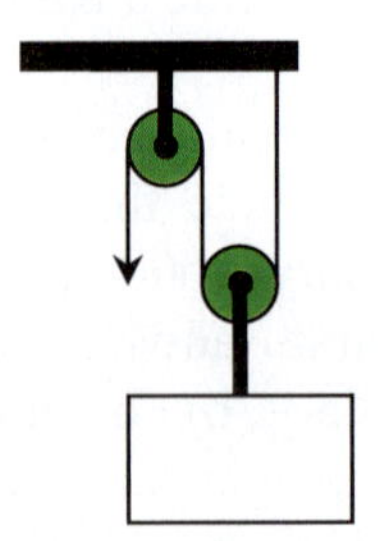

28. In the original movie *Flubber*, a mother accidentally drops her baby as she tries to manage the fridge and the baby. However, having a "flubberized" floor, the baby bounces without harm back up into her arms. a) What would the interaction time of the baby with the floor have to be so that the baby never experiences an average force more than his own weight? b) Modeling the floor surface where the baby falls as a spring of constant k, what maximum force does the baby experience? c) Using the results in a) and b) and given the baby's mass and the height of the floor as: $m = 5kg$, $h = 1.5m$, calculate the interaction time and the spring constant needed for the maximum force to not exceed the baby's weight.

29. Given that a proton is accelerated to a kinetic energy of 1 MeV, at what speed is it moving?

30. What is the potential energy of a simple pendulum as a function of angle? What is it in a small angle approximation?

31. A barrel goes over Niagara Falls, which is 176 feet high. Assuming it fell without touching the water and no friction, what speed would it have at the bottom? What energy would it have?

32. A bike starts at rest at the top of a hill of height h, travels down along the road into a smooth valley of height $h/4$, and then ascends to height $h/2$. How fast is it going at the top of this second hill? Neglect the energy of rotation in the tires in the bike and all frictional forces.

33. In order for a projectile of mass m fired from spring-loaded gun at height h aimed horizontally to hit a target at height h_t and distance d away, a) what compression is required for the spring of spring constant k? b) Suppose the next event requires you to shoot the bullet through a short tube orientated at angle θ to the horizontal, what spring compression is then required (assume the distance, d, is now chosen by the event organizers such that this is possible).

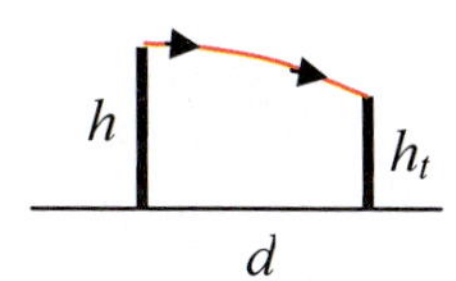

34. A horizontally driven conveyor belt has boxes of mass m dropped vertically on it every τ seconds. a) Calculate the average force necessary to keep it moving at a constant speed V. b) Calculate the average power it takes to maintain the belt moving at a constant speed, V c) Show how this comes about, assuming a frictional force of magnitude F_f operates to bring the boxes up to speed V of the conveyor belt and assuming the conveyor belt supplies whatever is necessary to keep its own speed maintained at V. d) Model the mass m being brought up to speed as an inelastic collision. In particular, assume the small mass m at rest is captured by a very much larger mass M body initially moving at speed V. Calculate the heat energy that must be expended for this to happen using conservation of momentum and energy. e) Using a Taylor's expansion show it reduces to our answer for $M \gg m$.

35. Consider a ball *parachuting* to the ground on a windless day. Why do heavier bodies reach higher terminal speeds? Assume frictional force is proportional to $\dot{y}$ or $\dot{y}^2$. Show weight and friction vectors.

36. To better understand the law of conservation of energy, consider a *one* dimensional case with a different definition of energy. In particular, take the kinetic energy to be: $K = \sum_i |\mathbf{v}_i| = \sum_i \frac{|p_i|}{m_i}$, so that the *sum* of the absolute value of the speeds of the bodies in an isolated system is conserved. This appears to be another way to keep activity under control, so that it never dies out or spins out of control. a) What force would correspond to such a definition of kinetic energy? b) Given this new definition of kinetic energy, what happens if we start by considering N particles *each* moving at a given speed, $\mathbf{v}$, as a system, and then a moment later consider the whole system as one body moving at the same speed $\mathbf{v}$? c) To make sense of this law, suppose that a change happens when the particles coalesce in some way to form one body. Consider consequences of two cases: with potential energy and without. Also, what effects would such a result have on activity in the universe? d) To avoid this sort of problem, what kind of general constraint would be put on a possible definition of kinetic energy for equal mass, equal momentum division of a body into parts (thus excluding the definition suggested in this problem)? Give the answer in words as well as mathematically *and* show that such a constraint applies to the real definition of kinetic energy which is given in the chapter.

37. In a similar fashion to the above problem suppose a one dimensional world in which we had conservation of kinetic energy, K, where $K = \sum_i |p_i| = \sum_i m_i |v_i|$, so that the *sum* of the absolute value of the momenta of the bodies in an isolated system is conserved. a) What force would correspond to such a definition of kinetic energy? b) This hypothetical definition of kinetic energy does not suffer from the defect of the one hypothesize in the previous problem; what defect does it have?

38. Suppose that all momenta were discrete so that one could only gain momentum in multiples of a certain smallest value, p_0. Such a discrete unit would be called a quantum of momentum. Explain how, using the total mass of the universe, $\underline{M_{max}}$, one might define a standard unit for energy that would somewhat simplify the thinking about energy.

39. Why is a large body elastically hitting N small balls one at a time affected differently than when it hits N small balls all at once? Calculate the two effects. Neglect the direct interaction of small balls with each other and take the mass, M, of the large body, and the mass, m, of the small balls to be such that one *cannot* assume $M >> Nm$, but do have: $M >> m$ and $M > Nm$, that is include the next level of approximation . Also, separately discuss the case in which $M >> Nm$ is true.

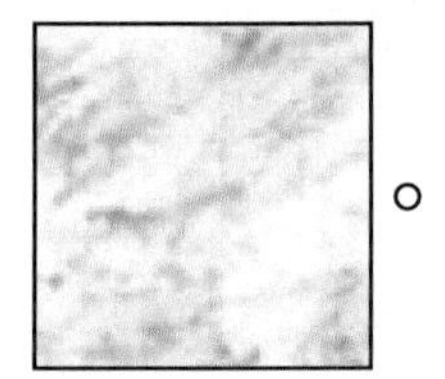

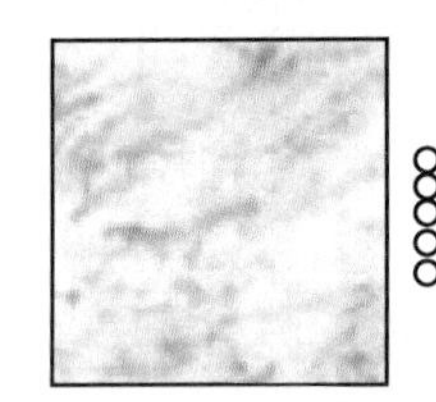

40. a) Briefly, what would the world be like if energy were not conserved? This answer gives an indication of the so called *final* cause of energy conservation. b) Generally, what causes energy to be conserved (so called *efficient* cause)?

41. In a related problem to the above, consider draining a large ball of mass M of its momentum, P, by having it elastically collide, in succession, with a series of individual small particles. a) First consider each such particle to have the same mass, m such that: $M >> m$. Calculate what the ball's speed will be after the ith interaction. Also, calculate the momentum imparted to the ith particle. b) Why does each successive particle take a little bit less momentum from the ball, thus preventing the particles from ever draining all the impetus out of the ball? c) What mass should successive particles have so that the same amount of momentum is removed from the ball for each collision?

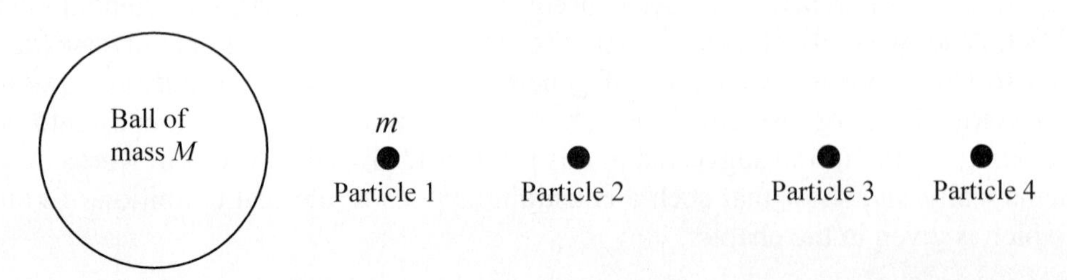

42. Suppose, as proposed in the chapter, one considers the effects of friction on a moving book. Start by setting a single book of mass m in motion at speed V. The book on the surface has coefficient of kinetic friction μ. a) How long will it take to stop? Suppose now one adds another book on top. Further, suppose the books are very sticky so the two books stay joined during the entire motion. b) How long will it take to stop now? c) What is the ratio of the frictional forces acting in each case?

43. Suppose one has to exert a force F to move a very heavy plywood wood box across a certain surface. The box has area A in contact with the floor and has mass M. Suppose we affix a piece of ¼ inch plywood of the same type, but of area $5A$, to the bottom of the box, what force will it now take to move the box? Assume a coefficient of kinetic friction: μ.. Explain.

44. In order to better understand work, energy and motion, consider the effect that a constant force acting on an body of mass m has after successive time intervals Δt seconds a part. a) In particular, write down dp, p, v, Δx of the body and the differential work done on the body $dW = v\,dp$ for each time: $t = 0,\ t = \Delta t,\ t = 2\Delta t, ... t = n\Delta t ... t = t_0$. Also, discuss the limit as Δt goes to zero. b) Why does the contribution to the kinetic energy, in a given interval of time, change with time? Comment on the relationship between speed and kinetic energy. c) Explain in light of the results of this problem, how mass, which is the receptivity to the action of the impetus, i.e. the source of uniform speed, is associated with *acceleration* in the Newton's second law formula: $F = ma$, so that one can write $m = \dfrac{F}{a}$? Discuss from both an empiriometric and physical point of view. *Hint*: Use the definition of F given in the first form of Newton's second law from which $F = ma$ is derived. Also use the definition p.

45. Explain how one gets from the differential definition of kinetic energy to the power formula $P = \vec{F} \cdot \vec{v}$. What does this mean physically?

(Problems continued on next page)

46. Consider two spherical balls of mass m, which have the great proportion of their mass concentrated at the center so that rotational motions can be neglected and which collide *completely elastically* with *no frictional forces* operating. The first ball is moving towards the second which is at rest as shown in the diagram below. The diagram also shows the moving ball at the point of impact as a dotted circle. Note the angles. a) Calculate what happens after the collision in the lab frame and in the center of mass frame. In the center of mass frame give a complete physical explanation of the behavior. Try this experiment in *Interactive Physics* (for instance, see 5.2-two balls colliding to give right angle.ip). b) How much energy does the system have in the lab frame? How much in the center of mass frame? Explain the difference.

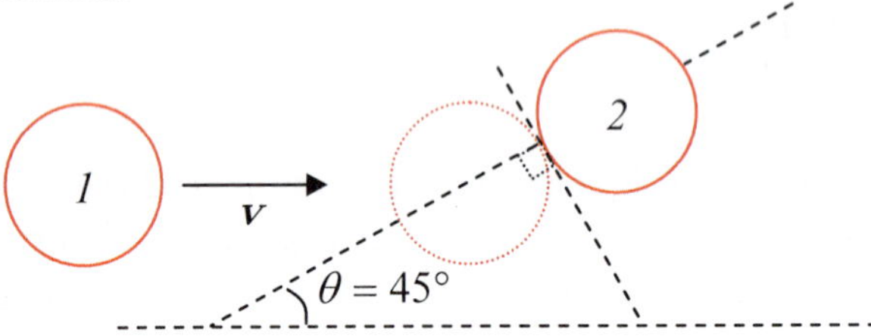

47. Show in general that $m_{earth} \gg m_{test}$ implies, in ordinary circumstances near a small region of the surface of the Earth such as discussed in section c. above (Examples of Conservative Forces) that $F_{earth\,on\,body} \gg F_{test\,mass\,on\,body}$

48. Show that in the limit of a body falling from rest at infinity towards the Earth, the final speed is: $v = \sqrt{\frac{2GM}{r}}$.

49. A body of mass m moving at speed V impacts a second body of mass M at rest. Assuming a one-dimensional inelastic collision in which amount Q of energy is dissipated during the collision, what are the speeds of the bodies after the collisions?

50. In advanced thinking about a manned trip to Mars, one needs to think about how to supply equipment for use during the time on Mars and for the return trip to Earth. Consider the possibility of a parachute drop that slows down the payload to a speed of V *m/s*, which then depends on a crushable material to cushion its crash. Supposing one encased the equipment in a crushable material that would crush about d meters upon impact on the surface. Given the surface area of the container is $S\ m^2$, the mass of the equipment plus its protective shell is m and the equipment can withstand $P_{max}\ N/m^2$, how much would the material have to compress in order for the equipment to arrive safely on the surface? Make simplifying assumptions as necessary.

51. Calculate the kinetic energy, K, needed to make a Higg's boson (assume it has a "mass-energy" of $200GeV$) from colliding a positron ($e+$) and an electron ($e-$) 1) when one particle is at rest and 2) when both particles are coming at each other at equal speed. Which is larger the total kinetic energy need in case one, K_1, or that of case 2, K_2? In this problem, use the Newtonian definition of kinetic energy $K = \frac{p^2}{2m}$ adding a rest mass potential energy term (called the rest mass-energy) according to $V_{rest} = mc^2$, so that $E = K + V_{rest}$.
Note the rest mass-energy of an electron: $2m_e c^2 = 511KeV = .511MeV$

52. Repeat the above calculation for making a Higgs boson using the special relativistic formula for energy instead of the Newtonian one. That is, use $E = \sqrt{p^2c^2 + m^2c^4}$ where $c \sim 3\times10^8 m/s$ to calculate the kinetic energy need in each case to produce a Higgs boson.

53. Show the following is true for all N:

$$E_{Lab} = \frac{1}{2}M\vec{v}_{cm}^{\,2} + \frac{1}{2}\sum_{i=1}^{i=N} m_i \vec{v}'^{\,2}_i \; with\, M = \sum_{i=1}^{i=N} m_i$$

54. Calculate the final velocity, v_{final}, of the single spring rocket in the text assuming its exhaust nozzle is as wide as the tank holding the balls.

55. Given that the receptivity to the action of the impetus starts at m_0 and changes, as predicted by special relativity, with the intensity of the impetus in the following way $m = \gamma m_0$ with $\gamma = \sqrt{1+\left(\frac{p}{m_0 c}\right)^2}$ and c is the speed of light, what is the kinetic energy of the body?

56. Physically, why does kinetic energy (rate of transfer of impetus) relate to pressure (force per area) in the way discussed in the text?

57. The room you are in can be modeled as a box of fixed temperature with many particles that are elastically colliding. a) Why does the energy (activity) available divide equally among translational (linear x, y and z directions) and rotational modes, so there is just as much energy moving particles back and forth as up and down or as rotating around the z-axis? In empiriometric language, why do the particles find each degree of freedom equally accessible? (Note that quantum mechanical considerations come in to play for rotational energies, but we ignore them here. This also applies to heavy versus light particles). If one has, for example, N_2 and O_2 which have different masses, energy will divide equally between those two masses. b) The energy per degree of freedom is approximately kT. How much energy is this in Joules?

58. Solar flares are a major danger to astronauts traveling to Mars. The flares eject high energy protons that can damage human tissues, causing increasing likelihood of cancer for instance. On Earth we are protected by our atmosphere and by the Earth's magnetic field which traps the particles in the Van Allen Radiation belt. A *rad* is equal to $10^{-5}\,J/g$; it is a unit of energy deposited per gram of material, such as human tissue. As the protons (or particles they cause in interacting with materials) move through human tissue, they deposit energy by breaking chemical bonds or ripping off electrons. This process damages the tissue. As a first cut solution, suppose one considers making a lead chamber that the astronauts would enter at the onset of a flare. Given the stopping power for *2 MeV* protons of is $dE/dx = 100 MeV\,cm^2/g$, approximately what thickness of lead shielding would be needed to stop such protons from getting into some protective chamber in the craft. The density of lead is: $\rho = 11.3\,g/cm^3$?

59. How much energy is required to get the space shuttle (2 million kilograms) from the radius of the Earth to the distance of the Martian orbit (notice that this means we assume no use of gravity assists and ignore the speed of the Earth and Mars)? Liquid methane (CH_4) fuel, density of 420 kg/m^3, releases 900 *kJ* of energy per mole. How much liquid methane would one need to get there?

60. An Estes *Hydrogen Fuel Rocket* is propelled by splitting water molecules by electrolysis ($H_2O \rightarrow H_2 + \frac{1}{2}O_2$) and the recombining them and using the resulting energy to launch itself upward. Assume that the rocket's cylindrical tank (*where: r ~ 1.1cm, h ~ 3cm)* of volume *V* is filled with hydrogen and oxygen (from electrolysis) at standard temperature and pressure before launch. Ignore air resistance and friction and assume that the mass of the empty rocket is *m ~ 25g*. All the fuel is burnt at launch, i.e. one can assume the distance traveled during the burn is small and neglect it. Take the energy available upon converting hydrogen and oxygen to water as: $2.4\times10^5\,J/mole$.

Assuming all the energy goes into the rocket, how high would the rocket go? The actual rocket only goes 100-200*ft*; give possible reasons for any discrepancy between this and your calculation. In particular, discuss the natural inefficiency of rockets in putting some available energy into kinetic energy of the fuel.

61. Neglecting aerodynamic effects, what is the maximum acceleration a car of mass $m = 4000kg$ can undergo assuming a maximum coefficient of friction of $\mu_{\max} = .9$? What if $\mu_{\max} = 1.7$? What if, through aerodynamic effects, one increases the effective weight by a factor of β?

62. In a problem of Chapter 1 (reproduced below), we saw that the amount of calories a $100kg$ man expends to get up to speed to a 3 *mph* walk is very small. Why does this not necessarily imply that walking is a bad way to lose weight? How many calories are burned walking a mile, assuming $\mu = .3$? (Recall $4180J = 1\,food\,Calorie$)

63. Suppose two particles moved by impeti of the same intensity and direction hit and are stopped in your skin. Assuming the particles are the same size and have the same external surface properties but have mass M and m, where $M \gg m$, which particle probably does more damage to your skin? Explain this physically by considering a simple constant force model of the skin surface reaction.

64. Look again at the problem on Newton's Cradle given in Chapter 3. Consider the 5 balls, each of mass *m*, to be in outer space far from the effects of gravity or any friction. a) Suppose one of the 5 balls is separated from the others and is initially moving at speed v toward the others (which are all in contact with each other but not stuck together). Explain why, out of all the infinite possible outcomes that conserve energy and momentum, only the following one occurs: The first ball comes to rest along side balls two, three, which stay at rest, while the fourth moves off at speed v. Impose the simplest allowable symmetry requirement on the situation (which results in constraints on nature of balls used). Explain in terms of forces and Newton's third law; use a spring model for the elastic surface of the balls. b) Do the same for 2 balls moving at 5 balls. c) In our model, physically, what is it that is responsible for maintaining the needed symmetry to get the effect characteristic of the Newton's cradle? *Hint*: consider if the springs are not stiff enough, not very springy d) Explain how it is that the total kinetic energy is not conserved while the interaction is happening? See (*5.3-Newton Cradle model short springs.ip* , *5.4-Newton Cradle model.ip*)

65. a) Calculate the curl and divergence for each of the force fields given below. b) Which of the force fields are conservative? c) For the non-conservative forces, compute a simple closed path line integral (of your choosing) to show how much work is used in completing that loop.

i) $6y\,\hat{y}$ ii) $x^2\hat{x} + xy\hat{y} + z^3\hat{z}$

66. Given the following potential energies, a) calculate the force associated with each:

i) $V(r) = \frac{A}{r^2}$ ii) $V(x) = F_0 x$

67. If one knows a force that is derived from a potential, is that enough information to determine whether or not it is conservative? If so, is it? Explain. Are all conservative forces derivable from a potential?

68. Consider a system of four particles that do not collide. The initial position and speeds *(x, v)* of each particle, in appropriate units, is given by: $\{(1,1),(2,2),(1,3),(2,3)\}$. Draw a diagram showing v versus x at this initial time and label each particle, connect the dots that represent each particle and calculate the area in phase space covered by this system of particles. Allow one unit of time to pass and draw and calculate the new area. Repeat this one more time. What conclusion do you draw?
Note: The conclusion arrived at here can be proven for momentum/position space of any number of particles that do not collide; in that case, the result is called Liouville's theorem.

69. Show that conservation of energy requires, for an incompressible fluid (density constant) that $\rho gh + P + \frac{\rho v^2}{2} = \text{constant}$, where P is the measured pressure, ρ the mass density and v is the speed of the flow, each specified at the same given point. The constant is the same for every point along the path of the pipe. This important law of fluid dynamics is called Bernoulli's principle. Hint: consider a small differential element of water and note that its energy can only be changed by gravity and the pressure differentials.

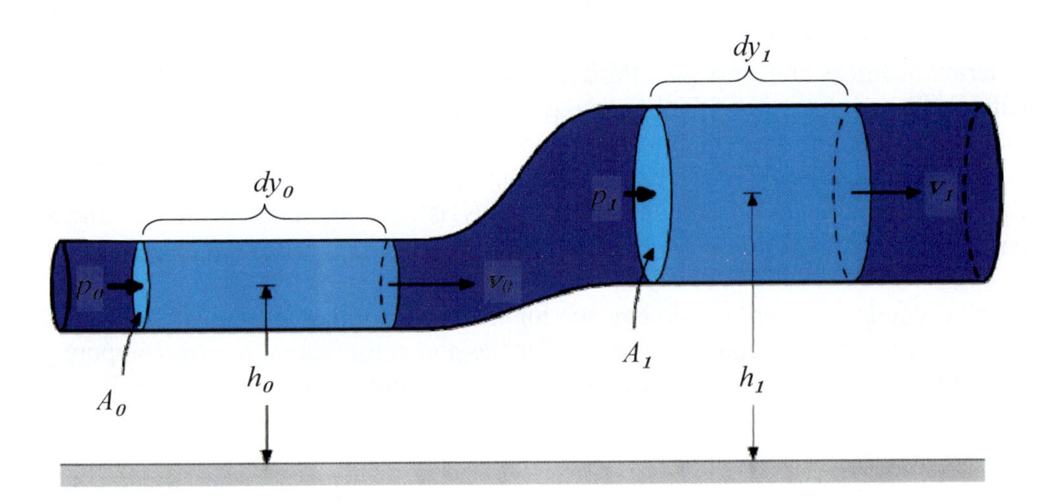

70. Suppose one would like to design and make a very basic hydraulic lift using water spraying vertically upward. A body of mass M is held up by water from a hose squirting vertically upward. Assume the water exits the hose at speed v_0 with a constant mass flow rate of $\frac{dm}{dt} = \dot{m}$. a) What is the maximum height that the object could possibly float (assuming such an equilibrium is possible) above the ground? Hint: Assume elastic collision to get maximum height.

Note that water is an (approximately) incompressible fluid (so its density remains the same) and water will tend to stick together up to a certain point.

71. Suppose you are sliding down a Martian hill in your spacesuit to get quickly back to the land rover. One starts at the top of the arbitrary shaped hill (given by the function $y(x)$) with negligible speed. a) Assuming a frictionless surface, at what point in your slide, i.e. what surface angle, do you leave the surface of the hill and begin "flying" in the thin Martian air. Assume a uniform acceleration of bodies on Mars of g. *Hint:* calculate the minimum acceleration needed to maintain the body moving along the curve of the hill and compare it to that provided by gravity. Note that the only component of the force that is relevant is the component perpendicular to the curve. b) What is the answer for the curve $y(x) = -\frac{1}{n}x^n$? c) What about for a circle? d) Explain why the parabolic shape is special. e) Solve this problem using the tangent ratio for the special case of the parabola *(n=2)*.

72. How big a hot air balloon is needed to carry a 100 *kg* man, if we heat air in a balloon to twice the outside temperature, which we take to be room temperature; that is, we heat the interior air from $300K \rightarrow 600K$? (note: this would probably burn the average material used in hot air balloons)*Hint:* use *PV=NRT*

73. To understand better how the helium balloon stays in the air consider the balloon as a very big, but light ball of mass M among much smaller air balls (molecules) each of mass m. In particular, consider M to be much less than the mass of the N air molecules that it displaces. (see 5.5-temperature of balls with gravity on with balloon.ip) in which the outside is modeled by 14 balls)

74. Consider the situation shown in the diagram below. Suppose the spring starts out compressed with all bodies motionless. Draw the forces acting on each body and explain in detail what happens to the center of mass of bodies *A* and *B*.

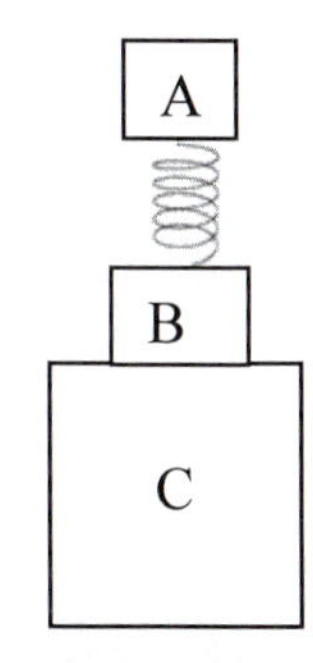

75. A rocket is pumped up with air to pressure P with pure nitrogen in volume V. The mass of the rocket when full is M and the ambient air pressure is P_0. a) Find the maximum rocket height. (ignore air resistance and friction). Use the ideal gas law and assume the pressure is increased while under constant temperature conditions to get the total work *stored*. Then, assume all this work is applied to raising the height of the rocket. b) Taking $P/P_0 = 2,\ M \sim .02kg, R \sim 8.3J/K, P_0 \sim 10^5 N/m^2$ the rocket's air "tank" has base a of $A \approx D^2 \sim (.03m)^2 = 10^{-3} m^2$ and a height of $h_{rocket} \approx .2m$, how high does the rocket go. c) The actual rocket goes only about 100*ft*; explain the discrepancy between this number and your answer.

Chapter VI

Harmonic Oscillators

Introduction

A child's swing goes up and then down and then up again. The pendulum on a grandfather's clock goes back and forth then back. These are all examples of motions that approximate the motion of a simple *harmonic oscillator*.

The simplest locomotion that can be studied is that of uniform linear velocity, (i.e., a fixed mass body with an unchanging impetus), then comes uniform linear acceleration (which implies the application of a constant force to a fixed mass body).

The motion of the *harmonic oscillator* is the *simplest* type of *bounded* motion that can be studied. It is a motion in one dimension. We will see shortly that uniform circular motion,[1] which is the simplest *two dimensional* bounded motion, can be thought of in terms of two harmonic oscillators. Because of its simplicity, the harmonic oscillator can be of great help in understanding the nature of force and impetus, especially in their quantitative aspects.[2] The harmonic oscillator has the simplest force variation, for its force varies only as a *linear* function of the distance.[3] Because harmonic motion is so elemental,

---

[1] Uniform circular motion results from uniform radial acceleration, which, in turn, results from a force of unchanging magnitude acting on a body always toward some fixed point (which thus is a central point).

[2] Here we use "quantitative" analogically, as we do throughout the text, to include any use of numbers, including, for instance, measurement of magnitudes of qualities.

[3] In two dimensions, Bertrand's theorem proves that the only central forces that have closed (i.e. come back to the same point) orbits for all possible bound particles are: 1) the inverse square force (i.e. force directed

it holds a place in the analysis of bounded locomotion (of parts and wholes), similar to that of the line in the analysis of complex curves in calculus. Namely, as the tangent line is the first approximation to a curve at a point, the harmonic oscillator is the first approximation of many complex bounded motions. We will see this below in our simple discussion of perturbation theory and, in related ways, Fourier analysis. The central importance of harmonic motion will become clearer as you advance in your study of physics, for example when you study quantum field theory.

Simple Oscillator

The simple harmonic oscillator neglects friction and assumes isolation from external forces. We will gradually add these effects back, but for now it's instructive to review what we know about the simple case.

Simple harmonic oscillators are circular in the sense that their motion repeats a given cycle over and over without diminishing. This latter is obviously another case of a mental construct of convenience, for real world physical oscillations eventually change; yet, as usual, we can learn a lot about such oscillations by leaving out those effects, and many systems can be approximately modeled by the simple harmonic oscillator.

Slingshot

The most obvious example of a repetitive motion is revolution in a uniform circular motion. Concretely, consider the ancient slingshot in which a heavy object was attached to a string twirled about and then released in the direction of the target, shown in Figure 6-1 below. Neglecting friction, the heavy object, say a rock, would continue in circular motion for reasons we will discuss in Chapter 7. Assume that the rock is in uniform circular motion for a period before it is fired.

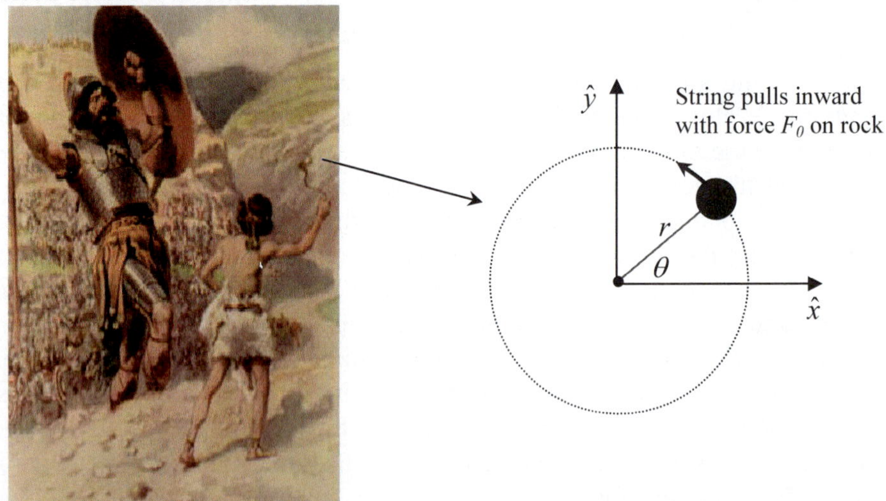

Figure 6-1: Ancient slingshot; slingshot is in uniform circular motion before being released at target.

---

towards a single point, varying as the inverse square of the distance from that point), such as the gravitational force discussed in Chapter 8, *and* 2) the linearly varying force of the harmonic oscillator.

Analyzing the motion briefly, we see that the force $\vec{F}_0$, which is some function of the radius r_0, is constant in the radial direction during this circular motion of radius r_0. Mathematically, we write: $\vec{F} = -F_0\,\hat{r}$ (note that, for a fixed mass body at a fixed radius, this implies uniform radial acceleration). Decomposing this by simple trigonometry (for details see Chapter 7), into Cartesian coordinates, we get:

6.1 $$F_x = -\cos\theta\; F_0 = -\frac{x}{r_0}F_0\,,\quad F_y = -\sin\theta\; F_0 = \frac{y}{r_0}F_0\,,$$

so that Newton's second law, in the x and y directions, $F_x = m\ddot{x}$ and $F_y = m\ddot{y}$, gives:

6.2 $$m\ddot{x} = -x\frac{F_0}{r_0} = -k\,x\,,\quad m\ddot{y} = -y\frac{F_0}{r_0} = -k\,y$$

Where k is a constant, not changing in time.

Notice the symmetry of these two equations-- i.e. they have exactly the same form—that results from the uniform circular motion. The general form of these equations: $m\ddot{x} = -k\,x$, is ***the harmonic oscillator equation***. Note the linear dependence of the force in equation 6.1, and the resulting linear dependence in the second derivative of the position.

Again, the two equations 6.2 together describe the uniform circular motion that we posited. By choosing the orientation of our coordinate systems (or our starting time) so that $x = 0 \;\; at \;\; t = 0$, we can describe the motion mathematically as:

6.3 $$x = A\sin\omega t,\;\; y = A\cos\omega t \quad \text{where } A = r_0 \;\; and \;\; \omega = \sqrt{\frac{k}{m}} = \sqrt{\frac{F_0}{m\,r_0}}$$

Each one of these then is a solution to the harmonic oscillator equation. Standard usage makes the first the icon or symbol for harmonic motion. Thus, the term *sine wave* primarily names $Sin\,\omega t$, but it also serves as the generic name for all waves of the form:

6.4 $$x = A\sin(\omega t + \phi)$$

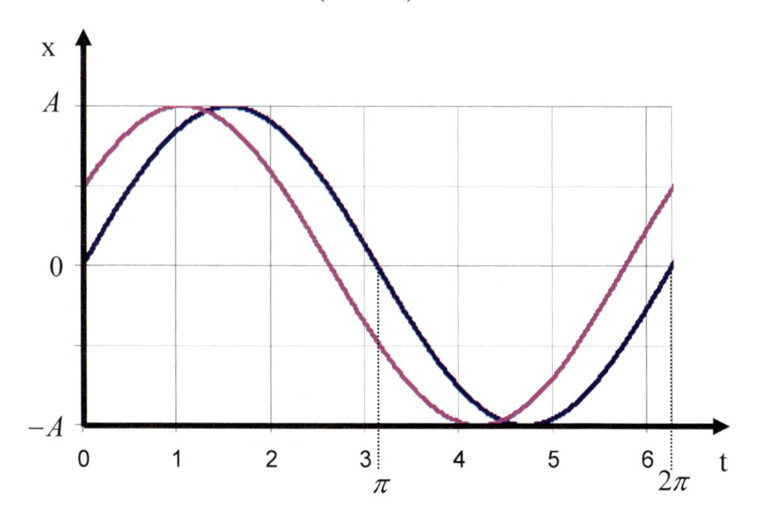

Figure 6-2: Blue curve is one cycle (from 0 to 2π) of $x = A\sin(\omega t + \phi)$, with $\omega = 1\; radian/\mathrm{s} \;\; and \;\; \phi = 0$, which means $T = 2\pi$ second. The pink curve shows that the wave shifts to left when ϕ is increased to .5 radians.

Figure 6-2 shows the prototypical sine wave in blue. A is called the *amplitude* of the harmonic oscillators motion. ϕ is the *phase angle*, which, as done above, is set to the value that fits given initial conditions. ω is called the *angular frequency*; it tells how many radians that the object in circular motion traverses per second. The *period*, T, is how long it takes for the object to complete one orbit; hence:

6.5 $$T = \frac{2\pi}{\omega}$$

From noticing the importance of circular motion in two dimensions or the importance of the simple harmonic motion in one dimension (which can analogically be thought of as two spatial dimensions), one can see the importance of the sine and cosine function in motion. Indeed, from the importance of the circle in geometry, one already knows the importance of the sine and cosine function and thus can glimpse where its analogical power comes from in describing motion. In fact, Jean Baptist Joseph Fourier (1768-1830) noted the primitive nature of sine and cosine, and as a result of his work, we now have a series parallel to the Taylor series, called the Fourier series:

6.6 $f(x) = \sum_{n=1}^{n=\infty} a_n \sin(n x)$ Where a_n are the constants analogous to those of the Taylor expansion discussed at the end of Chapter 2; the Taylor expansion around $x = 0$ written in a similar notation is:

6.7 $$f(x) = \sum_{n=1}^{n=\infty} A_n x^n \ where\ A_n = \frac{f^n(0)}{n!}$$

We will return to Fourier analysis, as it is called, after more discussion of the harmonic oscillator.

Spring Harmonic Oscillator

Perhaps the best way to picture a harmonic oscillator is as the spring with two masses that we have already discussed in Chapter 5, reproduced in Figure 6-5. However, we begin with the dynamically simpler case in which one mass is not allowed to move, which is just the limit of the latter with one mass "infinitely" large (note the being of reason created here). A single free mass under the action of a spring tied to an immovable wall is shown at successive times in Figure 6-3. Note the figure does *not* show a spring hanging from the ceiling, for this would bring gravity into play. Rather, the entire setup lies on a smooth floor, so that only the spring force is able to change the impetus of the ball.

As previously seen, springs have a force that can be modeled as proportional to the distance they are displaced from equilibrium. Namely:

6.8 $$\vec{F}_{spring} = -k\, x\, \hat{x}$$

Newton's second law then gives the equation found above for each dimension of uniform circular motion:

6.9 $$m\ddot{x} = -k\, x$$

This is obviously the harmonic oscillator equation with sine wave solution given above in equation 6.4, which is shown in orange in Figure 6-3.

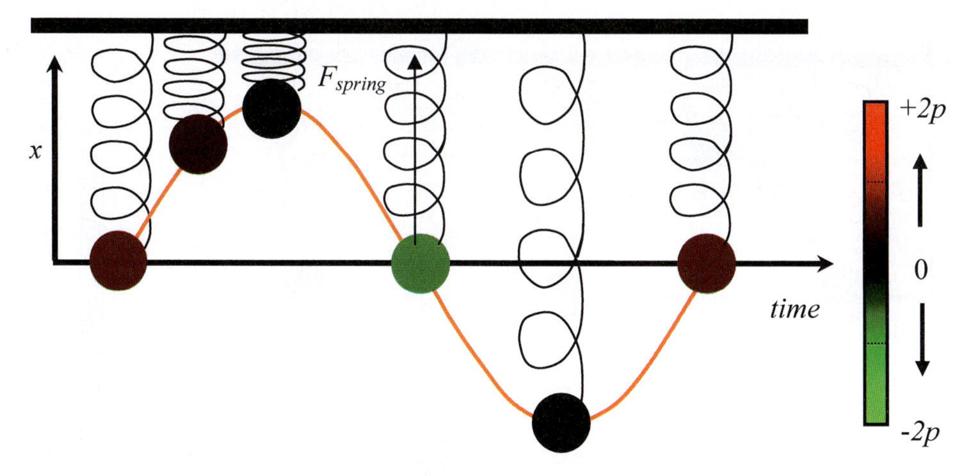

Figure 6-3: *Evolution of vibrating spring*. View looking down on a spring attached to a ball resting on a smooth floor; the opposite end of the spring is attached to a wall shown as a thick black line. Impetus acting to move a body along the x-direction is shown as red, while green is used for impetus acting in the opposite direction. For instance, the first snapshot of the motion shows a maroon colored ball, with the spring in unstretched neutral position; the color (maroon) of the shown intensity indicates that the ball's impetus tends in the $+x$ direction with an intensity, p (cf. scale on the right). The orange curve shows the sine wave path that the motion of the ball makes as time passes.

Energy of Harmonic Oscillator

To understand the harmonic oscillator, as with every system of moving bodies, understanding its kinetic and potential as well as total energy is crucial. The total energy determines what can happen, and it is conserved throughout isolated processes. As we saw in Chapter 5, the kinetic energy tells us the level of activity now and potential energy tells us how much activity *can* occur. As we also saw in Chapter 5, the total energy of the harmonic oscillator can be found by integrating equation 6.9 over x. Using E_0 as the constant of integration, we recall:

$$6.10 \qquad E_0 = K + V(x) \qquad \textit{with } K = \frac{1}{2} m v^2 \quad V(x) = \frac{1}{2} k x^2$$

Note that the total energy (i.e., the total amount of activity of the system --either actual or in the process of becoming actual), E, of the system is constant. Figure 6-4 below shows the values of the kinetic energy, K, and the potential energy, V at several moments in the oscillator's swing.

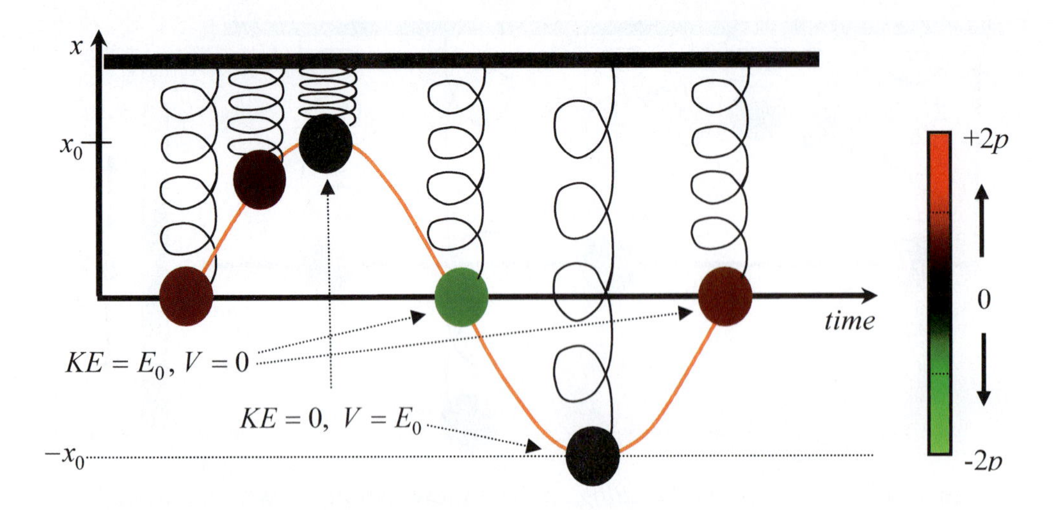

Figure 6-4: Kinetic and Potential energy throughout evolution of an harmonic oscillator. The total energy, $E = E_0 = 1/2m\mathbf{v}^2 + V(x) = 1/2m\mathbf{v}_0^2 = V(x_0)$, is constant, i.e. unchanging throughout time for the spring.

The kinetic energy is clearly a maximum when the spring is in the neutral position. The ball then moves along the x-axis under the impetus of intensity p, at speed $v_0 = p/m$. Gradually, the force of the spring builds up acting more strongly against the ball de-activating impetus in the ball until at the maximum spring compression (maximum potential energy), no impetus remains, i.e. $p = 0$. However, at this point, the force is highest so the rate of increase of momentum is maximum, so that by the time the ball is back again at $x = 0$ the ball has speed v_0 again but in the opposite direction.

Two mass Harmonic Oscillator

For two equal mass springs, defining directions as shown in Figure 6-5 below, we write the forces on the balls labeled A and B as:

6.11 $$F_{spring\to A} = -k(x_A + x_B) \qquad F_{spring\to B} = -k(x_A + x_B)$$

We also have:

6.12 $$F_{spring\to A} = F_{A\to spring} \qquad F_{spring\to B} = F_{B\to spring}$$

Note that defining $\hat{x}_A$ and $\hat{x}_B$ as opposite directions as done in Figure 6-5 ($\hat{x}_A = -\hat{x}_B$) helps in making clear the restoring nature of the spring force, but means that equations 6.12, which represent Newton's third law, appear without minus signs. This is fine, but should remind us that care needs to be taken in keeping track of directional definitions.

Equations 6.11 and 6.12 imply:

6.13 $$m\ddot{x}_A = -k\left(x_A + x_B\right)$$
$$m\ddot{x}_B = -k\left(x_A + x_B\right)$$

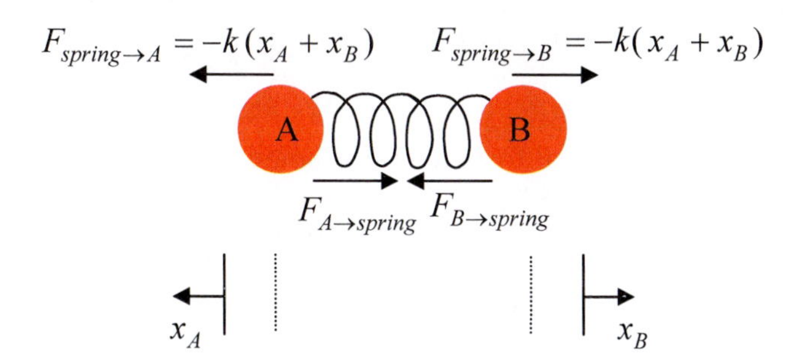

Figure 6-5: The spring and two balls is the prototypical harmonic oscillator. Shown are the forces acting on each ball as well as the back reaction force of the ball on the spring.

Adding the last two equations together and taking $x = x_A + x_B$ gives:

6.14 $2m\ddot{x} = -k\,x$

This equation implies an oscillation frequency or *resonance frequency* of : $\omega_0 = \frac{1}{\sqrt{2}}\sqrt{\frac{k}{m}}$, which continues undamped forever.

Interactive Simulation

See *6.1-spring with two masses.ip* *Interactive Physics* simulation to watch this interaction.

Damped Oscillator

Of course, actual harmonic oscillators have damping, i.e. effects that damp (inhibit) the motion. One common type of damping is proportional to speed. Such damping is commonly seen in fluid dynamics, for example, for a projectile moving through water when the speed is slow enough not to cause turbulence. We thus write:

6.15 $F_{damping} = -b\,v$

Such a damping law is also seen in electrical circuits such as shown in Figure 6-6.

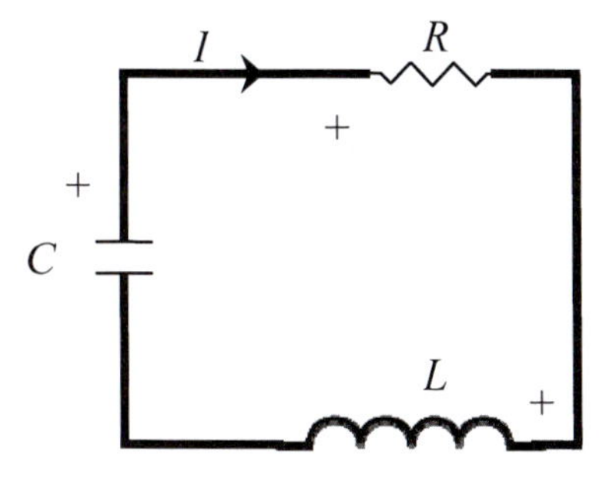

Figure 6-6: This electronic circuit is a harmonic oscillator with damping linearly proportional to the current, *I*. It has the equation of motion: $L\frac{d^2I}{dt^2} = -R\frac{dI}{dt} - \frac{I}{C}$. Comparison with equation 6.14 gives: $x \sim I$, $m \sim L$, $b \sim R$, $k \sim 1/C$. Resistance, *R*, to the flow of current, capacitance, *C*, for storage of charge, and inductance, *L*, will be studied in *electricity and magnetism*, the next course subject in the physics pedagogy. For now, it is only important to see that the similar equations arise in apparently very different circumstances, showing the analogical generality of the mathematics that comes from the fact that first property of all material things is extension.

The mechanical oscillator, for example the one mass spring oscillator, gives:

$$m\ddot{x} = F_{friction} + F_{spring} = -b\dot{x} - kx \tag{6.16}$$

We recast this equation in a standard form to define some important parameters:

$$\ddot{x} + \gamma\dot{x} + {\omega_0}^2 x = 0 \quad where \quad \gamma = \frac{b}{m} \quad and \quad \omega_0 = \sqrt{\frac{k}{m}} \tag{6.17}$$

The general solution to this equation for $\gamma < 2\omega_0$ is:

$$x(t) = Ae^{-\left(\frac{\gamma}{2}\right)t}\sin\left(\omega_{damp}t + \phi\right) \quad where \quad \omega_{damp} = \sqrt{{\omega_0}^2 - \frac{\gamma^2}{4}} \tag{6.18}$$

It is reserved as an end of chapter problem to plug into equation 6.17 to verify that this is indeed a solution. Again, this solution is only valid for $\gamma < 2\omega_0$, i.e. when the square root in equation 6.18 is real.

In graph of equation 6.18 shown in Figure 6-7, we see that the effect of the exponential is to make a *decaying* sine wave amplitude.

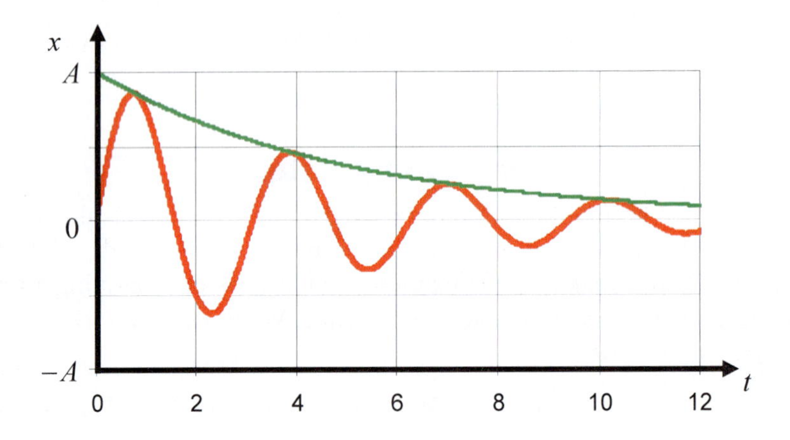

Figure 6-7: The red curve shows *decaying amplitude* sine wave characteristic of damped harmonic motion. As energy is dissipated to the surrounding environment, the energy of the oscillator is drained away. The green curve shows the exponential decay law of the amplitude.

The energy of a damped harmonic oscillator is:

$$\begin{aligned} E(t) &= KE(\dot{x}(t)) + V(x(t)) = \frac{1}{2}m\mathbf{v}(t)^2 + \frac{1}{2}kx(t)^2 \\ &= \frac{1}{2}A^2e^{-\gamma t}\left(k - \frac{1}{4}m\gamma^2\cos 2(\omega_d t + \phi) - \frac{1}{2}m\omega_d\gamma\sin 2(\omega_d t + \phi)\right) \end{aligned} \tag{6.19}$$

Note that the energy *of the oscillator* is no longer conserved, because the damping is caused by an external force, caused by a "damper," that de-intensifies the impetus in the oscillator, gradually decreasing its energy. By acting back on the damper in usual fashion described by Newton's third law, the oscillator activates impetus in the damper,[4] causing increased activity, i.e. higher energy, in the damper. We say, in a shorthand expression: energy is drained from the oscillator to the surroundings. Total energy is conserved but the energy of the oscillator alone is no longer conserved.

Assuming that the damping is small, so that there are many cycles before the amplitude drops significantly ($\omega_0 \gg \gamma/2$, we can use $\omega_d^{\,2} \approx \omega_0^{\,2} = \frac{k}{m}$ which gives:

6.20 $$E(t) \approx \frac{1}{2} k\, A^2 e^{-\gamma t}$$

Notice that this formula reduces to the same value as the undamped case *at* $t = 0$; such checking using already established results, or intuitive special cases, is important in verifying one's calculations. Now, the equation shows the energy drops by $1/e$ every $1/\gamma$ seconds. We thus define:

6.21 $$\tau = \frac{1}{\gamma}$$

as the *decay time* of the oscillator. Using the definition of γ, we get: $\tau = \frac{m}{b}$. Note that since it is energy, the measure of the level of activity, that is most important in interactions, it is energy that defines the characteristic decay time, *not the amplitude*. For completeness, note that the *decay of the amplitude*, using the $1/e$ fall-off definition, is:

6.22 $$T_{amplitude\ fall\ off} = \frac{2}{\gamma}$$

Q of an Oscillator

One needs light damping to get the above sinusoidal decay law. The very ability of an harmonic oscillator to oscillate is at stake if there is too much damping. For this and other reasons, it is important to quantify the level of damping. More specifically, we need to define a quality factor, Q, to rate a given oscillator, in such a way that the higher the Q, the better the oscillator is; i.e., it has less damping.

Now, since energy is the key parameter, we define the quality factor or Q of the oscillator as the fraction of *energy* lost in some characteristic time. But, what characteristic time do we use? We have two: the decay time and the inverse of the oscillation frequency (which specifies many "radians are traveled" in a unit of time). For a low loss oscillator,[5] the oscillation period is the period of the approximate repetition of the motion itself, as opposed to the decay of energy which relates to the *cause* of the motion, so we choose the oscillation frequency, ω_d.

Thus, given that the oscillator has energy E at a given time and loses say ΔE in the small period of time $\Delta t = 1/\omega_d$ around that given time, the fractional energy loss is

[4] For example, the damping force would arise from rubbing against the floor in the case of the spring oscillator described earlier or from pushing against the water in case of a body moving through water. In an actual spring, there are losses associated with internal stresses of the spring itself. The result of the damping is to heat up the spring in this latter case and the floor and water in the former cases.

[5] For higher loss oscillators, ω_d is still the choice that approximates the repetition of the motion of itself, for, while the high loss oscillator doesn't very closely repeat its path of motion, it does still cross the equilibrium point during every cycle.

$\Delta E / E$. However, this dimensionless number is smaller for less loss. So, we are interested in the inverse of this number, for we want the better oscillator (i.e., one that oscillates the longest) to have a higher quality factor. Hence, we quantify the "quality" of an oscillator, *Q, the quality factor,* as:[6]

6.23 $$Q = \frac{E}{\Delta E} = \frac{\text{Total energy of oscillator at given time}}{\Delta E \text{ lost in } 1/\omega_d \text{ sec around given time}} = \frac{E}{\left(\frac{1}{\omega_d}\right)\frac{dE}{dt}}$$

Thus, for the lightly damped case, we thus get:

6.24 $$Q \approx \omega_d \frac{E}{\gamma E} \approx \frac{\omega_0}{\gamma} = \omega_0 \tau$$

Thus, given two oscillators both with $Q >> 1$, we can compare the quality of the two oscillators by comparing the ratio of vibrational frequency ω_0 to damping factor, γ. Equation 6.24 can be understood as follows. When the approximate oscillation frequency, ω_0, is fixed and the energy decay time τ increases, it is clear that the oscillator oscillates longer, and Q is higher as it should be. Similarly, for fixed energy decay time, τ, as the oscillation gets faster, it implies more oscillations occur within the given decay time and Q is larger as it should be.

Musical Instruments as Harmonic Oscillators

Every musical instrument is somewhat like a damped harmonic oscillator. However, unlike a single damped harmonic oscillator, their activity is not confined to a band centered on one frequency, because they have various modes of oscillation. We can model them as many harmonic oscillators with frequencies that are each a multiple of a given *fundamental* frequency. The amplitude of each behaves in sine wave fashion. Since the forces are linear, the sum of all the oscillations can be considered to add the ordinary way like arithmetic. Thus, we can consider the sound of each musical instrument as approximated by a series of *independent* harmonic oscillators of frequencies that are just multiples of a given fundamental. The fundamental nature of the sine wave, seen again here, earlier led us to expand a function in terms of such fundamental frequencies. We called it the Fourier expansion (see equation 6.6).

The musical scale reflects this harmonic independence in its use of the octave or double frequencies, so for instance A4 (middle A) is 440 cycles per second ($\omega = 2763/s$), while A5, the "A" an octave above it is 880 cycles per second. Of course, the lowest available frequency depends on the characteristics of the particular instrument, but, in this case, must be a sub-multiple of 440. Musical sounds, like all physical things, have a base in extension, the first meaning of quantity, so it should again be no surprise to learn that the beauty we appreciate in music is strongly related to such mathematical underpinnings as harmonics. The common origin of the English words *harmonic* and *harmony* shows their deep interrelation. Pythagoras (b. 532BC) was the first to recognize the mathematics of

[6] An equivalent more common, though more opaque, definition is: $$Q = \frac{\text{Total energy of oscillator at given time}}{\text{energy lost per radian at given time}} = \frac{E}{\frac{dE}{d(\omega_d t)}}$$

music. This and other insights about the importance of number led Pythagoras to think all was number.

Each musical instrument playing the same note, say middle A, can be said to be generating the same fundamental frequency. Each musical instrument is distinguished by the harmonics, by the harmonic oscillations of frequencies which are integral multiples. Said another way, the coefficients, a_n , of the harmonics in the Fourier expansion (cf. equation 6.6) are different for different instruments. For a given note, an oboe would have one set of a_n terms, a flute a different set. Again, thinking of the sound as many independent oscillators is expressing a reality, but not the full reality. Indeed, we are reminded of this when we recall that the Fourier expansion requires an infinite number of such oscillators-- i.e. a completed infinity, thus a contradiction and thus a purely mental construct that cannot exist outside the mind. It is very like using an infinite number of infinitely small lines to make a curve; the ability to do so does not make a curve the same as a line.

Similarly, a musical instrument is in fact not simply a bunch of harmonic oscillators, but a complex interaction of bodies (such as a string drawn over a cavity) that will cause us to consider the *wave equation* which is a whole course of study in itself. Such a study, which usually covers waves and oscillations of many types, is obviously not the subject of this course. However, the concepts developed here are the generic subject matter required for going to that next step, so we provide the above outline to set the stage.

Indeed, we can add a little more back of what we've left out. In particular, musical instruments, like all real oscillators, are driven—e.g., plucked, blown or hit-- at one time or another, so we proceed to analyze the harmonic oscillator subject to an external force in addition to resistive damping.

Forced and Damped Harmonic Motion

We thus get:

$$\begin{aligned} F &= F_{spring} + F_{damping} + F_{external} \\ &= -k\,x - b\,\dot{x} + F_e \sin \omega t \\ m\,\ddot{x} &= \end{aligned} \qquad (6.25)$$

Remembering we can model any time-varying force applied to the oscillator by a Fourier sum of appropriately chosen harmonic amplitudes, we lose no generality by taking the external force to be a sine wave. Converting 6.25 to standard form, we get:

$$\ddot{x} + \gamma \dot{x} + \omega_0{}^2 x = \frac{F_e}{m} \sin \omega t \qquad (6.26)$$

A trial solution of $x = A \sin(\omega t + \phi)$ yields:

$$A = \frac{F_e}{m} \frac{1}{\sqrt{\left(\omega_0{}^2 - \omega^2\right)^2 + \left(\omega\gamma\right)^2}}$$

$$\phi = \arctan\left(\frac{\gamma\,\omega}{\omega^2 - \omega_0{}^2}\right) \qquad where: \gamma = \frac{b}{m} \quad \omega_0 = \sqrt{\frac{k}{m}} \qquad (6.27)$$

In order to understand these results, we re-write equation 6.27 so as to manifest *dimensionless* arguments:

$$A = \frac{F_e}{m\,\omega_0{}^2} \frac{1}{\sqrt{\left(1 - \left(\frac{\omega}{\omega_0}\right)^2\right)^2 + \left(\frac{\omega}{\omega_0}\right)^2 \left(\frac{\gamma}{\omega_0}\right)^2}} = A_0 \frac{1}{\sqrt{\left(1 - \left(\frac{\omega}{\omega_0}\right)^2\right)^2 + \left(\frac{\omega}{\omega_0}\right)^2 q^{-2}}}$$

$$\phi = \arctan\left(\frac{\omega}{\omega_0} q^{-1} \frac{1}{\left(\frac{\omega^2}{\omega_0{}^2} - 1\right)}\right) \qquad where: q \equiv \frac{\omega_0}{\gamma} \sim Q \;\; for\ light\ damping \qquad (6.28)$$

We can better understand these equations by plotting them as a function of $\frac{\omega}{\omega_0}$ as shown in Figure 6-8.

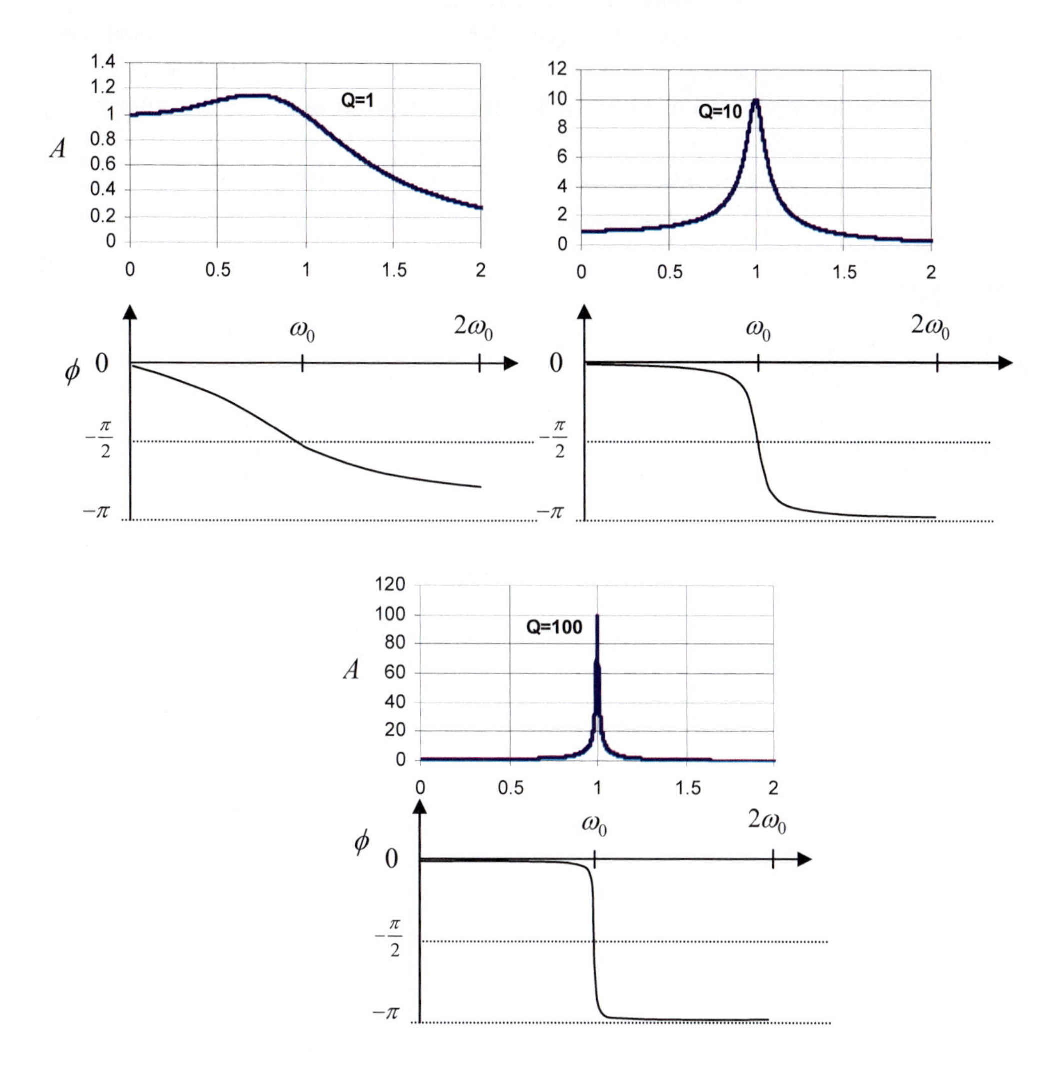

Figure 6-8: Plots of the *amplitude transfer function*, A / A_0, showing the amplitude and phase that results for various driving frequencies. *Top Left*: for $Q = 1$ system: plot of *amplitude* versus *frequency* $\omega_{0/}\omega_0$ and *phase* versus *frequency* under it. *Top Right:* Same plots for $Q = 10$ system. *Bottom:* $Q = 100$.

Notice how, as the damping decreases (Q larger), the resonance peak sharpens and the transition in angle occurs more quickly.

Simple Pendulum

To understand the forced oscillator, consider the familiar example of pushing a child on a swing. The swing is like the simple pendulum[7] shown in Figure 6-9. From that figure, we can write the differential equation for the motion in angle θ by introducing polar coordinates that are natural to the circular symmetry established by the fixed length string. The following summarizes the arguments to establish the form of the velocity and acceleration in polar coordinates.

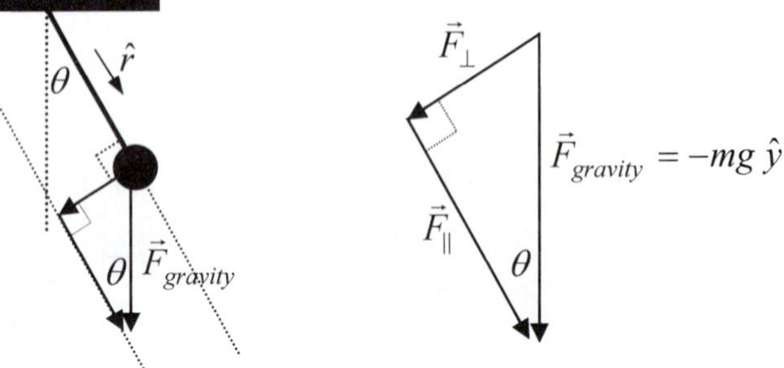

Figure 6-9: Swing as a *simple pendulum*. *Right:* zoomed view of decomposition of gravitational force $\vec{F}_{gravity}$ into components parallel to the suspension string, $\vec{F}_{\|} = F_{\|}\hat{r}$ and perpendicular to the string, $\vec{F}_{\perp} = -F_{\perp}\hat{\theta}$.

$$\vec{r} = r\,\hat{r}$$
$$\hat{r} = \cos\theta\,\hat{x} + \sin\theta\,\hat{y}$$
$$\hat{\theta} = -\sin\theta\,\hat{x} + \cos\theta\,\hat{y}$$
$$\Rightarrow \dot{\hat{r}} = \dot{\theta}\hat{\theta},\ \dot{\hat{\theta}} = -\dot{\theta}\,\hat{r}$$

6.29
$$\vec{v} = \frac{d\vec{r}}{dt} \equiv \dot{\vec{r}} = \dot{r}\,\hat{r} + r\,\dot{\hat{r}} \Rightarrow$$
$$= \dot{r}\,\hat{r} + r\dot{\theta}\,\hat{\theta}$$
$$\vec{a} \equiv \frac{d^2\vec{r}}{dt^2} \equiv \ddot{\vec{r}} = \ddot{r}\,\hat{r} + \dot{r}\,\dot{\hat{r}} + \dot{r}\,\dot{\hat{r}} + r\,\ddot{\hat{r}} = \ddot{r}\,\hat{r} + 2\dot{r}\,\dot{\theta}\,\hat{\theta} + r\ddot{\theta}\,\hat{\theta} - r\dot{\theta}^2\hat{r} \Rightarrow$$
$$= \left(\ddot{r} - r\dot{\theta}^2\right)\hat{r} + \left(r\,\ddot{\theta} + 2\dot{r}\dot{\theta}\right)\hat{\theta}$$

Now, forces along the radial direction must be balanced because the string is stiff in that direction. Hence, we only need consider the angular equation, which is the component perpendicular to the suspension string. It can be written, using $r = l$ *and* $\dot{r} = 0$:

6.30
$$F_\theta = m\,a_\theta = m\,l\,\ddot{\theta}$$

Finally, by substituting the force calculated with the aid of Figure 6-9, we get the equations of motion of a harmonic oscillator in θ:

6.31
$$-m\,g\sin\theta = m\,l\,\ddot{\theta} \quad \Rightarrow \ddot{\theta} = -\omega_0^{\,2}\,\theta \quad where \ \ \omega_0 = \sqrt{\frac{g}{l}}$$

---

[7] It is interesting to realize that Buridan (see box in Chapter 3) accounted for the swinging of the pendulum using his understanding of impetus.

Here we assume that the swing is only pushed through small angles so that the sine function can be approximated by the angle in radians. Note that, in this limit, the swing resonance frequency only depends on the length of the swing and gravitational pull of the earth, not on the mass of the child.

In this way, we see that our assertion that the swing is a harmonic oscillator is true at the right level of approximation. We can thus treat it using the full power of the equations developed earlier for forced damped harmonic oscillators (assuming a linear damping force) and graphed in Figure 6-8.

From those figures, we see that if we push the child at a frequency much less than the resonant speed of the swing very little amplitude is developed relative to what will result if we push near resonance. [8] One must push for awhile to establish steady state motion, for the *beginning* of the pushing amounts to a non-sinusoidal driving force; *idealized* sinusoids have no beginning. At slow speed, the motion also is in sync (phase) with the pushing. However at fast rates, the swing motion is exactly opposite that of the pushing. For example, if we push with $F_e \sin \omega t$, with $\omega \gg \omega_0$ the oscillator will move such that $\theta = -\theta_0 \sin \omega t$; so that when the driving force is $F_e \hat{\theta}$ (for instance at $\omega t = \pi / 2$), the position is $\theta = -\theta_0$, i.e. the driving force acts against gravity, while with $\omega \ll \omega_0$, it would be $\theta = \theta_0$.

To further our understanding, we now analyze the energy of the forced oscillator.

Energy of Sinusoidally Forced Damped Harmonic Oscillator

Calculating the steady state energy of the forced oscillator (as opposed to the energy after an excitation of an unforced oscillator given in equation 6.19) gives:

6.32 $$E(t) = \frac{1}{2} m\, v(t)^2 + \frac{1}{2} k\, x(t)^2 = \frac{1}{2} A(\omega)^2 \left(m\, \omega^2 \cos^2(\omega t + \phi) + k \sin^2(\omega t + \phi)\right)$$

Taking the time average of this, using the time average of the sine and cosine functions (see end of chapter problems), we get:

6.33 $$\langle E \rangle = \frac{1}{4} m\, A(\omega)^2 (\omega^2 + \omega_0{}^2)$$

Substituting for $A(\omega)$ from equation 6.27 gives:

6.34 $$\langle E(\omega) \rangle = \frac{1}{4} \frac{F_e^2}{m} \frac{\omega^2 + \omega_0{}^2}{\left(\omega_0{}^2 - \omega^2\right)^2 + (\omega\gamma)^2}$$
$$= \underbrace{\frac{F_e^2\, \omega_0{}^2}{4m\, \omega_0{}^4}}_{E_0} \frac{\left(\left(\frac{\omega}{\omega_0}\right)^2 + 1\right)}{\left(\left(1 - \left(\frac{\omega}{\omega_0}\right)^2\right)^2 + \left(\frac{1}{q} \frac{\omega}{\omega_0}\right)^2\right)} \qquad where\, q = \frac{\omega_0}{\gamma}$$

[8] Notice the kind of odd pushing we require here is a continual contact. If we push in the standard fashion, we wait for the swing to return and hence end up pushing largely at the resonance frequency.

When we take very light damping, so that $q \to Q >> 1$, the relation 6.34, is zero (i.e. $E \sim 0$) except near $\omega \sim \omega_0$. Thus, if we take $\omega \sim \omega_0$, except for the terms $\sim 1 - \frac{\omega}{\omega_0}$, we get (see end of chapter problem):

6.35 $$\frac{E_0}{2} \frac{1}{\left(\frac{\omega}{\omega_0} - 1\right)^2 + \frac{1}{4Q^2}} \qquad \textit{where } Q \sim \frac{\omega_0}{\gamma}$$

Plots of *energy versus frequency* using the simplified equation 6.36 are shown in Figure 6-10 for various Q values. This curve is called a *resonance curve* or a *Lorentzian.* Note, as with the amplitude curve, we have a resonance peak that gets sharper with higher Q.

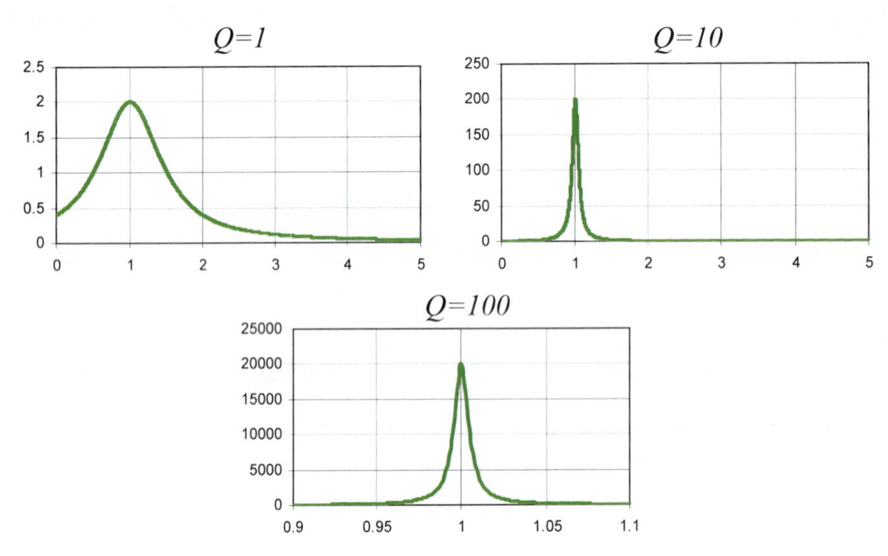

Figure 6-10: *Lorentzian energy versus frequency,* $\frac{\omega}{\omega_0}$, for three values of Q: 1, 10 and 100.

In order to speak about this curve's most obvious characteristic, its peak, specifically the width of that peak, we define the ***Full Width at Half Maximum.*** In the lightly damped case of equation 6.35, the peak is at $\omega = \omega_0$ with $E(\omega_0) = 2E_0Q^2$. And, to cut this value in half, the first term in the dominator must equal the second; thus, we obtain the frequencies ω_+ and ω_- that respectively bound the left and right side of the peak:

6.37 $$\omega_{\pm} = \omega_0 \left(1 \pm \frac{1}{2Q}\right)$$

Thus, the Full Width at Half Maximum (FWHM) is:

6.38 $$\Delta\omega = \omega_+ - \omega_- = \frac{\omega_0}{Q} \sim \gamma$$

One can verify the width formula for the particular *Q's* illustrated in Figure 6-10 above by reading off the halfway points. We also note that, because of the relation 6.38, one can write the quality factor, *Q*, for lightly damped oscillators as:

6.39 $$Q = \frac{\omega_0}{\Delta\omega}$$

This makes intuitive sense because the wider the fractional resonance the more frequencies there are in addition to the oscillatory frequency of the primary actors (e.g. bodies and spring), and, thus, the more the damping is succeeding in dissipating that oscillatory activity, hence, the lower the *Q*.

One can view an harmonic oscillator as a filter that is more selective for higher *Q*. Indeed, it is just this effect that is used to pick out a single radio station out of all the ones flying through the air. If one, for instance, wants to capture 96.1 FM, one must exclude all frequencies except those within a range around 96.1 *MHz*.[9] The better this filtering is done, the less crosstalk from other stations one will hear.

The oscillator's response to various driving frequencies is called its *frequency response*. However, it is also helpful to look at the response of the system to different actions in time. For example, suppose you strike an oscillator such as a drum head or a piano key, what happens? The response to a brief sharp impact is called the *impulse response* and is another sort of finger print for an oscillator. The impulse response for the damped harmonic oscillator looks similar to equation 6.18.[10]

An Example of Q

Suppose you strike the key for middle A ($440/s$) on a piano and let it vibrate; further suppose it drops off by $1/e$ in energy in about 1.8 seconds. Thus, $Q = \omega_0\tau = 2\pi(440)\tau \sim 5\times10^3$. (Also, using equation 6.21, we can get $\gamma = 1/\tau$, the damping coefficient). Notice the value of *Q* is in the thousands. Silicon micromechanical oscillators have been made with *Q*'s of 6×10^5. Quartz crystal oscillators used in radio frequency detection typically have *Q*'s from 10^4 to 10^6.

Expansion of Potential to Harmonic Term

We have now seen some of the many motions that harmonic oscillators are useful in describing. A general class of motions that can be analyzed in this way are those with potential energies that can be expanded in a Taylor's series. Notice we've already done such an expansion implicitly in our discussion of the simple pendulum when we approximated the $\sin\theta$ function by θ, for $\sin\theta = \sum_{n=1}^{\infty}\frac{(-1)^{n+1}}{n!}\theta^{2n-1}$ is the Taylor's expansion. More generally, given a potential, $V(r)$, one can generally expand it via Taylor giving:

6.40 $$V(r) = V(r_0) + V'(r_0)(r - r_0) + \frac{1}{2}V''(r_0)(r - r_0)^2 + \ldots$$

[9] The FM refers to frequency modulation, which means the voice and music is encoded in frequency variations.

[10] For an impulse, *δ(t)*, the response, for the simplest initial conditions, is $\sim e^{-t\gamma/2}\sin(\omega_d t)u(t)$, where *u(t)* is the unit step function.

The first term is constant, not contributing to the force, while the second term contributes only to a constant force, and thus is not relevant to locally bound motion. The third term is the harmonic term.[11] Note that we need this term to be positive to have bound motion. Geometrically, this means that the curve is concave upward. In terms of our analogical use of the calculus to describe the motion, this is equivalent to requiring the curve of *potential versus position* to be shaped like a valley or, as it is often called, a well. These are helpful analogies because whatever the actual force and body may be, if they are described by such a potential energy curve, they are in a way like a man on a bike in a valley. He can oscillate up and down the slopes of the valley with a fairly low Q unless the road and his bike are exceptionally good.

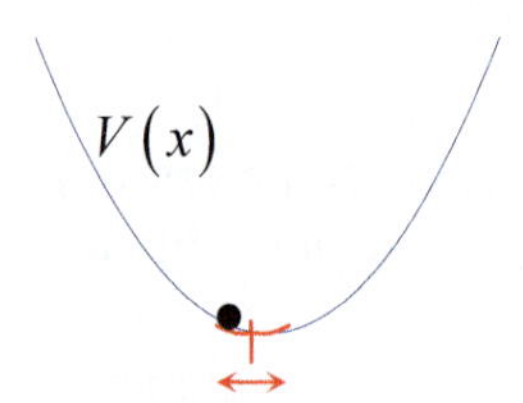

Mathematically, we say the second derivative of the potential energy determines the stability of the system. If $V''(r)$ is positive then the system will tend to restore itself under the influence of small perturbations. This is called a stable system.

[11] The fourth term would put us in the "anharmonic" regime, or perturbations to harmonic oscillator motion.

Summary

The simplest type of locomotion is uniform linear velocity (fixed intensity of impetus) of a fixed mass. A child's swing and the pendulum on a grandfather's clock approximate the simple back and forth motion of an *harmonic oscillator*, which is the simplest type of *bounded* motion. It is a one dimensional motion, but with a particular simple type of force, one that varies linearly with position $F = -Cx$, where C is a constant. Because of its simplicity, it holds an elemental place in the study of bounded motion. Its place is similar to that of tangent lines which can be used to build curves if one considers tangents of tangents etc., i.e. all the derivatives.

Circular motion can be divided into two harmonic oscillators one in the x-direction and one in the y-direction. An analysis of the force that causes the change in impetus needed to make circular motion gives the following equations:

$$m\ddot{x} = -x\frac{F_0}{r_0} = -k\,x, \quad m\ddot{y} = -y\frac{F_0}{r_0} = -k\,y$$

Each of these has the form of the *harmonic oscillator differential equation*: $m\ddot{u} = -ku$, the general solution of which is:

$$u = A\sin\left(\omega_0\, t + \phi\right) \quad \text{where } \omega_0 = \sqrt{\frac{k}{m}}$$

The frequency is $f = \omega_0 / \left(2\pi\right)$ and the period $T = 2\pi / \omega_0$.

As previously mentioned, the harmonic oscillator motion is elemental, so this sine wave motion can be profitably used to decompose more complex motions. Such a decomposition is done in the *Fourier Series*: $f(x) = \sum_{n=1}^{n=\infty} a_n \sin\left(n\,x\right)$, where the coefficients tell us how much weight to assign each of the multiples of the fundamental frequency (i.e. the so called harmonics of the fundamental). So, for instance, one gets $\sin x$, when: $a_1 = 1,\ a_{n\neq 1} = 0$.

The *idealized spring* has the simple harmonic force law, $F = -kx$, where k is called the spring constant. The simplest version is with two bodies, one attached to each side of the spring with one having so much more mass that its activity can be ignored; we say, via a being of reason, it has infinite mass. The movable body has energy $E = \frac{1}{2}m\mathbf{v}^2 + \frac{1}{2}kx^2$.

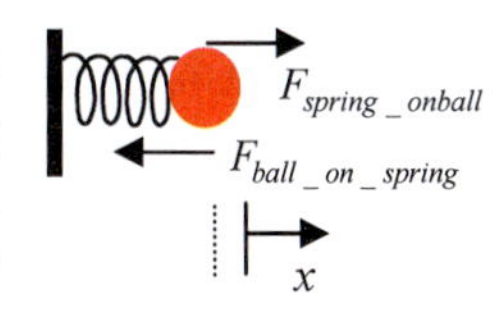

We include more of the real situation into our analysis by adding damping. Damping opposes the motion of the oscillation with, commonly, a force proportional to speed: $F_{damping} = -b\dot{x}$. This gives the equation for *damped harmonic motion*:

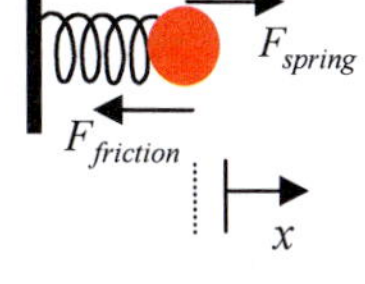

$$\ddot{x} + \gamma\dot{x} + {\omega_0}^2 x = 0 \quad where \quad \gamma = \frac{b}{m} \quad and \quad \omega_0 = \sqrt{\frac{k}{m}}$$

with general solution for $\omega_0 > \gamma/2$, (see end of chapter problems for other cases):

$$x(t) = Ae^{-\left(\frac{\gamma}{2}\right)t} \sin\left(\omega_{damp} t + \phi\right) \quad where \ \omega_{damp} = \sqrt{\omega_0^{\,2} - \frac{\gamma^2}{4}}$$

When an isolator is *lightly damped,* i.e. $\omega_0 \gg \gamma/2$), the energy is $E(t) = \frac{1}{2}kA^2 e^{-\gamma t}$, which gives the *decay time* of the oscillator, the amount of time it takes for the *energy* to drop by *1/e*, as $\tau = \frac{1}{\gamma} = \frac{m}{b}$. The decay time for the *amplitude* is 2τ.

In order to quantify the degree of damping of an oscillator, we define the "quality of an oscillator," called the quality factor or *Q*, which, for lightly damped oscillators, is:

$$Q = \frac{E}{\Delta E} = \frac{E}{\left(\frac{1}{\omega_d}\right)\frac{dE}{dt}} \approx \omega_d \frac{E}{\gamma E} \approx \frac{\omega_0}{\gamma} = \omega_0 \tau$$

The higher the *Q* the more cycles an oscillator will go through per oscillation, i.e. the better the oscillator oscillates.

We next include an external force to discuss *forced damped harmonic motion.* The equation is: $\ddot{x} + \gamma\dot{x} + \omega_0^{\,2} x = \frac{F_e}{m}\sin\omega t$. This has a solution $x = A\sin(\omega t + \phi)$, where:

$$A = A_0 \frac{1}{\sqrt{\left(1 - \left(\frac{\omega}{\omega_0}\right)^2\right)^2 + \left(\frac{\omega}{\omega_0}\right)^2 q^{-2}}} \qquad \phi = \arctan\left(\frac{\omega}{\omega_0} q^{-1} \frac{1}{\left(\frac{\omega^2}{\omega_0^{\,2}} - 1\right)}\right)$$

$$where: \gamma = \frac{b}{m}, \ \omega_0 = \sqrt{\frac{k}{m}} \qquad A_0 = \frac{F_e}{m\,\omega_0^{\,2}}, \ \ and \qquad q \equiv \frac{\omega_0}{\gamma} \sim Q \ \ for\ light\ damping$$

The aspects of the solution can be understood for various *Q* values by analysis of the graphs given in Figure 6-8.

Musical instruments can be modeled as many forced harmonic oscillators. In this analysis each instrument is distinguished by its Fourier decomposition so that the same note on two different instruments will have different decompositions.

The energy of a lightly damped sinusoidally forced harmonic oscillator can be approximated by the Lorentzian line shape:

$$E(\omega) = \frac{E_0}{2} \frac{1}{\left(\frac{\omega}{\omega_0} - 1\right)^2 + \frac{1}{4Q^2}} \qquad where\ Q \sim \frac{\omega_0}{\gamma}$$

Its characteristics are graphed for various *Q* in Figure 6-10. In particular, notice that the phase of the forced oscillator changes by 180 degrees when one moves through resonance. The width of the curve is partly characterized by the *F*ull *W*idth at *H*alf *M*aximum (FWHM) is: $\Delta\omega = \frac{\omega_0}{Q} \sim \gamma$, so $Q = \frac{\omega_0}{\Delta\omega}$.

By expanding the potential energy in a Taylor series, one can treat a small motion in an arbitrary potential as harmonic in the limit of small perturbations of the expanded variable. In particular, when the second derivative of the potential $V''(r_0)$ is positive, it is effectively the spring constant of a harmonic oscillator that oscillates arbitrarily close to the expanded variable at $r = r_0$.

Terms

Simple Harmonic Oscillator

- Harmonic oscillator equation
  - Spring Harmonic Oscillator
- Amplitude
- Phase Angle
- Angular Frequency
- Period

Damped Harmonic Oscillator

- Resonance Frequency
- Damping factor
- Q or quality factor of an oscillator
- Decay time
- Decay time of amplitude

Forced Harmonic Oscillator

- Lorentzian
- Full width at half maximum

Modeling and Harmonic Theory

- Fourier series
- Frequency response
- Impulse response
- Stable system

Mixed Problems

1. What value of Q for a damped oscillator gives the fastest return to equilibrium? This is called a *critically damped oscillator.*
Hint: consider a solution of the form $(A + Bt)e^{-\alpha t}$.

2. Consider an *over-damped oscillator* which is a damped harmonic oscillator with a very high damping so that $\gamma > 2\omega_0$, what does the motion look like in this case? Write the equation for the motion *and* graph it. Compare it with the *under-damped oscillator* (i.e., $\gamma < 2\omega_0$) discussed in the text.

3. In the gravity wave detector called LIGO, Laser Interferometric Gravity Wave Observatory, ground vibrations can be mistaken for gravity waves if the detector test masses are not sufficiently isolated from ground vibrations. Assume that one wishes to detect $1kHz$ waves from the inspiral of two black holes, and that one needs an isolation factor of 10^4 in amplitude at that frequency. a) Assuming one uses a system (shown below) with a spring of constant k and dash pot of damping constant b such that $F_{damp} = -b\mathbf{v}$ and taking $\gamma = 2\omega_0$ (called critical damping), what resonance frequency is needed and what value of k does it imply? b) At this resonance frequency and neglecting the mass of the stages, how many stages of such devices one on top of the other would it take to reduce the ground noise by 10^{12} in amplitude? c) Should one have any damping? Explain.

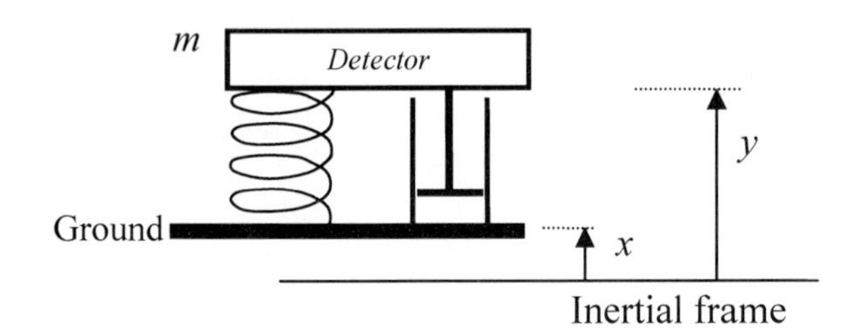

Hint: Given Δh_0 as the compression of a spring due to the weight of detector of mass m and h_0 is the unweighted equilibrium length of spring, we would like to know: $z = y - h_0$.

4. Define the *sine* of an angle as the ratio or comparison between the length of the opposite side and the length of the hypotenuse of a right triangle. Similarly, define the *cosine* as the ratio of the length of the opposite to that of the hypotenuse. a) Explain, how they are, thus, of necessity intrinsically related to the unit circle. b) What about a circle of arbitrary radius r_0?

5. $\sin\theta$, though fundamentally a ratio between the lengths of two sides of a right triangle, can be graphed to get a curve; i.e., we graph $y(x) = y_0 \sin x$. The resulting "sine wave" has greater curvature at the peaks and zero curvature at the zero crossing points. The first estimate of how much a curve bends ("curves") at a given point

can be obtained by analyzing how the tangent at points around the given point changes. That is, we look at the change of the first derivative, namely, the second derivative: $y''(x)$. In this particular case, note that the second derivative gives back the negative of the original function. Symbolically, we have $y''(x) = -x$. a) Now, moving from the static considerations of geometry (which is the study of the first property of material things, i.e., extension), as we have so many times, take the plot as representing time on the x-axis, and explain physically what the equations and the graphs mean. *Hint*: Take $x \to \omega t$.

6. Bertrand's theorem proves that, in two dimensions, only two potential energy functions can produce *stable,* closed trajectories (i.e. come back to the same point in a cyclic manner): an inverse radial potential such as the gravitational or electrostatic potential $V(r) = -\frac{C}{r}$, and the harmonic oscillator potential $V(r) = \frac{1}{2}kr^2$.

If one starts with a circular orbit that is maintained by force that has a potential: $V(r) = \frac{1}{3}kr^3$, show that, as long as there is no angular velocity, this forms a closed orbit. What is its angular speed? *Note:* in more advanced courses, one will show that a small perturbation can send it into a non-closed orbit.

7. The General Mobile Radio Service (GMRS) and Family Radio Service (FRS) band has channel 1 at $462.5625 MHz$ and channel 2 at $462.5875 MHz$. What is a rough estimate of the minimum Q of the filter that one must have to distinguish these frequencies?

8. Elementary particles such as the W boson survive only for a certain short time before "decaying" into other things. By an analogical extension of mass and energy called mass-energy, we say the mass-energy of the boson is about $80 GeV$, and in this dynamic decaying, we find there is a relation between how long the W boson survives and how accurately we can specify its mass-energy. This results in a width like the one found for the under-damped harmonic oscillator. By taking the $\omega_0 \to 80 GeV$ and the frequency width (which has been measured for the W boson) as $\Delta\omega \to 2 GeV$, where this later is an indication of the decay rate (the larger this width the faster the decay rate): a) What is the (analogical) quality factor, Q, for the W boson? b) Using the Lorentzian line width formula given in the text, recast it in the following way:

$$P(\omega) \propto \frac{\Gamma/2}{(\omega - \omega_0)^2 + (\Gamma/2)^2}$$ where $\Gamma = \gamma \to 2\text{GeV}, \omega_0 \to 80 GeV, \omega \to E$, graph the probability, $P(\omega)$ of seeing the particle of Energy E.

9. Expand $\sin(a+b)$ using the angle addition identity and show that:

$A\sin(\omega t + \phi) = B\sin\omega t + C\cos\omega t$

10. Calculate the average of $\sin^2(\omega t + \phi)$.

11. Calculate the average kinetic energy and average potential energy from equation 6.32 and use the result to prove equation 6.33.

12. Assuming the Earth and Mars both make perfectly circular orbits around the sun of radius r_{earth} and r_{mars} with periods of revolution T_{earth} and T_{mars}. Define a rectangular coordinate system such that the Earth and Mars are both on the x-axis at $t = 0$. a) Write the equation for the x and y coordinates of the Earth and Mars as a function of time. b) Using the standard polar coordinate, θ, give the angle of both the Earth and Mars at time t. What phase builds up for Mars when the earth returns to the x-axis?

13. Discuss *generally* how it is that the equations of damped harmonic oscillators can describe such radically different things as a spring oscillator with friction and mass, and a series electric circuit with capacitor, resistor and inductor such as shown in Figure 6-6. In general, why does it work? *Hint:* Think about how it can be that two apples plus two apples gives four apples is also true for oranges.

14. A microwave oven works by bombarding food with high frequency radio waves of 2.45 *GHz*. The food typically will have water molecules as part of its makeup or mixed in with it. Water molecules have strong electric dipole moments, that is, a charge configuration that can receive the action of these electromagnetic waves in such a way as to cause the molecules to oscillate back and forth dissipating energy in frictional heat. If we model this process as a forced damped harmonic oscillator with resonance frequency $f_0 \sim 20GHz$ and just above critical damping, so that $\gamma = 2\omega_0$. Taking the cross sectional area, σ, of energy that is blocked by the food (water) in the oven to be proportional to the amplitude given by $E(\omega)$ in equation 6.34, what is the *ratio* of the penetration depth $x_0(\omega) = \dfrac{1}{n\sigma(\omega)}$ at resonance $20GHz$ to that at $2.45\ GHz$. Note: n is the number density of water molecules which absorb the energy (eventually as thermal motion). Note this approximation could be the beginning of an experiment to test how good this harmonic model of water absorption of electromagnet energy in the microwave range (no pun intended) is.

15. Calculate the speed of the under-damped harmonic oscillator given in the text.

16. Expand the equation for 6.19 E(t) for a damped oscillator in powers of ω_0. Explain why the approximation used to get the approximate energy decay law in the next equation is correct.

17. Calculate the Q of the curves given in Figure 6-10 using the formula $Q=\Delta\omega/\omega_0$

18. Graph equation 6.34 for the energy as a function of frequency for various values of q and show that it has peaks (places were derivative is zero and second derivative is negative) on each side of the y-axis and that it has only one peak for $q < a$, for some a. Find a.

19. a) Find the Q of a child's swing using the following data. The swing with a child of mass m in it oscillates 5 times in 11.5 seconds. When a child is gently swinging (i.e. at low amplitude) the amplitude reduces to ½ its initial value in about 16 swings. b) The length of the swing is ~1.3m, what is the calculated period?

20. The following data was taken from a piano with the release pedal pressed so that the strings of the piano could vibrate freely. In particular, the A (440 Hz) key was hit and held for three separate data runs. The volume of the note dropped by about a factor of two in 4 seconds in the first run and 4 seconds and 5 seconds in the second and third. Note that when something sounds half as loud, it has 1/10 the received sound power, or in equal time periods, 1/10 the integrated energy. (Mathematically, for ratios greater than one, we can write: $L_2 / L_1 \approx 2^{(E_2/E_1)/10}$) a) What is the Q of the piano? b) Discuss the difference between intensity of the sound and the amplitude of the string vibration. Why would you guess the sound intensity would be related to energy?

21. Consider the Fourier expansion of a square wave such as shown below.

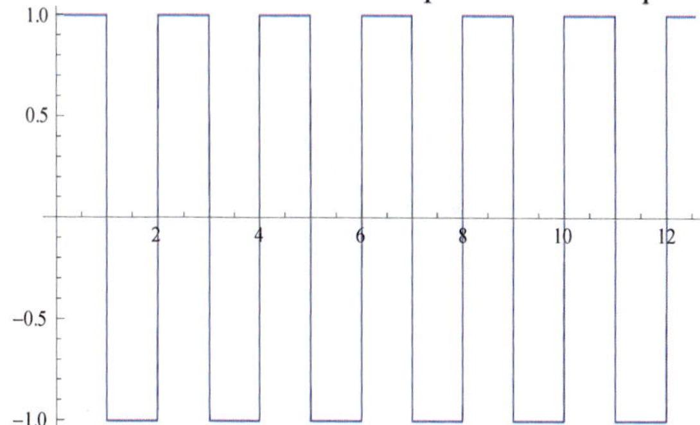

The first two non-zero terms of the expansion of this function are:

$$\frac{2}{\pi}\sin 2\pi t + \frac{2}{3\pi}\sin 6\pi t + ..$$

For one cycle, graphically add these two functions and thereby begin to show that adding them approaches the square wave. Explain why the term $\sin 4\pi t$ has a zero coefficient in the expansion, i.e. $a_2 = 0$.

Note when 100 terms of the Fourier series are added, we get a very close approximation. Note the over ringing at the jump points.

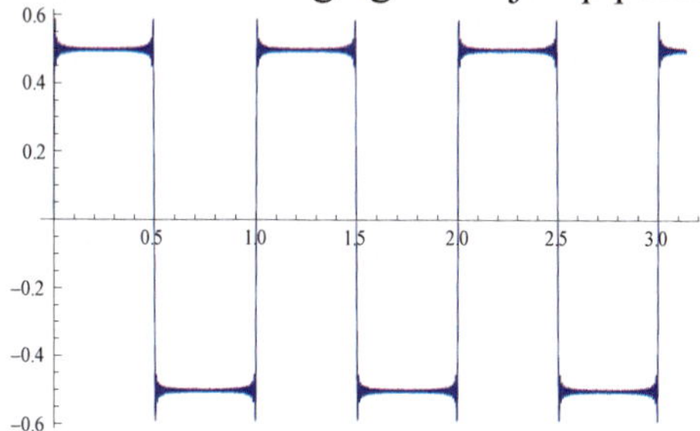

22. To bring home the key sense in which the decomposition of a curve into (sine and or cosine wave) frequencies is a being of reason, consider the following curve: Suppose this curve repeats every 2π so that we get a smooth curve over some portion of space. In what sense can this smooth curve be considered to be composed of the two curves sin[t] and sin[$2t$] given their sum is the (complete) Fourier expansion of the function for this curve and what sense is it not.

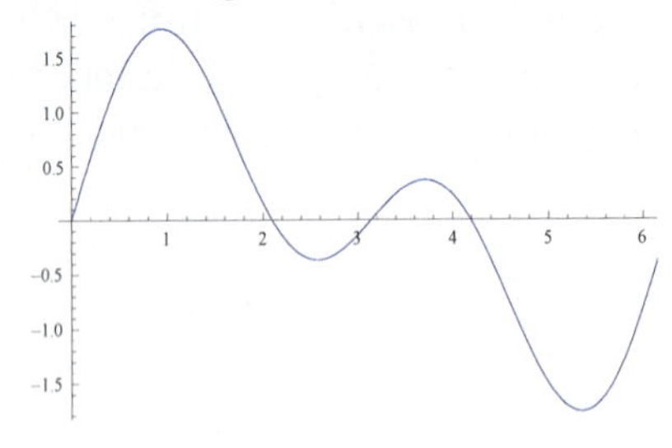

23. Show that the steps from equation 6.34 to 6.35 are justified:

24. Show that equation 6.18 is a solution to equation 6.17.

25. Describe what the three dimensional graph of the motion of a harmonic oscillator would look like. (place vs. time)

26. Consider two bodies of mass m in a coupled harmonic oscillator as shown. There are three springs of constant k and gravity is switched off. Write and solve the equations of motion. *Hint:* Think about the *two* simplest possible motions that these balls can undergo. These are the so called *normal modes* and can be found by examining the symmetry of the physical situation. There are *two* symmetrical types of motion: 1) the balls moving *together to the right* then *together to the left* and 2) the balls moving *away from each other* and then *back towards each other*. If a single normal mode is excited, it cannot excite any other normal mode; this is the meaning of "normal" (i.e. in the sense of perpendicular or orthogonal).

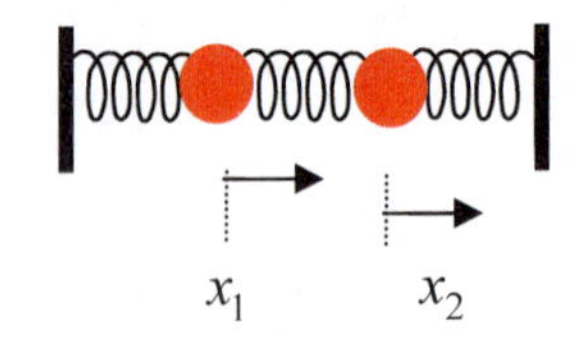

27. In quantum field theory, roughly speaking, one considers all of space as filled, like a mattress spring, with harmonic oscillators. We allow the harmonic oscillators to get infinitely small so that every point in space has an harmonic oscillator of certain amplitude; that is, we consider each oscillator to be a part of a field.

Given N coupled oscillators that can only vibrate in the vertical direction in a mattress, how many normal modes (that is, when one excites one of these modes alone none of the others will be excited later, no matter how long one may wait) are there? *Hint*: Consider the simple linear case of *2* coupled oscillators above and generalize from it. Notice, also, that each normal mode solution is a particular solution, $x_n(t)$ and that the general solution to the motion can be written as a weighted sum of the solutions; $x(t)=\sum_{n=1}^{M} c_n x_n$. Notice also that the number of ways that the oscillator can move (number of degrees of freedom) must be reflected in the solution.

Chapter VII

Angular Momentum

Introduction

We have seen the beginnings of what are the two most important concepts in modern physics: conservation of *energy* and *linear momentum*. We now investigate the third most important concept in physics, the conservation of *angular momentum.*

Whereas conservation of linear momentum means, loosely speaking that we never lose the tendency to move in a given linear dimension, angular momentum conservation means that we never lose whatever tendency there is to move around a point.

The Definition of Angular Momentum

A Free Point Particle

Take a small massive object moving uniformly in outer space far from any other objects that might be able to influence it. Turn off gravity and all other forces. Assume one other very small body is nearby as shown in Figure 7-1A. Take this as the origin of our Cartesian coordinate system, and, by a mental construct, take both small objects to be points as shown in Figure 7-1B. Draw a vector connecting the origin to the object, and we see that this vector is constantly changing. As the object moves from the far left to the far right, the angle that this *position vector* makes with the x-axis increases moving the vector through one half of a circle centered at the origin, i.e. 180 degrees.

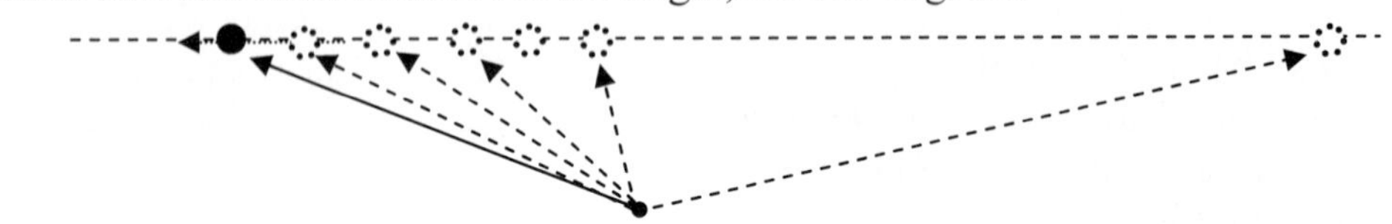

Figure 7-1A: The vector drawn between a small marker body at various points as it traverses its linear path from the right to the left. Dotted lines indicate past locations and their associated position vectors.

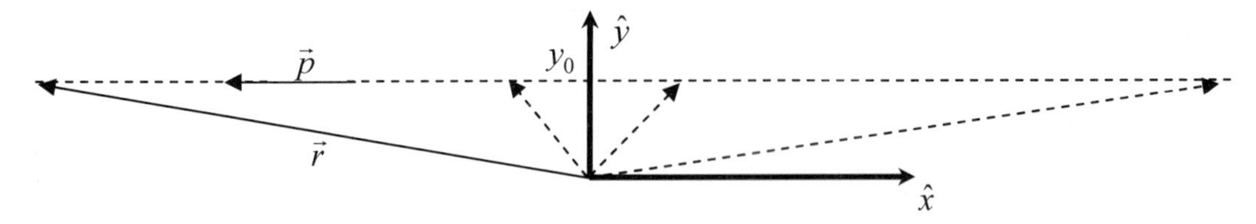

Figure 7-1B: The Cartesian plot of the movement of the ball relative to the chosen origin. Note how the position vector rotates around this origin.

We can thus call the motion of the large object relative to the marker point *angular motion*, and the correlative motion of the object's given intensity of impetus, *angular momentum*. That is, the momentum is, in a way, going around the point, so it is angular rather than linear. The farther the point is from the line of motion of the body, the larger the moment (lever) arm on which the momentum can act, and hence the more angular momentum we say the body has. Furthermore, for a given fixed distance from the origin, the more momentum the body has, the more angular momentum it has. Thus, we define the amount or measure of angular momentum, which we call L, as:

7.1 $$L = p \times (\text{distance to line of motion})$$

In our case above, we can write:

7.2 $$L = p\, y_0 ,$$

But what if it's moving in some other direction relative to our chosen axis? This is where all the work of Chapter 2 in constructing objects that capture the key quantitative aspects in a notation and set of rules comes in handy. Namely, we need to use vectors. We have momentum which, as we've seen, is a vector (the magnitude of the intensity of the impetus and its direction) and we have our position vector. We now need a way to write an equation for $\vec{L}$, which is a vector quantifying the rotation of the momentum around a point. This discussion should bring to mind the special analogical generalization of multiplication, the vector cross product (the "perpendicular" product discussed in Chapter 2). Recall that the cross product points in the direction of the axis along which one vector gets pushed into another, which is, within a minus sign defined by convention, the same as the rotational sense that we just discussed. Hence, we define L as:

$$\vec{L} = \vec{r} \times \vec{p} \quad (7.3)$$

As we have mentioned, like linear momentum, angular momentum is conserved. We say the quantity $\vec{L}$ (the measure of angular momentum) is conserved, and we write this conservation mathematically as:

$$\vec{L} = \sum_{i=1}^{N} \vec{L}_i$$

We consider N bodies in our system numbered consecutively, and the subscript i labels which body is under consideration.

Why is this quantity conserved? Why doesn't it change? It is clear why in the above example at least; there, the momentum is conserved and the perpendicular distance doesn't change. To understand in general why angular momentum is conserved, we have to dive more deeply into the meaning of angular momentum.

A Particle in Circular Motion

Obviously, since angular momentum concerns with motion about a point, a simpler example from the point of view of that motion, though more complicated from the point of view of mechanics, is rotation about a point, as shown in Figure 7-2 below.

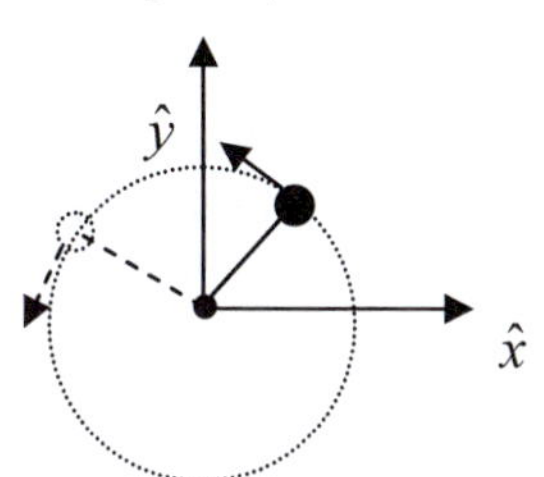

Figure 7-2: Rotation of a body around a point is the simplest example of angular motion; the motion is wholly angular.

For concreteness, suppose this experiment consists of a small massive object on a string swung around. Assume the mass of the string is negligible so that, as we saw in Chapter 4, its force is transmitted undiminished to the small object. Furthermore, we assume the object has established a steady state and rotates smoothly around the contact point, which we take to be a pin with a joint that allows the string to rotate freely with no resistive force, i.e., no friction.

As we know, without the string, the object would continue in uniform linear motion; however, with the string, a force is exerted which has the following form in cylindrical (or polar if we consider just two dimensions) coordinates:

$$\vec{F} = -F_0\,\hat{r} \quad (7.4)$$

With this, we can now check that the trajectory is indeed what we suppose based on such a force. Recall that we can break the motion into two independent dimensions, which we label by x and y as defined in Figure 7-2.[1] The origin is the pivot point for the string. Let the arrow from the origin to the place of the object at a given moment of interest define the x-direction. Similarly, the place where the object *next* intersects the perpendicular to the x-direction defines the y-direction. Thus, in rectangular coordinates the force is:

7.5
$$\vec{F} = -F_0 \cos(\omega t)\hat{x} - F_0 \sin(\omega t)\hat{y}$$
$$F_x = -F_0 \cos(\omega t), \quad F_y = -F_0 \sin(\omega t)$$

Hence, the equations that describe the motion which are given from Newton's second law become:

7.6
$$m\,\ddot{y} = -F_0 \cos(\omega t)$$
$$m\,\ddot{x} = -F_0 \sin(\omega t)$$

Integration yields:

7.7
$$x = r_0 \cos(\omega t)$$
$$y = r_0 \sin(\omega t)$$
$$where: r_0 = \frac{F_0}{m} \omega^2$$

This is the circular motion we expected.

What now is the momentum in the angular direction, i.e. the angular momentum? Using equation 7.3, we get $|L| = r\,p \; and \; \vec{L} = r\,p\,\hat{z}$, where $p = m\mathbf{v} = m\omega r$ and we take $\hat{z}$ to be out of the page--recall this means (looking down on the page) rotation in the counter-clockwise direction.

Notice something peculiar: the linear momentum of the ball, say in the x-direction, at any moment is constantly changing, though the angular momentum is constant. This is true for any direction one might pick. Pick one and see for yourself.

So, we can say total linear momentum changes but angular momentum doesn't. We say the linear momentum is not conserved, and the angular momentum is conserved. The linear momentum is not conserved because there is a force. We know by now that all forces are associated with bodies; in this case, it is the pin in the center which is anchored eventually to the Earth; so the unaccounted-for linear momentum goes into the Earth. Ultimately, the linear momentum is conserved when there is nothing to break the symmetry (see box on symmetry and footnote 1) of the underlying quantity.[2] In this case, once again excluding the Earth, linear momentum conservation is violated, because there is a force that breaks the general symmetry, in particular, linear symmetry that the quantity in that region has when not qualified by that force. Of course, not just any asymmetry will cause a violation of the conservation of linear momentum; it has to be one that can cause a force to

---

[1]The central object defines only a center of a circle, and the motion only defines a succession of places displaced in angle around this center in the plane; i.e. the natural parts that make up the plane in this case (radial and angular parts) are more commensurate with cylindrical coordinates than rectangular (Cartesian) coordinates. Despite this, a dissection of the space into rectangular slices is still possible by further specification. For instance, given a simple circular disk, we note its parts are naturally divided with cylindrical symmetry; however, the disk is by no means incompatible with many other types of divisions, but whatever it is, it must be specified. To see this, it may be helpful to imagine the circular path of the particle to be bounded by a very large rectangle, thus clearly "forcing" parts in the x and y directions to be established, or in mathematically terms, forcing a rectangular symmetry at the boundaries which can be extended inward.

[2] Recall this is the first property upon which all other are "built"

be applied to the body, i.e., something that can change the body's intensity of impetus; in fact, there is no active physical quality that we know of that does not do this (though some have more effect, others have much less). In addition to this, the force must be one that breaks the symmetry of the line, making one part of the line different from the next.

By analogy, we expect the conservation of angular momentum to be related to the symmetry around a point. This trivially applies to the first example of a free particle because it has no forces whatsoever. More to the point, in our current example, there are forces but there is no force acting around the point. The only force that acts is along the string. Indeed, strings can't act perpendicular to their length, because they have no stiffness in that direction.

Euclid

Five Postulates of Euclid:

1. To draw a straight line from any point to any point.
2. To produce a finite straight line continuously in a straight line.
3. To describe a circle with any center and radius.
4. All right angles equal one another.
5. (so-called *parallel line postulate*) If a straight line falling on two straight lines makes the interior angles on the same side less than two right angles, the two straight lines, if produced indefinitely, meet on that side on which are the angles less than the two right angles.

was born in 325 BC and died about 265 BC in Alexandria Egypt at the scholar's center (called a museum) founded by Aristotle's pupil Demetrius of Phaleron. The Museum of Alexandria boasts other important men of science, mathematics and engineering such as, Aristarchus, Archimedes, Eratosthenes and Hero.

Euclid's "*The Elements,*" is a classic on geometry that set the standard for rigor and clarity for many centuries. It contains its five postulates (see left), the last of which was successfully revised by among others F. Gauss, to discover non-Euclidean geometry. Even that new type of geometry takes its name from Euclid who first analyzed that most fundamental geometry that we get directly from our senses.

To illustrate the value of truth, not just for its practical use, but because that is what our minds seek and need and to manifest that there is no commensurability between knowledge and money, Euclid dealt with one of his students in the following way:

> A youth who had begun to read geometry with Euclid, when he had learnt the first proposition, inquired, "What do I get by learning these things?" So Euclid called a slave and said "Give him threepence, since he must make a gain out of what he learns."

Through this example, we begin to see a more quantitative way of treating the conservation of momentum, both angular and linear. Specifically, if we reflect the quality of force into quantity, for example, by writing a function or vector field, which represents the measure of the intensity of the force, dependent on position in space, we can say that the conservation of momentum is due to the linear symmetry of the space. In the case of angular momentum, the conservation is due to rotational symmetry. In fact, of course, the quality will change the space and thus the space will reflect the quality in some way;

remember that all physical qualities have a correlative in extension (continuous quantity, the first property).

Translational symmetry means if one moves an object, for instance, along the x-axis, it results in a situation indistinguishable from before the change. Rotational symmetry means, for instance, if one rotates a uniform sphere in place, it looks just the same as it did before in a rotational symmetric "space."

These sorts of symmetries are characteristic of Euclidean geometry which was analyzed rigorously by Euclid (see box) in the center founded at Alexandria by one of Aristotle's students.

Symmetry and Euclidean Geometry

Symmetry

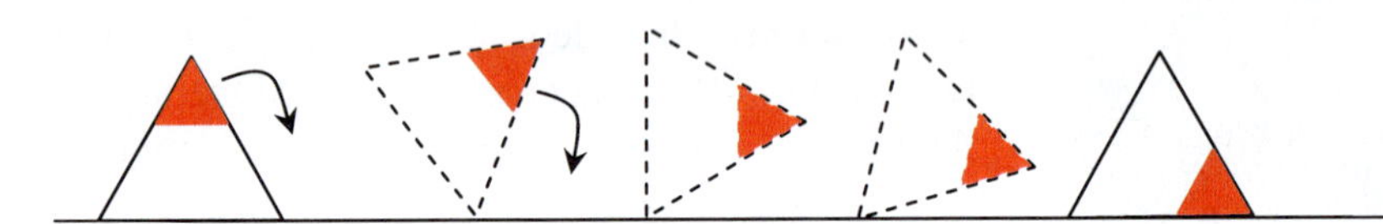

Discrete symmetry: Rotating this equilateral triangle by any multiple of 120 degrees (1/3 of one whole rotation) produces a triangle in an orientation that is indistinguishable at this level of abstraction (leaving out such things as the red tip drawn above for reference) from its initial orientation.

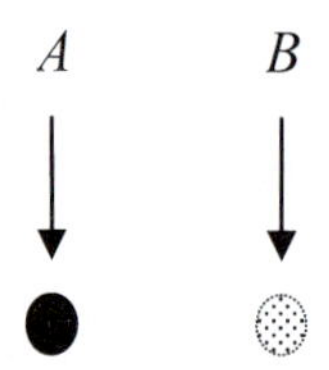

Left/right ***Translational*** **Symmetry**: if the boundaries are extended out of sight, I cannot tell the difference between when the ball is at position A or position B moved some distance to the right from that location. This is a continuous symmetry. This type of symmetry is associated with conservation of *linear* momentum.

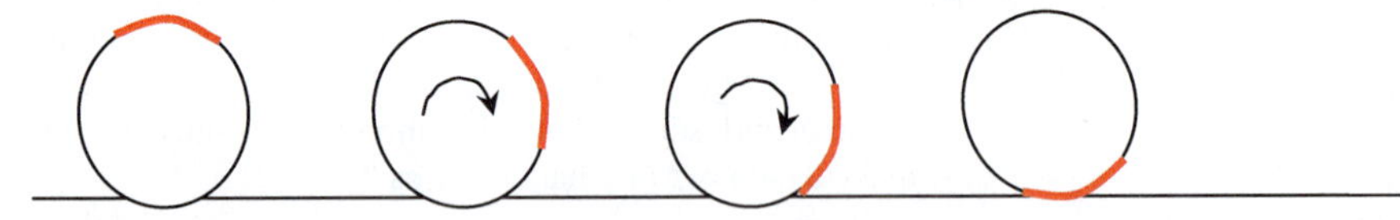

Rotational **Symmetry:** Rotating this circle about its center by *any* angle produces a circle in an orientation that is indistinguishable at this level of abstraction (i.e., leaving out such things as the red arc drawn above for reference) from its initial orientation. Note that such *qualities* as the red arc shown above are abstracted out by definition in mathematics generally and, in particular, is not included in the definition of a circle. This type of symmetry is associated with conservation of *angular* momentum.

A moment spent on Euclidean geometry will be helpful. Euclidean geometry is what one sees when one strips out most of the particulars in ones imagination, leaving only a sort of empty outer space like image, with one part not differentiated from another. For example, suppose you start with an image of the classroom, and then, in your imagination, eliminate all the desks, all the shelves and gradually all objects in sight, including all light and even the air which you obviously cannot see but can feel. You will then have a kind of "outer space world." Of course, our imagination is about things we've sensed, so we can't really eliminate all things we've sensed, we can only minimize the content, imagining a very dark area with very thin air etc., such as outer space. Our minds then can abstract the general properties of this image and lay down axioms to capture them, as Euclid did. This space is *homogeneous* and *isotropic* at every point.[3]

Homogeneous means everything looks the same no matter what direction we walk whether it be: up/down, left/right or back/forth. In our case, we are looking to see whether the potential energy changes as we move. In fact, if the interior of the space is homogeneous (in this sense) throughout and if we push the boundary of the space out of sight, we cannot tell the difference between one part and another. *Isotropic* at a point means when you stand at one point and rotate around and look around you will see the same thing in every direction. If we take a region to be homogeneous *to an arbitrarily small scale* and we again push the boundary out of sight or better still for this purpose, by a mental construct, imagine that it is pushed out to infinity," then *every point* is isotropic.[4]

As is clear from what we said earlier, homogeneity in a region results in conservation of linear momentum in that region, and isotropy at a point results in conservation of angular momentum at that point. To be precise, let's recall that homogeneity in a linear dimension (translational symmetry) simply means that each point is the same as the next in that given dimension, and homogeneity of a 3-dimensional region means having translational symmetry in all three dimensions in that region. Isotropy at a point means a rotation of the points on a small circle centered on that point leads to same physical situation (defined by the context). For example, suppose the circle was made up of small masses, i.e. small massive bodies, and there is an isotropic potential around the circle center, then each one would have the same potential energy before the rotation as after.

Conservation of Angular Momentum

Angular momentum can be considered as a separate conservation law in the sense that isotropy at a point doesn't require homogeneity near the point. Consider a bull's eye pattern. Further, we've already seen a case in which conservation of linear momentum was "broken" by a force, while angular momentum conservation was not.

Of course, homogeneity of the potential energy within a region, i.e. the condition that translation in *any* dimension doesn't change the potential energy (which simply means no forces!), *does* necessarily imply that there is rotational symmetry, i.e. no change in

---

[3] A space that is isotropic at every point is necessarily homogeneous (even in the analogical generalizations considered in cosmology). Isotropic at one point clearly doesn't guarantee homogeneity; think of sitting at the center of the lone planet (a spherical one) in the universe.

[4] Here when we use the word isotropic and homogeneous we refer to the limit of as small a size scale as we choose and we refer to simple Euclidean geometry. In cosmology, one averages over a certain size scale (operational assigning of common time can also come into play) and one considers analogical generalizations of Euclidean space.

potential due to rotation (no angular forces) so that angular momentum doesn't change.

As an example of the role of symmetry, consider the potential energy $V(\theta)$, associated with the force field $\vec{F}(\theta) = f(\theta)\hat{\theta}$ shown below in Figure 7-3a. The force field does not change as one moves perpendicular to the paper, i.e., it is independent of the z-dimension. This is also true if we move along a radial direction. However, if we cross the center along a diameter, V does change such that *force* switches directions. Figure 7-3b illustrates how an angular force implies a breaking of translational and rotational symmetry of the potential.

As a point of interest, note that there is a discontinuity in the potential near $\theta = 2\pi$ (where the potential drops from 2π to zero) because of the nature of this type of potential. This "kink" results in a (infinite) force directed contrary to those around the rest of any given circle, which is illustrated in Figure 7-3a by the long arrows. This example illustrates the fact that no angular force that increases or remains constant with angle can be represented by a potential function.[5] In fact, any force field that "curls" around a point cannot be written as the derivative of a potential function near that point.[6]

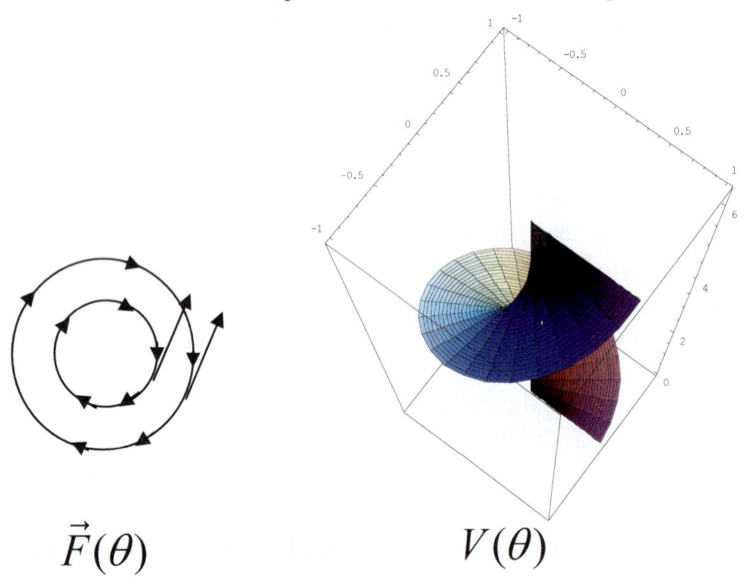

Figure 7-3: *(left)* **a.** Two dimensional illustration of a force $F(\theta)$ acting in an angular direction around a central point. It is associated with the potential $V(\theta)$ shown on the right. *(right)* **b.** The potential $V(\theta)$ associated with a force field with an angular directed force. The potential energy at a given point in the x-y plane is represented as the height along the z-axis above that point. In this way, high points have high potential energy, and so the figure can be thought of as a land terrain in which a ball would naturally tend to roll down the hills into the valleys. In this analogy, our potential is like a spiral children's slide. Note that, because of the nature of the potential, the child has to climb up after the ride

[5] More generally, this is true for any angular force that is a monotonic function of the angle.

[6] Mathematically, if $\vec{\nabla} \times \vec{F} \neq 0$, then $\vec{F} \neq \vec{\nabla} V$, for any function V.

A related question that deepens our understanding of angular momentum is: does angular momentum conservation follow from Newton's three laws (in simple Euclidean space)? Again, if we are precise about our meaning, the answer is yes. Any object (and this will include fields in their own way (see Chapter 8)) that exerts a force – including in a rotational manner – will have to experience a counter force acting on it in the opposite direction that will then result in conservation of linear momentum and conservation of angular momentum as well. There is a common objection that holds true only when one analyzes such forces leaving out the fields themselves.

The argument starts by considering the world to be made out of idealized point particles (a mental construct)[7]. Consider the effect of one such particle on another. In such a world, there is no distinction in direction, except directly along the line of the particles; hence, there can be no rotational forces, no forces trying to make one particle spin about the other. It is next pointed out that 20th century physicists have shown that elementary particles have spin, something that defines a direction to the particle and thus allows rotational symmetry to be broken, disallowing plain point elementary particles—at least all the elementary ones we know now have spin. Such spin *could* allow forces like those shown below, which satisfy Newton's second law and thus conservation of momentum, but not rotational symmetry and thus result in a violation of conservation of angular momentum. Notice how the forces tend to make the particles rotate in a counter-clockwise direction.

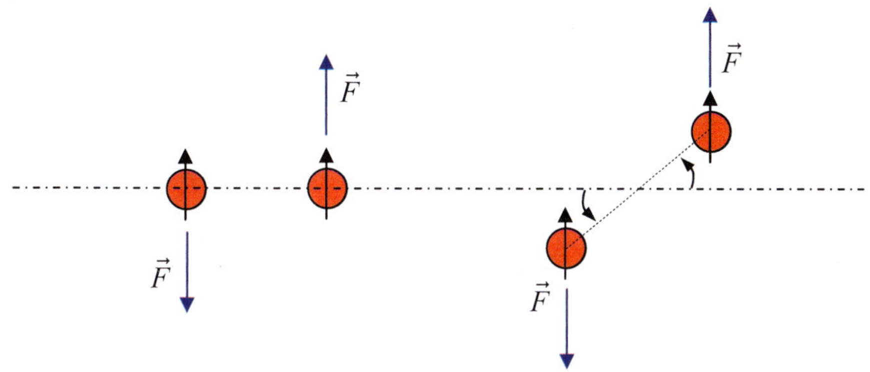

Figure 7-4: *(left)* **a.** Two particles with spin and a postulated force shown in blue. *(right)* **b.** As force acts particles move to the new positions shown after a certain time. Note how the effect of the force is to induce an angular motion or twirling about a common center.

However, notice that the forces seem to act over the distance with no intermediary.[8] More to our point, the spin has an effect; that's how we know about it, so it is mediated by something and that something, say it's a field or the like, is in fact what breaks the symmetry, so an analogical extension of Newton's third law would require it (the field) to experience an equal and opposite "force" from the particle, thus satisfying both conservation of linear and angular momentum.

---

[7] To leave the mental construct, consider small particles and don't consider the effect of their extension.

[8] While not impossible, this is not physical, for the primary property of physical things is their extension

Dumbbell in Rotation

To better understand the relation between linear and angular momentum, consider now a stiff very thin rod supporting heavy balls at each end. The center is affixed to a pivot as shown in Figure 7-5. Start the dumbbell shaped rod rotating around the pivot. Now, consider switching on gravity. In Figure 7-5, gravity is represented by a uniform constant vector field as shown in the figure. When the field comes on, no change occurs. Why? The symmetry of action around the pivot point for the dumbbell shaped rotator is such that gravity acting on the right ball trying to turn it clockwise is exactly countered by gravity acting on the left ball, which is trying to turn it counterclockwise. Hence, with respect to this rotator and field there is a rotational symmetry. In later courses, you will learn about a mathematical function called a Lagrangian.[9] The Lagrangian is written for a given physical system; for this system, one would immediately see that the Lagrangian is rotationally symmetric and thus would conserve angular momentum.

Of course, if the symmetry of the dumbbell was broken by making the ball on one end heavier than the other, then the dynamical rotational symmetry would also be broken, which implies the bar would rotate one way or the other. Yet, still there would be a symmetry in an average sort of way, because, every 360 degrees, it comes back to the same configuration. Not surprisingly, such average symmetries, which themselves result from other symmetries (in this case a discrete symmetry), lead to conservation of angular momentum in an average sense.

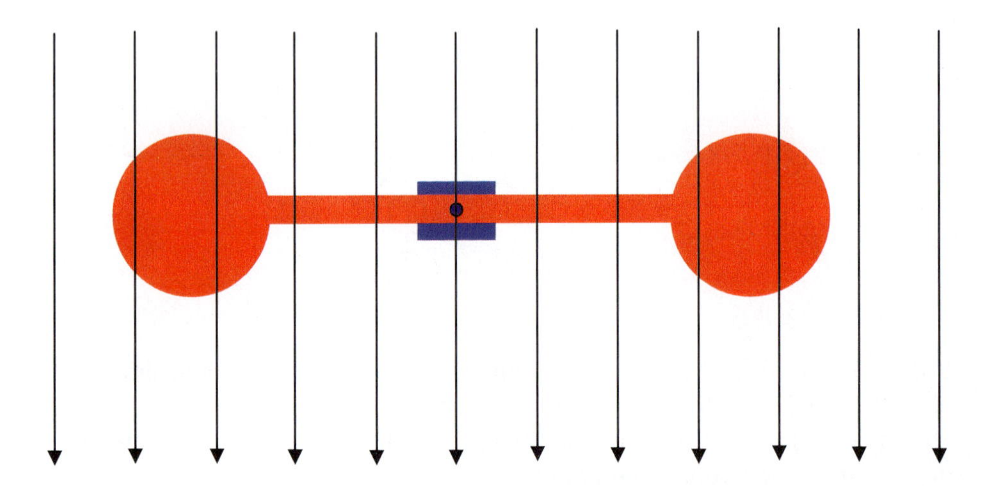

Figure 7-5: A dumbell, i.e. a rod with equal mass weights an each end, in a uniform field, affixed at center to a pivot protruding from the wall.

[9] It is related to the principle of virtual work discussed in Chapter 4.

Torque

We can see explicitly how conservation of angular momentum comes about by taking the time derivative of the equation for angular momentum:

7.8
$$\vec{L} = \vec{r} \times \vec{p}$$
$$\frac{d\vec{L}}{dt} = \frac{d\vec{r}}{dt} \times \vec{p} + \vec{r} \times \frac{d\vec{p}}{dt} = \vec{r} \times \frac{d\vec{p}}{dt}$$

Where, in the last equality, we used, $\vec{p} = m\frac{d\vec{r}}{dt}$ and assume m to be constant. Further, recalling that the change in momentum in a particle at a rate of $d\vec{p}/dt$ is caused by a force, $\vec{F}$ we get:

7.9
$$\frac{d\vec{L}}{dt} = \vec{r} \times \vec{F} \equiv \vec{\tau}$$

Noting the analogy with Newton's second law, we say that a torque, $\vec{\tau}$, causes a rate of change of angular momentum. When there are no torques, no forces trying to drive the body in a rotational type motion, $\vec{L}$ is constant (never changes); this is what we mean by conservation of angular momentum.

Equal and Opposite Forces Causing Conservation of Angular Momentum

An object whirling about on a string is an example of conservation of angular momentum which results from the force not being directed in an angular direction about the point under consideration. We now consider the case of a free body of mass m that is, for one moment, pushed around a given point by another body of mass m. Assuming initially that each body shares a speed v at a given radius from the point, we have, leaving off the direction and only considering magnitudes: $p = mv$; $L_1 = r\,p$ $L_2 = r\,p$. This yields a total of: $\left|\vec{L}_{total}\right| = L_1 + L_2 = 2r\,p$

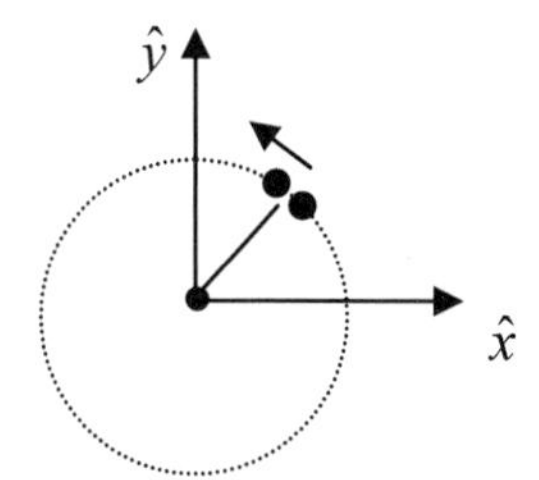

Figure 7-6: One mass m body momentarily pushing a second mass m body in along a tangent of a circle centered at the origin.

If *body 1* gives *body 2* an impulse of p, intensifying its current impetus (say through the release of a spring), then we get $L_1 = 2r\,p$. Now, as we say in Chapter five, because the force that caused this impulse (p) acts through a time, say T, then the opposing but equal in magnitude force that *body 2* exerts on *body 1* lasts that same amount of time, so that the impulse *body 1* receives is also p but in the opposite direction, thus conserving linear momentum. This then means $L_1' = 2r\,p \quad L_2' = 0 \; so \; L_{total}' = 2r\,p$, where the primed L

(L') indicates the angular momentum *after* the collision, as opposed to the symbol without the prime which means before the collision.

In general, we can see this mathematically by taking the derivative of:

7.10 $$\vec{L}_{total} = \vec{r} \times \vec{p}_1 + \vec{r} \times \vec{p}_2$$

to get:

7.11 $$\frac{d\vec{L}_{total}}{dt} = \vec{r} \times \frac{d\vec{p}_1}{dt} + \vec{r} \times \frac{d\vec{p}_2}{dt} = \vec{r} \times \vec{F}_1 + \vec{r} \times \vec{F}_2$$

And, Newton's third law, $\vec{F}_1 = -\vec{F}_2$, implies:

7.12 $$\frac{d\vec{L}_{total}}{dt} = 0$$
$$\vec{L}'_{total} = \vec{L}_{total} = constant$$

Why Do Some Angular Motions Take No Work?

Uniform circular motion, which is prototypical of angular motion, and thus of angular momentum, has an unusual aspect to it. It has changing velocity but not speed. This means, as we pointed out in Chapter 5, that no work is done because $\vec{F}$ is perpendicular to distance traveled for the entire path making $W = \oint \vec{F} \cdot d\vec{x}$. How can this be? It is paradoxical that the string does no work; after all, it *is* constantly changing the direction of the motion. To completely understand this behavior, we have to make clear what we mean by changing direction. To do this, consider polar coordinates[10] r, θ as shown below in context of ball held in a circular orbit by a string fixed to a point on the Earth. As we often do, we ignore the Earth's motion by taking its mass to be effectively infinite.

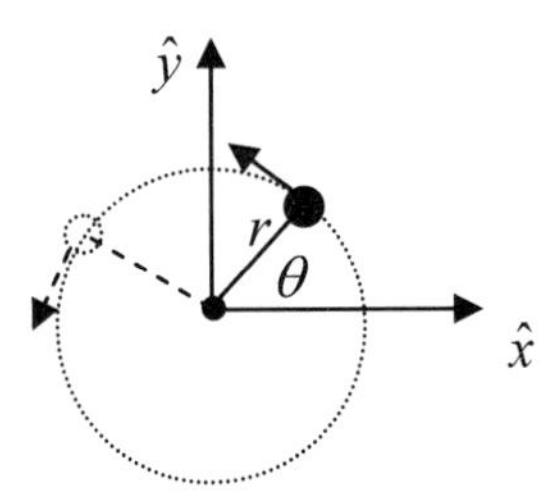

In these coordinates, $\vec{F} = -F_0 \hat{r}$, and the equations of motion are:

7.13 $$m a_r = m(\ddot{r} - r\dot{\theta}^2) = -F_0; \quad given\ r = R = \text{constant} \Rightarrow -mR\dot{\theta}^2 = -F_0$$
$$m a_\theta = m(r\ddot{\theta} + 2\dot{r}\dot{\theta}) = 0 \Rightarrow m r \ddot{\theta} = 0 \Rightarrow \dot{\theta} = \omega = \text{constant}$$

Thus: $W = \int \vec{F} \cdot d\vec{x} = \int_0^{2\pi} F_\theta R\, d\theta = 0$, since $F_\theta = 0$ In these coordinates, the ball is moving constantly in the $\hat{\theta}$-direction, which is a type of conserved momentum which we call *conservation of angular momentum*. The momentum in the $\hat{r}$-direction is also

---

[10] Also, see Chapters 2 and 6.

conserved, for the ball doesn't move at all in that direction. Notice that according to these coordinates, the ball doesn't change direction.

When we talk about constant speed but changing direction, we are talking about the primal directions that come from our simplest picture of space in which one imagines as little structure as possible and so begins by only considering lines. The most fundamental lines in two dimensions are perpendicular lines: for example, length and width. In these directions, using *x-y* labels for the rectangular coordinates, we saw earlier in this chapter that:

7.14
$$m\,\ddot{y} = -F_0 \cos(\omega t)$$
$$m\,\ddot{x} = -F_0 \sin(\omega t)$$

7.15
$$x = r_0 \cos(\omega t)$$
$$y = r_0 \sin(\omega t)$$
$$where: r_0 = \frac{F_0}{m}\ \omega^2$$

Notice that linear momentum is ***not*** conserved in either the x or y direction. In fact, we see that the net momentum in the $\hat{\theta}$ direction is conserved, because the momentum that was in x gets transferred to y in just the right way. Where then, speaking loosely, does the linear momentum go? We already saw it goes to the Earth. How? Again, the string exerts a force on the ball, the ball acts back on the string with an equal and opposite force, which is transmitted to the pin and, then, in turn, to the Earth. We can now qualitatively describe what happens in this motion.

During the first quarter of the cycle of the motion (from $\theta = 0\ to\ \theta = \frac{\pi}{2}$), the ball starts with a momentum of p in the y-direction, but ends with no y-momentum (see Figure 7-7); this is because work ***is*** done by the string on the ball in the y-direction, but that's not the whole story. The string exerts a force on the Earth in an equal and opposite direction during this time, causing the Earth to gain momentum p in the positive y-direction. Also, the ball starts the first quarter cycle with no x-momentum, but ends with an impetus of intensity p in the negative x direction, while the string forces activate an impetus of intensity p in the positive x direction, thus again maintaining conservation of momentum.

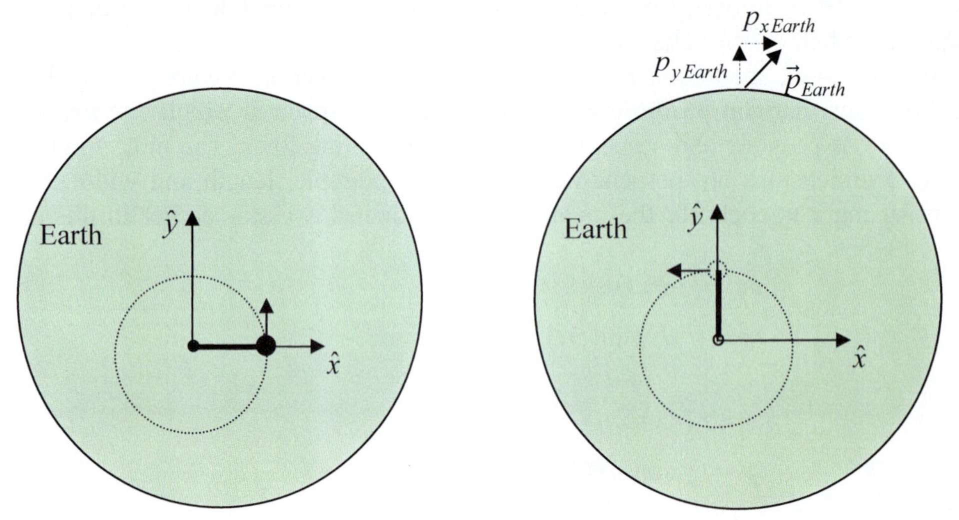

Figure 7-7: *(left)* ***a.*** Ball on a string moves under an impetus of intensity p directed in the positive y-direction, i.e. $\vec{p} = p\,\hat{y}$. The Earth is viewed in a rest frame in which initially it appears to have no impetus. *(right)* **b.** After a quarter cycle of the motion is complete, the ball moves with momentum p in the negative x-direction, $\vec{p} = -p\,\hat{x}$. Through the action of the string during the first quarter cycle of the motion, the Earth now has momentum p in the x *and* y direction: $\vec{p}_{Earth} = p\,\hat{x} + p\,\hat{y}$, keeping the net momentum constant at $\vec{p}_{net} = p\,\hat{y}$.

Similarly, from $\theta = \frac{\pi}{2}$ *to* $\theta = \pi$, the ball "exchanges" all of its negative x-directed impetus for negative y-directed impetus. During the second *half* of the rotation (from $\theta = \pi$ *to* $\theta = 2\pi$), the forces exactly reverse from the first half resulting in "draining" the momentum out of the Earth and dumping it back into the ball, so that the Earth and the ball are returned to their initial states when the ball returns to its initial position on the x-axis.[11]

We can see that the linear directions (e.g. length and width), which we've labeled as the x and y directions, have physical precedence because the ball would fly off, following the line of its instantaneous motion, if there were no string to pull it back. Again, fixed impetus would move the ball in a straight line; it changes direction because of the force of the string changing its impetus. In using cylindrical coordinates, we make solving our problem easier, because we incorporate two pieces of the physics of the situation into such coordinates: the impetus given the ball at any moment is in a given linear direction, ***and*** the impetus is constantly frustrated by the force of the string.

---

[11] Note the use of language. We cannot literally "drain" momentum as if it were water we could siphon out of a bucket. Momentum is the measure of the intensity of a quality. A quality is not a substance, such as a liquid, that can be literally removed, but we can talk in such a way by analogy, and we do in common speech. There is nothing wrong with this; indeed it can be a great help as long as we remember what we mean. Here we simply mean that the intensification of impetus in one body ("filling") is correlated with a de-intensification ("draining") in another. Such relations come from a profound symmetry of nature that we've illuminated to some degree but will become clearer as one advances in his understanding of physics.

Thus, the answer to the conundrum of how it is that circular motion avoids doing work is that it *only* does so with respect to specially chosen coordinates. In fact, when we respect the impetus' tendency to continue in a straight line and the distinctness of the perpendicular dimensions, we see work is done *within each linear dimension.* Recalling that work is defined as that which causes new (or destroys old) kinetic activity in a body, we say the work *defined in the special coordinates* is zero, because the string drains impetus from the ball in one linear dimension and fills it in the other in such a way that the kinetic energy of the ball in the $\hat{\theta}$-direction is constant (as is its momentum in that direction), assuming a frictionless system and an infinitely massive Earth.

For example, in the first quarter cycle of the motion (cf. Figure 7-7), we can see that work is done *within* each linear dimension. The Earth does work (via the string) on the ball in the negative x-direction, while the Earth does *negative* work on the ball in the negative y-direction (and the ball does positive work on the Earth in the positive y-direction) in such a way that whatever kinetic energy the ball gains in the negative x-direction, it loses in the positive y-direction, so that its *total* kinetic energy remains the same (again assuming an infinitely massive Earth in the absence of friction).

We now work several examples to better understand solving problems using angular momentum.

Example Problems

Use of Cylindrical Coordinates

Because of the importance of cylindrical coordinates (cf. Figure 7-8) for angular motion, let us reproduce and further explain the acceleration given in Chapter 6.

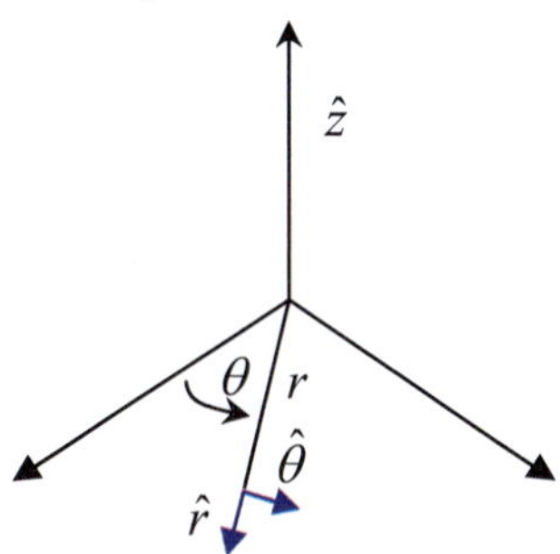

Figure 7-8: Cylindrical coordinates consist of regular polar coordinates along with the z-axis of rectangular coordinates. The directions used are thus: $\{\hat{r},\hat{\theta},\hat{z}\}$.

Ignoring the z component, which is just like the Cartesian z, we consider, as before, only the two polar dimensions, r and θ:

7.16 $$\vec{a} = \left(\ddot{r} - r\dot{\theta}^2\right)\hat{r} + \left(r\ddot{\theta} + 2\dot{r}\dot{\theta}\right)\hat{\theta}$$

The $\ddot{r} \equiv \dfrac{d^2r}{dt^2}$ term is simply the radial acceleration, very similar to an $\ddot{x}$. Because the two polar coordinates are, unlike the primordial x-y coordinates, somewhat tied up in each other, we have second term in the $\hat{r}$ direction. It is called the "centrifugal" term written as: $r\dot{\theta}^2$.

There are also two terms of the acceleration in the $\hat{\theta}$ direction. The first is the simple angular acceleration term, $r\ddot{\theta}$ which is still vaguely analogous to the $\ddot{x}$. The second term is the called the Coriolis term, $2\dot{r}\dot{\theta}$.

These polar acceleration terms can be better understood by imagining yourself on a spinning device, such as an old-fashioned record player or a merry-go-round. In particular, imagine that the disk spins at a rate $\omega = \dot{\theta}$ while you stand at a fixed point on it. On such a platform, bodies that are stationary in the lab frame (an inertial frame), say helium balloons hovering above the merry-go-round, appear to undergo an acceleration given by $-\vec{a}$ with $\dot{r} = 0,\ \ddot{r} = 0,\ \ddot{\theta} = 0$. Bodies that are moving at uniform speed in the inertial frame appear to undergo Coriolis acceleration as well. Thus, two of the four terms in equation 7.16 immediately come into play for those living in such a spinning world.

Indeed, as we will discuss in Chapter 9, on a manned trip to Mars in which gravity is simulated by centrifugal force, the terms of equation 7.16 will describe the effects that the astronaut crew will deal with everyday for nearly a year. The forces one would feel in such circumstances come from the so-called fictitious forces mentioned in Chapter 4. These "forces" appear only because we are in a frame that is accelerating in some way.

We can see this already in linear motion. For instance, in an accelerated car, a dropped ball is apparently accelerated backwards with force F equal to the balls mass times the car's acceleration; i.e. mathematically: $F_{fictitious} = m_{ball}a_{car}$. As the car accelerates, the front of the car accelerates away from the ball, the back towards it. To you, it appears as if the ball is accelerating at a_{car}, and since it has mass m_{ball}, a force of magnitude $m_{ball}a_{car}$ appears to be acting on it. Though in the frame of the car it looks like a force is pushing it back,[12] from the Earth's rest frame, we see that it's the car that's accelerating. It's really the car that will hit the ball (if the ball doesn't hit the floor first) --nothing is pushing the ball back! It is just that in the moving frame everything else accelerates. Notice that if we are in an accelerating frame, something must be accelerating us, and this is the real force we see once we allow ourselves to focus on it. In the case of the accelerating car, the force is the road pushing back against the tires which first pushed against it under the action of the engine.

On a merry-go-round, one is undergoing continual uniform rotation of angular speed ω. If you throw a ball from the center to someone on the edge, you can see the effects of using the position of people on the merry-go-round as reference points. (By the way, picking *a system of reference points* is what we mean when we say "pick a reference frame.") The ball will veer off; this is the Coriolis term.

We already have seen the centrifugal term in our earlier example. The ball moving in a circle was, in the frame rotating with the string, apparently pulled radially outward by a force $mr\dot{\theta}^2$ (see equation 7.16 above). The radial and angular acceleration terms are obviously not seen in uniform rotation, by the very definition of uniform rotation.

It is sometimes natural and practical to use an accelerating frame. In such cases, we can get insight into the problem as long as we are careful to remember that these forces are

[12] If it was you falling in the car, you wouldn't feel like there was force on you.

not really changing the momentum of the bodies but only apparently because I, the observer, am changing my momentum.

Many things in ordinary life are the result of the effects of such moving frames. Let's consider a couple.

What Drives a Hurricane?

In the hurricane that hit New Orleans on August 29th 2005, which is shown in Figure 7-10, there is something driving the wind inward; this is heat from the sun coming from the ocean; there must also be something that drives it to spin. What does? It must be pretty strong given the wind speed of hurricanes (see chart giving hurricane wind speed by category number below Figure 7-10). It is the Coriolis force; we live on a spinning planet. It's as if we were on an endless merry-go-round. So, we have centrifugal and Coriolis fictitious forces—remember it's really the Earth moving under the objects, but we can speak this way if we know what we're doing. Centrifugal force will not give us a hurricane spin, because it's not in an angular direction (we'll see it's weak as well); however, the Coriolis term will. In terms of vectors, the Coriolis term can be written:

7.17 $$\vec{a}_{coriolis} = -2\vec{\omega}_{earth} \times \vec{v} \ .$$

We plug in the Earth's rotation rate and take, say the latitude of New Orleans $\sim 30^\circ N$, giving:

$$-2\vec{\omega}_{earth} \times \vec{v} = -2\left(\omega_{earth} \sin\alpha_{latitude}\hat{z} \times (-v\,\hat{r})\right) = 2\omega_{earth}\, v \sin\alpha_{latitude}\left(-\hat{z} \times (-\hat{r})\right) = \omega_{earth} v\,\hat{\theta}$$

(Here we use full cylindrical coordinates (see Figure 7-8) with the z-axis perpendicular to the tangent of the Earth at the given latitude).

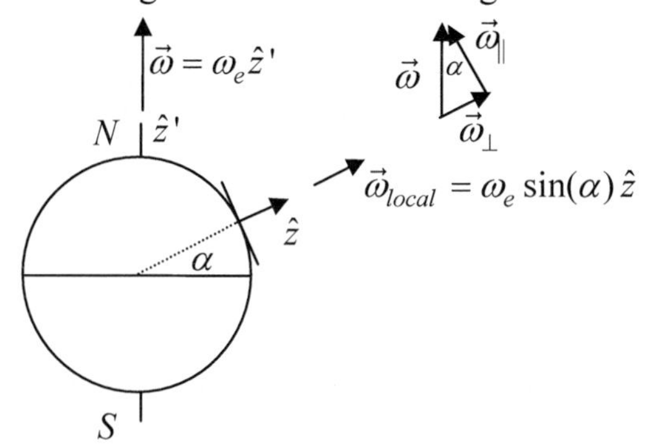

Figure 7-9: Local cylindrical coordinate system defined at latitude $\alpha \sim 30^\circ$. $\hat{z}$ points upward at the given spot; this means it is perpendicular to the local horizon (which is the set of lines that form a plane tangent to the surface).

If we allow this force to act on the scale of a day, which for a hurricane is an underestimate, but will nonetheless give us an order of magnitude result, we get:

7.18 $$v_{\theta\, result} = a_{Coriolis} T_{day} \approx \omega_{earth} T_{day} v = 2\pi\, v \approx 2\pi\, 20mph \approx 120mph,$$

which is certainly the order of magnitude of the wind speeds. By contrast the centrifugal acceleration is:

7.19 $$\vec{a}_{centrifugal} = -\vec{\omega}_{earth} \times (\vec{\omega}_{earth} \times \vec{r}) = -\omega^2{}_{earth} r\left(\hat{z} \times (\hat{z} \times \hat{r})\right) = \omega^2{}_{earth} r\,\hat{r}$$

giving:

7.20

$$v_{result\, cent.} = a_{centrifugal} T_{earth} = \omega^2{}_{earth} T_{earth} = 2\pi\,\omega_{earth} = 7.3x10^{-5}\, m/s = 1.6x10^{-4}\, mph$$

Figure 7-10: ***Hurricane Katrina: one of the largest storms to strike land in the US.*** It was the sixth-strongest Atlantic hurricane ever recorded and the third-strongest landfalling U.S. hurricane ever recorded. At its peak, it was a category 5 (see below) with sustained winds of 175 *mph*, but it dropped to a category 3 with sustained winds of 125 *mph* before landfall. [13]

Category[14]	Wind Speed Range	Possible Damage
1	75-95 *mph* (33-42 *m/s*)	Minimal damage to vegetation
2	96-110 *mph* (43-49 *m/s*)	Moderate house damage
3	111-130 *mph* (50-58 *m/s*)	Extensive damage to some structures
4	131-155 *mph* (59-69 *m/s*)	Extreme structural damage
5	>155 *mph* (>70 *m/s*)	Catastrophic building failure possible

Which Way Do Sinks and Toilets Drain?

Many have been told and think that this Coriolis force also drives the spin of water as it goes down the sink, tub and toilet drains. Is this so? There are even stories, from those who visit the equatorial regions, of tour guides who claim to demonstrate, by putting sinks on each side of the equator, that the Coriolis force causes draining water to spin counter-clockwise a couple yards to the north of the equator and clockwise a couple yards to the south. Whether the tour guides are unscrupulous and/or ignorant or whatever, it remains that the Coriolis force does not cause the spinning.

A look at equation 7.17 shows that the acceleration does change direction,[15] but it also shows that the magnitude vanishes on the equator and is very small near the equator. Indeed, we can shown that the magnitude is not, under normal conditions, great enough to cause the sink drain spin anywhere on the Earth.

Looking at equation 7.18 above for Coriolis term near New Orleans, we get:

$$\boldsymbol{v}_\theta \approx \omega_{earth} T_{day} \boldsymbol{v}_r \Rightarrow \frac{\boldsymbol{v}_\theta}{\boldsymbol{v}_r} \approx \omega_{earth} T \approx 1 .$$

[13] Wikipedia

[14] Data taken from physics fact book by Glenn Elert and his students.

[15] To see this, remember these accelerations apply to objects seen by a local observer at the given point on the Earth's surface.

We see that the ratio for which the radial to tangential velocities are comparable is for time scales on the order of a day. Hence, since water in a sink, for instance, leaves, in the order of minutes, the amount of "impulse" that can be applied to the water in a sink or tub or toilet is very much too small, because there is just not enough time for it to build up.

Angular Momentum Vectors

To better appreciate the nature of angular momentum, $\vec{L}$, consider the following examples.

Rotating Pendulum

The rotating pendulum shown in Figure 7-11 manifests the dependence of the angular momentum on the reference point. The pendulum consists of a ball hanging by a string attached to a rod at a point, *D*, which is above the center of the ball's *plane of rotation* and along the *axis of rotation*. We can show that for such a pendulum, the angular momentum about *C* (center) is conserved ($\vec{L}_C = \vec{r}_C \times m\vec{v}$), while the angular momentum about *D* ($\vec{L}_D = \vec{r}_D \times m\vec{v}$) is not.

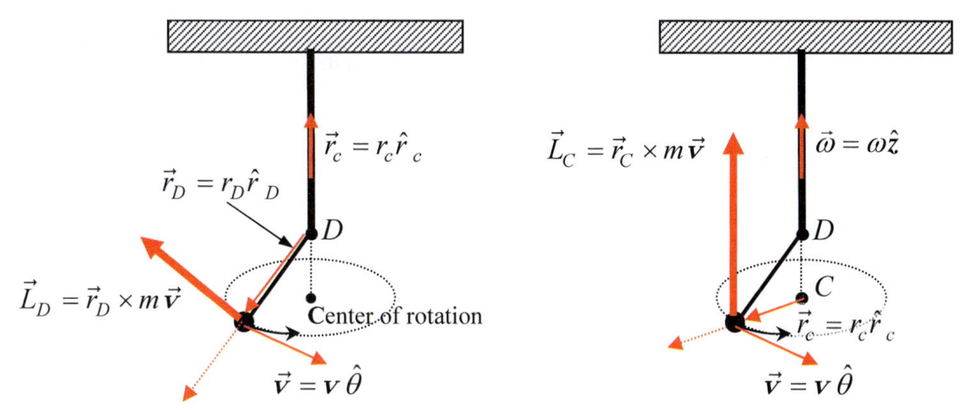

Figure 7-11: Two views of a pendulum with a hanging body that is rotating, rather than simply swinging back and forth. They illustrate that angular momentum depends on the chosen reference point. **a.** *(left)* The angular momentum about point *D* has an off vertical direction and is not conserved. **b.** *(right)* The angular momentum about point *C*, the center of rotation, is vertical and is conserved.

Notice that the component of $\vec{L}_D$ parallel to the rod, i.e. the vertical line which we label $\hat{z}$, doesn't change, but the component of $\vec{L}_D$ perpendicular to $\hat{z}$ rotates around $\hat{z}$ as the ball moves in its circular path. By contrast, $\vec{L}_C$ points along vertical ($\hat{z}$ direction) and does not change as the ball rotates. The angular momentum vectors at various positions along the orbit of the ball are shown in Figure 7-12a, which shows $\vec{L}_D$ and in Figure 7-12b, which shows $\vec{L}_C$.

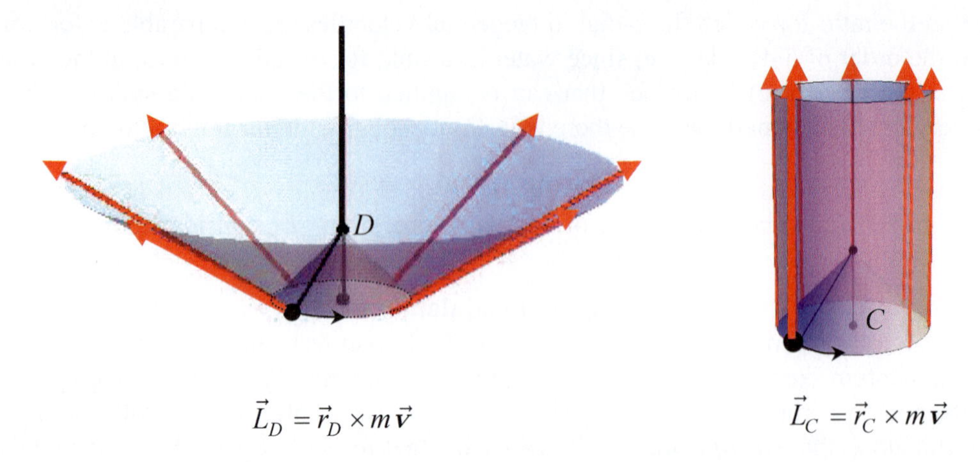

Figure 7-12: a. *(left)* The angular momentum about point D, $\vec{L}_D$, represented by *red arrows*, changes direction as the ball rotates around point C. **b.** *(right)* The angular momentum about point C, $\vec{L}_C$, represented by red upward pointing arrows does not change direction as the ball rotates.

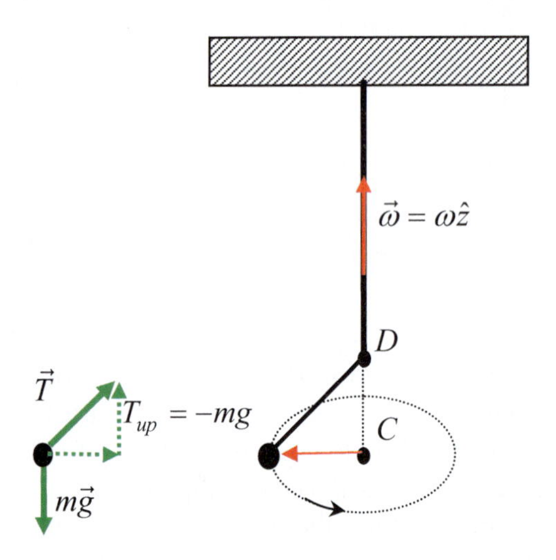

Figure 7-13: The forces of tension, $\vec{T}$, in the string and gravity, $m\vec{g}$, on the ball of mass m cancel in such a way as to leave only a radial force. This force causes a torque about point D but not C. One can verify this by applying the formula: $\vec{\tau} = \vec{r} \times \vec{T}_{radial}$.

Figure 7-13 shows the forces acting on the hanging ball. Given the circular motion, it is no surprise to learn that the force acting is radially directed. This radial force causes a torque at point D but not C, thus explaining the non-conservation of angular momentum of the first and the conservation of the second.

Bicycle

A bicycle has a complex angular momentum behavior, some of which is illustrated in Figure 7-14. Imagine trying to characterize the motion of the wheel using the center of the bike path instead of the center of the wheel. We will see later in this chapter that the spin/orbit theorem guarantees that the *orbit* of the bike and the *spin* of the tires can be treated separately, each with respect to their own preferred point of reference as shown below.

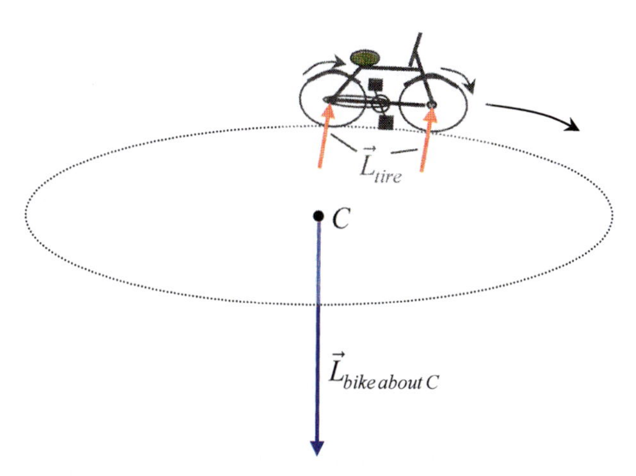

Figure 7-14: A bike tilts slightly inward as it moves in the circular path shown. The downward pointing *blue arrow* represents the angular momentum about the point C that results from the motion of the bike around the circular path. The angular momenta of the rotation of the bike's wheels are accounted for separately with the *red arrows*.

Skewed Rod

A rotating skewed rod such as shown in Figure 7-15 shows that the angular velocity $\vec{\omega}$ of an object is *not always parallel* to its angular momentum. The rod is welded at the pivot point C at some acute angle, θ, off the vertical axis. For simplicity in understanding, we take the bulk of the mass to be concentrated at the ends of the rod. In this case, the calculation of the angular momentum of the system is well approximated by that of the two end masses alone.

Figure 7-15: A skewed rod rotating at some angle θ, such that $0° < \theta < 90°$ as shown, implies that the angular momentum, $\vec{L}_C = \vec{r} \times m\vec{v}$, is not parallel to the angular speed, $\vec{\omega}$. Note that the red circle with the black dot in it refers to a vector pointing directly out of the page, that is, perpendicular to the plane of the page.

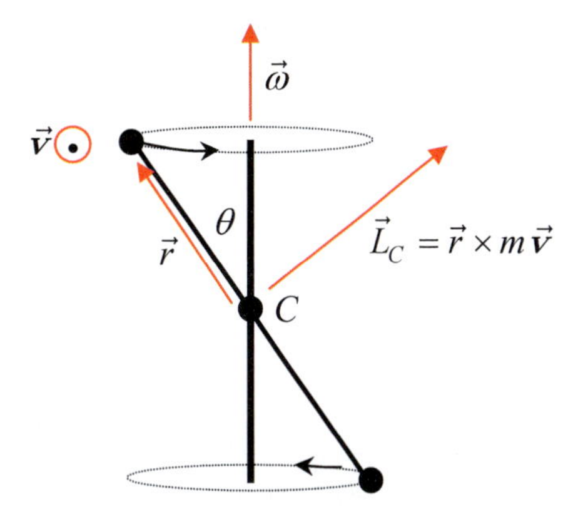

Also, note that the radial component of the angular momentum, $\vec{L}_C$, rotates around the vertical axis and thus is not conserved. The torque causing this change in angular momentum about C comes from a force that must ultimately be provided by the pivot at point C. This force is necessary to change the impetus in the end masses in such as way to keep them in a circular orbit (note the similarities here with the rotating pendulum). Absent such a force, the rod would fly outward and spin in the plane perpendicular to the vertical axis, lining up the angular momentum vector with the angular rotation axis as it is in a thrown twirling baton.

Rotation of Rigid Bodies

Consider an isolated system of massive bodies with interactions that have the required rotational symmetry so that total angular momentum is conserved. As we discussed earlier, this means that if one calculates the angular momentum for each body at one moment and adds them all up, that value will be exactly the same in the next moment and all moments before and after. Again, mathematically, we say:

7.21
$$\vec{L}_{Total} = \sum_{i=1}^{N} \vec{r}_i \times \vec{p}_i = \text{constant}$$

Now, instead of a system of separated bodies, consider a single rigid massive body. By definition, when one part of a rigid body is moved, the rest of it moves simultaneously (another idealization[16]). Also, to start, consider a *planar* rigid body; by a planar body, we mean a body that can be made by dragging a planar figure along the axis perpendicular to it. One such rigid body is shown in Figure 7-16a below.

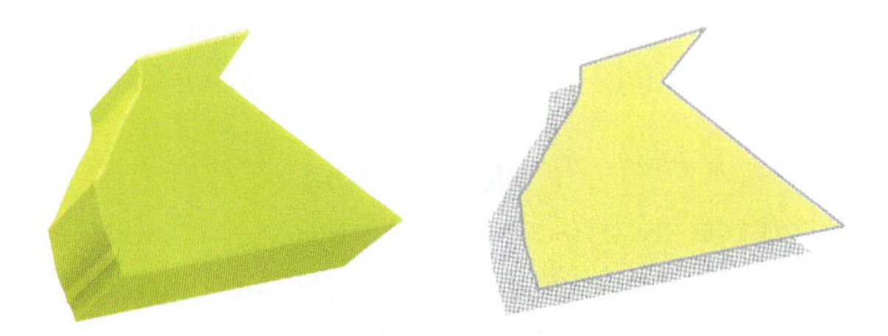

Figure 7-16: a. *(left)* Arbitrary shaped *planar* body. A planar solid is one that can be made by dragging a planar figure along its own normal axis, letting the motion of the edges define the boundary of the solid. **b.** *(right)* The left figure was made by dragging the shape on the right.

[16] We will see later that speeds of interaction from one part of a body to another or between bodies are limited, in a real sense, by the speed of light.

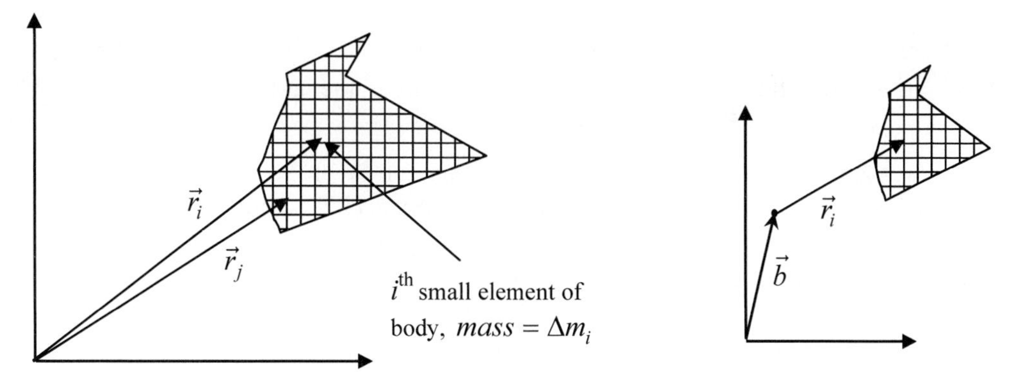

Figure 7-17: a. *(left)* A planar rigid body is composed of many small parts; we label the position of the ith element by $\vec{r}_i$. We can take the calculus limit of infinitesimally small parts (in this case we have mass *dm* at $\vec{r}_i$). **b.** *(right)* Angular momentum considered with respect to an arbitrary point can be obtained by shifting one's reference point by $\vec{b}$ from the origin.

Next, labeling the parts of the body by the index *i*, we can write an expression like equation 7.21 for the single body. We can then simplify the expression by using cylindrical coordinates (which, as is clear by now, is natural to angular momentum, because $\vec{L}$ (and $\vec{\omega}$) defines an axis).

In particular, consider a planar rigid body with parts that are arbitrarily small (see Figure 7-17a above). Taking the body to lie flat on the plane perpendicular to the z-axis, and assuming the body rotates about the z-axis, we can then add up the angular momentum of all the parts to get:

7.22 $$\vec{L}_{Total} = \sum_{i=1}^{N} \vec{r}_i \times \vec{p}_i = \sum_{i=1}^{N} \vec{r}_i \times m_i \vec{v}_i = \sum \vec{r}_i \times r_i \dot{\theta}_i \hat{\theta}_i m_i = \dot{\theta} \sum \vec{r}_i \times r_i \hat{\theta}_i m_i = \dot{\theta}\, \hat{z} \sum r_i^2 m_i$$

Here, as above, $\vec{r}_i$ is the radial vector from the point around which one wants to know the angular momentum. The second to the last equality follows because of what we mean when we say a rigid body rotates around a point (*exercise*: how is this different than the way a tilted ellipsoidal Earth revolves around the sun? *hint*: draw distinction between rotation around the center of mass of a body and other motions in which only the center of mass point moves, while the orientation of the body stays fixed in an inertial frame). Moving the body through an angle θ means that all parts of the body move through that same angle. Or mathematically, $\theta_i = \theta_j = \theta$; obviously, this equation remains true when we take an arbitrary number of time derivatives .[17]

In the continuum limit (taking the little pieces to zero size) we get:

7.23 $$\vec{L}_{Total} = \hat{z}\dot{\theta} \int r^2 \, dm = I \omega \hat{z}$$

[17] Note the choice of labeling for the coordinates (e.g. using $\hat{z}$) implies no loss of generality. This labeling could be applied to any physical situation in two dimensions. A generalization showing how to include a different reference point is shown in Figure 7-17b.

Here, we define I, the **moment of inertia** as: $I \equiv \int r^2 dm$. Given our assumptions and since, by definition, the angular rotation rate is along the axis of rotation, we have: $\vec{\omega} = \omega \hat{z}$; so we can write: $\vec{L}_{Total} = I\,\vec{\omega}$.

Now, the moment of inertia of any rigid body can be broken into components I_x, I_y, I_z such that I is a diagonal matrix with those components. This is most immediately evident for symmetric bodies, but, surprisingly, it is true for *any* rigid body. Still, in this text, we will focus on symmetric bodies because of their simplicity. Using the axes along which the reduction happens, that is along the *principal axes*, we can always write:

7.24
$$\begin{pmatrix} L_x & 0 & 0 \\ 0 & L_y & 0 \\ 0 & 0 & L_z \end{pmatrix} = \begin{pmatrix} I_x & 0 & 0 \\ 0 & I_y & 0 \\ 0 & 0 & I_z \end{pmatrix} \begin{pmatrix} \omega_x \\ \omega_y \\ \omega_z \end{pmatrix}$$

As we have already mentioned, the angular momentum associated with an arbitrary angular rotation of an arbitrarily shaped object will not always be in the same direction as the rotation vector, i.e. $\vec{L}$ is not always parallel to $\vec{\omega}$. In this text, we will define our axes along the principal axes, allowing us to treat L_x, L_y and L_z independently, i.e.:

7.25
$$L_x = I_x \omega_x \quad L_y = I_y \omega_y \quad L_z = I_z\,\omega_z$$

However, clearly, even with this simplification and the simplification of using symmetrical bodies, $\vec{L}$ is not necessarily parallel to $\vec{\omega}$. Indeed, only in special cases, for instance, when $\vec{\omega} = \omega_z \hat{z}$, will we get: $\vec{L} = I\vec{\omega}$.

Note the strong analogy with momentum: $\vec{p} = m\vec{v}$. Rigid bodies thus have linear momentum of their center of mass and rotational motion about their center of mass. Indeed, using center of mass coordinates, we can prove what we earlier called the spin/orbit theorem. Namely, the angular momentum of a body can be broken up into linear motion of the center of mass, rotational motion of the center of mass about a chosen origin ("orbit") and rotation about the center of mass ("spin").

Such a decomposition of motion will be familiar to those who follow NASA's many space projects. The Earth's motion is a simple example. The Earth (its center of mass moves) orbits around the sun. The Earth spins around its own axis (motion about center of mass). The Earth undergoes a linear motion through the cosmic background radiation (approximately $400\,km/s \sim 900{,}000\,mph$).

Angular Momentum and Center of Mass

To prove the spin/orbit theorem, we consider equation 7.21 with $\vec{r}_i = \vec{R} + \vec{r}_i\,'$, thereby introducing an offset of the reference point such as shown in Figure 7-17b, where $\vec{R}$ is the center of mass coordinate defined, as usual, as:

7.26
$$\vec{R} = \frac{\sum_{i=1}^{N} m_i \vec{r}_i}{\sum_{i}^{N} m_i}$$

Figure 7-18 shows the relevant vectors for our planar rigid body.

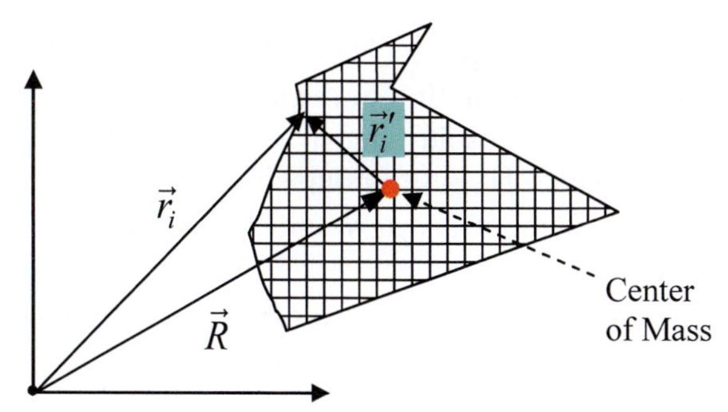

Figure 7-18: A body with a planar shape considered in its natural plane

Substituting, we obtain:

$$\vec{L}_{Total} = \sum_{i=1}^{N} \vec{r}_i \times \vec{p}_i = \sum_{i=1}^{N} \left(\vec{R} + \vec{r}_i\,'\right) \times m_i \frac{d\left(\vec{R} + \vec{r}_i\,'\right)}{dt}$$

7.27

$$= \sum_{i=1}^{N} m_i \vec{R} \times \dot{\vec{R}} + \sum_{i=1}^{N} m_i \vec{R} \times \dot{\vec{r}}_i\,' + \sum_{i=1}^{N} m_i \vec{r}_i\,' \times \dot{\vec{R}} + \sum_{i=1}^{N} m_i \vec{r}_i\,' \times \dot{\vec{r}}_i\,'$$

The second and third terms are each zero because the center of mass coordinates are by definition such that $\sum_{i=1}^{N} m_i \vec{r}_i\,' = 0$. That is, the very definition of the center of mass is the place where the sum of all the distances to the given element weighted by the mass of that element is zero; heuristically, it is the middle of all the mass.

Mathematically, the details of how the second and third terms sum to zero is seen in the following manipulations:

$$\sum_{i=1}^{N} m_i \vec{R} \times \dot{\vec{r}}_i\,' = \vec{R} \times \sum_{i=1}^{N} m_i \dot{\vec{r}}_i\,' = \vec{R} \times \sum_{i=1}^{N} m_i (\dot{\vec{r}}_i - \dot{\vec{R}}) = \vec{R} \times M(\dot{\vec{R}} - \dot{\vec{R}}) = 0$$

$$\sum_{i=1}^{N} m_i \vec{r}_i\,' \times \dot{\vec{R}} = \left(\sum_{i=1}^{N} m_i (\vec{r}_i - \vec{R})\right) \times \dot{\vec{R}} = M(\vec{R} - \vec{R}) \times \dot{\vec{R}} = 0$$

Hence, we are left with (using $\vec{V} \equiv \dot{\vec{R}}$):

7.28

$$\vec{L}_{Total} = \sum_{i=1}^{N} m_i \vec{R} \times \dot{\vec{R}} + \sum_{i=1}^{N} m_i \vec{r}_i\,' \times \dot{\vec{r}}_i\,' = \vec{R} \times M\vec{V} + \sum_{i=1}^{N} \vec{r}_i\,' \times m_i \dot{\vec{r}}_i\,'$$

$$= \vec{L}_{orbit} + \vec{L}_{cm}$$

Thus, we see that the angular momentum breaks into a piece describing the rotation of the object (monitored by its center of mass) around a point, which we call orbital angular momentum and a piece that corresponds to the angular momentum of the parts of the object around its own center of mass; this later we call spin. The spin is a uniform change of orientation, which recall is one of the nine categories.[18] The orbital momentum is related

[18] In fact, a change of orientation of this type also involves some change of place unless the object is symmetric about the axis of rotation. This is because the rotation of an irregular shaped object implies there is a protruding part that must occupy a different part of the region around it than it did before the rotation. Of course, in a collection of separated massive bodies, for which the spin/orbit theorem also applies, such distinctions apply only analogically.

to change of place, a different category. Orbital angular momentum, because it results from, at least instantaneously, linear motion of the center of mass, can be made to look as if it were zero, at least for a moment, as any linear speed can. However, the spin cannot be made to appear to go away except by putting ourselves in a rotating frame centered on the center of mass, in which, of course, we would feel other forces because we would necessarily be accelerating.

Also note that equation 7.28 is valid even when $\dot{V} \neq 0$, i.e. when the center of mass is accelerating, for we did not calculate it in the center of mass frame, but in the inertial lab frame.

By the same mathematics, we used earlier, we can show that there are, indeed, two types of moments of inertia (see end of chapter problem) and we write:

7.29
$$\begin{aligned}\vec{L}_{Total} &= \vec{R} \times M V + \sum_{i=1}^{N} m_i \vec{r}_i\,' \times \dot{\vec{r}}\,' \\ &= \vec{L}_{orbit} + \vec{L}_{cm} \\ &= \vec{L}_{orbit} + I_{cm}\,\omega_z\end{aligned}$$

In the last equation, we have assumed that the body can be separated into three independent rotation axes and that it is rotating along one of those axes. We have further taken the rotation to be along the z-direction, losing no generality by this simple act of naming. In order to make any use of these formulae, we need expressions for the moment of inertia. We now turn to a specific case. More moments are tabulated in the calculational techniques section at the end of the chapter.

Moment of Inertia of Rectangular Cross Section Body

A simple example that will give the idea of how these moments are calculated is a body with rectangular cross sections of dimensions a, b and c as shown in Figure 7-19 below. As usual in such calculations, we assume uniform mass density ρ.

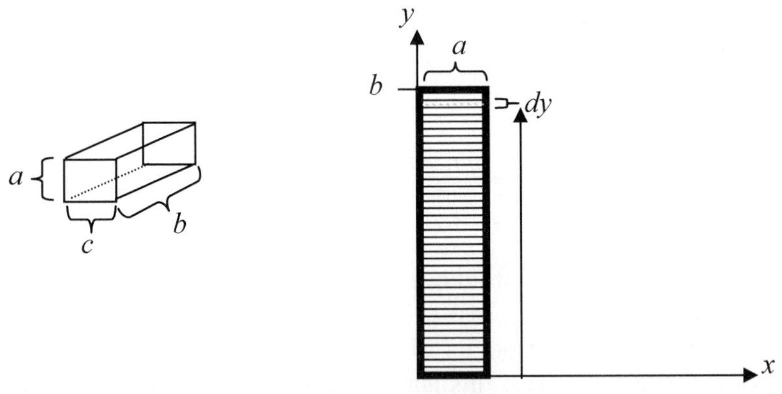

Figure 7-19: A rigid body of rectangular cross section (left), can be divided into parts as shown on the right.

Using the above diagram, we see each element has mass:[19]

7.30 $$dm = \rho c a\, dy,\ take\ \lambda \equiv \rho c a\ gives: dm = \lambda\, dy$$

The mass of the object is obviously: $M = \lambda b$. This means the moment of inertia about the x-axis I_x is:

7.31 $$I_x \equiv \int_0^b r^2 dm = \int_0^b y^2 \lambda\, dy = \frac{\lambda b^3}{3} = \frac{M b^2}{3}$$

The moment of inertia about a parallel axis, going say through a point A, displaced by an amount l from the center of mass can be written:

7.32 $$I_A = I_{cm} + M l^2$$

One can prove a general theorem to this effect, called the parallel axis theorem (see end of chapter problem 46). This theorem is directly related to the previously proved spin/orbit theorem. Note that the first term is the moment due to spin, and the second is the moment due to the center of mass motion around a point a distance l away, i.e. as if it were a point body of mass M, traveling around the point A. This is all a result of the special place that is the center of mass.

Using the parallel axis theorem, we calculate the moment of inertia about the x-axis for our bar (using $l = b/2$) is:

7.33 $$I_{x-cm} = \frac{M b^2}{12}$$

Where did Babe Ruth Most Likely Hit the Baseball?

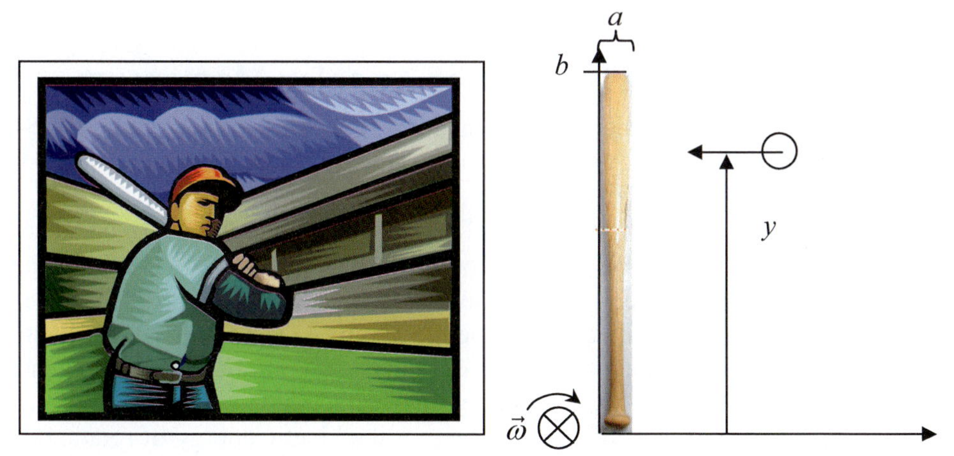

Figure 7-20: Babe Ruth was one of the best homerun hitters of all time. The left shows a coordinate system setup around a bat to calculate the ideal place for a batter to hit the ball. The inset circle with an "X" across it indicates that the rotation vector of the bat, $\vec{\omega}$, points directly into the page.

[19] The symbol λ is frequently used to denote the linear density of an object, in this case linear mass density, and it is measured in *kg* per *m* in mks system.

The man who hit 714 homeruns would likely be very efficient and not allow any backlash against his hands when he hit the ball. This means that, at the moment of the bat's impact with the ball, there was no need of a force applied by his hands on the bat; $\vec{F}_{hands\ on\ bat} = 0$.
Now, the linear impulse given by the ball to the bat is:
7.34

$$\text{Impulse} = \int F_{ball \to bat}\, dt = p^{bat}{}_{after} - p^{bat}{}_{before} = M_{bat} v_{after} - M_{bat} v_{before} = M_{bat} \frac{b}{2}(\omega_{after} - \omega_{before})$$

In the last equation, we have used the fact the speed of the center of mass of the bat can be written as the angular speed, ω, of the bat around the pivot point (i.e., where he holds the bat) times the distance to the center of mass, $b/2$.
Now using the equivalent of the impulse theorem for angular momentum:

7.35 $$\textit{Angular Impulse} = \left|\Delta\vec{L}\right| = \int \tau\, dt = \int \left|\vec{r} \times \vec{F}_{ball \to bat}\right| dt = y\,\text{Impulse} = y\, I_{bat}(\omega_{after} - \omega_{after})$$

Substitution of the linear impulse into the last equation gives:

7.36 $$y\, M_{bat} \frac{b}{2}(\omega_{after} - \omega_{before}) = I_{bat}(\omega_{after} - \omega_{before})$$

Thus,

7.37 $$y = \frac{I_{bat}}{M_{bat} l_{cm}} = \frac{I_{bat}}{M_{bat} b/2} = \frac{2 I_{bat}}{M_{bat} b} = \frac{2}{M_{bat} b} \frac{M_{bat}\, b^2}{3} = \frac{2}{3} b$$

Here, we have approximated the moment of inertia of the bat by the value of the rectangular cross sectional solid calculated above.

So, the right place to hit the ball is approximately two thirds the way up the bat.

Work-Energy Theorem for Angular Momentum

Having established the analog of momentum (the angular momentum), the analog of mass (the moment of inertia), and the analog of the impulse theorem, we introduce an analog of the work-energy theorem.

Remember the mass is the measure of the receptivity of the body to the action of the impetus. For a rigid body, the impetus coupled with the action of internal forces can maintain uniform rotation; we call the measure of this complex of impetus and internal forces *angular momentum*. The measure of the receptivity to the action of this complex of activities is still related to the mass but generally only in a complicated way dependent on its distribution in the body. This receptivity or, better said, resistance to rotation is measured by *the moment of inertia*. Thus, we expect the kinetic energy of rotation, like the linear kinetic energy, to be dependent on not just the active quality, the impetus, but also on the receptive one, the moment of inertia. To help in seeing the relations between linear and angular momentum see the summary of the analogies given in Table 7-1 below.

Table 7-1: *Comparison of parameters of Linear Motion and Angular Motion.*

Quality or Measure	Linear Motion	Rotational Motion of Rigid body
Cause of motion An active power (intensity and direction measure)	Impetus (***Momentum***, $\vec{p} = m\vec{v}$)	Impetus/ internal forces (***Angular Momentum***, $\vec{L} = I\,\vec{\omega}$)
Resistance to impetus a quality	***Mass***, *m*	***Moment of Inertia***, *I*
Measure of motion	***Speed***, $\vec{v} = \frac{d\vec{x}}{dt}$	***Angular velocity***, $\vec{\omega} = \frac{d\vec{\theta}}{dt}$
Measure of change of motion	***Acceleration***, $\vec{a} = d\vec{v}\,/\,dt$	***Angular acceleration***, $\vec{\alpha} = d\vec{\omega}\,/\,dt$
Cause of change of impetus[20] (intensity and direction measure)	***Force***, $(\vec{F} = \frac{d\vec{p}}{dt})$	***Torque***, $(\vec{\tau} = \frac{d\vec{L}}{dt})$
Measure of action of force	***Work***, $W_{lin} = \int \vec{F} \cdot d\vec{x}$	***Work***, $W_{rot} = \int \vec{\tau} \cdot d\vec{\theta}$
Measure of activity of motion	***Linear Kinetic Energy***, $KE_{linear} = \frac{1}{2} m\mathbf{v}^2$	***Rotational Kinetic Energy***, $KE_{rotation} = \frac{1}{2} I\,\omega^2$
Rate of Change of Activity	***Power*** $P_{linear} = \frac{dE_{linear}}{dt}$	***Power*** $P_{rot} = \frac{dE_{rot}}{dt}$
Associated Symmetry (complex quantitative and qualitative relation based in underlying Euclidean space)	***Linear Translation***	***Rotational***

Assuming, as usual, that the moment of inertia is diagonal so that rotations around the various axes do not couple, we take $\vec{\omega} = \omega_z \hat{z}$ as a stand-in for any of the three orthogonal directions. Thus, we can define the work in the case of circular motion in the following way, taking *I* to be constant:

$$7.38 \quad W_{rot} = \int \vec{\tau} \cdot d\vec{\theta} = \int \frac{d\vec{L}}{dt} \cdot d\vec{\theta} = \int I\, d\vec{\omega} \cdot \vec{\omega} = \frac{1}{2} I\, \omega_{after}{}^2 - \frac{1}{2} I\, \omega_{before}{}^2$$

When there are no torques present obviously: $W_{rot} = 0$, and thus the rotational kinetic energy, $\frac{1}{2} I \omega^2$, is constant, i.e. conserved.

---

[20] Note that, as we mentioned in Chapter 4, the word "force" is used both for the quality that accomplishes a change of impetus and the measure of the intensity of that quality. We have distinguished the impetus and the measure of its intensity, the momentum to make clear the distinction between a thing and that which measures it. Still, one can us, and we have been using, this verbal shorthand for that case as well, i.e. just talking about what happens to the momentum. This, as we've said so many times, is fine as long as we don't forget what we mean.

In a way similar to the linear case, we can define the potential energy function and define the mechanical rotational energy round a given point as:

7.39 $$E_{rot} = V_{rot} + KE_{rot}$$

We can show this is always constant (see end of chapter problem 43) when the potential term implicitly includes all torques acting on the chosen system.

Two Types of Rigid Body Activity

With this groundwork, we can show that any particle's kinetic activity, and thus the measure of that activity, the energy, can be split into activity of linear motion *of* the center of mass and activity of rotational motion *about* the center of mass. Mathematically, we can write this as:

7.40 $$KE_{external} = KE_{linear-CM} + KE_{rot-about\,CM}$$

The first term allows us to think of the given rigid body as a mass concentrated at the center of mass of the body. This latter is obviously a mental construct, existing in the mind, not in reality, but it is very useful because it encapsulates an important truth.

Examples

Bikes, Tops and Gyroscopes:

Why does a child's top, which is just simply a gyroscope, precess, that is spin, slowly around? How fast does it do so? Why does a top not fall over? The later answer will also partially answer why you can sit without falling on a bicycle when you're riding--and thus obviously the wheels are rotating---but not when it is still. We have the tools to answer these questions. Consider the top shown in the Figure 7-21 below.

Figure 7-21: A toy top rotates in two ways. The blue plastic disc spins in its metal frame, while the entire top slowly orbits around the point at the top of the red plastic holder.

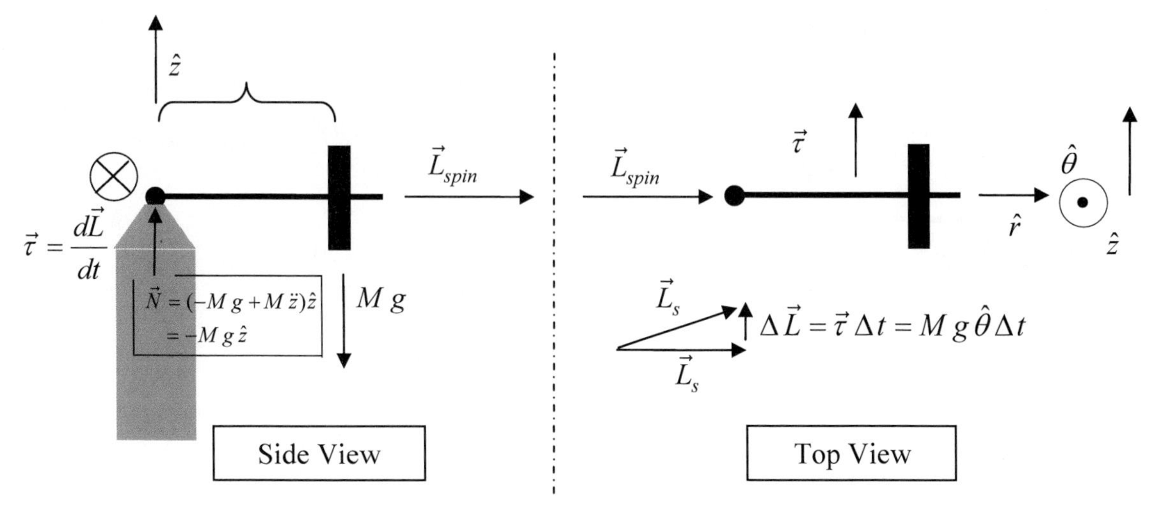

Figure 7-22: a. *(left)* Side view of a toy gyroscope. Notice that the normal force, $\vec{N}$, acting at the pivot point, includes a term that would be non-zero if the top were falling **b.** *(right)* Top view of the gyroscope detailing its angular momentum changes due to the action of the torque.

From our experience[21] that the top does not fall, we know that the force of gravity and the force of the object holding up the end of the gyro are equal and opposite (cf. Figure 7-22a above). Once we have such a condition, we have an answer of sorts, but one that needs more explanation. Before continuing with the answer, we stop to use the fact that with vertical forces balanced, the center of mass of the gyro cannot move in the *z*-direction.

Making use of cylindrical coordinates, we can quickly show that the gyroscope precesses at a rate of Ω:

First, note, using Figure 7-22b, that the torque acting on the gyroscope is:

7.41 $$\vec{\tau} = \vec{r} \times \vec{F} = l\,\hat{r} \times \left(Mg\,\hat{z}\right) = Mgl\,\hat{\theta}$$

Note how the torque adds to the spin vector in a way to cause it to rotate.
Thus we get:

$$\vec{L} = L_r\,\hat{r} \Rightarrow$$

7.42 $$\frac{d\vec{L}}{dt} = \dot{L}_r\,\hat{r} + L_r\dot{\theta}\,\hat{\theta} = L_{spin}\dot{\theta}\,\hat{\theta} = L_s\Omega = m\,g\,l$$

$$\Rightarrow \Omega = \frac{m\,g\,l}{L_s} = \frac{m\,g\,l}{I_{spin}\,\omega_{spin}}$$

Note here we have used the fact that there is no mechanism (in our idealized model) for the L_{spin}, the magnitude of $\vec{L}_{spin}$, to change.

To further see why the top rotates rather than falls, we look at a simple top with only two masses such as shown below. The figures below illustrate the situation under analysis; the file gyroscope2.avi at *www.iapweb.org/interactivephysics/pfr7*, which is an animation of the motion of this type of gyroscope, will also help in visualization.

---

[21] Generally, the little ups and downs that the top undergoes are too small to be seen.

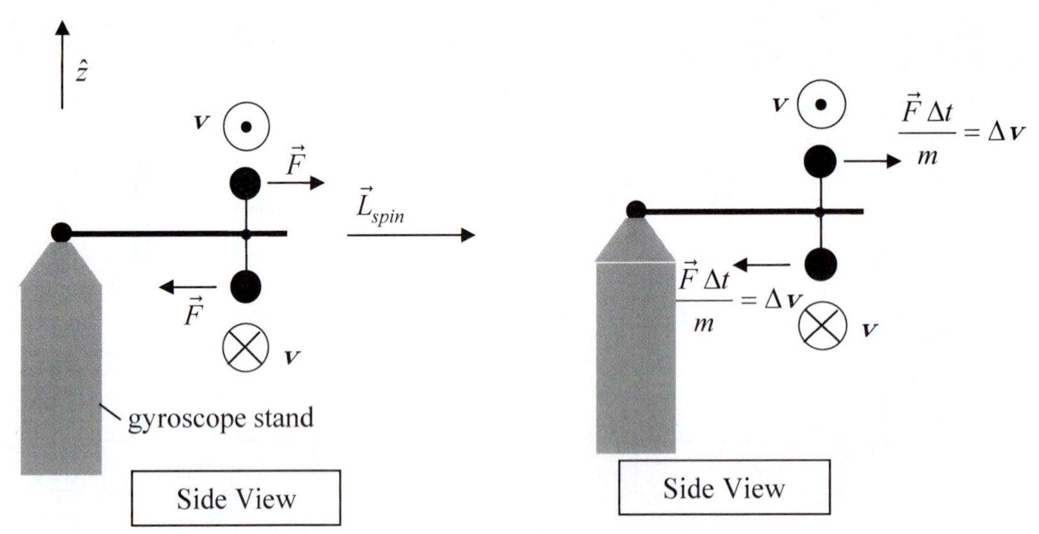

Figure 7-23: *Two Side Views* of a toy top help manifest why it does not fall. We initially suppose that the velocities of each ball have the same magnitude and opposing directions. Later we will realize that the center of mass of the dumbbell precesses uniformly, implying an instantaneous linear speed, $v_{procession} \ll v_{spin}$ (we also assume $\omega_s \gg \omega_p$); this implies, for the balls in the configuration shown, a velocity of $v_p - v_s$ for the top ball and a velocity of $v_p + v_s$ for the bottom ball at the same instant. This leaves $\left|v_p \pm v_s\right| \approx v_s$, allowing us to take the equal-speeds and opposite-directional velocities as an initial approximation.

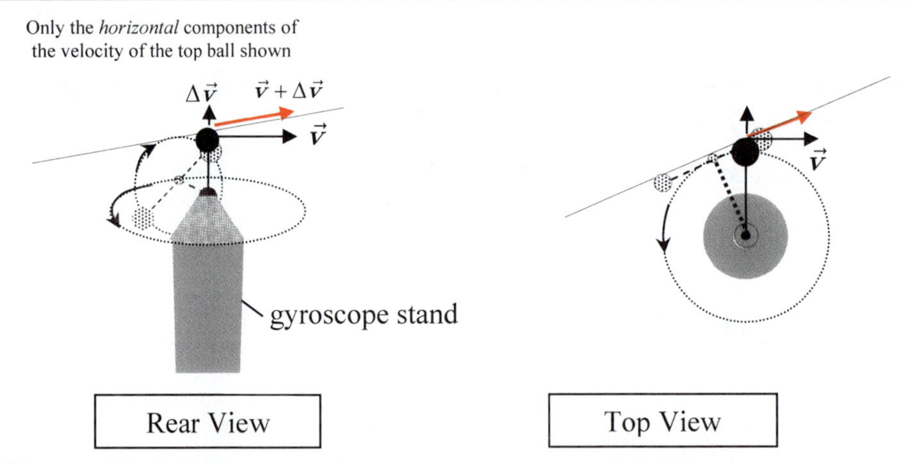

Figure 7-24: a. *(left)* Rear view, looking down a little on the top, showing the horizontal components of the velocity of the top ball. The solid black dumbbell represents the dumbbell at $t = 0$; the dotted dumbbell shows the dumbbell at the very short time later $t = \Delta t << 1/\omega_{spin}$. Notice the proportion between the distance traversed by the center of mass and that traversed by the ball is exaggerated to allow visualization. **b.** Top view, showing same two moments in time; i.e., the dotted ball again represents the dumbbell's position at $t = \Delta t$.

In the side views shown Figure 7-23, we see that, instead of a full gyroscope ("top"), we have a dumbbell, which is basically two spherical masses (balls) on a (massless) rod connected to a main (massless) shaft from which the top pivots. The rod and the shaft are here idealized transmitters of the various forces. Here we also assume that the speeds of the ball are such that the dumbbell's center of mass is not moving *and* that the dumbbell is not falling; our analysis below will then show that not falling is only consistent with a center of mass that precesses around the pivot. More than that, we can show that, given the right initial conditions, it will precess uniformly without falling.

Gravity pulls on the center of mass of the whole gyroscope (dumbbell plus main shaft), which is approximately at the center of the dumbbell; this, in turn, pushes the end of the main shaft onto the stand shown in solid gray in Figure 7-23 and Figure 7-24. This results (due to internal forces) in a force on each of the balls in the manner shown in the figures; these forces act to try to turn the gyro around its center of mass. This leads us to change the question we are asking.

In particular, since we already know the top cannot fall as long as the vertical forces are balanced, and we have already said they are, we change our question from "Why does the top not fall?" to "Why does it not rotate about the pivot point where the main shaft meets the stand." After all, it must rotate about the pivot if it is to fall. Indeed, it *cannot* fall *straight* downward, for the presence of the pivot means that the top is constrained (to the extent the pivot remains intact during the fall) to move in a circular orbit downward; the pivoted rod exerts horizontal forces as well as vertical ones.

Note that, in thinking about the gyroscope, it is important to remember that it is impetus that causes motion, not force; force causes a change in impetus, which causes a change in motion; this is why a ball with an upward impetus moves upward despite the downward force of gravity.

To begin to discover why, under certain initial conditions, the top does not pivot downward, look at the change in motion caused by the forces (which ultimately are induced by gravity) shown in Figure 7-23 during a very short period of time Δt, (i.e., $\Delta t << \dfrac{1}{\omega_{spin}}$, which also means $\Delta t << \dfrac{1}{\omega_{procession}}$ since we assume $\omega_s \gg \omega_p$). The velocity of the top ball *in the horizontal plane*[22] is shown both at $t = 0$ and at $t = \Delta t$. During the interval Δt, the forces cause a change in the top ball's speed that is very small compared to the speed from the already existing impetus of the *spin*. The rear and top views illustrated in Figure 7-24a shows how this force on the dumbbell changes the ball's velocity so as to cause a precession.[23] That is, the only way the ball can acquire a velocity with the directions required, without falling, [24] is for the ball to move in the way shown; to see this, examine the tangent line drawn in the rear and top views shown in Figure 7-24. (Note the interesting fact that the speed of the top ball of the dumbbells is opposite to the speed of precession).

---

[22] Note that we do not discuss, and the figure does not show an arrow for, the component of the spin speed that points downward, for it is not relevant to the precession.

[23] The bottom dumbbell moves in a corresponding way; you are encouraged to try to envision it for yourself.

[24] In fact, as one can see in the full calculation, if we allowed the initial conditions to be such that the top could fall, it would still have to precess over any finite time.

In short, the fact that the ball is moving so fast means that the little change in speed that the force is able to cause only acts to slightly change the angular momentum direction which is,[25] still in the spin direction, but now the spin direction is itself in a different direction, as shown in Figure 7-22. That is, the spin precesses.

Said another way, the torque, which we tend to think can only cause the top to rotate about the pivot point, causes a much richer behavior when the object under consideration is spinning fast. In that case, the torque changes the angular momentum by changing the impetus in the two balls. The net effect of these changes is that the instantaneous *magnitude* of the linear momentum of the center of mass of the dumbbell remains constant, while its direction continually changes, so that the dumbbell's center of mass executes a uniform circular motion. This then means that the top (dumbbell plus perpendicular lever arm), in steady state, must have already acquired an impetus that moves the dumbbell's center of mass tangentially. The role of the gravitational force, then, *in the non-falling steady state*, is to continually change the directional type of the center of mass impetus while maintaining the intensity (in each of the tangential directions) constant so as to cause uniform circular motion in *analogy* to the way a string pulls an object with tangential impetus into circular motion. We investigate this behavior in more detail in an end of chapter problem.

Again, physically, the force of gravity, via the structure of the top, increases the radially outward impetus of the top ball, causing the dumbbell to move into a plane of spin rotated from the initial one. The bottom ball has a complementary effect, resulting in a uniform precession, where the downward pull of gravity, via the structure of the top, continuously causes the requisite impeti changes.

Notice that throughout this analysis we have ignored the contribution to the angular momentum from the orbiting of the gyroscope. One can easily calculate the ratio between the spin angular momentum and the orbital angular momentum to see that this is a valid approximation.

To see in detail why the top doesn't fall, we would have to write down and solve the appropriate approximate equations with wider initial conditions. If you do so, you will see that, when a gyro is released at rest, it actually falls and starts bouncing up and down, i.e., nutating, while it precesses, but the friction quickly dampens the nutation out, leaving it in the steady state case described above. In a toy gyro, the dampening is so quick that one cannot even notice it.

Poynting-Robertson Effect

This final example concerns a particle of dust in orbit around the Sun such as shown in Figure 7-25. What will happen to the dust particle over time? In particular, what will happen to its angular momentum?

In order to answer these questions, we need a couple of completely new concepts. We need to know that light is not itself a massive body and thus cannot have impetus in the same sense as the things we have been discussing, but does "have" impetus in an *analogical* sense. In particular, it has the capacity to cause a certain impetus, and thus energy, in bodies that absorb it. The energy of light, E, is experimentally known to be related to its frequency, ν, according to the Planck law: $E = h\nu$. Furthermore, as we will

[25] This is an approximate statement which treats the angular momentum due to the precession of the top as small.

see in special relativity, when light is absorbed, it increases the mass of the body according to Einstein's famous equation: $m_{light} = E/c^2$. Indeed, analogically, we say light has mass; that is, light "has mass" in a way that has some relation to the way massive bodies have mass. The nature of this relation and the further meaning and implication of these new concepts is properly the subject of another course. However, it is important to begin to glimpse something of the analogical generality that physics uses which you will encounter in later studies. Beyond this, the concepts themselves, even at this simple level (though again they cannot be truly understood without further study), are important and are needed to understand the interesting problem at hand and many other aspects of our world.

We will analyze this problem in two frames: the sun fixed frame and an instantaneous rest frame of the dust.

In the sun fixed frame the situation is relatively simple once we put ourselves into an inertial frame in which the sun is at rest, i.e. we "turn off" all forces (including gravity[26]) that act on us and move at a uniform velocity matching of the sun (which we assume to be effectively infinite in mass so that the dust has no effect on it). We will also neglect (in both frames) the force of the light pressure on the dust, which pushes it radially outward (*exercise*: using the solution below, verify this latter approximation is valid.) We now start our analysis in the sun fixed frame.

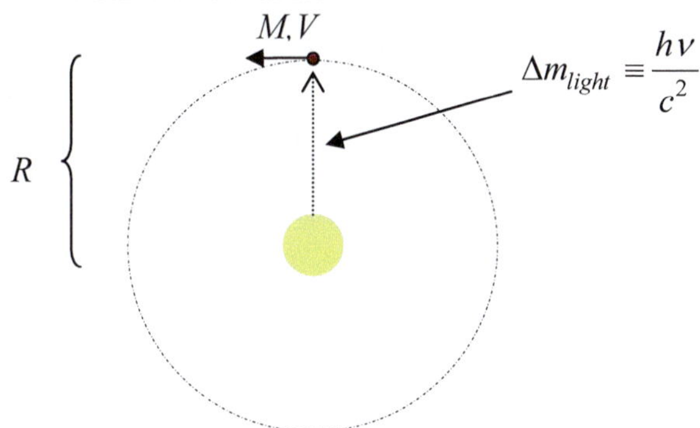

Figure 7-25: *In rest frame of Sun, shown is* a dust particle in orbit around Sun. In the figure, we see that the energy, $E = h\nu$, of one photon of light has, "converted" using Einstein's equation ($E = mc^2$), an "effective mass" of $\frac{E}{c^2}$.[27]

Sun Fixed Frame

Consider the angular momentum of the dust around the center of the sun, which is distance R away from the dust at a particular instant. The dust particle is taken to have mass M and move initially in a circular orbit with tangential speed V. Sunlight hits the dust,

---

[26] Turning off gravity's action is only a mental construct (being of reason), since there is no way to do this in the real world. We can account for gravity by simply considering ourselves doing all our measurements relative to an inertial frame very far from the Sun so that its gravitational pull will not affect, within the needed precision, our inertial frame. In both the sun frame and the instantaneous rest frame of the dust, one could instead use a *local* freefall frame as one's approximate inertial frame, dealing with gravity through the equivalence principle discussed in Chapter 8.

[27] See Chapter 10 for more on the analogical extension of the concept "energy" to include "mass."

and then is radiated outward from the dust again. At equilibrium, as much radiant energy as falls on the dust particle is re-emitted. Thus, we summarize (cf. Figure 7-25) what we know in three steps.

1. *Just before a light "ray" called a photon hits*:

7.43
$$L_{dust} = M\,V\,R$$
$$L_{light} = \Delta m_{light} \cdot 0 \cdot R$$

2. *Just after the photon hits*:

Maintaining an artificial distinction between dust and light even after the light is absorbed for consistency and convenience, conservation of angular momentum gives:

7.44a,b,c
$$L'_{dust} = M\,V'R$$
$$L'_{light} = \Delta m_{light} V'R$$
$$\Rightarrow L'_{total} = (M + \Delta m_{light})V'R$$

7.45
$$L'_{total} = L_{total} = L_{dust}$$
$$\Rightarrow V' = \frac{M}{M + \Delta m_{light}} V$$

NB: we use the prime to mean the value of a quantity after an event (in this case, the light being absorbed by the dust), two primes will indicate a quantities value after the event after it and so on.

Now, yellow light dominates the light emission of the sun. For a photon of yellow light, which is ***very*** loosely and analogically speaking the smallest piece of yellow light, using Einstein's relation between mass and energy and $E = h\nu$:

$$\Delta m_{light} = \frac{h\nu}{c^2} = \frac{h}{c\lambda} = \frac{h}{c\,5x10^{-7}} = 4.4x10^{-36}\,Kg$$

Now, taking the dust particle to be an approximately spherical shape about $D \sim .1mm$ thick, with a density, $\rho \sim 4gm/cm^3$ and using $V_{sphere} \sim \frac{D^3}{2}$, we get:

$$M = \rho V \sim 2\times10^{-6}\,gm = 2\times10^{-9}\,kg$$

Thus, $\Delta m_{light} << M$ and we can expand equation 7.45 in a series approximation to get:

7.46
$$V' = \frac{M}{M + \Delta m_{light}} V \sim \left(1 - \frac{\Delta m_{light}}{M}\right)V$$

Substituting this into equation 7.44c, we get:

7.47
$$L'_{dust} = MV'R \sim (M - \Delta m_{light})VR$$

Thus,

7.48
$$\Delta L_{dust} = L'_{dust} - L_{dust} = -\Delta m_{light} VR$$

3. *Just after the light re-radiates*. It re-radiates isotropically in the rest frame of the particle, so that there is no change in momentum. Why no momentum change? Because as much momentum goes forward as backward and the same can be said for every direction, for that is the definition of isotropic radiation. (*Exercise*: look at this isotropic radiation in sun frame and show how there is no force acting there either---see discussion later in this section). Thus, we can write

7.49
$$L''_{dust} \sim MV'R$$
$$L''_{light} = 0$$

Hence, since the angular momentum does not change any further with the re-emission of the light, i.e. $L''_{dust} = L'_{dust}$; we can write:

7.50
$$\frac{dL}{dt} = -\frac{dm_{light}}{dt} VR$$

Now, using the definition of the luminosity as energy per unit time, we note that the energy per unit area radiating from the sun is: $\frac{Luminosity_{sun}}{4\pi R^2}$.

Then, dividing by $h\nu$ gives the number of yellow photons emitted per unit area from the sun. To get the number of photons that are intercepted by the dust particle, we multiply by its cross sectional area, σ_{dust}. Finally, we multiply by the mass of each photon, $\frac{h\nu}{c^2}$ to get:

7.51
$$\frac{dm_{light}}{dt} = -\frac{Luminosity_{sun}}{h\nu\ 4\pi R^2} \sigma_{dust} \frac{h\nu}{c^2}$$

Thus giving:

7.52
$$\frac{dL}{dt} = -\frac{Luminosity_{sun}}{h\nu\ 4\pi R^2} \sigma_{dust} \frac{h\nu}{c^2} VR = -Luminosity_{sun} \frac{\sigma_{dust}}{4\pi R^2} \frac{1}{c^2} \frac{L}{M_{dust}}$$

This yields:

7.53
$$L = L_0 e^{-kt}$$

Where k is a constant (you are asked to calculate this constant in an end of chapter problem 32). Hence, the angular momentum decays exponential with time. Now, we analyze the same problem in the dust frame as seen in Figure 7-26 below.

Dust Fixed Frame

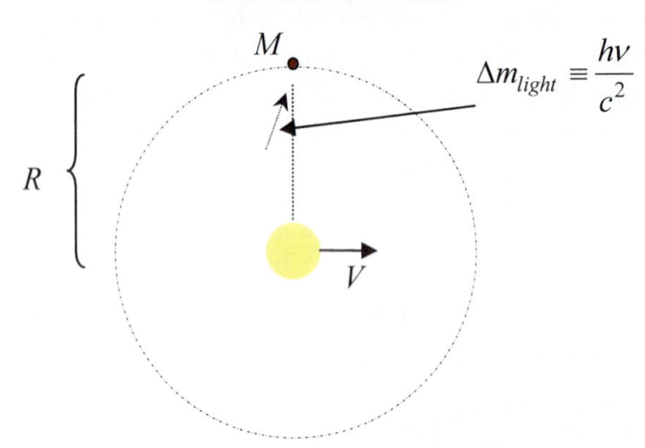

Figure 7-26: A dust particle in orbit around Sun *in an instantaneous rest frame of particle*. In the analysis in both frames, we make the Newtonian approximation, leaving aside the issues of special relativity to get a rough idea of the answer. In particular, this approximation shows up in this frame when we use Newtonian laws to add to the speed of light illustrated below.

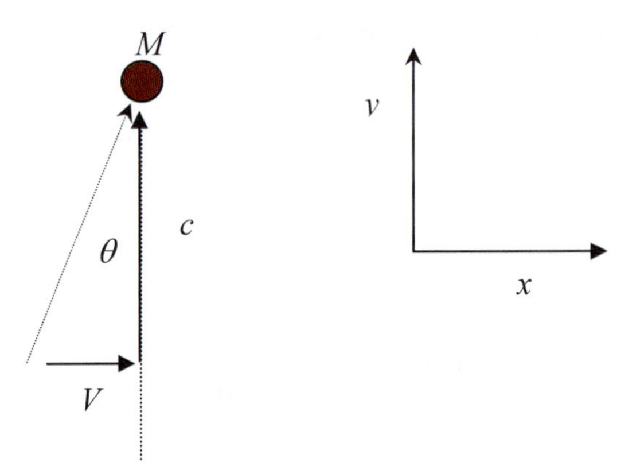

Figure 7-27: Brown dot is dust particle of mass M which moves at speed V in the sun fixed frame. In the instantaneous rest frame of the dust, the light appears to develop a component of speed that is off the radial direction. The path of the light as seen in the sun fixed frame is shown as a dotted line labeled c, the speed of light. The dotted line coming in at an angle θ to the radius is the apparent incoming direction of the light in the dust fixed frame.

In the instantaneous rest frame of the dust, we calculate the angular momentum of the dust around the sun. Since this inertial frame remains the instantaneous rest frame only to the degree the dust particle does not bend toward the sun. In particular, we take the rest frame to be valid only as long as the angle traversed during our analysis of the absorption and remission process is such that: $\Delta\theta \sim \frac{V\Delta t}{r_{earth-sun}} \ll 1$. This is not a problem, for we can define the angular momentum of the particle around the sun in such instantaneous rest frames at any point in the orbit, because of its symmetry. This viewing from instantaneous rest frames, though more complex, gives insights that are complementary to the Sun fixed frame.

To begin the analysis, we recall some useful facts:
In the sun frame:

7.54
$$\vec{V}^{Sun}_{dust} = V\hat{x}$$
$$\vec{V}^{Sun}_{light} = c\,\hat{y}$$

Thus, in the dust frame:

7.55
$$\vec{V}_{dust} = 0,\ \vec{V}_{light\ in\ dust\ frame} = \vec{V}^{Sun}_{light} - \vec{V}^{Sun}_{dust} = -V\hat{x} + c\hat{y}$$

In the dust frame, the light comes in at an angle of θ degrees from the radial line drawn from sun to the dust, cf. Figure 7-27. Note here, as in the treatment in the sun fixed frame, we are making assumptions so that we can use Newtonian framework; if we were to do this problem more completely we would use special relativity (Chapter 10). As it is, for the various steps, the current approach yields:

1. *Just before the photon strikes:*

7.56
$$L_{dust} = 0$$
$$L_{light} = -\Delta m_{light} \cdot V \cdot R$$
$$\Rightarrow L_{total} = -\Delta m_{light} V\,R$$

2,3 *The photon strikes the dust and re-radiates:*

The photon imparts angular momentum to the dust, giving it a velocity V' in the direction of the sun and adding Δm_{light} to the moving body which now "consists of" the light and the dust, following our artificial distinction between the light and the dust even when absorbed. The body next looses that Δm_{light} of its mass by *isotropic* emission of light, so that no net force, and hence no net torque, is exerted on the dust during this emission, giving $V'' = V'$ and $L_{dust}'' = L'_{dust}$; this also means, since the only two angular momenta are that of the dust and the light, that $L''_{light} = L'_{light}$. Thus, we have (still maintaining our artificial distinction between light and dust particle after the absorption):

7.57
$$L''_{dust} = L_{dust}' = M\,V'\,R$$
$$L''_{light} = L_{light}' = \Delta m_{light} V'\,R$$
$$\Rightarrow L''_{total} = M\,V'\,R + \Delta m_{light} V'\,R$$

Or, equivalently, we could simply write directly

$L''_{total} = L'_{\text{dust with absorbed light}} = \left(M + \Delta m_{light}\right)V'\,R$.

Using conservation of total angular momentum, $L_{total} = -\Delta m_{light} VR = L''_{total}$, we get:

7.58
$$V'' = V' = -\frac{\Delta m_{light} V}{\Delta m_{light} + M}$$

Thus, expanding as before in $\frac{\Delta m_{light}}{M}$:

$$L''_{dust} \sim -\Delta m_{light} V \; (1 - \frac{\Delta m_{light}}{M}) R \sim -\Delta m_{light} V \; R \tag{7.59}$$

Hence:

$$dL_{dust} = L''_{dust} - L_{dust} \sim -dm_{light} V \;\; R \tag{7.60}$$

And, from here, the argument proceeds as before, except we transform back to the sun fixed frame to get the decay law. In end of chapter problem 33, we explore the transformation laws for angular momentum and why the differential angular momentum is the same in both frames.

Understanding the Physics in the Two Frames

Notice that, in the dust frame, it is the torque applied by the off-angle absorption of the photon that causes the angular momentum change in the dust, whereas in the sun frame it was the increase in mass of the dust particle caused by the photon absorption. In this later case, loosely speaking, Δm_{light} had to share the impetus that was already moving the dust, i.e. the light and the dust had to come to the same speed, otherwise we would belie our premise that the light was absorbed.

In comparing the two rest frames above, it's helpful to use a simple analogy. Consider, as shown in Figure 7-28b below, a boy moving (in the ground frame) without friction in a wagon at speed v. He and his wagon are like the dust particle. Two balls, which are like the light in our analogy, fall vertically from overhead, i.e., they have no horizontal ground speed; the vertical motion is not relevant to our problem, so we only discuss the horizontal motion.

In his rest frame, the two balls come moving at speed v at him; he grabs them with a net and thus begins to move. In the ground frame, he is moving at speed v, but the balls are not moving; he grabs the balls which slow him down, for the impetus now has to move the wagon *and* the balls, and thus can only do so at a reduced speed.

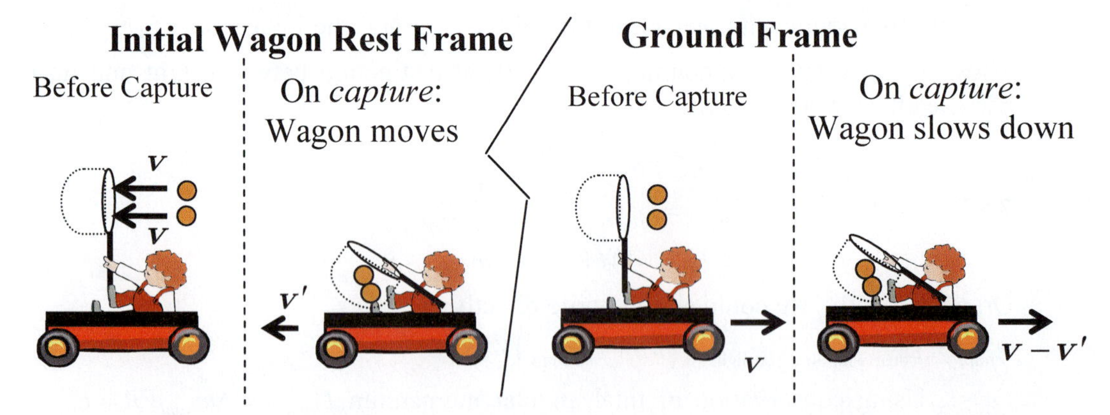

Figure 7-28: a. *(left)* From the point of view of the boy in the uniformly moving wagon, i.e. from the frame that moves uniformly to the right at speed v, the balls are flying at the boy at a speed v to the left. From this inertial frame, when the boy catches the balls, the wagon begins to move. **b.** *(right)* Seen from someone standing on the ground, the wagon is moving,

and the balls are not. However, once the boy captures the balls in his net, the car slows down as it maintains the same intensity of impetus, but now being received by a little bit larger mass, that of the boy, the wagon *and the balls*.

Next, as shown in Figure 7-29, the boy throws the balls out of a wagon moving at speed v. He throws one forward and the other backward at the same speed as seen in his frame. In comparing the figures for the two frames, one might think that a force acts on the wagon in the ground frame that doesn't appear in the moving frame. After all, in the moving (wagon) frame, the balls have the same speed (u), whereas in the ground frame, the wagon is emitting more energy forward than backward. Thus, in the ground frame, it *appears* that there is a momentum imbalance that will result in a net force on the wagon.

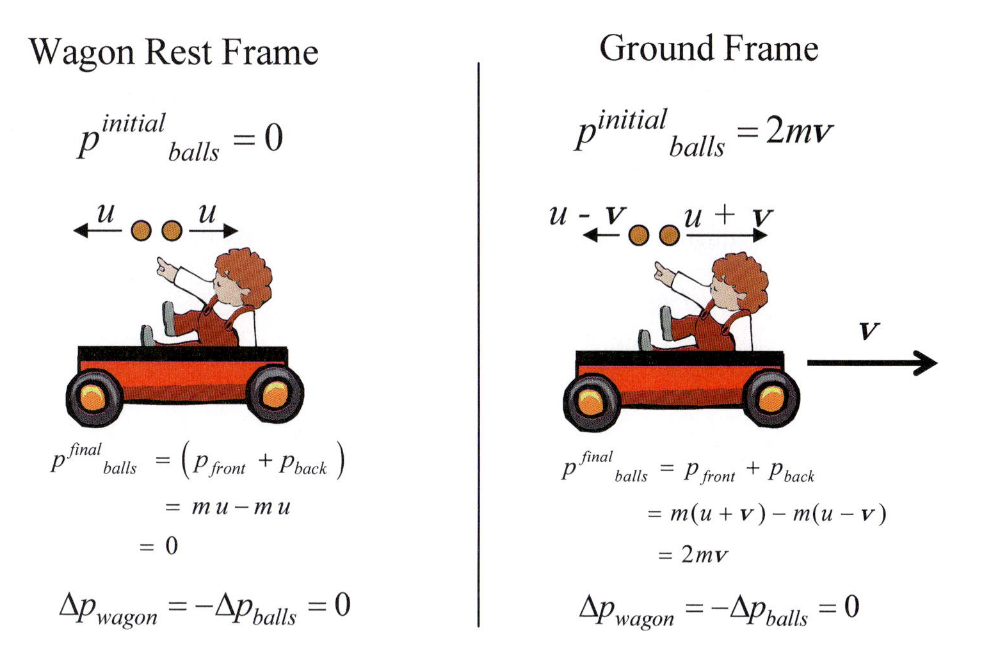

Figure 7-29: a. *(left)* In the wagon frame, the boy throws two balls, one forward and one backward at the same speed. **b.** *(right)* In ground frame, the same situation appears asymmetrical, i.e. the forward moving ball is slower than the backward moving one, The combinations of ground speeds shown is the only way for there to be no force on wagon. Note how the conservation of momentum works out in each frame. (Here, we exclude balls from the definition of the wagon).

However, this is only because we have forgotten that the *balls* initially have momentum in the v direction ($p_{balls} = 2mv$). Thus, in order for the balls to cause a net force on the wagon, the backward moving ball needs to be going faster than the forward moving ball by more or less than $2v$ (that is, assuming the balls do move, respectively, backward and forward in the *wagon's rest frame*).

For example, just dumping them over does nothing to the wagon speed; yet, look (cf. Figure 7-30) a momentum of $2mv$ comes off the wagon!

Figure 7-30: In the ground frame, the boy drops the two balls that have been sitting in the wagon. The wagon speed is unaffected.

Radiation Force

Once we know that as angular momentum decreases so does the radius (see problem at end of chapter), we have shown that a particle in orbit around the sun will gradually spiral inward. In so doing, we have assumed that the radiation force from the light from the sun can be neglected. We now calculate, very approximately, at what radius the force becomes such that it blows the particle way.

Using the equations above for the mass of a photon, we can write:

7.61 $$p_{photon} = \frac{h\nu}{c^2}c = \frac{h\nu}{c}$$

Then, the force due to Radiation pressure is:

7.62 $$F_{rad} = \frac{dp}{dt} = p_{photon}(\frac{\#\,photons}{\sec}) = \frac{h\nu}{c}\left(\frac{\frac{L_{sun}}{4\pi R^2}\pi r^2}{h\nu}\right) = \frac{L_{Sun}}{c}\left(\frac{r}{2R}\right)^2$$

Where r is the radius of the dust particle and L_{sun} is luminosity of the sun, ***not*** its angular momentum.

When the radiation pressure is greater than gravity the particle cannot stay in orbit and gets blown away, this happens when $F_{rad} = F_{gravity}$ thus:

7.63 $$L_{sun}\frac{r^2}{R^2}\frac{1}{4c} = \frac{GM_{sun}m_{dust}}{R^2} = \frac{GM_{sun}\frac{4\pi r^3\rho}{3}}{R^2}$$

Solving this equation then gives:

7.64 $$r = \frac{L_{sun}}{GM_{sun}}\frac{3}{16\pi c\rho}$$

Using:

$$L_{sun} = 4\times10^{26}\,Watts\;\; M_{sun} = 2\times10^{30}\,kg,\;\; \rho = 2g/cm^3 = 2\times10^3\,kg/m^3$$

$$G = 6.67\times10^{-11}\frac{m^3}{s^2kg},\;\; R = 1.5\times10^{11}\,m$$

We get:

$$r \sim .3\mu m \text{ or } D \sim .6\mu m$$

Summary

The three most important concepts of empiriometric physics are, when suitably generalized, the conservation of *energy*, *momentum* and *angular momentum*.

As conservation of momentum means that the *net* tendency of an isolated system of bodies to move in a given linear direction is never lost, conservation of angular momentum means the *net* tendency to move in an angular direction around a given point is never lost.

Angular momentum is defined as: $\vec{L} = \vec{r} \times \vec{p}$, where $\vec{r}$ is the displacement vector from the chosen origin to the body under consideration and $\vec{p}$ is the body's momentum. Since the cross product was defined in Chapter 2 so as to incorporate in our mathematical formalism the fact that two vectors naturally induce a rotational axis (a real fact that comes from the first property of all physical things), we make use of it to express the more fully physical fact that massive bodies move angularly around an offset point via their impetus.

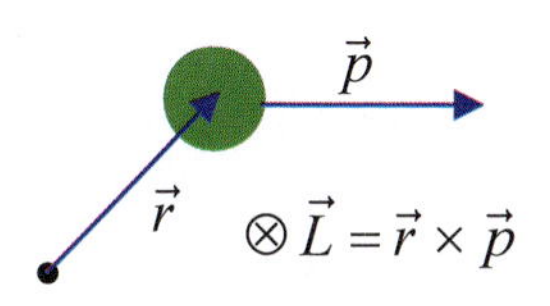

For the case of a freely moving body, it is clear that angular momentum is conserved because the linear momentum $\vec{p}$ is conserved and, by definition, the distance from a given point along the perpendicular to the direction of motion is unchanging. Since Newton's third law guarantees conservation of the *net* linear momentum at the point of interaction of massive bodies, angular momentum also must be conserved for a system of massive bodies subject to forces resulting only from their mutual interactions.

Conservation of angular momentum of a system is written:

$\vec{L}_{total} = \sum_{i=1}^{N} \vec{L}_i = \text{constant}$, where the N bodies in the system are consecutively numbered and the subscript i indicates which body.

Angular momentum does *not* require a new quality called angular impetus that moves things angularly similar to the way impetus moves things linearly. Angular momentum is the result of forces acting in concert with impetus, making a body turn. Conservation of angular momentum is the result of the symmetry of the acting forces, such as the string in the whirling ancient slingshot. As such, it is a derived law. Conservation of angular momentum is a consequence of the conservation of linear momentum *if* we require that a suitably general application of Newton's laws apply to all bodies massive and non-massive (i.e. things that can apply force such as fields; we will clarify this point further in Chapter 8). Nonetheless, we can still treat angular momentum as if it were the measure of a separate quality. Doing so allows us to ignore, in our usual way, aspects in which we are not interested, in turn, allowing us to focus on the angular symmetry.[28]

Understanding the symmetries of the physical world is essential to understanding conservation laws. The first property of all physical things, continuous extension, parts one next to another bereft of any qualities including, by a being of reason, the distinct shapes of the parts, has 6 types of symmetry: three linear or *translational symmetries* and three *rotational symmetries*. In this three dimensional mathematical space, called three dimensional Euclidean space, a given place on a line cannot be distinguished from any other place on the line; this is true for any of the three perpendicular lines that we usually label the *x, y,* and *z* axes. Furthermore, at any point in this unqualified space, a rotation about any one of the three axes cannot be distinguished from a different rotation about the same axes.

By assigning a force to some or every point such as we do with a force field, we can discuss the symmetry of the space in an analogical way. When, for example, there is a force in a given linear direction we say the translational symmetry is broken; left is different from right as seen in the change in the potential energy function with distance.[29] Because of the force, the break in the translational symmetry, momentum is not conserved.

However, when we include the source of the force (say a second massive body), there is an equal and opposite force acting on it, and we say the symmetry is maintained, because the amount of momentum caused in one direction is exactly balanced by that given to the other thus, in a more distant analogical way, preserving the original symmetry. In more advanced physics courses, such symmetries will show up clearly in our equations when we choose the appropriate coordinate system.

In analogy to momentum, which is changed by force, we say that the angular momentum is changed by torque. The torque, in analogy to Newton's second law, is written, $\vec{\tau} \equiv = \vec{r} \times \vec{F} = \frac{d\vec{L}}{dt}$. A torque, which changes angular momentum, results from a force acting at a distance from the chosen pivot point.

Uniform circular motion is the simplest type of angular motion. Analyzing it according to the natural coordinate system of polar coordinates, we say that no work is done in such motion. This is only true to the extent that we ignore the changing direction of the force and thus of the shifting of momentum from the *x* to the *y*-direction. Once we refocus on the more complete aspects of the physical reality that include the nature of the impetus to carry objects in a linear, not circular, direction, we are naturally drawn to realize the necessary role that the Earth's motion plays in the actual physical situation. Namely, the Earth does work via the string on the ball and vice versa. Still, this does not militate against the usefulness of focusing on the angular symmetry of the problem, which in fact is its most distinguishing feature.

---

[28] It is interesting to note that medieval scientists glossed over the distinction between angular and linear impetus in a way that is similar to the modern empiriometric treatment.

[29] Or, what is functionally equivalent, the qualification of the space by the forces changes the extension itself, which thus directly breaks the symmetries of the bare extension discussed above.

Cylindrical coordinates naturally arise in discussing angular momentum. In these coordinates the acceleration is: $\vec{a} = \left(\ddot{r} - r\dot{\theta}^2\right)\hat{r} + \left(r\,\ddot{\theta} + 2\dot{r}\dot{\theta}\right)\hat{\theta} + \ddot{z}\,\hat{z}$. In an accelerated frame, a mass m will experience a fictitious force, i.e. one that appears because of the observer's state of acceleration, not an actual force. This force is: $\vec{F}_{fict} = -m\vec{a}$. In a rotating frame, where cylindrical coordinates have an important place, this force consists of four important terms: the radial acceleration, $-m\ddot{r}$, the centrifugal force, $mr\dot{\theta}^2$, the angular acceleration, $-mr\ddot{\theta}$ and the Coriolis force, $-m\left(2\dot{r}\dot{\theta}\right)$. Of course, if the frame is accelerating along the z-axis we also have the standard rectangular acceleration: $-m\ddot{z}$.

Using cylindrical coordinates, for a planar body, we can write the angular momentum as: $\vec{L}_{Total} = I\omega\hat{z}$, where $I = \int r^2 dm$ is called the *moment of inertia.* Also, for bodies that are symmetrical about each of the three axes we can write: $L_x = I_x\omega_x$, $L_y = I_y\omega_y$, $L_z = I_z\omega_z$ calculating I_i using $I_i = \int r_i^2 dm$, where r_i is the distance perpendicular to the i-axis, namely: $r_x = \sqrt{y^2 + z^2}$, $r_y = \sqrt{x^2 + z^2}$, $r_z = \sqrt{x^2 + y^2}$. Often used moments of inertia along with helpful rules in deriving others are given in the calculational techniques section below.

The moment of inertia is analogically like the mass in the formula $\vec{L} = I\vec{\omega}$.For constant I, we can write an analogical equivalent to Newton's second law when one has constant mass, namely: $\vec{\tau} = I\vec{\alpha}$.

The angular momentum vector is not always parallel to the rotational axis. In general, we have: $\vec{L}_{Total} = I\vec{\omega}$, where I is a matrix, so that the directions do not have to be parallel. The use of a matrix is just a notation that allows us to compactly express a very complex relation; we will not generally need matrix manipulation in this text.

We can always break up the angular momentum of a body about an arbitrary point P into an orbital component and a spin component: $\vec{L}_{Total} = \vec{L}_{orbit} + \vec{L}_{cm}$. The *orbital angular momentum is the angular momentum of the center of mass* of the body about the point P and the *spin angular momentum is the angular momentum of the body about its center of mass*. Writing in terms of rotation about an axis we take to be the z-axis, we can write: $L_z = \underbrace{M\left(\vec{R} \times \vec{V}\right)_z}_{Orbital} + \underbrace{I_z\omega_z}_{Spin}$, where $\vec{V}$ *and* $\vec{R}$ are the location and velocity of the center of mass.

By analogy with the work-energy theorem of linear momentum, we can write the work-energy theorem for angular momentum, considering only rotation about the z-axis ($\vec{\omega} = \omega_z\hat{z}$). $W_{rot} = \int \vec{\tau} \cdot d\vec{\theta} = \frac{1}{2}I\omega_{after}{}^2 - \frac{1}{2}I\omega_{before}{}^2$

When there are no torques present, obviously $W_{rot} = 0$, and thus the rotational kinetic energy, $\frac{1}{2} I \omega^2$, is constant, i.e. conserved. We also obtain a conservation of mechanical energy law very like that which came out of the analysis of linear motion: $E = V + K$ is constant, where V is the angular potential energy.

The measure of kinetic energy can be split into the measure of the activity of linear motion *of* the center of mass and the activity of rotational motion *about* the center of mass. Mathematically, we can write this, for diagonal moments of inertia, as:

7.65
$$KE_{external} = KE_{linear\_CM} + KE_{rot-about\,CM}$$
$$= \frac{1}{2} M\vec{V}^2 + \frac{1}{2} I_{cm\_x}\omega_x^{\ 2} + \frac{1}{2} I_{cm\_y}\omega_y^{\ 2} + \frac{1}{2} I_{cm\_z}\omega_z^{\ 2}$$

Here $\vec{V}$ is the velocity of the center of mass. The first term allows us to think of the given rigid body as a mass concentrated at the center of mass of the body. In many problems, the conservation of translational and rotational energy can even be applied separately.

Conservation of *angular momentum* can be considered a separate conservation law from the conservation of *linear momentum*. The split between rotational and linear momentum is an important way of capturing key symmetries caused by the interaction of the impeti, forces and masses of the given system of bodies.

The analogy between the quantities (properties and their changes that are "quantified" or numbered by comparison to a unit) used in translational motion and rotational motion is summarized in Table 7-1. Again, for rotational motion, the generic qualities remain force, impetus and mass, it is only how they are specialized in particular physical situations that makes the rotational analysis relevant and thus fruitful.

Terms

Angular momentum
- Euclidean Geometry and first category of property of physical things (continuous or discrete quantity)
- Translational/rotational symmetry
  - Homogenous
  - Isotropic
- Conservation of angular momentum
- Simple angular acceleration
- Coriolis Force

Torque
- Moment of inertia
- Parallel axis theorem

Spin/orbit theorem
- Spin and orbital angular momentum

Luminosity

Calculation Techniques

Often used moments of inertia are given below with several helpful rules below it.

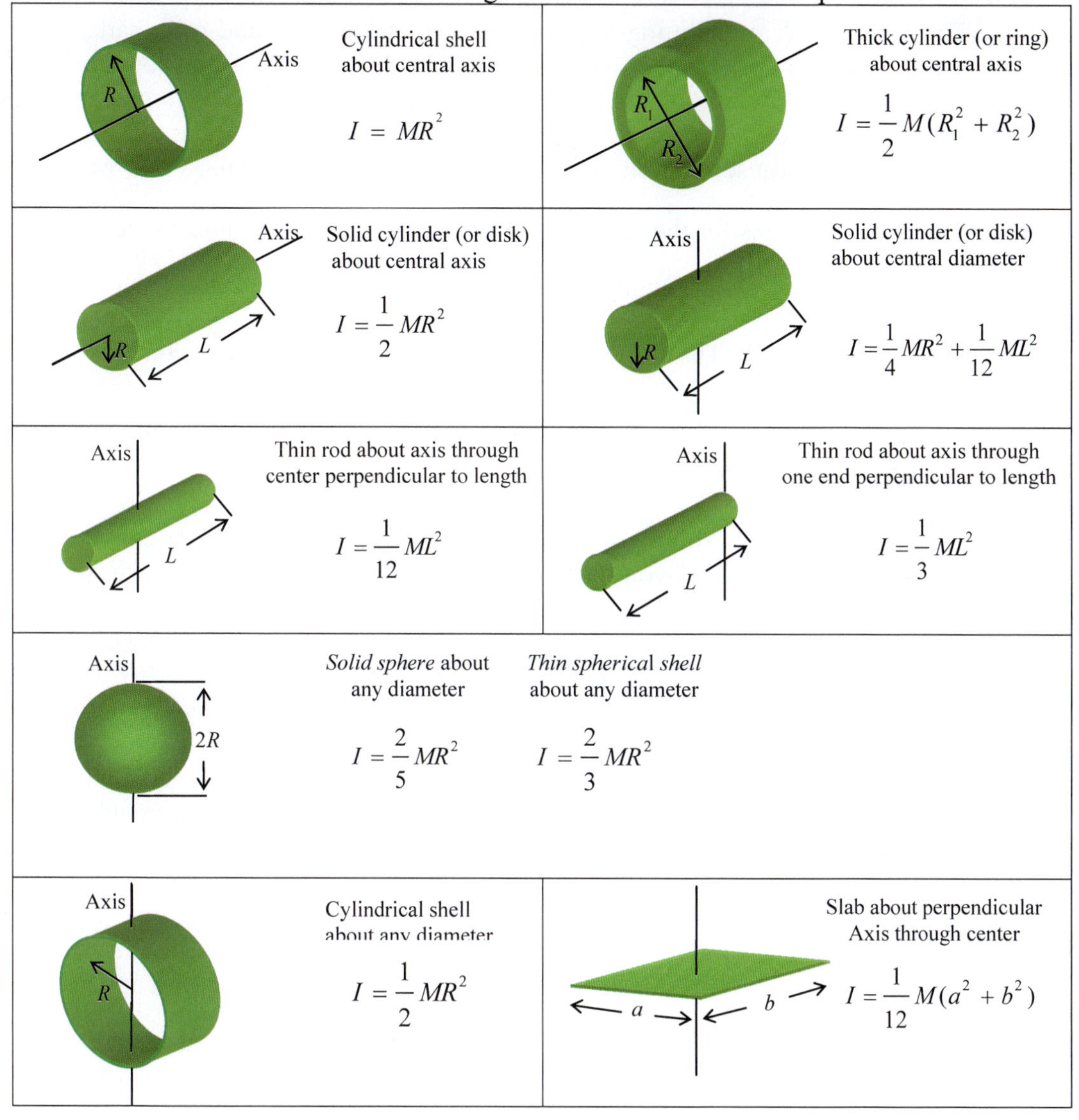

The moment of inertia about a parallel axis, going say through a point A, displaced by an amount l from the center of mass can be written: $I_A = I_{cm} + M\,l^2$

Perpendicular axis theorem: for a planar massive body, the moment of inertia about a given axis is equal to the sum of the moment of inertia about the two axes perpendicular to it that pass through the same point as it. Namely, using the figure to the right: $I_z = I_x + I_y$.

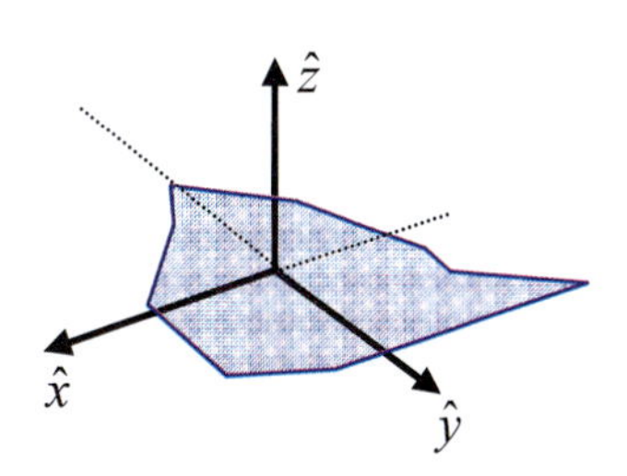

It can be helpful to analyze a system in two different frames to better understand the dynamics and to check your answer (as done in Poynting-Robertson calculation).

For a *rigid body*, in splitting angular momentum into spin and orbit parts, do not forget that the spin and orbit parts are only separable if one treats the spin about ***the center of mass*** and the orbital motion of ***the center of mass***. Furthermore, since torque about the center of mass cannot cause linear motion of the center of mass and force directly on the center of mass cannot cause rotation about the center of mass, we can, in that way, write separate $\vec{\tau} = \frac{d\vec{L}}{dt}$ equations for the spin and the orbital components. Thus, the total torque, that is the torque that changes the total momentum, can be written: $\vec{\tau} = \left(\sum_j \vec{r}_j' \times \vec{f}_j \right) + \left(\vec{R} \times \vec{F} \right)$. The torque about the center of mass for rotation axis aligned with the ith direction is: $\tau_i = I_i \alpha_i$ and is independent of the translational motion of the center of mass (it can even be accelerating).

When having trouble solving a problem, especially if you encounter contradictions in your thinking, remember that you may need to include more information, i.e. you may need to "switch more things on." For instance, in analyzing the behavior of a stick, you may begin by modeling it as a line; however, in order to understand the forces that act within it and on it you may have to add back its width and depth. Remember, beings of reason (pure mental constructs) are necessary in capturing aspects of a physical system in a formal mathematical system; nonetheless, the system, and the particular beings of reason that make it up, is not the thing itself and can thus, unlike reality itself, be contradictory if one is not careful in tracking what has been included and what has yet to be included.

Notation and Vocabulary Notes

Recall that primes can mean either "after an event" or "in the center of mass frame." We also will use primes for dummy variables in integration. As with any notation, the context determines the meaning.

Mixed Problems

1. Calculate the force acting on the dust particle in the Poynting-Robertson effect example.

2. Show that for an object in circular orbit that as the angular momentum decreases, so must the radius. *Hint:* use Kepler's law for orbiting a body of fixed mass: $\frac{T^2}{r^3} \sim constant$.

3. A man with an ancient slingshot has just finished whirling it up to speed and is now coasting (assume without friction) in a nearly circular motion. Draw a force diagram showing the chain of forces acting that finally ends in a force acting on the Earth, and explain how that force affects the Earth as the rock executes one orbit.

4. A pencil of length L and mass m falls from a vertical position with negligible initial rotation and no friction, a) what is the rotational speed, $\dot{\theta}$, as a function of angle? b)What is the rotational speed for a pencil with $L = .15m$

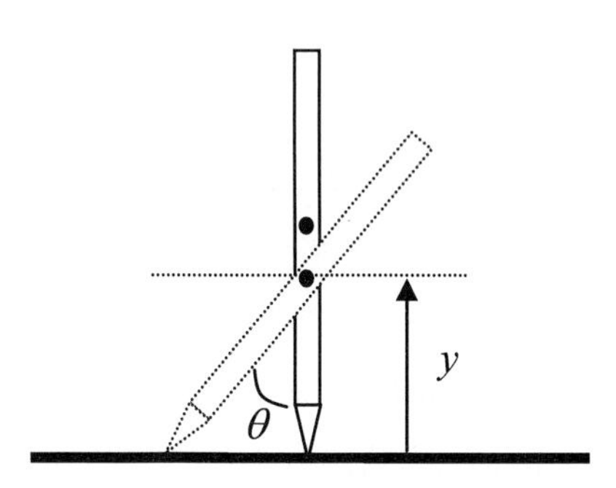

There is no friction, so there is no horizontal force acting on the center of mass

5. A pencil of length L of mass m starts in the vertical position and falls over with its point held in position (as opposed to the case described above) as it falls, a) what rotational speed does it attain upon hitting the table? b) what rotational speed does it attain if it has length of 15cm?

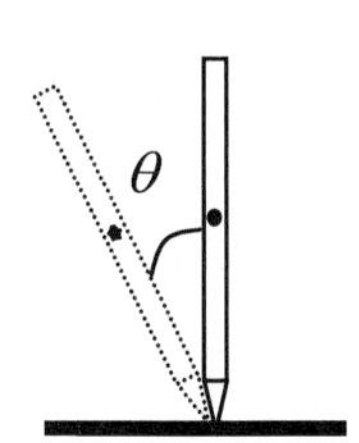

6. Assuming there is absolutely no wind and no table vibrations and the table is perfectly level and smooth, what is the maximum time that a pencil can be made to stand on its tip?

To solve this problem, you will need to know the following. There is a limit imposed by the natural vibrations of things. In particular, in quantum mechanics, a restriction on the ability to measure speed and momentum to arbitrary accuracy appears. Heuristically, if I try to locate something very precisely I need high frequency (longer wavelength) light to interact with the body I would like to view so that I can pinpoint it accurately; however, higher frequency light activates more momentum in the body, so in the process of learning about its position I lose my knowledge of its momentum. Similarly, if I use lower frequency light to determine the momentum accurately, I will not be able to determine its position accurately. Mathematically, this is written $\Delta p \Delta x \geq \hbar / 2$, where $\hbar$ is called Planck's constant. This relation is called the uncertainty principle. To solve the problem, follow these steps.

Step 1. Use the relations for force, work, energy, momentum, and position discussed in earlier chapters to hypothesize that: $\Delta E \Delta t \geq \hbar / 2$

Step 2. Using the small *angle* techniques of Chapter 6, write the energy of the system (it's a pendulum).

Step 3.

Take the average of this energy, i.e. $\langle E \rangle$. (note: the average is symbolized by the triangular brackets). With an eye towards using the uncertainty principle to take this average, substitute $\langle \theta^2 \rangle \rightarrow \langle \Delta\theta \rangle^2$ $\quad \langle \dot{\theta}^2 \rangle \rightarrow \langle \Delta\dot{\theta} \rangle^2$ in the average energy relation by using the fact that $\langle \Delta x \rangle^2 = \left\langle \left(x - \langle x \rangle \right)^2 \right\rangle = \langle x^2 \rangle - \langle x \rangle^2 = \langle x^2 \rangle$, where we take the average x to be zero, and the average rate of change of x to be zero because we want to stay near zero angle, and this is the best we can do in a jittery environment. This new energy is the energy due to fluctuation. In our problem, one wants the minimum of such energy. We take this to roughly correspond to the vibrational energy away from the zero energy point that QM requires; we take $\langle \Delta E \rangle \sim \langle E \rangle$.

Step 4.

Taking the best we can do as the equality in the uncertainty principle, use the methods of this chapter to write $\Delta p \, \Delta x \sim \hbar / 2$ in terms of the angle of offset, $\Delta\theta$, and its first derivative. Substitute this into the energy formula obtained in step 2 to write the energy solely in terms of the angle $\Delta\theta$.

Step 5. Find the angle that minimizes the energy, and use this and the step 1 result to find an approximate time that the pencil remains balanced (if only this "primal vibrational energy" were at work.) Also, the minimum energy calculated in the process is the minimal energy that an harmonic oscillator can possess and is very important in quantum mechanics especially at the very advanced level of quantum field theory.

7. Given the quantization of angular momentum, i.e. that the angular momentum must be integral units of $\hbar$, we can show that the electron in the hydrogen atom must take on discrete energy values. Assume the electron is in a circular orbit around the nucleus, and take the electron's potential energy function as: $-\frac{e^2}{r}$. Calculate the total energy of the electron and the radius of its orbit, showing explicitly what discrete values each can take.

8. Seeds that fall from Maple trees helicopter their way to the ground. Consider the toy helicopter, which is very much like these seeds, that can be made in the following way:

> Cut a rectangle out of ordinary notebook paper. Make it about 3" by 1.5". Cut about 2 inches down the middle into the longest part of the rectangle. Fold back the finger just cut to make a helicopter looking device. Attach a standard paper clip to the middle of the uncut edge oriented so that its long direction is aligned with the long part of the rectangle.

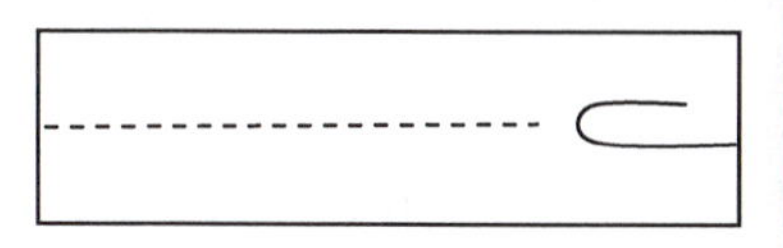

> Hold the paper clip with helicopter blades up and throw into the air.

Assuming the whole helicopter weighs ~10 *mg*, that it comes to a terminal velocity of V after falling about $h = .3m$ and that all the gravitational potential energy goes into rotational energy: a) Write the equation for the rotational frequency of the helicopter
b) Given, based on experiment, that once up to its terminal linear and rotational speed, it traverses 50 inches in about 1 second, how fast would it rotate? Make the above toy and check your estimate of its rotational speed as well as its terminal linear velocity and the height it falls before reaching terminal velocity. c) Where does energy that doesn't go into rotational or translational motion of the helicopter go? d) Assuming that a feather drops and does not spin, it will develop little terminal velocity, what happens to the rest of the energy. Note: to make a wooden helicopter: see
http://www.grc.nasa.gov/WWW/K-12/TRC/Aeronautics/Helicopter.html

9. Consider again the puck on ice sliding, without friction, toward the stick at speed v which was discussed in Chapter 3 and is shown below. The ball hits the stick at the very end of the stick; upon impact the ball attaches to the end, so that from that point on the stick and ball move together. The puck has negligible diameter and mass m, and the stick is assumed to be a thin rod which has mass M and length L. Take V to be the speed of the stick/ball system after the collision. Calculate the rotational speed of the stick/ball after the collision. Use the coordinate system (which is fixed with respect to the ice) and the labels shown below. (see also problem below)

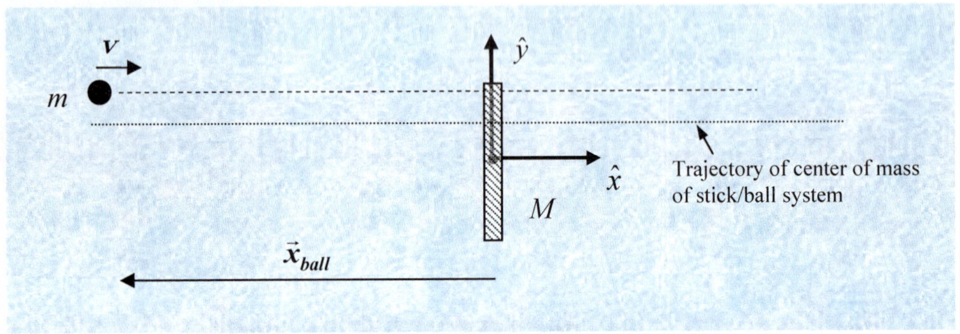

10. A puck hits a stick on the ice as described in the problem above, but this time it collides elastically. a) Describe the motion, giving equations for the linear and rotational velocities after the collision. b) Is any impetus activated in the y-direction (see the figure in above problem) and if so how does this occur? Play with this situation in *Interactive Physics* changing the ball mass and the elasticity of the collision.

11. Consider a horizontal rod of length L and mass m under the influence of uniform pull of gravity near the Earth's surface. The rod also has a force of magnitude mg applied upward to the right end. Given that the rod begins at rest, describe the resulting motion. In particular, what is the maximum rotational speed and at what angle from the initial horizontal does it occur? Give the result symbolically and then numerically, using $L = 1m,\ m = 10kg$. Verify your results in interactive physics program (see for instance, 7.1-dumbell falling off edge.ip.)

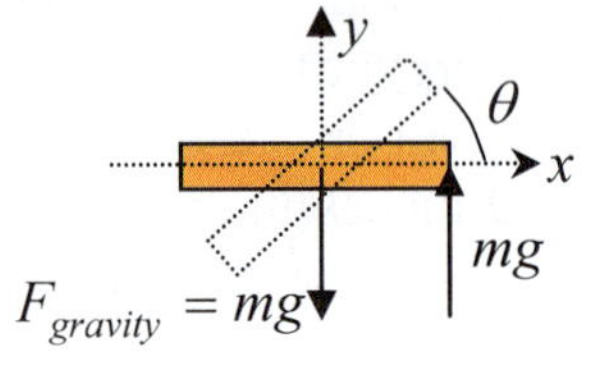

12. When a round toilet seat of outer radius R_2 and inner radius R_1 is up, it makes an angle of θ with the vertical. a) How big does θ need to be for the seat fall from a rest position? b) Given a seat at rest at an angle θ_0 larger than this minimum angle, what rotational speed, ω, does the seat reach by the time it hits the base of the toilet.

13. Many years ago, John Wheeler and his student Nobel laureate Richard Feynman once wondered what would happen if a rotating sprinkler were put in a pool of water and contrived so as to suck the water inward instead of expelling it. This thought experiment was meant to help answer the question: "What would happen if a sprinkler was operated 'in reverse,' sucking water in rather than spitting it out." a) What causes a normally operating sprinkler to rotate? b) What does happen in the "reverse" case ?

14. Two balls move at each other towards a head-on collision with no external forces acting. a) Explain how the angular momentum is conserved, not for each individual ball, but as a *sum* of the angular momentum of each ball. In particular, examine the case where the balls are confined to one dimension and one simple case where they are not. b) Explain in what sense (exactly) there is a rotational symmetry in this problem.

15. Any river that runs from North to South carries dirt from high latitudes to low latitudes, i.e. towards the equator. For a very long river, say one that went from the pole to the equator, carrying $\dot{m}$ kg/year of silt, what if anything does this action do to the length of the day?

16. Explain what role the little rotator on the back of some helicopters plays. Do a simple calculation to give an order of magnitude quantitative explanation of what would happen without the rotor.

17. a) Explain how one might use multiple gyroscopes to calculate *change in orientation* (roll, pitch and yaw as shown in the figure) of the spacecraft relative to a given inertial frame on the manned trip to Mars. Also, how can such an approach be used to determine one's *current orientation* without looking outside the craft? b) What else would be needed to track the craft's position as well as its orientation? Explain generally how to do it.

18. Calculate using knowledge of angular momentum, linear momentum and energy what you can say about how this pillar was knocked over? Was it likely done by a car or truck? The column is about 9' high including white top, 32'' wide and 32'' deep and its base puts the column ~1 *ft* off ground. Do so by calculating the minimum force required to unbalance the column and the minimum energy needed to knock it over. (*Note:* for simplicity, ignore this 1 foot base and assume the column rests flat on the ground.)

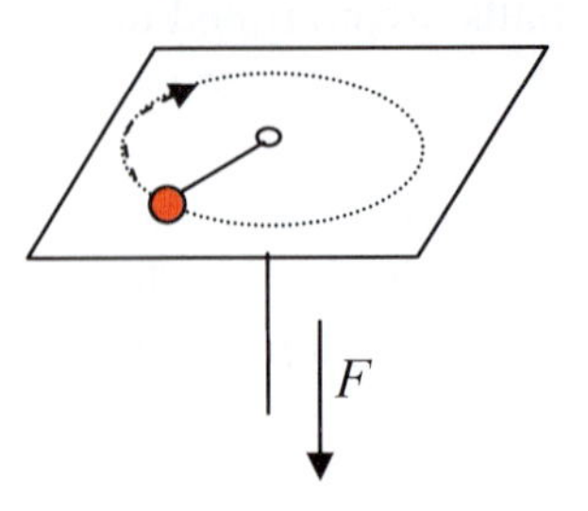

19. A ball is whirling without friction in a circle on a table as shown. The string is gradually pulled from the center, a) how much work is required to pull the ball from a circular orbit of radius r_0 into a smaller one of radius r_1? b) Why does it take any work, is not the force always perpendicular to the direction of motion?

20. a) Why does leaning one way when your riding a bike, cause you to turn the way you lean? Draw angular momentum diagram with the torque and resulting change in AM.
b) What happens to the linear momentum in the original direction and where does the linear momentum in the new perpendicular direction come from?

21. Suppose one "drops" a satellite from the orbit of Mars to the orbit of Earth in such a way that the force of gravity acting on the satellite turns a disc shaped flywheel, doing all of its work on the flywheel, which means negligible work on the linear motion of the satellite. What rotational speed does the flywheel of mass $m = 4\times10^5 kg$ and radius $R = 5m$ have, if one has negligible friction:

22. Astronauts on Mars may want to use its moons to verify their position using radio repeaters on the moons. To accomplish this for the Astronauts, we first need to know how far away the moons are. Suppose we measure the moon Phobos to take 3.3 hours to orbit, what is its approximate height above Mars? Deimos takes 30.2 hours; what is the height?

23. Supposing that hurricanes could form on Mars, how does the speed of winds that result from the Coriolis force on Mars compare to those on Earth for the time scale of a rotation?

24. Use the diagram and mathematics to show that, for a given position vector $\vec{r}$, the relationship between its rate of change in an inertial frame and a rotating frame is given by:

$$\left(\frac{d\vec{r}}{dt}\right)_{inertial} = \left(\frac{d\vec{r}}{dt}\right)_{rotation} + \vec{\Omega}\times\vec{r}.$$

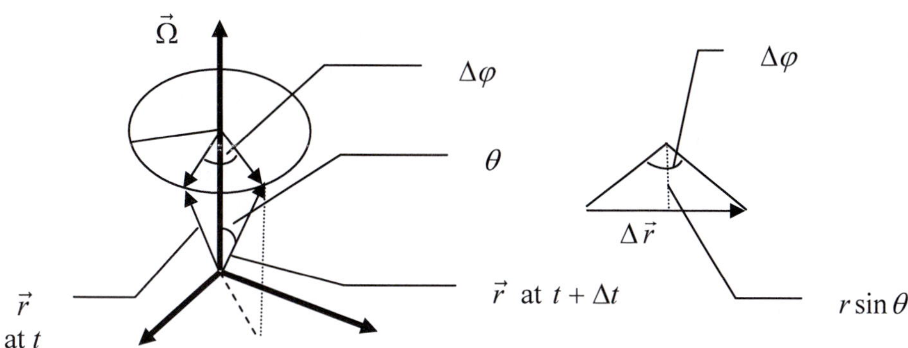

Fixed vector $\vec{r}$ changes by $\Delta\vec{r}$ as viewed from frame rotating at angular velocity $\vec{\Omega}$. Figure shows vector in rotating frame at times t *and* $t+\Delta t$

Note that, in this equation, the subscripts indicate in which frame one is measuring the position vector. Also, note that at any moment $\vec{r}_{rotation} = \vec{r}_{inertial}$, although they are measured in different coordinates (i.e. given different labels). However, this is not true for the vectors that are the derivative of each of these, because while the inertial coordinate directions stay the same, the rotational ones do not, for they are rotating.

As extra credit, you might try your hand at deriving the formula relating acceleration in the two frames, remembering to evaluate each derivative in one frame not two, to show:

$$a_{inertial} = \left(\frac{d^2\vec{r}}{dt^2}\right)_{inertial} = \left(\frac{d^2\vec{r}}{dt^2}\right)_{rotational} + 2\vec{\Omega}\times\left(\frac{d\vec{r}}{dt}\right)_{rotational} + \vec{\Omega}\times\left(\vec{\Omega}\times\vec{r}\right)$$

The latter two terms on the right hand side are the Coriolis and centrifugal terms discussed in the chapter.

25. The figure shows a ball of solder (silvery ball on wire) being heated in space by a soldering iron. As it is heated the flux (yellowish translucent ball on the left) emerges and spins around the solder ball as a elliptically shaped ball at a fast rate of rotation (see http://science.nasa.gov/headlines/y2004/16aug_solder.htm for full video). Make one simple conjecture about how this might happen.

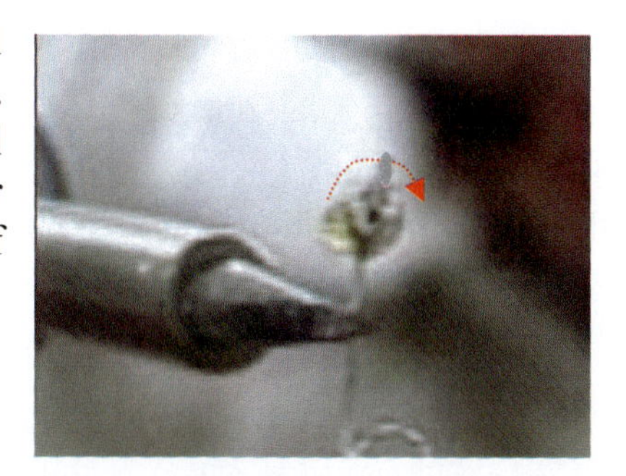

26. a) Explain generally how a teeter totter works. b) Using the same lever principles explain how a wrench works; in particular, what is the fulcrum? For a wrench of length 20 *cm* and a bolt of radius 2*mm*, what is the force advantage obtained. c) Given a 200 *kg* man puts all his weight on a .2 *m* long wrench, what torque can he produce?

27. Consider a ball of mass m attached to a (massless) string of length L which is pinned at one end to a much more massive body of mass M, such as the Earth. The ball whirls around the pinned point at an angular speed ω. a) By tracking the momentum in the x and y direction, show that for $m \ll M$, the kinetic energy of the small mass is approximately constant throughout its circular motion. b) If the Earth starts with no momentum when the ball is on the x-axis, how much kinetic energy does it have when it reaches the y-axis? c) Now, assuming a one can no longer neglect the mass of the Earth, describe the complete motion of the Earth and ball, including a) giving the translational and rotational *kinetic* energy when the ball is on the x-axis and when it is on the y-axis, and b) calculating the angular frequency of rotation of the system.

28. Given a top of diameter 6 *cm*, length from the center of mass to the end of the axis about 6 *cm,* mass of about 70 *g* and an observed precession frequency of $F = .5 rot/s$, calculate how fast the is top spinning. Can you think of a way to measure this in order to verify this prediction?

29. The dumbbell model of the gyroscope was explained heuristically in the text. Write down the equations for the motions for the top ball of the dumbbell shown in the figures below and use them to give a more quantitative account of the dumbbell analysis given in the text.

Hint: Using a rectangular coordinate system fixed in the lab frame with its origin at the pivot point, the velocity of the top ball of the dumbbell starting in the position shown below is:

$$\vec{v}^{Lab}_{top\_ball} = \left(\left(\mathbf{v}_s \cos\omega_s t - \mathbf{v}_p\right)\cos\omega_p t - r\omega_p \sin\omega_s t \sin\omega_p t\right)\hat{x} - \mathbf{v}_s \sin(\omega_s t)\hat{y}$$

$$+\left(\left(-\mathbf{v}_s \cos\omega_s t + \mathbf{v}_p\right)\sin\omega_p t - r\omega_p \sin\omega_s t \cos\omega_p t\right)\hat{z}$$

where $\mathbf{v}_s$ is the spin speed of the ball as seen in the center of mass frame; ω_s and ω_p are the angular speeds of spin and precession with $2\pi\omega_p << \omega_s$ Note the equation accounts for the rotation of the plane of spin of the dumbbell by writing the top balls position vector in the center of mass frame and then rotating that vector.

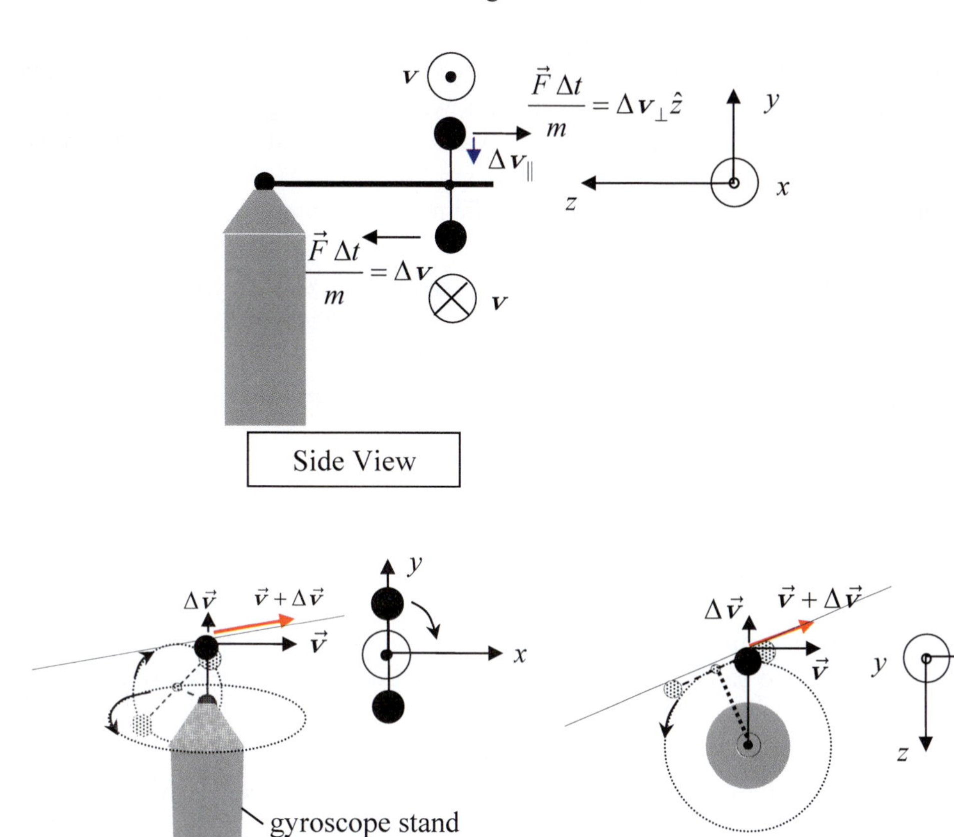

Note: Recall that to facilitate visualization, these figures exaggerate the motion of the precession with respect to the spin motion.

30. Is it possible for a body shaped and oriented as shown below to revolve on a pinned pivot without dipping or falling? Explain. (Note: unlike the gyroscope, this body does not spin around its center of mass)

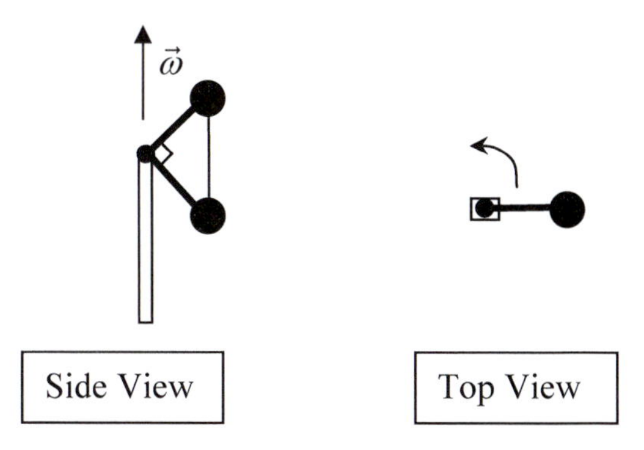

31. Hold one quarter fixed with the head oriented properly and put a second next to it with the same orientation and rotate it around the first allowing no sliding. a) What orientation does the second quarter have after a complete revolution? b) What if a dime is rotated around a quarter?
Note: a quarter has diameter 24mm and a dime has diameter 17.5mm.

32. Calculate the constant in equation 7.53.

33. In the text, we discussed calculation of the angular momentum for the Poynting-Robertson effect in two different frames. a) Explain what the difference between these two points of view is by transforming the angular momentum in the lab frame to the rest frame of the dust b) How can the same formula (viz. $dL_{dust} \sim -dm_{light} V R$) be relevant in both frames?

34. A man at rest on an initially motionless Earth starts running at speed v; what happens to the Earth? Estimate the effect.

35. A uniform sphere of mass M_{sphere} radius r_{sphere} can rotate around a fixed vertical axis on frictionless bearings. A massless cord passes around the equator of the sphere, over a disc-shaped pulley of moment of inertia I_{pulley} and radius r_{pulley} and is attached to a small object of mass M_{block}. There is no friction on the pulley's axle; the cord does not slip on the pulley. What is the speed of the object when it has fallen a distance h after being released from rest?

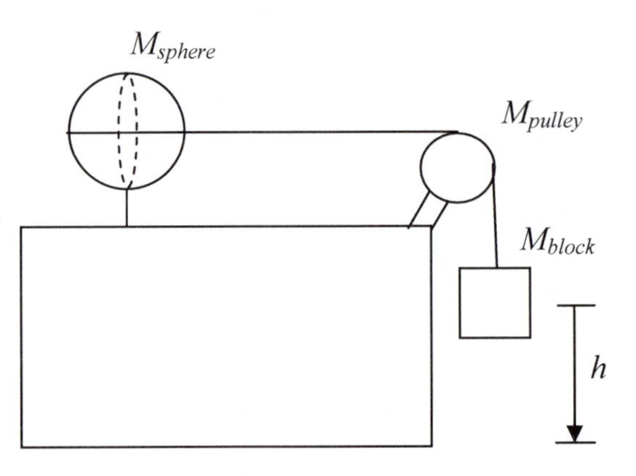

36. Consider a race car of mass m moving around a circular track of radius r at speed v, what minimum incline angle, θ, can the track have to just keep the car from climbing up the side of the track? Assume a coefficient of (static) sliding friction of μ, for the wheels on the pavement. Check your results with interactive physics (see, for instance, 7.3-car rounding circular track.ip).

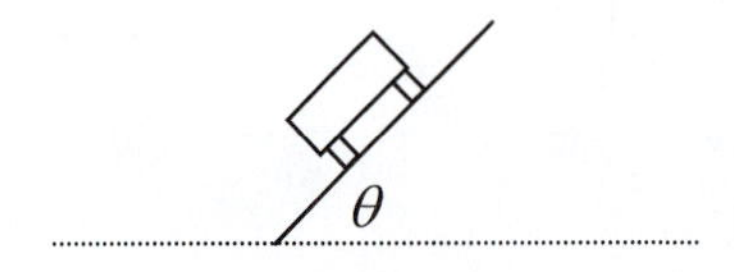

37. A solid wheel of mass m and radius r is connected to one end of a massless axle of length R. The wheel spins about the long axis of the axle in the horizontal plane with angular velocity ω_s. Gravity acts in the vertical direction. The axle is connected at the far end by a string of length L. The spinning wheel and axle precess about the vertical axis (through the point that connects the string with the ceiling) at angular velocity ω_p. Find the inclination angle α of the string from the vertical direction in terms of ω_s , R, r and g, assuming the horizontal displacement s of the string is much smaller than R and L.

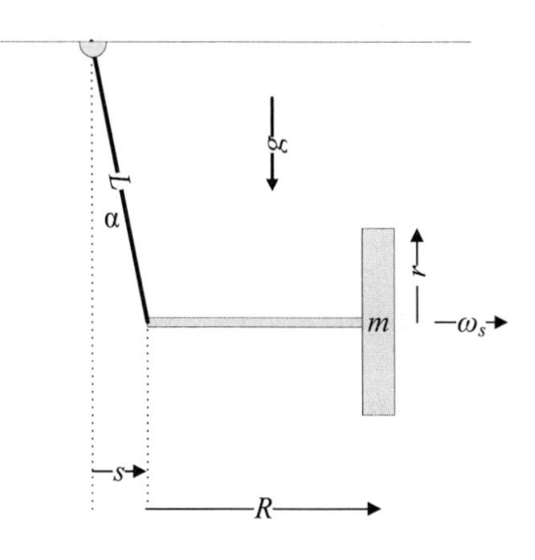

38. a) What force is needed and how much work is done in lifting a box of mass M a height h. b) Suppose one used a lever as illustrated below, what force and work are needed?

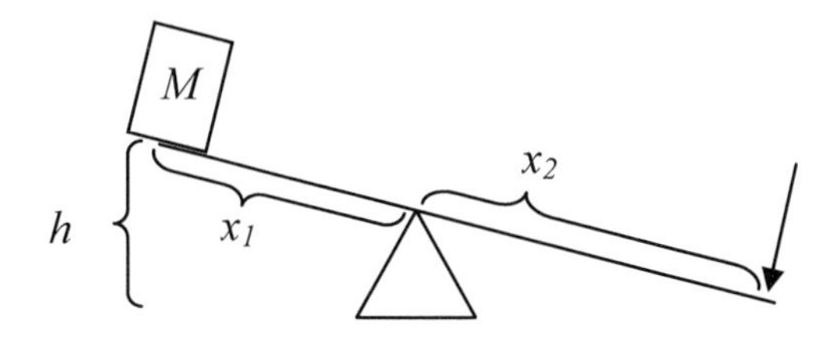

(Problems continued on next page)

39. Newton's second law passed another test [30]. In the test, a new limiting law to replace Newton's law, $F \sim a^2$ was proposed. In particular, $F = m\vec{a}\mu(a/a_0)$ where $\mu\left(a/a_0\right) = \frac{a/a_0}{\sqrt{1+\left(\frac{a}{a_0}\right)^2}}$. This law is referred to as MOND (the *m*odified *N*ewtonian *d*ynamics proposed by Milgorm).

The reason for such a proposal relates to Newton's second law implying unseen mass to account for the way materials rotate in galaxies. For galactic rotation curves to work out right, and thus avoid the need for unseen mass, the new formula needs: $a_0 \sim 1.2x10^{-10} m/s^2$. (NB: the natural scale for Hubble acceleration is $a_H = cH \sim 7\times10^{-10} m/s^2$).

Contrary to this hypothesis, the experimental test found "good agreement with Newton's law at accelerations as small as $5\times10^{-14} m/s^2$. The experiment used a torsion pendulum.

In simplest terms, a torsion pendulum is a strand of elastic material with a mass hung on its end under the influence of gravity. A slight twist of this stiff "string" along its long axis induces a restoring force proportional to the angle through which it is displaced. Thus, one gets an angular oscillation about the length of the "string."

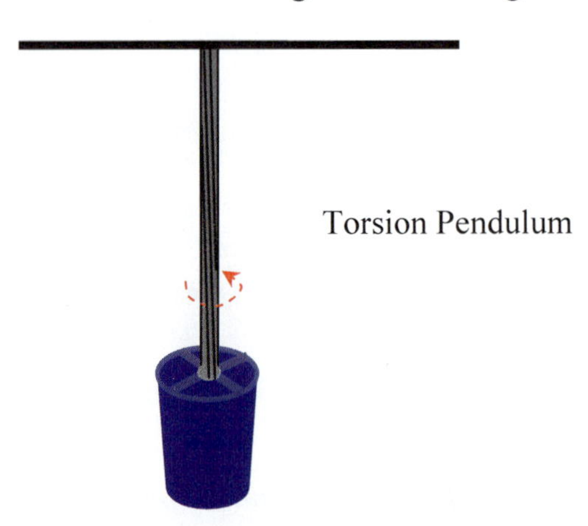

Show that, for a small enough amplitude that, if Newton's law is true (as opposed to MOND), the frequency should not depend on amplitude and give the form of the equations of motion for the MOND law as one approaches a_0.

Note: do not solve the problem exactly, use the fact that the angle of the torsion displacement is small and assume that $a >> a_0$, so that we approach the possible violation of Newton's second law. For simplicity assume a hollow cylindrical torsional pendulum as shown above. Also, to solve the equation approximately use a small-amplitude oscillatory test solution with frequency slightly off resonance and write a relation for the amount the frequency is detuned.

[30] See *Physical Review Letters* 98, *150801* (2007). See also *Physics Today*, June 2007, pg. 28.

40. A car of mass m moving at speed V hits an obstacle right at the edge of the frame as shown below. Assume that the car is a rectangular solid of uniform mass density and that, upon impact, it only pivots about the corner of the obstacle. If all the energy goes into tilting the car forward about the pivot, what is the maximum angle θ through which the car could tilt? Assume the car does not flip over. Also assume that the body of the car has height d, length L and width w, and that the car pivots at a height h below the horizontal plane of the center of mass of the car.

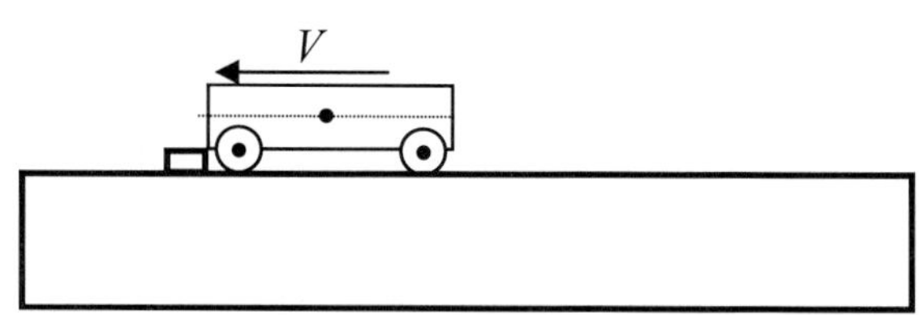

41. Show how to calculate the off-diagonal moments of inertia. Explicitly write the integral that defines the diagonal components.

42. By starting with the primary definition of kinetic energy, show that the kinetic energy of a rigid body can be spit into translational energy of the center of mass and rotational energy around the center of mass as given in equation 7.40.

43. Prove equation 7.39.

44. A yoyo of radius R with *no* initial impetus and which has mass m and moment of inertia I is allowed to fall a distance h freely under the influence of gravity while you hold its string fixed. What is the linear acceleration a, and angular acceleration α, of the yoyo as a function of time?

45. Why do we call centrifugal force, or any force that results from acceleration, a fictitious force, when such forces can, for example, kill an animal if they are high enough?

46. Prove the parallel axis theorem, equation 7.32.

Chapter VIII

Gravity

Introduction

Gravity and Impetus

In this chapter, we will concentrate on a particular type of force that acts on all massive bodies. Massive bodies, as we've seen, are bodies that have receptivity to the action of impetus; in particular, the mass of a body measures how much resistance to the action of the impetus that the body presents. Obviously, such bodies must be able to have impetus activated in them; that is, they must be able to be acted on by a force. Action and reception, like top and bottom go together so there must be a force to which massive bodies are naturally receptive. The one native to massive bodies *as such* is gravity, requiring only minimum further specification. There are others such as the electrical force. Still, while every massive body attracts every other, only bodies that also have another quality called charge interact by electric force.

Mass is a quality, indeed a characteristic quality of a class of bodies. Further, it seems that the more mass a body has, the more massive parts it has.[1] This is certainly true of a group of similar substances; for example, three marbles of the same type are more massive than two. It's also true for atoms. A uranium atom is much more massive than a hydrogen one. A hydrogen atom has a single proton, uranium has 96 protons and 146 neutrons, each comparable in mass to a proton. Though electrons are much less massive than nucleons, uranium also has many more electrons. Since it's true for atoms, it's also true for all ordinary substances such as animals and plants, which are composed of atoms; note that though these atoms are now *parts* of a new substance, they still have approximately the same mass as they did when they were their own individual substances. Thus, for instance, a squirrel has fewer massive parts than an elephant and a lead ball more than a rubber one of the same volume.

So, at least to this level,--indeed it seems to be generally true,[2] until we finally get to homogeneous parts or simple particles that are not divided[3]-- we can say that the mass[4] of a composite particle is also indicative of the number of massive parts. The greater the mass of a body, the greater the number of massive parts are in the entire body. Even the truth of this simple rule, which applies for ordinary bodies that are at least as complex as the nucleus of an atom, helps reveal how mass is a characteristic quality, revealing something essential about that body as a body. The distinction between massive and non-massive bodies is one of the most important distinctions in physics.

The more massive a body is the more gravity it has. Perhaps this is not so surprising if the above analysis is correct, because more parts means more sites for activity of and receptivity to gravity.

---

[1] Recall that most of the extent (size) of ordinary bodies comes from their *non*-massive parts. For example, in the atoms that compose everyday bodies, the massive parts make up only a tiny fraction of the size of the entire atom; "space" makes up the larger portion.

[2] More generally, we note that inertial mass is a quality of a body but always in relation to the surrounding space through which it travels. The mass of a body, i.e. its receptivity to the activity of the impetus (ability of the impetus to move the body at fixed speed), cannot really be thought of without at least an implicit reference to the space through which it moves. So, we can view the mass as indicative of the given massive body's relation to space (the plana discussed later), and thus we would expect the mass to be indicative of how force fields in the plana relate back to the body. A high inertial mass, for example, would indicate high "contact" between plana and body. This line of thought leads to the study of the vacuum including the hypothetical Higgs fields.

[3] Note, however, at one level, there is no apparent necessity to the rule: more mass, more massive parts. Obviously, the mass of a simple particle or a simple part of a complex particle is obviously not determined by the number of its actual parts since, being simple, it has none. Rather, its mass is determined by the qualitative state of the part. Furthermore, there seems to be no reason why, *in principle*, the *simplest* massive parts of one thing must have the same mass as those of another. Thus, *if it is possible*, a complex body with a few very heavy simple parts could have much more mass than one with many very light simple parts.

[4] In the language we begin to develop in Chapter 10, the component of the resistance to impetus that a body starts with at rest is called its rest mass. Analogical generalizations will be needed as we move to quantum mechanics and special relativity.

Newton showed that, assuming the gravitational force law, $\vec{F}_{gravity} = -\frac{G m_1 m_2}{r^2}\hat{r}$, where m_1 is say the mass of the sun and m_2 is the mass of a planet and $\vec{r}$ is the vector between them, one could explain the speed and orbits of the planets as they appear in the night sky.

Johannes Kepler, who gave mathematical form to those speeds and orbits, and Tycho Brahe (see history box), an extremely talented observer of the sky who provided the data for that analysis, were two of Newton's immediate predecessors. They and their contemporaries, as previously mentioned, ultimately trace their roots to the work of the men of the Middle Ages who, while building the science bequeathed them from Greece, successfully enculturated the respect for reason and for its power to understand the world. Again, a key fruit of that medieval work was the discovery of impetus, especially as articulated by John Buridan, who even noted that "gravity" increases impetus in falling bodies.[5]

In ordinary experience, we are somewhat familiar with the connection between mass and gravity. In particular, just from everyday experience, we know massive things, that is things that are generally hard to get moving, are also heavy, i.e., hard to hold above the ground. Indeed, this fact seems so clear that it can lead to confusion between these two aspects of massive bodies. ***Weight*** is the result of the force of gravity acting on a massive body, whereas ***mass***, in its first meaning, is the resistance of the body to the action of the impetus.

Newton's Law of Gravity

The Earth pulls on the body, but the body also pulls with equal force on the Earth. The more mass the more the body pulls. Again, mass is related to, not the same as, weight. Indeed, the word mass is used for three different aspects of a body. We can understand these aspects better by understanding Newton's equation for gravitational force. We rewrite it as follows, and apply Newton's second law to get:

8.1 $$\vec{F}_{gravity} = -G\frac{m_{A\_active}\, m_{B\_passive}}{r^2}\hat{r} = m_{B\_inertial}\,\ddot{\vec{r}}$$

Here $\vec{r}$ is a position vector pointing from one body, m_{A_active}, to another body, $m_{B\_passive}$, so that r is the distance between the two bodies. The first body generates an attraction on the second and vice-versa. This equation needs explaining in some detail.

For concreteness, consider a ball pulled by gravity to the Earth. Why does the ball fall to the Earth? The Earth has a power to cause an activity in space that is propagated through space to the ball, which in turn can activate impetus in the ball, i.e. give momentum to the ball. We call this power that permeates the region around a massive body the gravitational field. The receptivity of the ball to the action of the field reminds us that action and reception come together, and thus that we need both in the equation to specify the result of the action of a force (though sometimes only implicitly), namely a change of impetus which results in a change of velocity ($\ddot{\vec{r}}$). The deeper aspects of the nature of the

---

[5] Medieval and Early Modern Science Volume II, A.C. Crombie, Doubleday, 1959, pg 69. See also Physical Science in the Middle Ages, E. Grant, Cambridge University press, 1977, pg 53.

body, or of the complex of causes under consideration, sometimes allow us to include the effect of multiple aspects in a single measurement, and thus as a single variable in our equations. Gravity is an interesting such case as we will see shortly when we discuss the equivalence principle. [6]

Now, it takes time for the gravitational field to propagate to a given region around the Earth. If the Earth just came into existence, it would take, according to modern physics, $\frac{h}{c}$ seconds for the power of gravity to propagate to a height h above the Earth (see Figure 8-1 showing the field penetrating partially into space around Earth). Note that space is not simply "empty," there is now "gravity" in it. Gravity is a power (capacity) to activate impetus, a quality, which means it must be a quality of something. We say gravity is a property, in the category of quality, of "space." However, what we mean by "space" in the fully physical context is not the mathematical concept of space; that is, what we get when we leave behind all but extension, for extension cannot exist abstractly like that except in the mind. To make this distinction clear, we will use a different word for what we mean by the "space" here. That is, will use the word *plana* for that which gravity is a power of. *Plana* comes from the Latin word for field. Plana has extension, but it is not simply extension.

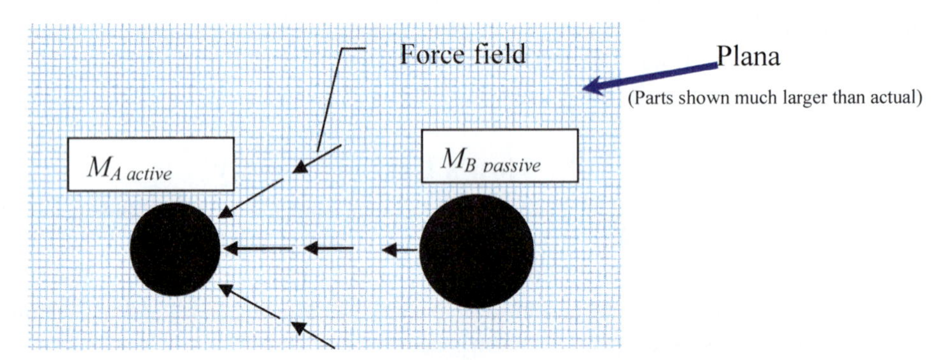

Figure 8-1: Body A acts on body B by acting through the space, which we also call the plana, Latin for field, because it receives the action of the field. The magnitude of the field (a quality capable of activating impetus in a body) generated by body A is proportional to M_{A_active} while body B's receptivity to the action of the field is proportional to $M_{B\_passive}$. In reality, it takes time for the field to propagate from body A to body B.

Notice we will sometimes still use the concept of space in its being of reason form, i.e. as if it could exist without matter instead of being simply the first property of all physical things, because such a construct allows us to concentrate attention on the extension itself and quantities that we relate to it. Like all such mental constructs, we will not only continue to use it, but will continue to make it more elaborate; again, as long as we don't forget what we are doing, this approach is a good and often even essential tool. In this case, we should not forget that properties (e.g., the power of gravity) can only exist in substances (e.g., plana).

[6] Also, the receptivity of "space" to have a field activated in it has been left implicit in the above equations, but obviously it does have such receptivity.

Finally, referring back to Figure 8-1 above, we can complete our explanation of physical meaning of equation 8.1. The field at some time in the past propagated from the Earth's surface (e.g., body *A*) into the adjacent part of the plana and then from one part of the plana to the next till it finally reached a region of height *h* above the ground. The power (ability to cause impetus) of that region of the plana, or "space" if you like (as long as you remember what that means), acts on the ball. The active mass, M_{A_active}, is a measure of the body's ability to produce a power of a certain strength quality (field) in the plana at a given point from the body, after accounting for the diminution that comes with traveling from the source to the given point. That power, i.e. that capacity of the plana to cause impetus, is called the gravitational field $\vec{g}$ of the Earth (body *A*) at that point. How receptive the body is to the action of the gravitational field is called the passive mass, $M_{A\_passive}$. Though not shown in the figure, body *B also* continually generates a gravitational field in the plana which acts on body *A*, so that we also have the same type of measures of the complex action and receptions of the plana and bodies for this reciprocal action; these are denoted by $M_{B\_active}, M_{B\_passive}, M_{B\_inertial}$.

Notice how different the new uses of the term mass are from the way we've used the term up till now. Up till now, we've meant *inertial* mass, the resistance to the action of the impetus. These various uses of the term mass point us back to the fact the inertial mass is a characteristic property revealing something deep about the nature of the body.

The depth is seen when we realize that the more inertial mass (the measure of the resistance to the action of the impetus) the more the active gravitational (the measure of the ability to activate the plana and to then activate impetus in another body) and passive gravitational mass (the measure of the ability of the body to receive the action of gravity from the plana). In fact, it can be shown, using experimental data (cf. end of chapter problem on the Eotvos experiment), that each is proportional to the other to high degree of accuracy.[7] That is, for every known body, the active, passive, and inertial masses are all proportional: i.e.: $m_a \propto m_b \propto m_c$ with the same constant of proportionality. Furthermore, by choosing our units appropriately we can make, for any body *X*,
$m_X \equiv m_{X\_active} = m_{X\_passive} = m_{X\_inertial}$ giving:

8.2
$$-G\frac{m_A m_B}{r^2}\hat{r} = m_B\ddot{\vec{r}} \quad \Rightarrow \quad \vec{G} \equiv -G\frac{m_A}{r^2}\hat{r} = \ddot{\vec{r}}$$

We could imagine that each type of substances could have different relative magnitudes of these three types of masses. It could be that some substances are distinguished by having different proportionalities between the various types of masses. Tin, for instance, could have different forms, for instance, one which has high inertial mass and low active gravitational mass, and another the reverse. These might be different isotopes (different numbers of neutrons) or an even more subtle difference. Yet, this has *not* been found to be the case.

Again, in equation 8.2 above, we use $\vec{G}$ to designate the strength and direction of the gravitational field, i.e. the power of the gravity activated in the plana to active impetus in a body at a given place. And, in order for the field to act, i.e. apply a force to the body,

---

[7] The proportionality is even more amazing than we've discussed so far, as it even holds for more general (analogous) use of the term mass that is important in advanced physics courses.

the body must have receptivity to such action, which is measured by the passive mass of the body it acts on. However, since the three different types of mass of the body are proportional in strength and, via our definition of their measure, can be represented by the same number, we can simply use the one word "mass" to mean all three of these without reference to the distinctions between them. So, by mass, we will now mean the complex of aspects of the body we discussed above, which are indeed closely related because they spring from the nature of massive bodies. Once more, thinking of them as the same doesn't make them the same, but by a mental construct (being of reason) we can carry around the latter truth without having to focus on the distinctions that we are purposely leaving for later thought.

Principle of Equivalence

In fact, going back to focus on the distinction between active and inertial mass was crucial in leading Einstein to the theory of general relativity. The equivalence between passive and inertial mass is called the principle of equivalence.

Einstein used the example of a falling man to illustrate the point. To understand the principle of equivalence, consider first instead of a falling man, a falling group of disconnected bodies, say round balls of different mass, that amount to about the size of a man. If we assume the balls are falling in a uniform gravitational field, that is, one that would apply the same force to a test mass of mass m at every point in the given region such as shown in Figure 8-2. By test mass, we simply mean a body of some known small mass, small enough that it doesn't significantly affect the gravitational field of the Earth.

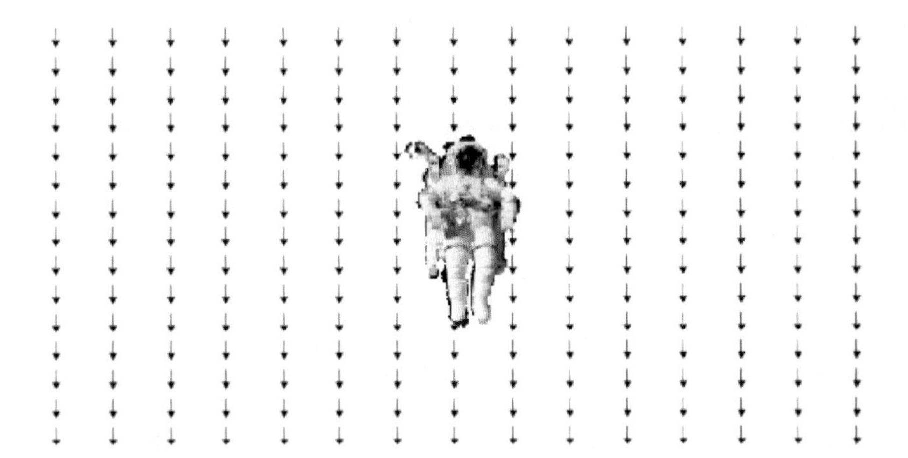

Figure 8-2: According to the equivalence principle, if one is in a frame attached to a test mass that is forever falling in a uniform gravitational field, such as illustrated above, such a frame is operationally indistinguishable from an inertial frame. Thus, all the equations of Newtonian physics apply in it as if it were an inertial frame (i.e., a uniformly moving frame with no gravitational field).

As the balls fall, the heavy ones have proportionally more force applied to them for they are more receptive to the action of the field,[8] but they also have proportionally more resistance to the action of the impetus activated by the force. For example, a ball of "mass" *2 m*, has twice the force acting on it, thus twice the *rate of increase* of intensity of impetus ($\frac{dp}{dt}$), but also twice the resistance to the action of the impetus, so that its speed *at every moment* is the same as a ball of "mass" *m*. That is, the two effects cancel as shown mathematically above in equation 8.2. Assuming all the balls start from rest, this means that they fall together in the same relative positions, not moving with respect to each other. The same would be true of the parts of the man. Neglecting air resistance as we have been, this means a man in a large free falling box would not feel any force on him. Though he's accelerating, all his parts accelerate together (see 8.1-falling man.ip). The principle of equivalence, the equivalence of the passive gravitational mass and the inertial mass, means that no part of the body is forced out of its equilibrium position. The group of bodies (or parts of the man's body) defines something very much like an inertial frame, except it is accelerating.[9]

Interactive Simulation

This is very different from the case of, say, a gentleman ice-skater applying a force to accelerate his lady partner across the ice. The force applied by the man is transmitted by pushing on the back of the woman. This force is transmitted to the front parts by compression of the parts before it. Even if the force were uniform and non-changing and were applied directly to each individual piece of the woman's body, each would be of different weight and this would mean that each part would respond with different speeds of motion. This, in turn, would lead to the parts separating from each other causing the various interconnecting parts to compress or expand, causing internal forces to be applied which would hold the body together, provided the applied force is small enough (see Interactive Physics 8.2-balls under equal force held together by damped spring.ip, 8.3-balls held together by damped spring in uniform force field.ip). As it is, the force *is* applied onto the back compressing the parts on the surface and transmitting the force to internal parts of the back. Internal forces also come into play (i.e. stresses within the body) to keep the body moving together at one speed at each moment.

The electric force is analogous to the gravitational force, but the analogy fails in so far as it clearly violates the equivalence principle. Charge generates the electric field, but there is no strong relationship between receptivity to the action of the electric field and the inertial mass, so that there is no cancellation of increase in resistance by increase in force. Mathematically, one easily sees the cancellation is absent:

8.3 $$\left|\vec{F}_{electric}\right| = \left|\frac{q_1 q_2}{r^2}\hat{r}\right| = \frac{q_1 q_2}{r^2} = m_2\,\ddot{r}$$

q_{1_active} $q_{2\_receptive}$, $m_{2\_receptive}$; r

---

[8] Again, more massive parts would explain this at some simple level, for, given equal receptivity in each part, more parts means more places to receive. Similarly, more parts mean the intensity of the impetus per unit mass is less, so that the body moves slower. Note such quantitative explanations, still require qualitative explanation which is more controlling, though we often notice and understand the latter first.

[9] To see this effect at your desk, take a rubber band (a fairly thin one is best) that tends to assume a circular shape (you might want to stretch to help its shape); when you hold it above the desk, it stretches to a long oval, but when you drop it, it assumes its natural circular shape.

Indeed, at some level, the electric force is the force driving the balls that are modeled in the *Interactive Physics* program mention above, as well as the force that keeps the parts of the human body together.

Of course, in our analysis of the falling man, we've neglected the action of the man's body on the Earth. The man also generates a gravity field in the plana that finally reaches the Earth and acts to give impetus to it at the same rate; i.e., it acts with the same force, as the Earth does on the man. The difference is that the Earth's mass is huge so that its speed does not change that much (cf. 8.4-ball falling towards earth.ip).

A problem now arises: how do we determine how fast the Earth is accelerating? What is our standard to measure from? All the massive things, that is, things of ordinary experience that we can sense will accelerate in this field. If one tries to use a very heavy body so that it will not move much in the field of the Earth, it now introduces a substantial field of its own, drastically changing the very situation you were trying to understand.

This is the problem which Einstein asked of Newtonian physics, namely where are our inertial frames, the frames that have objects moving at uniform speed that we have used so often. This led him eventually to general relativity, where local free falling frames, like those near a free falling man, substitute for Newtonian inertial frames. We will not go further into this complex but very interesting subject except to give a brief answer to where to find approximate inertial frames for the special case of an isolated body. Note that a uniform field is an idealization; no known real field is completely uniform over all distances. In the case of an isolated body, such as a planet or a solar system, for example, when one gets far enough away, the fields drop to arbitrarily small levels. Hence, one can measure with respect to objects at rest out there, where we have approximate inertial frames.

Center of Gravity

A further practical result of the equivalence principle occurs in the special case of uniform gravitational field. The center of gravitational mass is, because of the equivalence of the magnitude of the gravitational and inertial masses, the same as the center of inertial mass already defined. The equivalence makes little practical difference if the field is different at different places; however, when the field is uniform or, more accurately stated, can be approximated in some regime as uniform, the equivalence of the *center of gravity* and *center of* (inertial) *mass* points simplifies many problems, for example pendulum problems.

Key Simplifications

From this point on, having pointed out the simplification that can be made in equation 8.1 as well as its meaning, we reduce the equation for gravity to the simple form:

8.4 $$\vec{F}_{gravity} = -\frac{G\,m_1\,m_2}{r^2}\hat{r}\,,$$

Obviously, in our "everything but gravity turned off" universe, the force depends only on the masses of the bodies and the distance between them.

We also keep the simplifying approximation that the speed of gravitational field propagation is very much faster than anything else that's going on and in fact we make the mental construct that it is infinitely fast and thereby think of, for example, the Earth as acting directly on the other body and vice versa. As we saw in Chapter 4, this allows a simple interpretation of Newton's third law, for as we've seen, the force of gravity defined above works both ways yielding the same magnitude force but oppositely directed. Finally, because of this simplification, most of the time we will just treat the plana as the mental construct (being of reason) of empty Euclidean space as Newton did. This is not because the plana has no other effects but only because we are trying to concentrate on the most general properties we can without making the situation too complex. In the spirit of our approach, we switch as many things off as we can. The effects of interactions within the plana (or vacuum) are very small, so, though they might even end up being critical to the self consistency of the theory, we can safely leave them out if our goal is to elaborate only on those principles that actually make experimental predictions that we can measure at our chosen macroscopic level of accuracy and scale size. After all, reality still contains all that's needed and is consistent even when we take short cuts to understand certain aspects of it.

"Point" Particles and Central Force

To begin to apply the gravitational force law, consider two spherical shaped massive bodies. The gravitational fields of the two bodies activate the plana (vacuum) as shown in Figure 8-4. Each body acts to pull the other towards it. Mathematically, we can write the measure of the strength of the field at a point labeled by the vector $\vec{r}$ as:

8.5 $$\vec{G}_{Total} \equiv \frac{\vec{F}_{Total}}{m_{test}} = \vec{G}_1 + \vec{G}_2 = \frac{Gm_1}{(\vec{r}_1 - \vec{r})^2} \frac{\vec{r}_1 - \vec{r}}{|\vec{r}_1 - \vec{r}|} + \frac{Gm_2}{(\vec{r}_2 - \vec{r})^2} \frac{\vec{r}_2 - \vec{r}}{|\vec{r}_2 - \vec{r}|}$$

Here, $\vec{r}_1$ and $\vec{r}_2$ are the position vectors for body 1 and 2 that generate the field. Notice that, in this Newtonian approximation, the field of the two bodies interacts very simply by the addition of the measures of their field strengths. Each field attempts to generate the field strength that it would in the absence of the other, so that the field actually generated at a point is the simple linear *sum* of the measures of the strengths of two fields as they would have been generated without the other body. This ability to linearly add the field strength of each contributing body (e.g., $\vec{G}_1 + \vec{G}_2$) is known as the *principle of superposition.* It remains true, at our level of consideration, for any number of bodies, but, without benefit of experiment, one may have thought that the presence of one field would strongly retard (or enhance) the ability of other fields to be established in the plana. Experiment, however, verifies the principle of superposition, which we will use extensively.

To simplify our two sphere problem, take each body to be a distance of one unit from the origin along the x-axis, and take the center of the coordinate system to be at the origin. This means: $\vec{r}_1 = \hat{x} = -\vec{r}_2$. Furthermore, in Figure 8-3 and Figure 8-4, we consider only field points in the x-y plane, which is a two dimensional slice through the middle of the spheres.

Figure 8-3: Two spherical balls located at: $\vec{r}_1 = \hat{x}, \vec{r}_2 = -\hat{x}$. The gravitational field pulls along the directions: $\vec{r}_2 - \vec{r}$ and $\vec{r}_1 - \vec{r}$. The field produced in the plana for these balls is shown in Figure 8-4.

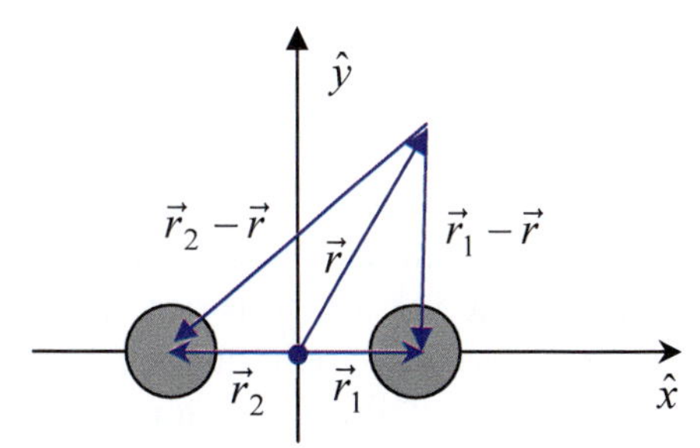

Each body acts along with the other to cause the field at any one point. So, in the middle between them, there is a point (note the simplifying idealization) at which the strength of each is such as to exactly impede the action of the power of the other. That is, there is no power available to act on a body at this point. Said another way, a body at this point will not be acted on by the gravitational force. In Figure 8-4, we mark this point of cancellation with a red dot.

To the right of the right body, the two fields both act in the same direction so they combine; they add so that the net force is bigger than it would be without the left body. Note that we are simply, following equation 8.5, summing the fields from the two bodies, because each ultimately contributes its part to a force on a body in the field. Of course, this kind of analysis could be applied to any number of bodies. Notice also that, except for ***not*** considering the field inside the radius of either body, we have taken each body as a point source.

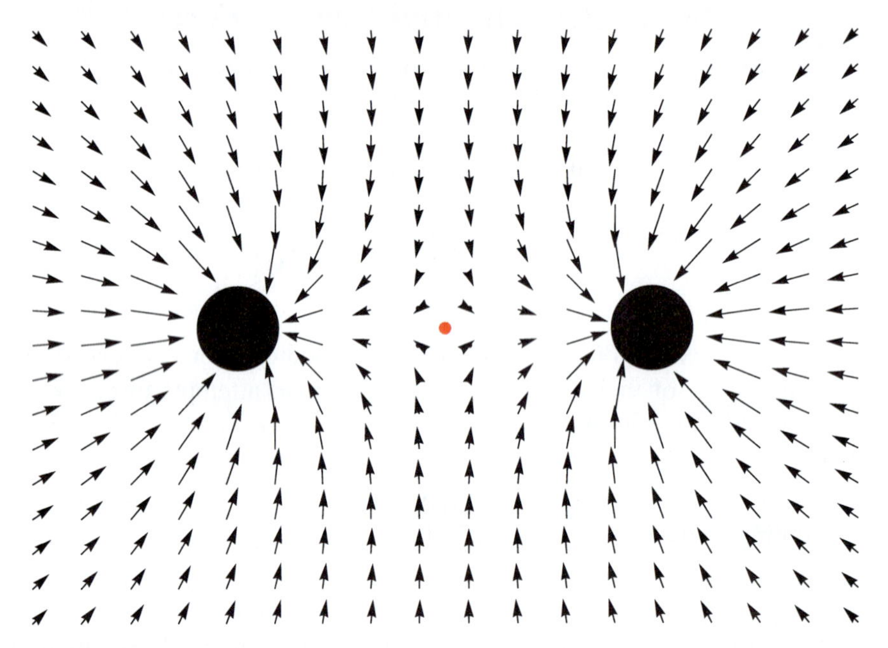

Figure 8-4: A two dimensional cross section (through the middle of the bodies) of the gravitational field due to two spherical masses. Note the gravitational field is zero at the point between them, shown by a red dot. Here the bodies are taken as point sources with the fields terminated at the surface of the body.

Indeed, the force law assumes a point from which to measure the displacement, $\vec{r}$. The correct and simplest way to interpret the force law is to consider it to refer to point particles, i.e. we take the bodies it references to be as small as we please with a certain fixed mass. Completing the limit towards an infinitesimally small body, we come up with a mental construct that is very helpful, "the point mass."

We can use the gravitational force law for extended bodies by integrating over the body. That is, we consider the body to be composed of small parts each of which contributes to the field at the point at which we would like to know the field; then, using the tools of calculus, we sum over all the contributions and take to the infinitesimal limit to get the field caused by the body at the chosen point. For a spherical body of uniform density, we can show that the net effect of the action of the body on the plana is the same as if the body were all condensed at its center point. That is, the mental construct of the point mass works for the *effect* of the field of a spherical body outside of its radius, and we can thus use it in the way already described above. This is more helpful than might be guessed since spheres are common in nature dominated by gravity, because objects that pull together naturally tend to form spherical shapes (it's a lower energy configuration). Because of this effect, for example, the sun and the planets are spherical to a good approximation.

Statics

The analysis that shows the spherical mass acts like a point mass is an example of *gravitational statics*, i.e. study of non-moving objects subject to gravity. Note that Gravitational statics, of necessity, involves a force (or forces) which counteracts gravity and so *is not only gravitational.* We will not discuss these other forces here, prominent among them the electric and magnetic forces, because detailed analysis of them falls outside the domain of Newtonian physics, but they are *implicitly* included otherwise the bodies would cave under the pressure of their own gravitational fields.

Dumbbell and Cylindrical Shell

To understand in detail the gravity field of spherical mass, we assume, as stated above, that it is composed of very many small parts, parts that are as small as we please, and, then, we sum over the contribution of each part. We will do this in steps. We start with the simpler configuration of a dumbbell with small masses (consider them point masses) at the ends as shown in Figure 8-5a and b below.

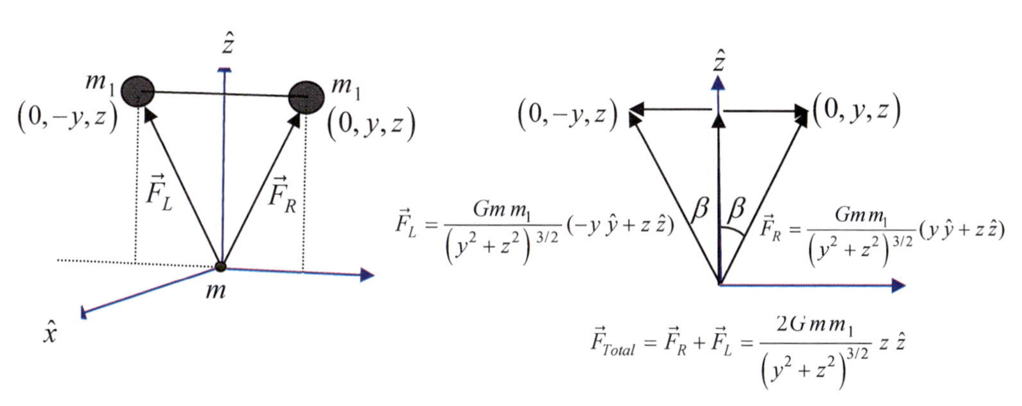

Figure 8-5: a. *(left)* Dumbbell attracts test body of mass, *m*. **b.** *(right)* Vector diagram and equations for calculating net force due to point masses at labeled locations.

The above figures show the equations for the attraction of the test body of mass, *m*, at the origin to each of the two masses of a dumbbell. The net force $\vec{F}_{Total}$ is also given there.

Next, we integrate over a uniform density cylindrical *shell* of very small thickness *t* and radius r_0, and height *2h*, in cylindrical coordinates, and calculate the net force that appears at a point outside the shell along the axis of the shell. Taking the origin as the viewpoint (the place we measure from) and $\vec{d}$ as the position vector of the center of the cylinder, we note that we can, by increasing or decreasing the magnitude of $\vec{d}$, d, change our viewpoint relative to the cylindrical shell, for example, moving inside or outside of it. Thus, the condition of being outside the shell can be written: $\vec{d} = d\,\hat{z}$ *where* $d > h$ with the relevant orientation and parameters defined in Figure 8-6 below.

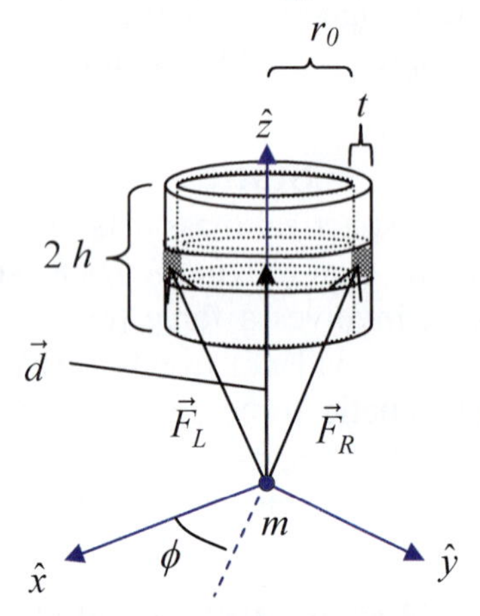

Figure 8-6: Calculation of the gravitational force of a hollow cylindrical shell on a body of mass, m, at the origin.

8.6
$$dF_z = \frac{2Gm\,z}{\left(r_0^2 + z^2\right)^{3/2}} \overbrace{\rho\, r_0\, d\phi\, dz\, t}^{dm}$$

$$F_z = \int_{d-h}^{d+h} \int_0^{\pi} \frac{2Gm\,z}{\left(r_0^2 + z^2\right)^{3/2}} \overbrace{\rho\, r_0\, d\phi\, dz\, t}^{dm}$$

8.7
$$= 2\pi\,\rho\, t\, r_0\, G m \left(\frac{1}{\sqrt{r_0^{\,2} + (d-h)^2}} - \frac{1}{\sqrt{r_0^{\,2} + (d+h)^2}} \right)$$

$$= \frac{G m M_{cylinder}}{2h} \left(\frac{1}{\sqrt{r_0^{\,2} + (d-h)^2}} - \frac{1}{\sqrt{r_0^{\,2} + (d+h)^2}} \right)$$

Where $M_{cylinder} = \rho\left(2\pi r_0 t\right) 2h$, for a shell of infinitesimal thickness t.

Spherical Mass

Now, assuming we have a ball of uniform density, ρ, we can calculate the force from the ball in the following way.

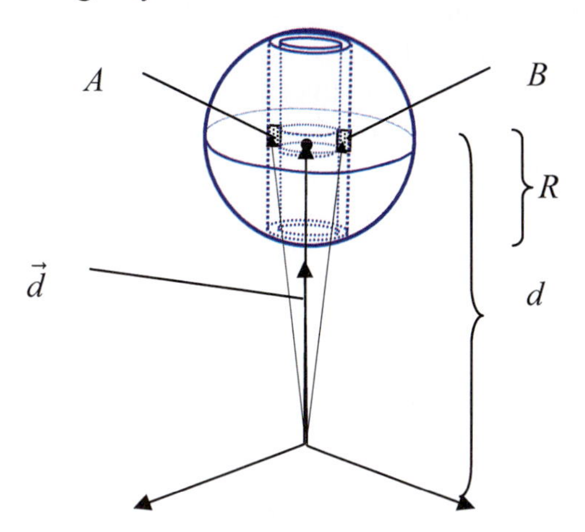

Figure 8-7: Drawing for calculating the gravitational force that a solid uniform density sphere exerts on a separate massive body at the origin.

As shown in Figure 8-7 above, we slice the ball into infinitesimal concentric cylinders and add up all the forces due to each such slice. By cutting the sphere into hollow cylindrical slices, we manifest the symmetry of the sphere that makes the field components of the parts sum as if they were simply emanating from along the axis through the center. Note the two forces coming from small regions *A* and *B*, obtained by slicing the hollow cylinder like a salami, cancel in the direction perpendicular to $\vec{d}$, the position vector of the viewpoint at which we wish to know the field. Adding up the cylindrical shells yields a net force (again directed only along the *z*-axis) of strength given by:

8.8
$$F^{ALL}{}_z = \sum_{all\ cylinders} F_z = 2\pi G m\rho \int_0^R \int_{d-\sqrt{R^2-r^2}}^{d+\sqrt{R^2-r^2}} \frac{r\ z}{\left(r^2+z^2\right)^{3/2}}\, dz\, dr$$

We have already done the integral over *z* in equation 8.7 except now $h = \sqrt{R^2 - r^2}$, $t \to dr$ and $r_0 \to r$ giving:

8.9
$$F^{All}{}_z = 2\pi G m\rho \int_0^R \frac{1}{\sqrt{d^2 + R^2 - 2d\sqrt{R^2 - r^2}}} - \frac{1}{\sqrt{d^2 + R^2 + 2d\sqrt{R^2 - r^2}}}\, r\, dr$$

Substituting $u^2 = R^2 - r^2$ gives: $rdr = -udu$ and then making the appropriate substitutions for each term in this remaining integral (exercise) gives the sought after result that the sphere acts as a point mass a distance *"d"* away:

8.10
$$F^{All}{}_z = \frac{G m}{d^2} \cdot \frac{4\pi \rho R^3}{3} = \frac{G m M_{sphere}}{d^2}$$

Notice that we have lost no generality by specifying the z-direction in the way that we have. Because of the symmetry of the sphere, no matter what point we pick in space for our origin, we will see the same situation as we describe above. Hence, we can always translate and rotate the given coordinate system so that the above relation is true for the point of measurement under consideration.

Using the result given in equation 8.7 for a cylindrical shell (take: $d = 0$), we can easily see that the field at the *center* of a cylindrical shell is zero. Again, since the cylinder pulls equally in all radial directions, only the z-direction remains as a possibility. Now, viewed from the center of the cylindrical shell, the shell can be thought of as consisting of two cylindrical shells of height h, one above the center and one below it. The *top cylindrical shell* causes an upward force on the test particle of mass m, the magnitude of which can be found by taking $h \to \frac{h}{2}$ and *then* $d = \frac{h}{2}$ in equation 8.7. Obviously, the force downward caused by the *bottom shell* is the same magnitude but opposite direction, so they cancel.

Likewise, the *center* of a *uniform solid cylinder*, being the result of adding many such concentric shells that contribute nothing to the field at the center, has no net gravitational force.

The fact that the field is zero at the center of uniform density spherical shell can be argued similarly. However, the spherical shell is more interesting; it has no gravitational field *anywhere* inside. We now show this in some detail, because understanding and following these types of calculations reveal much of the first property of physical things, which is their quantity. It is thus not surprising that both understanding and the ability to work such problems are important for progressing to more advanced physics.

It is clear that the center of the spherical shell has no net force, being attracted equally on all sides. However, this cancellation is not immediately obvious for an off-center viewpoint. It is now becomes clear why the choice of cylindrical coordinates for this problem is appropriate, for the only remaining obvious symmetry for an off-center viewpoint is the cylindrical symmetry around the axis through the *center* and the *viewpoint*. There is a hidden symmetry due to the $\frac{1}{r^2}$ fall off of the gravitational field that will couple with the spherical symmetry to make the appropriate cancellations, yielding a net zero force.

We show the off-set viewpoint in Figure 8-8a and b below; again, the viewpoint is offset, but inside the sphere. Notice the vector $\vec{d}$ no longer marks the center, but the location of our viewpoint. Remember we are dealing with a *hollow shell* not a solid sphere. The diagram carries the bulk of the needed information.

We can convey this diagram, at the risk of making you hungry, in the following way. Cutting perpendicular to any diameter, we first cut a slice, dz, of a solid sphere, yielding a pizza shaped piece of the sphere (see top view). We then imagine cutting this pizza into triangular slices in the usual way. Finally we consider *just* the crust, a ring, which has thickness $b / \sin\theta$, since we actually have a hollow sphere to start with and the center of the pizza, sadly, in our analogy is not present.

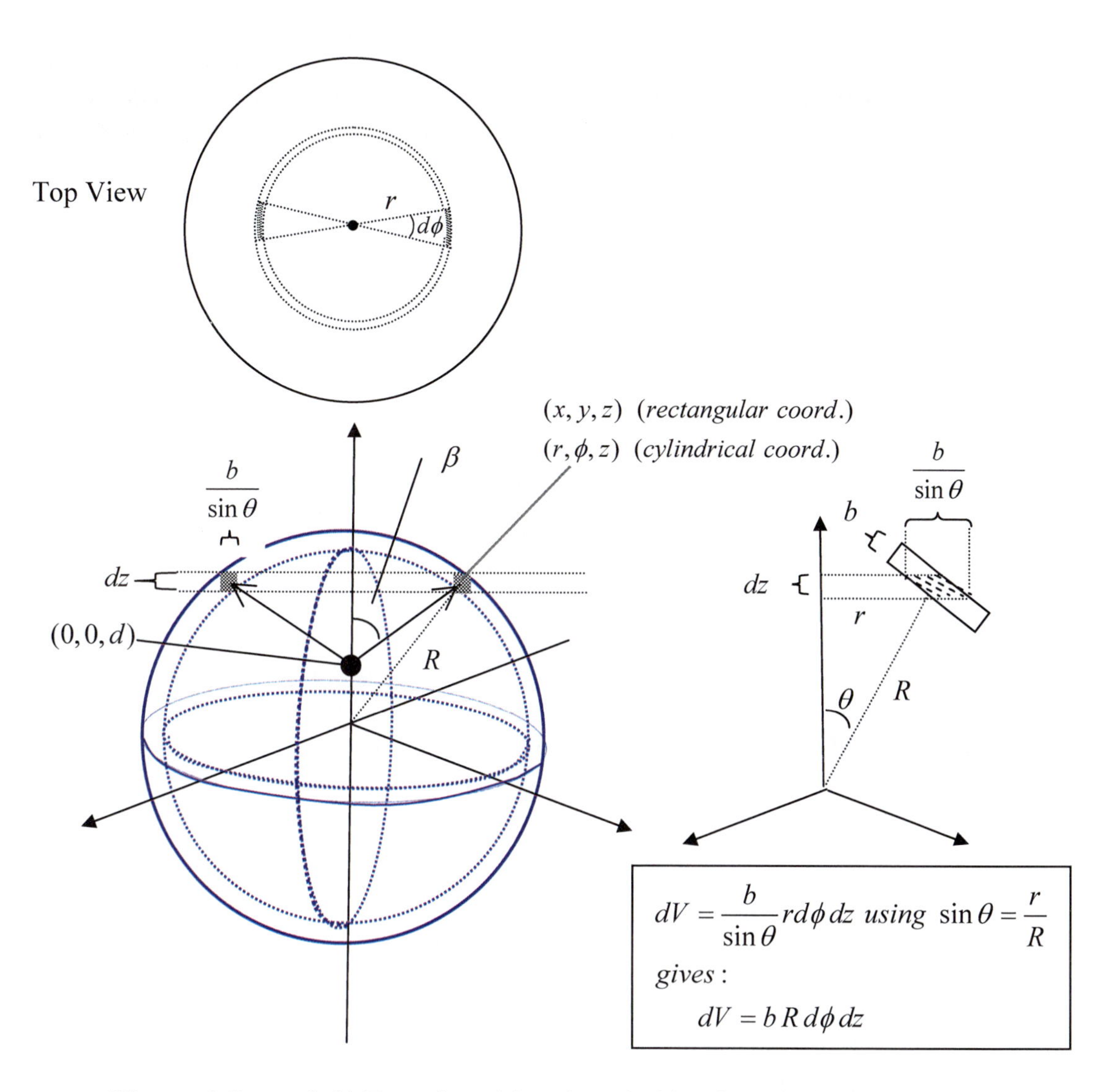

Figure 8-8: a. *(left)* Top view (above) and side view (below) of hollow spherical shell with a wedge cut out by dz and $d\phi$ action.. We calculate the field at the point shown as a black dot offset from the center. In particular, our viewpoint is at: $(0,0,d)$. **b.** *(right)* Blow-up of cross section of piece of ring cut out of the spherical shell.

We can check the analysis of the differential element given in Figure 8-8b by calculating the volume using it.

8.11 $$V = 2\int_{-R}^{R}\int_{0}^{\pi} b\, R d\phi dz = 4\pi R^2 b$$

Note that since we view from inside the sphere, $\left|\vec{d}\right| \equiv d < R$. To calculate the total force acting upward caused by the part of the sphere above $z = d$, shown in Figure 8-9, we get (not forgetting that r is a function of z):[10]

$$\begin{aligned}
F_{top} &= Gm\,2\pi R\rho\, b\int_d^R \frac{1}{r^2+(z-d)^2}\cos\beta\, dz \\
&= Gm\,2\pi R\rho\, b\int_d^R \frac{1}{r^2+(z-d)^2}\frac{z-d}{\sqrt{r^2+(z-d)^2}}\, dz \\
&= Gm\,2\pi R\rho\, b\int_d^R \frac{z-d}{\left(r^2+(z-d)^2\right)^{3/2}}\, dz \quad \text{using } r^2+z^2=R^2 \quad gives: \qquad (8.12)\\
&= Gm\,2\pi R\rho\, b\int_d^R \frac{z-d}{\left(R^2-2d\,z+d^2\right)^{3/2}}\, dz \\
&= \frac{GmM}{2d^2}\left(1-\sqrt{1-\left(\frac{d}{R}\right)^2}\right)
\end{aligned}$$

The cosine term in the first line projects out the component of the force in the z-direction. Note that, in our previous examples, the vector notation automatically introduced this term (it's what's responsible for the $3/2$ exponent).

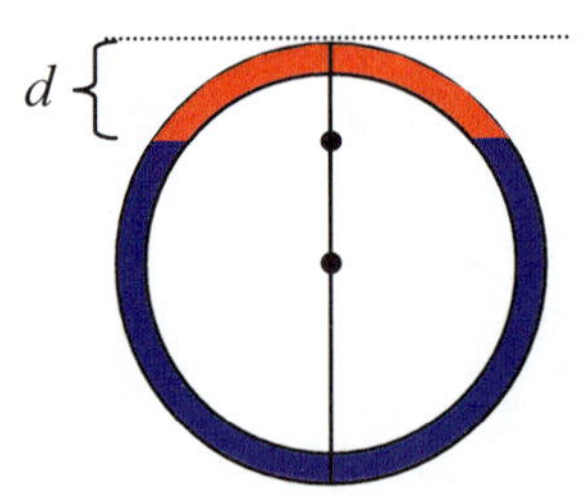

Figure 8-9: Cross section of the spherical mass shell that includes the viewpoint and the center of the shell. The portions that pull on the test particle upward are shown in blue and those portions that pull downward are shown in red.

One can show, by integrating z from $-R$ to d (see end of chapter problem), that the net force due to the bottom portion of the sphere (*blue* in Figure 8-9 above) is exactly the same magnitude as that due to the top (*red*) but oppositely directed. Thus, the two effects cancel out anywhere along the axis, implying no field at those points. Even more, because of the symmetry of the sphere, any point in the sphere can be made part of an axis through a diameter of the sphere. Hence, the argument is true of any point in the sphere, proving the field is zero, i.e. there is no field, through the inside of spherical shell.

---

[10] Note that one must be careful with which signs of the square roots one chooses. Using *Mathematica* to do the calculation without sufficient care, for example, can result in missing the correct sign at one point, thus resulting in an incorrect answer to the integral.

These sorts of symmetry insights are very important. They appear to simply be "tricks" because they short-circuit lots of work, but in fact, they are insights into the essential aspects of the reality relevant to the problem. In using them, one is just thinking more and doing less "turning the crank" of the algebra and calculus that carries so much of the information for us hidden within formalism.

In *all* the above example calculations, including the one in which we show that spherical bodies can be treated as if all the mass was acting from a point, we decompose the given body into arbitrarily small regions that act as point masses. Point mass idealizations are crucial in empiriometric physics.

Dynamics: Reduced Mass

Armed with our knowledge of point particles at rest, we now consider point particles in motion. At first, we see what we can say without the detail of the inverse square law.

Thus, consider an interaction due to a central (no angular dependence) force of unspecified radial dependence acting between particles: $\vec{F} = F(r)\hat{r}$, where $\hat{r}$ indicates the direction *from* the field generating particle, say *particle* 2, *to particle* 1 on which it acts. That is, *particle* 2 generates a field in the plana that exerts a repelling force of magnitude $F(r)$ (if $F(r)$ is negative it is attractive) on *particle* 1 and vice-versa. The equations that describe motion the interaction of two particles under the influence of such a force are quickly written in cylindrical coordinates using Newton's second law and Figure 8-10:

8.13
$$m_1 \frac{d^2\vec{r}_1}{dt^2} = -F(r)\hat{r}$$
$$m_2 \frac{d^2\vec{r}_2}{dt^2} = F(r)\hat{r}$$

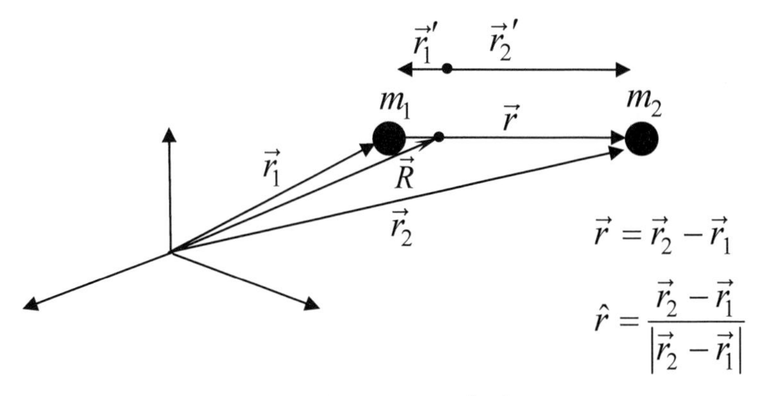

Figure 8-10: Position vectors in lab frame, $\vec{r}_1, \vec{r}_2$ and center of mass frame, $(\vec{r}\,'_1, \vec{r}\,'_2)$ for two bodies of mass m_1 and m_2 respectively.

8.14 $$\vec{R} = \frac{m_1 \vec{r}_1 + m_2 \vec{r}_2}{m_1 + m_2}$$

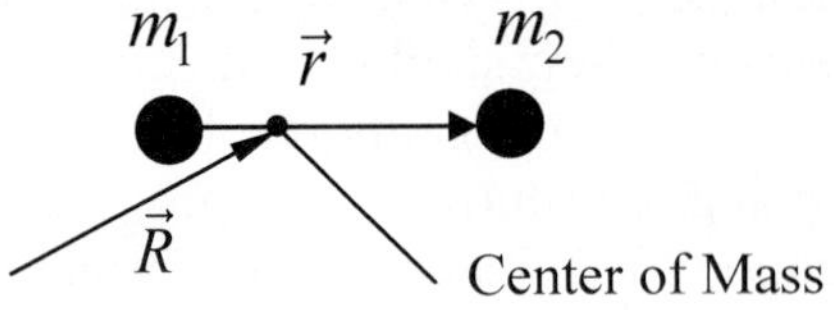

m_1 $\vec{r}$ m_2

$\vec{R}$ Center of Mass

Notice the definition of $\hat{r}$ used here: $\hat{r} = \dfrac{\vec{r}_2 - \vec{r}_1}{\left|\vec{r}_2 - \vec{r}_1\right|}$; for instance, taking *particle* 2 at the origin, gives: $\hat{r} = -\hat{r}_1$.

To express everything in terms of $\vec{r}$, we use equations 8.13 to write:

8.15 $$\begin{aligned} \ddot{\vec{r}} = \ddot{\vec{r}}_2 - \ddot{\vec{r}}_1 &= F(r)\left(\frac{1}{m_1} + \frac{1}{m_2}\right)\hat{r} \\ &= F(r)\left(\frac{m_2 + m_1}{m_1 m_2}\right)\hat{r} \end{aligned}$$

This gives a very simple equation for the motion:

8.16 $$\mu\,\ddot{\vec{r}} = \mu\left(\ddot{\vec{r}}_2 - \ddot{\vec{r}}_1\right) = F(r)\hat{r} \quad \text{where } \mu = \frac{m_1 m_2}{m_1 + m_2}$$

Note that this is also exactly the equation of motion for a single particle of mass μ traveling along a line under the influence of a force of magnitude *F(r)* in the $\hat{r}$ direction as illustrated in Figure 8-11. A little further thought reveals the origin of this simple model of the behavior of the two body central force interaction.

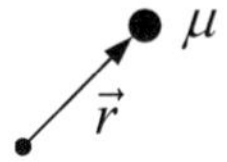

Figure 8-11: Two body central force interaction can be reduced to a problem of one body of "reduced" mass, $\mu = m_1 m_2 / (m_1 + m_2)$, located by a single position vector $\vec{r}$.

By Newton's first law, the center of mass of the two particles must move uniformly or be at rest, so the center of mass position can be ignored by moving (transforming) to the appropriate frame, say by getting in a truck moving at the right speed to be along side the particles. Now, the force is along the line between the particles, so the single direction $\hat{r}$ must be the only relevant one. By single direction, we ***do not*** mean that the motion in the inertial frame is confined to one dimension. We simply mean we are able to represent the general motion caused by this force by using just one variable or parameter (a single vector $\vec{r}$), called a single variable *parameterization* of the problem. This is an analogical use of the term variable for this "single" variable is a vector, which actually contains three variables itself (for instance, $\vec{r} = x\,\hat{x} + y\,\hat{y} + z\,\hat{z}$).[11] In summary, the forces must be equal and opposite and the relationship between the position vectors of the two particles must be functions of only the single vector $\vec{r}$ and the location of the center of mass. Hence, because we can eliminate the center of mass location, $\vec{R}$, there must be an equation such as 8.16 above that is only a function of $\vec{r}$.

To see this concretely, using equation 8.14 for $\vec{R}$ and $\vec{r} = \vec{r}_2 - \vec{r}_1$, we can write $\vec{r}_1$ and $\vec{r}_2$ in terms of $\vec{R}$ and $\vec{r}$ to get:

8.17
$$\vec{r}_1 = \vec{R} - \frac{m_2}{m_1 + m_2}\vec{r}$$
$$\vec{r}_2 = \vec{R} + \frac{m_1}{m_1 + m_2}\vec{r}$$

By substitution into the equations of motion and using the fact that there are no forces acting on the system so that $\ddot{\vec{R}} = 0$, we reproduce the one dimensional equations of motion.

Using the following equations for the center of mass coordinates, which we can read off from Figure 8-10:

8.18
$$\vec{r}_1' = \vec{r}_1 - \vec{R}$$
$$\vec{r}_2' = \vec{r}_2 - \vec{R}$$

simple substitution yields:

8.19
$$\vec{r}_1' = -\frac{m_2}{m_1 + m_2}\vec{r}$$
$$\vec{r}_2' = \frac{m_1}{m_1 + m_2}\vec{r}$$

Notice that the difference, $\vec{r}_2' - \vec{r}_1' = \vec{r}$ as clearly is seen in Figure 8-10 above.

Hence, putting the main results together, we have the equation of motion:

8.20 $$\mu\ddot{\vec{r}} = \mu\left(\ddot{\vec{r}}_1 - \ddot{\vec{r}}_2\right) = F(r)\hat{r} \text{ where } \mu = \frac{m_1 m_2}{m_1 + m_2} \text{ which we solve for } \vec{r}$$

and then we can get the equations of motion of each particle in the lab frame by substituting $\vec{r}$ into equations 8.17. μ is called the reduced mass for the obvious reason that this "parallel" combination of masses comes up in the reduced equation of motion.

---

[11] Again, such mental constructs are useful in focusing attention on a particular aspect of a problem in analogy to what we do naturally in everyday thinking.

Meaning of "Reduced" $\vec{L}$ and E

Looking at Figure 8-11, we see the natural angular momentum in the reduced problem is (using $\vec{v} = \dot{\vec{r}}$):

8.21

$$\vec{L} = \vec{r} \times \mu\vec{v} = (\vec{r}_1' - \vec{r}_2') \times \mu(\vec{v}_1' - \vec{v}_2') = \frac{m_1 m_2}{m_1 + m_2}\left(\vec{r}_1' \times \vec{v}_1' - \vec{r}_1' \times \vec{v}_2' - \vec{r}_2' \times \vec{v}_1' + \vec{r}_2' \times \vec{v}_2'\right)$$

$$= \left(m_1 - \frac{m_1^2}{m_1 + m_2}\right)\vec{r}_1' \times \vec{v}_1' + \left(m_2 - \frac{m_2^2}{m_1 + m_2}\right)\vec{r}_2' \times \vec{v}_2' - \frac{m_1 m_2}{m_1 + m_2}\left(\vec{r}_1' \times \vec{v}_2' + \vec{r}_2' \times \vec{v}_1'\right)$$

$$= m_1\vec{r}_1' \times \vec{v}_1' + m_2\vec{r}_2' \times \vec{v}_2' - \frac{m_1^2}{m_1 + m_2}\vec{r}_1' \times \vec{v}_1' - \frac{m_2^2}{m_1 + m_2}\vec{r}_2' \times \vec{v}_2' - \frac{m_1 m_2}{m_1 + m_2}\left(\vec{r}_1' \times \vec{v}_2' + \vec{r}_2' \times \vec{v}_1'\right)$$

Substitution of $\vec{r}_1' = -\frac{m_2}{m_1 + m_2}\vec{r} \quad \vec{r}_2' = \frac{m_1}{m_1 + m_2}\vec{r}$ into the last four terms shows that those four terms sum to zero, leaving:

8.22
$$\vec{L} = \vec{L}_1' + \vec{L}_2'$$

which simply says that the angular momentum naturally defined in the reduced problem is the sum of the angular momentum of each particle about the center of mass.

Similarly, for the energy, we can show (see end of chapter problem):

8.23

$$E = \frac{1}{2}\mu\vec{v} \cdot \vec{v} = \frac{1}{2}\frac{m_1 m_2}{m_1 + m_2}\left(\vec{v}_1' - \vec{v}_2'\right) \cdot \left(\vec{v}_1' - \vec{v}_2'\right) = \frac{1}{2}m_1\vec{v}_1' \cdot \vec{v}_1' + \frac{1}{2}m_2\vec{v}_2' \cdot \vec{v}_2'$$

Simplest Two Body Case

With this little information, and noting that it is common in many situations to find small mass bodies orbiting around very large mass ones (i.e. $m_1 >> m_2$), we get, in center of mass frame coordinates (i.e. coordinates natural to moving along with center of mass):

$$\vec{r}_1' = -\frac{m_2}{m_1 + m_2}\vec{r} = \frac{m_2}{m_1\left(1 + \frac{m_2}{m_1}\right)}\vec{r} \approx \frac{m_2}{m_1}\vec{r} \approx 0$$

8.24
$$\vec{r}_2' = \frac{m_1}{m_1 + m_2}\vec{r} = \frac{1}{1 + \frac{m_2}{m_1}}\vec{r} \approx \left(1 - \frac{m_2}{m_1}\right)\vec{r} \approx \vec{r}$$

$$\mu = \frac{m_1 m_2}{m_1 + m_2} = \frac{m_2}{1 + \frac{m_2}{m_1}} \approx m_2$$

(*Exercise*: show these expansions)

That is, the center of mass is approximately at m_1 and the particle m_2 moves around it. In short, we have, in the limit, a true one particle problem for m_2 under influence of the force field of m_1. This approximation is used very frequently in the study of orbits in outer space.

Gravity Assists

A simple use of this approximation is the gravity assist that is often used in getting spacecrafts to planets. NASA spacecrafts, like all crafts, are limited in how much fuel they can carry; thus, borrowing energy from (or dumping energy to) a planet can be essential to the success of a mission.

Take the case of using Venus to get a boost as shown in Figure 8-12 in the rest frame of the solar system, i.e. our "lab frame." We note that $M_{Venus} >> M_{craft}$, which gets us back, via the approximation above, to the one body problem of mass, $\mu = M_{craft}$. Consider the craft in the rest frame of the center of mass of the system. Figure 8-13a and b below show the relevant vectors. Viewing the situation in the center of mass frame allows us to easily see the relative increase in speed in the spacecraft.

In the center of mass frame, we know for a two body collision (see Chapter 5) that the speed of any one body before and after the "collision" must remain unchanged, but the direction can be different. Examination of Figure 8-12 reveals the craft must scatter in the clockwise direction shown because of the direction of gravity. Hence, by transforming back to the rest frame of the solar system (neglecting gravity of sun) as shown in the Figure 8-13b, we predict an *increase* in speed, and hence an *increase* in energy, of the craft as a result of its encounter with Venus. By similar arguments, we see from Figure 8-14a and b that the energy *decreases* for that different scenario. We can save lots of fuel by steering the craft into orbits that take advantage of these energy transfers with planets.

A very simple analogy will help in understanding the fundamental principle of gravity assists. Think of the interaction of the craft and the planet as like a collision in which we concern ourselves only with the state of the bodies before and after the interaction, not during the interaction itself. A long way before the craft's direction is reversed and a long way after the reversal occurs the interaction is negligible. With this in mind, one can compare the craft/planet interaction to throwing a super ball at a semi truck moving down the highway at some fast fixed speed. If the ball is thrown at an oncoming truck, it will bounce and return with much more speed than it was sent, while if it is thrown at a receding truck, it will return with less speed. In an end of chapter problem, you are asked to explain the gravity assist results in a physical way in some detail.

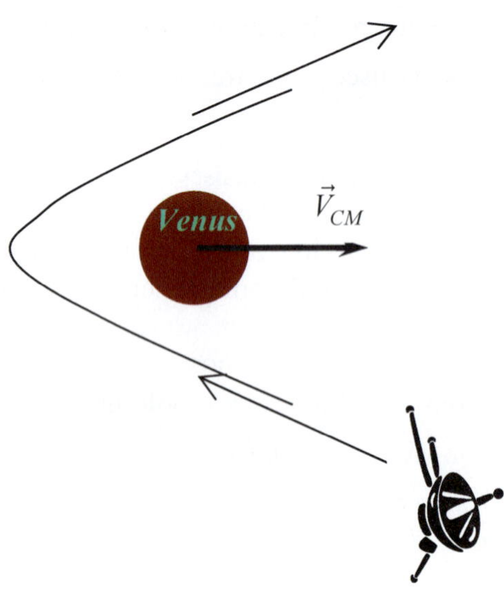

Figure 8-12: Hyperbolic orbit of spacecraft undergoing gravity assist (energy gain) from Venus.

Energy Gain Scenario

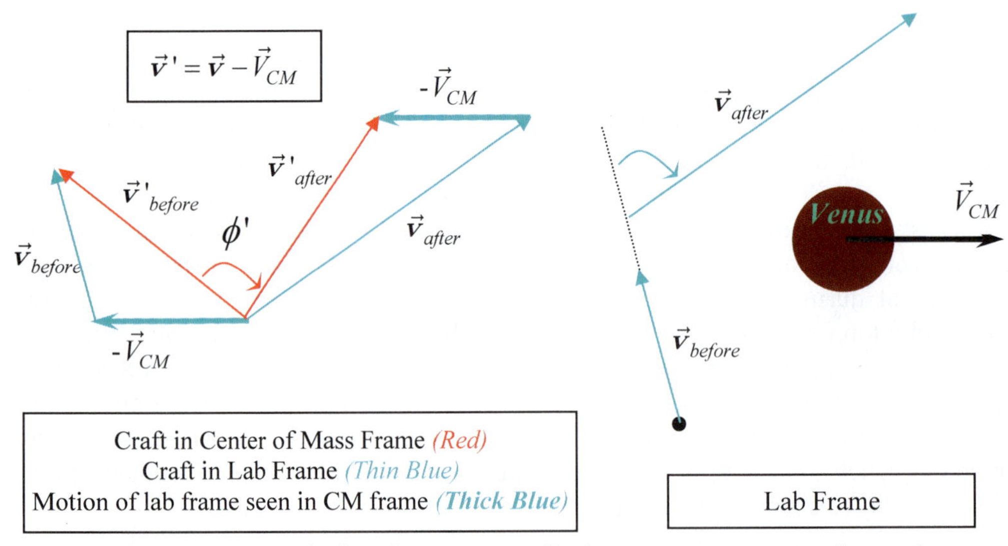

Figure 8-13: Analysis of a spacecraft that gets an energy boost by a "slingshot" effect around (behind) Venus (see Figure 8-12) **a.** *(left)* Graphical calculation of v_{after} (i.e., spacecraft speed after sling shot effect *in the lab frame*) from v'_{after} (i.e., spacecraft speed afterward *in the center of mass frame*) using $\vec{v}\,' = \vec{v} - \vec{V}_{CM}$. **b.** *(right)* The before and after velocity vectors of the spacecraft as seen in the lab frame. Note that when the spacecraft is ricocheted forward to move in the direction of Venus, as shown, it steals energy from Venus, while, in the reverse case, the spacecraft gives energy to Venus.

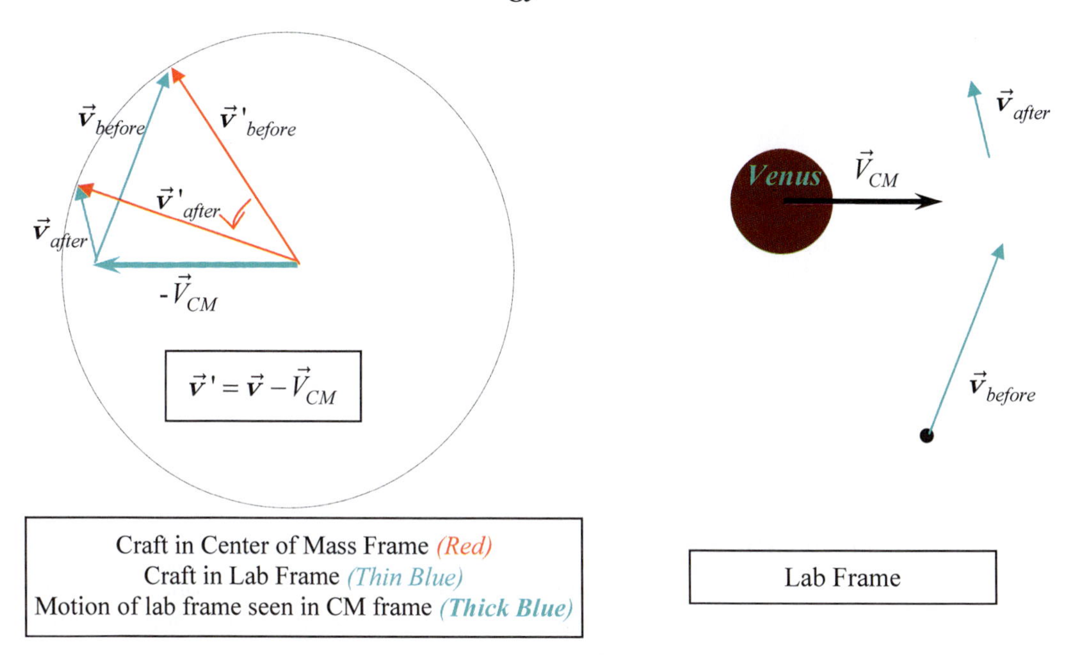

Figure 8-14: Analysis of a spacecraft that dumps energy by a "slingshot" effect around (in front) Venus (see Figure 8-12) **a.** *(left)* Graphical calculation of $\boldsymbol{v}_{after}$ (spacecraft speed after slingshot effect *in the lab frame*) from $\boldsymbol{v}'_{after}$ (spacecraft speed afterward *in the center of mass frame*) using $\vec{v}' = \vec{v} - \vec{V}_{CM}$. **b.** *(right)* The velocity vectors of the spacecraft as seen in the lab frame.

Solving the Reduced Problem

We can now attack the problem of the motion of two bodies under their own mutual gravitational interaction in its single parameter form. Using Newton's second law to write the equations in cylindrical coordinates gives the following components:[12]

$$
\begin{aligned}
r: &\quad \mu\left(\frac{d(\dot{r}\hat{r} + r\dot{\theta}\hat{\theta})}{dt}\right)_r = \mu(\ddot{r} - r\dot{\theta}^2) = F(r) \\
\theta: &\quad \mu\left(\frac{d(\dot{r}\hat{r} + r\dot{\theta}\hat{\theta})}{dt}\right)_\theta = \mu(r\ddot{\theta} + 2\dot{r}\dot{\theta}) = 0 \\
z: &\quad \mu\ddot{z} = 0
\end{aligned}
\tag{8.25}
$$

The *last* equation is easy to interpret. It means that either the body's location is fixed in the z direction, or it moves uniformly in the z direction. For convenience one can,

[12] We assume that the mass does not change.

by picking an appropriate inertial frame from which to view, take: $z=0$. Hence, ***given no external forces, the orbit that starts in the x-y plane will never leave it.***

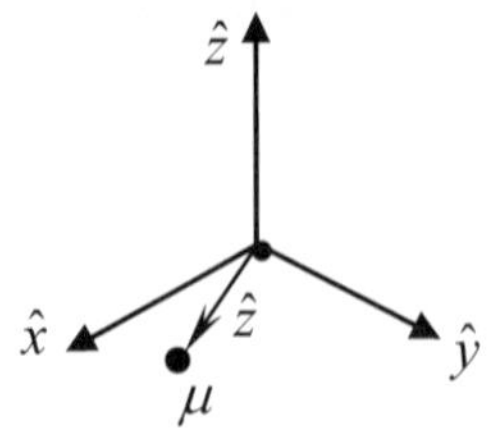

Using this fact and the definition of angular momentum about the center point of origin of the position vector $\vec{r}$ (which now lies in the *x-y* plane by definition) one gets: [13]

8.26
$$\frac{d\vec{L}}{dt}=\frac{d(\vec{r}\times\mu\vec{v})}{dt}=\frac{d(\mu r\,\mathbf{v}_\theta\hat{z})}{dt}$$

So, the second equation in 8.25 gives:

8.27
$$\left(\frac{d\vec{L}}{dt}\right)_z=\frac{d\left(\mu r^2\dot{\theta}\right)}{dt}=\mu r\left(2\dot{r}\dot{\theta}+r\ddot{\theta}\right)=0$$

Hence, ***the angular momentum,*** $\vec{L}=\mu r^2\dot{\theta}\,\hat{z}$***, never changes.***

Now, the *first* equation in 8.25 is the one that contains all the complication. To understand the equation, we integrate the equations of motion and use the work-energy theorem to write (the details are left as an end of chapter problem):

8.28
$$\begin{aligned}E&=\frac{1}{2}\mu v^2+V(r)=\frac{1}{2}\mu\dot{r}^2+\frac{1}{2}\mu r^2\dot{\theta}^2+V(r)\\&=\frac{1}{2}\mu\dot{r}^2+\underbrace{\frac{1}{2}\frac{L^2}{\mu r^2}+V(r)}_{V_{effective}}\end{aligned}$$

Notice that since $L\equiv\left|\vec{L}\right|$ is constant, the middle term can be treated as if it were a potential energy term and the object were simply moving in one dimension. Here again, we see the power of reflecting things into the mathematical domain. In this reading, the potential then is made up of two terms. Indeed, the first term can be seen to be simply the potential resulting from the centrifugal force which is an artifact of viewing things from a merry-go-round, i.e. a rotating frame, rather than an inertial one.

To see how to use this effective potential, consider the case where *V(r)* is identically zero. Concretely, consider the case of a baseball thrown at a massive catcher, such that $m_{ball} << M_{catcher}$. The ball misses the catcher's mitt (say this is his center of mass) by a distance *b*, called the impact parameter.

---

[13] We can also write: $\dot{\vec{L}}=\mu\vec{r}\times\left(\frac{d\vec{v}}{dt}\right)_\theta=0$ we leave it as an *exercise* to show this.

Figure 8-15: Catcher fielding an incoming wild pitch offset by impact parameter b from his mitt. In our scenario, the catcher doesn't even try to catch the pitch and the ball continues on by at uniform speed over his shoulder.

The angular momentum of the ball around the catcher's mitt is $\vec{L} = \vec{r} \times \vec{p}$. Mathematically, taking, for ease of calculation, $\mu = m_{ball} = \frac{1}{2}$, $L = 1$, we have $V_{eff}(r) = \frac{1}{r^2}$; this case is shown in Figure 8-16 below.

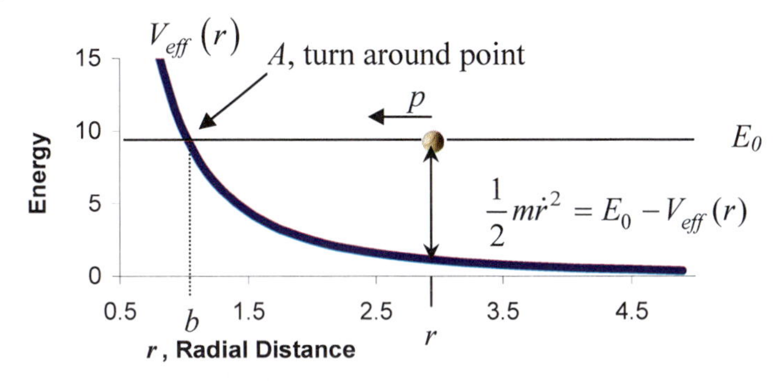

Figure 8-16: Plot of ball of momentum p moving towards the catcher. The ball approaches point A, "the turning" point, so called because if the potential were real rather than an "effective" potential, the ball would be forced to change directions at this point. In the case of the catcher, the radius decreases until the turning point where the radius begins to increase because the ball's closest approach occurs at $r = b$.

Using the equation for the kinetic energy, it is clear from the drawing that the rate of change of the radial distance is maximum far away and minimum close up. The radial distance is, by definition always positive; recall in polar coordinates, for instance, opposing directions are distinguished by a 180 degree difference in angle.

One can think of the curve in Figure 8-16 as a hill that a bicyclist is ascending, not by peddling but by sheer momentum. Namely, the bicyclist uses only previously acquired impetus, which is gradually dissipated by the action of gravity until point A is reached when all the momentum has been "converted" to potential energy. At this point the bike would begin to roll backwards down the hill, picking up momentum as it went. In this analogy, the bike/bicyclist system is like the baseball. However, the incompleteness of the analogy is revealed when we note that the ball obviously does not reverse direction when it reaches point A. Nonetheless, as in the case of the bike, the length of the radius vector from the catcher to the ball does increase, because though the ball does not reverse direction there, point A is its point of closest approach, after which, by definition, the radius vector must get bigger.

Note on Coordinates: The choice of coordinates is very important in both solving and getting insight into the import of the solution of any given physics problem. Generally, in digesting the problem to be solved, we need to decide which aspects of the given physical situation are important to the problem and which are not. Appropriately chosen coordinates are important in this regard. For example, angular coordinates can hide certain issues, such as the two different types of "turning points," but they can highlight other ones. For instance, they are needed to explain why planes look like they're going slower when they are coming at you than when they are moving perpendicular to your line of sight. (*Exercise*: show that the rate of change of the angle, θ, from which we deduce the apparent speed of the object (in the oncoming view, for instance, θ the angle subtended by a sphere of diameter L) moving at speed v, is: $\dot{\theta}_\perp = \frac{v}{r}$ for perpendicular motion and $\dot{\theta}_\parallel = \frac{L}{r}\frac{v}{r} = \left(\frac{L}{r}\right)\dot{\theta}_\perp$ for motion at or away from you. Notice this comparison is fundamentally related to the way we see the motion of an object moving towards us. It is different from the slightly more direct way that we see the motion of objects that move perpendicular to our line of sight.

Later, we will do a similar analysis with the particle in the presence of gravity's inverse square law.

For now, from the fact that E is constant, we can calculate the radius as it changes with time, at least implicitly, by solving for $\dot{r}$

8.29
$$\dot{r} = \sqrt{\frac{2}{\mu}}\sqrt{E - V_{eff}(r)}$$

and then for dt to get:

8.30
$$t = \sqrt{\frac{\mu}{2}}\int_{r(0)}^{r(t)} \frac{dr}{\sqrt{E - V_{eff}(r)}}$$

Later, we will solve for the angle dependence of the radius, $r(\theta)$, which will give the shape of the path that the particle follows, i.e. the orbital path.

Inverse Square Law

The gravitational law is central (directed towards a center point), obeys the equivalence principle, and varies as the inverse square of the distance. Before we investigate the general behavior of the bodies under this law, let us look at a few interesting cases.

Mail to "Down Under" via a Hole in the Earth

Assume its possible that we've drilled a hole down through the Earth from, say Bermuda island (Latitude 32.3N, Longitude 64.75W), which is just off the eastern coast of the US, through the center of the Earth to near Perth Australia (32S, 116E). The hole would be about 8000 miles long! See Figure 8-17 below.

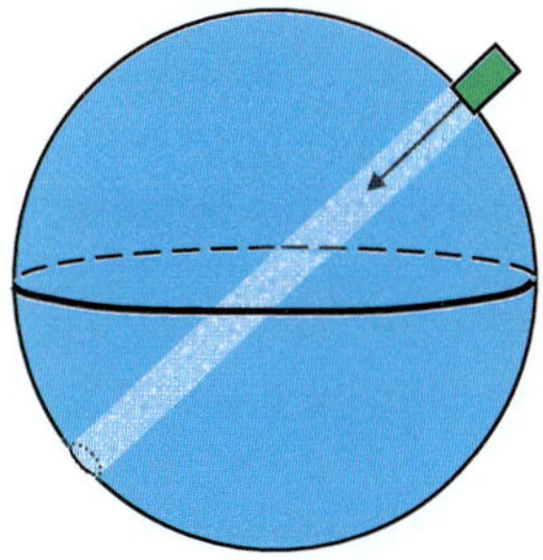

Figure 8-17: Earth with an 8000 mile "mail" hole drilled from Bermuda to Perth, Australia.

Let's say we drop a metal capsule down the hole in Bermuda while another student waits in Perth for the capsule; he hopes it will come out in the way such capsules do at bank drive-thrus. What would happen? Assuming the Earth has uniform density and that there is no friction, we can, with only a little thought, say quite a bit. As the capsule falls, it will continually gain kinetic energy as the gravity field acts to intensify its impetus, causing it to fall faster and faster. However, the field acting on the capsule decreases as the capsule approaches the center, until finally the field reverses and then gets stronger in its ability to de-intensify upward impetus as it moves away from the center. The momentum of the capsule thus reaches a maximum at the center and the capsule begins to slow down as the field acts to decrease the intensity of its impetus trying to pull it back to the center until finally there is no remaining impetus when it completes the mirror journey, arriving in Perth. Thus, the student at the other end can safely grab the capsule as it coasts to rest at about the same height above his ground as it was above ours when we released it.

How long would it take for our package to get to our friend in Perth? How fast would it be going at the center?

Before we answer, it is interesting to note that Bishop Oresme (d. 1382 AD, see history box in Chapter 2) had already discussed the possibility of dropping things through such a hole and the potential oscillation that he correctly thought would result.[14]

[14] Oresme's says: "I posit that the Earth is pierced clear through and that we can see through a great hole farther and farther right up to the other end where the antipodes [poles] would be if the whole of the Earth were inhabited; I say, first of all, that, if we dropped a stone through this hole, it would fall and pass beyond the center of the Earth, going straight on toward the other side for a certain limited distance, and that then it would turn back going beyond the center on this side of the Earth; afterward, it would fall back again, going beyond the center but not so far as before; it would go and come this way several times with a reduction of its

To answer the questions, we must first calculate how the field varies inside of a solid Earth. From our previous calculation in other sections, we already know that, at any given radius from the center, the net strength of the field caused by all the mass above the given radius is zero. That is, because the mass is spherically distributed, the fields generated by the various parts above the given radius cancel out so that there is no force applied to the capsule due to those parts. Furthermore, all the mass below that radius acts as if it were concentrated and acting from the center.

In particular, using the mass inside of the sphere of radius r, we get:

8.31 $$\vec{F} = -\frac{Gm(\frac{4}{3}\pi r^3 \rho)}{r^2}\hat{r} = -\frac{4\pi G m \rho}{3} r\hat{r}$$

Hence, we can immediately calculate the potential energy, for $r \le R_{earth} \approx 6378\,km$:

8.32 $$V(r) = -\int \vec{F} \cdot d\vec{r} = \int \frac{4\pi G m \rho}{3} r dr = \frac{2\pi G m \rho}{3} r^2$$

Note that here we have neglected the motion of the Earth (which in the exact treatment given above amounts to taking $\mu = mass\ of\ capsule$).

Now, using this expression for the potential energy, and assuming the capsule starts at rest, we get the total energy of the capsule just before release is: $E = \frac{2\pi G m \rho}{3} R_{earth}^{2}$.

This energy represents the capsule's *potential* for activity, i.e., potential for a certain rate of transfer of impetus. From this, we calculate (see problem) the maximum speed of the capsule is about 7900 *m/s* (18,000*mph*, note the speed of sound in the Earth is about 22,000*mph*, predicting about 22 minutes for any stomping noises[15] we might make to get through the Earth to Perth).

At any point during the fall we can write, using the principle of equivalence to divide out the mass:

8.33 $$\frac{E}{m} = \frac{1}{2}v^2 + \frac{2\pi G\rho}{3} r^2$$

Notice the similarity to the energy of a spring with a mass attached.
This energy equation implies, as discussed above, that the time to go from the surface to the center is:

8.34 $$T = \frac{1}{\sqrt{\frac{4\pi G\rho}{3}}} \left| \int_{R_e}^{0} \frac{1}{\sqrt{R_e^2 - r^2}} dr \right| = \frac{1}{4}\sqrt{\frac{3\pi}{G\rho}}$$

Here we have used the total energy calculated above. Notice that $\sqrt{G\rho}$ has units of frequency (1/*second*). Substituting for $\rho_{earth} = 5515\ kg/m^3$ ($\approx 5.5 gm/cm^3$ 5.5 times

---

reflex motions until finally it would come to rest at the center of the Earth...." Quoted by K. V. Magruder from *Le Livre du Ciel et du Monde* (Madison: University of Wisconsin Press, 1968), translated by Albert D.Menut, pg573.

[15] In reality, of course, damping effects would make our stomping much too weak and too much a part of the background noise to be detected. Earthquakes tremors, which cause sound in the Earth, can however be measured at great distances.

water) and using $G \approx 6.67 \times 10^{-11} \frac{m^3}{kg\, s^2}$, we calculate the time to get to the center, i.e. half the trip, as $T \approx 21min$, so that our Australian friend has to wait about 42 minutes for our capsule of mail to arrive.

A Baseball Pitched Around the World

Consider a slightly more complex motion; a baseball in orbit. Like the above, because the mass of the Earth is so much larger than the object and because the Earth is approximately a spherical mass distribution, we can take, to some good approximation, the center of the Earth to be the center of mass, and hence a fixed point in an inertial system.

Start by imagining that I hurl a baseball as fast as I can. In so doing, I generate in the baseball a certain intensity of impetus. As the ball flies, it is acted on by gravity in the plana (generated by the Earth), gravity changes the intensity of the impetus at a certain rate. As the ball falls, its momentum increases until it finally hits the ground. The resulting path of travel, shown in Figure 8-18, is an ellipse cut-off at the start and end points of the trajectory by the Earth's surface. Note the interior trajectory in Figure 8-18b is drawn assuming that all the mass of Earth is at its center.

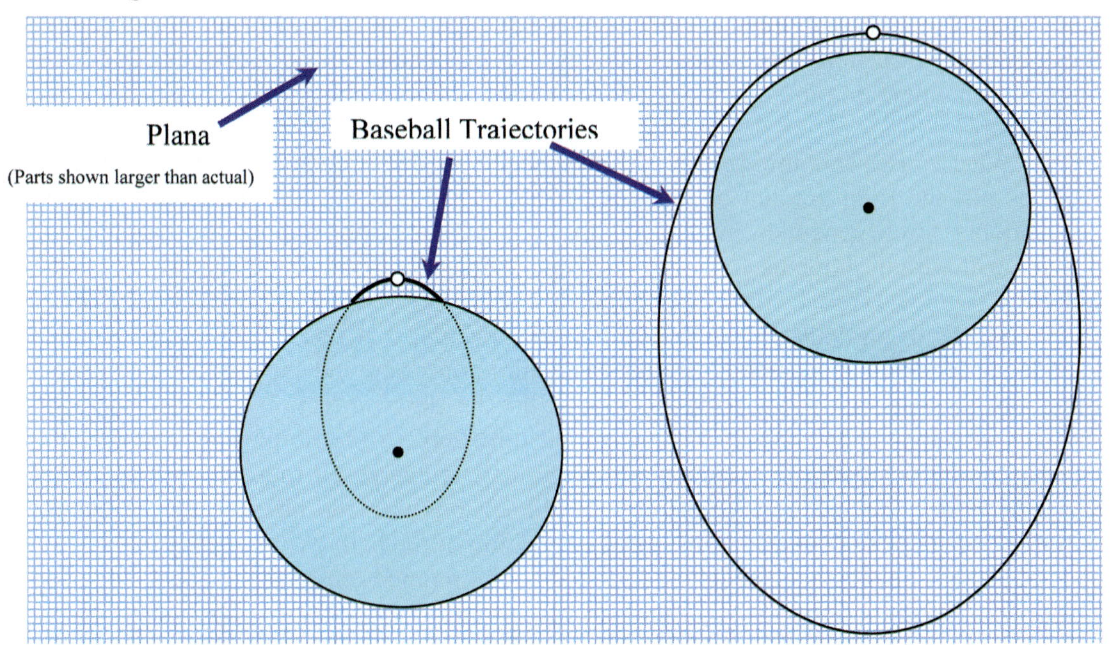

Figure 8-18: a. *Left side*: Trajectory of baseball cut-off by hitting Earth. The would-be trajectory (assuming Earth's mass is at the center) is shown as a dotted line. **b.** *Right side*: Trajectory of baseball is wide enough that it falls past the edges of the Earth.

Now, imagine that Superman hurls the ball as fast as he can. If he activates enough impetus in the ball, it will still fall but will now miss the edge of the Earth and thus continue to "fall" around the Earth. If it is thrown hard enough, it will settle into an orbit such that the perigee,[16] i.e. the closest approach to Earth, is the point of the throw and the furthest point, the apogee, is dependent on the speed of the throw.[17] If Superman had thrown the ball at the appropriate speed, it would have entered a circular orbit rather than an elliptical one. However, even if we start with an elliptical orbit, we can still make it circular.

The Circular Orbit

To see how, note that, for example, at the apogee, there is no component of the ball's velocity that is directed toward the Earth; all the velocity is tangential. We can use a rocket to increase this tangential velocity at apogee, thereby increasing the height at the perigee. What speed do we need to reach? Answering this question in a complete way takes us into an analysis of circular motion in a central gravitational field.

The speed required for a circular orbit depends on the height of the orbit. We take: ***h_{ap} = orbit height at apogee, R = radius of the Earth***; this makes the total radius from the focal point of ellipse (center of Earth) to the ball at apogee: $r_{ap} = R + h_{ap}$. Suppose we would like a circular orbit of this radius, i.e. $r = r_{ap}$. This means, as we discussed generically in Chapter 7: when the ball falls through the distance shown in Figure 8-19 (i.e., from 12 o'clock counterclockwise to 9 o'clock), it must lose all of its momentum in the horizontal direction. However, by the end of that same fall, it must gain the same amount of momentum (and thus speed) in the vertical direction that it had originally in the horizontal.[18]

What causes this interaction that then results in the circular orbit? Generally, the ball has just the right amount of impetus so that, starting at the top of Figure 8-19, the gravitational force generates impetus in the vertical direction and, as the ball "falls," gravity will "drain" impetus from the horizontal direction in the detailed way described below.

To see in particular how the circular orbit occurs, first recall that we can analyze each linear component separately, for the linear dimensions are independent. In the vertical direction, the ball starts out with no impetus in the down direction and ends with an impetus intensity (momentum) of $m \cdot v$ in that direction. In the horizontal direction, the exact opposite occurs as the momentum of the ball is decreased to zero from an initial value of $m \cdot v$. This symmetry between the dimensions is made possible by the central nature of the force and the initial impetus and position of the ball.

In each dimension, gravity acts to change the momentum over the given ¼ period of the orbit. By symmetry, i.e. since each quarter of uniform circular motion is the same as any other quarter, we see that this behavior repeats itself for each of the four cycles of the motion. We are now ready to calculate how much total change in momentum results from the activity of the gravitational force in the vertical direction during a quarter cycle and, using that result, get the speed of the circular motion.

We know that a circular orbit is most easily described in the following way:

---

[16] From Greek and Latin "peri" meaning near, and Greek "geo," meaning Earth

[17] Periapsis and apoapsis are respectively the *generic* terms for perigee and apogee; i.e. they are the terms which refer to locations in an orbit around an *unspecified* body.

[18] This must be true because the basic dynamics remains unchanged if I rotate the circle through 90 degrees.

$X = -r\sin\omega t$, $Y = r\cos\omega t$, where X and Y are, respectively, the vertical and horizontal coordinates of the baseball in the coordinate system shown below. Note that at time $t = 0$ the ball is at the top as shown in the Figure 8-19.

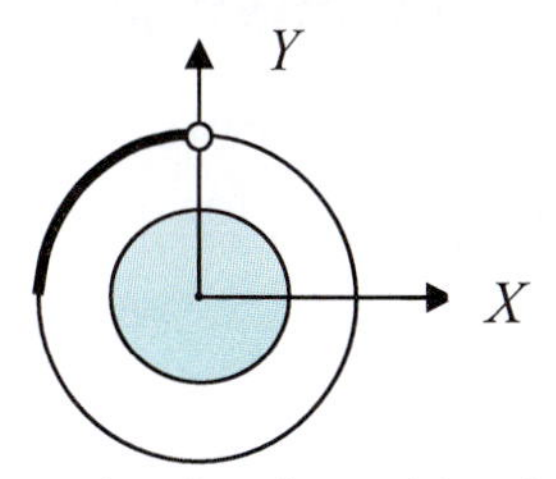

Figure 8-19: Baseball shown in circular orbit with an *x-y* coordinate system established at Earth's center. The dark part of the orbit is of interest in calculating the speed needed for circular orbit. It represents the ball falling from the position shown at 12 o'clock to a later position at 9 o'clock.

Using simple trigonometry, the Y component of the force can be written:

8.35
$$F_Y = -\frac{G\, m_{baseball}\, m_{earth}}{r^2}\cos\omega t = \frac{dp_Y}{dt}$$

We must thus integrate over time to get the net change in momentum, which must, as said above, be $m_{baseball}\,v$. Hence, taking the period, the time it takes the ball to complete one orbit, to be T, we have:

$$m_{baseball}v = \int_0^{\frac{T}{4}} \frac{G\, m_{baseball}\, m_{earth}}{r^2}\cos\omega t \; dt$$

8.36
$$= \frac{r}{v}\int_0^{\frac{\pi}{2}} \frac{G\, m_{baseball}\, m_{earth}}{r^2}\cos\theta \; d\theta$$

$$= \frac{1}{v}\frac{G\, m_{baseball}\, m_{earth}}{r}$$

Where we have taken: $t = \theta/\omega$; $T = 2\pi/\omega$, $\omega = v/r$ to get the second and third equations, Equations 8.36b and 8.36c.

Thus, the velocity needed at radius r to put the baseball in a circular orbit of the same radius r is:

8.37
$$v = \sqrt{\frac{G\, m_{earth}}{r}}$$

The first stage of space-flight is usually low Earth orbit, similar to the "launch" of our baseball. Thus, such a launch is the first step of our manned mission to Mars. Although President Bush's plan involves launching from a moon base, we may want to come from Earth sometimes.

To address the issues raised by the manned Mars mission that we tackle in Chapter 9, we need more understanding and more tools. Indeed, we now have enough understanding of how gravity works to begin our full out assault on the problem. The science that addresses the problems of space flight is called Astronautics. In the next section, we cover the basics of orbital mechanics.

Astronautics

Effective Potential

First, in Figure 8-20, we plot $V_{effective}$ for a central gravitational field. Recall that $E = \frac{1}{2} m \dot{r}^2 + V_{eff}(r)$, where:

8.38 $$V_{eff} = \frac{L^2}{2\mu r^2} + V_{gravity}(r) = \frac{L^2}{2\mu r^2} - \frac{G M m}{r}$$

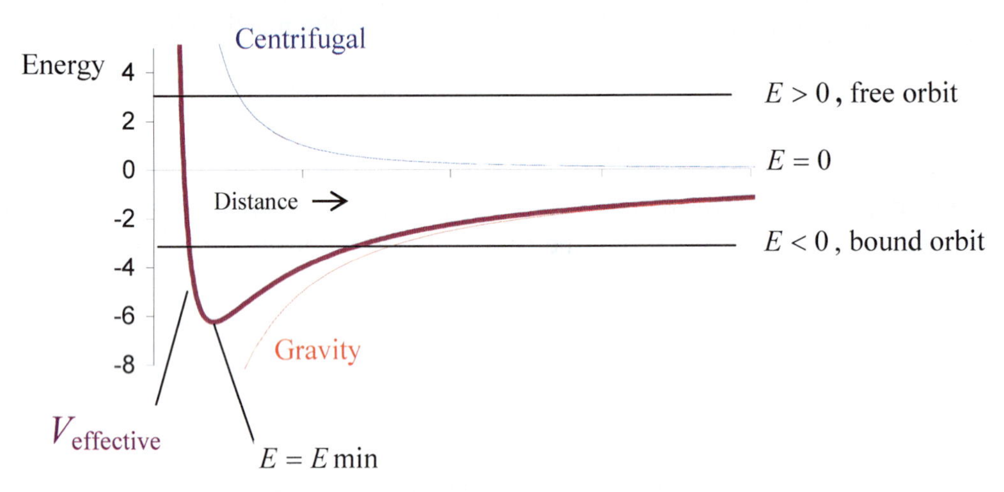

Figure 8-20: The effective potential $V_{effective}$ for a central gravitational field versus radial distance from the source point is shown in purple. The centrifugal component of the effective potential is shown in blue and the gravitational component is shown in red. Bodies with $E > 0, E = 0$ and $E < 0$ take on free orbits, a parabolic orbit and bound orbits, respectively.

The first thing one probably notices about this diagram is the minimum. Viewed as a landscape profile, there is a valley, a point which everything "tries" to roll to. Solving for the minimum, $V'(r) = 0$ implies $r_0 = \frac{L^2}{GMm^2}$ or $L = m\sqrt{G M r_0}$. There is only one distance that satisfies the energy minimum. This is the circular orbit.

There are four distinct cases of gravitational two body orbits:

1. ***Circular***, $E = E\min < 0$, constant radius, constant angular speed.
2. ***Elliptical***, $E\min < E < 0$, a bound orbit that traces an ellipse in one revolution. The satellite moves between a minimum and maximum distance, the apoapsis and periapsis.

3. ***Parabolic***, $E = 0$, just barely enough energy to escape. Satellite comes in from very far away, which by completion in a being of reason we call "infinity" then boomerang's around central object and proceeds back to infinity.
4. ***Hyperbolic***, $E > 0$, unbound or free orbit, like 3.

Finally, there is also special case in which $L = 0$, for which the bodies collide head on.

These descriptions are of general importance because they come from the solution of a very basic differential equation which, in turn, relates to a very basic type of motion. For instance, an equation of the same form appears in general relativity when certain approximations are made. In fact, that equation of general relativity describes the universe as a whole (again, under certain approximations and averages), which is the standard theory used in cosmology.

Starting with a circular orbit (*case I*) and adding *only a little* energy, we can show that we get an ellipse, i.e. *case II*. Using the techniques of Chapter 6, we expand the potential as:

8.39
$$V_{eff}(r) = V(r_0) + V'(r_0)(r - r_0) + \frac{1}{2}V''(r_0)(r - r_0)^2 + \dots.$$

The second term is zero because the circular orbit is minimum energy, while the first term is determined by the particular radius chosen and, moreover, being constant, doesn't show up in the force. Hence, the determining term is the last. It is simply the harmonic oscillator term, giving:

8.40
$$\omega = \sqrt{\frac{V''(r_0)}{m}} = \frac{G^2 M^2 m^3}{L^3} = \frac{L}{m r_0^2}$$

In the last equality, we eliminate G and M expressing ω only in terms of L and r_0.
We write the radial equations of motion as:

8.41
$$m\ddot{r} = -\frac{dV(r)}{dr} \approx -V''(r_0)(r - r_0)$$

This means we can write the solution for the orbit as:

8.42
$$r = r_0 + \delta r \cos \omega t$$

Where δr is the small amplitude oscillation we introduce by adding energy; we choose the cosine to give a convenient initial condition. Is this equation that of an ellipse? The equation for an ellipse, as can easily be verified by plotting it, is:

8.43
$$r = \frac{A}{1 + \varepsilon \cos \theta} \quad where\ A = a(1 - \varepsilon^2)$$

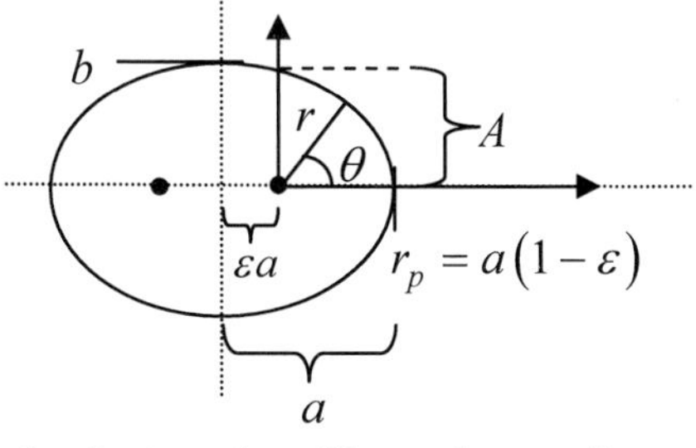

b, r, θ, A, εa, $r_p = a(1 - \varepsilon)$, a

A is called the semi-latus rectum and is the distance from the foci to the ellipse along a line perpendicular to the long axis of the ellipse. As an aside, note that this equation is actually the general equation for the conic sections discussed later.

Johannes Kepler was born in AD 1571, in Weil der Stadt, Württemburg, in southwest Germany, and died in 1630. He was an astronomer whose Christian (Lutheran) convictions sprinkled his work. He also did important work in optics. Kepler's interest in astronomy started young when his mother took him outside to see the great comet of 1577. Later, in 1604, he saw a supernova, a rare event about which he wrote.

The world class observer Tycho Brahe, who was responsible for so much of the most accurate sky data to his time, invited Kepler to Prague to become his assistant, and Kepler moved in 1600. With access to Brahe's data and intensive thought and study, Kepler made his most important discoveries. His introduction of elliptical orbits was a pivotal event in physics as it paved the way for Newton's work connecting motion in the heavens with motion on Earth. Kepler's decision to use ellipses came by analyzing Brahe's data on the orbit of Mars, which is, as is relevant for our manned trip to mars, much more elliptical than that of Earth.

Kepler had some suggestive thoughts on gravity but apparently did not understand the role of momentum. On gravity he says in his *Astronomia Nova:* "*If two stones were placed anywhere in space near to each other, and outside the reach of force of (other bodies), then they would come together...at an intermediate point, each approaching the other in proportion to the other's mass.*"[1] In the mid 1600's, the famed Italian Catholic astronomer Giovanni Cassini (who realized the moons of Jupiter obey Kepler's laws) and Jesuit collaborators tested Kepler's idea that the planets moved in ellipses[2] in the solar observatory which the Pope had ordered be part of the Basilica San Petronio. Kepler's saw that even simple mathematics can capture key aspects of nature; this is seen especially in his "Kepler's laws."

Kepler also published a book on the date of the conception of Christ titled *De Vero Anno quo Aeternus Dei Filius Humanam Naturam in Utero Benedictae Virginis Mariae Assumpsit* (Concerning the True Year in which the Son of God assumed a Human Nature in the Uterus of the Blessed Virgin Mary) in which he hypothesized about the star of Bethlehem.

Kepler's motivation in his work can be summarized by a couple of quotes. He said:

> I was merely thinking God's thoughts after Him. Since we astronomers are priests of the highest God in regard to the book of nature, it benefits us to be thoughtful, not of the glory of our minds, but rather, above all else, of the glory of God.[3]
>
> I wanted to be a theologian; for a long time I was unhappy. Now, behold, God is praised by my work even in astronomy.[4]

[1] http://galileoandeinstein.physics.virginia.edu/1995/lectures/morekepl.html
[2] http://cseligman.com/text/history/cassini.htm
[3] Quoted in *Kepler* by Max Casper, Dover, 1993, pg 88 ISBN 0-486-67605-6
[4] Letter to Michael Maestlin, 3 October 1595. *Johannes Kepler gesammelte Werke*, ed. M Caspar et al., Munich, 1937, **13**, 40. referenced by http://www-groups.dcs.st-and.ac.uk/~history/Quotations/Kepler.html under the name of J.V. Field.. Also quoted in *The Composition of Kepler's Astronomia Nova* by James Robert Voelkel, Princeton Univ. Press ISBN 0-691-00738-1, pg 32.

Since we started with a circle, it is clear that $\varepsilon \ll 1$ ($\varepsilon \sim 0$) and $A \sim r_0$, so that we have (think about why):

8.44 $$r \approx \frac{r_0}{1+\varepsilon\cos\theta} \approx r_0\left(1-\varepsilon\cos\theta\right)$$

Which, taking $\varepsilon = -\delta r / r_0$, is simply the same as equation 8.42 with θ being swept out at an angular speed ω, so that the $\theta = \omega t$. Hence, the new orbit, after perturbation, is an ellipse. We will now present all the most general facts about orbits, including that all bound orbits are ellipses (lines have $L = 0$, $\varepsilon = 1$ and circles ($\varepsilon = 0$) can be considered degenerate cases of ellipses). Studying Kepler's laws of planetary motion will lead us into the salient facts.

Kepler's Laws

Johannes Kepler (1571-1630) (see history box) working with the data collected by Tycho Brahe came up with the crucial idea that orbital paths did not have to be circles but could be ellipses and indeed are accurately described by ellipses.
Kepler came up with three laws:

1) The orbit of each planet is an ellipse with the Sun at one of the foci
2) Equal areas are swept out in equal time.
3) T^3 / a^3 is the same for all the planets around the sun (or any one given body), where a is semi-major axis of the orbit and T is the orbital period.

Law I: Ellipses et al

The first law is easily generalized into a key law in orbital mechanics that breaks the orbits into the four types, called conic sections that include the ellipse and circle as the bound cases. We can show this key fact of orbital mechanics in two ways.

First, the most direct way, as we've already seen, is to solve equation 8.28 for dr / dt and substitute the inverse square law potential of gravity to get:

8.45 $$\frac{dr}{dt} = \sqrt{2}\sqrt{\frac{E}{\mu} - \frac{L^2}{2\mu^2 r^2} - \frac{GM\,m}{r\,\mu}} = \sqrt{2}\sqrt{\mathcal{E} - \frac{\ell^2}{2r^2} - \frac{GM\,m}{r\,\mu}}$$

Where, in the second equation, we've used the energy and angular momentum divided by the reduced mass, which we label respectively as $\mathcal{E}$ and ℓ (we call them the specific energy and momentum) and which do not change. We, then, use the conservation of angular momentum already shown (cf. equation 8.27, using $\ell \equiv L / \mu = r^2\dot{\theta}$) to write:

8.46 $$\frac{d\theta}{dt} = \frac{\ell^2}{r^2}$$

Dividing the above two equations gives an equation for $\frac{dr}{d\theta}$ in terms of constants of the motion, $\mathcal{E}$ and ℓ that can be solved (exercise) to give the equation for $r(\theta)$ which has the general form of a conic section which we now derive using a second method.

The second way uses the power of vector notation and insights in working with vectors. We first write Newton's second law:

8.47
$$\ddot{\vec{r}} = -\frac{GM\,m}{r^2\mu}\hat{r} = -\frac{G'}{r^2}\hat{r} \qquad \text{where: } G' \equiv \frac{GMm}{\mu}$$
$$\ddot{\vec{r}} \times \vec{\ell} = -\frac{G'}{r^2}\hat{r} \times \vec{\ell}$$

In the last equation, we cross by $\vec{\ell} \equiv \vec{L}/\mu$. That equation can be re-written using conservation of angular momentum:

8.48
$$\frac{d\left(\dot{\vec{r}} \times \vec{\ell}\right)}{dt} = -\frac{G'}{r^2}\hat{r} \times \vec{\ell} = -\frac{G'}{r^2}\hat{r} \times \left(\vec{r} \times \dot{\vec{r}}\right)$$

To solve this we need to answer the question: the time derivative of what gives:

8.49
$$-\frac{G'}{r^2}\hat{r} \times \left(\vec{r} \times \dot{\vec{r}}\right) = \frac{G'}{r^2}\left(\left(\vec{r} \cdot \hat{r}\right)\dot{\vec{r}} - \left(\dot{\vec{r}} \cdot \hat{r}\right)\vec{r}\right) = \frac{G'}{r^2}\left(r\dot{\vec{r}} - \dot{r}\vec{r}\right),$$

where the second equality follows from the vector identity[19] $\vec{a} \times \left(\vec{b} \times \vec{c}\right) = \left(\vec{c} \cdot \vec{a}\right)\vec{b} - \left(\vec{b} \cdot \vec{a}\right)\vec{c}$ and the last equation follows from fact that the $\hat{r}$ component of $\dot{\vec{r}}$ is $\dot{r}$. Trying a simple guess, $\hat{r} = \frac{\vec{r}}{r}$, which will obviously give the needed $1/r^2$ term portion as well as a second term, we get: $\frac{d\left(\frac{\vec{r}}{r}\right)}{dt} = \frac{\dot{\vec{r}}}{r} - \frac{\vec{r}}{r^2}\dot{r}$. So that we finally get:

8.50
$$\frac{d\left(\dot{\vec{r}} \times \vec{\ell}\right)}{dt} = G'\frac{d\hat{r}}{dt}$$
$$\Downarrow$$

8.51
$$\dot{\vec{r}} \times \vec{\ell} = G'\hat{r} + \vec{P}_{eriapsis}$$

To get an equation for r as a function of the angle swept by the orbit, we dot by $\vec{r}$ to get:

8.52
$$G'r + \vec{P} \cdot \vec{r} = \vec{r} \cdot \left(\dot{\vec{r}} \times \vec{\ell}\right) = \vec{\ell} \cdot \left(\vec{r} \times \dot{\vec{r}}\right) = \ell^2$$

Here we've used a vector identity to help us express the right hand side solely in terms of the constant of the motion ℓ. We now solve for r, getting the general form of the orbit in terms of the cylindrical coordinate, θ, measured from the x-axis, which is defined by the vector $\vec{P}$. $\vec{P}$ is, in turn, the direction from the periapsis focus to the nearest orbit approach. We get:

8.53a,b
$$\boxed{r = \frac{\frac{\ell^2}{G'}}{1 + \frac{\mathrm{P}}{G'}\cos\theta} = \frac{A}{1 + \varepsilon\cos\theta}}$$

[19] The student is asked to prove this vector identity and others in the end of chapter problems of Chapter 2.

This is the general form for conic sections we sought (and which we proved was approximately true for a special case in equation 8.44) with $A \equiv a(1-\varepsilon^2)$ and $P \equiv \left|\vec{P}\right| = \varepsilon G'$. Comparing the general form of a conic section given in equation 8.43 to the trajectory given in equation 8.53a gives:

8.54a,b $$a\left(1-\varepsilon^2\right) = \ell^2 / G' \text{ and } \varepsilon = P/G'$$

Still, we wish to find ε in terms of the specific energy and momentum, $\mathcal{E}$ and ℓ, and we'd also like $\mathcal{E}$ in terms of key orbital parameters like a. To obtain these relations, we need the following simple analysis.

Examination of the equation 8.53b, shows that closest approach, i.e. the periapsis, ($\theta = 0$) is at radius:

8.55 $$r_p = \frac{a(1-\varepsilon^2)}{(1+\varepsilon)} = a(1-\varepsilon).$$

Similarly, the radius of the orbit at apoapsis (at $\theta = \pi$) is:

8.56 $$r_a = \frac{a\left(1-\varepsilon^2\right)}{(1-\varepsilon)} = a\left(1+\varepsilon\right)$$

Considering the specific energy at the periapsis (where $\dot{r} = 0$), $\mathcal{E} = \frac{\ell^2}{2r_p^{\,2}} - \frac{G'}{r_p}$ and using the equation 8.55 for r_p and equation 8.54a for ℓ^2 we get:

8.57 $$\boxed{\mathcal{E} = -\frac{G'}{2a}}$$

Solving 8.54a for the eccentricity, ε and substituting $a = -\frac{G'}{2\mathcal{E}}$ gives:

8.58 $$\boxed{\varepsilon = \sqrt{1 + \frac{2\mathcal{E}\ell^2}{G'^2}}}$$

recall: $G' \equiv \frac{GMm}{\mu}$

For such an orbit we also have $\ell = \mathbf{v}_a r_a = \mathbf{v}_p r_p$, because in both apoapsis and periapsis, $\mathbf{v}$ is perpendicular to r.

For given energy and angular momentum and masses, we have the eccentricity and semi-major axis, a, given by equations 8.58 and 8.57, which are valid for all four types of orbits. Given these parameters, we describe the various types of orbits in more detail.

1) The ***circular orbit*** we already understand very well (it has $\varepsilon = 0$).
2) The ***elliptical orbit*** looks as shown ($0 < \varepsilon < 1$).

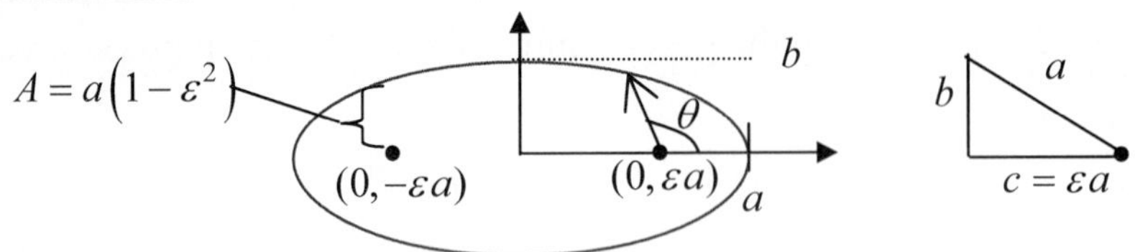

$A \equiv$ semilatus rectum. $a^2 = b^2 + c^2$
Redrawing so the focus is at the origin, as it is for the standard equation for the ellipse given in equation 8.53 and reproduced below, we get:

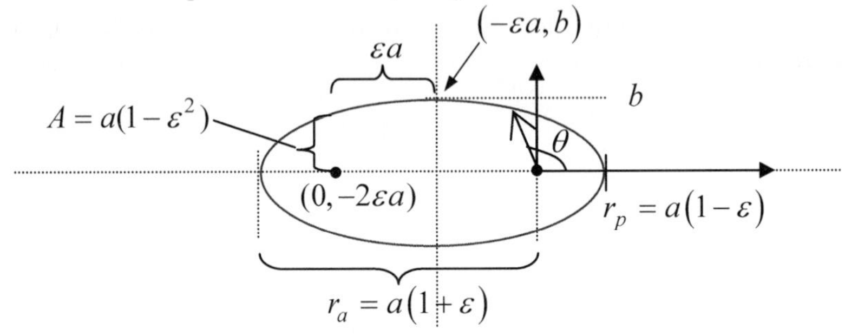

$$r(\theta) = \frac{a(1-\varepsilon^2)}{1+\varepsilon\cos\theta}, \text{ so that: } r(0) = r_p = a(1-\varepsilon),\ r\left(\frac{\pi}{2}\right) = A = a(1-\varepsilon^2),$$
$$r(\pi) = r_a = a(1+\varepsilon)$$

An ellipse can be defined as *the locus of points in a plane such that sum of the distances to each of two foci is constant* ($= 2a$). In the problems, you are asked to create an ellipse using this property, a string and two nails.

3) The ***parabolic orbit*** looks, excepting the degenerate case of a line, as shown ($A = \frac{l^2}{G'}$ $\varepsilon = 1, a = \infty$):

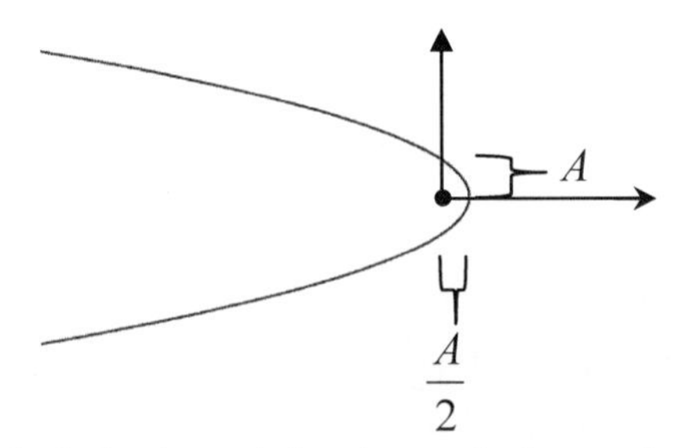

Note that the focus of the parabola is the origin. A parabola can be defined as *the locus of points equidistant from the focus and a line called the directrix*, which is perpendicular to the line of symmetry of the parabola. In the problems, you are asked to create a parabola using this property, a string and a "T square,"

4) Finally, ***the hyperbola*** ($\varepsilon > 1$) has two branches shown in blue. The solid blue curve here is meant to be compared to the orbits above. The red lines indicate the asymptotic flight path.

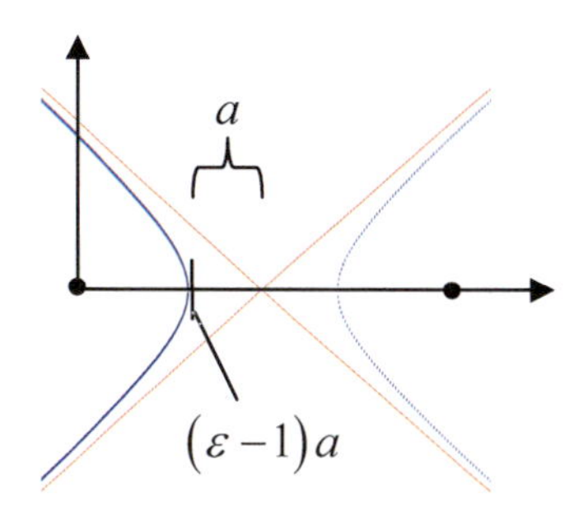

An hyperbola can be defined as *the locus of points in a plane such that the difference between the distances to the two foci is a constant* ($= 2a$)*, i.e. the same for every point.* In the problems, you are asked to devise a method for mechanically drawing an hyperbola using this property.

These orbital shapes have also been called *conic sections* because they result from the intersection of a plane with the two cones, as shown in Figure 8-21.

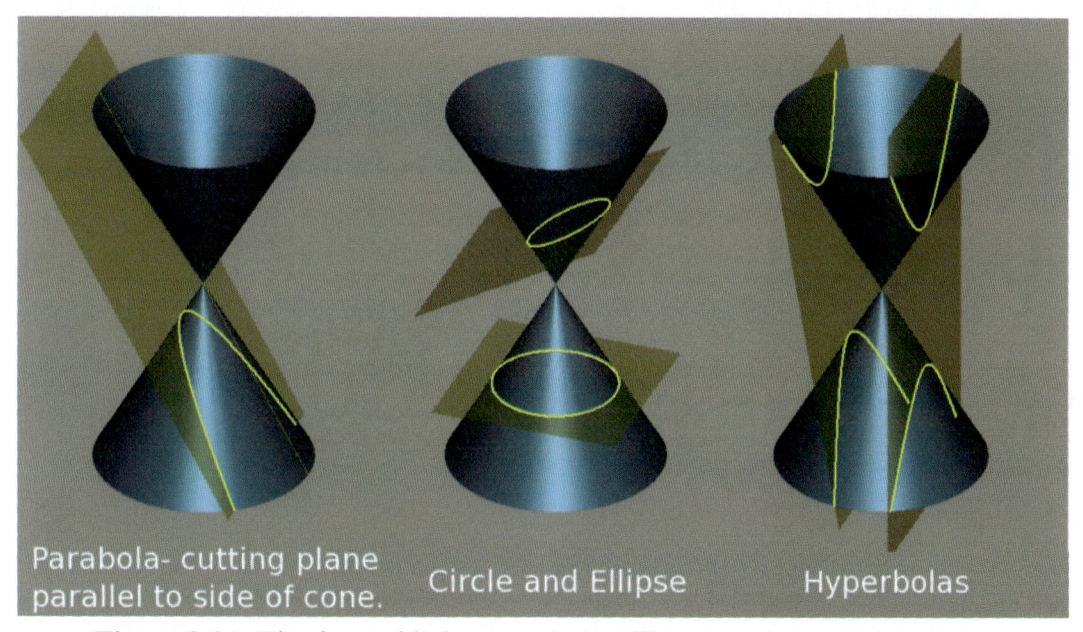

Figure 8-21: The four orbital types, circle, ellipse, parabola and hyperbola can be formed by allowing a plane to intersect to two right circular conical surfaces as shown. Duk on English Wikipedia has granted Gnu Free documentation License.

Law II: Equal Area in Equal Time

That the rate of area swept is constant follows trivially from conservation of angular momentum and understanding of the nature of angles.

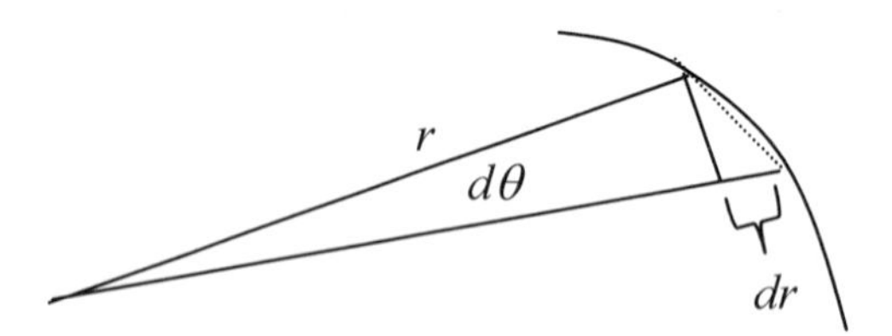

The area swept out, dA, by a movement through an angle $d\theta$ is:

8.59 $$dA = \frac{1}{2} r\, r d\theta + \frac{1}{2} r d\theta dr \rightarrow \frac{1}{2} r^2 d\theta \quad \text{because } d\theta dr \sim d\theta^2 \text{ is second order}$$

$\Rightarrow$

8.60 $$\frac{dA}{dt} = \frac{1}{2} r^2 \dot{\theta} = \frac{1}{2}\frac{L}{\mu} = \frac{1}{2}\ell = \text{constant}$$

Thus, Kepler's second law is an indirect statement of the conservation of angular momentum. Hence, it is true for all central forces, not just inverse square ones, because all such forces conserve angular momentum.

Law III: T^2 / a^3 Same for All Objects Orbiting Around a Common Body (Sun)

Integrating equation 8.60 to get the total area of the ellipse (which can be shown to be $\pi a b$) and the total period, T and using equation 8.54a again for ℓ^2 gives:

8.61 $$\pi a b = \frac{T}{2}\sqrt{a G'\left(1-\varepsilon^2\right)}$$

Now, by noting (using the second figure under the elliptical orbit section above) that at $(x, y) = (\varepsilon a, b) \Rightarrow r = \sqrt{\varepsilon^2 a^2 + b^2}$ and $\cos\theta = -\frac{\varepsilon a}{\sqrt{b^2 + \varepsilon^2 a^2}}$ and by plugging into the general ellipse equation (equation 8.53b) we get: $b = a\sqrt{1-\varepsilon^2}$. Substituting this into equation 8.61 and squaring each side yields.

8.62 $$\boxed{\frac{T^2}{a^3} = \frac{4\pi^2}{G'}}$$

Here we see that *Kepler's 3rd Law is only approximately valid* because $G' = G\left(M + m\right)$, and so there is a dependence on the planets mass but it is very small since $m / M \ll 1$.

Now that we have the basic principles of gravity and celestial mechanics we can move on to some of the details in problems and finally to Chapter 9 in which we will advance further our understanding of the principles by looking at orbit transfers and other issues to lay out our trip to Mars.

Summary

Since massive bodies, by definition, have the quality of mass, receptivity to the action of impetus, they must have the ability to have impetus. Thus, there must be some force that can activate impetus in any one of them. This generic force is gravity. All massive bodies are receptive to the action of gravity; indeed all massive bodies also cause gravity, so that all massive bodies are attracted to each other, try to move towards each other.

Each massive body acts on the media surrounding it, generating a field in it that can act on other massive bodies to pull them towards it. We call the media between massive bodies *plana*, from the Latin word for field, since it is receptive of the fields, that is fields can exist as qualities of it making it to be this way rather than that. The plana is usually called *space* or *vacuum*; both of these names can be used if one realizes that space does not mean extension without substance and vacuum does not mean *nothing*. Extension is a property (the first property), and nothing, taken literally as no-thing, is a being of reason. Plana seems to exist both as part of substances (virtual plana) and between substances, but, in usual fashion, we ignore this distinction to concentrate on the generic forces and their effects. Thus, things with physical aspects can be summarized in the chart below. Though the lightly shaded area is not studied in detail in this text, we touched on the important distinctions in it in Chapter 1, and all we say in this text applies in an analogically general sense to those substances as well.

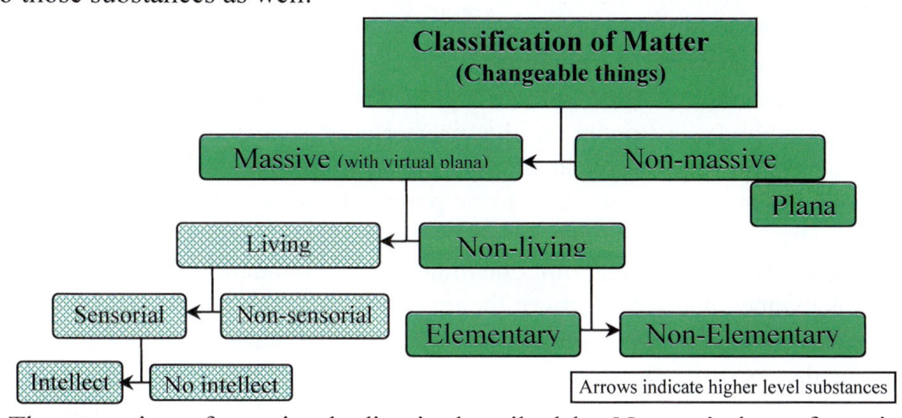

The attraction of massive bodies is described by Newton's law of gravity which relates the measures of the various aspects as: $\vec{F}_{gravity} = -G\dfrac{m_{A\_active}\, m_{B\_passive}}{r^2}\hat{r}$, where $\vec{r}$ is the position vector from point body A to point body B. Note that gravity is a *central force* with strength that decreases *inversely with the square* of the distance. It works in the following way. Body A generates a field proportional to its *active gravitational mass*, m_{A_active} in parts of the plana near it; those parts of the plana then activate parts of the plana near them and those, in turn, activate the ones next to them and on and on. At some point the field reaches body B and acts on it according to its receptivity measured by $m_{B\_passive}$ and called its *passive gravitational mass.* Newton's second law then gives the acceleration of body B as: $\vec{F}_{gravity} = m_{inertial}\ddot{\vec{r}}$. Thus, we have three types of mass. Experiment reveals that for any given mass the active, passive and inertial mass are proportional, so we can write our equations using just $m = m_{active} = m_{passive} = m_{ineritial}$ by defining our relations such that the proportionality between the three types of mass is one. Of course, the three qualities remain distinct, but we drop the distinction for our formula and calculations in order to focus on the sense in which they are unified. Thus, in this way we speak of "mass" without reference to type. Thus, in some way, the essence of (inertial) massive bodies includes the passive and active gravitational mass. Because of its special importance, the equivalence of the quantities (the measures of these qualities) of passive gravitational mass and inertial mass, $m_{B\_passive} = m_{B\_active}$ is called the *principle of equivalence*.

When massive bodies A and B are fixed, they generate the field in the plana shown below. The effect of each of the two bodies on the plana is additive so one can write the field at given arbitrary point, $\vec{r}$, as:

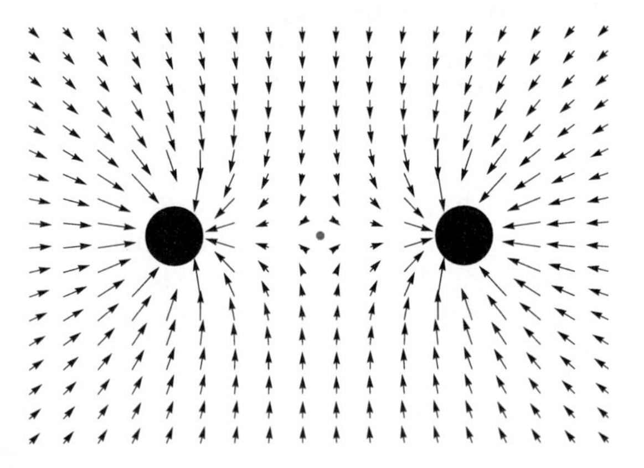

$$\vec{G}_{Total} \equiv \vec{G}_A + \vec{G}_B$$

$$= \frac{Gm_1}{(\vec{r}_A - \vec{r})^2}\frac{\vec{r}_A - \vec{r}}{|\vec{r}_A - \vec{r}|} + \frac{Gm_2}{(\vec{r}_B - \vec{r})^2}\frac{\vec{r}_B - \vec{r}}{|\vec{r}_B - \vec{r}|}$$

Indeed, in our Newtonian approximation, the field of the any number of massive bodies interacts very simply by the addition of the measures of their field strengths. This ability to linearly add the field of each contributing body is known as the *principle of superposition.* Because of this principle, the field of any body can be determined by treating it as composed of many small parts, which in the limit we take to be points. Thus, by adding the contributions of each of these parts in the limit, i.e. taking the integral over the contribution of each part, we can determine the field at a given point due to the given extended body.

In this manner, a key property of the inverse square law field is revealed. We prove that a spherical shell of mass in empty space has no gravitational field inside of it, and outside of it, the field it generates can be treated as if it came from a point mass at its center. Thus, the gravitational field *external* to a solid sphere of mass M centered on a point P generates the same field as (and thus can be treated as) a point mass of mass M located at P.

Because of the central and inverse square nature of the force of gravity, one can reduce the analysis of the motion of the two body problem to a one body problem where the relative motion is given by the vector: $\vec{r} = \vec{r}_2 - \vec{r}_1$, giving the equation of motion: $\mu\ddot{\vec{r}} = F(r)\hat{r}$ where $\mu = \frac{m_1 m_2}{m_1 + m_2}$ is called the *reduced mass*. The center of mass of the system comes naturally into play in this reduced formalism and is located at: $\vec{R} = \frac{m_1\vec{r}_1 + m_2\vec{r}_2}{m_1 + m_2}$. The equations of the masses in the center of mass frame are: $\vec{r}_1' = -\dfrac{m_2}{m_1 + m_2}\vec{r}$, $\vec{r}_2' = \dfrac{m_1}{m_1 + m_2}\vec{r}$, where the primes indicate center of mass coordinates. The angular momentum and energy are also conveniently written in the reduced formalism and they have the following simple interpretations in the center of mass frame. $\vec{L} = \vec{r} \times \mu\vec{v} = \vec{L}_1' + \vec{L}_2'$, $E = \frac{1}{2}\mu\vec{v}\cdot\vec{v} = \frac{1}{2}m_1\vec{v}_1'\cdot\vec{v}_1' + \frac{1}{2}m_2\vec{v}_2'\cdot\vec{v}_2' = K'_1 + K'_2$.

When one mass is larger than another (i.e. $m_2 \ll m_1$), we have : $\vec{r}_1' \approx 0$ $\vec{r}_2' \approx \vec{r}$, $\mu \approx m_2$, so that the center of mass is approximately at body one and one can approximate the motion as being due to body one changing the impetus of body two and treat body one as fixed in position; this greatly simplifies orbital calculations.

Using this approximation for a spacecraft's orbit near a planet, we can show that one can get gravity assists from planets that may be helpful on our trip to Mars. Traveling in front of the moving planet gives a velocity decrement and traveling behind it gives a velocity increment to the spacecraft.

Solving the reduced problem immediately shows that the orbit will always remain in a given plane. Also, the angular momentum, *L*, about the axis of rotation of two bodies is conserved, since there is no force applied perpendicular to the radius vector to the orbiting body. This gives a simple relation for the energy using cylindrical coordinates: $E = \frac{1}{2}\mu\dot{r}^2 + \underbrace{\frac{1}{2}\frac{L^2}{\mu r^2} + V(r)}_{V_{effective}}$. Properly speaking, the second term is a kinetic energy term, but mathematically it is a function of radius rather than derivatives of radius because of the constancy of *L*, so that formally (i.e. in the mathematical formalism), one can treat it as part of the potential energy.

Using conservation of energy and solving the above equation gives the time it takes the body to move from r_0 to r: $t = \sqrt{\frac{\mu}{2}}\int_{r_0}^{r}\frac{dr}{\sqrt{E - V_{eff}(r)}}$

For the case $V_{eff} = \dfrac{L^2}{2\mu r^2} + V_{gravity}(r) = \dfrac{L^2}{2\mu r^2} - \dfrac{GMm}{r}$. When $L = 0$, the potential reduces to the standard potential energy of a point particle of mass *M* acting on a point mass *m*. Under the general potential, V_{eff}, whose components are graphed below, there are four possible orbits with the eccentricity of the orbits, ε also shown.

1. ***Circular***: ($E = E\min < 0$, $\varepsilon = 0$) of radius *r*, central mass *m*, $v = \sqrt{\dfrac{Gm}{r}}$

2. ***Elliptical***, $E\min < E < 0$, $0 < \varepsilon < 1$,moves between a minimum and maximum distance, the apoapsis, r_a and periapsis, r_p.

3. ***Parabolic***, $E = 0, \varepsilon = 1$, just barely enough energy to escape. Satellite comes in from very far away, which by completion in a being of reason we call "infinity" then boomerang's around central object and proceeds back to infinity.

4. ***Hyperbolic***, $E > 0\ \varepsilon > 1$

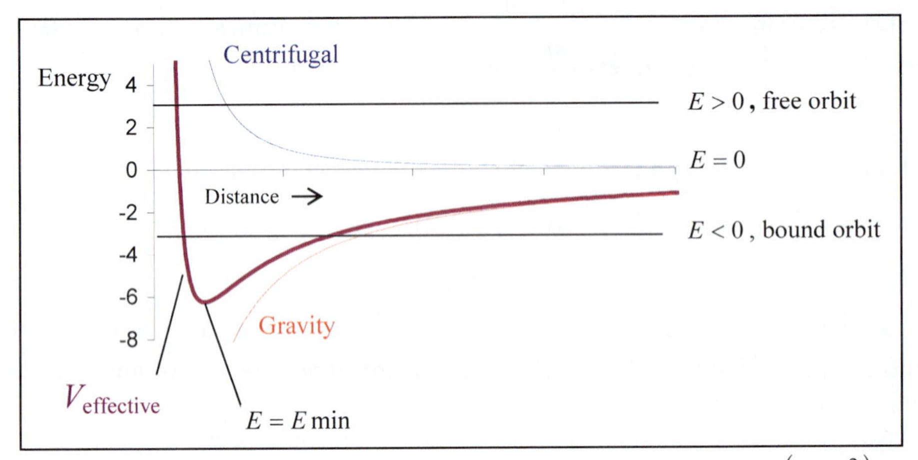

The equation for conic sections of various eccentricities, ε is: $r(\theta) = \dfrac{a\left(1-\varepsilon^2\right)}{1+\varepsilon\cos\theta}$. In terms of key orbital parameters the equations can be written.

$$r = \frac{\frac{\ell^2}{G'}}{1+\frac{P}{G'}\cos\theta} = \frac{A}{1+\varepsilon\cos\theta},\quad \ell \equiv L/\mu = r^2\dot{\theta},\ \varepsilon = \sqrt{1+\frac{2\mathcal{E}\ell^2}{G'^2}},\quad \mathcal{E} = -\frac{G'}{2a},$$

$G' \equiv \dfrac{GMm}{\mu}$, $A \equiv a(1-\varepsilon^2)$, $P \equiv \left|\vec{P}\right| = \varepsilon G'$, where $\mathcal{E} = E/\mu$ and $\ell = L/\mu$ are, respectively, the *specific* energy and *specific* angular momentum. The ellipse, the standard bound orbit, is shown below with the key geometric parameters labeled.

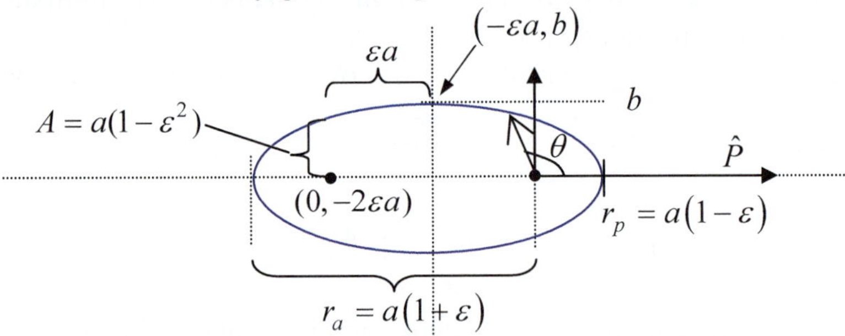

Kepler's three laws summarize important information about planetary behavior. The first law says that orbit of each planet is an ellipse with the Sun at one of the foci. The second, a result of conservation of angular momentum, is that the radius vector from the focus to the planet sweeps out equal areas in equal time. The third law states that the T^3/a^3 is a constant ($\dfrac{4\pi^2}{G'}$) for all planets in orbit around the sun, where T is the period of the orbit and a is the length of the semi-major axis.

Terms

Gravitational Field

Plana

Mass: Three types: Inertial mass

Active gravitational mass

Passive gravitational mass

Rest mass

Principle of equivalence

Center of Gravity

Gravitational Statics

Point Mass

Reduced Mass

Reduced angular momentum

Reduced energy

Specific Energy and Angular Momentum, $\mathcal{E} = E/\mu$ and $\ell = L/\mu$

Astronautics

Orbits

Effective potential

Perigee, Apogee (generally, apoapsis and periapsis)

Semi-latus rectum

Calculation Techniques

Approximate the speed of propagation of gravitational fields to be much faster than any other time scale in the given problem, indeed, take the speed to be infinite (a being of reason).

To get the field of an extended body, use the superposition principle and calculus to treat it as composed of an infinite number of point particles. Mathematically, we write: $\vec{G}_{net} = \int_{Body} \frac{G}{|\vec{r}' - \vec{r}|^2} \frac{\vec{r}' - \vec{r}}{|\vec{r}' - \vec{r}|} dm(\vec{r}\,')$

Sum over all parts of mass *dm* of extended body

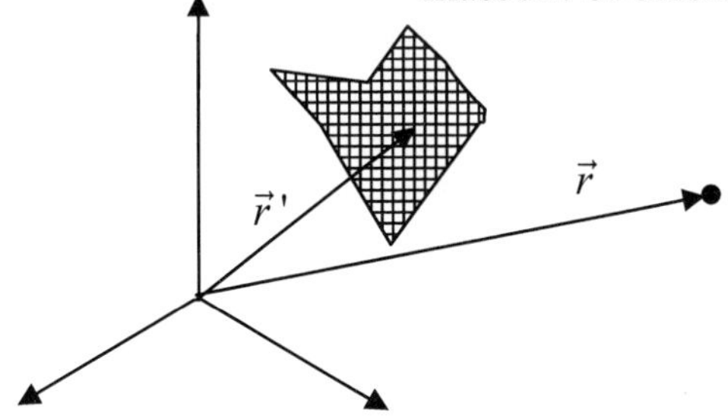

Mixed Problems

1. Calculate how much a 100kg man weighs on each of the visible planets, assuming they each had solid surfaces on which to stand: Mercury, Venus, Earth, Mars, Jupiter, and Saturn. Do the same for: our moon, and Titan and Io of Jupiter.

2. Calculate, using Newton's law of universal gravitation, the weight of a 100kg man at the a) Earth's surface b)10 miles above c) 100 miles above d) 1000 miles above e) 10,000 miles above f) the distance to the moon above the surface.

3. Calculate and sketch the field produced in the plana by the following bodies. a) point of mass m b) an infinite line of uniform linear mass density λ c) an infinite plane of uniform surface mass density σ.
Hint: use symmetry

4. Show that equation 8.28 follows from equation 8.25.

5. The Eotvos experiment was designed to test the Equivalence Principle. It consists of a dumbbell hung from a wire and was originally done in the Earth's field with a rotating apparatus, such as shown below (note that since the sun is so far away, take the two forces as parallel).

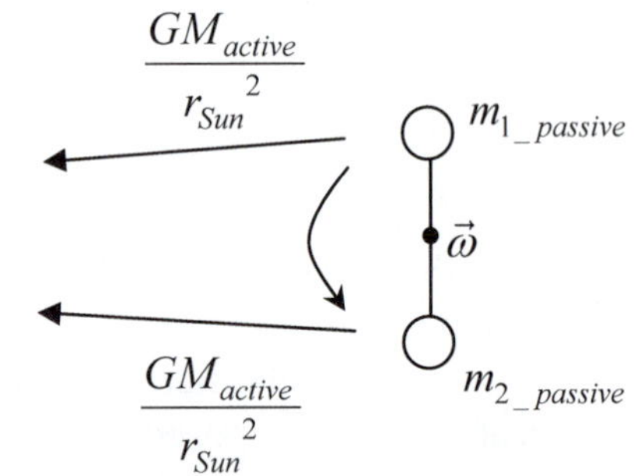

The R.H. Dicke version of the experiment[20] used the rotation of the Earth and did not introduce any further rotation. To analyze a simple version of the experiment, consider the dumbbell hanging over the North Pole. It is hanging from a stiff wire fixed to the overhead structure so that the wire twists as the Earth turns, thus turning the dumbbell with it.

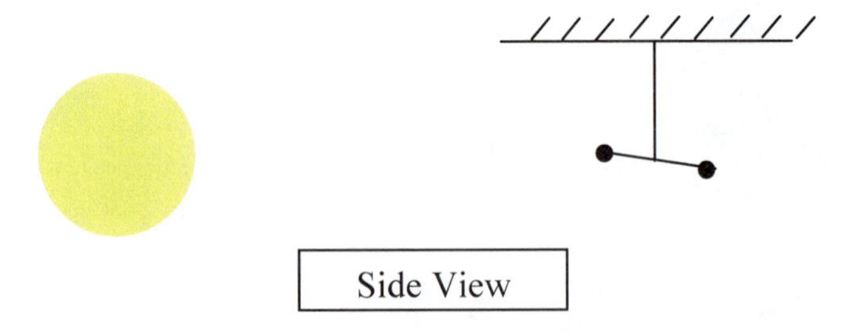

Side View

[20] P. G. Roll, R. Krotkov and R. H. Dicke, Ann. Phys. (NY) 26, 442 (1967)

a. Calculate the torque around the suspension on the pendulum in the falling frame of the Earth by calculating the forces on each body. First, do the calculation ignoring the sunward tilt of the wire, then include that tilt by using the vector identity:

$$(a+b)\cdot\left((c\times a)+(d\times b)\right)=c\cdot\left(a\times(a+b)\right)+d\cdot\left(b\times(a+b)\right)$$

to show that the component of the torque along the wire is: $\tau_w = \dfrac{(\vec{r}_2-\vec{r}_1)\cdot(\vec{F}_1\times\vec{F}_2)}{\left|\vec{F}_1+\vec{F}_2\right|}$.

b. Express this torque in terms of the *Eotvos* ratio: $\eta \equiv \left|\dfrac{2(a_1-a_2)}{a_1+a_2}\right|$, (note these accelerations are the ones seen in inertial frames (i.e., far from all gravity)), and show that the torque can be written: $\tau \sim d\,\eta\, m\, g_{sun}$. Where d is the distance between the dumbbells and m is the approximate mass of each dumbbell and g_{sun} is the acceleration due to the gravity of the sun at the orbit of the Earth.

Hint: There will be a fictitious centrifugal force that will on the average balance out for Earth-pendulum system but generally will not for the dumbbell masses if there is a difference between the passive and inertial mass, say for different material. Dicke used Gold and Aluminum, on the postulate that two different substances might mean two different ratios of m_p / m_i.

6. Show that Newton's third law leads to the equivalence of active and passive mass, thus (coupled with the Eotvos type experiments which show—at the appropriate level of approximation—the equivalence of passive and inertial mass) establishing the proportionality between all 3 types of mass measures.

7. Calculate the total change in the acceleration due to gravity at the surface of the Earth, g, over the height of $100km$ (definition of outer space).

(Problems continued on next page)

8. Air is on the Earth because it has mass, being composed of molecules, so that gravity acts to pull it to the Earth. Thus, because air is compressible, it is denser near the surface and less dense higher up. a) Using the following simplifying assumptions, derive relations for the variation of the Earth's atmospheric pressure *and* density as a function of radial distance from surface. b) Taking the atmosphere to end when the density drops below *1/e* of its surface value, what would you estimate the atmospheric height to be? c) Given this general behavior of the atmosphere, explain the following: a group was riding in a car traveling up into the mountains, at a certain point a loud pop was heard; later investigation revealed a bag of potato chips had popped (this is true story).

Assume that the atmosphere varies only with altitude, so that the column of air above any given point on the surface is the same at every place on the surface. Also, consider a column of air that is narrow enough that we can neglect the curvature and treat it as simply a right cylinder of air that starts at the surface with base of area A and extends to some height h. By integrating over the whole Earth, we get an area $S_e = 4\pi r_e^2$ (yielding a total atmospheric volume, $V_0 = S_e h$). Assume that the total number of particles in the air is N_0 and take m_p = {the mass of a mole of each particle of air, which is 78% N_2, 21%O_2} ~ mass of N_2 per mole ~28*g/mole*. Further assume that atmosphere has the same temperature (iso-thermal approximation) throughout the column. Assume ideal gas law: $P = nRT$, where R ~8.3 *J/K/mole*, and $n = N/V = moles/m^3$ so that $n_{avg} = N_0 / V_0$.

9. Given a helium balloon that can just lift a mass of $m_0 = 100g$ off the ground. a) Describe its behavior when it is in a car which is accelerating at a rate of $g = 9.8$ m/s^2 carrying a mass of $m = 70.7g$? b) Suppose the car is accelerating at 2g?

10. Calculate the escape velocity of the moon, that is, the speed at which one must leave its surface in order to escape the pull of the moon.

11. If we define a Newtonian black hole as a body with a gravitational field so strong that even a massive body moving at the speed of light cannot escape to infinity, what radius and mass must such a black hole have?

12. Take the role of Isaac Newton, and deduce the law of the acceleration due to gravity from Kepler's laws. We know the acceleration is purely radial, because the angular component of the acceleration must be zero given the form of Kepler's second law. Next, a) Use Kepler's first law, the equation for an ellipse ($r(\theta) = \dfrac{a(1-\varepsilon^2)}{1+\varepsilon\cos\theta}$), and second law (which gives $\dot{\theta} = \dfrac{l}{r^2}$, where l is the specific angular momentum) to deduce the form of the gravitational acceleration. (*Hint:* use $a_r = \ddot{r} - r\dot{\theta}^2$ and use the second law and the first law to write the equations for the needed derivatives; also use, $u = 1/r$, a useful transformation in orbital mechanics) b) Use Kepler's second law to deduce the constant and thus write the law of gravitation.

13. The flat "rotational curve" for our spiral galaxy, which is like the one shown at the right, has been used to argue for a violation of Newton's law of gravitation. However, most argue for dark matter, i.e., matter not visible in the electromagnetic spectrum. The *rotation curve* is the plot of the speed of visible mass in the galaxy (stars and any other visible matter such as gas) as a function of the distance from the center, for our galaxy, the Milky Way. The matter distribution that one deduces from the rotation curve does not mach the matter seen; hence the discussion of dark matter.

Given Newton's law of gravity and making the simplifying assumption that the galactic mass is distributed in cylindrical shells of uniform *angular* density, and neglecting the contribution of the field at a given radius by mass outside that radius, what mass density profile (density as a function of radius) does this constant speed profile imply?

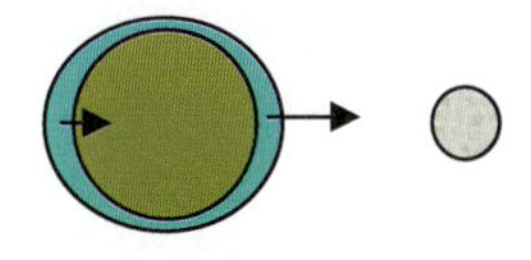

14. The tides on the Earth are caused by the *difference* in the force of gravity on the water nearest to the moon (and sun) versus that furthest from the moon. Give an estimate of this field differential, (i.e. force per unit mass of water) for the moon.

15. Calculate the surface gravity of Earth using the orbital period of the moon or other satellite around the Earth (assume a circular orbit for the moon).

16. Using $\vec{P} = \dot{\vec{r}} \times \vec{\ell} - G'\hat{r}$ from equation 8.51: a) Write an equation for the eccentricity in terms of the speed and distance at pericenter, v_p and r_p. b) Derive the equation, which we will use in Chapter 9, for the speed at the pericenter, v_p, in terms of r_p and r_a and v_a c) Use conservation of angular momentum to obtain a relation for v_p in terms of r_p and r_a alone.

(Problems continued on next page)

17. Suppose one makes a tunnel under the Earth on the line from Chicago (42N, 88W) to Boston (42N, 71W) for a tram to cut traffic in that heavy corridor. The distance between the cities is 714 miles as the crow flies and 840 miles along the road. How long would it take for a round trip if one simply lets the tram fall back and forth down the hole without friction?
In the figure below, the tunnel is shown larger to make it visible, and the Earth is shown rotated to align with convenient axes.

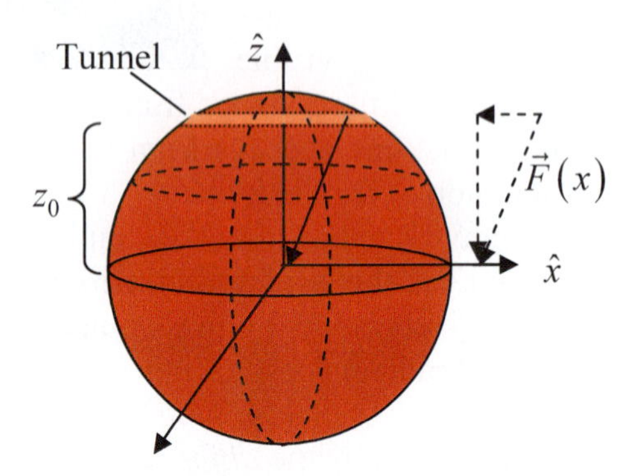

18. Suppose, at some point in the future, there is a thriving community of people on Mars. a) Further suppose, they drill a hole from pole to pole on Mars to transport equipment and materials between the two halves of the planet; how long would a one way trip along this hole take, using only the force of gravity, assuming no friction. Compare this with the speed of a car traveling along the surface. b) Consider two colonies on Mars, one of which is 5 degrees north of the equator and the other is 5 degrees south of the equator both on the same line of longitude (see the figure in the above problem). How long would it take for the two colonies to transport people and things through a hole on the chord that connects the two colonies using only the force of gravity, neglecting friction? Compare this with the speed of car travel on the surface.

19. Bishop Oresme argued in the Middle Ages that "assuming … and that a man were at the center standing straight with his head in one direction from the center and his feet on the other, I say that such a man would have his head and also his feet on top or upwards, and he would be no more in a lying than in a standing position, nor facing downward more than upward." Explain this qualitatively and quantitatively in terms of the gravitational field near the center.

20. Using conservation of angular momentum and conservation of energy at the periapsis and apoapsis and the relations: $r_p = a(1-\varepsilon)$, $r_a = a(1+\varepsilon)$, prove Kepler's third law in the following way. a) Calculate the ratio of the speeds at periapsis and apoapsis, $\dfrac{v_p}{v_a}$.
b) Calculate the speed at the periapsis in terms of *only* the semi-major axis, the masses, the gravitational constant, G and the eccentricity. c) Using this last result, calculate the rate that area is swept from the foci to the orbiting body, $\dfrac{dA}{dt}$ in terms of the same variables.

(this problem continues on next page)

d) Take T to be the period of the elliptical orbit and use the swept out area just derived to prove Kepler's third law.

21. Explain physically what happens during a gravity assist by using the model of a ball hitting a semi truck given in the text. Assume a one dimensional idealization in which the craft comes in along the x-axis toward the origin and then returns along the positive x-axis out to infinity. Also, show how to calculate the approximate value of the assist in terms of velocity increment given by the interaction of the craft with a planet. Prove that the one dimensional idealization sets an approximate upper bound on the speed increment in the full two dimensional case. (To better understand the effect in terms of a collision analogy use the Interactive Physics model 4.1-collision of masses with springs.ip --you will need to play with the parameters to get some precision).

22. Voyager II was launched in August of 1977 on a path to leave the solar system and as of February 2008 was still functioning, while a full 85 AU from the sun. Using the one dimensional collision assumption discussed above, calculate an upper limit on the gravity assist (i.e. give a limit on the maximum possible increment in speed in *km/s*) that could have been given to Voyager II by each of the planets below. Compare those results to the values in the graph below. Note that the goal of the flyby of Neptune was not to gain further speed but to get close to Neptune's moon Triton.

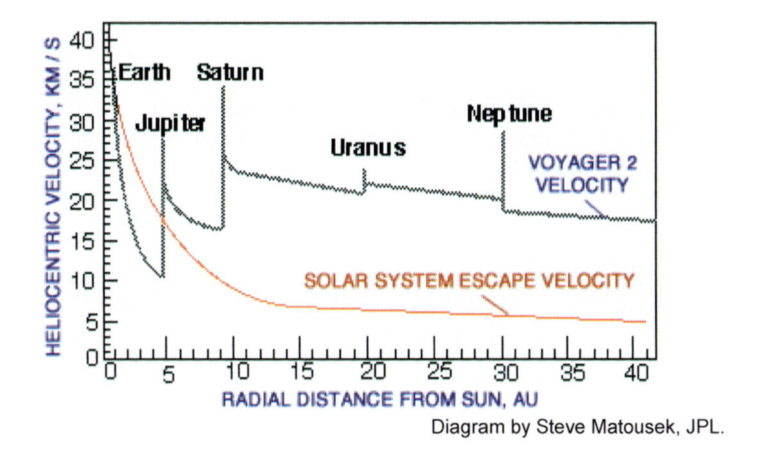

Diagram by Steve Matousek, JPL.

(Problems continued on next page)

23. Show that a spacecraft coming in from infinity towards a planet will be deflected by an angle 2θ, where $\theta = \tan^{-1}\frac{\frac{GM}{b}}{v^2}$, where b is the impact parameter and the initial speed at infinity as illustrated below. These kinds of "scattering" problems are of general importance in modern physics.

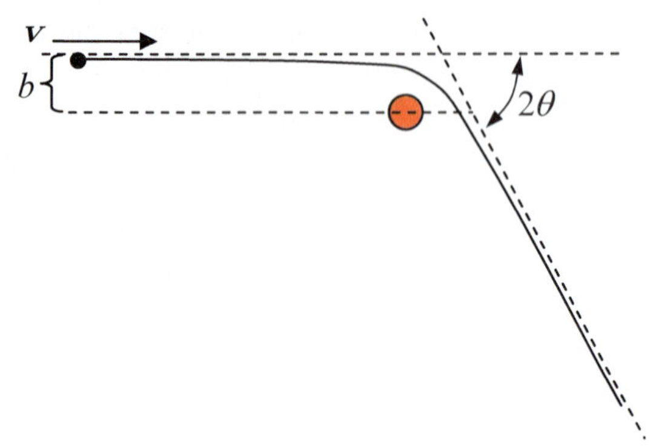

24. Satellite TV dishes point at one place in the sky, because they are pointing at a satellite that transmits the TV information, which is in geo-stationary (more loosely called geosynchronous) orbit. This means the satellite is in an orbit that exactly follows the rotation of the Earth. Calculate the radius of the equatorial orbit that is needed to accomplish this type of orbit.

25. Assuming the Space shuttle orbits at $300km$ radius above the Earth, a) how long does it take to go around the Earth once? b) Assuming an object could orbit right at the surface of the Earth, how long would it take to complete one orbit?

26. Why are the GPS satellites not in geo-stationary orbits? Given that GPS satellites are in a $12h$ orbit, what is their altitude? For circular orbits, what is the length of the semi-major axis of their orbits?

27. A rope of length L hanging from the ceiling has a uniform mass density λ; calculate the tension as a function of the distance from the hanging end of the string, $T(y)$.

(See *Interactive Physics* 8.6-tether hanging in uniform gravity.ip)

28. Some futurists have suggested that a tether hanging from a super space station in the sky is a good way to make space accessible. An elevator or other type transfer vehicle could then transfer people from some fixed station on the ground into space. In this problem, we investigate this proposal to discover whether a tether from Earth to orbiting satellite (the space station) is feasible. We do this by making some rough estimates. a) Calculate the tension in the tether as a function of distance and, making the appropriate approximations (do not attempt to solve the full equations), give the force the tether exerts on the satellite. b) Assuming that one can use a tether made of carbon nanotubes will the tether support its own weight? *Note:* carbon nanotubes have a density of $1300kg/m^3$ and a

tensile strength, the maximum force per unit area in tension that it can stand before it breaks, of $6.3 \times 10^{10} N/m^2$.

c) Consider the same project for a tether to Mars. Describe in *broad* outline how that might work assuming the orbital plane of Earth and Mars are exactly the same?

To get a feel for tethers in space, see *Interactive Physics* models below and try your own experiments. 8.7a-tether-dumbell model.ip
8.7b-tether-11 element model.ip
8.7c-tether- longer tether smaller earth radius.ip

29. It costs roughly ½ billion dollars for a shuttle to get into space, and given *SpaceShipOne*, the winner of the X-prize contest to get men into space *twice* within two weeks, costs $100,000. Assuming the tether hanging from an asteroid in geostationary orbit (see above problems), discussed above, were feasible, approximately how much would it cost per trip for energy. Assume the elevator or crawler that takes the ship up the wire is "plugged" into an electrical power grid which charges $.1 per kilowatt-hour. Assume the ship weighs 20,000 pounds.

30. Show that the popularly acclaimed "whatever goes up must come down" rule mentioned in Chapter 5 can be violated in a simple way by a ball that does not hit the Earth. Choose an appropriate throwing angle and calculate the speed it must be thrown at to miss the Earth. Try it in *Interactive Physics.*

31. a) Show, by integrating z from $-R$ to d, that the net force acting downward due to the bottom portion of the sphere (blue) is exactly the same magnitude as that acting upward due to the top (red), which is given in equation 8.12. b) Use this result to calculate the force pulling downward at the center of the spherical shell due to *only* the bottom part of the shell.

32. Prove: $E = \frac{1}{2}\mu\vec{v}\cdot\vec{v} = \frac{1}{2}m_1\vec{v}_1{}'\cdot\vec{v}_1{}' + \frac{1}{2}m_2\vec{v}_2{}'\cdot\vec{v}_2{}'$

33. The author of this book has given relativity colloquia in which he discusses the equivalence principle using a rubber band in free fall. Explain how a rubber band might be so used. (It may also be instructive to play with *Interactive Physics*: 8.8-Einstein equivalence principle toy.ip)

34. In the dropping mail through the center of the Earth problem in the text, the energy before release is $E = \frac{2\pi G m \rho}{3} R_{earth}{}^2$. From this, calculate the maximum speed of the capsule.

35. Recall the formula for the specific energy of an object falling through a hole drilled directly through a sphere of uniform mass density ρ is: $E = \frac{m}{2}v^2 + \frac{2\pi G m\rho}{3}r^2$. Using the analogy of this energy with the harmonic oscillator, calculate the period of one cycle of oscillation of such a falling object, and compare the result with the time calculated in the text.

36. Using spherical coordinates, give a general method for finding opposite points of the Earth and then show Bermuda island (Latitude 32.3N, Longitude 64.75W), and Perth Australia (32S, 116E) are about opposite each other.

37. Is it possible, in principle, neglecting all resistance and assuming Earth is a perfect sphere with no obstacles on it, to throw a ball from $h = 5\,ft$ off the ground such that the point of throw is the apogee rather than the perigee of the resulting orbit and prove your answer.

38. The definition of the ellipse, parabola and hyperbola are, as given in the text:

An *ellipse* can be defined as *the locus of points in a plane such that sum of the distances to each of two foci is constant* ($= 2a$).

A *parabola* can be defined as *the locus of points equidistant from the focus and a line called the directrix*, which is perpendicular to the line of symmetry of the parabola.

An *hyperbola* can be defined as *the locus of points in a plane such that the difference between the distances to the two foci is a constant* ($= 2a$), *i.e. the same for every point.*

a) An ellipse can be created by using two nails separated by a certain distance and a piece of string tied to make a closed loop as shown below. Carry out this construction and explain how it makes a figure that satisfies the definition of an ellipse.

As we move the nails (foci) further apart the ellipse drawn gets more oval.

b) Create a parabola using this property, a string and a "T square." Explain how this construct works to make the figure described by the definition of a parabola.

c) Create hyperbola using a straight edge and some string. As for the above, explain your construct.

39. Contrary to the approximation that we make nearly everywhere else in this text, when a massive body of charge q_1 is introduced suddenly into a new region, it actually takes a while for the electric field it generates to travel through "space" to any other body. That is, the field, the quality of the plana ("space") that can eventually apply a force on a second charge must first be activated *by* charge 1 in the part of space nearest to charge 1. Then, the field must be activated in the next part of space etcetera until it finally reaches charge 2. In short, the field has to "travel" through the parts of the vacuum in a finite time that we do *not* ignore. a) Assume that there is another body of charge q_2 in the region that has been there forever when the first charge is suddenly created; explain what happens to Newton's third law for massive bodies? b) Assume now that both charge 1 and 2 are *both* introduced suddenly into a region with no other fields present. Given the equation of electric interaction in equation 8.3, discuss what conditions must be placed on the speed of travel of the field in the two directions, if Newton's third law is, in some sense, not violated for massive bodies. Why do we say in some sense?

40. Tycho Brahe told Kepler to concentrate on Mars. Tabulate the eccentricities of Mars and the other planets that are visible with the naked eye. a) Leaving out Mercury, imagine drawing the ellipses corresponding to their orbits, could you determine visually from a drawing that they were ellipses rather than circles?

Kepler later was reported to have said.

"*By the study of the orbit of Mars, we must either arrive at the secrets of astronomy or forever remain in ignorance of them*"

(Problems continued on next page)

41. When stars get old, they eject material such as shown in these planetary nebulae (the name arises because, through small aperture telescopes, they look somewhat like the gas giant planets such as Uranus). Stars in the mass range that do this type of ejection become white dwarf stars, which are supported against gravitational collapse by the so-called electron degeneracy pressure. Neglecting the energy of the nebula, how much gravitational potential energy has been released from the sun when it reaches the white dwarf stage? Assume 50% of its mass, $M_{sun} = 2\times10^{30}\,kg$, is lost to the nebula and the radius goes from $r_i = 7\times10^{8}\,m$ to $r_f = 9\times10^{6}\,m$.

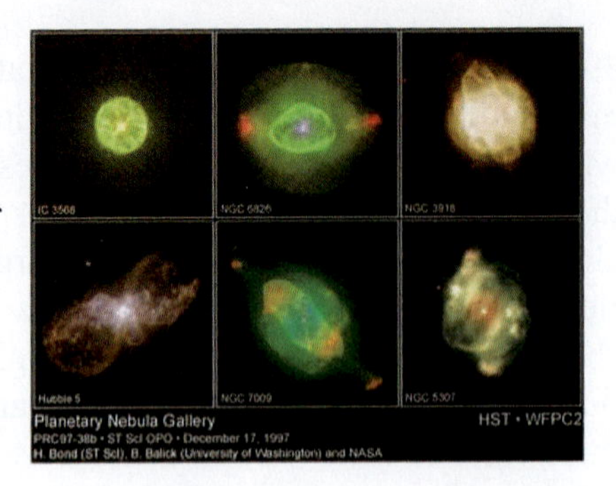

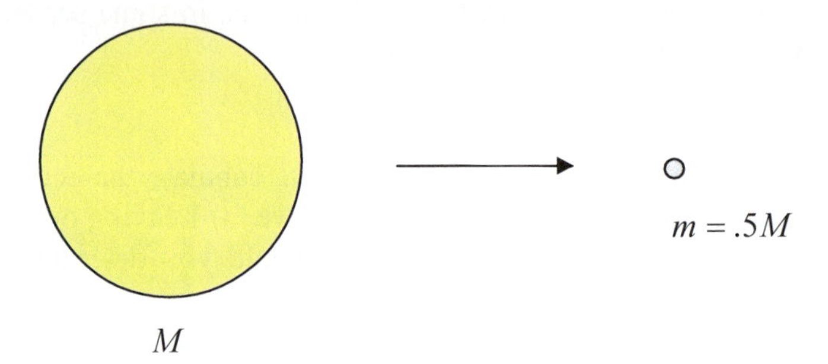

42. The Cavendish experiment measured the force between two everyday sized balls. Look up the Cavendish experimental setup, draw the experiment and explain briefly, but quantitatively, how the experiment can be used to measure the gravitational constant, G.

43. Explain why the form of the equations for the electric and gravitational forces, in a sense, automatically satisfies Newton's third law, implying something about the nature of these fundamental forces.

NASA, ESA, The Hubble Heritage Team (STScI/AURA), J. Bell (Cornell University) and M. Wolff (Space Science Institute)

Chapter IX

The Manned Trip to Mars

Introduction

The Mission

Though Mars is one of the most easily visible of the planets to the unaided eye, it is still only a small reddish-white dot in the sky. To the eye aided by the insight of the mind and the technology the mind conceives, we see that Mars is much more. Size, including apparent size, bears little indication of the object's importance. Indeed, that little planet, barely distinguishable from many stars, turns out to be so much more than its size that, in a statement reminiscent of the heavy commitment made by President John F. Kennedy to send men to the moon, on January 14th 2004 AD, President George W. Bush announced a plan for a manned mission to Mars.[1]

By the end of that month, many unmanned vehicles had already been to Mars (see Figure 9-1 for a list), including the highly successful Mars Rovers: *Spirit* and *Opportunity* (see Figure 9-3). In thinking about a trip to Mars, as for any major undertaking, one must tap deeply into past experience and understanding. Experimental physicists know that reality is complex and that neglected effects can end up delaying or even canceling an experiment. Obviously, the stakes are high for such a trip for human lives are at risk. The first task in developing our plan to get to Mars is to be clear about the immediate purpose, the reason we want to go to Mars, for that is the ultimate driver of our choices of how to go.

Mission[2]	**Launch**	**Termination**	**Objective**[3]	**Country**
Mariner 4	11/28/64	12/21/67	Flyby (7/15/65)	USA
Mariner 6	2/25/69	8/69	Flyby (7/31/69)	USA
Mariner 7	3/27/69	8/69	Flyby (8/5/69)	USA
Mariner 9	5/30/71	5/72	Orbiter (arrives 11/14/71)	USA
Viking 1	8/20/75	8/17/80	Orbiter (6/19/76)	USA
		11/13/82	Lander (7/20/76)	USA
Viking 2	9/9/75	7/25/78	Orbiter	USA
		4/11/80	Lander (9/3/76)	USA
Mars Global Surveyor	11/7/96	11/5/2006	Orbiter	USA
Mars Pathfinder	12/4/96	9/27/97	Lander / rover	USA
2001 Mars Odyssey	4/7/01	Currently operational	Orbiter	USA
Mars Express Orbiter	6/2/032	Currently operational	Orbiter	ESA
Spirit rover	6/10/03	Currently operational	Rover	USA
Opportunity rover	7/7/03	Currently operational	Rover	USA
Mars Reconnaissance Orbiter	8/12/05	Currently operational	Orbiter	USA

Figure 9-1: List of successful unmanned missions to Mars [4]

[1] http://www.usatoday.com/news/science/2004-01-14-bush-space_x.htm

[2] To show how fraught with problems and the few nations who are able to even try as well as to get an idea of success rate note that: (per Wikipedia): the USSR had 12 failures, and 5 partial successes (counting when orbiters associated with landers succeeded but landers did not), Russia (after fall of USSR) had 1 failure. US had 5 failures (includes the two *Mars Polar* crafts as one mission as above). Japan and the European Space Agency (ESA) each had one failure.

[3] Per *The Case for Mars* by Zubrin, Touchstone, NY 1997

The purpose of the mission is to understand more about Mars through direct exploration. Why Mars? Mars has much to offer in its remarkable differences from Earth as well as its similarities. Appendix III is a compilation of facts about Mars. Mars is a solid planet like Earth and unlike the planets beyond it.[5] Mars looks like parts of the Utah desert as seen in the image below (Figure 9-2):

Figure 9-2: Mars near *Columbia Hills* (in Gusev crater) shown from Mars Rover Spirit on July 15, 2005.[6] These hills are dedicated to the astronauts who died in the space shuttle Columbia.[7] At the top of Husband Hill, the tallest hill which is named after the ship's commander, Rick Husband, movies were captured of a bevy of dust devils. These dust devils were responsible for keeping the dust off the solar panels that power the rovers. In the pre-flight worst case analysis, engineers and scientists originally thought that Martian dust would eventually accumulate on the panels ending that rover's mission. The rovers were designed to last just 90 sols (Martian days), but they have lasted from the time of landing in January 2004 AD to the moment of this writing (June 2007), which is over 1000 sols.

Mars does have an atmosphere, though its pressure is 1/100 that of Earth. Mars does have water, though it is not normally in liquid form on the surface (see section titled *Time on Mars*). Mars as far as we know doesn't have life now, but may have in the past. If it did, it was most likely simple microbial life. In any case, Mars' similarities to Earth make it attractive because this increases the chances that it can give us insight into the richest arena (because of its life, including its plants, animals and, at the summit, us) of the physical

[4] http://en.wikipedia.org/wiki/Exploration_of_Mars#Launch_windows

[5] The planets are, in order from the Sun: Mercury, Venus, Earth, Mars, Jupiter, Saturn, Uranus, Neptune and Pluto, which can be memorized by the following sentence: **M**an **V**aliantly **E**mbarks on **M**agnificent **J**ourneys into **S**pace to **U**nlock **N**ew **P**hysics. Though Pluto is not technically a planet, we include it because of its longstanding importance in learning about the planets, including having been considered one until 2006.

[6] http://marsrovers.nasa.gov/spotlight/spirit/20050713.html

[7] Columbia is a feminine of Columbus often used poetically to designate the United States as in the previously popular song *Columbia the Gem of the Ocean*.

universe we know so far, the Earth.[8] Still, both the differences and similarities are helpful in increasing our understanding. Sometimes we have to see something in a different context before we really understand it or even notice it.

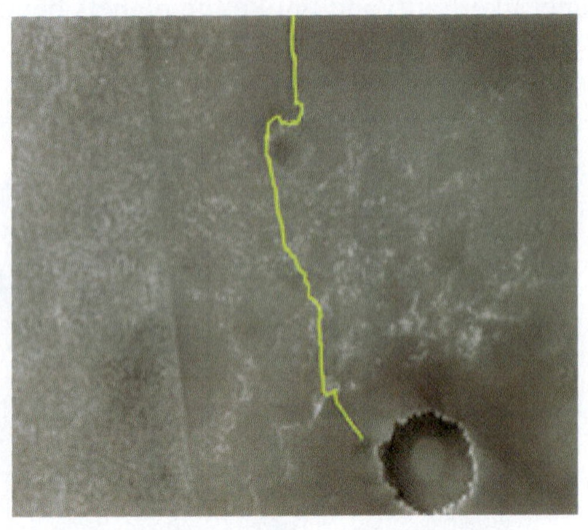

Figure 9-3: Path of rovers on Mars for *about* first 1000 sols of their missions. *Left*: Spirit, *Right:* Opportunity

We can divide our journey into three parts: *getting there*, *time on Mars* and the *return to Earth.* This rough division can serve us as long as we don't make it absolute, for obviously many issues cross such boundaries.

Getting To and Returning from Mars

Getting to Mars involves launching from the ground, speeding through space, and landing on Martian ground. The return trip is similar. Since the principles are fundamentally the same, we will only calculate the outgoing trip and leave the return trip details as problems at the end of the chapter. In particular, in a simple analysis, transit to Mars can be thought of as six phases:

1. Rocket launches vertically, bends over to optimal elliptical transfer orbit.
2. Inject into circular low Earth orbit.
3. Inject into an orbit that sends it on an intercept path with Mars where trajectory corrections are applied at various points along the way.
4. Inject into circular orbit around Mars.
5. Enters elliptical orbit by aero-braking, and then soft lands on Mars.

1-Rocket Launch

The first phase, the rocket launch, is conditioned by the other phases. So, at this point, all we need for it is the generic rocket equation which we recall from Chapter 4:

9.1 $$\Delta V = V_{final} - V_{initial} = \boldsymbol{v}_{exhaust} \ln\left(\frac{M_{initial}}{M_{final}}\right) - g\,t$$

[8] Of course, this does not mean that other areas of the universe do not give us insights that cannot be had on Earth. For example, one would likely never learn about the reality of the black holes, and the principles behind them, by looking at the Earth alone. However, once one does understand them; one has a deeper understanding of the generic principles operating in the universe, including on the Earth.

We also note chemical rockets have many advantages over other possible technologies at the point of this writing. Chemical rockets will require fuel and an oxidant such as the hydrogen-oxygen main engines of the Space Shuttle. An example of such a rocket is shown in Figure 9-4 below.

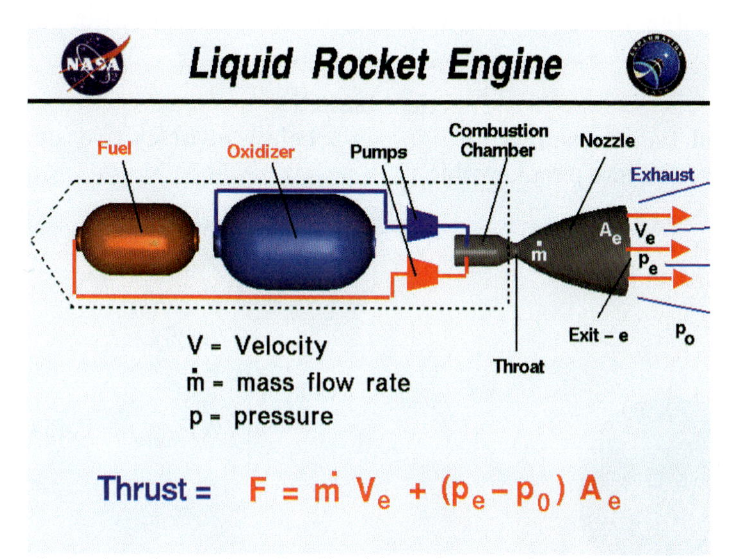

Figure 9-4: A schematic of a liquid rocket engine that produces a thrust based on fast particles generated in the combustion chamber by chemical exothermic reactions.

In problems at the end this chapter, we look at various types of rockets and use the principles that we have already established in earlier chapters to determine what type and how much fuel that we need.

2-Injected into Circular Low Earth Orbit

If there were no atmosphere, a Hohmann transfer elliptical orbit (see Figure 9-6b) would expend the least energy in transferring the craft to a low Earth orbit. For such a transfer, the rocket launches tangent to the Earth at sufficient speed so that the point of launch at the Earth's surface becomes the periapsis (perigee) of an elliptical orbit, and the point of the radius of the low Earth orbit opposite the launch point becomes the apoapsis (apogee). We would then fire rockets at the apogee and establish the rocket in the low Earth circular orbit; hence, the spacecraft would only follow half the elliptical trajectory of the Hohmann orbit.

In fact, we do not do this, for the Earth's dense atmosphere causes significant drag, resulting in large energy expenditures for more horizontal launches (cf. Figure 9-6). Thus, in such cases, a more vertically directed orbit (Figure 9-5) that goes through less atmosphere is energetically favorable, expending less fuel. Still, the calculation of the Hohmann orbit in the absence of atmosphere can give us a lower bound on the required transfer energy and calculation of the vertical transfer energy (also ignoring atmospheric drag) will give us an idea of how much atmospheric drag we should be willing to tolerate.

We leave the more complex problem of working out the optimal orbit for the problems at the end of the chapter. The trajectory of such an orbit is called a *synergy curve* and is computed so that the rate of kinetic energy lost *by gravity* is equal to that lost *by air resistance*. Other insertion optimization issues not related to energy expenditure, such as structural integrity and astronaut comfort and safety, are also discussed in the problems. Now, we turn to analyzing the two orbits of interest: the vertical transfer and the Hohmann orbit transfer.

Vertical launch

The vertical transfer shown in Figure 9-5 below involves two steps: the vertical assent then the insertion into circular orbit.

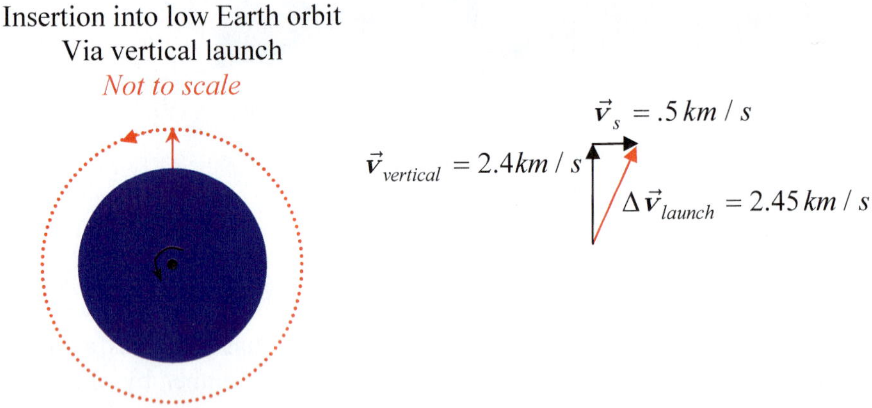

Insertion into low Earth orbit
Via vertical launch
Not to scale

$\vec{v}_s = .5\,km/s$

$\vec{v}_{vertical} = 2.4 km/s$

$\Delta\vec{v}_{launch} = 2.45\,km/s$

Figure 9-5: Rocket is put in circular orbit by firing vertically until proper altitude is reached, and then the rocket is fired to give it the velocity needed to enter a circular orbit at that altitude. This maneuver lowers the atmospheric drag relative to a Hohmann transfer. However, it does so at the expense of not being able to use the Earth's field or Earth's orbital motion, indeed causing one to work against both as shown in the text.

The speed needed for vertical assent is simply found by using the potential energy difference between the ground and the low Earth orbit. We get:

9.2
$$v_{vertical} = \sqrt{2\left(\frac{GM_{earth}}{r_e} - \frac{GM_{earth}}{r_e + h}\right)} \qquad note: r_{circle} = r_e + h$$
$$\sim \sqrt{\frac{2GM_{earth}}{r_e}}\sqrt{\frac{h}{r_e}} \sim \sqrt{2gh} \sim \sqrt{2\cdot 9.8\,m/s^2 \cdot 300 km} = 2.4\,km/s$$

Here, we assume that the field of gravity doesn't change much over the altitude from ground to the low Earth orbit (of say $300km$). In this approximation, $\sqrt{2gh}$ is the speed with which a ball must be thrown upwards in order for it to reach a peak height of r_c under the influence of a constant gravity field, g, acting on the ball.

For the circular orbit, we calculate the speed needed by equating centrifugal and centripetal force, giving:

$$\mathbf{v}_c = \sqrt{\frac{GM_{earth}}{h + r_e}} \sim \sqrt{\frac{GM_{earth}}{r_e}} \left(\sqrt{1 - \frac{h}{r_e}} \right) \sim 7.7\, km/s \tag{9.3}$$

Now, we need to incorporate one further piece of information. The Earth is rotating, at speed $\mathbf{v}_{surface} = \frac{2\pi r_e}{1\,day} \sim \frac{6.28 \cdot 6400\, km}{24 hours} \frac{1\, hour}{3600} = .47\, km/s$, so the initial launch thrust must have a component that: **1)** deactivates the impetus that is moving the rocket horizontally along with the surface of the Earth and **2)** establishes the needed impetus in the vertical direction. As illustrated in Figure 9-5 above, this means the net incremental $\Delta\mathbf{v}$ needed is:

$$\Delta\mathbf{v} = \sqrt{2.4^2 + .5^2}\, km/s = 2.45 km/s\,. \tag{9.4}$$

In summary, the *first burn* (launch) requires $\Delta\mathbf{v} = 2.45 km/s$, the *second burn* (injection to circular orbit) requires $\Delta\mathbf{v} = 7.7 km/s$.

Hohmann Transfer to Circular Orbit

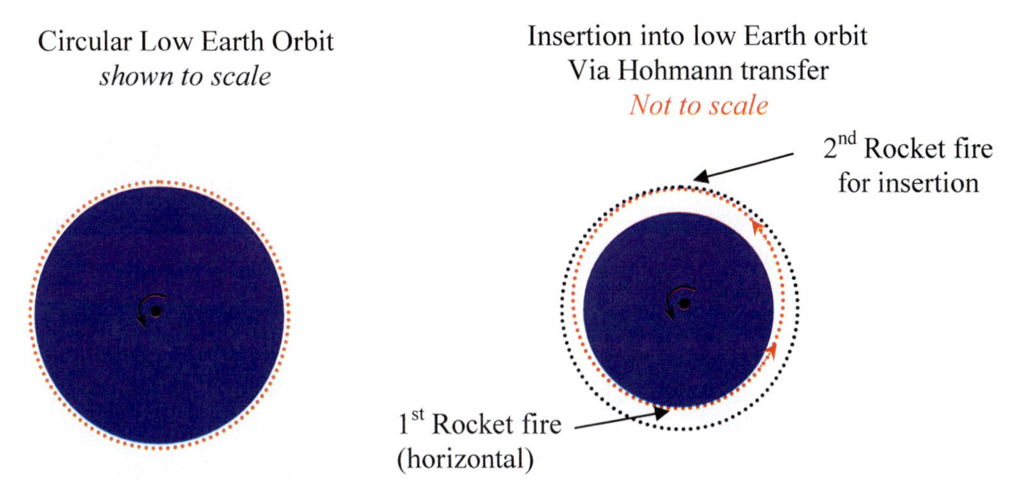

Figure 9-6: a. *(left)*: Low Earth Orbit (LEO) shown to scale. **b.** *(right)*: Hohmann transfer orbit from surface of Earth to tangent to LEO. Both views are as seen from observer directly and far above the North Pole. Notice the large amount of atmosphere the rocket must traverse in such a transfer orbit.

To calculate the $\Delta\mathbf{v}$'s for the Hohmann transfer, we first apply conservation of angular momentum. Namely, the angular momentum of the spacecraft is the same at apogee and perigee, which implies:

$$\mathbf{v}_a r_a = \mathbf{v}_p r_p \tag{9.5}$$

Next, conservation of energy implies:

9.6 $$-\frac{GM}{r_p}+\frac{1}{2}m v_p^{\;2}=-\frac{GM}{r_a}++\frac{1}{2}m v_a^{\;2}$$

Or,

9.7 $$2GM\left(\frac{1}{r_p}-\frac{1}{r_a}\right)=v_p^{\;2}-v_a^{\;2}$$

Substituting equation, 9.5 into 9.7 gives:

9.8 $$v_p=\sqrt{\frac{2GM_{earth}}{r_p+r_a}\frac{r_a}{r_p}}$$

Using: $r_p=6400km$, $M_{earth}=6.0x10^{24}\,kg$, $r_a=r_{circle}=h+r_e=6700km$
$G=6.67\times10^{-11}\,N\,m^2\,/\,kg^2$ gives: $v_p=8\,km\,/\,s$.

Now, the speed of the rotation of the Earth is, from above, $v_{surface}\sim.47\,km\,/\,s$, so, since both speeds are in the same line, the total speed required at launch is:
$\Delta v_{launch}=8\,km\,/\,s-.5\,km\,/\,s=7.5\,km\,/\,s$

At the apogee of the orbit, we then attain a speed of: $v_a=\frac{r_p}{r_a}v_p\sim7.2km\,/\,s$; so, to get up to $v_c=7.7km\,/\,s$ given in equation 9.3, we must apply an increment of only .5*km/s*.

Hence, for the *Hohmann transfer*, the *first* rocket burn (launch) must cause $\Delta v=7.5km\,/\,s$, while the *second* burn (orbit insertion) must cause $\Delta v=.5km\,/\,s$; this compares with 2.45 and 7.7 respectively for the above *vertical launch*. Not only are the latter speed increments *both* larger, but the Δv is larger for the second burn than the first in the vertical launch. This means we have to carry the fuel against the force of gravity for longer in the vertical launch. This is a particular reason for a more generally applicable rule that less fuel is expended when a given velocity increment is applied at lower altitude. Again, this is because the rocket sheds the fuel needed for the increment in velocity before climbing further out of the gravity field well, rather than afterwards.

A fundamental rule of astronautics

The most general statement of this rule of astronautics, sometimes called the fundamental rule of astronautics, is: *for minimum energy expenditure apply braking and launching thrusts (Δv's) as close to the center of attraction as possible*. Like all such rules one must be very careful in applying it. Understanding a thing is the proper way to know how to exercise care in its use. To understand the general reason for this rule, we need the relation between the kinetic energy and the speed; in particular we are interested in the relation among their differentials:

9.9 $$\Delta KE=m\,v\,\Delta v$$

Note the larger the v the more a Δv is effective in changing the kinetic energy. To understand its import, consider two cases: a craft in an elliptical orbit and a craft landing vertically.

In the elliptical orbit, the craft moves most swiftly when it is at its periapsis. Thus, application of a $\Delta \boldsymbol{v}$ tangent to the orbit at perigee will change the craft's kinetic energy the most, namely by $m\boldsymbol{v}_{peri}\Delta\boldsymbol{v}$, whereas application of the same $\Delta\boldsymbol{v}$ at the apoapsis would result in only a change of $m\boldsymbol{v}_{apo}\Delta\boldsymbol{v} = \left(\frac{r_p}{r_a}\right) m\boldsymbol{v}_{peri}\Delta\boldsymbol{v}$.[9]

For the case of landing vertically, we see that if the rocket is fired to decrease the craft's speed when the craft is some altitude above the ground, the gravitational field will again speed the craft up as it falls to the ground from that altitude. This means we would need a second burn at some point. Furthermore, since, in such a case, both the first and second burns are applied when the craft is going at slower speeds than for a single burn nearer the ground, the thrust induced $\Delta\boldsymbol{v}$'s are less effective in changing the kinetic energy.

Taking off is similar; getting up to speed all at once allows us to increase the intensity of an already large impetus, giving those last incremental increases of momentum the ability to get further, because that high intensity of impetus carries it further through the field. The gravitational force field, which we write as $\vec{F}(\vec{r})$, will gradually "drain" the momentum; thus the more momentum, for a given fixed mass, the faster it will go through the regions of high field intensity (i.e., it craft has more speed), thus lessening the field's effect on the craft's final speed (relative to if we tried, instead, to give a little at the beginning and a "remedial" amount at some intermediate point). Again, while the work done is *only* dependent on the radial distance through the field (recall $W = \int \vec{F}\cdot d\vec{x}$ and, for gravity: $\vec{F} = -\vec{\nabla}\phi$), the momentum change is dependent on the time during which the force acts: $\Delta\vec{p} = \int \vec{F}dt$ (see end of chapter problem).

Notice the *time symmetry* of the solutions between take off and landing. Time symmetry refers to the following fact. Consider a craft launched to space from Earth. The craft will have to climb out of Earth's gravitational potential well, and thus have its momentum decreased as it goes until "at infinity" (notice the "being of reason" way of speaking), it moves at some asymptotic speed called the *hyperbolic excess speed,* $\boldsymbol{v}_h$.[10] Now, if we roll the video of this craft's trip backwards, the craft appears to be falling towards the Earth initially at a speed $\boldsymbol{v}_h$ and then gradually moves faster and faster as it falls. Interestingly, a craft that actually started falling in from "infinity" with a speed $\boldsymbol{v}_h$ would, as the craft's impetus intensified due to the action of the Earth's field,[11] increase in speed in the manner seen in such a reverse video. In a certain way (if we keep in mind what we really mean), we can say running time backward is the same as running time forward. This is a statement of time symmetry. Of course, time symmetry shows up in our equations. In particular, in time symmetric equations, we can replace t with $-t$ and still have the same equations (*exercise:* show this for Newton's 2nd law).

[9] This doesn't mean there are not reasons for firing at apoapsis. Such firings are just not the lowest energy maneuvers of the type we are describing. To crash on the surface, for instance, we could use an apoapsis firing after entering a highly elliptical orbit.

[10] Note the escape speed for analogical reasons is called the *parabolic speed.*

[11] More precisely, as the plana activated by the Earth's field acted on the craft.

This time symmetry can be seen "operating" in the two mentioned reasons for firing rockets close to the center of attraction: 1) Firing the rockets so that the efficacy of firing at different craft speeds (not velocities, i.e. whether ingoing or outgoing) is accounted for 2) Firing the rockets so that fuel weight is best utilized. For instance, consider how time symmetry arises in reason number two. For take-off, the mass ejected at launch to get the rocket to speed obviously no longer needs to be brought through the gravitational field as it would if we waited till we get into space to do part of the burn. And, for landing, holding onto the mass till the end makes all the fuel mass available for braking the craft (payload), not the fuel. Just as in the launch case there is no point in carrying the *fuel* mass against the field, in the landing case, there is no point in spending fuel in countering the action of the field in speeding up the fuel, which is what we do if we fire the rocket before its fall is complete. (True, we slow the craft as well, but we can do that more effectively when it is moving faster.) Indeed, even if *all* the fuel is burned high above the surface (which insures the craft will crash, not land), the speed of the exhausted fuel will be less than it would be if the burn happened just before the surface; i.e. we would have effectively slowed down the fuel. Of course, we do not carry fuel in order to slow it down or just for the sake of carrying it but to use it for stopping or accelerating the craft (with instruments, people and/or cargo). Such arguments are of great value in revealing the physics of a situation and should also be used as entry points to quantitative analysis that will reveal even more under thoughtful reflection.

Now, such time symmetry disappears when we include dissipative effects such as aero-braking. Because of such effects, ordinary life is not time symmetric. For example, we don't expect a cup to suddenly begin sliding across the table at increasing speed, though we do expect that if we slide a cup across a table, it will gradually slow down, finally coming to a stop.

Dissipative actions such as friction result in *time dependent* forces that destroy time invariance (time symmetry). Similarly, if the Earth's field, say increased with time, we would, generally speaking, lose time reversible dynamics for space flight. Or, if impetus itself were to "wear out" we would also violate time reversal invariance. To see this latter, consider a system of free bodies in the absence of any forces in which each body starts with particular impetus activated in it (called the initial conditions) at say 9:00AM; obviously, the bodies will move in a certain way. If at noon, we were to instantly reverse the directional tendency of the impetus of all the bodies (while keeping the magnitude of their intensities the same), each body would retrace its original path *at the same speed* but in the reverse direction until they finally arrived again at the same location with the same speeds that they had at 9:00AM. By contrast, this would not happen if (say) impetus were such that its strength increased with time. In that case, the balls would be going very much faster than they were initially and would, for instance in a simple no-collision situation, get back to the initial spatial configuration much faster than they left it and with different speeds. Thus reversing the initial conditions at some point would not result in a retracing of the same system states; it would not be like running the movie backwards. In other words, we would lose time symmetry.

Still, for flight in space in which we only consider the action of gravity of relatively fixed bodies (or if we don't have relative fixity, if we include, at some level, the action of all bodies involved), [12] arguments for landing can be turned around and used for braking. In other words, we often consider such time symmetric situations in which we assume the forces at play are *approximately* time independent.

A Word about Approximations

We pause for a paragraph to focus on this important idea of making suitable approximations that is so important in physics. As we've seen before, we often consider such limited situations even though they may never really be completely true. In doing so, we leave out causes and/or effects that may actually be necessary to the existence of the very substances we are considering.[13] But, this selective leaving out of certain things allows us to concentrate on the dominant causes and effects in the given situations. We choose to do this because, in a given situation, the magnitude of the measured effects seen (and hence the causes) are largely controlled by the causes we choose to consider, not by those we leave out. Indeed, as is usual in physics, the calculations we have done so far in this chapter have ignored causes that exercised less significant effects on the motion.[14] In those and other cases, we can always return later and add the causes/effects that were not as important in the given situation to get a better and more complete understanding as well as a better quantitative approximation to the variables we are interested in.[13] For anyone wishing to learn physics, it is of paramount importance to gain expertise in discerning what effects are dominant in a problem and what are secondary. Working through problems thoughtfully is probably the best way to develop that expertise.

Hohmann versus Other Transfers

The analysis of "the fundamental rule of astronautics" gives us insight into why, neglecting atmospheric drag (which is not negligible for the Earth, but is for other places), the Hohmann transfer requires much less energy than the vertical transfer. It is left as an end of chapter problem to prove that, under certain conditions, the Hohmann orbit is the least energy transfer orbit.

Circular Orbit Once Attained

Now, taking into account drag, we finally establish ourselves in a low Earth orbit by the synergistic path which is a combination of vertical followed by horizontal flight (Hohmann). Once we've established low Earth circular orbit, we need to know the time for the satellite to complete one orbit. The period depends on the orbit height which we take to be approximately the height of a shuttle mission. Figure 9-7 shows shuttle mission STS-73 (Space Transportation System, mission 73) and some of its orbital parameters; its orbit height, h, is about $270km$. Ignoring the mass of the satellite, as we have been doing, we get:

9.10
$$\frac{T^2}{(R+h)^3} = \frac{4\pi^2}{GM}$$

---

[12] Gravitational radiation can "carry away" momentum, and thus cause a dissipative effect; we here exclude this small effect on the approximation grounds discussed shortly.

[13] For example, in the case of atoms, initially no one was aware of "spin" since it did not have any large effect on measured spectra at the time. Thus, it was not included in the calculation of spectra, and even today is not included for rough calculations. Yet, without the Pauli Exclusion Principle, which comes from the spin ½ nature of the electron, the atom would *not even be stable*, and thus no spectra would be possible. Later, the smaller effects of spin on spectra were calculated and measured experimentally.

[14] We also focus on a given time and distance scale.

Using: $\frac{GM}{R^2} \equiv g$, $R = 6400km$, $h = 300km$ *implies* $R + h \sim R$ yields a period of:

$$9.11 \qquad T = \frac{2\pi}{\sqrt{GM}}(R+h)\sqrt{R+h} \sim \frac{2\pi}{\sqrt{\frac{GM}{R^2}}}\sqrt{R} = 2\pi\sqrt{\frac{R}{g}} = \frac{2\pi}{\sqrt{\frac{g}{R}}} = \frac{2\pi}{\omega_{sat}}$$

So the formula is like that of a pendulum of length approximately the size of the Earth giving $\omega_{sat} = 1.2\times10^{-3}\sec^{-1}$ which implies $T \sim 5100$ seconds ~ 1.4 hours.

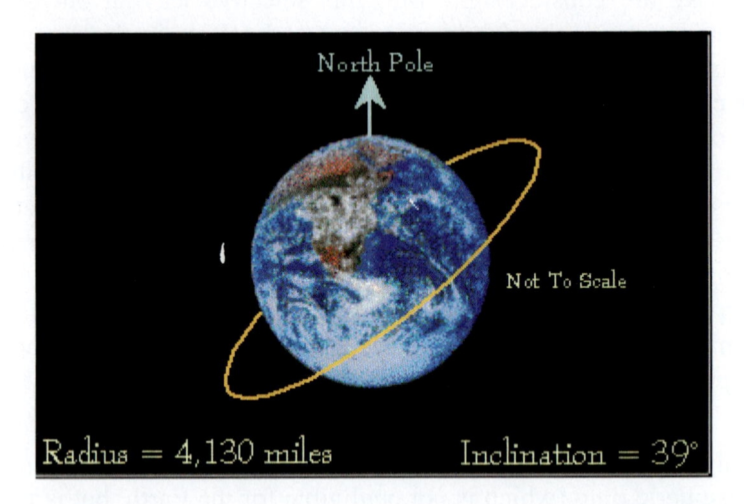

Figure 9-7: A typical shuttle orbit appears as shown; the parameters given were the ones actually used for STS-73.[15] The approximate values of three key shuttle orbit parameters are: $Radius = 6,647.782km\ (4,130mi)$, $orbital\ inclination = 39°$. $Velocity = 7,747\ m/s\ (17,330\ mph)$

3- Transit from Earth to Mars

To get to Mars from our low Earth circular orbit, we consider the low energy Hohmann orbit shown in Figure 9-8 below.

As for the case attaining Earth orbit from the ground, the orbit of the spacecraft around the Sun that takes the craft from Earth to Mars is called its transfer orbit or trajectory. Constructing such a transfer orbit is like throwing a football to a moving receiver; the quarterback has to lead him by just the right amount. But the transfer is more than this; one has to account for the radial pull of gravity, first of the Sun then of all the planets the craft gets near. The opportunity to launch a spacecraft on a transfer orbit to Mars occurs about once every 25 months.

[15] From: http://liftoff.msfc.nasa.gov/academy/rocket_sci/orbmech/orbit/orb_pic/orb2i.html

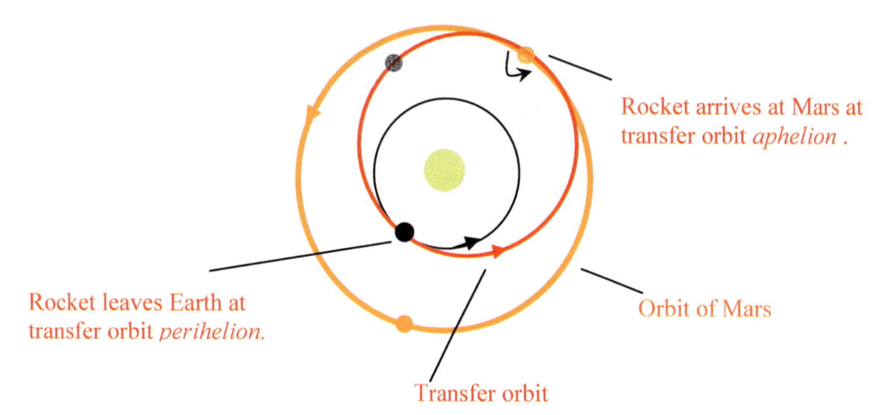

Figure 9-8: Hohmann transfer orbit viewed from the north celestial pole. In such a view, the orbits of Mars and Earth both revolve *counterclockwise* around the Sun, *and* they *also* each rotate *counterclockwise* around their own axis, being tilted by about 23.5 and 25.2 degrees from the Earth and Martian orbital plane, respectively. We approximate the Martian orbit as circular; this approximation is only very rough for Mars, so we need to include, at some point, its elliptical nature (it has $eccentricity = .0935$ or 6 times that of Earth).

Hohmann Transfer Neglecting the Earth's Field

To obtain the speed needed to get to Mars by Hohmann transfer orbit, we rewrite equation 9.8, which gives the periapsis speed for orbits around the Earth, for an orbit around the Sun, obtaining:[16]

9.12
$$v_p = \sqrt{\frac{2GM_{Sun}}{r_p + r_a}\frac{r_a}{r_p}}$$
$$= 32.75\, km/s \quad (Earth - Mars\ Hohman\ perihelion\, speed)$$

In the second line, we have used the parameters given in Figure 9-11 for Mars, but we have assumed a circular orbit for Mars ($\varepsilon \sim .09$) and Earth (this latter is good approximation, $\varepsilon \sim .02$). In particular, in this equation we take:

$r_a = r_{mars} \equiv \dfrac{r_{a\_Mars} + r_{p\_Mars}}{2} \approx 228 \times 10^9 m,\ r_p = r_e = 150 \times 10^9 m$ and $M_{sun} = 2.0 \times 10^{30} kg$. Thus, 32.75 *km/s* is the speed needed to get to Mars from Earth's radius. However, the Earth is moving in orbit around the Sun in the direction of Mars, so we already have some of the relative speed needed;

Interactive Simulation

[16] One can calculate values for an *Interactive Physics* simulation in following way. We do simulation using $r_{earth} = 3m$, $G = m^3/(kg\, s^2)$, $M_{sun} = 10kg$, $r_{mars} = 4.5m$ gives $\omega = 1$ or $T = \frac{2\pi}{\omega} \sim 6\sec$, which means $r_p = 3m$ and $r_a = 4.5m$ giving $v_p = 3.464 m/s$ (see simulation 9.1-Earth to Mars- Hohmann.ip) time scale is: 1 *earth year* $\sim$ 6 *second* ; 2 *months* $\sim$ 1 *second*

the impetus of the Earth is already carrying us in the right direction at (using equation 9.3 for circular orbit speed):

9.13 $$v_c = \sqrt{\frac{GM_{sun}}{r_{earth-sun}}} \sim 29.8\, km/s\,.$$

Thus, *neglecting* the energy needed to escape Earth's pull in order to get to Mars on a Hohmann transfer orbit, our rocket needs an increase of speed of:

9.14 $$v_{without\, earth's\, field} = v_{wef} = v_p - v_c = 32.75 - 29.8\, km/s = 2.95\, km/s$$

Incorporating the Field of the Earth

Of course, we now need to include the significant contribution of Earth's gravity. The speed needed on the ground to escape the Earth's field (i.e. to get very far away) is, using conservation of energy:

9.15 $$v_{escape} = \sqrt{\frac{2GM}{r}}$$

Note the factor of two in this equation not present in the equation for the circular orbit speed of the Earth. Substitution of Earth parameters into the escape velocity equation 9.15 gives 11.2 *km/sec* or 7 *mi/sec*. The question now is: if we were launching from a non-rotating Earth or other fixed point in orbit around the Sun at the Earth's position, at what speed must our rocket launch? In other words, how do we use these two speeds: v_{wef} and v_{escape} to get the speed increment that would be needed for a launch from a non-rotating Earth?

To do this sort of calculation, we recall that gravity is a conservative force which means the path an object takes doesn't matter but only the difference in potential energy (which in this case is a function of the radius only) between the beginning and ending points. Furthermore, it is helpful to note that the contribution of potential energy from two bodies is simply the sum of that contributed by each. Hence, we can calculate the change in kinetic energy due to each separately. The last thing we need is the principle of equivalence that guarantees that if we are freely falling under the influence of a gravitational field that we can neglect the action of that field and proceed as if, for calculation purposes, there were no field.

In particular, we know what we need to get to Mars accounting *only* for the Sun's field (i.e. the field that the Sun activates throughout the solar system plana). The speed we would need at launch to act against the Sun's field and have the requisite speed at apoapsis if the Earth had no gravity is 2.95 *km/s*. Using the equivalence principle by assuming the Sun's field doesn't change much over the most significant range of the Earth's field and assuming the Sun's field doesn't (nonlinearly) affect the Earth's field we can do the calculation of the escape speed as we have above, namely as if it were taking place in an inertial frame with just the Earth's field acting on the craft. Then, considering the rocket as departing from a non-rotating Earth, we find, substituting into equation 9.15 as already done above, that the speed needed to move against Earth's gravity and just come to rest (have zero speed with respect to the Earth) "at infinity" is 11.2 *km/s*. Now, the total *kinetic energy* we need is the sum of the kinetic energy needed to overcome the Sun's field and that needed to overcome the Earth's field. This, in turn, means, using the proportionality of kinetic energy to the square of the speed, the total speed we need, accounting for the Earth's field, to get to an orbit the radius of Mars leaving from a non-rotating Earth is:

9.16 $$v_{total} = \sqrt{v_{wef}^{2} + v_{escape}^{2}} = \sqrt{\frac{GM_{sun}}{r_{earth-sun}}\left(\sqrt{\frac{2r_{m-s}}{r_{e-s}+r_{m-s}}}-1\right)^{2} + \frac{2GM_{earth}}{r_{earth}}}$$

$$\sim \sqrt{2.95^2 + 11.2^2}\, km/s = 11.6\, km/s$$

Incorporating Speed of Circular Orbit around Earth

However, we are still not done, because we start not from a non-rotating Earth but from a circular orbit which already has us moving in the right direction. So, taking away the speed of the circular orbit given by equation 9.3, namely 7.7 *km/s*, we get:[17]

9.17 $$v_{launch} = 11.6 - 7.7 = 3.9 km/s$$

(*Exercise*: calculate circular speed for an object orbiting right at Earth's surface; what then would v_{launch} to Mars be?). Finally, this is the launch speed needed (we say it is the Δv required) to insert the spacecraft (from the circular orbit) into an orbit with an *effective* speed at perihelion (Earth's radius) of about $v_p = 32.75\, km/s$, so that, neglecting Mars' field, the speed at the aphelion (when spacecraft is at Mars' radius in its Hohmann elliptical orbit) is:

9.18 $$v_a = \frac{r_p}{r_a} v_p = \frac{150km}{228km} 32.75 km/s = 21.5\, km/s$$

4 and 5: Injection into Circular Orbit and Landing on Mars

Now, Mars is moving at a speed: $v_{Mars} = \dfrac{2\pi \cdot 227.92 \times 10^6 km}{686.98\, day\, 86{,}400\, s/day} = 24.1 km/s$. If there were no Mars' field, and no atmosphere on Mars, we simply would speed up by about 2.6 *km/s* right at the moment to reach the surface. However, Mars both generates its own gravitational field and has "air" (95% CO_2, the rest mostly N_2 and Ar).

Incorporating the Martian Field

Since there is a field, speeding up would only require us to make an even more severe braking maneuver. This problem is very similar to the one we tackled at the end of the last section in which we incorporated the gravitational field of the Earth after accounting for that of the Sun. As before, the potential energy is converted into kinetic energy, namely the field gradually increases the impetus of the spacecraft in such a way that:

9.19 $$v_{final}^{2} - v_{initial}^{2} = \frac{2GM_{mars}}{r_{mars}} = v^{2}_{escape\ mars}$$

The last equality converts the potential energy to its equivalent kinetic energy written in terms of the escape velocity. Concretely, $M_{mars} = .642 \times 10^{24} kg$ $r_{mars} = 3400 km$ giving:[18]

---

[17] Notice the multiple approximations we have made. Among others, we have neglected the action of the Earth and Mars on the Sun, the craft on all bodies (a very small effect), the field of the planets after a certain distance (because for instance we take the distance to the craft at Mars (from Earth) to be "infinite" (this has the effect of leaving out the time dependence of the fields that results from their motion), the solar winds acting on the craft, the light pressure acting on the craft. *Exercise*: list other effects that we've neglected.

9.20 $$v_{escape\,Mars} = \sqrt{\frac{2GM_{mars}}{r_{mars}}} = 5.0km/s$$

To further emphasize and understand the causes at play, consider two scenarios in the rest frame of Mars. First, assume the craft is far away from Mars, moving away at the already mentioned speed of 2.6*km/s*. Further, suppose that the craft reached this state by being launched from Mars with enough impetus that after the draining action of Mars' field as it escaped Mars' gravity it still maintained a 2.6*km/s* motion. Second, consider the Hohmann transfer from Earth to Mars. Neglecting the effect of the Martian field, the craft would arrive *at* Mars with a relative velocity such that the craft would fall behind at a rate of 2.6*km/s*. This is just like the final state of the first situation, except now the craft is near Mars and thus must have already fallen through the gravity field of Mars, though we *did not* account for it yet. Clearly, two sources of fields are changing the impetus: that of the Sun and that of Mars and need to be accounted for.

How do we get the requisite velocity increment that accounts for both? Knowing that the craft's speed at arrival at the Martian orbit but very far from Mars would be $-2.6km/s$ and knowing the first scenario describes a craft that gets to that state from launch, we note that the required speed to attain such a state is the launch speed from the first scenario. This launch speed accounts for the Martian gravity, while the final speed ($-2.6km/s$) includes the accounting for the Sun's gravity. Thus, the speed of launch of the first scenario equals the final speed of a craft arriving at the surface of Mars after coming from Earth (under the given conditions) against the Sun's field and passing through the Martian field. Of course, the spacecraft does not actually arrive far from Mars with a recessional velocity (in which case it would never get there). Once we incorporate the Martian gravity acting during the transfer, we see, by time symmetry applied to the first scenario, that the craft actually reaches the effective infinity point of Mars' field with a speed towards Mars of 2.6*km/s.*

Finally, recall from Chapter 5 that the square of the speed is proportional to the rate of transfer of the impetus, and the work done by the fields is proportional to this rate of transfer. That is, though the speed is not conserved, energy is conserved,[19] and, hence, we can find the total speed the craft will have at the surface (neglecting now only the atmospheric drag) due to the action of the two relevant fields (on the craft's impetus) by adding the square of the escape speed and the incoming speed (2.6*km/s*). Or, said another way, the final speed at the surface is given by solving equation 9.19 for v_{final} using $v_{initial} = 2.6km/s$.

9.21 $$\Delta v = v_{final} = \sqrt{v_{escape\,mars}^{\;2} + v_{\text{incoming}}^{\;2}} = \sqrt{5^2 + 2.6^2}\,km/s = 5.6km/s$$

This then, neglecting the Martian atmosphere, is the speed increment that must be applied right before contacting the Martian surface in order to land softly.

More on Time Symmetry and Approximation

Before continuing the landing discussion, we focus for a moment on the particular time symmetry mentioned above. That is, the dynamics is the same whether we let the craft

---

[18] Notice Mars is about 1/10 the mass of Earth, and about ½ the radius, *exercise*: what does this say about Martian density relative to Earth? How do you explain the difference? What effect does each have on the gravity?

[19] The proportionality is replaced by equality after measures and constants are appropriately defined.

fall from far away with a beginning speed of $2.6\,km/s$ finally crashing on the ground at 5.6 *km/s*, or the craft is launched at 5.6 *km/s* and finally left moving away at a speed 2.6 *km/s* when it gets infinitely far away from Mars. The latter is the same as running the former backwards. This time symmetry is true when the planet is in an inertial frame. As we have mentioned, it is only in an inertial system in an approximate sense, for the planet is actually moving in the field of the Sun (and, much less importantly, in the field of the craft) and the craft is moving in the field of the Sun *and* the planet. To get a better approximation to our transit to Mars, these effects need to be included, as do the components of the relative motion of the craft and Mars that have been left out, including the elliptical nature of the Martian orbit. These are just some effects that still need to be accounted for. Thus, as we've emphasized, because it's often missed by students, the above numbers are approximate, and thoughtful approximation is key to properly understanding and doing physics. And, again a deeper point in the same vein is that though the effects we neglect (and thus the causes) aren't "big", the causes may be necessary nonetheless for the whole of reality to be as it is.[20] Because of the many variables, we can expect that, even after including many details, that we will need to do midcourse corrections as other NASA missions to Mars have done.

Accomplishing the Requisite Velocity Decrement for Landing

We can accomplish our decrement in Δv of 5.6*km/s* by a few rocket thrusts coupled with aero-braking, using the atmosphere to circularize the orbit and then a parachute and some low thrust rockets to finish a soft landing. In particular, a circular orbit at low altitude around Mars has a speed:

9.22
$$v_{c\,Mars} \sim \sqrt{\frac{GM_{Mars}}{r_{Mars}}} \sim 3.6 km/s$$

From this number, we can calculate the required Δv for putting the craft into low Mars orbit. Namely, using the difference in square velocities, we get:

9.23
$$\Delta v_{to\,attain\,circular} = 4.3\,km/s$$

Once we attain this circular orbit, say by scraping the atmosphere, we can descend to the surface via the thrusting maneuvers illustrated in Figure 9-9. The rocket firing at the key point in the orbit (largely determined by desired landing site) establishes an apoapsis for a highly eccentric orbit that will bite deeply into the Martian atmosphere. Once the craft is slowed sufficiently, parachutes can open, easing the craft down until it is near the ground. At that point thrusters fire to softly rest the craft and its inhabitants on the Martian surface. This then completes the journey to Mars in outline, yet we have still to consider important details such how long it will take for the Hohmann transfers and what it would take to get us there faster than a Hohmann transfer orbit.

[20] For instance, small effects can accumulate over time and affect whether Earth's seasons or even whether the orbit is there long enough for the formation of life.

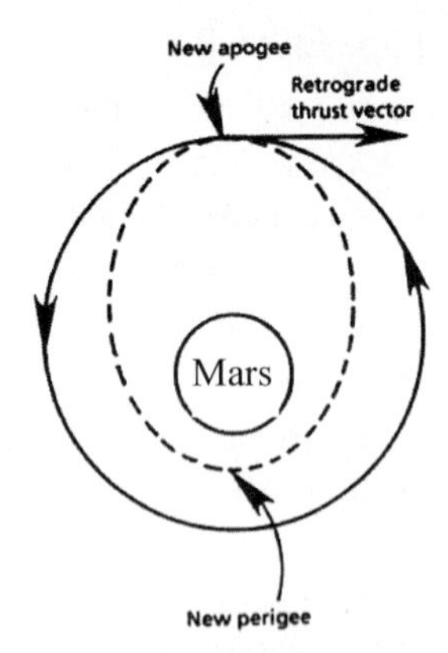

Figure 9-9: Moving into highly elliptical orbit to get far enough into the Martian atmosphere for aero-braking to begin sequence for landing on Mars.

Back Along the Way

So, returning now to our journey to Mars, we note that it is critical to our mission design to know how long the trip to and from Mars will take. If we try to get there too fast we may have to carry too much propellant, making the cost prohibitive. If we take too long, many other problems are introduced. Within this set of problems, two categories stand out: 1. psychological and spiritual issues confronted by all explorers, such as the astronauts being cooped-up and isolated for longer than may be healthy and 2. physical health issues, such as those related to food and water, but in this category, two stand out: radiation exposure and lack of gravity for extended periods. Radiation exposure from cosmic rays and solar flairs both need to be considered. It is known from *Skylab*, *International Space Station* and other studies that human bone density suffers severely from extended micro-gravity environments. Shorter time in transit helps both problems.

From the point of view of orbital mechanics, this means we need to know the time it takes for our craft to move through a certain angle which tracks the motion relative to a given point. There are two angles that can do this tracking; they are called the *eccentric anomaly*, *E,* and the *true anomaly*, θ, shown in Figure 9-10 below.

To get the time, we need Kepler's *third law* that equal areas of the ellipse are swept out in equal time, thus we can write:

9.24 $$\frac{\Delta t}{Area\,swept} = \frac{t}{A_1} = \frac{T}{\pi ab} = \frac{2\pi\sqrt{\dfrac{a^3}{G'}}}{\pi\,a\,b}$$ (where, we take $t = 0$ when at the periapsis)

We need a couple of facts relating the circumscribed circle to the elliptical trajectory shown in Figure 9-10b below. The equation for an ellipse and a circle are, respectively:

9.25 $$\frac{x_e^{\;2}}{a^2} + \frac{y_e^{\;2}}{b^2} = 1 \qquad \frac{x_c^{\;2}}{a^2} + \frac{y_c^{\;2}}{a^2} = 1$$

The circle is simply an ellipse with $b = a$.

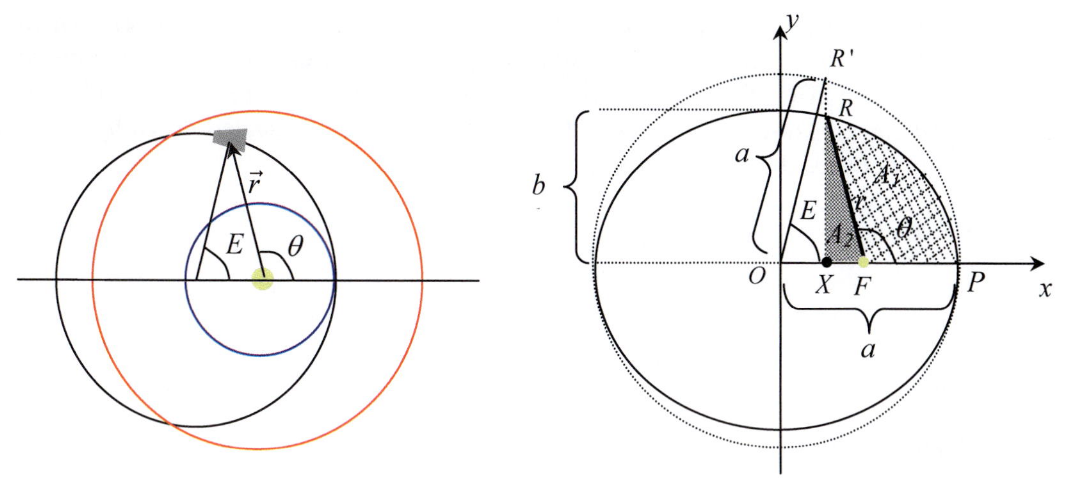

Figure 9-10: a. (*left*): locating craft with θ angle (*true anomaly*) and E angle (*eccentric anomaly*). Trajectory of craft, Mars and Earth are shown, respectively, in black, red and blue. **b.** (*right*): The craft's elliptical trajectory is shown, again, by a solid black line. That trajectory has a dotted circle of radius "a" drawn around it. "a" is the semi-major axis of the ellipse measured from the center to the periapsis. The right focal point, which is the approximate location of the Sun, is labeled F; the periapsis is P. The area swept by θ is shown lightly cross-hatched, and its measure is labeled A_1.

We can calculate the ratio of the vertical components of each by noting:

9.26
$$y_e = b\sqrt{1-\frac{x_e^{\,2}}{a^2}} = \frac{b}{a}\sqrt{a^2 - x_e^{\,2}}$$
$$y_c = \sqrt{a^2 - x_c^{\,2}}$$

At a fixed x location, i.e. $x = x_c = x_e$, we thus get:

9.27
$$\frac{y_e}{y_c} = \frac{b}{a}$$

Now, from the Figure 9-10b, we note that:

9.28
$$A_1 = Area\ of\ RXP - A_2$$

A_2 is the area of a triangle with base $XF = FO - XO$, where $FO = a\varepsilon$ (*exercise:* show this from definition of eccentricity) and, $XO = a\cos E$, so that:

9.29
$$XF = a\left(\varepsilon - \cos E\right)$$

and the altitude:

9.30
$$RX = \frac{b}{a}XR' = \frac{b}{a}a\sin E$$

because of the above calculated vertical scaling factor given by y_c / y_e. Hence, we get:

9.31
$$A_2 = \frac{1}{2}a\,b\sin E\left(\varepsilon - \cos E\right)$$

We need the area of *RXP*, the area bounded by the *ellipse* and the lines *XP* and *XR*, which has the same base (namely the line *XP*) as the area $R'XP$, the area bounded by the *circle* and the lines *XP* and XR'. Since area is proportional to the length of either perpendicular dimension and since I know the area bounded by the ellipse is just scaled-down by b/a, the relationship between these two areas is:

9.32 $$A_{RXP} = \frac{b}{a} A_{R'XP}$$

Now:

9.33 $$A_{R'XP} = \text{the area of } R'XP = \{\text{sector area}\} - \{\text{area of triangle } R'OX\}$$

The sector area is given generally by $\dfrac{\theta r^2}{2}$ which reduces to πr^2 for the full circle (i.e. $\theta = 2\pi$). Thus, we have:

9.34 $$A_{R'OP} = \{sector\ area\} = \frac{1}{2} E a^2 \quad and\ A_{R'OX} = \frac{1}{2} a^2 \sin E \cos E$$

giving:

9.35 $$A_{R'XP} = \frac{a^2}{2}\left(E - \sin E \cos E\right)$$

So that, substituting into 9.28 and 9.31 to get A_1 in terms of *E* gives:

9.36 $$A_1 = \frac{ab}{2}\left(E - \varepsilon \sin E\right)$$

Substituting into 9.24 gives the desired equation as a function of the eccentric anomaly:

9.37 $$\boxed{t = \sqrt{\frac{a^3}{G'}}\left(E - \varepsilon \sin E\right)}$$

However, the true anomaly, θ, is the angle of interest, so we still need a relation that gives $E(\theta)$. This also can be gleaned from Figure 9-10b above (*exercise*) by noting that $XF = r\cos(\pi - \theta)$ so that:

9.38 $$\cos E = \frac{a\varepsilon + r\cos\theta}{a}$$

To eliminate $r(\theta)$ so that we get a simple function of θ on the right hand side, we recall the equation for an elliptical orbit from the end of Chapter 8:

9.39 $$\boxed{r(\theta) = \frac{a\left(1-\varepsilon^2\right)}{\left(1+\varepsilon\cos\theta\right)},}$$

Combining this with equation 9.38, we get the needed relation:

9.40 $$\boxed{\cos E = \frac{\varepsilon + \cos\theta}{1 + \varepsilon\cos\theta}}$$

Now, we would like to calculate how much fuel will be expended to get to Mars if we attempt to get there faster than the minimum energy Hohmann orbit (Figure 9-11). Such a transfer orbit is shown in Figure 9-12 and is called a type I transfer.[21] Given the desired time of transit, to determine such an orbit, we only need to calculate $\mathbf{v}_p$, the speed at the perihelion.[22] For such a transfer, we will start with $r_p = r_{earth}$, so the only parameter left to determine is the speed at perihelion. Transit time can be written as a function $t(\varepsilon, a)$, given fixed $G' \sim GM_{sun}$. Thus, to use equations 9.37 through 9.40, we need the functions: $\varepsilon(\mathbf{v}_p, r_p)$ *and* $a(\mathbf{v}_p, r_p)$. Using the equations at the end of Chapter 8, we obtain the required results:

9.41 $$a = \frac{1}{2}\left(r_a + r_p\right) = \frac{1}{2}\left(\frac{A}{1-\varepsilon} + \frac{A}{1+\varepsilon}\right) = \frac{A}{1-\varepsilon^2} = \frac{r_p^{\,2}\mathbf{v}_p^{\,2}/G'}{1-\varepsilon^2} \qquad \text{using } A = \frac{l^2}{G'}$$

9.42 $$\varepsilon = \frac{\left|\vec{P}\right|}{G'} = \frac{r_p \mathbf{v}_p^{\,2}}{G'} - 1$$

Substituting the previously calculated *Hohmann* value for the periapsis speed, i.e., $\mathbf{v}_p = 32.7 km/s$, into the appropriate equations, we obtain a transit time from Earth to Mars of about 258 days (see end of chapter problem).

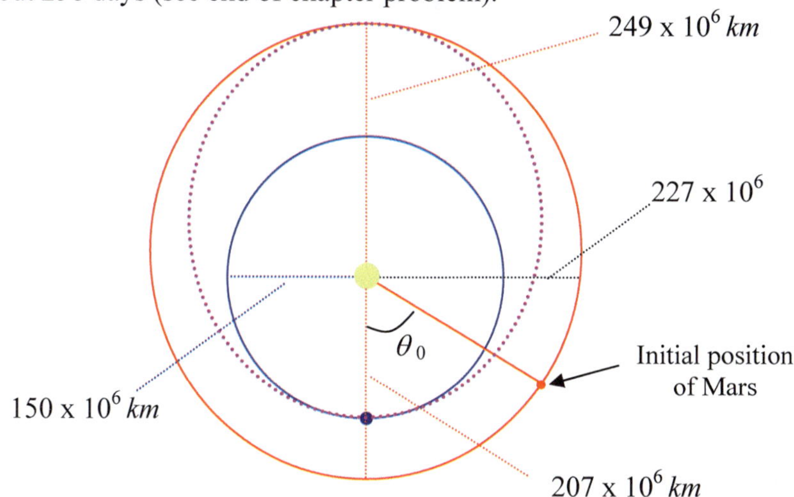

Figure 9-11: Hohmann transfer orbit to Mars shown in purple. In the calculations in the text, we neglect the elliptical shape of the Martian orbit, which is exaggerated for clarity above, and treat it as approximately circular. By fixing r_a and r_p for the transfer orbit, we fix ℓ and *P*, the two constants in the equation. Such a transfer is also called conjunction transfer because Mars is in "conjunction" in the following sense. If we were to stay at the position in space where Earth was when the craft left, then, when the craft arrived at Mars, it would look, from that position, as if Mars (and, thus, the

[21] Technically, a type I transfer takes the craft to Mars in less than 180 degrees around the Sun, while a type II does so by covering greater than 180 degrees.

[22] Since the constants *P* and ℓ used in Chapter 8 in the equation for the elliptical orbit are determined for all time, once we choose r_p and $\mathbf{v}_p$, such choice determines a unique ellipse. The choice of transit time amounts to a choice of $\mathbf{v}_p$.

craft) and the Sun were in the same place in the "sky;" namely, they would look like they had a coming together or a "con" "junction." This is contrasted with opposition missions that result in Mars and the Sun appearing to be at opposite ends of the "sky" or in opposition. This terminology is common in observational astronomy.[23]

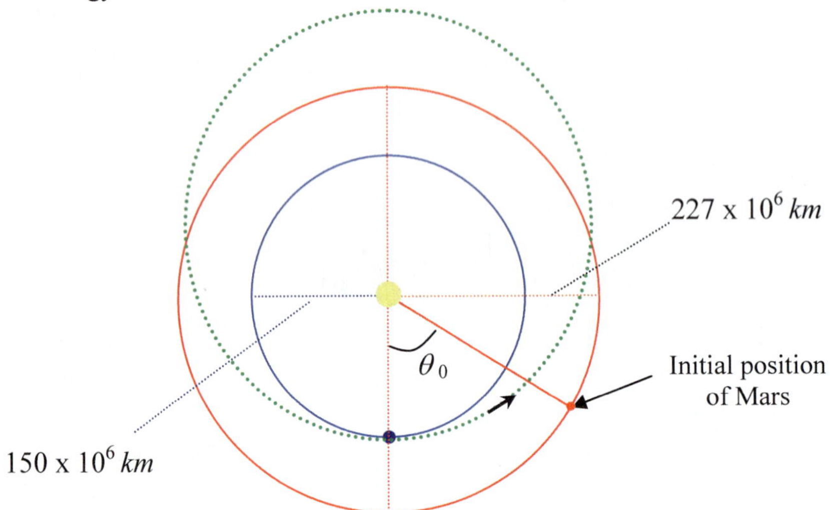

Figure 9-12: Type I orbit shown in green (a type II would take longer than the Hohmann transfer). The Mars Rover *Spirit* launched on June 10, 2003 and arrived on January 4th 2004 into a trajectory like this. Both *Spirit* and *Opportunity* used type I transfer trajectories to get to Mars. Note that Martian orbit is approximated as a circular orbit around the Sun, instead of an elliptical orbit with the Sun at one focus as shown in Figure 9-11.

To do the reverse calculation and obtain the v_p needed for a 180 day (about 6 months) journey to Mars, which is seen to be optimum by some, we continue to assume, for simplicity (but will eventually need to better our approximation), that the Martian orbit is a circular orbit of mean radius: $r_m = 2.28 \times 10^{11}\, m$.

Even with this assumption, the problem cannot be solved analytically in the way one can solve the quadratic equation, but it can be solved by various numerical techniques. Because many problems in physics end up in this state, numerical methods are important. What's more, some physics is done by getting "a feel" for the equations first using numerical trials. In our case, we note that if the orbit is not significantly modified, the 180 day trip should be about 180/258 of the full conjunction trip, namely the craft will have gone an angle of:

9.43
$$\theta_{180} \sim \frac{180}{258}\pi \sim 2.2$$

---

[23] From the figure, we see semi-major axis $a = 228\ million\ km$, $r_p = A/(1+e)$ $r_a = A/(1-e)$ where $A = semi\text{-}latus\ rectum$ (see Chapter 8) implies $\varepsilon \sim .09$ (as tables verify). $a = r_p + c$ implies $c \sim 21\, million\ kilometers$, and, thus, $a^2 = c^2 + b^2$ implies $b \sim 227\ million\ kilometers$

Substituting this into equation 9.37 using 9.40, using $v_p \sim 32.75 km/s$ from Hohmann transfer (see equation 9.12) and $r_p \sim r_{earth}$ $(\Rightarrow \varepsilon \sim .21)$ gives a transit time of 151 days. This smaller value is to be expected because the craft moves faster nearer to the periapsis so that it will need to go further in θ to cover the 180 days than it would if it were in a uniformly moving (i.e. circular) orbit. With this starting point, we can find the required orbit (including θ and v_p) by writing a program (see end of chapter problem) that increments values of v_p until we find one where the crossing occurs such that transit time is about 180 days. Using Mathematica, for example, we find that $\theta = 2.40831\ (E \sim 2.58)$ and $v_p = 32754.3 + 510\ m/s \sim 33.3\ km/s$. In the same manner, we can make a graph showing how much Δv we need as we try to reduce our trip time; this graph is shown in Figure 9-13 below.

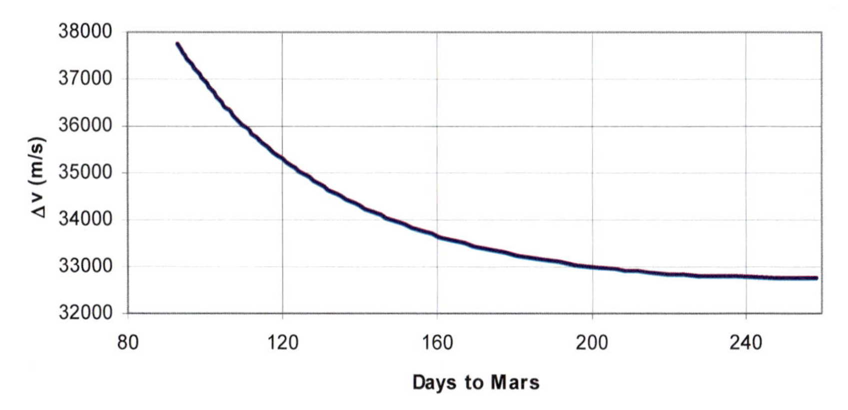

Figure 9-13: Plot of transit time to Mars versus Δv of spacecraft needed to get there from fixed point at a distance of 1 AU (distance from Earth to Sun, 93 million miles or 8 light minutes) from the Sun. Shorter than ~ 200 days, the cost of increasing Δv goes up dramatically, being relatively flat before this. In the text, we showed that Δv of 32.75*km/s* would get us to Mars in about 258 days.

How long will we stay on Mars if we use the 180 day type I transfer ("180 day fast conjunction transfer") both there and back? In this plan, we must pick a launch date at which Mars *leads* Earth by about 44 degrees, so that during 180 days while Earth travels about 178 degrees around the Sun, Mars travels 94 degrees to meet the spacecraft (which has by that time has itself traversed a 138 degree angle around the Sun) just arriving at the Martian orbital distance. This leaves Mars, at the arrival of the spacecraft, *behind* Earth by 40 degrees. For the return trip, Mars must again *lead* by 40 degrees. We thus have to wait another 607 days (see end of chapter problem) for another such 40 degree separation to give us a launch window for a 180 day trip back. Hence, we take 6 months to travel to Mars, we have 1.7 years on Mars and it takes another 6 months to get back making for a total trip time of 2.7 years. But what happens if we cannot insert into an orbit around Mars or we cannot land?

A safer option is a so called free-return orbit. In such an orbit, gravity naturally takes the craft back to Earth; that is, the period of the transfer orbit is such that when the craft returns to the Earth's orbital distance, it finds the Earth there again. Apollo 13 had that very problem, and the men survived because they early-on were able to switch to a free return orbit. It costs more fuel initially because it requires: $v_p = 34.9 km/s$, but it gets one to Mars in only about 126 days. This would leave us more time on Mars to explore (see end of chapter problem to explore how long). If we decide to abort the Mars orbit and landing, it will take us 2 years to get back (see end of chapter problem). As long as we've brought plenty of provisions, we should be able to survive to get home safely.[24]

If we think we can survive a three year free return (i.e. a return based on a 1.5 year period elliptical transfer orbit), such a mission is closer to the Hohmann (least energy option) (see end of chapter problem). However, because of its long free return time, such a mission seems more suited for unmanned trips carrying cargo that precede the human activity on the surface. Before leaving these different transfer orbits, it will be helpful for the student to experiment with the *Interactive Physics* software models: *9.4-Earth to Mars 3 year free hohman free return-- circular martian orbit mars at 44 degrees .ip* and *9.5-Earth to Mars 1.5 year Hohmann orbit to produce 3 year free return with earth.ip* as well as the ones mentioned in footnote 24.

Interactive Simulation

Cargo to Mars

As the Mars Direct Plan[25] recommends, it would be a good idea to have cargo already on the surface before the astronauts arrive, so such a free-return orbit might be more helpful for such cargo. As per Mars Direct, the cargo could arrive in 6 months, and even begin doing things on Mars.

For example, carrying all the fuel for the return trip to Mars would be prohibitive. It would be helpful to send advanced equipment and materials to begin manufacturing fuel before the astronauts even get there; as a precaution we could even wait till it's all made before we launch the astronauts. Also, the advanced unmanned mission could transport a surface transportation vehicle and a large habitat to provide extra living space for the astronauts when they arrive. The extra living space could also be part of building a base for long term exploration; after all, real exploration takes long term concerted effort not a simple one time landing as important as that may be in proving technology. Furthermore, robots equipped with complex sensor and data collection/transmission capacity as well as the ability to leave radio markers at a few key sites, thus inform the crew (before they arrived or even launched) about their future living environment and prepare the way for navigating to a good close-by landing site. The Mars Direct program requires all these particulars.

Artificial Gravity

By picking a fast trajectory to Mars, we have reduced our radiation worries. Indeed, according to *Case for Mars*,[3] cancer risks are only slightly increased. The other key concern mentioned earlier was degradation of bone density due to lack of gravity. This can be mostly (though not completely) solved by creating artificial gravity for the trip there and back. Different type vehicles have been suggested. *Case for Mars* suggests the last-stage of

---

[24] View the simulations for this, the Hohmann and the 180 day flight above, which are respectively (*9.3-Earth to Mars 2 year free return.ip* and *9.1-Earth to Mars- Hohmann.ip* and *9.2-Earth to Mars type I about 180 d --martian circular orbit.ip*).

[25] The "Mars Direct Plan" is championed by Bob Zubran, who was in charge of the team at Martin Marietta (now Lockheed-Martin) that came up with it in 1990 (see his already mentioned book *The Case for Mars*).

the Ares launch vehicle be used as a counterweight for a centrifugal force type of artificial gravity. The upper-stage would be attached with a tether, so that when it is jettisoned, it can be used to set up a spacecraft upper-stage spin shown in Figure 9-14. It is left as an end of chapter problem similar to those done in Chapter 7 to calculate the force, including all components, Coriolis as well as "downward" acting components, for those standing in the craft with the given parameters. More will be said about the issues we confront on the way to Mars in other problems; for now, we are ready to pick a launch date and arrival times.

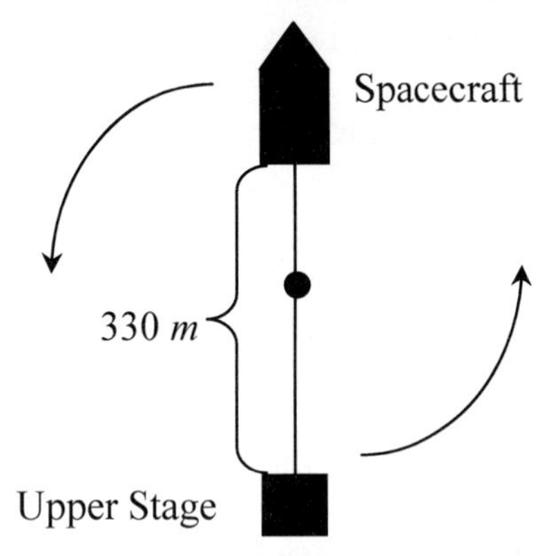

Figure 9-14: The burnt-out upper-stage of the launch vehicle and spacecraft itself spin around their center of mass at 2 *rpm* for the cruising time to and from Mars, creating a simulated gravity that is about the same strength as gravity on Mars, i.e. about $\frac{3}{8}g$.

Launch Date

Choosing the 2 year free return means we need Mars to *lead* Earth by roughly 40 degrees (which is roughly the same as needed for the 3 year free return and hence also roughly the 180 day fast conjunction). To calculate possible launch windows for such a transfer given knowledge of the current position of Mars and Earth, we need to write down equations for the angular position of each planet and then use these to find when Mars and Earth are in the correct relative position.[26] Indeed, it would be helpful to use such a calculation to make a slide-rule like tool[27] that we can read off the relative position of Mars at any given date (see problem at end of chapter).

Launching in June of 2020, for instance, gives us a Mars lead angle of about 43 degrees. The craft will then arrive in about January 2021 AD Earth time if we use the three year free return (about same as 180 day fast conjunction). For the two year free return, it would arrive at Mars on October 2020 AD. We will see shortly, when we discuss the Martian calendar, that this date translates to about (depending on launch day) Januarius 1074 ADM while the launch date, June 2020 AD, is around Nonus 1074 ADM.

Time on Mars

Once on Mars, there is much scientific research that needs to be done, but before this can happen the astronauts need to stay alive and, at some point, to return to Earth. We have to bring most of our food, although some can be grown on board in fluids (so called hydroponic gardening). We know that there is water on Mars; for instance the North polar cap is water ice (the south polar cap has, in addition to CO_2 ice, some water, cf. Figure 9-15) there are even water clouds and some evidence for underground water reservoirs (see

[26] Note that the further in the future that we try to use such equations, the more accurate approximations we will need, because small discrepancies compound over time.
[27] Such as done by Zubrin, *op. cit.* footnote 3.

Figure 9-16). Indeed, just counting the ice and permafrost, there is enough water to cover a smooth Mars thick with water.

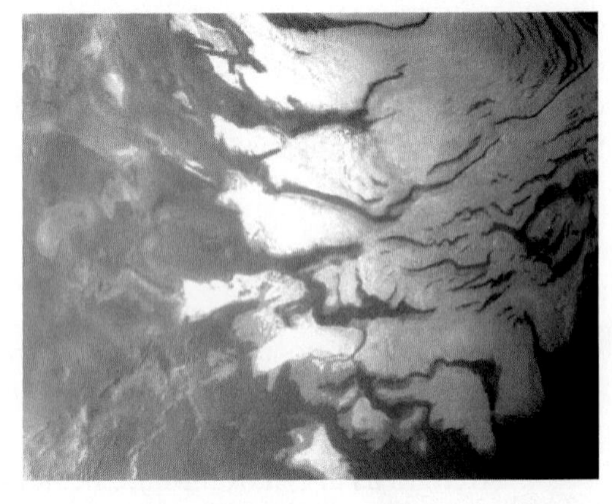

Figure 9-15: A partial view of the Martian south polar ice cap, taken Feb. 2004 by ESA's *Mars Express* orbiter, from an altitude of 269*km*. In the visible range shown here, we mostly see CO_2 ice. However, in January 2007, *Mars Express* deduced that steep slopes known as 'scarps', made almost entirely of water ice, fall away from the polar cap to the surrounding plains, and the permafrost. The North Pole by contrast is almost exclusively water ice. This confirmed other studies showing water ice on the Martian South Pole. In short, water ice, whether mixed into the dirt (such as permafrost) or ice on the surface, is plentiful on Mars.

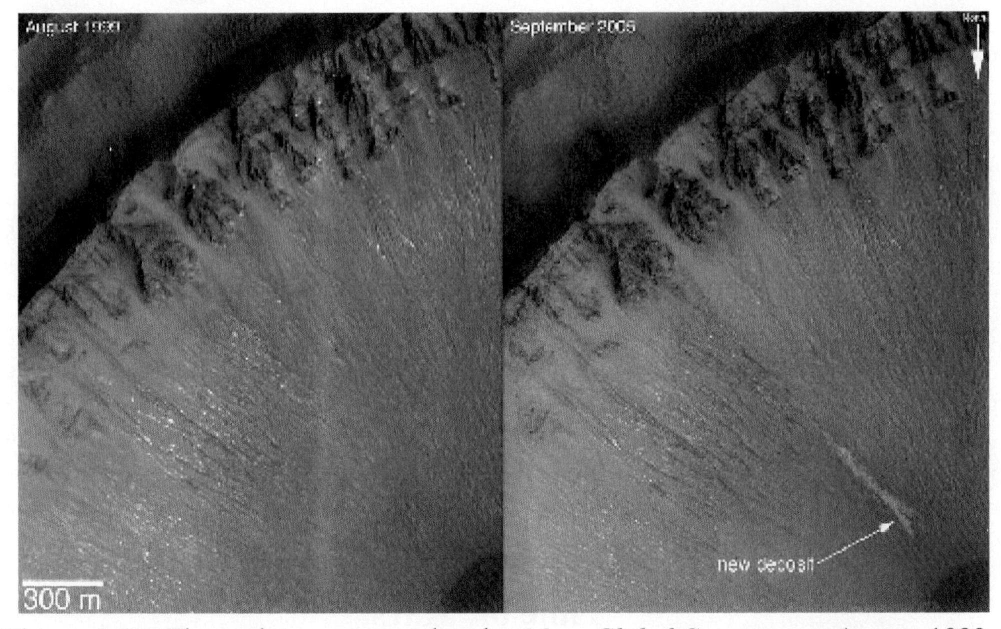

Figure 9-16: These pictures were taken by *Mars Global Surveyor* on August 1999 (*Left*) and September 2005 (*Right*). The latter shows a new feature which appeared by February 2004. The reports say that gullies likely carried five to ten "swimming pools" worth of salty water down the slopes of the craters. The surface temperature is very cold so the water would likely quickly freeze or boil (water boils at $10°C$ on Mars!, for it can only be liquid from[28] about 0 to $10°C$ given the low pressure) upon reaching the surface. The explanation of squirting underground water for the new feature is as of early 2007 questioned by some; more investigation is needed.

[28] Adding salt, which is likely on Mars, extends this range.

How practically accessible the water is remains an open question. However, we can make water if we bring the lightest element (hydrogen) with us (in liquid form); in particular, there is plenty CO_2 in the air that, by chemical reaction with hydrogen, will form water and methane, which, as we discuss next, meshes with the need for fuel. In the long term, one should investigate pulling water out of the air and soil.

Once we satisfy the need for food, water and protection from the elements, the next most important issue is getting back. To do this, we need to have a way of manufacturing fuel. If we were to send liquid hydrogen as part of our cargo mission, we could use the following reaction to make methane fuel (CH_4):

9.44 $CO_2 + 4H_2 \rightarrow CH_4 + 2H_2O$

Methane/Oxygen engines could then be used for rocketing home as well as for internal combustion engines for exploration vehicles on the surface. Oxygen could be obtain by the following reaction: $2H_2O \rightarrow 2H_2 + O_2$.

There are many more activities to consider on the surface for the first and follow-on missions; we leave thinking about the rest to the problems at the end of the chapter and to your imagination. We move on to consider possibilities that some have suggested.

Permanent Settlements and Colonies on Mars?

Two significant differences between Mars and Earth are the extremely cold temperature and the very thin atmosphere, lacking oxygen. *Spirit*, which landed in Jan. 2004 in Gusev crater in the southern tropical region in its summer, recorded, on one January day, a $40^\circ F$ high and $5^\circ F$ low, but the temperature dropped to $-130^\circ F$ at night. The temperature on Mars can get as cold as $-250^\circ F$ (on the poles in the winter) and as warm as $70^\circ F$; Mars has an average temperature of $-80^\circ F$.

To increase and stabilize the temperature to make it more habitable for plant, animal and human life, some have suggested that Mars' temperature might be changed by heating up the south pole thereby sublimating the CO_2 ice into gas that can go into thickening the atmosphere and increasing the temperature through the greenhouse effect of the CO_2 which acts to "trap heat" by not allowing infrared radiation to escape as easily. The greenhouse heating, they go on to explain, would then increase the evaporation rate of CO_2 causing a further increase in temperature, thereby causing further melting and thus, through this positive feedback, the temperature and pressure of the planet could be significantly raised.

If this really would work and *if* the soil is as fertile as some have claimed then plants could be grown and oxygen would then begin to fill the air (thus needing more greenhouse gases to keep the planet from cooling again). Such a process, which has been dubbed "terraforming," would take many centuries; indeed, it has not been proved in any real detail, and we simply don't know enough about Mars yet to make definitive statements. Still, the point here is that one cannot know until one asks the questions.

In the *Case for Mars*, Zubran asks the question "what happens next?" and lays out a very interesting plan for the colonization of Mars. What according to him would people do on Mars? Mine raw materials for one. He points to the possibility of rich resources on Mars itself and in the nearby asteroid belts. Mining the asteroid belts is much easier from Mars than from Earth, as one can see by comparing the Δv needed from Mars versus what one would need from Earth. Of course, eventually one wants, at least as far as can be seen clearly at this point, the mined material for use on Earth, but one needs an outpost to supply food and other materials.

This then brings in a whole host of questions, most of which eventually involve physics questions, though many initially spring from only simple human needs. For example, if even only a few small community settlements are built, the whole spectrum of human political, social and religious needs will have to be met if a healthy Martian life is to be established. There are many analogies here with the European discovery and colonization of America. Columbus's first journey was exploratory, but Columbus put priests on his *second* voyage. Social organization began quickly. Much later, the Mayflower Compact served as initial political documents. In a certain way, even more so than the early colonies, these early Martian settlers would need to establish such cultural institutions, because their physical problems will be *qualitatively* different from those in the "old world." America, for instance, had plenty of air, gravity and temperatures comparable to that of Europe, whereas Mars is deficient in those areas.

To really maintain and grow a nascent population, families must be accommodated thus affecting mission design and even science goals (which can be expanded with increasing presence). Spaceship redesign and even different transfer trajectories with different travel times might be necessary as families start to immigrate. The long journeys will certainly make for long periods of interrupted relationships that may make it desirable to begin family (as opposed to individual) transportation sooner rather than latter. These considerations bring us to a key point. The distance to Mars is unbridgeable in a way that no place on Earth is or can be. Namely, the finite speed of light means we *cannot* really converse with our friends and family on Earth. The distances will bring back the need for communication more like the now rare art of letter writing than phone conversation. It will force a relationship with people on the home planet which, despite our increasing technological sophistication, will indeed be much like those early American settlers separated from their friends and family who had to rely solely on letters. Instant voice communication will not be possible, because the speed of light will take 22 minutes when Mars is at its furthest (Solar conjunction) and 6 minutes when Mars is at its closest (Solar opposition) (cf. Figure 9-11). To see this, imagine any conversation in which after you finish talking and it takes 40 minutes (or at best 11 minutes) for the response to come? Will you even remember the comment you made?

Martian Calendar

Finally, whatever our activity on Mars, whether a permanent or sporadic manned presence or even just unmanned robotic missions, our increased activity on Mars makes it necessary to keep track of events on Mars *according to time on Mars*. In 2007, when the rovers beamed back their latest images, temperatures and other findings, the times were simply given in terms of Earth dates and times-- the one exception was the use of the "sol" which is a Martian day (midnight to midnight or Sun up to Sun up). When we hear that it's -30 degrees or that there's a massive dust storm on Mars,[29] we'd like to know whether it is the equivalent of our January or our July.

---

[29] It is thought that the dust storms on Mars are caused by the sublimating CO_2 in the southern hemisphere carrying dust into the air. This is why there are dust storms in the Southern summer, which is the northern winter.

Recall that time is the measure of motion, one motion measured by a simpler motion or a more primary causal motion. The motion that drives other motions is the key motion by which to measure. For example, in the current understanding of the universe, we date events from the big bang, because all events are driven by the progression of the big bang at the most global level (we might call it the universal driver). The expansion of the universe, for instance, was responsible for the cooling of the primeval matter that then was able to form the first hydrogen. Similarly, on Earth, we measure events according to days, years and months, for the rotation and revolution of Earth and the revolution of the moon around the Earth drive, at the planetary level, the motions on Earth. Everything from the weather, tides, plant life cycles, animal sleep and wake cycles, to when we go to work is driven by one or more of these motions.

As the amount of Martian data (about the past and present) and the number of scientific events that occur on Mars accumulate, the need to have a time tag (date) that manifest the state of Mars at the given moment becomes more critical. Again, it is whether its winter or summer on Mars, or whether its day or night on Mars, that usually matters most for an event on Mars. Furthermore, the need to correlate Martian events with Earth events becomes increasingly important as joint histories are built. Even times before our initial visits will be important.

Making a Martian Calendar

Now, Mars has a tilt just like Earth causing seasons very similar to Earth. Mars also has a day that is about the same length as that of Earth (24.65979 *h* or 24 *h* 39 *min*, 35.244 *s* or 88,775.244 *s*).[30] This means if we keep our definition of a second and hour, we don't get very nice clocks, so we will use hours and minutes and seconds dividing the Martian day, a sol, to give 24 Martian hours in a sol, 60 minutes in an hour, and 60 seconds in a minute. This will make the Martian second about 1.02749125 (it is actually this number that is given in reference cited in footnote 30) longer than an Earth second.[31] However, Mars' year is almost twice that of Earth's (1.88 Earth years) and Mars' orbit is elliptical. The result is that the Martian northern autumn is the shortest season, followed by the winter, followed by the summer, with spring being the longest season. The temperatures are moderated a little in the northern seasons, because, for example, the Sun is closer in the fall and winter. And, the seasons are exaggerated in the southern hemisphere.

If we were to divide the Martian tropical year (the motion of Mars around the Sun from, for instance, one winter solstice to the next) into twelve months, we could not keep 30 sols in a month, but must have closer to 60. Related to this, if we divide the year by only 12, we then don't have fine divisions for distinguishing the various parts of the seasons. This is especially problematic on Mars since the length of the four seasons are not the same, increasing the need for such delineation; the more varying features, the finer the unit that is needed and the month is the unit that we use for such distinctions. For example, we might say, "it's warm for January this year." Furthermore, we'd like to keep the current

[30] Allison and McEwen, *A post-Pathfinder evaluation of areocentric solar coordinates with improved timing recipes for Mars seasonal/diurnal climate studies* Planetary and Space Science 48 (2000), 1999 pgs 215-235

[31] Since the second is a standard unit of measure (so many oscillations of an atomic spectral line frequency), and it is by design close to the Earth second, though still not identical with it (a little like the lunar month's relationship the calendar month), we may want to call this Martian second differently to avoid confusion: for instance, a secondm, abbreviated as secm. The names of the other units (hours, days, years, months) are less important as they are not used in this more absolute fashion.

naming of the months so that something of our already assimilated understanding about seasons can be carried over to Mars, thus making the Martian calendar more accessible to all and facilitating translation between the two calendars.

Thus, we settle on 24 months to get a fine division which is about the same as our own lunar division of a month. Mars' moons travel too fast to be useful time scales, yet the Earth and its moon are visible from Mars with the naked eye. Figure 9-17 shows a view of the Earth and the moon as seen by *Mars Global Surveyor*.

Figure 9-17: Earth and Moon as viewed from Mars by *Mars Global Surveyor*. The Earth and Moon would be visible from Mars with the naked eye.

Also, since our biology and that of other living things on Earth have been influenced by this lunar variation, the lunar (month) scale will be of enduring value to us. In order to maintain some continuity of our intuition, we also establish a larger division which consists of two such months, thus dividing the year into 12 parts (whose parts are 28 day months), that we call a *mensis*, Latin for month. We use the names of the Earth months, January through December, to name each mensis. We name the months that make up each of these menses after the masculine and feminine names for those months. So, for example, the mensis of January is composed of *Januaria* followed by *Januarius*. April is composed of Aprilia and Aprilius. The Latinized names follow the Roman tradition from which we inherit our calendar and the Latin names of geographical places on Mars. The whole calendar is illustrated in Figure 9-18 below. There are 668.5921 sols in a tropical year.[32] We divide the sols among the months as shown in Figure 9-18.[33] Notice the ladies are first except for the first month of the year which is Mars, because it is the name of the planet. The year begins with Mars and ends with Martia, thus symbolically pointing to the unity of the year by bracketing it with the masculine mensis at the beginning of the year looking towards the feminine counterpart of that mensis at the end of the year, with its "family" of

[32] This is the average of four numbers 668.5906 sol, 668.5880 sol, 668.5940 sol, and 668.5958 sol , which are respectively, the number of sols in between successive vernal equinoxes, summer solstices, autumnal equinoxes and winter solstices per http://www.giss.nasa.gov/tools/mars24/help/notes.html. The average value is consistent with sol/day ratio 1.02749125 given earlier from reference in footnote 30.

[33] It takes 175,000 years for the tilt axis to precess around in a circle once. For Earth, it takes 26,000 years for Earth.

all twins in-between. **An interactive software (a Java applet) implementation** of the calendar which gives the relation between the current Earth calendar and our Mars calendar can be found at www.iapweb.org; you are encouraged to use it to find historical dates and even your birth date in Mars time.

As on Earth, the 12 menses are numbered by 1 through 12, so that the mensis of April is number 3, the mensis of September is number 7. Indeed, because on Mars, Mars is now the first month as it was for the ancient Romans, September, October, November and December which mean 7, 8, 9 and 10 respectively have, once again, real meaning in our Martian calendar. We distinguish the two months *within* a mensis using an "*m*" for the masculine month and "*f*" for feminine month. So, for example, we'd write Julia 23, 1000 ADM as 5f/23/1000 but Julius 23, 1000 ADM as 5m/23/1000.[34] Now, one labels the mensis of March 1 *and* 13 for it is both the first month and the thirteenth. More to the point, its first half, Martia, is the last month of the year, so it can be written 13f, but its last part, Mars, is the first month of the year so can be written 1m; this leaves 13m and 1f as possible designators for Mars and Martia respectively that don't appear to be very useful, but could be used.

The calendar starts when on Earth it is January 1, 1 AD, symbolically set as the birth (or conception) of Christ.[35] If we set the date on Mars at this moment to Mars 1, 1 ADM, which means AD according to Mars time,[36] then it turns out we can make Januarius 1st the winter solstice, which will become the astronomical marking point for the calendar. This makes all the other seasons fall in months similar to what they do on Earth (see Figure 9-19).

Seasons drive everyday life as well as holidays, which have, throughout history, been in accord with seasons. On Earth, the vernal equinox is important, for example in the calculation of Easter. It was the drifting of the seasons relative to the Julian Calendar that led Pope Gregory XIII in1582 to adopt a new calendar (our current calendar), with the addition of leap year rules. We also need a system of leap year rules to prevent a shift due to the fact that the number of sols does not evenly go into a tropical year. The rule we adopt is that leap years are years that are odd numbered or multiples of 10, except for centuries not divisible by 500. Leap years have 669 days and the rest of the years have 668 days (see end of chapter problem). The leap year rule is applied to BCM years by subtracting one from the BCM year and applying the rules; zero (BCM 1), which is, strictly speaking, neither even nor odd, is by definition not a leap year.[37]

---

[34] In more primitive beginnings of the computer age, we would have to worry about using letters rather than numbers in such notation, for in those times, programming was hard and memory was short. Making programs that convert between the numbering of the months from 1-24 to {1f,2f,2m,3f,3m,4f,4m….13f} is easy now and will be even easier in the future.

[35] In the *logic* of the current calendar, the first day of the year, i.e., New Years Day, marks the birth of Christ. This can be easily seen if one considers the abbreviation AD and BC at the center of the calendar. On December 31st, 1 BC just before midnight it is BC, before Christ, whereas the moment after midnight it is January 1, 1 AD, i.e., Anno Domini, in the year of our Lord; i.e., Jesus is now considered to have arrived. Note also Christmas is, according to long Christian tradition, 12 days long (recall the song), with New Years Day about in the middle. The symbolism of the calendar is not affected by the fact that historians have been fairly certain for a long time that the monk that originally derived the calendar was likely in error on the birth date of Christ.

[36] More completely, ADM means Anno Domini secundum temporem *Martianum.* Similarly, BC (Before Christ) will be written BCM, meaning *Before Christ according to time on Mars*.

[37] This is defined to make the Martian Winter Solstice of 1BCM to fall on Januaria 1 BCM.

Northern Hemisphere: Spring, Summer, Autumn and Winter

Mars	28	**Octavia**	27
		Octavius	28
Aprilia	28	**Nona**	28
Aprilius	28	**Nonus**	28
Maia	28	**Decana**	28
Maius	28	**Decanus**	28
Juno	28	**Januaria**	27
Junus	28	**Januarius**	28
Julia	27	**Februaria**	28
Julius	28	**Februarius**	28
Augusta	28	**Martia**	27/28
Augustus	28		
Septima	28		
Septimus	28		
		Total sols	**668/9**

Figure 9-18: List of months (28 sols (Martian days)) grouped as menses (pairs of months) making up a 668 sol year (669 sols for leap year). The menses are named January to December in the obvious way, so that, for instance, January consists of Januaria/Januarius. The spring equinox is $\sim$ Mars 23 and the winter Solstice is: $\sim$ Januaria 1. Green, red, brown and black are, respectively spring, summer, fall and winter months; a month is assigned the color of the season that starts in it.[38] Note that the feminine month of a mensis in which a season change occurs (March (Martia), July (Julia), January (Januaria), October (Octavia)) has 27 sols, except on leap years when Martia has 28 sols.

[38] The suggested abbreviates for the months are: MarM, AprF/AprM, MayF/MayM, JunF/JunM, JulF/JulM, AugF/AugM, SeptF/SeptM, OctF/OctM, NovF/NovM, DecF/DecM, JanF/JanM, FebF/FebM, MarF and the Mensus having the same abbreviations as the terrestrial months, Jan, Feb., Mar…

Event on Mars	*Martian* Winter Solstice (dur. of winter)	**Beginning of *Martian* Calendar**	*Martian* Spring Equinox (dur. of spring)	*Martian* Summer Solstice (dur. of summer)	*Martian* Fall Equinox (duration of fall)
Earth date	8/12/ 1BC[39] (164 days)	**1/1/ 1 AD** (142 days from Martian Winter Solstice)	1/23/ 1AD (201 days)	8/12/ 1AD (177 days)	2/5/ 2 AD (145 days)
Mars date	Januaria 1, 1BCM (159.6 sols)	**Mars 1, 1ADM** (138.2 sols from Martian Winter Solstice)	Mars 23, 1ADM (195.6 sols)	Julia 22, 1 ADM (172.3 sols)	Octavia 27, 1ADM (141.1 sols)

Figure 9-19:[40] Beginning and duration of various seasons on Mars in Gregorian Calendar and Martian Calendar.[41] The first Martian winter solstice of the first millennium is on 6/30/2AD[42], which is Januaria 2, 1 ADM.

When to celebrate holidays? Earth holidays happened on Earth, so it seems reasonable that they should track Earth dates, while as Mars develops its history, people on Earth should mark events in its history, say the first Martian constitution, according to the Mars calendar. This is already done on Earth; for example, those in the southern hemisphere, celebrate Christmas according to the northern hemisphere's tradition where it originated, even thought it is not the winter solstice in the southern hemisphere but summer. Christmas on Mars would thus move around the year and would be celebrated twice during the Martian year.

Children born on Mars would similarly celebrate their birthdays according to time on Mars. Children on Mars might want to celebrate "mirror birthdays" since the Martian year is twice as long as a terrestrial one. Consider two girls born at the same time, one on Earth the other on Mars. When the Earth girl is 10 years old, the Martian girl will be about 5.5 Martian years old, just about celebrating her *mirror* 5th birthday and a ½ Martian year away from her sixth. Indeed, because of Mars' elliptical orbit, the season of the mirror will depend on what year the births happen. Again, time is the measure of one motion by another motion. We see the motions by which one chooses to measure will cast what you are measuring in different ways because it is a different comparison, a different relation. This relativity comes up in empiriometric physics all the time.

---

[39] Approximately at 1900 hours UT

[40] For reference, the Julian days (measured from noon January 1, 4713 BC) for the Earth date column are, respectively: 1721282, 1721424, 1721446, 1721647, 1721824. Also, the exact sols for the Mars date column are, respectively: 159.61206482, 138.20069027, 195.62210384, 172.26424069, 141.12042317 sols. The later are obtained by: # sols = # days/1.02749125.

[41] Note, for reference, the Earth's winter solstice of 1 B.C. is 12/22/1 BC.

[42] It is about ~1721969.25 Julian day. Note our calendar starts the day at midnight, whereas Julian days are counted from noon, thus we have to subtract 12 hours (.5 days) from that reckoning to get standard Julian reckoning. This only means for Julian days we carry around .5 day link for midnight UT time.

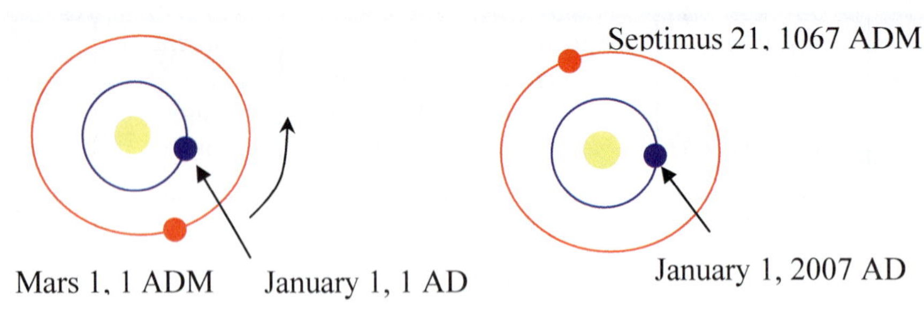

Figure 9-20: a. *(left)*: The planets' relative positions at the start of both calendars; note that the calendar on Earth (blue) begins on *Jan. 1 AD* while on Mars (red) it begins on *Mars 1 1ADM*. **b.** *(right)*: The relative location of Mars and Earth on January 1, 2007 AD. Note that on that date, the Earth is in winter while Mars is ending summer; but for January 1, 1 AD, the Earth has just begun winter, while Mars is soon to end winter.

Calendar as Icon of the Importance of Physics in Ordinary Life

As an example of the usage of the calendar consider the following: when it was *January 1 2007 AD* on Earth, it was *Septimus 21, 1067 ADM* on Mars. In Figure 9-20 we show the position of Earth and Mars in the solar system for this point in time (as well as for the start of the calendar year). We see here the year is approximately half that of Earth's because of the difference in the orbits. To understand and make a good calendar requires a real understanding of celestial mechanics. In making the calendar, we've also seen the flip side. Namely, physics doesn't exist in a vacuum, but is part of our lives. It is a true knowledge of the world that demands to be integrated with all the rest of our understanding, not left as a technical specialty on a separate shelf from one's real life but integrated like all true thinking into the whole of our understanding *and* doing.

We've seen that our knowledge of the principles of Newtonian physics is crucial in planning a manned trip to Mars, but physics is part of the larger whole. The calendar reveals *both* how intertwined the nature of the planets and their interactions are to everything we do, from thinking about birthdays to dreams of terra-forming Mars.

Conclusion

With the conclusion of this chapter, we have seen powerful principles of Newtonian physics and how to obtain quantitative answers to complex problems in the real life situations in which orbits, as well as walking across the room and dropping objects, came into play. The crown of the success of Newtonian physics is that it reveals something of the generic unity of nature of the planets and Sun and the bodies on Earth. They are all bodies, i.e. things can change, but they are bodies that exercise forces that activate impetus in other bodies, and they, themselves, have the capacity to have impetus activated in them. In the appropriate domain, all bodies that we know of, in Newton's phrase, maintain their state of rest or uniform motion unless acted on by an outside force. Without that simple insight, first articulated by Medieval men such as Philoponus and Buridan, modern science and all the things we discuss in this chapter would not be understood, let alone discussed and done. Impetus has many more implications than we have discussed explicitly up to this point. Chapter 10, our final chapter, brings in an important such implication and in the process shows us something more of the power of the insight.

Summary of Equations

Rocket equation: $\Delta V = V_{final} - V_{initial} = \boldsymbol{v}_{exhaust} \ln\left(\frac{M_{initial}}{M_{final}}\right) - g\,t$

Speed of a body of mass m in a *circular orbit* at height h above heavy body with mass $M \gg m$ and radius r.

$$\boldsymbol{v}_c = \sqrt{\frac{GM}{h + r_e}} \sim \sqrt{\frac{GM}{r}}\left(\sqrt{1 - \frac{h}{r}}\right) \text{ for } h \ll r$$

Periapsis speed, $\boldsymbol{v}_p$, needed to insert a body into an elliptical orbit with apoapsis r_a and periapsis r_p around a much more massive body of mass M. We used this equation for calculating Hohmann transfer speeds.

$$\boldsymbol{v}_p = \sqrt{\frac{2GM}{r_p + r_a}\frac{r_a}{r_p}}$$

Hohmann transfer orbit to from Earth to Mars is shown below:

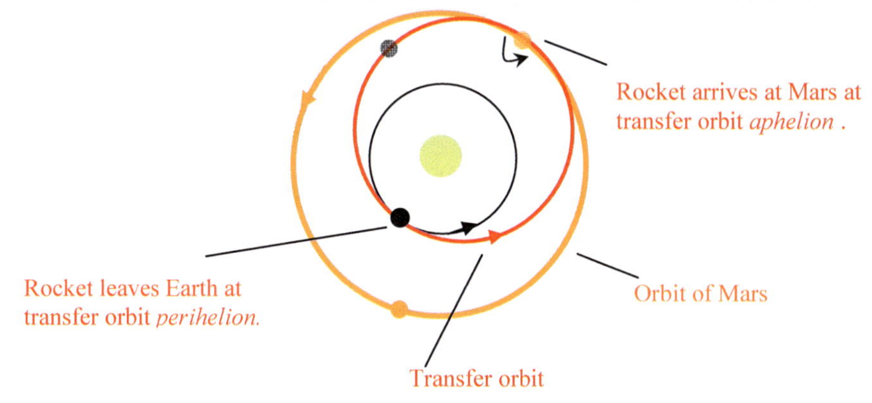

Fundamental rule of astronautics: *for minimum energy expenditure apply braking and launching thrusts ($\Delta \boldsymbol{v}$'s) as close to the center of attraction as possible.*

Orbital period for body of mass m orbiting in elliptical orbit of semi-major axis a around body of mass $M \gg m$:

$$T = 2\pi\sqrt{\frac{\mu}{GMm}}a^{3/2} \sim 2\pi\sqrt{\frac{1}{GM}}a^{3/2}$$

Total speed needed for insertion into Hohmann transfer orbit *from* Earth (or any other planet) *to* an outer planet (such as Mars), accounting (approximately) for the gravity of the Earth and Sun and the speed of the Earth orbiting around the Sun but not other motions or fields (in particular, without consideration of the gravity of destination planet):

$$\mathbf{v}_{total} = \sqrt{\mathbf{v}_{wef}^{\;2} + \mathbf{v}_{escape}^{\;2}} = \sqrt{\frac{GM_{sun}}{r_{earth-sun}}\left(\sqrt{\frac{2r_{planet-sun}}{r_{earth-sun}+r_{planet-s}}}-1\right)^2 + \frac{2GM_{earth}}{r_{earth}}}$$

Where $\mathbf{v}_{wef}$ is the speed needed for the insertion into the orbit without the *Earth*'s field and r_{earth} is the radius of the Earth and $r_{A\text{-}B}$ is the orbital radius from A to B. In the chapter, we use the values obtained by assuming approximately circular orbits.

Equation for an elliptical orbit in terms of *true anomaly*, θ, defined in figure below :[43]

$$r(\theta) = \frac{a\left(1-\varepsilon^2\right)}{\left(1+\varepsilon\cos\theta\right)}$$

Where:

$$a = \frac{1}{2}\left(r_a + r_p\right) = \frac{1}{2}\left(\frac{A}{1-\varepsilon} + \frac{A}{1+\varepsilon}\right) = \frac{A}{1-\varepsilon^2} = \frac{r_p^{\;2}\mathbf{v}_p^{\;2}/G'}{1-\varepsilon^2} \qquad \text{using } A = \frac{l^2}{G'}$$

$$\boxed{\Rightarrow \varepsilon = \frac{\left|\vec{P}\right|}{G'} = \frac{r_p\mathbf{v}_p^{\;2}}{G'} - 1}$$

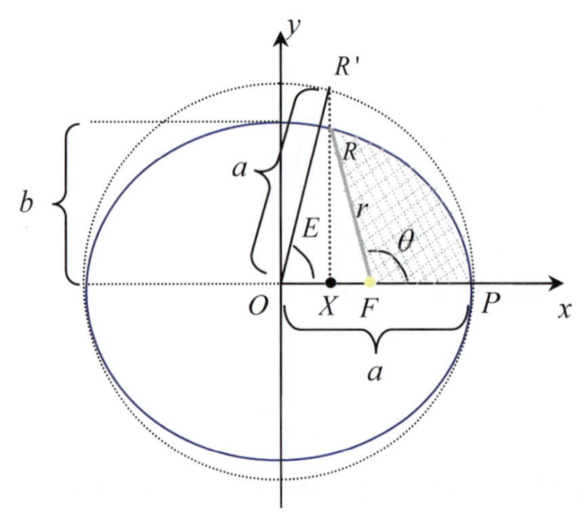

The time it takes for a satellite in an elliptical orbit to move through an angle, E (*eccentric anomaly)* defined above is given by:

$$t = \sqrt{\frac{a^3}{G'}}\left(E - \varepsilon\sin E\right), \text{ with } \cos(E) = \frac{\varepsilon + \cos\theta}{1+\varepsilon\cos\theta}$$

Where the eccentric anomaly, E, is obtained from the true anomaly, θ, using the equation above on the right.

[43]What we here call A is sometimes called p.

Mixed Problems

1. Using the same approach as the chapter, what trajectory and what fuel burn, in terms of Δv, is needed for each step of our *return* trip from Mars?

2. Explain why the increment in speed given by a rocket firing is a good unit to use for determining what trajectories are most efficient. In particular, what is its connection with fuel usage for a) one firing scenario b) two rocket firings at separated points in the trajectory (note: we *do not* consider here the multistage case, i.e. in which part of the weight of the rocket is dropped after a burn is complete)?

3. Consider an *n*-stage rocket in which each stage is identical to the others and which are fired in succession the moment the previous stage burns out. Assuming the spacecraft is launched far from all fields, so that no external forces act on it, what speed (called the terminal speed) will the spacecraft have after all stages have been fired and detached? *Hint:* Decompose the mass of the spacecraft into three components: $n\,m_{stage} + n\,m_{fuel} + M_{craft}$, where m_{stage} is the mass of each stage, not including its fuel, i.e. the mass of the support structure for the stage; m_{fuel} is the mass of the fuel in each stage, and $M_{payload}$ is the mass of the "payload," i.e. that which one wishes to get into transport.

4. Using the results of the above problem, determine how many stages a multistage model rocket (using the standard engines available at stores) would need for its last stage to reach a height of 100 *km* (sometimes defined to be edge of space). Generally assume the validity of the multistage rocket equation (including neglecting the field of the Earth in that equation), and simply calculate how many stages it would take to attain the launch speed needed to reach a $h \sim 100km$ above the Earth's surface: $v = \sqrt{2gh} \sim 1.4km/s$.

Below is a chart listing the total impulses for various types of model rocket engines. Assume a D12-0 rocket for each stage with initial total weight 42.2*g* of which propellant's weight is 24.93*g*. To solve the problem (and in case, you would like to use your results to stage a real model rocket for extended flight experiments), we also note the labeling of model rockets works as follows. For example, given a D12-3, the first symbol indicates the class; the second is the average thrust, 12*N* in this case, and the last is the delay time (3 seconds) before the charge blows the parachute out or for us drops the last stage. We assume no delay time.[44] Recall: $1\,B = 1\,Ns$

Class	Total Impulse	Class	Total Impulse
A	1.26-2.50 B	I	320.01-640.00 B
B	2.51-5.00 B	J	640.01-1280.00 B
C	5.01-10.00 B	K	1,280.01-2,560.00 B
D	10.01-20.00 B	L	2,560.01-5,120.00 B
E	20.01-40.00 B	M	5,120.01-10,240.00 B
F	40.01-80.00 B	N	10,240.01-20,480.00 B
G	80.01-160.00 B	O	20,480.01-40,960.00 B
H	160.01-320.00 B	Note: above 160 requires certification and permits	

[44] Specs from http://en.wikipedia.org/wiki/Model_rocket_motor_classification

5. On February 21, 2008, a failed US spy satellite in low Earth orbit was shot down because of its near full complement of toxic fuel and its imminent uncontrolled orbital decay into the atmosphere. Describe how one could shoot this satellite down so that it would fall more or less straight to the ground, neglecting the atmosphere (which is actually not a good approximation).

6. Types of propulsion are usefully classified by their **specific impulse**, I_{sp} which, in analogy to specific energy angular momentum, is the impulse per unit mass of propellant ($Buridan/kg = m/s$). Below is a chart of the specific impulses for various types of rocket engines. Note that, since a given mass, *m*, is often measured as a force on the Earth, via the action of the Earth's surface gravity (*mg*), we can also express I_{sp} in units of seconds, because impulse is measured in *Buridan* (or *Ns*) and weight is measured in Buridan/s (or *N*). Given the increment in velocity for insertion into a Hohmann orbit to Mars, calculate the mass ratio needed for spacecrafts with the following propulsion systems.

Engine[45]	**Effective Exhaust Speed (m/s)**	I_{sp} **(s)**	**Energy/kg (MJ/kg)**
Solid rocket	2500	250	3.0
Bipropellant liquid rocket (e.g. H_2+O_2)	4400	450	9.7
Nuclear Thermal rocket	8800	900	
Ion thruster	29 000	3000	430
VASIMR (max) (Plasma ejection, *theoretical*)	290 000	30 000	43 000
Light	300,000,000	31,000,000	

7. Assuming the use of bipropellant liquid fuel, and assuming a single stage which finishes firing close to the ground, about how much fuel in kilograms is needed to carry a cargo (payload) of 28.6 tonnes (1 tonne=1000*kg*) outside the influence of Earth's gravitational field?

8. Suppose we use the methane that we have made on Mars (*Note:* burning methane, CH_4+O_2, yields a specific impulse of 380*s*) to enter a Hohmann orbit to return home to Earth. If our craft (without fuel) has a mass of 25,000*kg*, how much fuel will be needed?

[45] Data from http://en.wikipedia.org/wiki/Specific_impulse

9. A craft needs to climb out of a uniform field gravitational "well" of height h_0 and come to rest just outside the well. Ignoring the duration of the rocket burns:
a) Demonstrate by calculation that we expend less energy in giving the rocket (payload, not fuel) all the requisite impetus, and thus speed, at launch rather than giving part at launch and part when the initial impetus is drained and the rocket comes to rest some fraction h/h_0 of the way through the trip b) Demonstrate that one minimizes the sum of the increments of speed needed when all the fuel is used at launch; said another way, show that the first burn is the most effective. Explain the key *physical* principle involved in each case. c) Given that the craft enters the uniform gravitational field at speed v_1, calculate the impulse given to the craft by the field in rising to a height h. Using this equation, show that the distance traversed by the craft is independent of the initial mass, thus, once again, manifesting the principle of equivalence. Also, show quantitatively, using simple analysis, how the craft goes further with higher v_1.

10. Consider a trip to the Moon from a circular low Earth orbit via a Hohmann transfer orbit to a circular low altitude orbit around the moon then finally a Hohmann transfer to the lunar surface. What increments in velocity are needed (and when are they needed?) to accomplish this mission?

11. Consider a trip from the surface of the *Moon* to *Mars* via a Hohmann transfer orbit to the Martian orbit to a circular low altitude orbit around Mars. a) What increments in velocity are needed to accomplish this mission? b) Using this result and the result from the previous problem construct a trip from Earth to Mars via the moon and compare the required rocket firings to that for a direct trip from Earth to Mars. Some have suggested a trip to the Moon on the way to Mars would save money. Which requires less fuel?

12. For the trip to Mars, to calculate the increment in speed needed to land on the surface of Mars from the Hohmann orbit, we used the root of the sum of the squares of the escape speed of Mars and the speed a craft would have given a Hohmann transfer orbit to Mars excluding the field of Mars. Expound further on why this works physically.

13. Consider mining the asteroids in the belt between Mars and Jupiter. In particular, there are four asteroids in the belt that border on being classified as planets: Ceres, Vesta, Pallas and Hygia. Ceres, shown to right, has been reclassified as a dwarf planet; it is the largest member of the belt. Compare the velocity increments needed to get to Ceres from Mars and from Earth. Assume all relevant bodies lie in the same plane of orbit around the Sun, that they all have circular orbits and that the radius of Ceres orbit is about $4\times10^8\,km$.

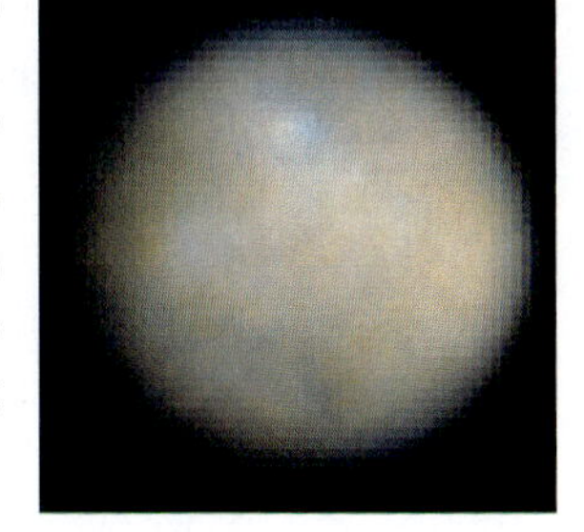

14. Is it easier to get to the Moon from Mars or from Earth? Show by approximate calculation of the needed Δv's.

15. Suppose a weather satellite is in an elliptical orbit around Mars. Suppose further that one needs to increase its kinetic energy at periapsis and one can only generate a Δv increment in speed. Is it better to apply this change at the periapsis or at the apoapsis? In answering, calculate how much the periapsis (also called pericenter) speed and kinetic energy change when a Δv (as a result of a rocket firing) occurs at the apoapsis of the orbit. Also, explain the result in physical terms.

16. a) How much $\Delta v = v_1$ is needed to move from the surface of Earth r_e to a radius r_2 from the center of the Earth? b) How much is needed to escape to infinity from r_2? c) How does this compare with the Δv needed for escape from the surface?

17. Comment on the idea from the text about communication with Martian friends and family from Earth. Specifically, why will the distances involved bring back the need for something like letter writing?

18. Consider the single thrust and double thrust maneuver shown. At what speed will the double thrust maneuver require less fuel than the single? In particular, calculate Δv for each case for a spacecraft that leaves a circular orbit to reach "infinity" with a speed v_∞.

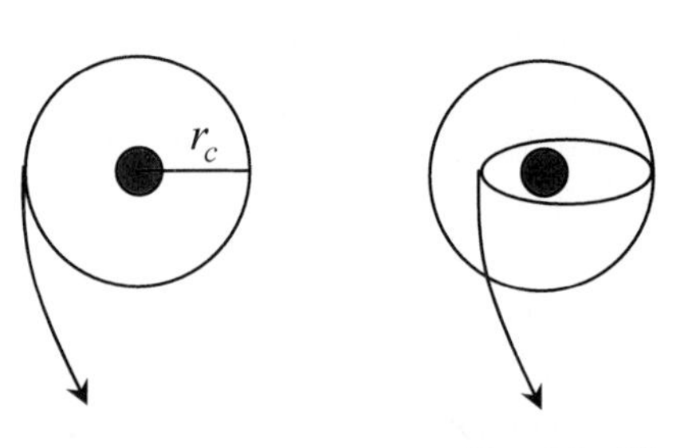

19. A communications satellite is orbiting Mars. Assume an astronaut repair crew (or a robotic repair craft) takes a vertical flight to the satellite from the surface of Mars. Calculate the equation for the time of flight for the repair mission using the following steps. Start with the following equation: $v = \sqrt{2GM\left(\frac{1}{r} - \frac{1}{r_{sat}}\right)}$ and explain its origin (note: v is the radial velocity of the repair craft at radius r; M is the mass of Mars, and r and r_{sat} are, respectively, the current (i.e. after time t) radial distance (from Mars' center) to the repair mission craft and the radial distance to the satellite). Then explain how one writes: $t = \frac{1}{\sqrt{2GM}} \int_{R_m}^{r_{sat}} \frac{dr}{\sqrt{\frac{1}{r} - \frac{1}{r_{sat}}}}$. Where R_m is the radius of Mars. Finally, use the substitution:

$$r = r_{sat} sin^2 \frac{\Phi}{2}.$$

20. Use the data on Mars to determine in what constellation the Sun will be for each of the Martian months.

21. When we modeled the trips to Mars in *Interactive Physics*, we wanted Mars and Earth to complete rotations on the order of seconds not years. Explain quantitatively how to scale

solar system times to a time that can be used in an *Interactive Physics* model of the situation and show it works for one *IP* example. Also, give an equation for scaling speeds.

22. In the chapter, we discussed the need for "artificial gravity" on the trip to Mars; below we reproduce the figure. a) Calculate all the forces the astronauts feel on the trip to Mars, assuming it spins at a rate of 2 *rpm*. b) Suppose you run as fast as you can across a 7m long room in the direction perpendicular to the tether, what force will act on you? c) Suppose you move "up" ladder as fast as you can, what force acts on you?

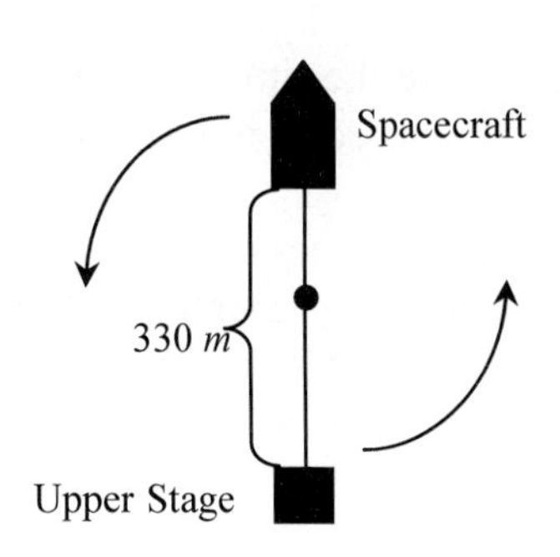

Note: assume the mass of the spacecraft, *m,* is three times more massive than the upper stage.

23. In general terms, how might one maneuver a rotating craft such as the one suggested in the chapter for generating artificial gravity?

24. Prove, as mentioned in the chapter, that the round trip to Mars that uses 180 day long type I orbits for both the trip to Mars and the trip back to Earth requires roughly 607 days stay on Mars.

25. Calculate how often the opportunity to put a spacecraft into a Hohmann transfer orbit from Earth to Mars occurs. Verify your result in *Interactive Physics*.

26. Neglecting the gravity of the Earth, calculate the Δv needed at periapsis, the angle traversed during the transit time and the transit time for following type I transfer orbits from Earth to Mars: a) Hohmann transfer orbit b) 180 day type I transfer time c) for the 126 day, 2 year free return transfer time. c) A three year free return type I transfer that is close to Hohmann (it's very close), i.e. based on a 1.5 year period elliptical transfer orbit? d) a three year free return type one orbit, i.e. based on 3 year orbit. Experiment with these orbits in *Interactive Physics*. How do your results compare?

27. Using the results of the previous problem, calculate how much Mars must lead (or lag) Earth at launch so that the craft will intercept Mars on its type I trajectory for each of the following: a) 3 year free return based on 1.5 year period transfer orbit. b) 180 type I transfer orbit c) 2 year free return orbit

28. Give a general equation for calculating the length of the semi-major axis of a type I transfer orbit that gives a free return to Earth after T years. Discuss how this could be used when T there is an integer N such that NT is an integer?

29. Using the results of the above calculate the time one needs to stay on Mars if a 126 day type I transfer (a 2 year free return orbit) is used for the outgoing and return trips.

30. What speed is needed to leave Mars to return to Earth on a Hohmann orbit and on a 180 day type I transfer orbit? How long will it take?

31. Calculate the time it takes to do Hohmann transfer from Earth to low circular orbit of 300km described in this chapter.

32. To get an intuitive idea of how much the eccentricity of the Martian orbit matters in our trip to Mars, use Interactive Physics or another gravity modeling software application to show that for the 2 year free return orbit there is little difference between the following two cases in the time it takes the craft to reach the Martian orbit: i) a craft launch that takes place when the Earth is lined up with the Martian perihelion and ii) one lined up with the Martian aphelion;

33. Construct a tool that can give relative positions of Mars and Earth. Use a circular orbit for Earth and an elliptical one for Mars. Mark off month intervals on Earth and on Mars" orbit. Start them so that they show the position of Earth and Mars at January 1, 2000. Show how this can be used to calculate launch and landing dates. Put the Martian months and the Earth months on the tool so that it can be used to find out what Martian month and season (color code the seasons as in the text) it is on Mars at any given date on Earth. Use it to find what season it is now on Mars and check it with the software implementation of the Mars calendar at iapweb.org.

34. Discuss how calendars are influenced by the religious thought in the most general sense of the "cult" of a culture, the shared understandings and beliefs about the world and the point of life. If Columbus, for example, were to land on Mars, what calendar might he choose? Why did the Romans choose the one they did? Why did the Jews choose theirs? Babylonians?

35. Another possible danger to astronauts is from meteors. On the right is a picture of a "new" 25*m* wide crater discovered on Mars at the end of 2006. On Earth, we are protected somewhat by a thick atmosphere. Indeed, crater formation on Mars has been seen by orbiting satellites: estimate the chances that an astronaut might get hit by a meteor forming a crater, using a reported rate of 3 per year and assuming all the craters are about 25*m* in diameter.

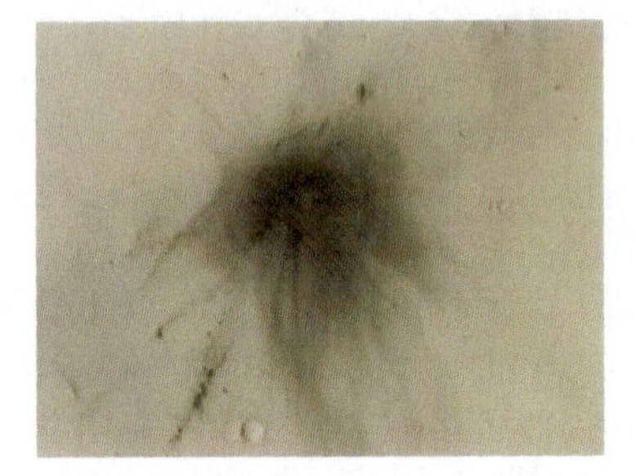

36. Suppose at some distant time in the future, engineers succeed in establishing an atmospheric pressure of 5 psi (about one third of the pressure on Earth) on Mars by sublimation of CO_2 ice, which they suppose is enough atmosphere to grow plants and allow men to walk around without special pressure suits, using only scuba like breathing apparatus for their O_2. a) Using the methods similar to the air pressure variation problem in Chapter 8, calculate how many molecules of CO_2 would be needed to increase the pressure on Mars to this level. b) Use this calculation to also obtain the approximate number of air molecules on Earth. c) Explain briefly what causes atmospheric pressure both on Mars and Earth and then explain how this could be used to approximate the number of CO_2 molecules needed to create the 5psi on Mars.

Photograph of a Martian sunset taken by Spirit at Gusev crater, May 19th, 2005.

37. We would like to understand something of the heating we will experience from the Sun on our trip to Mars. Thermal design is an important part of spacecraft success. Toward this end, consider the influence of the Sun on Mars and Earth. The amount of light (electromagnetic radiation) that reaches the Earth from the Sun (before entering the atmosphere) is about $1300 W/m^2$ (the so-called *solar constant* at the Earth). a) Given that the energy is conserved and spreads out evenly, how much energy per unit time per unit surface area reaches Mars? b) Given that a black body (which we assume that Mars and Earth are for this calculation) with average temperature of T measured in Kelvin, the energy per unit area radiated is: $S = \sigma T^4$, where the Stefan–Boltzmann's constant is: $\sigma = 5.67 \times 10^{-8} \frac{W}{m^2} \frac{1}{K^4}$. Knowing that, the incoming energy must balance the outgoing energy, otherwise the Earth would constantly get hotter or colder, and assuming the only energy source is the Sun and the only way of dissipating heat is through radiative heating described by the previous equation, what temperature would the Earth be if it were not rotating, and it were an insulator so heat on one side did not transfer to the other? Now, assume the Earth comes to some average temperature because, for example, of its rotation and convective heat transfer. c) How about Mars for each case?

38. a) Calculate how much energy would be needed to manufacture CH_4 on Mars in the way discussed in the text. b) How much for O_2? c) Given the solar constant, the amount of light (electromagnetic radiation) reaching Mars is $590 W / m^2$ how big of a 10 % efficiency solar array would be needed? d) Discuss how one could do this using other sources of energy: e.g.: wind or thermal energy from volcanic underground heating of water, estimate relevant parameters.

39. Suppose there were 686.91237 sols in a year ....what leap year scheme would you recommend? Keep it simple and explain your answer.

40. Mars's tilt axis changes from 15° to 35° (or perhaps even more) over a 124,000-year cycle, whereas Earth shifts from 21.5° to 24.5° over 41,000 years. What effect does such a shift in the axis have on the seasons?

41. Given the sidereal year on Mars is 686.980 days and the tropical year is 686.973 a) calculate (roughly) what the precession of the tilt axis of Mars is. b) Generally, what causes the precession of planetary rotation axes? Give a rough everyday analogy. c) Explain what effect such precession has on the stars one sees on Mars and on Earth.

42. Argue that the Hohmann orbit transfer from one circular orbit to another is the least energy two-impulse transfer orbit.

43. Look up the Interplanetary Transport Network and bi-elliptic transfer (right). Explain generally and briefly their relation to the Hohmann transfer.

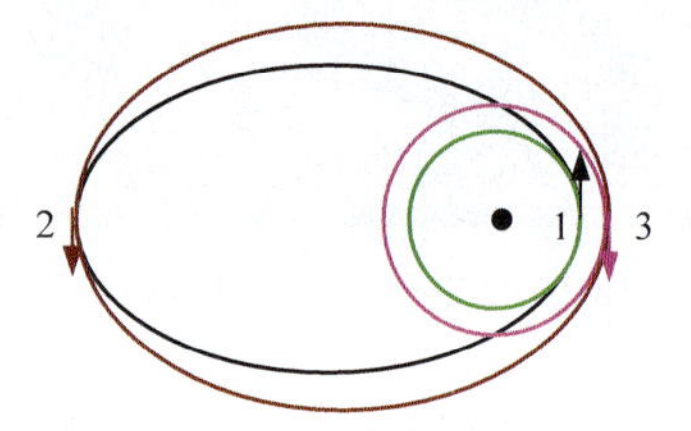

44. List the approximations we have made in our calculations of relevant parameters for our mission to Mars in the chapter. What is it that we finally mean by "approximation?"

45. a) Using the leap year prescription in the text calculate how much error accumulates in the calendar after 10 years, after 100 years, 500, 1000, 3000 years. b) How do the errors in the Mars Calendar compare to those in the Gregorian Calendar for the Earth?

Chapter X

Introduction to Special Relativity

Preview

Special Relativity is an important topic with many new concepts to digest. To help illuminate these fascinating ideas, we give an overview of the path we will follow in unpacking the beauty of the world uncovered in and by them.

We first reiterate the importance of impetus and mass, which have been central topics of the text up to this point. In particular, the centrality of inertial frames is brought to the foreground of our attention. Because of the nature of impetus, any experiment done in one inertial frame will give exactly the same result in another; this is referred to as Galilean relativity, and is also described as the invariance of the laws of physics under transformation between inertial frames. Galilean relativity is only true to the extent one can approximate the maximum speed of communications between bodies as infinite. Infinite speed means that events happening in a distant place can be synchronized with events happening locally.

Once we realize that the speed of communications between bodies is finite, we will see that other changes in bodies must occur if we are to retain frame invariance. Furthermore, we will see that if there is not a maximum speed beyond which no body can be measured to move, effects can appear, to some observers, before their causes in ways that a maximum speed limit can prevent. We will call this speed faster than which no object can be measured to go, **the maximum *apparent* speed**. Since light travels at this maximum apparent speed, we will also call this maximum apparent speed, the speed of light, c. The existence of a maximum measured speed is the second ingredient needed for special relativity. The first ingredient is the frame invariance of the laws of physics, which is backed by direct and indirect experimental evidence.

To make precise our analysis of the nature and meaning of this invariance in the context of a maximum apparent speed limit, we introduce the concept of houses in which experiments are done and compared to other houses. *A house* generally means any inertial frame restricted to a region; thus we speak of either a *closed* house or an *open* house. A closed house, the first meaning of *house*, gets its generic usefulness from the existence of boundaries around physical substances. Substances, and machines which are modeled after them, are important in our analysis for the following reason.

At the current level of abstraction, i.e. including only those aspects of things related directly to impetus, substances act and behave in uniform motion just as they do at rest; we call this ***the integrity principle***. This is not the case for accelerated motion. Substances that have impetus change in ways which cause the substance's integrity of operation to be maintained. We formalize this fact in ***the principle of internal relativity***, that the laws of physics are the same in the interior of every house and that the intrinsic state of a house is only a function of its speed. With this principle, coupled with the existence of a maximum speed to guard against violations of apparent causality, we can then deduce ***the full relativity principle***, Einstein's special relativity, that any experiment done in one inertial frame will yield the same in any other frame and that the speed of light is the same in all frames. An *inertial frame* in special relativity is a system of rulers and clocks moving at uniform speed. We speak of an *observer doing experiments in a frame*, when he uses rulers and clocks that move with him in the frame to do all his measurements. As with Galilean relativity, special relativity can be summarized by its transformation laws. The special relativistic transformation laws can be shown to follow from the full relativity principle and the existence of a finite maximum apparent speed.

Before demonstrating in detail how the principles of special relativity arise from the integrity of substances in uniform motion, we establish the physical sense of the constancy of the speed of light in all directions and inertial frames, which seems to be a logical impossibility. First, we introduce *an operational definition of simultaneity* which gives a practical rule for focusing attention on motions within a given frame. We do this by considering as the motion of sound, which we take as analogous to light, on a moving flatcar. We show that by setting our clocks in the moving frame with a certain delay as a function of position, we can make the speed of sound in the forward and backward directions the same. Even so, we still do not obtain the same speed as the rest frame.

Before resolving this difficulty, we introduce two other apparent difficulties with re-establishing frame invariance. We liken these effects to the wind on the flatcar. Experimentally, bodies that move relative to the frame of the observer contract in length *and* increase in mass. We also explain how the later is equivalent to a *time dilation*. Next, we introduce the radar method of incorporating operational simultaneity. In radar, we send out a signal, noting the time of transmission; then we wait for the signal to reflect off the target and return and we note the reception time as well. With those two times, we can calculate the equations for the time and location of the reflection event. These equations are then written as time and position in the moving frame as functions of time and position in the rest frame, i.e., the ground (the frame which we take to be at absolute rest). In other words, by using the radar method we deduce transformation laws between the two frames. Once we apply the time dilation factor to these equations, we see that they are the transformation laws of special relativity, which naturally give us the constant speed of light in all directions and all inertial frames. Proper interpretation and understanding of these equations however still remains.

With the foundations of the physical understanding of how we can have a constant speed of light and, more generally, how one can have frame invariance with a maximum speed limit, we now turn to develop the points left behind earlier.

First, we prove that internal relativity and a maximum apparent speed limit leads to the principles of special relativity. Next, we show (in a section that can be skipped in a less advanced course or a first read) how to develop the full transformation laws from maximum apparent speed and a symmetry principle. Finally, we lay out the full transformation laws, i.e. the so called Lorentz transformations, and their inverses.

Second, we return to the open/closed house distinction to discuss the role plana plays in the propagation of light. The distinction between virtual plana, i.e. plana that is a part of a substance, such as yourself, and free plana is used and discussed. Finally, we briefly summarize our part one results.

In the second part, we move to understanding the meaning of the transformation laws by using them to predict various phenomena such as length contraction and time dilation. We also point out the following, which can be predicted by a simple application of the full relativity principle. Clocks (or rods) in the moving frame will appear to tick slower (contract) for observers in the rest (unprimed) frame, while clocks (or rods) in the rest frame will appear to tick slower (contract) for those observing in the moving frame. To illuminate the physical issues involved in this symmetry, we discuss the so called twin paradox.

In this part, having already noted the consequences of the mass increase with speed for the rate of transfer of impetus, i.e. the energy, we move to discuss the analogical generalizations of mass and energy that flow naturally from special relativity. We also introduce a generalized version of velocity, momentum and force. With the more developed formalism, we return to further our understanding of the nature of light and its propagation. Lastly, we discuss what is meant by the convertibility of mass and energy.

Part I: The Fundamentals

Introduction

Up till now, we've covered very general principles of mechanics, leaving out much in order to say what we could at a very simple level. Of course, much more is known than we're able to cover in this text and still more is yet to be discovered. Yet, there is another level that we can reach given our current understanding. This huge step up the ladder of understanding of mechanics was first codified by Albert Einstein in 1905.

It begins, as every discovery does, with further observation and a more complete digestion of what is already understood. For the latter, Einstein began with *Galilean* relativity, which we touched on earlier in Chapter 3.

a) Galilean Relativity

Einstein noted, as did medieval scientists such as Buridan and their successors such as Galileo, that a group of objects moving together at the same speed is a special kind of system. Its members act as if they were at rest one relative to another. Galileo, using the ship analogy as did Buridan and Oresme[1], describes this phenomenon in detail:

> "Shut yourself up with some friend in the main cabin below decks on some large ship, and have with you there some flies, butterflies, and other small flying animals. Have a large bowl of water with some fish in it; hang up a bottle that empties drop by drop into a wide vessel beneath it. With the ship standing still, observe carefully how the little animals fly with equal speed to all sides of the cabin. The fish swim indifferently in all directions; the drops fall into the vessel beneath; and, in throwing something to your friend, you need throw it no more strongly in one direction than another, the distances being equal; jumping with your feet together, you pass equal spaces in every direction.

[1] Oresme says, for instance, "If air were enclosed in a moving ship, it would seem to the person situated in this air that it was not moved." *Book of the Heavens,* Book II chapter 25, from Grant, *A Source Book of Medieval Science*," page 505, Harvard, 1974

When you have observed all these things carefully (though there is no doubt that when the ship is standing still everything must happen in this way), have the ship proceed with any speed you like, so long as the motion is uniform and not fluctuating this way and that. You will discover not the least change in all the effects named, nor could you tell from any of them whether the ship was moving or standing still. In jumping, you will pass on the floor the same spaces as before, nor will you make larger jumps toward the stern than toward the prow even though the ship is moving quite rapidly, despite the fact that during the time that you are in the air the floor under you will be going in a direction opposite to your jump. In throwing something to your companion, you will need no more force to get it to him whether he is in the direction of the bow or the stern, with yourself situated opposite. The droplets will fall as before into the vessel beneath without dropping toward the stern, although while the drops are in the air the ship runs many spans. The fish in their water will swim toward the front of their bowl with no more effort than toward the back, and will go with equal ease to bait placed anywhere around the edges of the bowl. Finally the butterflies and flies will continue their flights indifferently toward every side, nor will it ever happen that they are concentrated toward the stern, as if tired out from keeping up with the course of the ship, from which they will have been separated during long intervals by keeping themselves in the air. And if smoke is made by burning some incense, it will be seen going up in the form of a little cloud, remaining still and moving no more toward one side than the other. The cause of all these correspondences of effects is the fact that the ship's motion is common to all the things contained in it, and to the air also. That is why I said you should be below decks; for if this took place above in the open air, which would not follow the course of the ship, more or less noticeable differences would be seen in some of the effects noted."[2]

Dialogue Concerning the Two Chief World Systems

In short, a system-- such as a spaceship or the Earth-- which is moving at a uniform speed has a certain impetus that is divided among the constituents in such a way to cause the uniform speed. Still, the system can just as well, for observations within the system, be thought of as *not* having that impetus. But, you may ask, what sense does it make to talk of the uniform motion of a system composed of bodies moving in many directions at many speeds? We really are talking about an average kind of motion, the motion of the center of mass. This is exactly why center of mass coordinates are so valuable. Those coordinates re-label things in such a way that we can concentrate on the bodies *in* the system under consideration, yet we always have handy their relation to the outside world if needed. And, again, we can only do so because of the property of impetus that keeps things moving at uniform speed.

Moving to view the system in a frame other than the center of mass effectively adds momentum to the entire system. Remember that more impetus means more speed and thus faster transfer of the impetus across the room, as well as the more impetus itself that the body "carries" with it. Mathematically, we write:

$$P_{Lab} = MV_{cm} + \sum_i m_i \mathbf{v}'_i \qquad E_{Lab} = \frac{1}{2} M V_{cm}^{\ 2} + \frac{1}{2} \sum_i m_i \mathbf{v}'^{\,2}_i \quad with\, M = \sum_i m_i$$

10.1

$$P' = \sum_i m_i \mathbf{v}'_i = 0 \qquad E' = \sum_i \frac{1}{2} m_i \mathbf{v}'^{\,2}_i$$

Here the prime refers to the quantity as measured in the center of mass frame.

[2] Translated by Stillman Drake,by Stillman Drake, University of California Press, 1953, pp. 186 - 187 (Second Day)

We see that the activity that's added is simply that due to the impetus $P_{CM} = MV_{cm}$ that moves the entire system at a speed V_{cm}. As seen in the top energy equation, this activity is measured by the energy in the same way as the activity as any other impetus is.

We thus leave behind the fact that the system may or may not have a certain impetus and treat the system as if it (the center of mass of the system) were at rest. This is the Galilean relativity we've already discussed. Relativity here means motion is considered in this relative manner only.

Indeed, we often go a step further to make thinking about the relations of interest in the system easier and treat the relations as if they were able to exist as such. It's after this final step that we have a mental construct that can be made part of a formal system. This is something we do to carry around what we know in a convenient package with symbols that keeps our concentration on the subject at hand. And as we've said so many times, our leaving out something doesn't make that something not exist.[3] Yet, none of this should come up, for, by now, we know what we are doing.

b) Time and Maximum Speed

Still there is a problem stemming from other things we've left out. We have treated time and space as if they existed independently of physical bodies; both are properties of physical things and so can't exist without them.

Time is a succession of "nows," which is measured by motion (recall it is a one of the nine categories of properties of physical substances). To understand motion, we need to understand how various things interact. Interaction by contact is one way, but things in the universe are separated and contact is infrequent and not the main way things interact. How do things interact at a distance? The two most obvious answers are by gravity and by light. Gravity actually takes a certain amount of time to move through the plana from one body to another. Light takes a certain amount of time to travel, which was known during Einstein's life. These things don't happen in no time. In fact, in all known physical domains, no information has been measured traveling faster than the speed of light in a vacuum. In the simplified mechanics we've set up, we've made the idealization that things communicate at an infinite speed. This did just what we wanted, that is simplified things and made it very easy to know what measure to take for motion, i.e. time, in each frame. In particular, in the infinite speed idealization, any motion in one frame can easily be measured by motion in another frame because the motions are in immediate communication.

---

[3] After all, reality is not only relation. It's not *only* the relation between the particles of a system and the relation of the system to the other systems that matters. We want to put the full reality together at some point and remember that the systems themselves are realities. Relation is, after all, *just one* of the 9 categories of properties *and* it's a weak one for it depends on the real *intrinsic* properties of quantity and quality, of being extended in a certain way with certain actual qualities, for it's only between things with intrinsic existence that we can have relations. Also, reality is not both something and not something at the same time and in the same way. Something is either moving or not. It is either actually being moved by impetus (or some other agent or quality) or it is receiving the impetus of another to move across it. For example, recalling a fundamental principle that we have already learned, if I move my hand across the table in such a way that my hand gets more impetus so as to move over the table, the table is receiving the action of my hand motion; the table is not moving in that regard; though it is moving as part of the Earth.

Infinite speed, of course, is one of those mental constructs (beings of reason), one of those things that can only exist in the mind, that helps artificially seal off our theory for concentrating on what remains. It is now time to fill back in some of what we left out. Special relativity does this and has been verified in its domain of applicability directly via experiment and indirectly via theory that has stood the test of experiment,

Special relativity arises when one realizes there is no physical infinite speed communications. The implications of a finite speed of interaction are somewhat surprising especially when coupled with the central feature of Galilean relativity, the equivalence (in a sense to be clarified) of all inertial frames. The two key implications are:

1) There is a maximum apparent speed,[4] that is a maximum speed that any body or effect can appear (be measured) to travel.

2) This speed appears the same in every inertial frame.

These are directly related to the preferred status that impetus conveys to massive bodies in uniform motion (constant speed). That is, when such bodies are closed off from interactions with the rest of the universe, the interactions within that system will appear the same as that within any other identically configured system that is moving uniformly at any other speed. Once we establish arguments for a maximum apparent (measured) speed faster than which no interaction can travel, we then conjecture, via the relativity principle, that this maximum is the same in every frame. Ultimately, of course, the arguments come from and must be verified in the physical world through direct sensorial contact or by reasoning from such contact. Experiments of all types do verify, at the appropriate level, the conclusions we reach.

Einstein postulated that the maximum speed was realized only by light-like phenomena, i.e. by those fastest mediators of interaction between distant objects. This postulate makes sense because if the speed of propagation of these actions, such as light, were slower than the fastest massive bodies, then they would cease to be a remedy for lack of contact between massive bodies. Contact between massive bodies could occur sooner than the light could mediate an interaction. In the next subsection, we will think more about how light might travel in accomplishing this mediation and, in the process, demonstrate the relative need for an apparent maximum speed.

Apparent Reversal of Causality

With what we currently know, it is probably not possible to argue that a maximum apparent speed is an absolute necessity. However, we will argue that it is a relative necessity. More specifically, we argue the natural order of causes and their effects cannot be maintained without a maximum apparent speed.

To do this, we introduce the two possible modes of light travel. The light, or whatever mediator, may travel **I)** independent of, *not intrinsically linked* to, the media of travel or **II)** *linked* to the media. That is, it may travel in *bullet-like* fashion or *sound-like* fashion. In either case, we run into trouble if there is not a maximum speed limit.

We start by considering the following arrangement of a man shooting a bear that is attacking someone (see Figure 10-1). Consider the two ways the actions of the shooter and bear can be mediated to the observer.

---

[4] By maximum apparent speed, we mean that there is, in principle, a natural limit to the apparent speed of bodies and interactions among bodies, ***not*** simply that there is something that happens to be moving faster than anything else in the universe, but nonetheless could, if "pushed" a little more, appear to move faster.

Case I: Take light to travel in a ***ballistic*** fashion at a finite speed c, and allow the bear to be moving at the shooter (and you) at a speed v_{bear}, while you and the shooter remain at rest. The word "ballistic" here refers to the hypothesis that the light comes off objects with its speed c added to the speed of the object, i.e., $c + v_{object}$. This behavior is called ballistic because it is ***like*** the speed of a bullet coming out of a gun, which equals the speed it would have were the gun at rest plus the actual speed of the gun, as illustrated in Figure 10-1. If light is like this, if the bear is moving fast enough, and if the distances are right, the faster moving light emitted from the dead bear will get to you before the light from the bullet and the shot gun. You will see the bear being shot dead before you see the gun being fired.

Figure 10-1: *Case Ia:* We hypothesize that light travels in ballistic fashion in which its speed, c, adds to the speed of source. You and the shooter are at rest; the bear runs at you at speed v_{bear}. Yellow arrows show the "movement" of light.

Figure 10-2: *Case IIa:* We hypothesize that light travels in sound-like fashion. You and the shooter are at rest; the bear runs at you at speed v_{bear}. Yellow arrows show the "movement" of light.

Case IIa: Assume light travels in a ***sound-like*** fashion, i.e. that it always moves at some finite speed c *m/s* relative to the medium which it travels in, say air, and allow the bear to be moving at the shooter (and you) at a speed: $v_{bear} > c$. See Figure 10-2. If the speed is fast enough the bear will pass up the light emitted at the time of the fatal bullet hitting the bear, so that the *dead* bear will fly past you *before* you see the bear getting hit by the bullet that killed him. Again, you will see the effect before the cause!

Case IIb: Take the light to travel in ***sound-like*** fashion and take the *shooter and bear* to be moving together through the air at a speed $v > c$ toward you. See Figure 10-3. Distances can be chosen (see end of chapter problem) such that, from your "at rest in the air" vantage point, you will see first the bear dying and then the bullet being shot from the gun; in fact, you will see the bullet move from the bear and go *backward* into the gun. Again, there is an appearance of backward causality.

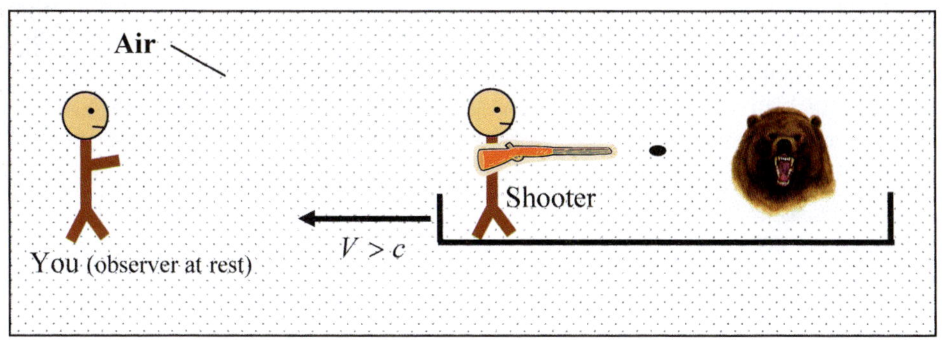

Figure 10-3: *Case IIb*: We hypothesize that light travels in sound-like fashion. You are at rest. Shooter and bear are moving at you at speed V.

A couple issues may now come to mind. First, obviously sound has the very problem that we would like to avoid. However, remember we use sound only as an analogy to light; that analogy fails finally because sound is not the fastest possible communications. In short, what is a problem with sound may be mitigated by light,[5] if light travels at the maximum speed.

Second, you may have already noted that: for stationary bodies, whenever an observed event, say a bullet killing a bear, is between its cause and the observer, since the event is closer to the observer than the cause, the event will be seen before its cause (*exercise*: give an example). This apparent reversal of causality arises from the very nature of extension and could only be overcome by an infinitely fast speed of light travel that we know does not exist. However, the reversal of apparent causality due to the motion of the various bodies, i.e. the type we just discussed, is another deeper loss of apparent causality.[6] A universe that has this type of apparent reversals of causality would certainly be much less organized than one without it.[7] Remember we are talking about the primary means of the universe coordinating its activity in one part with another. Thus, we now investigate the consequences of a maximum apparent speed.

c) Introduction to Internal Relativity and Special Relativity

i) Frames and "Houses"

To see the key consequence of the existence of a maximum speed coupled with the fundamental insight of Galilean relativity, we consider the universe as viewed from various isolated regions that move uniformly. We call such regions ***houses***, for they are like independent communities of particles that, yet, are still part of the whole universe. Houses are distinct from frames in that though a house, like a frame, moves at a constant speed, it

[5] As we will discuss, light is often used to signify the fastest possible measured speed. Thus, more precisely, we can say, the problem could be mitigated by some other body or effect making use of the potential to move as close to that speed as we like or even at the speed of light in non-massive cases.

[6] Notice that we have not yet shown that there is a way out of this motion-induced apparent causality, just that without a maximum speed there is no way out under our generic (simplifying) assumptions.

[7] It would, for example, rule out the possibility of a motion, such as the big bang driving all motions in this more orderly way. Note that, though the straightforward meaning of this statement will be changed when we switch on gravity and consider general relativity, its new meaning will still be valid analogically.

is confined to a certain region and things inside the house of primary interest cannot see outside or generally feel influence from outside the house. Houses *inside the house of primary interest* are allowed to look outside of themselves as long as their view stays within the primary house, and the effect of such interaction on any house is assumed to be minimal so as to not affect its overall state. That is, we exclude effects that would change the unique status conveyed by the impetus to inertial frames. Experimentally, this means we find (or create) arenas where such effects are small. In allowing some interaction of parts, a house models (in the most abstract way) a physical system, including substances such as animals, devices such as computers, or any collection of physical things. When we pick a system, we concentrate on its parts and their motion, such as blood flowing through veins and molecules moving in blood, leaving out of consideration the environment of the system.

Generally a house is a confined region of a frame, whether it is an open house or a closed house. The first meaning of a house is a *closed* house, which is based on the existence of boundaries to a substance and their complex parts. A closed house does not allow plana (e.g., the cosmic background radiation-filled vacuum)[8] to flow through it. By contrast an *open* house is a frame confined to a certain region that does allow exterior plana to move by, for instance, as the house moves. Note that we are not attempting here to give definitive statements about plana (vacuum), the study of which properly belongs to another course such as Quantum Field Theory, but only discuss the apparent possibilities. Indeed, special relativity will finally leave the actual state of the plana out of account, and will replace it with the mental construct (being of reason) of "empty space" creating a powerful system for studying the generic physical aspects that remain.[9]

Frames cover the entire universe and consist of a system of clocks and measuring sticks (rulers) traveling at a fixed speed; coordinate systems label the places and times in a given frame. Picture a giant platform the size of the universe moving at fixed speed with observers, each equipped with a ruler and a clock, stationed at every point. Note that platforms are usually two-dimensional surfaces; to really cover the whole universe, we would need a three-dimensional lattice work, such as shown in Figure 10-4. From a frame, one can view any event in the universe (again, with causes such as gravity "switched off", or so small as not to affect our measurements at a given level of approximation); thus, we speak of an event as viewed from a given frame. Note that a house is a frame with certain restrictions on it; in particular, to get a house, a given frame is confined to a region in which regions of other frames, i.e. internal houses, can also be set up. Note that for an open house these internal distinctions are not as important.

---

[8] As analogy for the difference between an open house and a closed house, one might consider the difference between an enclosed metal box moving with respect to the cosmic background radiation and an open platform moving through the cosmic background radiation. The analogy fails because ultimately we exclude any effects that destroy the generic symmetry we seek to describe, and the cosmic background radiation would clearly do so.

[9] The "empty space" being of reason is not just actual extension without the other accidents and with no substance in which to inhere; it also leaves out of account any details about space, by keeping the extension in potency with respect to its actual parts (for instance, as to whether it is divided or not divided), allowing the possible comparison (measurement) of lengths in the given frame to determine the distances as needed, which, in turn, specifies place as a number in some coordinate grid as discussed in Chapter 2. Again, this allows our attention to be focused on the relations of interest for understanding impetus and its effect at a generic level.

Figure 10-4: Three dimensional lattice structure useful in thinking about a frame. Frames cover the entire universe and consist of a system of clocks and measuring sticks (rulers) traveling at a fixed speed (coordinate systems label the places and times in a given frame).

ii) Maximum Speed in Various Houses

Now, "turning on" only what is related to impetus and its effects in the way we have discussed, consider a house *A*, moving at some uniform speed. It, in turn, has houses inside of it, among which are houses *B* and *C*. In order to avoid a violation of apparent causality, we have already seen that there must be a maximum speed limit in house *A*. The same maximum apparent speed limit must apply to house *B* and *C* as seen from house *A*; otherwise, we again violate apparent causality. If, for example, the maximum possible apparent speed of light traveling in *B* were lower than that in *A*, then the order of events in *B* can be overshadowed by communications occurring between the objects via light moving in *A*. On the other hand, the maximum apparent speed of light traveling in *B* cannot be higher than *A*, even neglecting the fact that *B* is technically a part of *A* which already makes it impossible, for then it would mean *B* could circumvent apparent causality in *A*. Similar arguments apply to viewing from any internal house, say viewing from *B* ***or*** *C*, to any other house (see end of chapter problem). This implies that viewing *from* any given house (and thus frame as well) one must see the same maximum apparent speed in every house, but, at this point, the maximum speed seen *from* one house need not be the same as that viewed from another. Thus, the primary house may see the maximum apparent speed as *c*, whereas a moving house inside of it may see it as c'.

To establish a mathematical notation, we label houses by an index *i*, where c_i is the maximum apparent speed (speed of light as we will call it) seen *in* the ith house as viewed *from* the primary house (house *A*). Whereas, c_i' is the maximum apparent speed seen in the ith house, seen *from* another house (say the 1st house); or generally *from* the nth house, the maximum apparent speed seen *in* the ith house is written: $c^{(n)}{}_i$. In terms of this notation, using what we have said above, the maximum apparent speeds seen in the ith house from any one house (or frame) must be equal, we get: $c \equiv c_i$ and $c' \equiv c_i'$ and $c'' \equiv c_i''$ etc. for all i. Or written in total generality, we require: $c^{(n)} \equiv c_i^{(n)}$ for all values of i (i.e. all houses) for all values of n (i.e. as seen from all houses). Again, note that, at this point, it is not required that the maximum apparent speed be the same as viewed from all houses. Namely, at this point, there is *not* a number, $c_{\max}$ such that $c_{\max} \equiv c^{(n)}$ for all n.

iii) Uniqueness of Inertial Motion

Next, as we have hinted at, we want to make use of the fact that Galilean relativity suggests that the nature of impetus and how its action is received (i.e. mass) is such that each system (e.g. house) moving at a fixed speed behaves, in important ways, the same as the system moving at a different fixed speed. (We mean this only within the confines of what has been included in our analysis, i.e. what things have been "turned on"). In short, a system's behavior, as viewed by one moving along with it, i.e. as viewed from its own rest frame as we say, is the same no matter how fast it goes. *This is important, because this means the integrity of the interrelation of the parts of a substance, and thus of the substance itself, is maintained while it is moving uniformly, no matter how fast it moves.* We call this ***the integrity principle.*** For instance, this means an animal's internal workings continue in the same manner whether the animal is in a stationary car or in a uniformly moving car. We can imagine this could have been otherwise. Indeed, such *is not* the case for *accelerated* motion; too much acceleration will kill an animal and destroy a machine. For this very reason, all NASA rockets must meet certain *maximum* acceleration requirements, and all NASA payloads (for example, scientific and communication instruments) have to be specially designed to withstand certain minimum acceleration requirements.[10] Whereas, there are *no similar speed requirements* for the payloads, since fixed speed leaves the internal parts of a system working in the same relative state as they were when they were not moving.

iv) Combining Maximum Speed and Relativity

Now, in order to bring home the principle that each house (i.e. system, such as an animal or a machine) should maintain the same integrity it has when at rest when it is moving uniformly, we introduce ***the internal relativity principle***. The internal relativity principle states that no experiment done *within* a given house can distinguish one house from another and that only the absolute speed of a given house determines whatever differences there may be between houses.[11] This basically is a restatement of the integrity principle that we argued for above from the existence of impetus. We will see that this principle (with its high level of abstraction, which includes turning off all but basic things) makes it *impossible* to determine what is absolutely at rest and what is not. Internal relativity (including the fact that absolute speed alone determines the state of a house), as we will show, implies a full relativity principle.

This full principle, called ***the* relativity principle**, states that there are no detectable differences between houses even when one looks at effects outside of the house. In other words, viewed from inside a particular house, an experiment *in that house* will give the same results when placed *outside of that house,* as long as it continues to move at the same

[10] Vibration requirements can be considered a subset of this.

[11] Note some real differences may not be manifested by their effects at the current level of abstraction. We will see, for example, plana behaves such as to cancel any of its effects at this level of approximation.

speed relative to the house. Once we have shown that no effects of the plana show up at our level of abstraction, we will then formulate *the relativity principle* simply as: whatever experiment one does in one *frame*, will yield the same results in any other uniformly moving *frame*. Again, such relativity obtains because all objects of the uniformly moving frame, including the clocks and measuring sticks, move under the action of an unchanging impetus on a fixed rest mass in such a way that the effect is the same, as indicated by their equal uniform speed. A more massive body needs more impetus to be affected in the same way as a less massive body; if a pencil and a brick are given the same impetus, the pencil is more affected, as revealed by its faster motion. With the full relativity principle, as with Galilean relativity, one cannot determine the absolute motion of any given house; one can only tell relative motion. Unlike in Galilean relativity, in special relativity, there is a maximum apparent speed limit.

These, then, are the key principles necessary to expound Einstein's theory of special relativity: 1) there is a maximum apparent (measured) speed limit and 2) the physics one measures in any experiment is invariant under change of inertial frames. A key deduction from these principles is the invariance of the speed of light.

In order to really understand these principles and the meaning of the constancy of the speed of light, we need to investigate the propagation of light in a general way. We will do this, and then complete the arguments that indicate that the speed of light is measured to be *c,* a universal constant in all inertial reference frames, and then deduce the transformation laws that take an observer from one moving frame to another. As the transformation laws will be the key mathematical form of the theory of special relativity, we, first, briefly introduce the much simpler transformation laws of Galilean relativity.

Transformation Laws of Galilean Relativity

In Galilean relativity, the transformation rules are simple. Mathematically, transformation rules are a set of equations that relate the place of bodies at each moment in time to the apparent place and time of the body as seen by an observer moving in a certain way. We are not necessarily transforming from the actual situation at rest to that perceived by a moving observer. It could be a transformation between the perceptions (measurements) of ***two moving*** observers. In Galilean mechanics (as opposed to Newtonian mechanics), we do not distinguish an absolute rest frame, so the transformation is always between two observers; the word "rest" is then used only relative to a given frame.

In our mathematical treatment, we, of course, use coordinates, such as the Cartesian x, y and z to label the place of a body and the variable t to label the time it is there. The system of labeling for discussing coordinate transformation is shown below in Figure 10-5. Since a measurement involves a change in a body such as the emission of light that later registers on a detector, the formalism's coordinates most properly refer to such changes, or *events*; in other words, (t, x, y, z) refers to the time and position at which an event occurs in a given frame. (Note here that our focus on events, i.e., change, should not lead us to forget that change presupposes substances that change.)

In Galilean (and Newtonian) mechanics, time is the same in every frame, giving equation 10.2a below, and the spatial dimensions are the same except for the axis that the frame is moving along. Notice we choose to call this axis the x-axis; such a configuration is called ***standard configuration***. Standard naming includes: calling the first frame the "lab frame" or the unprimed frame or Σ and calling the second frame the "moving frame" or the primed frame or Σ' frame.

10.2a-d

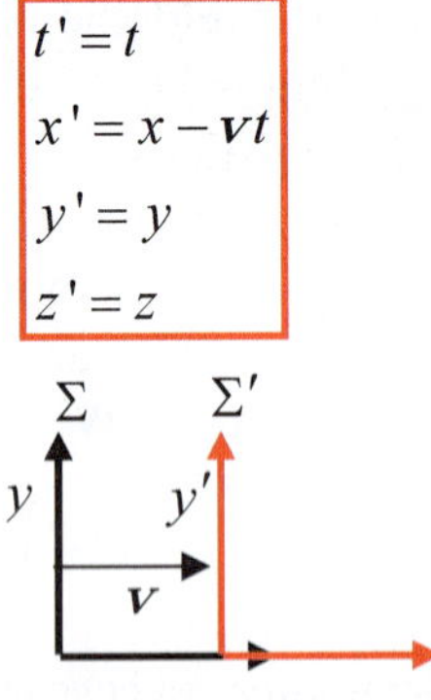

Figure 10-5: *Standard Configuration* with coordinate transformation between the *lab frame*, Σ, and *moving frame*, also called the primed frame Σ'. We define x' and t' so that $x = x' = 0$ and $t' = 0$ at the moment $t = 0$.

Introduction to Transformation Laws of Special Relativity

a) What do We Mean by Constant Speed of Light?

Before we can write the transformation laws for special relativity, we need to investigate how we can have a maximum speed that is the same in every frame, and, more to the point, what we mean by it.

i) Light

First, it's important to recall that the interaction between separated bodies is controlled by those mediators that have the most effective combination of strength and fastest travel speed. So for example, sound mediates between separated bodies but it's not the fastest such mediator and is, also, linked to the speed of light.[12] In the region around a strong magnet, the speed of the propagation of the magnetic field is the same as that of gravity but the magnetic field (which you'll see in future courses is related to light in its most general meaning) is much stronger. In both cases, light, as the carrier of new information about the electromagnetic field, would be the controlling influence.

Secondly, we need to say a couple words on language. Since light is the fastest evident example of interaction between separated bodies, *we will usually use the word light to refer to any action of one body through some distance in the plana onto another body that travels at the fastest possible speed.* Keep in mind though this doesn't mean they are the same thing. If God were to create (out of nothing) a flashlight in the on position, its gravitational field would move through the plana at the same speed as the light would. Still, gravity and light are not the same thing as could be seen by the radically different way that each interacts with bodies in their path. In taking advantage of this way of speaking as well as a summary mode of description, instead of saying "the maximum apparent speed of communication between bodies is independent of choice of uniformly moving frame," we will often simply say "the speed of light is constant or invariant."

---

[12] The speed of sound is conditioned on reactions that happen at the speed of light, so, for instance, if the speed of light was suddenly made much slower the speed of sound would be affected.

ii) Propagation of Light

To understand how odd it is and what we mean when we say that light travels at the same apparent speed no matter where it is measured, consider a simple analogy using sound instead of light.[13] Refer to Figure 10-6 below of a railway car on the ground. For simplicity, we assume the ground under consideration is truly at rest.

Think of two men, conveniently named Front and Back, on a flatcar that is part of the swiftly moving train. Assume it is dark and that they can only communicate by sound. Each yells out a quick "Hi" to the other. We want a *quick* "Hi" so we don't worry about how much the flatcar moves during the time of speech.[14] Back's message takes longer to get to Front than the reverse, because as the sound travels to Front, Front is moving away and so the distance the sound must travel to Front is increased. By contrast, Front's "Hi" takes much less time because the distance that the sound has to travel is decreasing all the time that the sound moves through the air toward Back. Hence, someone measuring the speed of sound on the train finds different speeds of sound in forward and backward directions. The measurements would proceed as follows in our thought experiment.

The observer on the train ignores the movement of the train, including the heavy wind he feels. Concentrating on the platform as if it were a completely closed system immune from outside influence, he times how long the sound takes to move over a standard ruler. He finds the sound moving toward the front is traveling slower than it would if the car were at rest and sound traveling to the back is traveling faster. Whereas, an observer on the ground that measures the time it takes for the sound to cross a standard ruler finds that the sound travels at the same speed in both directions.

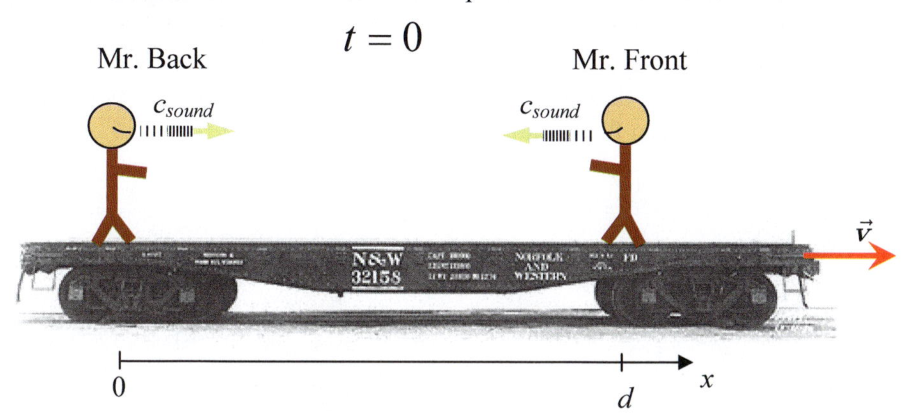

Figure 10-6: Mr. Front and Mr. Back a distance d apart: they speak at the same actual time $(t = 0)$ on the ground.

iii) Operational Simultaneity and Time Delay

Obviously, as we expect with sound, we don't have a constant speed of propagation, not the least because it is an asymmetric situation and thus not even constant

[13] To circumvent any possible confusion, note (as hinted at earlier) that sound is dependent on light (i.e. when the molecules that carry the sound interact they do so by forces (largely electrical) that travel at the speed of light) so we will not (indeed, cannot) short circuit in any of what we're going to say by this analogy, but only do what we have in mind, which is to prepare the way.

[14] If the time of speech were too long, we would be forced to note that the first part of the speech would be excited in one region of the air and the next part in another and so on.

within the frame. However, we can remedy this. To do so, we have to consider what we mean by simultaneity. Simply said, simultaneity means two things exist at the same time. We might say "the couple sat at the table at the same time" or "they sat at the table simultaneously." We often are interested in whether two events, that is two happenings or changes began at the same time, i.e. were simultaneous. But how do we determine this? This is Einstein's insightful question. If we had two identical things separated by a distance, and there were no qualities to distinguish one region of the distance from another, then we expect that an action of each *would be simultaneous if the effect of each action reached the center at the same time.* This is an operational definition of simultaneity.

Now, in the spirit of discussing the closed system aspect of the flat car, that is, of ignoring the motion of the flatcar and just considering things with respect to the car itself, we can define simultaneity in the above operational way. This will not, of course, make the events truly simultaneous any more than ignoring the system's uniform motion makes it not moving. Still, for our intent, and solely for that narrow domain of discussing events relative to the platform, this definition is not simply a convention; it allows us to focus attention on a *particular* reality of the moving platform. Namely, the nature of the platform's impetus that keeps it moving uniformly, keeping the parts at rest with respect to each other, makes it seem as if all its parts were actually at rest. By defining simultaneity on the platform in this manner that is natural to the platform in a very important way, we will leverage further insight.

To accomplish the definition, we put a listener at the middle point between the two men. Front and Back could then adjust the time they give their greetings until the listener hears them at the same time, i.e. the sound from Front and Back reach him at the same time.[15] In so doing, it is obvious that we will find that Front must delay his "Hi" compared to Back. This, in turn, means that the definition of time in the moving platform will be shifted by a certain amount as a function of distance in the moving frame.

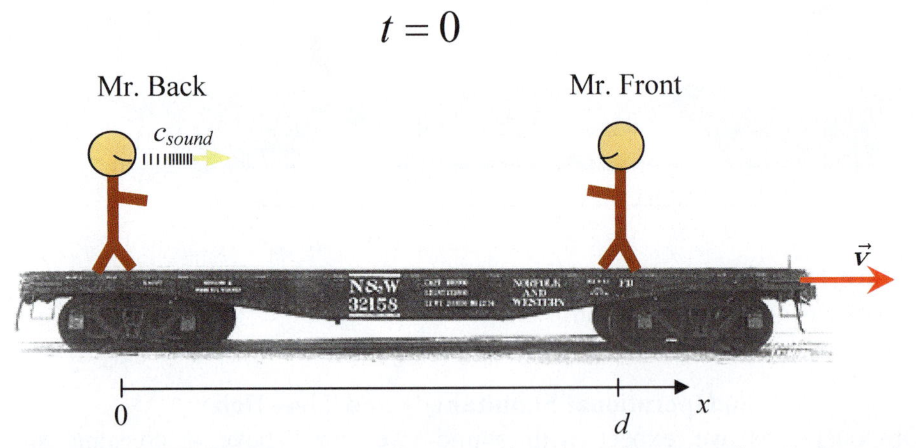

Figure 10-7: Two men a distance d apart: they speak at the same *operationally-defined* time in the moving frame. At time $t = 0$ shown above Mr. Front has yet to speak.

[15] A group of experimenters fixed alongside the tracks also makes recordings of the events of sound emanating from the man and being received at the middle.

In particular, taking Back and Front to be, respectively, at $x=0$ and $x=d$ at $t=0$, the emission times we need for apparent simultaneity in the moving frame are, using standard notation:[16]

10.3
$$t_{x'=0}=0$$
$$t_{x'=d}=T \quad where\ T=\frac{\beta}{1-\beta^2}\frac{d}{c} \qquad \text{where, } \beta \equiv \frac{\mathbf{v}}{c}$$

We obtain this in the following way. First:

10.4
$$x_{middle}=\frac{d}{2}+\mathbf{v}\,t \qquad x_{Sound\ from\ Back}=c\,t_1 \qquad x_{Sound\ from\ Front}=d+\mathbf{v}\,T-c(t_2-T)$$

Now, if $T=0$ the "hi"s get to the middle of the car at *different* times according to:

10.5
$$t_1=\frac{\frac{d}{2}}{c-\mathbf{v}} \qquad t_2=\frac{\frac{d}{2}}{c+\mathbf{v}}$$

On the other hand, we need $T=\frac{\beta}{1-\beta^2}\frac{d}{c}$ to get the sound to *meet* in the middle, which happens at:

10.6
$$t_{meet}=t_1=t_2=\frac{d}{2(c-\mathbf{v})}$$

Thus, the point of meeting in the lab frame is:

10.7
$$x_{meet}=\frac{d}{2}+\mathbf{v}\,t_{meet}=\frac{d/2}{1-\beta}$$

Mr. Back speaks at $t=0$, and thus his "hi" arrives at the middle of the platform at $t=t_m$. Mr. Front speaks later at $t=T$ but his "hi" arrives at the middle at that same time t_m. To define time in the moving frame such that the "hi"s are always timed to appear to occur at the same time $t'=0$, we subtract off the delay T giving:

10.8
$$t'=t-T=t-\frac{\beta}{1-\beta^2}\frac{x'}{c}$$

This equation, as written, is somewhat general; it tells us how to synchronize clocks in the moving frame so that the "hi" said at x' will reach the middle point, $x'/2$, at the same moment as the "hi" said at $x'=0$. For example, this equation tells us that if the "hi"s are spoken at: 1) $t=t,\ x'=0$ 2) $t=t+T,\ x'=d$, they are operationally simultaneous at $t'=t$.

Obviously, the spatial pair to this equation is:

10.9
$$x'=x-\mathbf{v}\,t$$

Substituting this equation into equation 10.8 gives:

10.10
$$t'=\frac{t}{1-\beta^2}-\frac{\beta}{1-\beta^2}\frac{x}{c}=\gamma^2\left(t-\beta\frac{x}{c}\right)$$

[16] In standard notation, recall we define x' and t' so that $x=x'=0$ and $t=t'=0$ at the moment $t=0$.

where $\gamma \equiv \frac{1}{\sqrt{1-\beta^2}}$.

Notice that, according to this equation, the time in the moving frame does not tick at the same rate as the time in the rest frame. In addition to being shifted it is also scaled by the factor $\frac{1}{1-\beta^2}$.

To better understand this scaling factor, consider a rest frame clock (i.e. a clock is fixed at some $x = x_0$) that emits equal interval sound pulses, for example, τ seconds apart. How far apart will these pulses appear to an observer in the moving frame, say stationed at $x' = 0$, in the moving frame? It depends on the direction from which one approaches the clock. Remember events that are operational simultaneous in one frame are not so in another. An simple calculation shows that the periods of clocks that one is approaching are contracted by $\frac{1}{1-\beta}$, whereas the periods of those that one is receding from are stretched by $\frac{1}{1+\beta}$, giving an average of $\frac{1}{1-\beta^2}$.

Thus, our current transformation laws from time and space coordinates in the lab frame to time and space coordinates in the moving frame are given by equations 10.10 and 10.9. In fact, equation 10.9 is not the fully consistent (per operational simultaneity) spatial pair, as we will see later. In any case, *we need to be careful in interpreting the transformation laws because, in them, we have linked time with position. This mixing will indeed be the hallmark of special relativity.*

The space-time mixing results from simple facts that need to be recalled in understanding the equations. For instance, $x = 0$ and $x = d$ *only* coincide with Back and Front respectively at $t = 0$; at later times, they correspond to different parts of the train according to $x' = x - vt$. Also, notice that we have, by convention, set $t' = 0$ at $t = 0$ and, moreover, that we can only synchronize t and t' at one spatial point on the platform. We clearly see, as we expect from our definitions, that $t = constant$, say $t = 0$, i.e. simultaneity in ground frame, does not yield simultaneity in the flatcar frame, for $t' \neq constant$ when $t = constant$. Again, the most important thing to notice here is that the *time is a function of distance.*

iv) The Wind Problem

Hence, we've solved the speed asymmetry issue; it is not a problem if we stick to the spirit of considering the moving frame as a closed system and measure things accordingly. To accomplish this frame centered view, we must synchronize our moving-frame clocks at all points in the moving frame. This is done by adjusting the clocks so that pulses that are released at the same time (according to them) from any two of them will arrive at the center point between them. With this operational definition, we have made one big step towards special relativity.

But, we still have problems. In our analogy using sound on the flatcar, although now the speeds are symmetric just as they are in the rest frame, the speed is not the same as it is in the rest frame; in particular:

10.11

$$c'_{sound} = \frac{\Delta x'}{\Delta t'} = \frac{\Delta x - v\,\Delta t}{\frac{1}{1-\beta^2}\left(\Delta t - \beta\frac{\Delta x}{c}\right)} = \frac{\frac{\Delta x}{\Delta t} - v}{\frac{1}{1-\beta^2}\left(1 - \beta\frac{\Delta x/\Delta t}{c}\right)}$$

$$= \frac{\frac{\Delta x}{\Delta t} - v}{\frac{1}{1-\beta^2}\left(1 - \beta\frac{\Delta x/\Delta t}{c}\right)} = \frac{c(\pm 1 - \beta)}{\frac{1}{1-\beta^2}(1 \mp \beta)} = \pm\frac{c}{\gamma^2}$$

(where, in the last line, we substituted $\frac{\Delta x}{\Delta t} \to \pm c$, and we use "+" for forward and "–" for backward moving sound).

In addition, we still have a rushing wind that we cannot avoid. In the case of light, we have an *apparently* somewhat similar "wind" problem; the mass increases with speed.

b) Mass Variation and Time Dilation

To see the problem, we return to the primary motivation for the relativity of frames in the first place, impetus. Bodies have this capacity to keep moving after the action of a force has ceased. This means bodies can be at rest with respect to each other, and thus appear to be at rest, even when they are moving. This is part of the reason why inside a uniformly moving ship, we can sometimes forget the ship is moving. But there's more; if we push on an object in the ship, it appears to respond the same whether or not the ship is moving. Furthermore, the form of the energy and momentum conservation equations is the same in the moving frame. This is because the impetus is additive or linear. Adding momentum p to an object adds $\Delta v = \frac{p}{m}$ speed to the body, no matter how fast it's already going. Because of this fact, the energy and momentum conservation equations, ($\sum \frac{1}{2} m v_i^2$ and $\sum m v_i$) apply in the moving frame in the way we have previously discussed, though the particular numbers are different, depending on whether the extra center of mass terms due to the motion of the system (see equations 10.1) are added in or not. However, as we "turn on" more effects, we find that the linearity of speed increase with momentum is only approximately true at low speeds and greatly violated at higher speeds.

Figure 10-8: Aerial view of Fermilab in Batavia, Illinois. In the photo, one sees part of two circular accelerator rings. The equipment at Fermilab is capable of accelerating protons (clockwise) and anti-protons (counterclockwise) to 0.980 *TeV* each. The two large rings are the last part of a chain of accelerators that bring the particles up to speed to be collided. The closest ring brings the particles up to 150 *GeV*, the other large ring brings the particle up to its maximum speed. As the particle's speed increases, its mass increases.

In particular, we find by experiment-- for example, in particle accelerators such as shown in Figure 10-8-- that the inertial mass increases as the particle's speed increases. It is as if the particle were fighting against a wind when it is moving. Of course, this is only an analogy to wind; it is not exactly the same as wind, only like it. As in any investigation, we move from a vague knowledge of a thing to more and more specific knowledge. Though, for instance, we know the nine categories of properties, we must experiment to learn more about the specific properties.[17]

In this case, it is known to great precision that the mass increases by $\gamma = \dfrac{1}{\sqrt{1-\beta^2}}$, i.e. $m = \gamma m_0$ where m_0 is called the rest mass. This means that, though things move together, they are harder to move faster when they are already moving. As we will see later, this changes our equations for energy and momentum. Thus, it seems our relativity of motion breaks down.

To recover it and to better understand the place of mass variation, we need to again take seriously the idea of the system of particles moving together as a closed system, i.e. leaving out the influence and the comparison with the outside.

Time Dilation Due to Mass Variation

Once we consider the frame without reference to the outside we have, by definition, no standards of comparison by which to say it's *harder* to move objects within the frame. As long as all the masses increase in the same way, there will be nothing within the frame to reveal that it is moving. Of course, when we do allow ourselves to transition between the given frame and the rest frame, we will see that when a particular body, say an electron at rest in a moving frame, is given an impulse, i.e. a certain increase in momentum, say Δp, its speed *increases less* than does an electron that is initially at rest in the lab frame.[18] Now, time is the measure of motion, so time in each frame is defined by a standard of motion, which we call a clock. The clocks are moving slower in the moving frame since the same Δp causes less motion there, so time moves slower in the moving frame. Experiments, such as the muon decay discussed in an end of chapter problem, have verified that time does slow down in moving frames. This slowing down of time in a moving frame is called ***time dilation.***

This is important. Take a moment to think a little about it. Making a measurement implies that we have a unit by which we measure. To say, for instance, we determined that it took six seconds to get the ice cream out of the freezer implies there is a motion that we take as our unit, e.g. the swinging of a pendulum or the movement of light from one place to another or the motion of a hand on a stopwatch. We then determine how many of these motions span the motion we that we seek to measure. For instance, we watch the second hand traverse six tick marks during the motion of the ice cream from the freezer shelf to the table.

[17] The further study of the nature of mass and the mass increase, i.e. decrease in the receptivity to the action of the impetus, is the subject of other courses; neither is understood yet at a deep level.

[18] The Δp is measured in the lab frame, which we take to be an absolute rest frame, but it can, empiriometrically, equally well be taken to be any other inertial frame that we wish, as we will see later. However, only in the hypothetical actual rest frame, will the Δp measured be the actual net intensity of the impetus. We say hypothetical because general relativity will cause us to further refine our thinking on frames.

So, in the case of two reference frames, since the motions, say of stopwatches, in each of the two frames proceed at different rates, we can measure time by the standard of motion in the moving frame *or* in the stationary frame. Of course, the moving standard is natural to the moving frame, and the stationary standard is natural to the stationary frame. The moving standard is *dilated* (slower), as is everything in the moving frame, relative to the lab (stationary) frame.

To be precise about this effect, consider the following thought experiment. Consider two universes[19] that we will call the unprimed universe, U, and the primed universe, U'. In the first universe, the mass does not change with speed, while in the second, like our universe, the mass increases with speed according to $m = \gamma m_0$. Suppose that each universe has identical rulers of length L. Furthermore, the time is measured by the following type of clock. Assume a known intensity of impetus p is given to a body with known mass m_0. Indeed, to avoid having to introduce separate standards into our universes, we could define the unit of length, the unit of intensity of impetus and the unit of mass to have these magnitudes. In any case, we define the unit of time to be the time it takes for the body to traverse the length of the ruler. To keep the clock working after passing the end of this ruler, we posit rulers lined up one after another along the *x*-axis in both directions as far as we can imagine. Suppose the body is equipped with some electronics so that when the body reaches the end of a ruler it counts and keeps a running tally of how many rulers it has traversed. In other words, reaching the end of a ruler is like the ticking of a digital watch.

Now, we compare time standards in the two universes to discover exactly how much time dilation, we have because of the mass increase.

In U, the time, T, between ticks is calculated in the following way:

The speed of the body is: $v = p / m_0$,

So that: $T = L / v = L m_0 / p$

In U', the time, T', between ticks is calculated in a similar way:

The speed of the body is: $v = p / (\gamma m_0)$,

So that: $T' = L / v = L \gamma m_0 / p$

Thus, the relationship between the time standards in the two universes is:
$T' = \gamma T$. Be careful in interpreting this expression. It is fundamentally different than the similar looking equation, which we will write below, that compares the number of ticks (standard units) that occur in one universe with those in another. $T' = \gamma T$ relates the *definition* of the standards in the two universes, which can then be used to write the latter type of relations. It implies that γ ticks in the unprimed universe must occur to get 1 tick in the primed universe. Note that the clock in the universe with mass increase runs slower, as it must. We now recast this result to write the relation in the more standard way. Namely, note that the passage of γ units of time in U means only 1 unit of time passes in U'. Hence, we write time dilation formula as:

10.12 $$\Delta t = \gamma \, \Delta t'$$

[19] Or two appropriately set up parts of our universe. Note that an end of chapter problem addresses the question of time dilation and mass change in a way that makes more direct use of reference frames.

Since very slow moving bodies such as those in a frame arbitrarily close to absolute rest are like U and moving frames are like U', we can say that, with respect to the mass,[20] time slows down in moving frames according to this formula.[21]

Though time dilation helps us understand the moving frame as whole, it obviously introduces problems for our larger effort to understand how we can get an invariant speed of light. Before we can resolve that issue, one other effect needs to be studied.

c) Length Contraction

We need to incorporate one more change that happens as we move faster. Remember that quality and quantity are correlated, so we don't expect to find a quality, e.g. impetus intensity, changing without a change in quantity as well. In fact, experiment has shown that the length of objects contract in the direction of travel according to $L = L_0 / \gamma$ where L_0 is called the proper length. Figure 10-9 illustrates how rods of given length are shorter in the moving frame than in the stationary frame. In the case shown, it takes two moving rods to equal the length of one rod at rest. Note how the parts of the rod shrink proportionally as well.

Length contraction and time dilation together will allow us to write the transformation laws between one "closed frame," corresponding to the point of view of a group of objects moving at uniform speed, and another. We will later see that length contraction and time dilation play complementary roles in the constancy of the measured speed of light. Indeed, we can already somewhat see how it will play out by noting that the slower motions make things take longer to move, but the length contraction of rods mean there is correspondingly less distance that needs to be moved across.

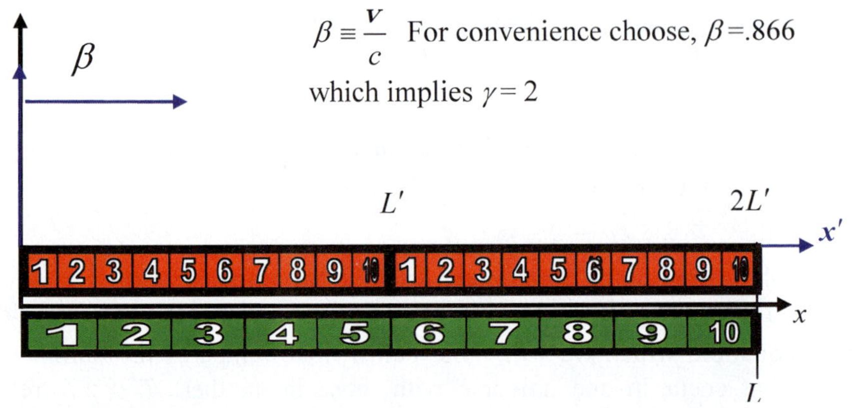

Figure 10-9: Notice how the parts of the top moving (red) rods are smaller than those of the fixed (green) rod, so that it takes two rods in the moving frame to make one rod in the fixed frame. $L' = L/\gamma$ or $L = \gamma L' = 2L'$

[20] We will need to incorporate of other aspects of bodies before this formula can have any general use.

[21] For those familiar with special relativity, this is not yet the time dilation formula as used there, because it is relating the time between an absolute rest frame (bodies with no impetus) and one that has some fixed (actual) impetus, whereas empiriometric special relativity prescinds from any such distinction and, indeed, could not make any sense of such a distinction.

d) Simultaneity in Special Relativity

We are now ready to show, using our operational definition of simultaneity, how we can incorporate the mass increase (in the form of time dilation) and length contraction to get the full transformation laws (for one spatial dimension) *from the rest frame to the moving frame*. Once we couple the derivation of this section with the discussion of the plana later in Part I and the interpretation of the laws given in Part II, we will have in place the physical underpinnings[22] of the laws, i.e. the natures that the laws describe. In the sections that immediately follow this one, we will discuss first principles[23] that explain these laws.

To begin, we use the "radar" method. Radar works by sending a radio signal toward an object at a certain time. The radio signal travels to the object, bounces off it and returns to the point of origin where it is received and the time of return is recorded. Knowing the time of transmission and time of return, one can calculate how far away the object is, as well as the time the object was "seen," i.e. when the radio wave reflected off of it. We can do the same thing if we are moving. Of course, normally, we would correct for the fact that we are moving, but if we choose to look from our closed system perspective, we can choose to not account for the effect of movement directly but just to define everything consistently within our frame. Again, this is basically an application of the operational definition of simultaneity.

To see how it works, consider the following particular case which consists of the sequence of events shown in Figure 10-10*a-d* below (for the figures, we assume $d < ct_{deer}$). It's a moonless night. 1) You step outside your moving railway car onto the running board on the side of the train. 2) You shine your flashlight at a moving object in the distance (*d* away) *alongside* the tracks ahead of the train. 3) The light reaches the object and reflects back. 4) The light returns, revealing a deer frozen in its tracks by the mesmerizing light.

We determine the time and coordinates of the event of the light striking the deer (event # 3) as given by the radar method, which, of course, will be different from the values: $t = t_{deer}$, $x = d$ in the rest frame; this can, of course, then be applied to other t, x. The four figures, Figure 10-10a-d, illustrate how to calculate the emission and reception time of the radar beam (t_E and t_R, respectively) in terms of t_{deer}, d and v. We do all calculations in the lab frame and deduce the times and distances seen in the moving frame.

---

[22] Technically, this is called the efficient causality.
[23] Technically, this is called the final causality.

Event 1

Step Outside at $x=0, t=0$ $(x'=0, t'=0)$

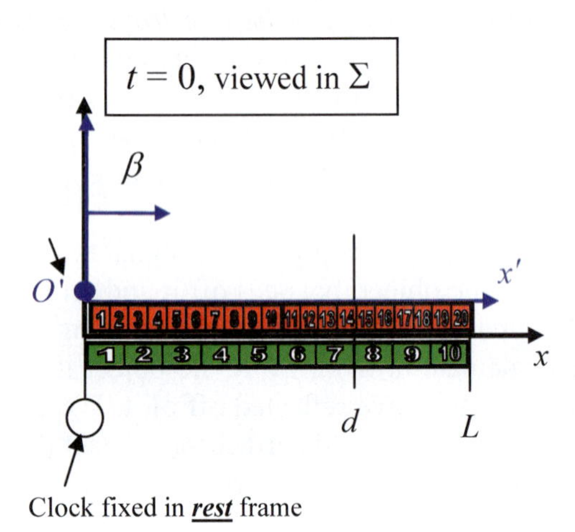

Event 2

Flashlight beam emitted from O' at $(t=t_E,\ x=b=\mathbf{v}t_E)$

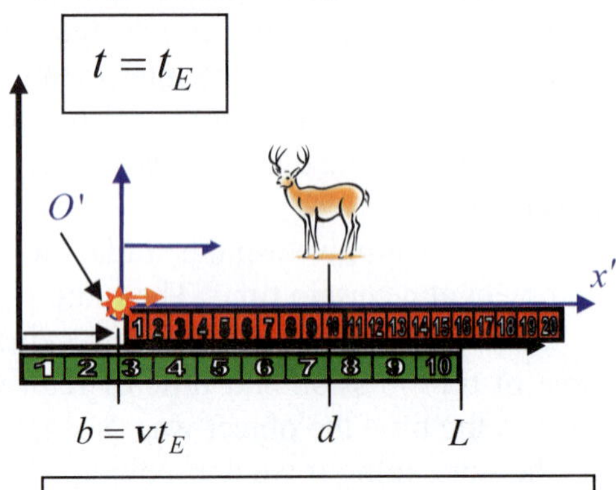

$x_{o'}=\mathbf{v}t$ (general eqn. for O' in Σ)

$x_\gamma=\mathbf{v}t_E+c(t-t_E)$

Event 3

Light Arrives at Deer and reflects off at $(t=t_{dear}, x=d)$

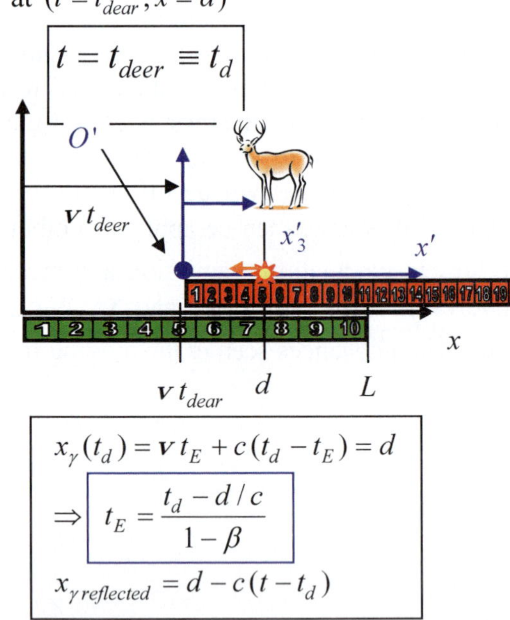

Event 4

Light Received at O' at $t=t_R, x=\mathbf{v}t_R$

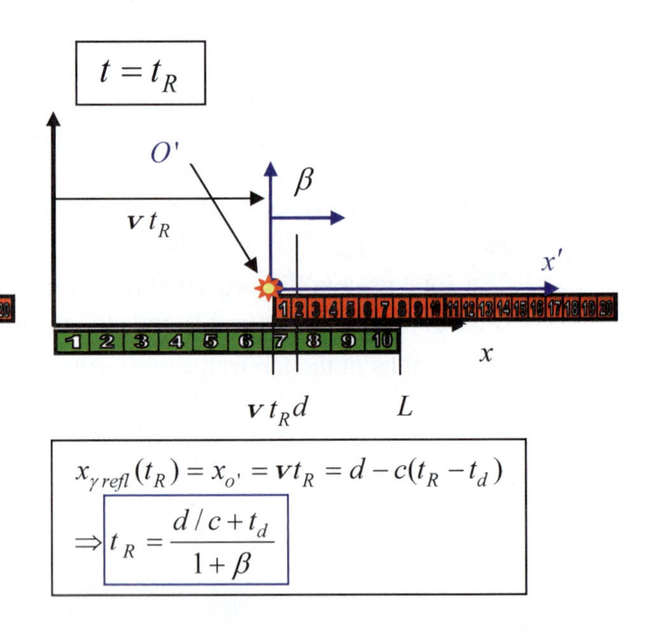

Figure 10-10: *Radar tracking of a deer* a. *<u>Event 1</u> (top left)* Observer steps outside railway car at time $t=0$ **b.** *<u>Event 2</u> (top right)* Observer shines flashlight alongside train toward deer alongside track at $t=t_{Emission}\equiv t_E$ **c.** *<u>Event 3</u> (bottom left)* Light strikes deer and reflects off it at $t=t_{deer}\equiv t_d$. The results are used to calculate the emission time t_E in the box below the image **d.** *<u>Event 4</u> (bottom right)* Observer sees light at $t=t_{received}\equiv t_R$. This reception time is calculated in the box below the image.

With the help of Figure 10-11 below (cf. also Figure 10-12), we write the equations for position, t and x deduced from the radar method.

10.13
$$t = \frac{t_E + t_R}{2}\ , \quad x = \frac{c(t_R - t_E)}{2}$$

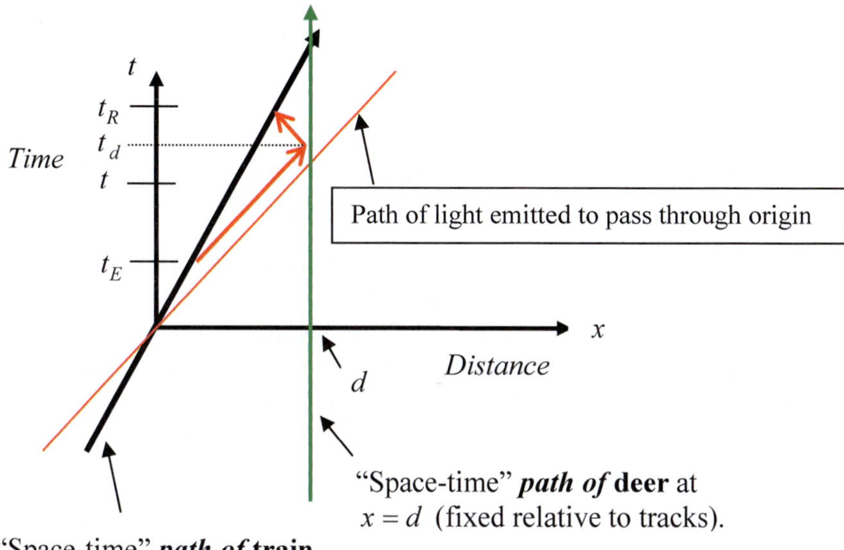

Figure 10-11: Radar method of determining distances is illustrated on a "space-time" diagram. This diagram shows events as seen in lab frame. It shows the positions as a function of time for all objects as viewed from the moving train using Earth-based clocks and rulers (so we are not yet at special relativity). The lower red arrow indicates the "space-time" path of light emitted towards the deer at time t_E. The upper red arrow indicates the path of the light that reflects off the deer at time t_R. The green and black line, respectively, indicates the path of the deer and train. Note the assigned time for the event of the light hitting the deer is labeled t and is offset from the actual time in the rest frame, but shifted time is part of what is needed to keep operational simultaneity. Note that the standard convention for such a diagram puts *time* on the vertical axis, even though the independent variable is, by other conventions, put on the horizontal; this is a further indication of the "mixing" of space and time.

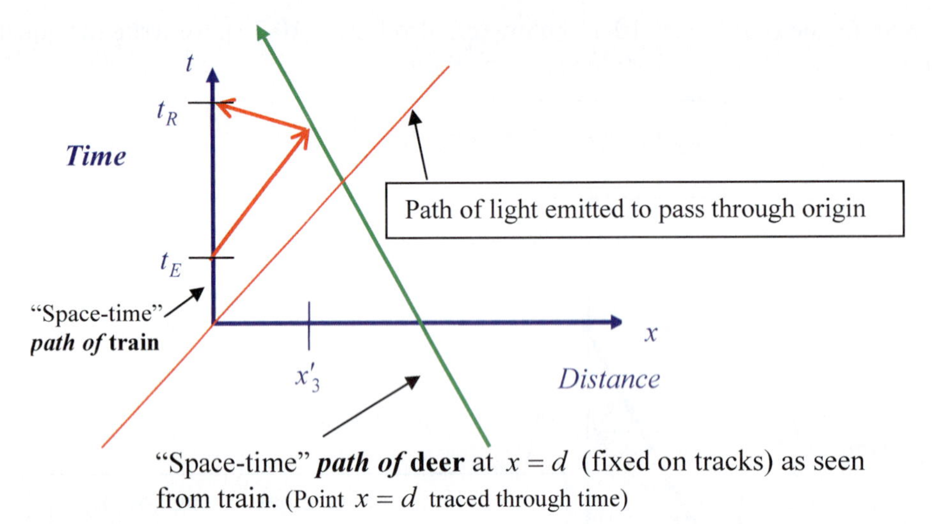

Figure 10-12: Same space-time diagram shown in Figure 10-11, but here referred to moving frame using ***Galilean*** transformation equations. The train observer sits at $x' = 0$, and so traces out the y-axis in this space-time plot. Note that the outgoing and reflected light beams (red arrows) do not have the same apparent speed c, showing *such Galilean transformation laws do not square with observed experiments*. The slopes of the red arrows are exaggerated to illustrate the effect.

Substituting the values for t_E and t_R given in Figure 10-10c and Figure 10-10d into equations 10.13, we get the following for the time and location of the deer in the moving frame:

10.14a,b
$$t = \gamma^2 (t_d - \beta \frac{d}{c})$$
$$x = \gamma^2 (d - \mathbf{v}\, t_d)$$

Note that 10.14a is the same as we obtained in equation 10.10.

However, we must account for the time dilation in the moving frame. The clocks we are using up till now are *fixed in the moving frame*, but we have not accounted for the difference in motion between rest and moving frame clocks. Lab frame clocks go faster (for every moment in lab frame only $1/\gamma$ passes in the moving frame), so we must divide by γ. It is easiest to see how the time dilation changes the transformation laws by applying the division by γ to equations 10.13 and then noting that the algebra yields:

10.15a,b
$$\boxed{\begin{aligned} t' &= \gamma (t_d - \beta \frac{d}{c}) \\ x' &= \gamma (d - \mathbf{v}\, t_d) \end{aligned}}$$

These are also the predictions of the Special Relativity transformation laws if we take $t_d \to t$ and $d \to x$. Notice these equations would be of no use were it not for the fact we have length contraction, because they would not correspond to what we measure and so would just be a kind of mathematical manipulation of what we already know with no further penetration into the physics. As mentioned, we will give the full interpretation of these equations in Part II of this chapter.

With an understanding now of what we mean by light traveling at a constant speed and about time keeping in moving frames, we can now deepen our understanding of special relativity and the transformation laws (equations 10.15a and 10.15b) that characterize it. We do so by carefully examining the core principles[24] and then deriving those laws in a more complete way, including explicitly handling both transformations from a given frame to a second frame in standard configuration *and* the reverse.

From Internal Relativity, Impetus and Maximum Speed to Special Relativity

First, we establish the principles of special relativity from maximum apparent speed and the internal relativity principle, including the nature of the impetus making speed the sole potential indicator of the nature of a house.

By internal relativity, every house must obviously see the same maximum apparent speed within it. However, it implies even more; it, as we have mentioned, implies, with the existence of such a maximum speed, the full relativity principle.

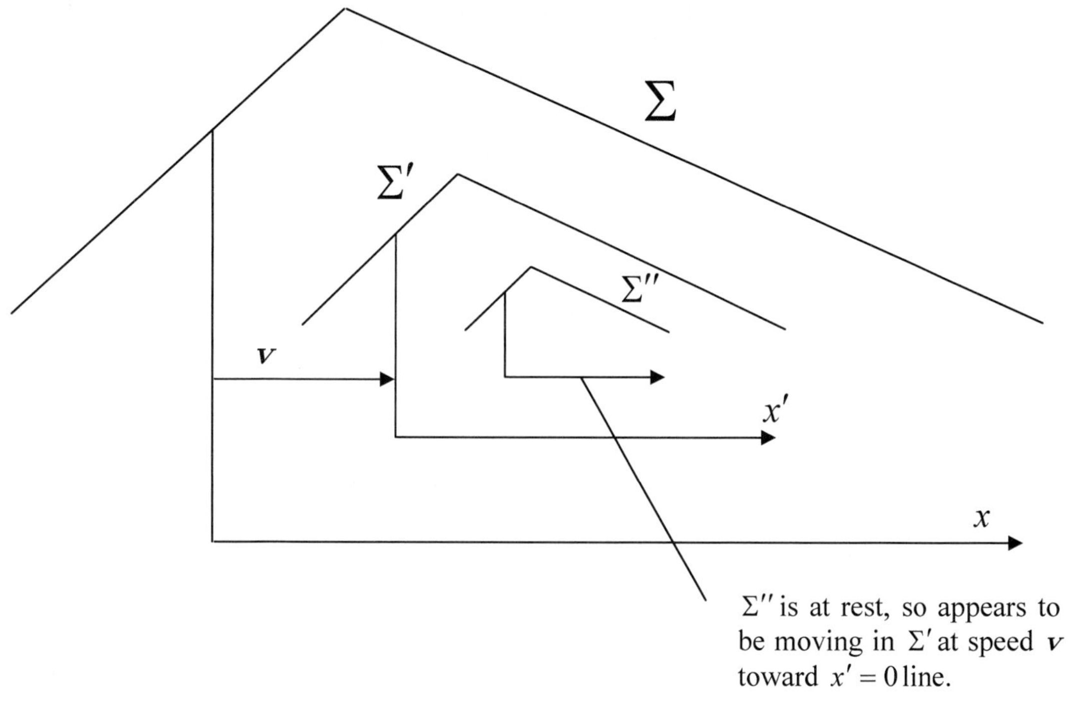

Figure 10-13: Houses are a helpful analogy for understanding the internal relativity principle. Each house can be considered to be isolated in the sense that we do not allow the interactions between houses to change the overall state of a house; we do not look outside the primary house, which is Σ in this case. Here we have three houses: Σ and Σ'' are at rest, while Σ' is moving.

[24] Final causes.

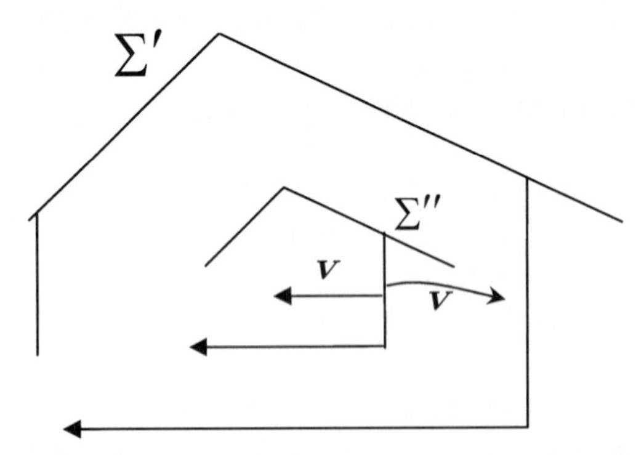

Figure 10-14: Changing origin and direction of axes in Σ' and Σ'' to reveal internal relational similarity with Σ and Σ'. Another way to think of this is to say this is the view from the other end of both houses; it is natural there to have the *x*-coordinate axis defined in the opposite sense. The bent arrow indicates how fast Σ' appears to move away when viewed from Σ''.

Start with the situation shown above in Figure 10-13. The house Σ, which is at rest, contains a moving house Σ' which, in turn, has Σ'' inside of it, which is also at rest. By the internal relativity principle, all experiments done in Σ must yield the same results as when done in Σ' and Σ''. Because the state of a house can only be a function of the speed and both Σ'' and Σ are not moving, they must be in the same state. Thus, when we observe experiments done in Σ' from Σ, we should see the same results as when we observe experiments in Σ' from Σ''. That is, we are *effectively* looking from the same house at the same thing.[25] Using the notation in which $A \rightarrow B$ means we look from house *A* to house *B*, this argument can be written:

10.16 Since $\Sigma'' = \Sigma$, $\Sigma \rightarrow \Sigma'$ is same as $\Sigma'' \rightarrow \Sigma'$

Now, by internal relativity, all experiments done and viewed in Σ' should give the same result as those done and viewed in Σ; i.e., each house will appear the same as the rest house and thus as every other house. Thus, when one confines one's view to Σ', Σ'' will appear to be the moving house. Indeed, by moving to the other end of the Σ' and Σ'' houses, reversing the coordinate axis and applying the rest of the arguments given in the section below titled *Deriving the Coefficients of the Laws,* we show in that section that, in Σ', it will look as if Σ'' is *moving away at speed* v as shown in Figure 10-14. Thus, jumping into the apparently moving house, Σ'', and looking back into Σ' should be exactly the same as looking from Σ' back to Σ. Taking advantage of our notation, we write this as:

10.17 Internal relativity implies $\Sigma'' \rightarrow \Sigma'$ should be same as $\Sigma' \rightarrow \Sigma$

Using 10.16, we can replace the first relation with $\Sigma \rightarrow \Sigma'$ giving:

10.18 $\Sigma \rightarrow \Sigma'$ should be same as $\Sigma' \rightarrow \Sigma$

In words, looking from the rest house (Σ) to the moving house (Σ') should be the same as looking back from the second to the first. In other words, there is symmetry between the two.

---

[25] The relative placement might be different, but it could be so within the given house as well.

Next, we apply the internal relativity principle one last time to see that the Σ house could just as well be moving. If we allow Σ to be moving at some arbitrary speed, the arguments still go through, since they only depended on Σ and Σ'' being in the same state and being indistinguishable from rest, not on being at rest. Thus, we can leave out any mention of what house is actually at rest and what is not, and we have the full relativity principle.

These arguments go through for both open and closed houses. Furthermore, since the speed of a house, whether it is opened or closed, determines the state of the interior of that house, it makes no difference whether we look at the same outside world from an open house or a closed one. Also, note that the environment outside the house *is* the same for both an open and a closed house, because we suppose that the closure of the interior house does not affect the state of the exterior house; only the speed of the exterior house determines its state. This amounts to a statement about the inertness of the exterior plana at our level of approximation. In fact, of course, things happen in the plana, for which special relativity will be a backdrop, such as the pair creation and annihilation discussed at the end of this chapter. In other words, an open house moving at a given speed sees, both inside and outside, just what a closed house moving at that speed would see. The closed or open specification results in no difference in the effects we see at this level of abstraction. Thus, we can open up all the houses and cease all considerations of plana and simply say that the laws of physics are inertial frame invariant. That is, *any* experiment will give the same result no matter what inertial frame it is carried out in. In the section below titled *Inside a Sealed Frame*, we will discuss the hypothetical changes in the plana (vacuum) that can account for the similarity of open and closed houses at the level of special relativity. In any case, in special relativity, we leave out the plana, for it becomes a non sequitur as we focus only on the *relations* of motions, lengths, masses and other aspects *within* a given inertial frame and the transformations laws that give the relations between any two such frames.

Interestingly, the internal relativity principle which restricts the results of measurement *inside* then restricts what can happen *outside*, including the speed measured looking outside. This happens because all systems (including substances) have parts (note analogical meaning of parts here) so that inside each system is a sort of microcosm of the outside. Indeed, again, this internal relativity principle really, coupled with the general need for a maximum speed, yields the full relativity principle and thus special relativity.

SR Transformations from Maximum Speed and a Relativity Principle

a) Overview: From Principles to Transformation Laws

To establish the laws, we need to establish the linearity of the transformation laws mathematically, which we will do shortly. Next, we could use the full relativity principle, that any experiment done in a given inertial frame yields the same result when done in any other inertial frame, and the need for a maximum apparent speed to argue that there must be the same maximum apparent speed in every frame, i.e., the invariance of the speed of light, c. Confining ourselves to one spatial dimension, we could then write the equation for the propagation of light, c, in one frame, and demand that it also propagate at the same speed in any other frame that is moving uniformly relative to the first. This is standard practice, and we will do it as an end of chapter problem. However, deeper physical insight into the meaning of the relativity principle and the law of maximum speed as well as the transformation laws themselves comes via the different approach that we will take.

In this approach, we note that integrity of moving systems could (apparently) be somewhat maintained by a less stringent condition on the similarity between inertial frames. Note, in this discussion, we will find it helpful to return to our house analogy.

In particular, we will start with a weak form of house "indistinguishability" or similarity that requires only that the maximum apparent speed in each house be below a certain value as seen from any house. In other words, we require there to be a maximum speed of say c_{Max} that bounds the max apparent speed as seen in any house; this means that any given house A might have a lower (or equal) actual maximum apparent speed, c_A, so that in that house (which includes views from the exterior house into interior houses, to exclude the apparent a-causal behavior) no measurable communication could appear to go faster than $c_A \leq c_{Max}$. Mathematically, as we've said, for all houses labeled by index i, $c_i = c, c_i^{(n)} = c^{(n)}$, but $c^{(n)} \leq c_{Max}$ for all n, where n represents the view of the primary house from a given house.

This is a reasonable restriction because without it any given house may have a maximum apparent speed, but it can be arbitrarily high, making huge differences in the sort of activity that can occur from one house to the next. Objects in some houses would be able to go, for instance, across the length of the universe in a flash, whereas in other houses it would take billions of years to cross the same apparent distance. In other words, we severely violate the integrity of the system (house) we are trying to establish, because it would be able to behave so differently within its own interactions depending on how fast it was going. For similar reasons, after we have given this upper bound, we would want, if we were to continue with this same line of thought alone, to restrict the lower bound as well, thus, creating a band for the maximum apparent speed. Instead, we stay with the given postulate by itself.

This postulate of a global maximum apparent speed will get us most of the coefficients of the equations of transformation between houses. However, as we expected, we must add a further assumption to get the rest of the way. At this point, we will hypothesize a symmetrical relation between the apparent relative motions of any two houses (and hence frames). That is, if we view a house to be moving away at speed v in a lab house (i.e. the house which we take as our initial rest house), then the lab house viewed from that moving house will also appear to be moving away at speed v. With our experience of working with defining operational simultaneity by shifting time, we can see that such a speed symmetry is not necessarily a given. To understand something of the nature of this symmetry postulate and of the internal relativity principle, we next deduce the symmetry by application of the internal relativity principle.

We then see that the symmetry of the apparent speed in both directions allows the completion of the form of transformation law, thus showing that those postulates are axiomatically equivalent to the full relativity principle and the existence of a maximum apparent speed. The result is the full transformation equations. These transformation laws are the same for any house whether closed or open, because the arguments that determine the coefficients apply to both; hence we can, as before, cease to talk about houses and use only the concept of frame. In sum, by imposing a maximum global apparent speed, a frame speed symmetry and operational simultaneity in defining time in each house, we establish the transformation laws of special relativity.

Once we've established the transformation laws using the principles of this section, we will then put empiriometric special relativity on its principled and formalized footing, including the proper context for the law of the invariance of the speed of light. We now move to the first step of establishing the transformation laws: establishing the linearity of the laws.

b) Linearity of the Transformation Laws

We have already laid out the basics of why the mathematical transformations that describe the connection between one house (and thus frame) and another must be linear. Namely, the impetus keeps things moving and so while one house can "run" slower than (and shifted from) another, there is no reason why identical types of motions in each house should not be uniformly moving relative to each other. Or, said positively, something moving uniformly in one inertial house (or frame) must do so in another.

Let's make this argument mathematically explicit. The key fact of the inertial house (frame), i.e. that the momentum, and hence the motion of particles not under the influence of external forces[26] is constant, means the acceleration of any such particle is zero. In particular, for the rest (Σ) house, $\frac{d^2x}{dt^2}=0$, where we use unprimed coordinates for the rest house. For choice clocks on the particle (called proper clocks), we can write $\frac{d^2x}{d\tau^2}=0$, where since they move uniformly with respect to each other $d\tau \propto dt$.

The unprimed house coordinates, x, are considered a function of the moving house coordinates and vice versa. For example, $x(t',x',y',z')$ *and* $x'(t,x,y,z)$; in the latter function, for instance, inserting the values for the time and place in the rest house will yield its x-position in the primed house.

To transform small changes in the numbers representing place and time in the rest house (unprimed coordinates) to those of the moving house (primed coordinates) and vice-versa, we recall:

10.19a-d

$$dx'^0 = \frac{\partial x'^0}{\partial t}dt + \frac{\partial x'^0}{\partial x}dx + \frac{\partial x'^0}{\partial y}dy + \frac{\partial x'^0}{\partial z}dz$$

$$dx'^1 = \frac{\partial x'^1}{\partial t}dt + \frac{\partial x'^1}{\partial x}dx + \frac{\partial x'^1}{\partial y}dy + \frac{\partial x'^1}{\partial z}dz$$

$$dx'^2 = \frac{\partial x'^2}{\partial t}dt + \frac{\partial x'^2}{\partial x}dx + \frac{\partial x'^2}{\partial y}dy + \frac{\partial x'^2}{\partial z}dz$$

$$dx'^3 = \frac{\partial x'^3}{\partial t}dt + \frac{\partial x'^3}{\partial x}dx + \frac{\partial x'^3}{\partial y}dy + \frac{\partial x'^3}{\partial z}dz$$

[26] Recall that fictitious forces which are seen in accelerated frames (such as centrifugal force) are due ultimately to some force acting on the system (for example, in the frame of the ball swinging around in a circle on end of a rope, it is the tension of the rope that causes the acceleration.)

Where the primed coordinate variables are labeled by four indices, 0, 1, 2 and 3 as: $x^0 = t, x^1 = x, x^2 = y, x^3 = z$. Note that "$x$" is the base of the symbol for time as well as space, codifying the mixing of space and time.

After "dividing" the equations 10.19a-d by $d\tau$, where τ is proper time, we can write, using the symbolic index ν to denote *0, 1, 2, 3*:

10.20
$$\frac{dx'^{\nu}}{d\tau} = \frac{\partial x'^{\nu}}{\partial t}\frac{dt}{d\tau} + \frac{\partial x'^{\nu}}{\partial x}\frac{dx}{d\tau} + \frac{\partial x'^{\nu}}{\partial y}\frac{dy}{d\tau} + \frac{\partial x'^{\nu}}{\partial z}\frac{dz}{d\tau} = \sum_{\lambda=0}^{\lambda=3}\frac{\partial x'^{\nu}}{\partial x^{\lambda}}\frac{dx^{\lambda}}{d\tau}$$

Differentiating this expression gives:

10.21
$$\frac{d^2 x'^{\nu}}{d\tau^2} = \sum_{\nu=0}^{3}\left(\frac{\partial^2 x'^{\nu}}{\partial x^{\lambda}\,\partial x^{\sigma}}\frac{dx^{\sigma}}{d\tau}\right)\frac{dx^{\lambda}}{d\tau} + \sum_{\nu=0}^{3}\frac{\partial x'^{\nu}}{\partial x^{\lambda}}\frac{d^2 x^{\lambda}}{d\tau^2}$$

Because when no forces are applied, no particle will be accelerating in either inertial house (i.e., $\frac{d^2x}{dt^2} = 0$ and $\frac{d^2x'}{dt^2} = 0$), the first term on the right hand side is zero. Hence, since this must remain true for all speeds of particles ($\frac{dx^{\lambda}}{d\tau}$),

10.22
$$\frac{\partial^2 x'^{\nu}}{\partial x^{\lambda}\,\partial x^{\sigma}} = 0$$

This, in turn, implies:

10.23
$$x'^{\nu} = \sum_{\nu=0}^{3} a^{\nu}{}_{\mu}\, x^{\mu} \text{ where } a^{\nu}{}_{\mu} \text{ is a matrix of constants}$$ [27]

(*Exercise:* check that this yields the vanishing partials above). Now, we will return to considering just one dimension in space, because this dimension can be chosen to be any one of the three (x, y or z), and thus no generality is lost for our current discussions.

Hence we have a two dimensional transformation in matrix notation:

10.24
$$\begin{pmatrix} t' \\ x' \end{pmatrix} = \begin{pmatrix} a_{00} & a_{01} \\ a_{10} & a_{11} \end{pmatrix}\begin{pmatrix} t \\ x \end{pmatrix}$$

Here we've dropped the superscript/subscript combination on the "a" for convenience; this combination will be important in advanced courses. In terms of individual equations we have:

10.25
$$\begin{aligned} t' &= a_{00}\, t + a_{01}\, x \\ x' &= a_{10}\, t + a_{11}\, x \end{aligned}$$

At this point, as mentioned, we could apply the postulate of the relativity principle and the postulate of the constancy of the speed of light in all frames to deduce the transformation laws of equations: 10.38 and 10.39. However, as mentioned, this is left as a problem which students are strongly encouraged to do. However, in order to give advanced students an opportunity to think more about the origin of the different constants, we approach it from a somewhat more physical point of view. We introduce two hypotheses: a

---

[27] Note that Λ is the standard symbol for these constants; we use "a" here to avoid introducing any more unfamiliar ground than is necessary.

global maximum speed hypothesis and a postulate of symmetrical apparent speed of house (frame) motion. We also show how the latter is deduced from the internal relativity postulate, thus opening that postulate up more. *In a less advanced course, this section can be skipped without losing the flow of the chapter*. However, if it is, the student that has a grasp of the basics of the chapter is encouraged to return and read it later for the chance it provides to fertilize and stir one's thinking about the basic issues of the chapter.

c) Deriving the Coefficients of the Laws

The constants in the second equation of 10.25 can be very quickly derived. We note that by definition of standard configuration and Figure 10-15:

10.26 $$\{\text{distance from } x'=0 \text{ to point in } \Sigma' \text{ in units of } \Sigma\} = x - \mathbf{v}t$$

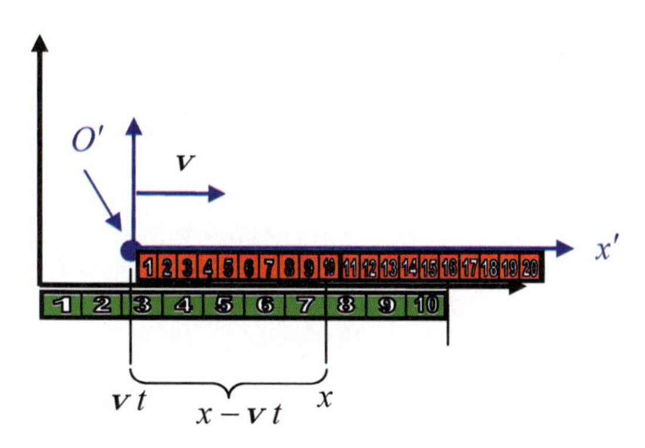

Figure 10-15: A house (frame) (Σ') moving at speed $\mathbf{v}$ and lab house (frame) (Σ) shown in standard configuration after the passage of t seconds of time. Note the distance from O' to some other point in Σ' given in terms of the label appropriate to Σ, x so as to facilitate writing the measure of the distance in units natural to the lab house (frame) (Σ).

As we recall and is evident from the figure above, we must multiply the left hand side by γ (which is not yet, for this calculation, the particular function of speed we assigned it earlier but simply an unknown) to convert to units of the moving house (or frame) giving:

10.27 $$x' = \gamma(x - \mathbf{v}t)$$

Here γ is a positive constant that we will determine later, and $\mathbf{v}$ is the speed of the moving house. For the remaining constants, we know that we want: $a_0 \equiv a_{00} > 0$, for when x is fixed, we don't want to have events in Σ' to be oppositely ordered from Σ;[28] and $a_1 \equiv a_{01} < 0$, for we want the time to be shifted in the way we've discussed ($t - T_{delay}$) so as to make the operational law of simultaneity work.

Solving for t and x in terms of t' and x' gives:

10.28 $$t = \frac{t'\gamma - a_1 x'}{(a_0 + a_1\mathbf{v})\gamma}$$
$$x = \frac{t'\mathbf{v}\gamma + a_0 x'}{(a_0 + a_1\mathbf{v})\gamma}$$

[28] If it were negative, one would have $t' = -t$.

i) Using a Global Maximum Bound to Maximum Apparent Speed

Now, imposing a global bound in the following way will give us a_0 and a_1. First, recall the steps that get us to this bound. We needed to put limits on measured speeds in order to insure the apparent inversion of causality that we discussed does not occur. We also found each house must see a maximum apparent speed, but that maximum may be different from one house to the next. Then, later, to enforce some house to house similarity, we argued for a maximum bound, c_{Max}, to these maximum apparent speeds. This least upper bound to the maximum apparent speeds at which particles can communicate is called *global* because it applies to all houses and hence all frames.

In this section, for mathematical convenience, we define our units such that this fastest speed measurable in any house is unity, i.e. $c_{Max} = 1$.

We can write the speed of any particle moving uniformly in a given house using the coordinates t, x (t', x') to mark the time and place of the particle as it travels through the unprimed (primed) house. We get:

10.29 $$u' \equiv \frac{dx'}{dt'} = \frac{\gamma(dx - \mathbf{v}\,dt)}{a_{00}dt + a_{01}dx} = \frac{\gamma\left(\frac{dx}{dt} - \mathbf{v}\right)}{a_0 + a_1\frac{dx}{dt}} = \frac{\gamma(u - \mathbf{v})}{a_0 + a_1 u}$$

where $u \equiv \frac{dx}{dt}$

and, we can write the equivalent for dx/dt:

10.30 $$u \equiv \frac{dx}{dt} = \frac{\gamma\mathbf{v}\,dt' + a_0\,dx'}{dt'\gamma - a_1 dx'} = \frac{\gamma\mathbf{v} + a_0\frac{dx'}{dt'}}{\gamma - a_1\frac{dx'}{dt'}} = \frac{\gamma\mathbf{v} + a_0 u'}{\gamma - a_1 u'}$$

where $u' \equiv \frac{dx'}{dt'}$

Recalling the condition set above, $a_0 > 0$ and $a_1 < 0$ and using the above equations and the maximum speed postulate, we can argue as follows.

First, consider a particle to be traveling at the fastest speed possible in the negative direction, as seen from the rest house (which can be any house within a given primary house), i.e. $u = -1$. Taking continuity as a reasonable assumption, this speed must transform to the fastest speed seen from the standard *moving* house so that slower speeds can transform to slower speeds and that, generally, the order of speeds is maintained.

Now, one might argue that there is something special about objects in a moving closed house, that is, objects moving in plana that moves with the house. We could say, for example, that objects viewed *from* an interior moving house are only seen to be able to reach the maximum speed when moving within the given internal house; in the primary house, they are seen to be moving slower. In this way, all speeds outside the house would transform to some speed a finite amount lower than the speed possible within the given house. However, such a situation would violate apparent causality. Recall that though the maximum apparent speed may vary when we change our *viewpoint* from one house to another, the maximum speed *seen from a given house* must be the same everywhere no matter where we look, including looking into other uniformly moving houses.

Hence, the maximum negative speed seen from the rest house must transform to the maximum speed seen from the moving house, though it need not be the same value. Then, using the fact that we cannot measure a speed of faster than c_{Max}, i.e. unity, seen from any house, including the primed house (frame), i.e. $|u'| \le 1$; the latter gives:[29]

10.31-1. $$u=-1,\ |u'|\le 1 \quad \Rightarrow \quad |u'| = \left|\frac{\gamma(-1-\boldsymbol{v})}{a_0 - a_1}\right| \le 1 \quad \Rightarrow \quad \gamma(1+\boldsymbol{v}) \le a_0 - a_1$$

Next, by similar arguments for the positive direction, and using the fact that the maximum apparent speed must be the same in positive and negative direction to preserve operational simultaneity, we consider $u = 1$ to get:

10.31-2 $$u=1,\ |u'|\le 1 \quad \Rightarrow \quad |u'| = \left|\frac{\gamma(1-\boldsymbol{v})}{a_0 + a_1}\right| \le 1 \quad \Rightarrow \quad \gamma(1-\boldsymbol{v}) \le a_0 + a_1$$

Now, consider a particle measured in the *moving house* (or frame) which is going unity speed in the positive direction (as measured in that house), i.e. $u' = 1$, and applying the same argument as above including the fact that unity is the fastest, we can measure it to go in the rest house, i.e. $|u| \le 1$ gives:

10.31-3 $$u' = 1, |u| \le 1 \quad \Rightarrow$$

$$|u| = \left|\frac{\gamma\boldsymbol{v} + a_0}{\gamma - a_1}\right| \le 1 \ \Rightarrow \gamma\boldsymbol{v} + a_0 \le \gamma - a_1 \quad \Rightarrow \quad a_1 + a_0 \le \gamma(1-\boldsymbol{v})$$

Lastly, we consider a particle moving in the negative x direction, i.e. $u' = -1$ and use $|u| \le 1$ to get:

10.31-4 $$u' = -1, |u| \le 1 \quad \Rightarrow$$

$$|u| = \left|\frac{\gamma\boldsymbol{v} - a_0}{\gamma + a_1}\right| \le 1 \ \Rightarrow -\gamma\boldsymbol{v} + a_0 \le \gamma + a_1 \quad \Rightarrow \ a_0 - a_1 \le \gamma(\boldsymbol{v}+1)$$

Since $a \le b \le a$ implies $a = b$, the final relations of equations 10.31-1 and 10.31-4 imply: $a_0 - a_1 = \gamma(1+\boldsymbol{v})$

Similarly, the relations 10.31-2 and 10.31-3 imply: $a_0 + a_1 = \gamma(1-\boldsymbol{v})$. Together these two equations give: $a_0 = \gamma$ and $a_1 = -\boldsymbol{v}\gamma$ (*Exercise*: verify that all of the relations, including the dual[30] of each of the above: 10.31-1 to 10.31-4)

---

[29] Note that all inequalities of the form $|u'| \le 1$ are actually two relations, i.e. $-1 \le u' \le 1$. In the four such inequalities in this section, we pick the one of each pair that is most productive in finding the solution--- the most restrictive one, when we know a prior, or the one not yet used. In the end, the solution must satisfy ***all*** relations, and we can come back and verify this once we get the solution.

[30] Each of the absolute value relations is actually a pair of relations.

We thus have the following transformation laws:

$$t' = \gamma\, t - \gamma\, v\, x$$
10.32
$$x' = -\gamma\, v\, t + \gamma\, x$$

and their inverse:

$$t = \frac{1}{\gamma\left(1 - v^2\right)}\left(t' + v\, x'\right)$$
10.33
$$x = \frac{1}{\gamma\left(1 - v^2\right)}\left(v\, t' + x'\right)$$

We have now determined all constants except γ. We've done this without the constancy of light requirement, though our arguments were initially driven by seeking how to accomplish it. As already mentioned, we postpone discussing the constancy of light, so as to force further exploration of the concepts. So, how do we find γ? We might think it time to bring in a second dimension, say the y-direction. Indeed, the form of γ is intimately related to the second direction as can be seen by an analysis of the famous Michelson-Morley experiment which we discuss in an end of chapter problem. The second dimension does impose tight constraints on γ, however it does not give the full form of γ if one uses only the maximum speed constraint (see end of chapter problem). To determine γ, we need the symmetrical relation between the relative motion of the frames that we previously discussed. We could simply postulate this symmetry; however, instead we take the occasion to further understand the Internal Relativity Principle, from which we can deduce the requisite speed symmetry.

ii) Revisiting the "Internal" Relativity Principle: Deduction of Relative Frame Speed Symmetry

In particular, we now return to think more about the unique system of houses (thus, frames) that is set up because of the impetus. Impetus gives the moving body, through its reception by the mass, a sort of secondary state that is closely analogous to rest for activities within the body, similarly so for a system of bodies considered from a center of mass perspective. This state is reflected in the velocity, the result of the impetus action in a particular massive body. Following this line of thinking, we, as already mentioned, begin to set up a formalism that will capture the effect of this privileged power of bodies. For concreteness, we return to our house analogy. Referring to Figure 10-16, which we reproduce from our earlier discussion with a helpful addition, we recall our setup. Consider a house, which is at rest, indeed we could even say that it is the universe rest frame; call this house Σ. Consider a smaller house moving at speed v within Σ; call this house Σ'. Lastly, consider a third, smaller house inside Σ'; call this house Σ''.

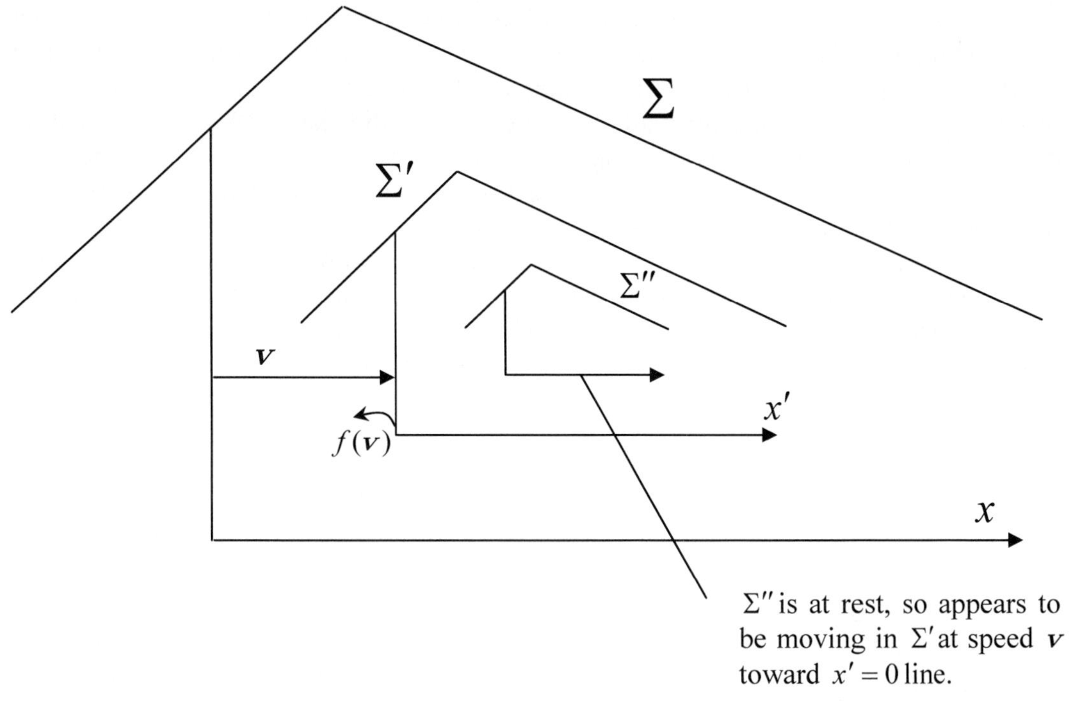

Figure 10-16: Houses are a helpful analogy for a region (closed or open) of a frame. Each house can be considered to be isolated in the sense that we do not allow anything outside to affect overall the state of the inside, though the inside can be viewed by the outside. Here we have three houses: Σ and Σ'' are at rest while Σ' is moving.

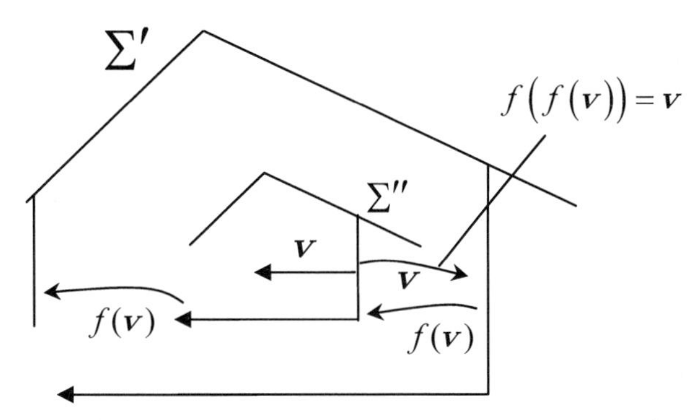

Figure 10-17: Changing origin and direction of axes in Σ' and Σ'' to reveal internal relational similarity with Σ and Σ'. Another way to think of this is to say this is the view from the other end of both houses; it is natural there to have the x-coordinate axis defined in the opposite sense.

We continue to assume that each house contains only empty space (non-interacting plana), so that there is nothing to distinguish one part of the house from another. Note that any bodies we may later put in will have the physics we discuss in this chapter as a generic background waiting further specification, which we can call the background space-time.

In this spirit of turning everything "off" but the most general causes, consider that the only thing that affects the activities within a given house is the impetus it has (which is indicated by its motion relative to Σ). In other words, for a closed house, we assume the walls of the house protect it against all other effects (even the plana is "locked in") and, in the case of an open house, we assume the effects in some way cancel so as not to matter at our generic level of consideration. If indeed the effect of the impetus is complete in "elevating" the state of uniform motion to give the body in motion an ability to maintain this special integrity, then confining ourselves to activities within the house should prove fruitful. Indeed, in this case, we should not be able to notice any change because the qualitative and quantitative change of state of the "house" and all objects in it change together. If, for instance, all things in the entire universe shrunk by half, we would not know it. There may, however, be evident a difference between houses of different impeti.

As previously, we will only be looking at the *x*-axis. The other axis is drawn only to show enclosure of housing. We will consider *y* and *z* axes in the problems.

From this reasoning, looking from Σ'' to Σ and vice versa should be the same, for they both are at rest, having no impetus. (Of course, at this point, this is strictly true only when one looks at Σ from a place where the floor of Σ'' contacts Σ so that there is no effect from propagation through Σ'.) Further, we can show that looking from Σ' to Σ should be the same as looking from Σ'' to Σ'. First, we assume, taking seriously the integrity given by the impetus, that each house looks just like the house at rest (Σ) as long, as one confines one's view to within the given house. That is, we now adopt the internal relativity principle. Then, looking from Σ to Σ', one sees Σ' moving at speed *v*; while looking back from Σ' (still within the primary house, Σ) to Σ, one will see Σ moving at a speed that can only be some function of v, $f(v)$. Looking from inside Σ', we see Σ'' moving away at a speed $f(v)$, for it is a rest house just as is Σ. Now, locking oneself within Σ', not looking outside of it (making it the effective primary house and an effective rest house (frame)), seeing Σ'' moving at speed $f(v)$ implies, using the rule just established in the Σ house (which must be the same in every house), when looking from Σ'' to Σ', one should see $f(f(v))$. And, from within Σ'', one should see Σ' moving at *v*, since Σ'' is the rest house, effectively the same as Σ. Hence, $f(f(v)) = v$ which means that $f(v) = v$ (exercise: why?[31]). This means that whether looking from Σ'' to Σ' or from Σ' to Σ'' (and also Σ to Σ' or from Σ' to Σ) the speed of the opposing house looks to be v. Now, all the arguments go through if Σ (with Σ'' moving the same speed) is not the actual rest house; thus any house *A* looking at the motion of house *B* sees the same speed of motion as the reverse. This proves the house speed symmetry we needed.

Mathematically, this means that since the Σ' house proceeds away at speed v so as to give:

10.34 $$x' = \gamma(x - vt)$$

then the Σ house must appear to proceed in the negative direction away at speed v giving:[32]

10.35 $$x = \gamma(x' + vt')$$

[31] Hint: what action when performed twice on any number gives still the same number?

[32] Notice that we are *not* switching our definition of coordinates as we did with Σ'' before this point.

iii) Applying Postulate of Speed Symmetry between Frames

Solving 10.35 for t', then substituting for x' using 10.34 gives:

10.36 $$t' = \frac{x}{\mathbf{v}\gamma} - \frac{x'}{\mathbf{v}} = \frac{x}{\mathbf{v}\gamma} - \frac{\gamma x}{\mathbf{v}} + \gamma t = \left(\frac{1}{\gamma} - \gamma\right)\frac{x}{\mathbf{v}} + \gamma t$$

Equating the coefficient of x in this equation with that of the transformation equation from equation 10.32a, i.e., $t' = \gamma(t - \mathbf{v}\,x)$ gives:

10.37 $$-\mathbf{v}\gamma = \frac{1}{\mathbf{v}}\left(\frac{1}{\gamma} - \gamma\right) \quad \Rightarrow \boxed{\gamma = \frac{1}{\sqrt{1-\mathbf{v}^2}}}$$

Here, we have ignored the negative solution to avoid negative length such as might come from an application of equation 10.35 giving: $\Delta x = \gamma\,\Delta x'$.

d) Special Relativity Transformation Laws

Thus, we've established the special relativity transformations. Transitioning to more practical units, we take $\beta = \mathbf{v}/c$ yielding (with γ as defined above):[33]

10.38a-d $$\boxed{\begin{aligned} t' &= \gamma\, t - \gamma\,\beta \frac{x}{c} \\ x' &= -\gamma\,\mathbf{v}\; t + \gamma\, x \\ y' &= y \\ z' &= z \end{aligned}}$$ and their inverse: **10.39a-d** $$\boxed{\begin{aligned} t &= \gamma\; t' + \gamma\,\beta\,\frac{x'}{c} \\ x &= \gamma\,\mathbf{v}\; t' + \gamma\, x' \\ y &= y' \\ z &= z' \end{aligned}}$$

(where: by symmetry, z is same as y and $\beta = \mathbf{v}/c$, $\gamma = 1/\sqrt{1-\beta^2}$ and c is the speed of light.)

With the form of the γ factor now established, we will now see that the fastest speed is realized by light, not by massive bodies. In particular, we've seen that the time dilation arises from the mass increase and other changes that result from the quality of impetus in the moving body. The mass increase is thus now seen to be, in a different way, a result of the relativity principle.

As we try to push a mass faster, it will always get harder to give it that next increment of speed, and we will never be able to accelerate it to the speed of light, though we can, in principle, get as close as we'd like. In particular, the student is asked to show in an end of chapter problem that the kinetic energy of a body at moving at speed $\mathbf{v}$, assuming the mass changes as γm_0, is:

10.40 $$E = K \equiv \int \mathbf{v}\,dp = \int_0^{\mathbf{v}} \mathbf{v}\; d(\gamma m_0 \mathbf{v}) = (\gamma - 1)\, m_0 c^2$$

So, it takes an infinite amount of energy to make a massive body go at the speed of light.

This means our thought experiments (such as in the section above) involving communications need to be considered either in the limit approaching the speed of light or as light rather than particles. This, of course, doesn't change the validity of the remaining reasoning and conclusions.

Now, we notice that our argument for the transformation laws did not depend on knowing the true rest frame; it was by definition, looking at only the relations between

---

[33] Note the inverses can be obtained from the initial equations by negating all *explicit* $\mathbf{v}$ dependences, i.e. taking $\mathbf{v} \to -\mathbf{v}$.

frames. This means, given the symmetrical nature of the equations, that each frame can be treated on equal footing, leaving out the real distinction between having impetus and not. The equations account for what we've chosen to focus on. We've deliberately focused on the similarity of each moving frame to the rest frame and finally, thus, to each other, and, to do this, we left out all those influences that would cause a difference. More precisely, our equations indicate how to transfer measurements in one frame to those of another in those situations in which effects that would break the needed symmetries (such as gravity) are small enough to be neglected.

Thus, it is convenient in thinking about special relativity to think only in terms of inertial frames and their relations to one another, thereby leaving out the complication of specifying a fixed rest frame, which would detract our attention from what's the same about the frames. In closing the system off in this way, we create a mental construct at the "outer edges" of the facts we're formalizing.

In such beings of reason, we, metaphorically, close the line upon itself to make a system so that we can follow around within our system without having to run into any dead ends as we pursue the things within our system.[34]

Hence, for example, sometimes rod *A* will be considered smaller than rod *B* sometimes larger, because we will be correlating appearances. Obviously we can't really mean that a rod is both smaller and larger than another rod. Back behind the appearances, we've captured reality in the equations; it's the reality we've discussed, not all of reality but, in a certain way, a general piece of it. Before concluding our discussion of the fundamentals of the coordinate transformations of Special Relativity, we need to address one more issue.

Inside a Sealed Frame

There is one loose end in our discussion. We have switched from houses to frames in our discussion. Can we do this? An important difference that we have often drawn between a house and a frame is that the plana moves along with the house, but does not necessarily move in a frame. It's the massive (i.e., having mass) measurement instruments and the system of clocks we setup near those instruments that determine our frame. We make measurements from some such setup that moves uniformly in space and we then say we have measured *an event* in that frame. Now, experiment has verified that it makes no difference whether the frame is encased or not; so, within the limits of such experiments, it makes no difference whether we encase the plana or not. Therefore, given that one allows everyone to look everywhere, one can then leave behind the idea of "houses" and just use the less restrictive idea of frames, as we do in special relativity. Still, one might like to better understand these "houses" or sealed frames.

You may have noticed that since the house frames are inside closed walls, the light doesn't have to catch up as the sound on the flatcar did. Putting up walls turns the flatcar into a box car and the wind thus ceases. Still, experiments shows that the transformation laws of special relativity are valid to high accuracy. Despite the breakdown of the analogy, we have to explain the facts.

---

[34] Again, such an attitude, as convenient and helpful as it is, doesn't belie the fact that we took the line from amongst a larger space in which it exists.

Concretely, for instance, we know that light coming from a flashlight in the rest frame entering a moving body, say through a thin window like entrance into an empty space, still looks like it is going at the speed of light relative to the ground, even though the light is moving through the space trapped in the body. This effect is shown in Figure 10-18.

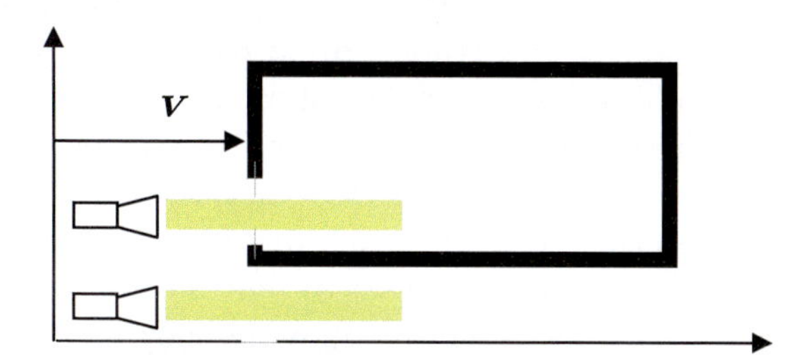

Figure 10-18: Light travels the same speed as seen from the ground, even when it enters the moving body above.

Most generally, we can explain this by pointing out that the moving body (of which the plana inside is considered to be a part) has a different impetus/mass combination (reflected by its velocity) and thus responds to light differently than a body at rest, indeed responds in such a way that the light keeps its initial motion, i.e. keeps up with the ground. It does so even when the light originates from within the box. In fact, even in the first case, the light interacts with the massive boundary layer (shown as thin window above) before entering the internal plana, where it is absorbed and reemitted. Hence, that light originates in the box in the limited sense that it is re-emitted by a moving part *somewhat like* a flashlight inside the box would. In that case, some combination of the way it is emitted and the nature of the plana inside determine the light to move in such a way as to keep up with light on the ground.

By contrast, in the case of sound catching up to the boxcar in our train, the sound goes through the boxcar wall, and inside moves at a speed c_{sound} through the air in the car, so that the sound speed with respect to the ground is that speed plus the speed of the train, i.e. $v + c_{sound}$. Hence, there is no equivalent effect for sound. This is because, though sound does ultimately depend on light, its major cause is the propagation of massive particles (molecules), not light, and thus it behaves differently. Though sound propagation does depend ultimately on interactions that occur at the speed of light, this dependency does not significantly affect the behavior of the sound.

Again, the fact that the moving body has impetus (that is how it moves) and the rest system does not is a real difference (a difference in quality) affecting the speed of light in the moving body in such a way that the light keeps pace with light moving in the rest system (e.g., body) emitted at the same time (see figure above). That is, there is a sense in which the light is traveling at a slower speed, $v - c$,[35] for it is traveling through the parts

[35] We have various measures at hand. In this case, we are measuring distances and times ***within the moving frame*** with distance and time units *from the rest frame*. All these express real facts about the situation; which ones we choose depends on what aspect we are concentrating on. Typically, we can capture most by using measurement units within each given frame and then use transformations to move between the frames.

of the plana of the moving body slower than it would in a body at rest. After all, it is moving in the parts of a plana that we are taking to be moving with the moving body.[36]

To explain, at the current level of abstraction, the difference in the way light travels in the moving body versus the rest body, we simply emphasize that a real change takes place in acquiring the quality (second property) of impetus and as always, there is a quantitative change (change in the first property, i.e. change in size and/or number of parts, see Appendices I and II for a general discussion) in the body that mirrors the qualitative change. To go much further would be a study in the nature of light and the plana. We leave the specifics for later courses; however, we will speculate a little about their more general aspects in a moment.

To better understand the effect, consider the sequence of events depicted in Figure 10-19 and Figure 10-20 in which two pairs of pulses of light are emitted at each other. One pair of such pulses is emitted so that they enter the moving body immediately; the other is emitted so that they travel underneath the moving body. Consider the inside of the moving body to be some single substance, not free plana (i.e., not a vacuum), but less interactive with light than water, in which light travels near c; it may be considered to be virtual plana, like found in most massive substances (in which case we can consider its speed to be c), for instance inside an atom. Notice that, for a given duration of time (measured by rest frame clocks), the number of parts traversed by the right moving pulse is much less than that traversed by the left moving one, and both are different than the number of parts crossed in the rest frame. Again, we account for this by the change in the state of the bodies in the moving frame; for example, because the virtual plana shrinks, light moves differently in it. The initial state of the light, e.g. moving to the left or right, also affects the travel in the moving plana. The right moving light interacts in a certain way with the window as the window moves *away from* it. The left moving light acts in a different way as it moves *towards* its entry window. The net effect is what we see.

Now, we change our problem slightly and consider the moving body to contain a vacuum, i.e. plain space (plana) that is not part of a substance but is a separate substance.[37] This free plana may be very much like the virtual plana. However, it may even be more inert and thus may not move along with the moving body. It may be stretched at the front and compressed at the back,[38] comparable to the way we currently think dark energy stretches space.[39] Such an immobile plana recreates the effect of "wind."

---

[36] In fact, if we measure using the parts in the moving frame using units of length of the moving frame (recall they are contracted) and time standards from the rest frame, the speed is $c\sqrt{\dfrac{1-\beta}{1+\beta}}$

[37] Substances, like a plant, for instance, are true wholes, rather than mere conglomerations, like a handful of marbles.

[38] For instance, it may be very stretchy and thus only require stretching and/or compression (formation of new parts), of the plana right near the boundary between the plana and the body wall. New plana parts would be pulled out of front parts of the plana and the back parts would be absorbed into back parts of the plana, all without necessarily having an effect on the middle. The expansion of the space of the universe since the big bang would be an example of this kind of effect. The plana substance could "feel" the effect of these changes, and be a source of its ability to change through the activity in the vacuum described, for example, by Quantum Field Theory. These speculations need to be left as such for now.

[39] Another possibility is that the plana is absorbed into the back and pulled out of the front end of the container rather than from the free plana inside the container. Remember most if not all massive objects are composed of virtual plana, most of the extension of atoms is due to virtual plana (plana that has been made part of the atom).

This latter would explain the vertical speed which we would see from the ground: $\boldsymbol{v}_{vertical} = \frac{c}{\gamma}$ (along with horizontal speed of $\boldsymbol{v}$ of the moving frame gives, $u_{light} = \sqrt{\left(\frac{c}{\gamma}\right)^2 + \boldsymbol{v}^2} = c$) (*Exercise*: show this fact.). For the virtual plana that we previously described, the effect is caused by the change in state of plana due to having rightward impetus, a property of the whole body,[40] creating a qualitative equivalent of the wind, partially reflected by, for instance, the reduction (by a factor of γ) in horizontal size of the parts of the plana that the light is traversing.[41]

[40] Recall that most physical substances have non-massive parts that are part of their nature; these parts do not interact as much, nonetheless, the electric and magnetic fields (and the internal energy associated with them), for example, between the massive parts are real properties of the virtual plana of the substances.

[41] Rightward impetus might, for example, cause horizontal groupings of the parts of the plana that make the left side bunch up while the right side spread. Such structure would be the quantitative reflection of a qualitative effect of making it easier for light to move to the left (against the impetus) in analogy to the way a force finds it easier to change the motion against the impetus action, i.e. slow it down. In this scenario, light would travel vertically along the minimal parts, but horizontal propagation would be dominated by the above groupings.

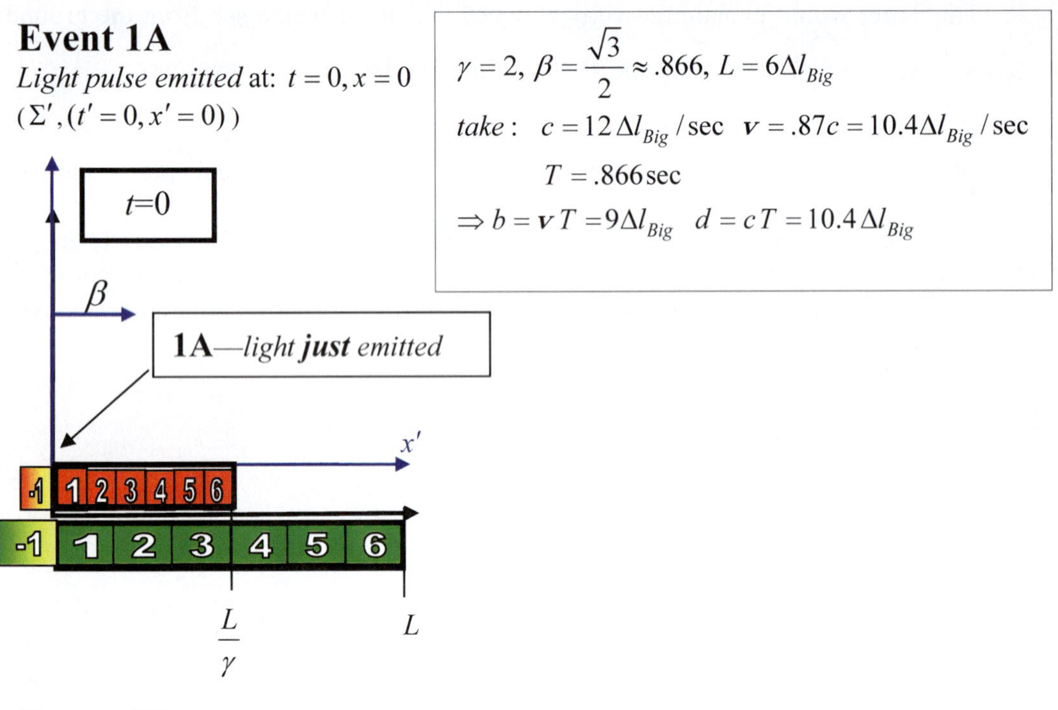

Figure 10-19: The first two events of a three-event experiment seen from lab (rest) frame, Σ. **a.** (*top*) First light pulse emitted; it is emitted to right at $t=0, x=0\,(\Sigma', (t'=0, x'=0))$ **b.** (*bottom*) Second light pulse emitted; it is emitted to left at $t=T, x=\gamma L$ $(\Sigma'\ (t'=0, x'=L))$ Various parameters assumed are given at top of diagram.

Event 2

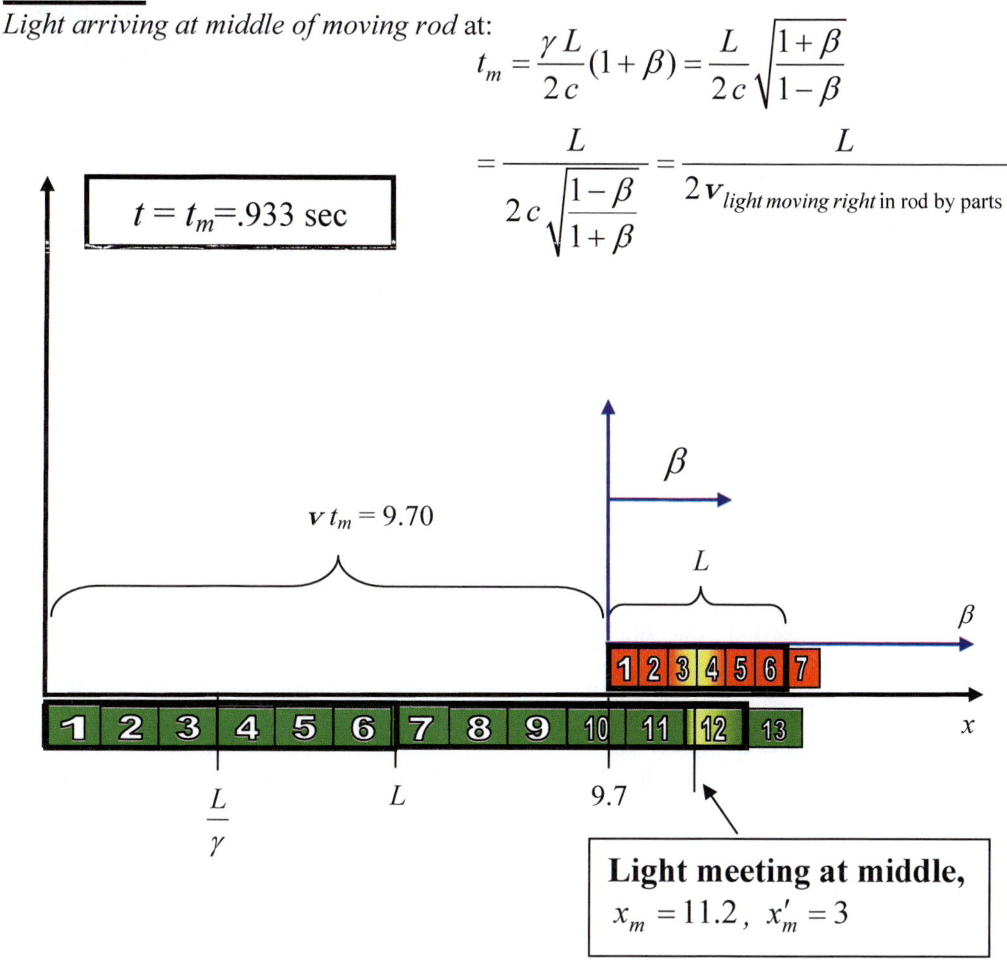

Figure 10-20: The last event of a three-event experiment seen from lab (rest) frame. Here the light meets in the middle of the moving rod at time t_m. Above is shown the (not measured in the frame, but deduced) speed of travel of the light moving to the right in the red (moving) rod in units of the red rods (which are shrunk) and stationary clocks, i.e. its actual speed moving through the red rod, as opposed to its apparent speed (supposing we could know the absolute rest frame), which we get to be: $v_{right} = 3\Delta l_{small} / .933\,\text{sec} = 3.2\Delta l_{small} / \text{sec}$. See end of chapter problem for full calculation of $v_{right\_\text{in rod by parts}}$ and $v_{left\_\text{in rod by parts}}$. We can also calculate that speed in absolute rod units, which is probably the most significant ontological speed; this gives $v_{left} = c + v$, $v_{right} = c - v$ (see problems). Within the standard system of relativity, we would say these are speeds based on mixed frames and thus dismiss them as not important, which is the case *within* the system; they are important for placing the system in the full physical context and recognizing what we have left out and what we have included.[42]

---

[42] Notice the analogical definition of place, because here we are talking about *a quality* (light) activating one part and then the next of the red bar, whereas place first means the location of *a substance* in its environment. Thus, technically, we should talk of two motions: the motion of the red rod and the motion of the quality's activity (which is technically called *alteration*) in the red rod.

Summary of Part I

We have seen that Galilean relativity hinges on the unique status of inertial frames. We pointed out that the similarity of moving frames to each other and to rest frames comes about because of the nature of the impetus, which preserves the integrity of the behavior of uniformly moving bodies in a deeper way than one might at first expect. We have also seen (at the level of our abstraction) that because there is no effect or body that travels at an infinite speed as Galilean relativity assumes, it makes sense that there should be a speed limit, a maximum apparent speed faster than which no physical communication can occur. If there were no such cosmic speed limit, then we would have a somewhat less unified, less intelligible, and less simple universe in which certain uniformly moving observers would be affected by causes before they are affected by some of the effects of those causes; i.e. they would experience inverted causality for those events.

We formalized the observed integrity principle, i.e., that moving systems (in particular substances from which we model other systems) behave in uniform motion just as they do at rest. We do this through the internal relativity principle, which states that the laws of physics are the same in the interior of every "house" and that the intrinsic state of a house is only a function of its speed. The internal relativity principle implies that we can drop the distinction between open and closed houses and just speak of inertial frames. Further, since houses internal to a given primary house are microcosms of the exterior of the primary house, we can via the internal relativity principle, prove a full relativity principle.

Indeed, the integrity principle in the form of the internal relativity principle coupled with the maximum apparent speed is shown to give the principles of special relativity: 1) the laws of physics are the same in every inertial frame 2) the speed of light is constant. From these one can establish the transformation laws of special relativity.

We showed that the physical realities that these laws describe include: the increase of mass with speed, the contraction of length with speed and the naturalness of operational simultaneity in an inertial frame. The mass increase with speed can be considered as a principle of time dilation, if we confine ourselves to a given frame. At the abstraction level that considers only effects related to impetus and using only standards (units) of time and length moving uniformly at the speed of the frame, i.e. confining our measurements to within the frame, we find that it is natural to synchronize clocks in just the way we would if the frame were at rest. Through the use of the radar method, which incorporates this operational simultaneity, and the application of the time dilation factor, we get the equations of transformation from rest to the moving frame; which in Part II we will see do specify the correct length contraction and preserve the symmetries of special relativity.

The impetus is the qualitative cause[43] of these facts. The more impetus a body has the more mass, i.e. the more resistance to the action of the impetus. The more impetus, the shorter a body becomes. Impetus also affects the way light travels in the plana such that the measured value of the speed is the same in every direction and in every inertial frame. Because of this, we leave out the environment (plana) of the effects and massive bodies under consideration, effectively allowing that environment to be in any state that does not contradict the predictions of special relativity. Indeed, within empiriometric special relativity, we leave out all mention of plana.

---

[43] Called the efficient cause.

Note that because we cannot distinguish between absolute rest and motion in special relativity, that distinction is dropped in favor of purely relative motion, and indeed makes no sense within *empiriometric* special relativity.

If we keep in mind the context outlined above (i.e., what we mean and all that we've left out), special relativity boils down to one principle; the results of an experiments performed in a given inertial frames will be the same if performed in any other inertial frame. From this, taking light (the fastest means of communication) to have a maximum apparent speed to avoid apparent reversals of causality, we can deduce that the speed of light is constant in all frames. And, again, from this, one can deduce the transformation laws of special relativity (called the Lorentz transformations) as is done in an end of chapter problem.

In this way, special relativity becomes a general principle that, if true, and all current evidence overwhelmingly supports it, must apply to every physical theory. Empiriometrically, we say any given theory (i.e. its equations) must be invariant under the special relativistic transformations. In the sense described, it then becomes the "background space-time" for all physical theories within its domain of applicability.

To understand the transformations in this sense means to understand the proper empiriometric interpretation of the mathematics. We now turn to that task.

Part II: Understanding the Transformation Laws and Related Dynamics

Simple Use of the Transformations

a) Length Contraction

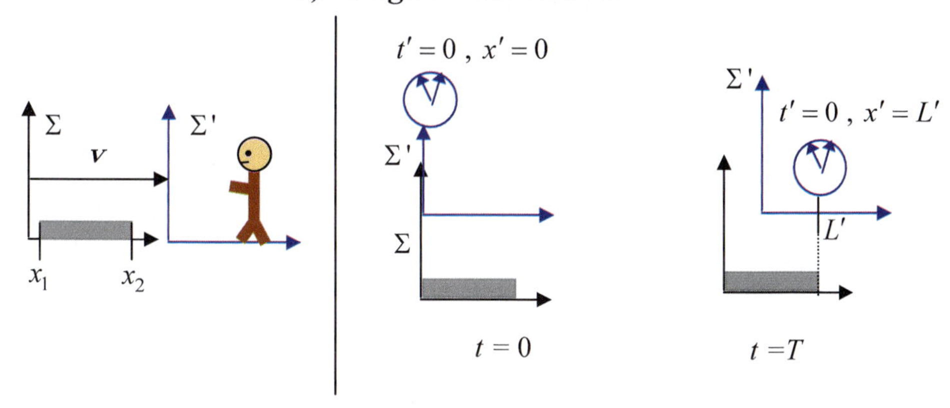

Figure 10-21: Meaning of "length contraction" *(Left)* The convention defining the standard relation between Σ and Σ′ is shown. The moving frame, labeled Σ′, moves at speed *v* away from the lab (rest) frame Σ with ruler (rod) shown at rest in Σ. (*Right*) Measurement of gray rod from the moving frame. The measurement uses synchronized clocks in moving (Σ′) frame to determine when to make the measurement at each end; in this case, the measurement is made at $t'=0$. The time in the rest (lab) frame for the measurements of the two ends is, respectively, $t=0$ and $t=T$. *The rod in the lab frame appears shorter when viewed from the moving frame.*

Given a rod of length $\Delta x = L$ in the rest frame (Σ), as shown in Figure 10-21, how long does it look from the moving frame (Σ')?

To answer, we must fill out more clearly what we are asking. In the true rest frame (where it actually, not just apparently, is at rest), the measurement of the rod reveals its true length, because the time we use is truly simultaneous, reflecting the actual state of the parts of the rod. For instance, we measure the ends of the rod at the points $x = 0$ and $x = L$ at $t = 0$. In the moving frame, we have defined time through the operational simultaneity; so, if we measure the location of the front of the rod and the back of the rod to be: $x' = 0$ and $x' = L'$ at $t' = 0$; those two measurements will not occur at a single (actual) time "*t*," and thus we will, in reality, be allowing the rod to move between the two measurements. Nonetheless, as we've pointed out, this operational definition captures important aspects of reality, so we use it and remember what we mean. So, to answer the question of how long it looks in the moving frame, we look at the transformation laws, which incorporate our definition of simultaneity.

The special relativity transformations (equations 10.38 and 10.39) give: $\Delta x = \gamma \mathbf{v}\, \Delta t' + \gamma\, \Delta x'$, where $\Delta x' = L'$, is the distance between the two events of viewing the rod at each end point as measured in the moving frame (using time and length units appropriate to it) at one time say $t' = 0$ (which means $\Delta t' = 0$), and $\Delta x = L$ is the length of the same rod as seen in the rest frame and Δt is the time lapse between events. Substituting $\Delta t' = 0$ and $\Delta x = L$, $\Delta x = \gamma \mathbf{v}\, \Delta t' + \gamma\, \Delta x'$ becomes:

10.41 $$\Delta x = L = \gamma L' \quad or \quad L' = \frac{L}{\gamma}$$

This means that it appears that the rod in the rest frame is *contracted* by γ. It appears γ times shorter than rods in the moving frame. Said another way, it appears that it takes γ rods in the rest frame to make one in the moving frame. The time lapse in the rest frame (actual lapse) between measurements is $T \equiv \Delta t = \gamma\, \Delta t' + \gamma\beta \Delta x'/c = \gamma\, \beta\, L'/c$ (note: this is the delay time required to obtain operational simultaneity—see end of chapter problem). Thus, the rod actually moves by $\mathbf{v}T$ between measurements, though we take no note of this quantity in the moving frame because of the way we define simultaneity.

Now, consider the reverse situation shown in Figure 10-22: a rod of the same type has been accelerated up to the moving frame and an observer measures it from the rest frame. The lapse time between the events of measuring the end points in the rest frame (actual time) is zero ($\Delta t = 0$). Substituting $\Delta t = 0$ and $\Delta x' = L'$ into equation 10.38b gives:

10.42 $$L' = \Delta x' = \gamma\, \Delta x = \gamma L \quad or \quad L = \frac{L'}{\gamma}.$$

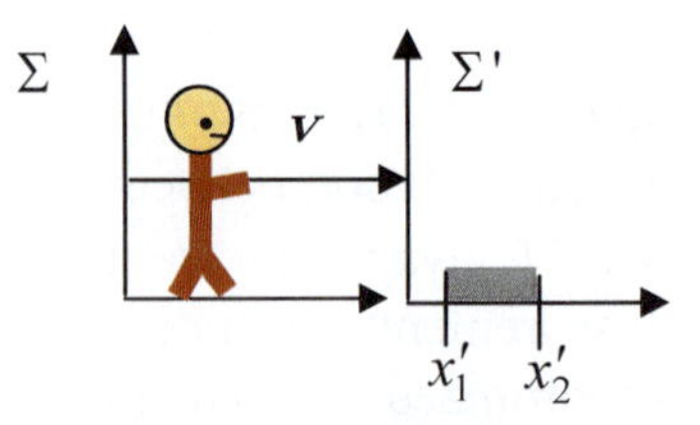

Figure 10-22: Viewing a rod in the moving frame from the lab (rest) frame. *The rod in the moving frame appears shorter in the absolute rest frame.* The rod actually does shrink. However, in special relativity, we assume we do not know, and in fact that it does not matter,[44] what is absolutely at rest and speak only of the relation between frames.

---

[44] As always leaving something out and treating as if it does not matter does not make it not exist and not matter; it only means we are focusing on what is of relevance for the discussion at hand and ignoring what is not.

Hence, we see exactly the reverse of the first situation (compare Eqs. 10.41 and 10.42). The rod in the moving frame looks shrunk now and the rods at rest look longer. It takes many moving frame rods to make one at rest rod. The rods actually do shrink, because this time the rod does not move from underneath it between when we measure the different ends.

Notice that the symbol L' has different physical meanings in the two cases. The prime always indicates that the measurement occurred in the primed frame. Thus, in the first case, L' meant the length *of an at rest rod* measured in the moving frame. In the second case, L' meant the length of *a moving rod* as measured in the moving frame, and L meant the length of the moving rod as measured in the rest frame. In short, the second case is about a moving rod; the first is about an at rest rod.

In fact, including a physical rod, in a way, violates our symmetry and, to that extent, makes one frame privileged over the other; after all, one "has" the rod of interest the other does not. It is only in that frame that the rod appears at rest.[45] Again, because of the integrity that the rod has in its uniformly moving frame (or true rest frame), we call the frame in which an object appears to be at rest its proper frame and refer to the time standard in that frame as its *proper time* and its length in that frame as its *proper length.* Thus, when the rod is at rest in the lab frame, L is its proper length; whereas when the rod is at rest in the moving frame, L' is its proper length. As we now see, time dilation involves similar issues.

b) Time Dilation

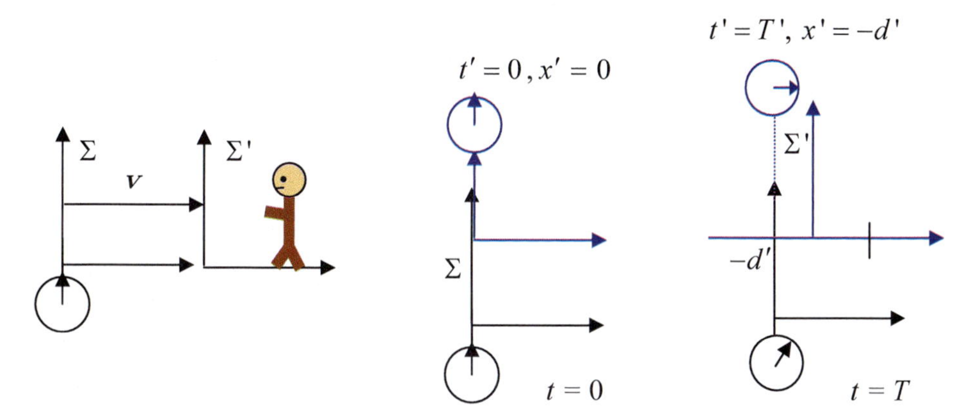

Figure 10-23: (*left*) A clock at rest in the lab (rest) frame, Σ, at $x = 0$ is viewed by an observer in the moving frame, Σ'. (*right*) We measure the time that elapses on the clock at rest in the lab frame, so that the position in the moving frame changes during the period of time T that lapses in the lab frame. *The clock in the rest frame appears to be moving slower when viewed from the moving frame.*

[45] The violation can be overcome by inserting a rod in the same relative place in the other frame, so the principle of our system still stands. In any case, our assumptions in the formalism are general and not meant to meet the needs of further specification but only do what general principles do and that is make general statements that leave room for many specific incarnations.

Given a clock in the rest frame fixed at $x = 0$ that shows T seconds elapse, how much time goes by in the moving frame? The two relevant events are two successive readings of the clock: one at $x = 0$, $t = 0$ the second at $x = 0, t = T$; using SR transformation equation 10.38a, we get:

10.43 $$T' \equiv \Delta t' = \gamma\, \Delta t - \gamma\beta\Delta x / c = \gamma\, \mathrm{T}$$

This means it looks like more ticks go by in the moving frame (where the observer is) than in the rest (lab) frame. For instance, if 1 second went by in the rest frame, it looks like γ seconds go by in the moving frame. So, the moving frame sees the rest frame clocks as moving slower.

Now, take the reverse situation shown in Figure 10-24. Given a clock in the moving frame fixed at $x' = 0$ on which $t' = T'$ seconds is allowed to elapse.

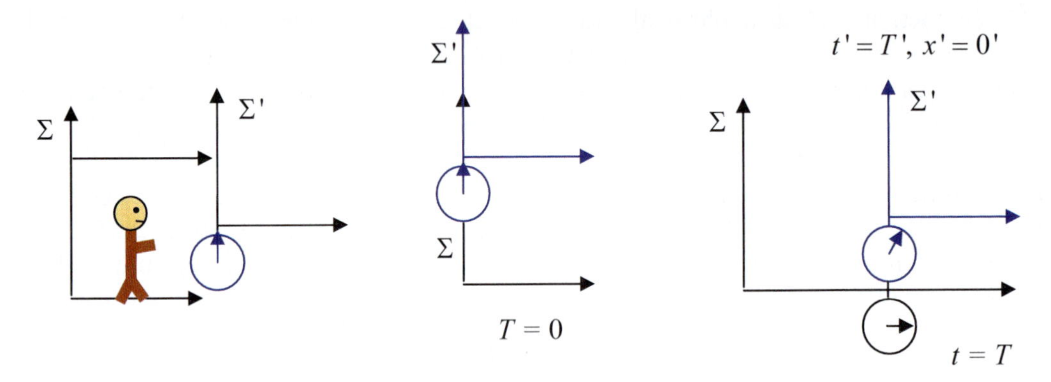

Figure 10-24: (*left*) Clock at rest fixed at the origin of the moving frame is viewed by an observer in the lab frame. (*right*) The time on the moving clock is read (measured) after a certain lapse of time; this means the first time reading is done at one place in the lab frame and the second is done at a different place as shown here.

The clock in the moving frame appears to be moving slower than the rest frame clocks when viewed from the rest frame. In fact, time does slow down as one moves though we usually do not know[46] what is actually moving and what is only apparently moving. Still, such *relations* between frames remain true.

This means, using the SR transformation equation 10.39b, with $\Delta x' = 0$ $\Delta t' = T'$,

10.44 $$T \equiv \Delta t = \gamma\, \Delta t' + \gamma\, \beta\, \Delta x'/c = \gamma T'.$$

This means that, when watching a clock in the moving frame from the rest frame, more ticks go by in the rest frame than the moving frame. For one second in the moving frame, γ seconds appear to go by in rest frame. Thus, the rest frame clocks appear to be going faster than the moving ones, which is the exact opposite of the first case.

Notice that we have the very symmetry we've strived for. Looking from one frame to the other always makes rods look shorter and times look slower in the other frame. To each frame, the other looks like it is the moving frame with slower clocks and shorter rulers (see end of chapter problem).

[46] This is true with respect to all of a given object's locomotion; for instance, though the Earth is moving around the sun, we usually do not include this when we say: he moved and he did not.

At this point, one can begin to glimpse the internal consistency of relativity. We have left out anything that might distinguish one frame from another and thus focused on the sameness among uniformly moving frames and the rest frame. In so doing, it now looks hollow to continue to single out a rest frame. And, within the system it is indeed so. Hence, as we've said, we simply stop making the distinction between rest and motion and thus between moving fast and faster except relative one to another. In this way, we keep the relational aspect that will allow us to piece the isolated views of reality back together. We say, still again, leaving something out doesn't make it not there, so it doesn't mean there is no rest frame and doesn't make time and space cease to be distinct. Though it does show us that distinguishing what's at rest and what's moving is harder than we initially might have thought.

c) Twin paradox

Forgetting the former point can lead to contradictions. A famous such apparent contradiction is the case of the twin paradox. In this paradox, one of the identical twins leaves on a trip to go to a planet (say the star Upsilon, in the Andromeda constellation, which is known to have planets around it) that, for argument's sake, is at rest with respect to the Earth and is 50 light years away. This means it takes 50 years for light to reach there in the rest frame of the Earth. This twin is moving at .9806*c*. The other twin stays home. Ignoring gravity and other such effects, what is the age of each twin when the traveler returns?

If we forget about the distinction between rest and motion, then there is a real problem. Simplistically applying the time dilation analysis above, one might think that the stay-at-home must conclude the traveler is younger and the traveler must conclude the stay-at-home is younger. In particular, it looks as if the stay-at-home will see $2\times\frac{50}{.9806}=102\,years$ pass by on Earth but only see $2T\sqrt{1-.9806^2}=2\left(\frac{50}{.9806}\right).1960=20$ years pass for the traveling twin *and* the traveler would get back and see just the opposite, i.e. that he was 102 years old and his brother was only 20 years old. Of course, he cannot be *both* 102 years old and 20 years old.

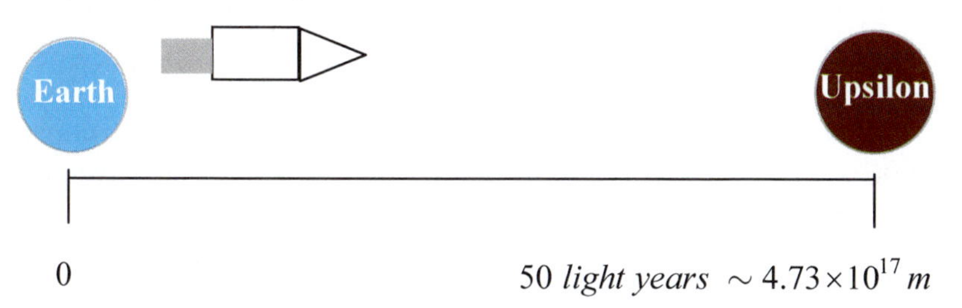

Figure 10-25: One of two identical twins leaves Earth at the origin for a round trip to Upsilon, which is 50 light years away. The other twin remains at home. Which twin is actually older when the travel twin returns?

Returning to the complete physical situation will resolve the problem. In so doing, we see that one frame actually has the Earth and the planet at rest the other does not. Further, it is the traveling twin that changes his impetus. Indeed, the traveling twin accelerates as he leaves, as he reverses course to return home, and as he lands when he arrives back at Earth;

i.e. he changes his impetus three times. Of course, within the formalism, we could simply say it's the acceleration that distinguishes the two, because our identification has been between uniformly moving frames and did not include accelerating frames. Thus, the actual acceleration of the rocket frame distinguishes it from the Earth frame. This is true as far as it goes.

However, we want to account for the whole situation. The acceleration can be reduced to a small part of the trip and in the limit can become arbitrarily small. The real difference results because one frame is, during the whole trip, in a qualitatively different state than the other. In a rarefied world where the only things in the universe are the two planets, the rocket and the plana, it's the planets that most affect the surroundings, i.e. the plana and the rocket.

Of course, because of the nature of the impetus, the whole system, planets and rocket, could be moving uniformly through a larger universe and all the analysis of who's older would remain true. To better understand the nature of special relativity, which seems counterintuitive because of all we have left out, it's important to go through this latter case.

The View from a Third Frame

Assume that the planets are moving together *to the left* through a larger universe which we now take to be the true rest frame. The traveler must then slow down his leftward movement in order to allow Upsilon to catch up to him, then he must speed up again to a faster than earlier leftward movement to catch up to the Earth again; finally he must slow down to match the speed of the Earth to land. In this case, with respect to the universe rest frame (which in this contrived situation gives absolute time), his clocks run less slow for the first part of the trip and slower for the trip home, while the stay-at-home is experiencing an intermediate rate of time passage.

Specifically, assume that the planets move at a speed .9806 with respect to the universe rest frame. This makes the distance between the two planets 9.8 light years, which is what is seen in the universe rest frame. Further, assume that the rocket reduces its own impetus to zero in allowing Upsilon to catch up to it. Hence, in the rest frame of the universe (and the rocket), the rocket takes $9.8/.9806 = 10$ years to get to Upsilon. On its return trip, the rocket is going .9806 c in the frame of the planets. Using the transformation laws for velocity, the speed seen in the rest frame is $\frac{.9806+.9806}{1+.9806\times.9806} = .9998$, giving a time dilation factor of .0196. The speed that the rocket overtakes the Earth in the universe frame is thus $.9998-.9806 = .0192$, which means it takes 510.4 years in the universe frame for the rocket to catch up to the Earth, which starts out 9.8 light years away. However, the universe observer sees time-dilated clocks in the rocket frame, which means he sees the rocket clocks register a total of 10 years during the duration of that part of the trip (assuming he lives the 510.4 years that pass in his frame!). So, the universe frame agrees with everyone else about how much time goes by in the rocket frame for the full round trip journey (i.e. a total of 20 years).

In the universe frame, it takes 10 years for the rocket to get to Upsilon and 510.4 years to return or a total of 520.4 years for the trip. Meanwhile, the time passage for stay-at-home twin according to observers in the universe rest frame is $520.4\sqrt{1-.9806^2} = 102$ years, again in agreement with all other reckonings.

Figure 10-26: The twin paradox viewed from a third frame that is moving with respect to the Earth and the spaceship.

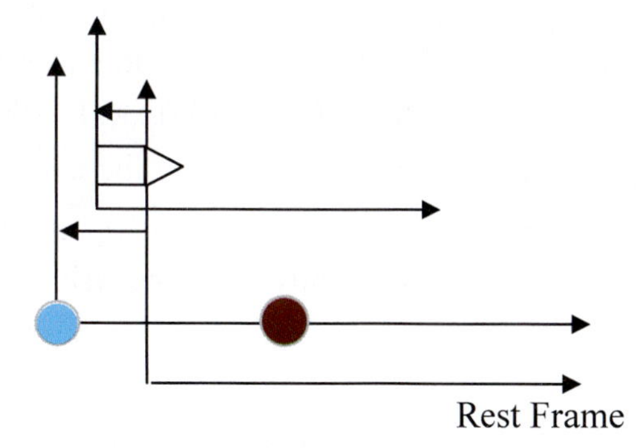

What is it that could make such profound distinction between different uniformly moving frames? At a deep level, as we have seen, it's the impetus and how it is received in the body. To answer how the various frames fit together as an integrated whole, rather than what appears to be a matter of choice within the SR formalism, we'd have to begin "turning on" other causes, not the least of which is gravity. In fact, once we turn on gravity, we realize how much of an idealization "an inertial frame" really is. As we mentioned in Chapter 8, the equivalence principle is key. In general relativity, the inertial frames will only exist in the limit, at a point in space-time. General relativity adds more of what we've left off, but that is for another course. Still, it also, like all good formal theories, "cuts the line, and closes the loop" at some point.

Dynamics

a) Generalization of Mass and Energy

i) "Convertibility" of Mass and Energy

For special relativity, we have seen that the kinetic energy of a particle is not $1/2m\mathbf{v}^2$ but is $KE = (\gamma - 1) m_0 c^2$. This fact arises from the variation of mass with speed, according to $m = \gamma m_0$ where m_0 is called the rest mass because it is the mass at very small speeds: $\mathbf{v} \ll c$. Einstein made the bold prediction that mass and energy were convertible. This means that particles can interact and lose some of their mass and have the amount of kinetic activity increased. Of course, a passive property is not an active quality; what we mean is that massive particles, particles which have essential properties of attracting and being attracted by other massive particles and of being able to receive the quality of impetus, have another essential property. When such particles interact with each other and with other things, they have the capacity to change substantially into, for example, parts of the plana with a certain activity. For example, an electron and a positron at rest can fall together and cause a burst of light (a type called gamma rays) in two directions. Light, in turn, has a certain capacity to act on massive bodies and make those bodies move. For example, when light heats a surface, it increases the motion of the molecules on the surface. Thus, light has a certain activity (or potential for activity), a certain energy in the wide sense, that we can measure by how much kinetic activity it produces. Further, since massive particles can be converted to non-massive plana and cause activity in the plana, we can, *loosely speaking*, say the mass is converted to energy. Indeed, we can analogically extend our concept of energy to include not only the activity of light but also the amount of massive material in the body that caused it, because it is related to the potential for activity.

Further, we find that the activity produced is proportional to the mass. The more massive the particle is, the more activity that is available. An electron and a positron can produce $2\times m_{electron}c^2 = 16.4\times10^{-14}\,J = 1.02\,MeV$ of energy.

In this way, we can define the energy to be the sum of the kinetic energy and the rest mass, i.e., analogous to $E = KE + V$, we get: $E = (\gamma - 1)m_0c^2 + m_0c^2 = \gamma m_0 c^2$. That is, energy in special relativity can be written:

10.45 $$E = \gamma m_0 c^2$$

When the body is not moving $\gamma = 1$, and we reproduce the famous formula $E = mc^2$ that we just used above. With this definition and the relativistic definition of momentum, an important relation can be proven:

10.46 $$E^2 = (pc)^2 + (m_0 c^2)^2$$

So, massive bodies have the essential aspect of causing gravitation in the plana, receiving the action of gravitational fields, of receiving and resisting the activity of impetus and of being converted to non-massive plana with a certain activity level (energy). The inertial mass, gravitational passive and active mass and the activity level caused by the destruction of the massive body are all proportional to the inertial mass. This makes sense at a generic level, for, as already mentioned, a higher inertial mass implies a higher contact with the plana, which implies a greater ability to affect the plana by any action.

ii) Types of Inertial Mass

Indeed, in discussing special relativity, we have also deepened our understanding of inertial mass itself. In particular, the increase of inertial mass with speed reveals that it has at least two basic components; one component is related to how fast the body is moving through the plana; the other is the receptivity to the action of the impetus asymptotically near rest. The latter, called the rest mass, is the starting point from which resistance to motion in the plana builds and, in that sense, the rest mass can be said to be the intrinsic receptivity, because it abstracts from any effect of the impetus on the receptivity. As we mentioned in Chapter 8, for any ordinary substance that is at least as complex as an atom, the more mass the substance has, the greater the number of massive parts, and thus the larger the initial resistance (the rest mass) is. Mass within a body can be, and as far as we know now always is, a combination of the two types. For example, the nucleus of an atom is made of quarks that move around at high speed. The net *rest* mass of the hydrogen nucleus (a proton) is 1 *GeV*, the mass of the two up quarks is .008 *GeV* and one down quark is .008 *GeV* giving a total of .016 *GeV*.[47] The remainder of the rest mass is related to the motion and potential energy in the strong force holding it together. The former is a resistance of the motion of the whole related to the motion of the parts, and thus analogically like the resistance of the motion of the whole in the plana, whereas the latter has no such direct analogy. It is a resistance of the motion of the whole that is due to force that holds the whole together.

It is important to note these two distinct contributions to inertial mass, because much of the resistance to the action of impetus that we see in things around us is apparently

---

[47] Things are more complicated than this but again, a more detailed discussion is part of the interesting discoveries of future physics courses. For example, the concept of mass is extended analogically further when we get to quantum mechanics that is so important in this realm.

due to their internal motions caused by internal impetus of the parts. Indeed, that component of the mass is directly proportional to a body's internal energy.

b) Momentum Transformation Laws

We have established the relationship (especially as manifested in the transformation laws) between two frames in terms of the space and time coordinates of each. We've labeled those coordinates by a common symbol, x^μ and indeed call them *space-time* coordinates, considering each coordinate as on equal footing in the formalism.[48]

Having such concise mathematical treatment of the concepts allows us to calculate and think in an economical way that minimizes errors. It leaps us into new territory without having to quite know how we got there. Of course, we should always go back at some point and explore where we've leaped from and how we got there, for that is the only way to get a clear and complete physical understanding. Still, it's indispensable to have the leaping tool.

We would like to extend the relativity formalism that we have developed so far to include the measure of the direction and intensity of the impetus, *the momentum,* and the related measure of the activity, *the energy*. We already know the momentum is:

10.47
$$\vec{p} = m\frac{d\vec{x}}{dt} = \gamma\, m_0 \frac{d\vec{x}}{dt} = m_0 \frac{d\vec{x}}{d\tau}$$

Here τ is the proper time and $\vec{x}$ is a vector given in the natural coordinate in some observer's frame.

The last equality makes use of the fact that the time associated with the instantaneous rest frame of the particle, called the proper time of the particle, doesn't require a γ factor. That is, that the proper time is a natural measure. This makes sense because the interactions of the body under consideration with its environment, e.g. its interaction with the plana, are, in one way, naturally considered with respect to its special state of uniform motion (which is determined by its impetus and mass). This is the physical reason why proper time is important, and why it will still be important in our formalism in which we have formally left out all but what is related to the most general aspects of the motion of bodies.

Obviously, $d\tau$ will not change as the observer moves from frame to frame. However, $d\vec{x}$ will. Indeed, we know the transformation for $d\vec{x}$ if we incorporate t as well. We use our space-time coordinates with its concise notation:

x^μ *for* $\mu \in \{0,1,2,3\}$ *such that* : $x^0 = t,\ x^1 = x,\ x^2 = y,\ x^3 = z$ and write:[49]

10.48
$$p^\mu = m_0 \frac{dx^\mu}{d\tau}$$

---

[48] A key fact that we've established is the constancy of the apparent speed of light in all uniformly moving frames. In coordinates this means, in one dimension that $x^2 - (ct)^2$ is the same in all frames; we say it is *invariant under coordinate transformation*. In general relativity, we even call this, in differential form, proper distance, *ds*, and in analogy to Pythagorean theorem we write: $ds^2 = dx^2 - c^2dt^2$. In SR, we can write s^2.

[49] One can distinguish a three-vector from a four-vector by using a Latin superscript, say *i*, which runs from 1 to 3, instead of a Greek superscript, say μ, which runs from 1 to 4. We can also use lowercase "*p*," for three-momentum and uppercase for four-momentum.

Notice how x^{μ} takes the place of the vector $\vec{x}$; indeed x^{μ} is a generalization of a vector called a tensor; we will simply call it a four-vector. More importantly, we have added one more component to $\vec{p}$; we know p^1, p^2, p^3 are the *x, y* and *z* components. p^0 is the *t*-component which equals γm_0, which is the special relativistic energy. Thus, in matrix notation we can write:

10.49
$$p^{\mu} = \begin{pmatrix} E \\ p^x \\ p^y \\ p^z \end{pmatrix}$$

Such a 4-vector describes the energy and momentum in a given frame. Note energy is a measure of the activity due to the impetus. As we've said, action and reception go together like top and bottom, so we need both the acting quality (impetus) and the receiving quality (mass) to understand the situation. The 4-momentum p^{μ} is a mathematical "entity" that concisely packs both the measure of the active quality (the ***momentum***) and the measure of the activity that results (the ***energy***), which thus incorporates the receptive quality. We can simply write the same measures in moving (primed) frame coordinates as p'^{μ}, but how do we determine its values in terms of p^{μ}?

We can transform easily, assuming standard configuration, using (setting: $c = 1$):

10.50
$$dx'^{\mu} = \begin{pmatrix} dx'^0 \\ dx'^1 \\ dx'^2 \\ dx'^3 \end{pmatrix} = \begin{pmatrix} \gamma & -\gamma \mathbf{v} & 0 & 0 \\ -\gamma \mathbf{v} & \gamma & 0 & 0 \\ 0 & 0 & 1 & 0 \\ 0 & 0 & 0 & 1 \end{pmatrix} \cdot \begin{pmatrix} dx^0 \\ dx^1 \\ dx^2 \\ dx^3 \end{pmatrix}$$

Thus, we get:

10.51
$$\begin{pmatrix} E' \\ p'^x \\ p'^y \\ p'^z \end{pmatrix} = \begin{pmatrix} \gamma & -\gamma \mathbf{v} & 0 & 0 \\ -\gamma \mathbf{v} & \gamma & 0 & 0 \\ 0 & 0 & 1 & 0 \\ 0 & 0 & 0 & 1 \end{pmatrix} \cdot \begin{pmatrix} E \\ p^x \\ p^y \\ p^z \end{pmatrix}$$

In particular, we have: $E' = \gamma E - \gamma v p^x$, $p'^x = -\gamma \mathbf{v} E + \gamma p^x$. These relations capture the way the momentum and associated energy of a given particle appear from the various uniformly moving frames. Taking a simple case, if the particle is at rest, $E = m_0$, and we move to a frame moving at speed *u*, we get $E' = \gamma_u m_0$ and $p' = -m_0 \gamma_u u$ as we should (where $\gamma_u = 1/\sqrt{1-(u/c)^2}$).

If we consider a particle of mass m_0 moving at speed $-u$, so that it has exactly the above values of momentum and energy, then in a frame moving at speed **v**, we get (*exercise*)

10.52
$$E' = \gamma(\gamma_u m_0) + \gamma \mathbf{v}(\gamma_u m_0 u) = m_0 \gamma \gamma_u (1 + u\mathbf{v}).$$

It's helpful to define a relativistic 4-velocity:

10.53 $$U^{\mu}=\frac{dx^{\mu}}{d\tau}=\begin{pmatrix}\gamma\\ \gamma u_x\\ \gamma u_y\\ \gamma u_z\end{pmatrix}$$

Thus, in the lab frame, a particle moving at speed $-u$ has 4-velocity of

10.54 $$U^{\mu}=\gamma_u\begin{pmatrix}1\\ -u\end{pmatrix}$$

(Where we've suppressed the y and z components, since we take them to be zero)

In the moving frame, we get:

10.55 $$U'^{\mu}=\gamma_{u'}\begin{pmatrix}1\\ u'\end{pmatrix}=\gamma_u\gamma\begin{pmatrix}1+u\mathbf{v}\\ -u-\mathbf{v}\end{pmatrix}$$

Thus, $\gamma_{u'}=\gamma_u\gamma(1+u\mathbf{v})$ which implies using the second line, that $u'=-\frac{u+\mathbf{v}}{1+u\mathbf{v}}$, giving a velocity transformation formula. It also shows that equation 10.52 is $E'=m_0\gamma_{u'}$ as it should be.

We also can define force as:

10.56 $$\vec{f}=\frac{dp^i}{dt}=\frac{d(\gamma m_0\vec{v})}{dt}=\gamma^3 m_0\vec{a}$$

Where, in the last line, we've assumed the force is applied parallel to the speed, and the last equality assumes m_0 is constant.

The 4-force is defined as:[50]

10.57 $$F^{\mu}=\frac{dp^{\mu}}{d\tau}=\gamma\frac{dp^{\mu}}{dt}=\gamma\begin{pmatrix}\frac{dE}{dt}\\ f_x\\ f_y\\ f_z\end{pmatrix}$$

Obviously, this transforms in the same way as the coordinates, because dp^{μ} does and because $d\tau$ is invariant under the transformation, as mentioned above. Note that the transformation matrix that generally takes any one of these entities from one frame to another is called the *Lorentz transform* and is usually symbolized by the capital form of the Greek letter lambda, i.e., Λ.

Now, considering a particle accelerated along the direction of its velocity, we can verify some of our previous analysis. Starting with the definition of 4-acceleration, $A^{\mu}=dU^{\mu}/d\tau$, and transforming to the frame in which the particle (which is moving at speed $\mathbf{v}$) appears to be momentarily at rest, i.e., $a'_x=A'^x=\Lambda^x{}_{\lambda}A^{\lambda}$, we can show (see end of chapter problem) that $a'_x=\gamma^3 a_x$ (note: lower case "a" represents the non-relativistic 3-

[50] A force that has constant m_0 is called a pure force; one that has constant speed is called heat-like.

acceleration, e.g. $a_x = \frac{d^2x}{dt^2}$). Hence, the non-relativistic force equation obtained in this way is the same as above. This shows that the transformation equations are just as they should be to make the somewhat complexly changing force (which can be seen from the hypothesized absolute rest frame) look just as if the particle were just beginning to be accelerated from rest in the instantaneous moving frame. This, in turn, shows again the formalism's strength in capturing the special integrity of bodies in uniform motion.

c) Light and the Relativistic Doppler Shift

We now would like to understand how light fits in with our new transformation laws and more general definitions for energy and mass. Light is extremely important as the means by which distant objects "communicate." There are other such intermediates by which distant bodies interact, such as gravity, but here we re-specify to just consider light. A key aspect of light is its wavelike behavior, which is characterized by its frequency and its wavelength in analogy to a water wave on the beach. Consider an isotropically radiating source that is at rest and emits light of frequency ν. Consider what it looks like from a moving frame shown in standard configuration in Figure 10-27.

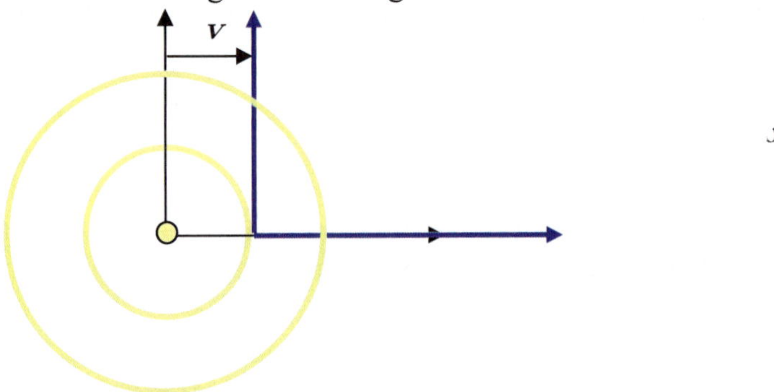

Figure 10-27: A light source at the origin isotropically emits light of frequency ν. Shown are two crests of the outward propagating light wave. We observe the light pulses from the standard moving frame.

The movement of the primed frame axis and the emission of the light is timed so that the crest of the first wave occurs at $(t = 0, x = 0), (t' = 0, x' = 0)$. The equation for the first crest's location on the x-axis is: $x_{\gamma 1} = ct$, the second crest is: $x_{\gamma 2} = c(t - T)$, where $T \equiv \frac{1}{\nu}$ is the period of the wave. The equation for the origin, O', is: $x_{O'} = vt$, which gives $\Delta t = \frac{cT}{c - v}$ for the time it takes between pulses received at O' as viewed in the rest frame. Accounting for the slower passing of time in the moving frame by dividing by γ gives

$$10.58 \quad \Delta t' = \frac{\Delta t}{\gamma} = \frac{\sqrt{1-\beta^2}}{(1-\beta)} T = \sqrt{\frac{1+\beta}{1-\beta}} T$$

(*exercise*: check the first equality with SR transformations) which means:

10.59
$$\nu' = \sqrt{\frac{1-\beta}{1+\beta}}\,\nu$$

The Doppler shift for an observer moving toward, rather than away from, the source can easily be calculated by taking $\beta \to -\beta$ (*exercise*: calculate the Doppler shift in the classical limit $\beta << 1$). Further, an application of the principle of relativity shows us that a moving source viewed from a fixed frame should give the same formula.

We have already seen that light has the power to cause activity within certain massive bodies, i.e., impetus in its parts, which is related to heat in a surface, for instance. Experiment also verifies that light has the capacity to activate impetus in a particle to move it as a whole, not just its parts; indeed all experiment indicate that it does. Thus, we can speak analogically of light as having momentum, since it has the potential to cause momentum.

Now, looking for a direct analogical way to ascribe mass to light leads to assigning it no rest mass, i.e. no intrinsic receptivity to the action of impetus, because, as we already saw: $m = \gamma m_0$ and $\gamma \to \infty$ *as* $\mathbf{v} \to c$, so the only way to avoid associating an infinite mass and momentum with light is to take $m_0 = 0$. But, then, how do we assign momentum to light? Again, looking for direct quantitative analogies within our formalism, we have

10.60
$$E^2 = p^2c^2 + (m_0c^2)^2 \Rightarrow E = pc\,.$$

This means the 4-momentum of a light pulse traveling in the *x*-direction is:

10.61
$$p^{\mu}{}_{light} = \begin{pmatrix} E_\gamma \\ \frac{E_\gamma}{c} \\ 0 \\ 0 \end{pmatrix}$$

So that in the moving frame we see:

10.62
$$p'^{\mu}_{light} = \begin{pmatrix} E'_\gamma \\ \frac{E'_\gamma}{c} \\ 0 \\ 0 \end{pmatrix} = \sqrt{\frac{1-\beta}{1+\beta}} \begin{pmatrix} E_\gamma \\ \frac{E_\gamma}{c} \\ 0 \\ 0 \end{pmatrix}$$

A profound fact can now be seen using the relations for the energy and frequency in different frames (*exercise*): namely,

10.63
$$E \propto \nu \quad or \quad E = h\nu$$

The value of the constant *h*, called Planck's constant, cannot be determined by special relativistic arguments. However, from quantum mechanics, we know the constant is $h = 6.63 \times 10^{-34}\,J\,s$. Quantum mechanics was developed independently of special relativity, so we see here some deep aspects of reality coming to the fore from very different approaches to it.

Notice how a well thought out formalism that captures certain key truths of the physical reality not only allows us to better understand those truths, but also allows us to jump ahead.

Still, we'd like to go back and see more deeply what we're saying. All this seems to confirm what we've often implicitly been assuming. That is, it appears that light is a property of the plana, rather than a body itself. If it is not a body, it cannot have rest mass in the first sense of the word. Furthermore, at our level of abstraction,[51] even free plana itself has no inertial mass. Hence, zero is a good number for the measure of the (analogical) rest mass of light. However, light does affect the mass of a body that absorbs it, usually by simply increasing the motion of the internal parts of the body. And it does seem to have the ability to activate impetus in another body. Something cannot give what it doesn't have at least the power to give. So, it makes sense to talk, by an analogical extension of the meaning of impetus, about the intensity of the impetus of the light and its measure, the momentum. We have to remember, since we're saying light is not a body,[52] that the impetus we're discussing is the ability of the plana (activated by the electromagnetic field), *to cause* impetus, rather than an actual existing impetus such as that found in a moving massive body. This is why it's an *analogical* extension of the meaning of momentum. Keeping in mind what we mean, we can, and will, say the light "has momentum."

Further, light is not accelerated up to speed, so we have no factor of ½ or squares in the relation between energy and momentum. Light only travels at one apparent speed, *c*, so that the rate of transfer of the ability (or power) to activate a certain amount of impetus (our *further* extended definition of energy) is $p \cdot c$.

d) The Connection between Light and Mass

i) An Argument from Gravity

Lastly, we explore the physical origin of the relation of the mass to energy, helping us see the connections between momentum, energy and mass. First, consider two particles at rest which annihilate each other becoming part of the plana and causing activity (light) in the plana, such as the electron and positron pair annihilation mentioned earlier. Suppose the annihilation results in the emission of two pulses of light in opposing directions. The moment before the pair annihilation a test particle, a small body of low mass in the gravitational field of the pair, has a gravitational potential energy due to the mass of the pair given by:

$$PE = -\frac{G(2m_e)m_{test}}{r} \quad (10.64)$$

(where m_e is the mass of an electron which is the same as that of a positron and r is the distance from the pair to the test particle the moment before annihilation)

The pair also has a rest mass ("energy") that gets "converted" to electromagnetic energy. However, if we do not assign an effective mass of some kind to the light that results, then energy is instantly created after the annihilation, because the negative potential energy component (equation 10.64) of the total energy disappears. Recall we thought conservation of energy was important, because it kept a mean between runaway activity and a quick

[51] That is, we include only what we have explicitly "turned on."

[52] Deeper discussions of light have to wait for electricity and magnetism, quantum mechanics, field theory and other more advanced courses

ceasing of all activity. Hence, this argument indicates that we should assign, in some analogous way, an active gravitational mass, m_e, to each pulse of light (assuming each pulse is identical except for being emitted in opposite directions). This, in turn, argues for the light pulses in some analogous way having the same inertial mass. This whole line of thought is indeed incorporated in some way in general relativity where, empiriometrically, energy generates gravity. Notice again the use of analogical concepts of mass and energy increases as we turn on more effects, incorporating more of reality into our formalism.

ii) An Argument from Integrity of Inertial Frames

We start, this time, with a single particle, say an atom, emitting two pulses of light in opposing directions. Figure 10-28 shows the event along with calculations of the momentum and energy in the rest and moving frame before and after the emission event. In these calculations, we use our first derivation of the relativistic Doppler shift. Our key principle here is that we do not want there to appear to be conservation of energy in one inertial frame and not in another. Observing from a moving frame introduces the appearance of impetus in bodies that have none and changes the apparent intensity of impetus of those with impetus; in order to maintain the integrity given in the internal relativity principle, interactions must occur in such a way as to take into account these changes in appearance. With this in mind, we begin our analysis.

First, using Figure 10-28, we note that conservation of momentum in the moving frame implies that there should be a mass change of the atom directly proportional to the *total* energy of the light emitted, $h\nu$, i.e. the total activity level of the light (i.e. potential for it to cause energy through imparting impetus to a particle). Mathematically, we write:

10.65 $$\delta m c^2 = h\nu_0, \qquad \text{where } \delta m \equiv m_0{}^{before} - m_o{}^{after} = m_0 - m_o{}^{after}$$

i.e., fundamentally Einstein's formula: $E = mc^2$. Note that this argument does not imply that *all* mass can be converted to energy; the argument assumes it. However, Einstein did assert it and experiment has backed his prediction to this date.

Further analysis of such atomic photon pair emission (cf. Figure 10-28) is helpful in understanding the mass-energy relation. In this analysis, we will see more clearly how the mass is an indicator of potential to cause activity in the plana in the form of light. First, we will lay out the form of the energy before and after *in each frame*, then we will analyze conservation of energy in each frame. For convenience, we below set $c = 1$.

<u>*Before*</u> the atom emits light, *in the rest frame* (Σ) it has mass m_0 giving:

10.66 $$K_{before} = 0,\ E = V_{before},$$

Here V is, at this point, some unidentified type of potential energy.

In the moving frame, Σ', the atom appears to have kinetic energy, giving:

10.67 $$E'_{before} = (\gamma - 1)m_0 + V'_{before}, \text{ because } K'_{before} = (\gamma - 1)m_0$$

<u>*After*</u> the atom emits the light, *in the rest frame*, we have:

10.68 $$E_{after} = h\nu_0 + V_{after}, \text{ because } K_{after} = h\nu_0$$

While in *the moving frame*, Σ', we have (*exercise*: show this result):

10.69 $$E'_{after} = \gamma h\nu_0 + (\gamma - 1)m^{after}{}_0 + V'_{after}, \text{ because } K'_{after} = \gamma h\nu_0 + (\gamma - 1)m^{after}{}_0$$

Conservation of energy in the rest frame thus gives:

<u>In Σ:</u>

$$\Delta E = E_{after} - E_{before} = h\nu_0 + \Delta V = 0$$

This means that energy must have been taken from the body, and possibly the plana, in the form of change of potential energy:

10.70 $$\Delta V = -h\nu_0$$

Having already seen, from conservation of momentum, that this results in a decrease of mass $\delta m = h\nu_0$, we take:

10.71 $$V_{before} = m_0 \text{ and } V_{after} = m_0 - h\nu_0$$

so that if we try to make $h\nu_0 > m_0$ the potential energy goes negative. In fact, classically this can never happen but it can quantum mechanically.

Now, conservation of energy in the moving frame yields:

In Σ':

10.72 $$\Delta E' = \Delta V' - (\gamma - 1)\delta m + \gamma h\nu_0 = 0$$

(Notice the minus sign is introduced because of the definition of δm). This equation implies: $-\Delta V' = -(\gamma - 1)\delta m + \gamma h\nu_0$

Applying $\delta m = h\nu_0$, gives:

10.73 $$\Delta V' = -h\nu_0$$

Thus, the change in potential energy is the same in both frames (compare 10.73 with 10.70). Thus, we set:

10.74 $$V'_{before} = m_0 \text{ and } V'_{after} = m^{after}{}_0 = m_0 - \delta m = m_0 - h\nu_0$$

giving:

10.75 $$\Delta V' = V'_{after} - V'_{before} = -\delta m = -h\nu_0$$

We could add a constant to both V'_{after} and V'_{before} without changing the result. However, then we would not get the correct result when $\beta = 0$, i.e. $V'_{before} = V_{before} = m_0$.[53]

Once we have noted this, we see that V does not change when one moves from one frame to another (cf. equations 10.71 and 10.74). This V is a special type of potential energy having to do with the essential nature of the body. It shows up still in moving frames because of the special nature of the proper rest frame, that is, the frame in which the object under consideration appears at rest.

Thus we have, using equation 10.67, $V'_{before} = m_0$, our definition of momentum, and ceasing to take $c = 1$:

10.76 $$E = \gamma m_0 c^2$$
$$\vec{P} = \gamma m_0 \vec{V}$$

Using these equations, we see clearly again that $E^2 - p^2c^2 = m_0{}^2c^4$, which is an invariant like $x^2 - c^2t^2$, thus allowing us to glimpse why the same transformations apply to energy and momentum as they do to time and space.

[53] We argue that adding a velocity-dependent term would be to effectively add another term dealing with the activity, not the potential for activity, and we have already handled the activity, i.e. the kinetic energy, separately.

	Before Emission	After Emission
Σ	m_0	$\frac{1}{2}h\nu$ $m^{after}{}_0$ $\frac{1}{2}h\nu$
$\hat{x}$		
Momentum	$\vec{P}=0$	$\vec{P}=-\frac{1}{2}h\nu_0\hat{x}+\frac{1}{2}h\nu_0\hat{x}=0$
Energy	$E=m_0$	$E=\frac{1}{2}h\nu_0+\frac{1}{2}h\nu_0+m^{after}{}_0=m_0$
Σ'	V $\gamma\, m_0$	$\frac{1}{2}h\nu_L$ Σ $\gamma\, m_0{}^{after}$ $\frac{1}{2}h\nu_R$
		$\nu_L=\sqrt{\frac{1+\beta}{1-\beta}}\nu_0$ $\nu_R=\sqrt{\frac{1-\beta}{1+\beta}}\nu_0$
Momentum	$\vec{P}'=-\gamma m_0 V\,\hat{x}$	$\vec{P}'=-\gamma m^{after}{}_0 V\,\hat{x}+\frac{1}{2}\frac{h\nu_0}{c}\left(-\sqrt{\frac{1+\beta}{1-\beta}}+\sqrt{\frac{1-\beta}{1+\beta}}\right)\hat{x}$

Conservation of Momentum: $-\gamma m_0 V=-\gamma m^{after}{}_0 V-\gamma\,\beta\frac{h\nu_0}{c}$

$$\Rightarrow\ \gamma\ \delta m V=\gamma\frac{V}{c}\frac{h\nu_0}{c}\ \Rightarrow \boxed{\delta m\, c^2=h\nu_0}$$

$$where\,\delta m\equiv m_0-m^{after}{}_0$$

Energy	$E'_{Before}=\gamma m_0$	$E'_{after}=\gamma m^{after}{}_0+\frac{1}{2}h\nu_0\left(\sqrt{\frac{1+\beta}{1-\beta}}+\sqrt{\frac{1-\beta}{1+\beta}}\right)$

Conservation of Energy implies: $\gamma m_0=\gamma\, m^{after}{}_0+h\nu_0\frac{1}{\sqrt{1-\beta^2}}$

Figure 10-28: (*Top*) Massive Body emits light pulses in opposing directions. Momentum and energy are shown before and after in the rest frame (Σ) of the particle. (*Bottom*) Same events shown in the moving frame Σ'.

Synopsis

Part I: The Fundamentals

Special relativity can be considered to arise from the observation of Galilean relativity and the finite speed of light.

Galileo, and Buridan (and others) before him in a different way, already recognized the unique status of inertial frames. They recognized that things in an enclosed vehicle, such as a ship or a plane, at some level, look the same whether the vehicle is moving uniformly at any given speed or is at rest.

An *inertial frame* is a system of rulers and clocks laid out, for instance, in a three dimensional grid pattern such as illustrated in Figure 10-4, moving at uniform speed. We speak of an *observer*[54] *doing experiments in a frame*, when he uses rulers and clocks (which are synchronized according to operational simultaneity) that move with him in the frame to do all measurements. Common experience, as well as careful experiments, indicates "*the equivalence of inertial frames,*" which means an experiment done in a given uniformly moving frame will give the same result if done in any other uniformly moving frame. This is generically true, but therefore only true within the level of approximation consistent within our level of abstraction, i.e. consistent with what we have left out; e.g., we may, for example, exclude effects such as air motion by enclosing the frame.

In other words, because of the nature of impetus, for any system (e.g., a single contiguous body[55] or a group of bodies in an enclosure) moving at a uniform speed $\boldsymbol{v}$, we can ignore, at some level, the impetus that moves the center of mass of the system and treat the parts of the system as if each had only the impetus their apparent speed would require.[56] In fact, by speaking of the momentum, the measure of the impetus, using the units (standards) *within* the enclosed frame, we accomplish this in a very natural way. As we have the seen in earlier chapters, ignoring the center of mass motion allows us to focus on the interactions in the system under consideration and not concern ourselves with the motion of the system as a whole.

The equivalence of inertial frames, i.e. the idea that the fundamental (generic) laws of physics are the same in every uniformly moving frame, is called *Galilean relativity*. Again, in saying this, we exclude all but what pertains to the impetus at the simplest level[57] so as to get at the most general understanding possible. Indeed, Galilean relativity "turns off" the finite speed of communications between bodies, assuming an infinite speed of communications.

---

[54] Or a data recorder later examined by an experimenter

[55] This could be a single substance or a group of substances that are in contact, say stuck together.

[56] We could, instead, only ignore, as we will later, the impetus that moves the center of mass *of the frame* (observer, rulers, clocks etc) and consider only motions relative to it. However, at this point where we consider enclosed frames, it is most natural to consider systems whose center of mass is moving at the speed of the frame, in which case, we ignore the center of mass motion of the whole system. If we do not do this, the enclosed system will eventually change its speed when collisions within it cause the transfer of impeti. For example, consider a 50,000 pound spaceship moving at speed $\boldsymbol{v}$ which has a 50,000 pound rock in the middle that is at rest. When the moving wall of the craft hits the rock, say inelastically, the craft will slow to half its original speed.

[57] This, of course, includes "turning off" gravity, which, when we bring back, will lead us, as it first did Einstein, into general relativity.

Bodies typically interact by contact. Fast communication through the plana (free space) is a kind of remedial way for bodies not in contact to interact and so is very important. Since light (electromagnetic radiation) is the chief means of such communication among bodies of everyday experience (and even for the parts within bodies), we often refer to the fastest such communications of all such types as the "speed of light." Experiments show that the speed of light is faster than the speed of any massive body, which thus allows it to act in the remedial fashion described.

As soon as one drops the idealization that the speed of light is infinite, one is confronted with problems with both the relativity of inertial frames and with the way causality is seen in various frames. These problems are resolved in special relativity.

If there is not a maximum apparent speed, i.e. a bound to the speed that can be seen (measured) within any given frame, then, we can show that effects can appear before their causes in a way that would not occur just from the natural distance separation. In other words, *if there is no maximum apparent speed, there can be violations of apparent causality* introduced by uniform motions. A universe that violates apparent causality is a more disordered universe than one which does not, so we hypothesize a maximum speed limit. Of course, this line of reasoning is only important because, in fact, experiments all point to the existence of such a speed limit.

In order to discuss the maximal speed issue and other issues involved in relativity, we introduce the concept of *house. A house* generally means any inertial frame restricted to a region; thus we speak of either a *closed* house or an *open* house. A closed house, the first meaning of *house*, whose generic usefulness derives from the existence of boundaries around physical substances, is a frame that is confined to a region by walls that keep plana inside, so that there are no plana winds; by contrast open houses do not restrict plana flow.

It is especially useful to consider the structure of the interior of closed houses as follows. A given primary house may have many houses inside of it, nested in various ways. Inside the primary house, one does not look out, nor does the outside affect the generic state of the interior. An interior house can look and be affected by the outside, as long as it does not affect its overall state.[58] That is, as long as the impetus of the house is not affected and *only* that impetus and how it is received determine the generic dynamics[59] within the house.[60] In this way, we allow ourselves to treat, at an abstract level, the interior of substances, for instance animals, which have organs that contain flowing (e.g. blood) and

[58] This is another example of a being of reason created by completing a limit. In this instance, we consider the case in which the effect of the changes caused by bodies outside the house are as small as we please so that we can consider them negligible, i.e. non-existent, for our purposes. In fact, of course, the causes we are discussing are intrinsically linked together in the physical world of substances (for instance, impetus changes accompany other changes). Still, we can for example, consider a very heavy house so that small changes in its impetus do not affect its speed at our chosen level of accuracy.

[59] This includes, for instance, the generic changes in the virtual plana in a closed house, the generic length contraction and mass increase of bodies with speed that result, along with operational simultaneity, in the measured constancy of the speed of light from house to house.

[60] In this way, as we will see, special relativity becomes the background space-time for all interactions at our current level of abstraction. We leave out, most notably, gravity which is intimately related to the impetus. We do this even though the natures of many of the things involved, such as atoms, gravitate as does even light. However, because such gravitational interactions are weak in many actual physical situations, we can often leave them out. Of course, even a little bit of gravity generically destroys the global special relativity we are discussing; however, the results of special relativity approximate the real cases when (as they often are) such effects are small enough.

fixed cells that contain organelles themselves, or any interacting system (if we introduce open houses as well) whether it be a substance or group of substances.

Now, still considering only the direct effects of impetus, leaving out causes and effects that would specialize beyond this, such as gravity, we hypothesize an ***integrity principle***. Namely, we notice that, with all other causes and effects "turned off," the effect of the special nature of inertial frames implies that substances, such as men and animals will function, act and live in uniform motion *just as* they do when they are at rest. That is, in contrast to the case of accelerated motion, the integrity of the substance is maintained unchanged in the case of uniform motion at any speed.

We incorporate this integrity principle in a general way into our thinking through ***the internal relativity principle***, that the laws of physics are the same in the interior of every house and that the intrinsic state of a house is only a function of its speed. With this principle coupled with the existence of a maximum speed to guard against violations of apparent causality, we can deduce ***the full relativity principle***, Einstein's *special relativity*, namely:

1) *All inertial frames are equivalent* in the sense that any experiment done in one frame will give the exact same result if done any other frame.[61]
2) *There is a maximum speed limit.* (No physical effect can travel faster than the speed of light.)

In a frame that is at rest in which we leave out all other effects but that of impetus, there is no reason for light to travel faster in one direction than another. Hence, a glowing spherical ball, for instance, puts out light that moves away uniformly in all spatial directions. Hence, this implies, applying the two postulates of special relativity, that *the speed of light is a constant* in every direction in every frame. That is, the expanding spherical wave front of light would be the same in every frame. For instance, an observer viewing the expanding wave front from a spaceship moving half the speed of light would still see a spherically expanding wave front.

Special relativity, as well as Galilean relativity, can be specified by their transformation laws. These laws are the equations that predict when and where an object will be seen in one frame, called the primed frame, given when and where it appears in the second frame, called the unprimed frame. The primed and unprimed frames are assumed to be in the *standard configuration* shown below.

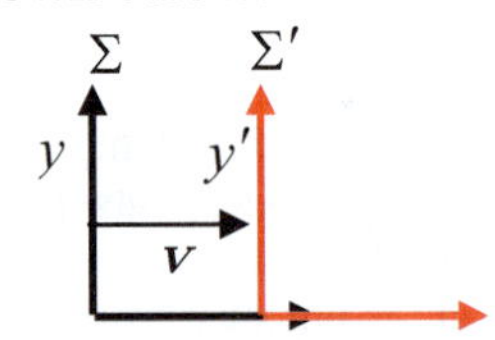

The Galilean transformation laws are: $t' = t,\ x' = x - vt,\ y' = y,\ z' = z$.

In such transformation laws, when we say *seen,* we mean *measured* in that frame, for example, by the radar method in which we send out a light signal and time its return, to determine the time and location of an event. By event, we mean a change of some kind,

[61] Note that we do not directly treat the transformations between arbitrary inertial frames, but only between those in standard configuration. These are, however, covered implicitly by the homogeneity at the level of abstraction (leaving out all but impetus and its most evident effects when gravity is weak). In particular, the rotational and translational symmetry of space as well as the "translational" symmetry of time, lead to the Poincaré transformations, which can be used to explicitly finish the demonstration.

such as the emission of a pulse of light. Of course, our focus on events (changes) should not cause us to forget that all objects are actually something before they change. We now turn to the transformation laws for special relativity.

Before *deriving* these laws, we confront the problem of what it could possibly mean for the speed of light to be constant in every frame. To answer the dilemma, we consider two men, one on each end of a moving flatcar and use sound as an analogy for light. By introducing a time synchronization rule appropriate to the flatcar in the moving air, we can make their "Hi"s reach the midpoint between them at the same time. This is an *operational definition of simultaneity,* because when the train is at rest, the two "Hi"s would be said to have occurred at the same time if and only if they reached the midpoint at the same time. Of course, this functional definition is ***not*** the same as true simultaneity, which is the existence of two things or motions at one moment (or one "now"). However, we would like to focus on the system as a whole, avoiding its overall motion, and such an operational definition reveals these real aspects of the system. Now, this operational definition results in a measured speed of sound which is the same in the forward and reverse direction. However, it still has the lingering effect from the wind that the speed measured on the flatcar is less than the speed of sound in still air.

A similar problem would occur in the plana, except that the impetus changes the mass and the length of all objects in the frame. Now, the effect of a body's impetus on the body itself is indicated by the speed alone, since the speed is a result of, and thus incorporates, both the action (impetus) and the reception (mass) of the body. Experiment shows that the faster a given body moves the more resistance to the action of the impetus it has, i.e. the greater the mass it has; in particular: $m = \dfrac{m_0}{\sqrt{1-\beta^2}}$, where m_0 is called *the rest mass*. The increased mass of the bodies results in slower motion of those bodies, which in turn, means time runs slower in this regard (non-massive motions will be seen to be proportionally slower as well); this is called *time dilation.* Similarly, the faster the body, i.e., the more its impetus is affecting the body, the more its length is reduced; this effect is called *length contraction:* $L = L_0\sqrt{1-\beta^2}$, where L_0 is the length at rest or the so-called *proper length.* Note that, in general, all qualities condition the first property of quantity or extension of a body; length contraction under the action of impetus is a particularly clear example of this.

Hence, impetus acts in such a way as to maintain the integrity of the system as if it were at rest. Using the radar method, which implicitly incorporates operational simultaneity, to determine times and distances, we can, formulating the description in the rest frame and then including the effect of time dilation and length contraction, deduce the special relativistic transformation laws from the rest frame to the moving frame. Thus, the constancy of the measured speed of light is accomplished by the use of operational simultaneity, i.e. the definition of time most useful in the frame itself, and the fact of time dilation and length contraction.

Having shown what we mean by the constancy of the speed of light, how it can be realized and what the form of the transformation laws is, we deduce the full relativity principle of special relativity from the internal relativity principle (including the fact that the properties of a given closed house can, at most, only be a function of its absolute speed) and the existence of a maximum apparent speed limit. To do this we must ultimately include open houses, i.e., frames that are limited in extent but which allow passage of the plana through their interior. Thus, by internal relativity, experiments done in open houses are required to give the same result as those done in closed houses. Note that closed houses are important in the argument, for with open houses alone one would never address the possibility of "trapped" plana.[62] Such open houses could represent machines with parts that are connected by open regions over which the exterior plana can move. Thus, the internal relativity principle requires that the internal workings (or any experiment performed "inside") of such machines behave *exactly* the same in uniform motion as they do at rest or in any closed house. Recall these statements all presuppose the requisite level of generality in which we leave out all but what is relevant to the impetus and its action.

Internal relativity leads to the full relativity principle needed for special relativity because of the nature of substances (or by analogy, any system of bodies) modeled by primary houses, which have parts inside that can also have parts inside that are each, in microcosm, like those in which it is nested. Using this fact, as expressed in the internal relativity principle, that each house is, in an important way, a microcosm of the whole allows us to extend the analysis of the inside to the outside of the house. This turning inside out to incorporate the whole universe is completed once we explicitly allow for open houses.

We could now derive the transformation laws directly from the special relativistic postulates as is usually done. However, to give further insight into the role of apparent causality and the internal relativity principle, we derive the transformation law using these latter two postulates.

We start by using the nature of impetus to move bodies uniformly to show that the transformation law[63] must be of the form: $\begin{matrix} t' = a_{00}\,t + a_{01}\,x \\ x' = a_{10}\,t + a_{11}\,x \end{matrix}$, i.e. linear in the space and time coordinates. Next, we use length contraction and the properties of unqualified space, e.g. its uniformity, to get the following: $a_{10} = -\gamma \mathbf{v}$, $a_{11} = \gamma$, leaving a_{00}, a_{01} and γ unknown. We use, respectively, time ordering and the need for a time *delay* in operational simultaneity to argue that: $a_{00} > 0$, $a_{01} < 0$.

To find a_{00} and a_{01}, we further analyze the conditions under which apparent causality is violated with the existence of a maximum apparent speed. We note that, to avoid a violation of apparent causality, within a given house *A*, there must be a maximum apparent speed. There also must be one as seen from any house *B*. Indeed from any house looking into any other house there must be a maximum apparent speed limit or one of the houses could be used to short circuit apparent causality for the other. In our notation, we thus have the following conditions on the apparent maximum speed of light: $c^{(n)} \equiv c_i^{(n)}$ for all values of *i* (i.e. all houses) for all values of *n* (i.e. as seen from all houses (and thus frames)).

---

[62] Also, closed houses motivate the internal relativity principle which is part of the intrinsic purpose, technically called intrinsic final causality, of the full relativity principle, namely to facilitate, at our level of abstraction, the ability of substances to transverse the universe in uniform motion while maintaining their internal activity unchanged by the fact of motion. Thus, we have called this maintenance an integrity principle.

[63] Note that by using open and closed houses, we can show that only one set of transformations is needed.

In order to keep one house interior from being too radically different from another without yet invoking internal relativity, we require a further condition that none of these apparent maximum speeds exceed a certain least upper bound, c_{Max}; mathematically, we write: $c_i^{(n)} \leq c_{Max} \equiv 1$. Using this restriction, we get two coefficients: $a_{00} = \gamma$, $a_{01} = -\mathbf{v}\gamma$.

Next, we use internal relativity to show the symmetry of the apparent motion between the rest and moving houses (and thus frames) in standard configuration. That is, if an observer is within a primary house *A* and sees a house *B* moving at speed $\mathbf{v}$ within that house, then if he moves to within house *B* to look back at house *A*, he will see house *A* moving at speed $\mathbf{v}$. Thus, functionally either house could be treated as at rest, depending on which is more appropriate to the physical situation to be analyzed.

This symmetry then allows us to calculate the gamma factor, which shows up in the mass increase and time dilation with speed. Thus, we have the full special relativistic transformation laws, the so called, *Lorentz transformations*:

$$\begin{aligned} t' &= \gamma\, t - \gamma\,\beta \frac{x}{c} \\ x' &= -\gamma\, \mathbf{v}\, t + \gamma\, x \\ y' &= y \\ z' &= z \end{aligned} \qquad \text{and their inverses} \qquad \begin{aligned} t &= \gamma\, t' + \gamma\,\beta \frac{x'}{c} \\ x &= \gamma\, \mathbf{v}\, t' + \gamma\, x' \\ y &= y' \\ z &= z' \end{aligned}$$

Where $\beta = \mathbf{v}/c$, $\gamma = 1/\sqrt{1-\beta^2}$ and c is the speed of light. Note these laws reduce to the Galilean transformations in the limit of $c \to \infty$, i.e. where there are speeds of communication that take no time to traverse a distance.

The internal relativity principle and thus special relativity itself leads to the consideration of light movement in moving plana versus non-moving plana. Virtual plana (as opposed to free plana), i.e. plana that is part of a massive substance, will change in the presence of impetus moving that substance. In particular, under the action of impetus, the speed of light in the virtual plana will change, and the extension (length) of the virtual plana will decrease according to the law of length contraction. Free plana is very inert and may stretch and compress so as to create an effective "plana wind" even inside frames. Here, we refrain from specifying exactly how free and virtual plana behave, only mentioning the possibilities because environment is one of the necessary categories of properties of physical things. The study of vacuum (plana), for instance, in Quantum Electrodynamics, is a fascinating topic. However, special relativity itself is able to ignore the plana because of its behavior at this level of abstraction, as is made evident by the identical behavior of open and closed houses. The plana behaves, at the appropriate level of abstraction, in just such a way so that the transformation laws of special relativity are true. Indeed, it makes no sense *within the confines of what is included in this theory* to continue to incorporate plana; instead we substitute the being of reason of empty space when working in the empiriometric theory.

Leaving out all but what pertains to the impetus and its action and then dropping the plana, thus results in a generic description that appears to precede all physical theories that specify further without violating the level of generality with which we started. In this way, special relativity is called the background space-time. When gravity, which because of its nature is directly connected with impetus and its action, is added back in, we then get a different background space-time.

Again, experiment is the final arbitrator in determining whether a general theory is correct or not. The theories we have been discussing, Newtonian physics and Special Relativity, have met the test of experiment in many direct and perhaps even more indirect ways. We should be careful with the word "theory" as we have used it here, for here the word theory does not carry the kind of doubt that it usually does in common speech, for instance, as when someone says "his theory about the situation is... ."

PART II

Understanding the Transformation Laws and Related Dynamics

The transformation laws show that a meter stick at rest in the lab (unprimed) frame looks shorter when viewed from a moving (primed) frame. And, one at rest in the moving frame looks shorter when viewed from the lab frame. Similarly, from the lab frame, moving frame clocks look to be moving slower than lab clocks, and from the moving frame, lab frame clocks look to be going slower. This state of appearance is a result of actual length contraction and time dilation viewed with the rulers, motions and clock synchronization best suited to the given frame. As the frame in which a ruler or clock is at rest is the most natural frame for the given state of that ruler or clock, we call the length and time the ruler and clock measure the *proper length* and *proper time.* These measurements will be the same, as we have seen through the internal relativity principle, in every frame including the rest frame.

The twin paradox investigates the issue of time dilation. One twin takes a round trip journey to a distant planet while the other remains at home; who is actually older, since each sees the other as aging less during the bulk of the trip? Fundamentally, the paradox results from forgetting that one has left out the actual state of the impetus of the relevant bodies. In fact, we do not even need to bring back the full actual state of the impetus, but only the fact that in any particular frame, the spacecraft has greater speed (thus more effect from the action of the impetus than the planets) for a long enough time[64] for its time dilation to be greater than the planets. Furthermore, we also note that though in special relativity, we leave out the effects of the actual bodies breaking the symmetry of space, they nonetheless do pick out unique rest frames (which will have import in general relativity when account is taken of gravity).

Empiriometrically, the key effect shows up in the *acceleration* changing the length of the traveling twin's trajectory through space-time. More physically, the traveling twin alone accelerates and decelerates. During either the acceleration or the deceleration (it looks to the rest frame of the Earth that it is during the acceleration) the actual speed and hence impetus increases, so that, for a time, he has greater time dilation than the planets. Of course, for another part of the time, he *may* have less impetus[64] and thus less time dilation than the planets; it is just that his time dilation is much greater during the faster time and so more than makes up for the time gained during the slower part of the trip.

Again empiriometrically, the special theory of relativity leaves out the actual value of the impetus, for in so far as what we have included in the theory, i.e. what we have not "turned off," it does not make sense. Such mental constructs (beings of reason) help to

[64] The details here depend on which inertial frame is the actual rest frame, and this, in turn, depends on the state of the entire universe, which we may not know (hence, a key advantage of leaving actual rest frames out of account—i.e. abstracting it out (in the general, not the first use, of that term)—the way it is in special relativity.

allow us to move around in the theory without running into loose ends because of what we have left out of account. With more experimental data and more analysis, we later include more in the theory, but these are subjects for other study.

Lastly, we discuss the generalizations of mass and energy that result from the effects of impetus (including the change of the mass and length with speed) that are described by special relativity. For a massive body with rest mass, m_0, the kinetic energy, the rate of transfer of the impetus, $dK = \mathbf{v}dp$ is given by, $K = (\gamma - 1)m_0c^2$. The increasing resistance to the action of the impetus with speed means that the more impetus that is activated in a body, the more impetus it takes to get the body to the next increment of speed; this is reflected in the energy which includes the speed which each "piece" of momentum is given in the body in its coming up to its final speed. Note that the expression for the energy diverges at the speed of light, which means it takes an infinite amount of energy to bring a massive body to the speed of light! Indeed, the amount of impetus needed is infinite since the mass is infinite. This reveals the non-massive nature of light.

Now, it turns out that particle/antiparticle pairs can annihilate and activate electromagnetic radiation in the plana.[65] The total energy of a particle, including this potency to be converted to activity in the plana, is: $E = \gamma m_0c^2$, giving the famous equation $E = m_0c^2$ in its proper rest frame. This activity in the plana is the "activity" or the energy of light moving from one part to the next. However, light, which is related to a quality or qualities in the plana,[66] has activity in a different (analogical) sense, for it is not a massive body moving from place to place, but it does have the potency to cause a certain impetus in bodies or parts of bodies and thus activity of massive bodies. Its energy is measured by its capacity to cause such motions.

In summary, massive bodies have the essential aspect of causing gravitation in the plana, receiving the action of gravitational fields, of receiving and resisting the activity of impetus and of being converted to non-massive plana with a certain activity level (energy). The inertial mass, gravitational passive and active mass and the activity level caused by the destruction of the massive body are all proportional to the inertial mass as might be expected, given the closer contact with the plana this latter tends to imply.

There are two types of inertial mass; the rest mass, and the augmentation of the rest mass that results from motion. In special relativity, it is only the relative speed that matters.

Momentum is also naturally generalized through the use of the differential proper time $d\tau$: $\vec{p} = m_0\dfrac{d\vec{x}}{d\tau}$. Notice here that if the body is moving, the left hand side of this equation requires a mixture of units between frames. However, writing in terms of time and spatial measures natural to the given frame, by explicitly including the mass variation, we see this is just our ordinary definition of momentum: $\vec{p} = m\dfrac{d\vec{x}}{dt} = \gamma\, m_0\dfrac{d\vec{x}}{dt} = m_0\dfrac{d\vec{x}}{d\tau}$. This equation is then generalized to four-dimensions, treating time as the fourth dimension in

---

[65] Furthermore, the quantum mechanical activity of the plana can cause part of the plana to, for example, form "virtual" electron-positron pairs for a brief time. Interestingly, they can be kept from falling back into the plana by the gravitational field of a black hole.

[66] Quantum mechanics indicates that this quality (or qualities) is quantized, and that light comes in discrete units called photons.

the analogical way already discussed. In our space-time coordinates notation, we use: x^μ with $\mu \in \{0,1,2,3\}$ such that : $x^0 = t, x^1 = x, x^2 = y, x^3 = z$ and write:[67] $p^\mu = m_0 \dfrac{dx^\mu}{d\tau}$.

p^μ is called the energy-momentum 4-vector, or simply the momentum 4-vector; it is a type of *tensor*. p^0 is the "*t*-component" which equals γm_0, the special relativistic energy. Thus, in matrix notation we write: $p^\mu = \begin{pmatrix} E \\ p^x \\ p^y \\ p^z \end{pmatrix}$. The related four velocity is:

$$U^\mu = \frac{dx^\mu}{d\tau} = \begin{pmatrix} \gamma \\ \gamma u_x \\ \gamma u_y \\ \gamma u_z \end{pmatrix}.$$

By definition all 4-vectors transform according to the Lorentz transformation laws given above.

The 4-momentum concisely expresses both the active qualities, through the measures of the impetus in the three orthogonal directions (spatial components: p_x, p_y, p_z) and the net effect of the receptive qualities through the measure of the activity that results, the energy (i.e., the time component: $p_0 \equiv p_t$).

The energy transforms, under the application of the Lorentz transformations, just like the time. Or in matrix form, transforming from unprimed to primed frame in standard configuration: $\begin{pmatrix} E' \\ p'^x \\ p'^y \\ p'^z \end{pmatrix} = \begin{pmatrix} \gamma & -\gamma \mathbf{v} & 0 & 0 \\ -\gamma \mathbf{v} & \gamma & 0 & 0 \\ 0 & 0 & 1 & 0 \\ 0 & 0 & 0 & 1 \end{pmatrix} \cdot \begin{pmatrix} E \\ p^x \\ p^y \\ p^z \end{pmatrix}$

where we take the speed of light to be: $c = 1$.

Using a similar line of argument, we get the law of relativistic addition of parallel speeds. Namely, given standard configuration, if a body is measured to be moving at a speed u in the unprimed frame, then it will be measured with speed u' in the primed frame according to: $u' = \dfrac{u - \mathbf{v}}{1 - \dfrac{u\mathbf{v}}{c^2}}$. Note that, for instance, when $u \to c, u' \to c$; so in keeping with our discussion of maximal apparent (measured) speed, the speed never exceeds c.

Using arguments similar to those we used to write the 4-momentum, we get the 4-force:

[67] Latin indices, such as *i* are used when we are discussing only the three spatial variables and Greek letters to include all four.

$$F^{\mu}=\frac{dp^{\mu}}{d\tau}=\gamma\frac{dp^{\mu}}{dt}=\gamma\begin{pmatrix}\frac{dE}{dt}\\ f_x\\ f_y\\ f_z\end{pmatrix}$$

The behavior of light can also be treated using the 4-momentum vector. Comparison of frequency changes from one frame to another with the corresponding energy changes using the transformation laws gives: $E = h\nu$, where, within relativity, h is an unknown constant, but is known as Planck's constant in quantum mechanics. This convergence of results from very different arenas of physics reveals both the unit of reality and the power of such mathematical formalism[68] in probing it.

Light is analogically said to have impetus according to the impetus it can give massive bodies. The state of the plana changes as "the electromagnetic field", a topic for the next course in the physics curriculum, activates one part then the next, and we say that this traveling "wave" (which is traveling at the speed of light) is the light. Now, since light is either moving at speed c or not at all, to define its energy (in the extended sense we already discussed), we do not need to track the speed at which each "piece" of impetus was "added," since its "impetus" was all added at once at speed c. Hence, we say the energy of the light is simply: $E = pc$, given p is the effective momentum of the light "packet" or photon.

Finally, we show that the connection between mass and light manifests the interactive consistency of the natures of light and massive bodies. Namely, we show that the convertibility of mass to energy (more accurately, the proportionality of the mass[69] to the potency of a given massive body to substantially change, become part of the plana and, in the process, cause activity in the plana) is crucially related to conservation of energy. Namely, without this relationship, conservation of energy would *not* obtain in every frame (given conservation of momentum), violating the integrity principle expressed in internal relativity; furthermore, we would even lose conservation of energy, using our Newtonian gravity approximation, in the rest frame of a particle/antiparticle pair annihilation.

In summary, special relativity, including the transformation laws that mathematically characterize it, follows from the integrity principle and the existence of a maximum apparent (measured) speed. The first principle proceeds from the nature of physical substances which are such as to enable them to maintain their proper operation and, hence their integrity, during uniform motion, thus allowing them to transit across the regions of a large universe such as ours. Further, as long as there is a finite maximum speed of communications between physical bodies, the only way to maintain the integrity at the level discussed, in particular, so as not to have violations of apparent causality, there must be maximum apparent speed of the type we have discussed.

---

[68] More generally, such systems in which reality is imported via axioms are powerful in predicting results as they facilitate avoiding errors in reasoning and help in rooting errors out when they occur. The feedback to experiment then constantly causes one to refine a theoretical system to include more information about reality or even purge it of erroneous information or disprove it completely in some cases.

[69] Note again the analogically general usage of the word mass.

These principles, of course, still await further specification, but once these principles are grasped you are ready for the next level of specification. This includes exciting and deep topics such as general relativity and even areas of reality which we do not yet understand; perhaps you will discover these answers for us. You should hone your technical skills in understanding and using equations to solve problems, for this will put you in contact with the principles of the field in a profound way.

Mixed Problems

1. Explain how it is possible to take $c = 1$ without violating the very physical realities, we are trying to understand. Support your analysis with examples.

2. In the chapter, we discussed the case shown in Figure 10-3 in which we assumed a Newtonian world and hypothesized that light behaves in a sound like fashion. We said that apparent causality would be violated if the distances were chosen correctly. Specify a set of distances and speeds for which this is case. Also, measuring from the time that the bullet is actually fired, give the time when the observers sees: a) the bear is hit by the bullet b) the bullet is half way c) the man fires the gun.

3. Given primary house A with internal houses B and C, each moving at some arbitrary uniform speed, show that having a lower maximum apparent speed in house B can violate apparent causality.

4. Draw a t versus x space-time diagram, called a Minkowski diagram, in which t is plotted on the vertical and x on the horizontal axis; take the speed light to have a slope of one. On it show the following:
a) The path of a ray of light traveling rightward from $-\infty$
b) The path of a ray of light traveling leftward from $+\infty$
c) Show the region that represents events that *cannot* causally connected to the event at the origin.
d) Show the region that represents events that *can be* causally connected to the event at the origin.

5. Starting with two frames in standard configuration. Draw clocks in each frame that are spaced one unit apart as seen in the unprimed frame. Given the speed of the primed frame is $\beta = \sqrt{3}/2 \Rightarrow \gamma = 2$, write the values of the clocks seen at a) $t = 0$ and b) $t = 8/\sqrt{3}$.

6. A man carrying a long 2×4 board parallel to the ground runs at a speed of $u=\frac{\sqrt{3}}{2}c$ relative to the ground. He runs into a barn that has a proper length of 6 feet. Given that the 2×4 has a proper length of 10 feet and that the barn has front and back doors which are both slammed shut the moment the back end of the 2×4 enters the barn: a) Is the 2×4 in the barn when the doors close? Does it fit? b) Explain. In particular, what happens when the 2×4 hits the back end of the barn?
Note: consider this problem *first* from a more fully physical point of view and *then* from the point of view of empiriometric special relativity in which we leave out any notion of actual impetus or absolute rest.

7. What constraint is placed on γ by considering motion in y direction from a) Σ to Σ' and b) Σ' to Σ ?

8. Given 1) the full relativity principle that the laws of physics are the same in every inertial frame and that 2) there is a maximum apparent speed, the speed of light, deduce the transformation laws of special relativity in the following way. Assume the linearity of the transformation law, which we proved in the chapter.
a) Explain why the speed of light must be the same in every frame.
b) Set up two frames in a three dimensional standard configuration and use the symmetry of space to write the transformation law for the y and z directions.
c) Assume a light source at the origin of a primed and unprimed coordinate system in standard configuration emits light uniformly in all directions. In the unprimed frame, the rest frame of the light source, write down the equation, in terms of x, y, z and t ,for the propagation of wave-front of that light as it spreads into 3 dimensional space.
d) Write the equation for that wave-front as seen in the moving frame (primed frame).
e) Using c) and d) write down an invariance relation. Use the symmetry of space (in the same way as done in part b) to eliminate any unknowns in the invariance relationship.
f) Focusing on the x, t part of the invariance, eliminate the minus sign in the invariant function by introducing, *i,* the square root of minus one.
g) Note the only way to generate new coordinates is through rotations. Use the following equations for rotations in two dimensions to write the Lorentz transforms: $x'=x\cos\theta+y\sin\theta$, $y'=-x\sin\theta+y\cos\theta$.

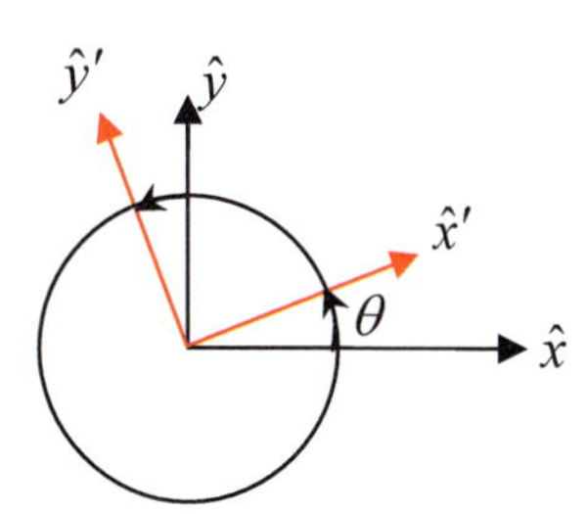

Note: the follow trigonometry relation will be helpful in developing the transformation laws for the x, t coordinates: $1+\tan^2\theta=\sec^2\theta$.

9. The Michelson-Morely experiment tested for the existence of a mechanical type substance, which was called "ether," that would carry light in exactly the same way that air carries sound. The experiment measured the speed along the direction of the Earth's motion around the sun and perpendicular to it. It used the so-called Michelson interferometer diagramed below. A rudimentary explanation of the interferometer follows. The laser sends out a light signal that is split at the half silvered mirror, *S*, half going to mirror A half going to mirror B. Since the distances to A and B are each *L*, if there are no difference in the way the light travels to the two mirrors the light should bounce off and return to S and reach the detector, D at the same time. Now, assume the interferometer is moving, and that light *always* travels at a speed *c* through the plana. Explain what we would expect a) in Newtonian physics 2) given our knowledge of special relativity. Look up the experimental results including modern versions of the experiment and explain which theory is supported by the evidence. Does the experiment deal with time dilation?

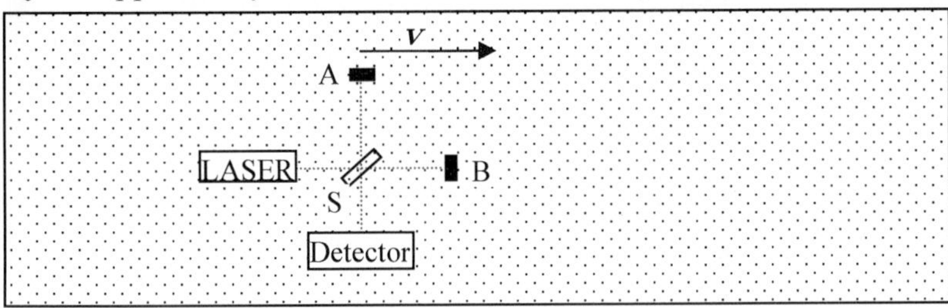

10. Given the total energy is $E = \gamma m_0 c^2$,use the relativistic definition of momentum to prove $E^2 = (pc)^2 + (m_0 c^2)^2$.

11. a) Calculate the trajectory as seen in the lab frame of a particle that is undergoing uniform acceleration of rate *a* in its own rest frame. b) Suppose we choose this acceleration to be *g*, the surface gravity of the Earth, so as to simulate gravity on our trip, how long does it take to get to *Alpha Centuri*, which is approximately 4 light years away? Give the proper time elapsed on the ship and the time elapsed on Earth.

Hint: Use the inertial frame that is instantaneously at rest with the particle and consider the differential increase in speed per time *in that frame* needed to accelerate the particle at a uniform rate. Then, use the velocity addition formula to deduce its speed as a function of time in the lab frame.

12. Check that equations 10.8 and 10.10 give the correct time delay needed to create the appearance of simultaneity in the moving frame in the way discussed in that section. In particular, show that the equations give: $t' = 0$, for the first events of Mr. Front and Back saying "Hi" when a) Mr. Front speaks at $t = 0$ b) when he speaks at $t = \tau$.

13. Suppose Mr. Front and Mr. Back repeat the experiment explained in the text, but instead of sound, they send out short pulses of light. The calculation of the delay time needed for operational simultaneity proceeds in the same way if we assume light travels in a media such as plana. Now, assume that Front and Back also throw baseballs at each other at the same speed, *u*, but opposite directions as seen in the flatcar's rest frame (i.e. using the natural time synchronization in the moving frame and taking account of length contraction and time dilation; in other words, a full special relativistic treatment). Show that the use of the delay time calculated for light, $T = \frac{\gamma^2 L}{c}\beta$, also means the baseballs will reach the middle of the flatcar at the same time. Take the moving flatcar to have length *L*, as measured in the ground frame; do all calculations in the ground frame and, for simplicity, assume $u < \mathbf{v}$.

14. Expand $KE = (\gamma - 1)mc^2$ to show it reduces in the limit to ½ $m\mathbf{v}^2$

15. Show mathematically how invariance of $t^2 - x^2$ is similar to invariance of $E^2 - p^2$.

16. a) Calculate the rocket equation for propellant that exits at relativistic speeds, i.e. where $\mathbf{v} << c$ is not a good approximation. What is the result when one uses light as a propellant? b) Suppose a rocket is powered by a light beam of power P_0 seen in the rest frame of the rocket. Give an implicit expression for β as a function of time in the initial rest frame.

17. To understand the analogies involved in extending to 4 dimensions and in including time as a dimension answer the following questions:

a) Describe a sphere in *4D* space-time, i.e. $x^2 + y^2 + z^2 - t^2 = d^2$

b) Describe the equivalent of a cube, i.e. the figure with faces that are squares with sides that do not cross dimensions and have the same length, in *4D* space-time?

18. How fast does a particle of rest mass m_0 move if its kinetic energy equals its rest mass energy?

19. Show that a photon with 4-momentum $p^\mu = \begin{pmatrix} h\nu \\ h\nu \\ 0 \\ 0 \end{pmatrix}$, cannot decay into an electron/positron pair.

20. A radar signal is emitted along the x-axis at time t_E from an antenna at the origin towards an object. The signal bounces directly backwards of the object, retracing its path back to the antenna. Assuming the signal is received at time t_R, and assuming the signal always travels at speed c, show that one deduces the following for the time and position of the object.

$$t = \frac{t_E + t_R}{2}, \quad x = \frac{c(t_R - t_E)}{2}$$

21. Muons are elementary particles that have a half-life of $1.56\mu s$, in their own rest frames. This means that if I have a certain population of these particles at any one moment, then after $1.56\mu s$ only half of that original population will remain, the rest having been destroyed. They are said to have "decayed." Assume: 1) The half life describes the only mode of decay during the transit to the surface. 2) There is a population of one billion at 10 *km* altitude each with a downward speed of about .98*c*. a) How many will survive the journey to hit the surface? b) Compare this result with the non-relativistic result.

22. Using the special relativistic transformation laws: a) calculate, the delay time as seen by an observer in an hypothetical absolute rest (unprimed) frame between two events that appear simultaneous in the moving frame in standard configuration. b) How does this result relate to the time delay, given in equation 10.3, found in the experiment with the two men on the flatcar.

23. In the problem illustrated in Figure 10-10, we found the location and time of an event (the event of light reflecting off a deer) using the radar method.

Suppose we allow the light to strike the deer and begin its trip back to you at the moment you walk out of the car. Assume you are at the back end of the car and that at the moment you walk out, as illustrated in the figure, you (and thus the back end of the car) pass the origin. This means the front of the car, which is a distance d away, is at the location of the deer at this same moment. To accomplish this timing in our formalism requires that we take $t_d = 0$. Deduce, using this setup, the time delay needed in the moving frame to make the arrival of the back of the car at $x = 0$ actually simultaneous with the arrival of the front of the car at $x = d$, as it must be given the length of the car is d. Show this value is the same as that predicted by the operational simultaneity calculation of the men on the flatcar given in Equation 10.3.

Note: In keeping with these two examples (i.e. the deer near the passenger car and the two men on the flatcar), do all calculations in a Newtonian framework (non-relativistically).

24. Draw a space-time (Minkowski) diagram for the unprimed frame (lab frame). On the diagram, draw in red the line that represents an outgoing ray of light that passes through the origin. Draw lines that represent constant time and constant position for a moving frame in standard configuration.

25. Use $dW = Fdx$ to show $K = (\gamma - 1)m_0 c^2$.

26. Consider a rod moving at speed v through plana that is at (absolute) rest. Setup the coordinate axes in standard configuration as shown in Figure 10-20. a) Calculate the actual time (which is measured in the frame at rest) that it takes each pulse to move half way through the rod. Also, calculate the speed of light in the rod using the extension of the parts of the moving rod (which are contracted) and the actual time as standards. Note the "mixed" (use of prime and unprimed standards) units. In following the light in the moving rod, rather than the transit of light relative to the rest frame coordinate axis, there are two different transit times and two different speeds to calculate; that of the *right moving* and the *left moving* light. b) Calculate the two speeds using the extension of an unmoving (uncontracted) rod as a standard. c) Why is this result useful in a general sense but *not* for empiriometric special relativity? Recall that in empiriometric special relativity, the speed of light is constant in every direction and as measured *within* any frame.

27. Given the transformation laws calculated from the deer near the passenger train problem, i.e. Equations 10.15a and b, show that the reverse transform gives the standard inverse transforms.

28. Using the special relativistic transformation law from a primed frame to an unprimed frame moving at speed v in standard configuration:
a) Construct (trivially) the transform to a double primed frame moving speed v'.
b) Construct the transformation law between Σ' and Σ'', reproducing the special relativistic law of velocity addition along the direction of motion.

29. The proper time is maximized by the uniform motion represented by a straight line on a $t - x$ plot, called a Minkowski diagram). Illustrate this by calculating the proper time for the case of a body that moves uniformly from one place to the next and for one that moves first at one speed, then at another in making the same journey. Draw Minkowski diagram to explain your result.

30. Using the acceleration 4-vector, show that for any given linear direction, which we label x that: $a'_x = \gamma^3 a_x$, where a'_x is the acceleration as seen in the moving frame in which the particle is instantaneously rest, and a_x is the acceleration seen in a lab (unprimed) frame in standard configuration with this moving frame, from which the particle appears to move at speed v. *Hint:* Use the invariance of the magnitude of a 4-vector; that is, for a vector V^μ, $(V^0)^2 - (V^1)^2 - (V^1)^2 - (V^1)^2$ is invariant under Lorentz transformations.

31. Calculate the Doppler shift in the classical limit $\beta << 1$.)

32. Write down the transformation law for the energy of a photon, a packet of light. Compare the transformation law for the frequency and deduce the relationship between frequency and energy.

33. Consider a man traveling at speed c/γ^2 in an absolute rest frame. How would he appear in a frame in which rods are shrunk by γ and clocks run slow by γ? Couple this with the experiment on the train (which uses operational simultaneity) to show how the speed of light, which travels in the plana, appears to be constant.

34. In the chapter, we discussed how mass increase with speed ($m = \gamma m_0$) can be thought of as time dilation with speed ($\Delta t = \gamma \Delta t'$), since the mass increase corresponds to a slower motion for a given momentum. In the discussion, we considered two universes, one with and without mass increase. In this problem, we consider again this issue, but this time we more directly incorporate the very frames that we are trying to understand.

In working this problem, recall that the problem is not yet a special relativistic problem, so frames are not only relative as we finally make them for empiriometric ease. In short, in this problem, we do not yet speak of relative motion only, but still incorporate actual impetus and actual (absolute) motion.

Consider a body, A, of mass m_0 in a frame that is at rest; that is, suppose the clocks and rulers used in this frame *actually* have no impetus. Of course, in reality, the clocks must have some motion in order to be clocks. In particular, suppose the uniform linear motion of body A is our clock. To be a clock in the frame at rest, its motion is assumed to be an arbitrarily small value u. Accordingly, we give it an arbitrarily small intensity of impetus p. Also, since, to compare times between frames, we need such a clock in the moving frame we choose this motion to be along the direction perpendicular to the motion of the moving frame for the following reason. By picking the perpendicular direction, say the y-direction, we minimize the effect the motion of our clock has on what we are investigating, i.e., the effect of the mass increase in slowing motion. (It's somewhat like having our motion take place in the "other universe" of the orthogonal direction.)

a) Given a ruler of length L laid down along the y-axis, calculate how long it takes body A, under the action of the impetus of intensity p, to traverse the length of the ruler.
b) Consider now body A moving along with a moving frame at a uniform speed $\mathbf{v}$ in standard configuration. Assume body A is given an intensity of impetus, p, in the y-direction, so that it moves along the y'-direction at speed u'. Assume the speed of the frame is much faster than the speed of body A: $u' << \mathbf{v}$. How long does body A now take to traverse the ruler of length L at rest along the y' axis?
c) Give the relation between the standards and deduce the relation between times in the two frames.
d) *Optional*: Given body A has a fixed speed $\mathbf{v}$ in the x-direction, write the equation giving the change of mass of the body that results from a small increment of momentum in the y-direction. Compare this to the relation of the change in mass of body A from a small change of momentum applied in the x-direction, with no y-velocity, and show that the effect is much less in the former case.

35. Using special relativity, a) give a simple example of how two objects can appear to move away from each other at the speed of light. b) Explain why this is not a violation of the postulate that no object be measured to travel faster than light. c) Can the speed violation in your example be used to violate apparent causality? d) Look up tachyons and discuss what their existence might imply.

36. Given two identical rods, A and B. A is at rest in the unprimed frame (say the lab frame) and B is at rest in the primed frame in standard configuration with it. Using the transformation laws, show that A appears shorter than B when viewed from the primed frame, and B appears shorter than A when viewed from the primed frame.

Appendix I

Substance and the Nine Categories of Properties

Section I: Discussion of Demonstrations

The types of demonstration of relative internal necessity that we present below are qualitatively different from the standard mathematical proof. This means they can be followed more intuitively. However, it doesn't mean they're less certain. Rather they are actually more certain, because they are primary. How so? The things that we will discuss below are seen directly or reasoned to from things seen directly and so will be the material out of which the axioms of standard mathematics are constructed.[1] Thus, mathematical type proofs will ultimately stand on many of the insights found below.

Substance and Property

The first thing we see is that there are things there. "There is an is" as the renowned author G.K. Chesterton said. That is, we see that there are things that have their own existence. In univocal terminology, i.e. having only one meaning, we give this the name *substance*. So, ***substance*** *is that which has its own existence.* We also see that those things that have their own existence are distinguished from those things that exist in them, which we call properties. Thus, *a* ***property*** *is that which exists in another.* For example, we say the squirrel is brown. The substance is the squirrel; the property is brown.

In empiriometric physics, we will often ignore the question of what is a substance, a part of a substance, or a group of substances, in order to focus in on some generic common aspect of the things under consideration. This, of course, doesn't make such distinctions unreal or even unnecessary; thus, we should be sure to reintroduce those questions later.

Change and the Nine Categories

In addition to seeing that things are there and that they have properties, we also see right away, through our various senses (we touch, hear, see, smell and taste), that all the things around us can and do change. A rock when *struck* makes a sound. A rock when *illuminated* can be seen. The rock is changed by being struck and being illuminated. We also notice that everything we sense has each of the nine categories of properties. Each has qualitative properties such as shape and directly sensible ***qualities*** such as color and hardness. Through these qualities we see each thing is extended, one part outside of the next, which we call ***quantity***. Each thing is ***related*** in some way to other things. For example, this copper bar is equal in length to that, or that is more round than this, or this ***caused*** that (this latter is another important principle, though not a category of property). Each thing we touch can act on another; its action must be received by another; the roles of

[1] Thus, it is incumbent upon us to use only those things that we know directly and to not bring in things that come later, such as conclusions from the specialized sciences. Even more, we are only talking about those things we see directly through our senses, so, in this appendix laying down the physical basics, we are not yet to the point of talking about the mind.

each can be reversed so that everything has ***action*** and ***reception***. Each thing has a ***place*** relative to other substances[2] touching it that define its location in the universe. Each thing has an ***orientation*** in that place. Each substance is in contact with substances that are, in turn, in contact with still other substances; these form its ***environment***. All this happens at a given point in ***time***, at a given now.

We see that these nine categories of properties of substances that we sense are all intimately related to the changeability of physical substances. Their changeability is their most generic essential aspect somehow necessarily related to the nine categories of properties. However, we only see this confusedly. To be science, it must be analyzed carefully. In other words, through conscious analyzing of these things that we see, we can understand why these categories are necessary for something to be changeable, i.e. to be one of the things that we know exists directly through our senses.

Capacity to be Changed
The Essential Characteristic of Physical Substances And How the Nine Categories Follow

Again, we recognize the changeability of things around us. In other words, we quickly see that the things that we sense are <u>actually</u> something but are <u>potentially</u> something else; that is, each can change properties or even cease to be and be replaced by something else. A baseball flies across a field. An apple ripens from green to red, but is still the particular part of the tree that we call an apple; we call such a change a *change of property*. A tree catches fire and turns to ashes; this is a *change of substance*.

Generically, again, we see that the things we sense are something, but they can become something else. Indeed, in the sense of touch, which puts us in direct physical *contact* with things, it is clearly through change that we sense what we sense. I touch something cold and it cools my hand. I sense the coldness (a property) of the substance(s) which caused my hand to get cold. Or again, if I touch something small, like a cold penny, I sense the penny gradually warming up (change) under the influence of my warm fingers.

Locomotion and Quantity

Now, the most immediately obvious and basic way things change is by moving, i.e. locomotion. I move my hand to the cold substance(s) or it moves to me.

What is locomotion? A substance is actually somewhere and a moment later it is somewhere else. Say a sea lion (a substance) needs to move to the other side of a given boundary, for instance, from the ocean onto the shore. In order to move across that boundary, the substance must have at least two parts in the direction of its motion. That is, it must have extension, parts outside of each other, or what we'll call *continuous quantity*. One part of the substance must cross a boundary while a second part remains behind. To just "jump" across the boundary, so that at one moment it is on this side and the next it's on the other, would be for it to be annihilated and then recreated on the other side. For what is preserved during such a "jumping" process? Nothing. This is not what we see and it

[2] We don't know the nature of these yet. They may be very tenuous like what we call "space" (which later in the text we will call plana).

doesn't make any sense. We see things moving and changing, not completely disappearing and then reappearing.[3]

Matter

In other words, physical change, including locomotion, is a *process* in which something is preserved throughout. A thing changes color or grows or dies, in each case something is preserved through the process. In the first case, the substance remains throughout; in the second, the original substance, say an animal, is gone, having changed into various dead organic substances (carbon compounds). We call the generic ingredient that perseveres through change *matter;* note that this is a univocal use of the word *matter* and is more generic than the one that we will often use in empiriometric physics.

In empiriometric physics, we will mean many different things by matter, but, in Newtonian mechanics, it will most often simply mean physical things of an unspecified kind. In the Newtonian context, we usually prefer this meaning because specification would lead us to concentrate on things that we deliberately wish to leave out of consideration. At other times in empiriometric physics, it will mean massive particles (as opposed to massless particles) or particles versus fields, which diverges from the most general definition given above. In short, we will *not* use the word *matter* univocally in empiriometric physics, but instead use it analogically in various ways. The primary analogate, off of which those meanings are based, is the one given above, i.e. that which is preserved through a change. This definition, which is the one used by Aristotle, is univocal (as, by definition, a primary analogate must be) and it is the most generic.

Now, in any change, we obviously don't have just *matter*, that which is preserved during the change. For in a change, there is something that changes, i.e. is not preserved. In change, something (a substance(s) or a property(s)) is actually one thing then becomes another. Generically, we call both that which a thing changes *from* and *into* a *form*. So that by which a thing actually is at a given moment is its *form*. For instance, we can say the apple changes *form* from a small round object to a big round object. Or in the sci-fi realm, one might say the shape-shifter changed form. "*Form*," in our more general use of the word, can be size or shape, for objects change size and shape, but it can also be color or hardness or whatever actuality we are discussing. Shape shifters, of course, do all the above change of properties. However, *form*, in this general sense, *can also* mean all of what it actually is. In this way, we say the chicken died, and Colonel Sanders ate it. So the form of the chicken was lost and replaced by the form of the Colonel. Material things thus are something (their form) but are also potentially something else (matter); they must have both, otherwise they couldn't change in the way they do. From this analysis, two important questions are brought to mind: What is it? and What is it made out of? These are questions that everyone asks from a very young age. We have now seen, at the most generic level, what physical things are; they are substances that can change.

Causality

As an aside from our main discussion of change and the categories, we also note that when something changes, something must cause the change. It's my touching the glass that makes it move or warms it. Something cannot come from nothing. From this we ask a

[3] Furthermore, as we'll discuss later, in the primary type of change (the most manifest one), change of property, the substance undergoes change for a noticeable period of time; it isn't instantaneous. There is a period during which I can say it is changing. The termination of change is often quick. The hit baseball moving through the sky can be quickly stopped by an expert glove.

third question: What caused it? That changes must have a cause is simply called *the principle of causality*. We make heavy use of this principle implicitly in nearly all we do in physics. There is also a very important fourth question: Towards what does it naturally tend? [4]

Objections to the Idea of Physical Change

Now, you may have an objection or, at least, a question about change that has become increasingly bothersome as our conclusions have multiplied. How do we know that there isn't a radically different type of change, i.e. not the type of change described above that we understand first through our senses? You may suggest that, "It may not have extension, but some other actuality that is preserved during this different type of 'motion'."

First, and most importantly, this new idea of motion would then, of necessity, have to be understood by analogy to that which we already know, namely, the change we understand through what's given us through the senses. Hence, we first need to understand this ordinary first type of change as clearly as possible. In other words, whether there is such a motion or not, we need to finish the analysis of the type of motion we know about directly.

Second, this other type of change may exist, but we must decide whether or not it exists through argument based on what we know directly; until we actually know it exists, we're not even sure the notion is not contradictory in someway.[5]

In brief, we are seeing that if the thing is not extended, has no parts one outside of the next, we have no real idea at this point what it could be, because we have not sensed anything like that. If there are such things we have to build our understanding of them based on what we do see.

But, how do we know that there aren't things composing what we see that are not changeable? For example, couldn't there be a constituent of some of the things (e.g., an elementary particle) that we see that is not changeable? No, this would imply effects that we don't see. For example, such a constituent must be able to move (once subsumed into a thing), because if it couldn't, we'd see things that cannot be moved and we don't. Furthermore, as constituents, they've already shown an ability to be changed for they've somehow become part of the thing that they constitute.

Now, you may further ask, how do we know there aren't things that can never move, but can undergo other types of change? For instance, say a change of color. First, even if we grant that it is impossible for it to move, it would still remain true that other substances must be able to move past it; it would be, after all, part of the universe. This, in turn, means it must be extended, because for a substance to move past it, rather than first be on one side of it than the other (that is, be annihilated and recreated on the other side), it must have some width, which first means at least two parts outside of each other. Second, this is *not* the type of substances we see; all that is commensurate with our senses (i.e. that we can walk near and touch) can be moved, at least, in principle, and always by breaking a

---

[4] For instance, moving objects, due to their momentum (impetus), tend to keep moving in a certain linear direction. Or, as another example, in the absence of other forces, two massive bodies tend to fall together due to their gravitational nature.

[5] For those unwise enough to not trust the immediate proof from first principles given here, Gödel's theorem waits around the corner to say that no system of ideas can be proven self consistent within a system.

piece off.[6] Even when something is not in practice movable as whole, we see that it is in principle movable given enough force.[7]

Change Occurs over Time

Further, that which takes place in no time has not undergone physical change in the primary sense, i.e. change of property. Change of property is primary because it is the most manifest and always precedes a change of substance. In the primary sense, the *change* itself only exists when the object is in the process of transitioning from one state to another. That is, change is the actualization of the potential which is ongoing, not before it's started or after it's stopped.[8] We all know this fact; for instance, a ball is moving *only* while it's actually heading from one point to another. It's only moving *while* it's reducing the potential places it could be to actual places that it is. If there is no process, there can be said to be no change. To flash from one state directly to another would be a different type of change of property than we see directly.

There do appear to be exceptions to this. Water heated in a pot, for example, boils at a given temperature not before. However, *this is the end or termination of a change*, namely the process of heating the water up. Also, the act of boiling is itself a process; bubbles form at the bottom and rise to the top; they don't "jump" over the intervening extension of the water.

Light seems to be the one real exception of things that we see directly; it seems to move from the light bulb, for instance, to the ground in *no* time. So, it seems the activity of light is instantaneous and never undergoes the process we call change; it apparently never changed in our primary sense. It appears we need to introduce an analogical type of change. It does look like this, but we also have experience of things going so fast that we can't visually keep up with them, for example, the turning of a wheel on a wagon, bicycle or car. Thus, with no further information than what is given directly to our senses, we should not conclude that light, which is so integrally a part of all we see around us, is so radically different than everything else. Aristotle and others made the mistake of thinking light was infinitely fast based on the good evidence of what could be directly seen. The evidence was, indeed, good, but, of itself, not enough to close the case. Using only direct measurement, such a statement of infinite speed can only get more and more probable, not certain, because the possibility that it's simply out of reach of current measurement capacity remains. Of course, we *now* know that light does travel at a finite speed,[9] so it

---

[6] When we incorporate information from the specialized sciences, the only thing that we see that may be like this is what we call "empty space," which is an idealization, even empiriometrically, because it is actually always "filled" with fields and virtual particles. Later in the text we will call this plana, which is Latin indicating that it is the receiver of the fields.

[7] Furthermore, we see, because of their extension, physical things can be cut up indefinitely. No doubt, at some point, there may be a limit of a sort; for instance, there may not be enough energy in the whole universe available to make the split. Thus, at this point, we might say this is as small as one can get. However, even here, there is a different sense in which one has not reached an, in principle, limit because this limit does not have to do with the nature of dividing the quantitative parts themselves (intrinsically), but limits imposed by the larger universe of which it's a part.

[8] The technical definition of change, using the primal terms of actuality and potentiality, *is the act of the potential as such.*

[9] During the 1670's AD, the Danish astronomer Ole Roemer made extremely careful observations of Jupiter's moon Io (which orbits Jupiter once every 1.769138 days) and found that its predicted appearance (from behind Jupiter) varied depending on the time of the year. In other words, it depended on Jupiter's distance from the earth. Jupiter orbits the sun about once every 12 years, so it doesn't move much during an earth

does undergo change in this primary way and thus there is no longer any reason, on its account, to introduce another analogous type of change.[10]

Quantity Summary

Thus, we start with change that we see, which, as we've said is a *process, not instantaneous*, and through which *something is preserved.* We've seen that all physical substances, that is the changeable sorts of things we know with our senses (that, for instance, move or can be moved), must be extended, i.e. have parts one outside of the next only sharing a common boundary. The first property of all material things is thus extension or what we'll call **quantity**; it's from here we get our initial ideas of geometry and number.[11]

Intrinsic Properties
Quantity
Quality
Relational Properties
Relation
Action, Reception
Place, Environment, Orientation
Time

Figure 1: The Categories of Properties

Quality

In order for something to be changeable, it must be something first, something must unify it into an individual thing; it must have some ***shape*** and some ***powers***, i.e. it must have properties in the ***category of quality***. In particular, it must, at least, have the power (a quality) to act and receive action, for that is what changing and being changed means.[12] If I move two things next to each other, the minimum we need is that they can resist, to some

year. Roemer realized that it was the finite travel time for light to get from Jupiter to earth that caused the discrepancies. From understanding the nature of the delays, he calculated the speed of light for the first time, though only roughly.

[10] Light suggests an analogy between locomotion of a substance and qualities that propagate their influence. For example, we will learn about gravity fields that propagate their influence through space in chapter 8. The quality in one part activates a quality in the next part and in this way moves across space.

[11] Once we've seen that there is extension, parts outside of each other, we can talk more of the actual minimal parts of an extended substance. Each of these parts is extended one relative to the next and, in a *secondary* sense, extended within themselves. Herein is a critical point. We only have the secondary sense of extended, once we've first seen the first sense. As cannot be overemphasized, we must always start with what we know and deduce from there. We see actual parts outside of each other; for instance, the parts of a certain grassy field, such as the bare spot and the surrounding weeds. From that primary sense of extension, we deduce the minimal parts and their extension in the secondary sense. We know it is an essential element to all material substances to have parts outside of each other; only then can we analyze what this means. Indeed, we find the secondary mode of extension comes up in a substance once the first is established. Once we've seen the dependent nature of minimal parts one on another, we can start thinking more deeply about the second mode of extension. This would lead us into very interesting territory, but it diverts us too far off our path. We only hasten to recall that the minimal parts are dividable; they just aren't actually divided (otherwise, they wouldn't be minimal).

[12] More detail is given on this in section II.

extent, going through one another; if not they'd cease to have the parts outside each other, as we have already discussed.[13] The qualities we sense are classified in the subcategory of powers called *sensible powers*.

Relation, Action, and Reception

In this same analysis, several other categories came out. This should be no surprise, for in the real world (as opposed to the way they can exist in our minds) the categories exist altogether; we cannot leave one out without losing the real. Of course, the distinctions we see are real and this is part of what we investigate. In doing the analysis, we are really just unburying all the interconnections among the different distinct categories and thus seeing how they all fit together to make the whole we see.

The category of action and reception are explicitly in our discussion of quality. Consider, for concreteness, a substance (a monkey) that uses its power to act, for instance to pick up a glass of water. The substance is now exercising a property in the category of action. The glass is receiving the ***action***; this is the category of ***reception***. Notice here we've implicitly invoked another category, ***relation*** as we must, for all the categories after quality involve relations to other substances rather than intrinsic properties (see figure 1 above). In discussing, action and reception, we implicitly invoke the relations of causality. The other subcategories of relations, i.e. equality and similarity, follow from the fact that there must be substances with different but related quantities and qualities in order for them to be able to interact with, i.e. change, each other.

Everything we know directly through our senses has the ability to act, *action*, as well as be acted on, *reception*. We know this because if it did not have the ability to act, it could not act on us and thus we would sense them and thus not ***directly*** know of their existence. Again, there may be things that cannot act, but we don't know them directly.

Place, Orientation, and Environment

In talking about locomotion, the category of place and environment were also needed. For a substance, to move it must first be in one ***place***, that is one location defined by the innermost immobile boundary of the surrounding bodies. This, like the other categories, can be understood even more deeply; place is complex, as described in Chapter 3, because something can be at rest with respect to the immediately surrounding bodies, say a fish in a bowl, but the whole bowl can be moving in a larger environment.

An object could not move if there were no other place for it to move. There must be an ***environment,*** other substance(s) beyond it, for it to move. We see many substances interacting around any given substance; for instance, a bear in a forest has many substances around it: plants, animals and inanimate things that make up its environment.

Now, in a given place an object can be rotated. The particular relation of a substance's parts within a given place is called ***orientation.*** It is most clear that orientation is a distinct univocal category, if one thinks of a spherical substance. In this case, rotating it around any direction doesn't affect the environment or the place of the object.

---

[13] We need to be very careful not to specify meaning solely to a particular case. Thus, though it is important to start with things we can touch and move, we should not confine the principles to be exactly as seen in those cases. More generically, examples are necessary, but they necessarily concretize the principle and thus can be misleading if this is not kept in mind.

Time

The last category has also been implicit in what we've already said. Everything exists at a given moment. Each such moment is lost and followed by another as change occurs. Indeed, change is clearly linked with time, for it is in time that change happens. ***Time*** is the measure of motion. It is one motion measured by another, either a causal motion measuring one that is caused, such as the relative motion of the sun (years and days) measuring the lifespan of an animal or it is a simple regular motion measuring irregular motion, such as a stopwatch measuring the motion of a racer running the 50 yard dash. The category of time is necessary because all motions in the universe are related in some way to all others.

Section II: Some Axiomatization

In the previous section, we gave demonstrations in paragraph form in order to stay as close to our primary understanding as possible. However, putting these insights into a more axiomatic form also will be helpful. Again, the goal here is to reveal more clearly and definitely, what we've already seen in a vaguer and less specific way. We need to be careful to not push for too much clarity and specificity but only the amount that can be had with the given information. Note further that we are starting with only the most generic information, that which we know directly through our senses.

Definition 1: A *physical substance* is a whole that exists of itself, and its characteristic property is its changeability.

Axiom 1 (self evident): That which exists of itself, substance, is, in principle, before its properties, including its parts.

> **Corollary**: We must establish facts about substances first, and then proceed to understand the interrelation of the parts, for, by the above axiom, whatever is in the parts is ultimately due to the substance. Said another way, we can only *understand* the interaction of the parts by analogy to the interaction of substances.

Definition 2a: Process means something that occurs through time rather than instantaneously.
Definition 2b: Change is the process of reducing potentiality to act, during which something is preserved.
Definition 3: Extension means having one part outside of another.[14]

Axiom 2: Two types of change

> **2a.** Change of property: during such change a substance is preserved but a property changes.
>
> > 2a_1. *quantity*: growth or diminution
> > 2a_2 *quality*: alteration: intensity variation, change of quality
> > 2a_3 change of *place*
>
> *Note:* Change of orientation, in so far as it's purely orientation and does not also involving place, has to do with parts of a thing not whole.
>
> **2b.** Substantial change is not change in the first sense but in a different sense. It is the termination of one or more property changes; one substance is lost and another comes about, but the potentiality for both substances to be remains. Discussion

---

[14] One might try to introduce minimal parts here instead of "parts outside of each other," (or too quickly before the latter is understood), but this would be a mistake because minimal parts only come about once we have actual extension. A substance that didn't have actual extension would not be anything (that we can think or talk about), so we cannot then talk about its parts. The thing has to come first then its components; in the same way, the consequences of a principle cannot be discussed without the principle.

relative to substantial change will follow the main discussion of change of property, which is particularly needed in this text.
Note: Substantial change is a fact we see. Namely, the changes of property of any substance can be of a type or reach a level at which the substance loses its integrity and is replaced by one or more others.

Axiom 3: Things cannot be and not be at the same time and in the same way.
Axiom 4: (Causality) Something (a new actuality) cannot come from nothing.
Axiom 5: It's only *through* contact with the given physical substance that it can be acted on.

Theorem One

All physical substances must have extension (***quantity***).

Proof:

Assume a substance has no parts outside of each other, i.e. no extension, Show it cannot change and therefore is not a physical substance.

I. *Change of property*

1) Show that it cannot undergo the most obvious type of motion, locomotion (2a_3):

Most fundamentally, if a substance does not have extension, it can not be in place, i.e. it could not be in contact with other bodies; thus it could not be situated among the realm of extended bodies. Lacking the capacity for place, it could not change place, i.e. move.

We can manifest more of what this means by attempting to understand locomotion in this context.

If the body doesn't have two parts then it is either on one side or the other, never on the way. Thus, it does not move, because there is no process.

Correlative principle: To speak of locomotion we must have an extended ***environment*** for the object to move in, for it to move it must start and end in two different ***places***.

2) Show that it cannot undergo change in quality (2a_2): (see theorem 1b below):

Two possible ways change in quality can be caused:

a) **External** (activated by another substance): with no extension, there is no contact. For example, consider a rectangle; it cannot make contact with another rectangular body if it itself has no width. Without contact, there is no way for such a physical substance to act or be acted on by physical things.

b) **Internal** (actuated by powers within itself). If this is to occur the substance must have at least two parts outside of each other, i.e. extension, so one can act on the other in, for instance, a sort of oscillatory fashion. (Again, all this goes back to what we see. If this is to happen other ways, we'd have to have proof of it and then it would not be the type of things we see around us but only like it in some way). Note that change within a substance such as this is only understood *after* change caused by one substance acting on another; indeed, change among the parts of a substance is modeled after such change.

3) Show that it cannot undergo change in quantity (2a_1):

Obviously, it cannot grow or diminish because this would involve further extension, and we said it had none.

II. *Change of substance*

Show that it cannot undergo substantial change (2a_1):

There is no way a non-extended substance can change substance, because it cannot have contact, so nothing could act on it.

Exploration: Again, to explore more, we leave the lack of contact out of consideration. If a non-extended substance were to change substance, such an event could ***not*** be the termination of a change of property, because, by the above arguments, it cannot undergo such a change. At one moment, the thing would be one substance; no change (i.e., no process) would happen; it would just suddenly be another.

Further, there would be nothing preserved during the change of substance. Without extension, there would be nothing that could be said to remain through the "change." It would be the annihilation of one thing, and the creation of another. It all, again, comes back to: we must start with what we actually see.

Theorems on Quantity

Theorem 1a. *Quantity is the first property of physical things.*

Proof:

Quantity and quality are intrinsic properties, i.e. properties belonging to the substance itself, not relational properties (see figure 1). The other properties are relational, requiring the intrinsic properties as well as extrinsic substances. Hence, quantity and quality come before the others.

Quantity (extension) is limited by shape (quality) and thus cannot be understood without that quality. However, quantity can be conceived without the more robust subcategory of quality called power, including the sensible powers. We can see extension of itself doesn't require sensible qualities by starting with a body, say a wall, and leaving out, by an act of the mind, all its properties but its extension, i.e. its parts outside of each other. In our thought, though not in our images, extension can be considered separately from qualities in the category of power (e.g., color). However, none of the other physical properties can be understood without quantity. Every quality we see is based in quantity; for instance, one cannot consider the color green without considering it as existing in an extended surface.

Further, quantity is that which mediates the potentiality for the substance to be other than it is (either by substantial or property change). Namely, because it has parts outside of each other, it can, in principle, be taken apart and put together differently.

In this way, we can say quantity is the base for the others, i.e. that upon which the others stand (from the point of view of potentiality).[15] Further, since changeability, i.e.

[15] Again, quantity and quality are the only intrinsic properties and thus come before the other categories of properties, which require them. They are co-relative, i.e. mutually dependent. Quantity is first with regard to potentiality, while quality is first with regard to specification (actuality).

potential for change, is the essential generic attribute of a physical substance, this makes quantity its first property.

Theorem 1b. *For every completely specified quantity, there is, at most, one specific set of qualities.*
Proof:

Because a substance is a whole, i.e. exists as one thing, not many, the various properties of a substance must be all interrelated. Whatever isn't interrelated in some way is not part of the substance. As shown in theorem 1a, the first property of all material things is extension (quantity), on which all the others are "built". By quantity we do not simply mean the size of the whole, but also the size, number and mutual arrangement of all the actual parts of the whole. We note that quality and quantity are correlative, mutually dependent, with the potential to be this quality or that one "growing out of" quantity.

Namely, quality cannot change without quantity also changing. To understand the meaning of this, suppose the contrary. If there were two different sets of qualities associated with the same quantity, then whatever the differences between the sets are would be independent of quantity, belying the fact that *all* quality is directly based in quantity. More specifically, whatever potentialities that a substance has are mediated through quantity.

We can show this dependency in the following way. Because of quantity (i.e., parts outside of each other), a body can be rearranged or taken apart and put back together differently. When I see a green surface, for example, the green *of itself* doesn't bespeak of any potentiality to be red or even a different shade of green; in fact, "adding green" not only doesn't add any further potentiality but it specifies the surface. By contrast, the fact that there is a right and a left side of the surface, manifest the substance's potential to be ripped in half. [16] Indeed, there is no other way we see for the thing to be other than it is, than by changing its parts (this is not to say the parts are simply the same as what it is).[17]

Now, because all the potentiality of a substance to be other than it is comes through the quantity, once we specify the quantity there is no further potentiality to specify.[18] Hence, we know that, at most, only one set of qualities can be associated with it. (Note we don't, by this fact, know what the qualities are-- or even if there is a quality possible within

[16] You may say that our senses do not see this radical dependence of quality on quantity but a lesser dependence, for a red disk doesn't seem any less possible than a blue disk. However, this is forgetting that we don't see all the parts of the disk. We see that the red disk is truly extended, and, no matter how close we look, each part we examine is itself extended and thus has further parts, one outside the next. In this way, we come to *understand* that the thing has parts that are too small for us to see. In other words, the quantitative specification is not simply the boundary, which we usually focus on in our imagination, but the whole quantitative arrangement of parts, even those too small to be seen. Because we don't usually notice this, we usually consider the actual division, size and arrangement of parts of the disk to be arbitrary, but, if fact, they must be specific. Thus, we realize that the disk's color only *appears* somewhat independent of the quantity, and, in fact, (through the arguments in this theorem) the color must be limited by, though ***not*** simply reducible to, the particular arrangement of parts. In short, the parts, which are too small to be seen, must be different in a red and green disk.

[17] Taking an analogous example from modern physics, an atom has a particular set of lines of emission across the electromagnetic spectrum. A different set of lines always implies different parts and different arrangements that come with different types of atoms as well as within a given atom.

[18] Note, this is not to deny the *potentiality* to be divided (or united) always present in quantity, but only to point out that once its actually specified in one way it is that way and not the others.

the contingencies of this world for the given specific quantitative state). In this way, we glimpse more of the reason for the fecundity of the empiriometric method, namely, that quantity says something implicit about quality.

This, again, means that if a quality changes, so must the quantity. That is, parts of the body undergoing alteration must decrease or increase in size and/or number (that's all quantity can do as quantity). If the parts of a substance decrease or increase in this way, it's likely that the parts must move within the substance and/or within the environment. This means that any given change (locomotion, alteration and growth (diminution)) will either require or likely require locomotion in the proper sense or locomotion of parts.

Note that the interrelation of quantity and quality discussed above is true for those things that we know directly through our senses, but we may find that there are other types of things in which the qualities are not like this. Still, we must start with what we know does exist, then let the evidence and reasoning take us where it will.

Theorem Two:

A physical substance must have ***qualities****.*

Proof:

1) Without a power (quality) to receive the action of another body or the power (quality) to change a property of itself, a body could not change.

Further Specification (auxiliary axiom):

All the physical things that we know directly (i.e., are commensurate with our senses so that we can move near them and touch them) can move. Thus, a physical substance must have a power (a quality) to *receive* the action of another body and/or the power (a quality) to move itself.

Furthermore, any substance that moves must have a power (quality) to *actively displace* what's in front of it (even if it's just "space,"). If it did not do so, it would not be able to keep its own existence as it moved; in other words, it wouldn't really move.

The moving substance can displace the medium of travel in only two ways; it could deal with the medium externally or internally. First, it could have the power (quality) to push, to *some* extent, other substances (or parts of a substance) out of the way. Second, it could assimilate what's in front to itself as it moves. A rudimentary analogy would be a man eating his way out of a mass of very springy cotton candy. In both cases, the body must have active qualities: in the first, the power to push, in the second the power to assimilate.

Exploration:

Suppose there is a physical substance (such as "space" [19]) that cannot move, but can be moved *through* by another.

Such a substance must have the ability to have its parts moved in some way. The body may be such that it accepts a "push" from the moving body, causing each of the contacting parts to push (which needs a quality) against the parts they contact

[19] This is not something we see directly. The closest we see is air, which we know is moveable. It takes analysis to come to the idea of "space" which we will call plana.

etc., ***or*** they may have qualities such as to allow the other substance to assimilate its parts. If it can be assimilated, it must have active qualities that can inter-act and can, in some way, become a real part of the *moving* body.[20]

2) Without a quality to unify the "parts outside of each other" into a whole, it would not be a whole; this implies, at least, a need for ***figure*** (shape) and, in what we see directly, a power to maintain its integrity as a whole under activity on it from the outside.
Specific Clarifications:

The various types of change require some obvious qualities:

2a_1 *Diminution* and *growth* must have shape that limits extension at any moment.

2a_2 *Alteration*: must have quality to change it

2a_3 *Locomotion* must have shape to have place.

Theorem Three

All physical substances have **relations** *with other substances.*

Proof:

Since they have extension, they have a certain number of parts, that number is greater, less than or equal to that of others, which are relations. They can have more or less power; more and less are relations. They can cause motion in another or receive motion from another; each of these is a relation as well. This establishes all subcategories of relation.

Theorem Four

All physical substances have properties of ***reception*** and ***action***.

Proof:

Since they are changeable they must have the ability to be changed, and while they are being changed, they have a property in the category of ***reception***.

Since a physical substance can cause change (e.g., because of its extension, it must either have the capacity to resist the motion of a substance or for the substance to move through it and thus must, at least, have the power to re-*act* in the way described in the *exploration* section of theorem two—see also: theorem 4a below and the auxiliary axiom of theorem two), when they do so they have a property in the category of ***action***.

Note: We can see that some physical things must have active qualities (and thus action) by noting that there would not be change if this were not the case.

Theorem 4a All physical substances that we know directly have the power to change another.

Proof:

We would not know it directly unless it acted on our sensorial powers.

---

[20] If the parts of the medium of travel were not able to be pushed out of the way (including by compression) and if it were completely un-reactive so that it could not become part of the substance, then it must simply disappear. Following this extremely odd supposition, if it did just disappear, then it would not be changing in our primary sense, because obviously nothing is preserved during such a change. This would clearly be a radically different type of change that we don't have any direct evidence for.

Theorem Five

All physical substances have a ***place****,* ***orientation****, and* ***environment****.*

Proof:

We've seen the need for place and environment in understanding locomotion (in proving theorem one), and that all the things that are commensurate with the senses can be moved. Further, alteration in things we see only occurs when contact occurs, i.e. through locomotion from the environment; for example, the apple ripens from green to red only as long as new nutrients and water are continually given to it.[21]

Generally, substances acting one on another in an environment (and thus with place and orientation) are what causes changes of all sorts to begin and to end, as well as causes the changes in the changes that we see.

Orientation follows on place; that is, its parts must be arranged in a certain way with respect to the contacting bodies (again even if this be "space filled with weak fields").

Theorem Six

All physical substances exist in ***time****.*

Proof:

Change can only be understood as happening in time. It proceeds from point A to point B through some duration of time. Time is the measure of motion, but motion proceeds through a continuum of "nows." They are inextricably linked.[22]

Again, the above discussion is about the primal realities that we see directly and which are inextricably linked. In such discussions, we bring out more clearly the mutual necessities that we saw before only confusedly. Also, we recall one final time, when confronted with reasoning or evidence that demand we assert the existence of things different from the physical things we see directly, we can only make the argument that they exist using those things and the principles that we see in them. Further, we can only understand those things, whatever they might be, by analogy to physical things, which we see directly. This is why physics, the study of physical things, is so very important.

Finally, we stop here in our analysis, though we have only said very general things, things that hardly restrict at all. Yet, this was our goal. In the text and in future physics courses, you will learn more of the specifics of our world; this is important, for reality is concrete not general. Also, much more depth can be had even in the generic type

---

[21] Note, in this and the previous sentence, we needed to introduce extra axioms concerning movement and alteration in bodies we see. We could also just introduce the simple fact that there are multiple substances. To argue straight from changeability, we have to deal with the possibility of alteration without any attending locomotion. To do so, we can conceive of a substance that acts one part on another, and which then acts back on the first proceeding back and forth in an oscillatory fashion ad infinitum. Such a substance would not need contact of external substances to change in quality. However, it would *never* be able to not change (even in principle because, by definition, it would have to be the whole universe). Being able to do so is implicit in the idea of changeability. It's the difference between: *can* do something and *must* do something. In other words, it would not be a *changeable* substance but *necessarily changing* substance. To make stopping the change possible, we need environment and place (and hence orientation).

[22] In this way, the potentiality in time proceeds, ultimately, from extension.

discussions above. For example, many of the above points on substance have counterparts when we consider the changes within a substance, specifically with respect its parts. Such information will be important in further study of physics. We find, for instance, as you may know, that inside an atom, the light emission occurs because of locomotion of parts.

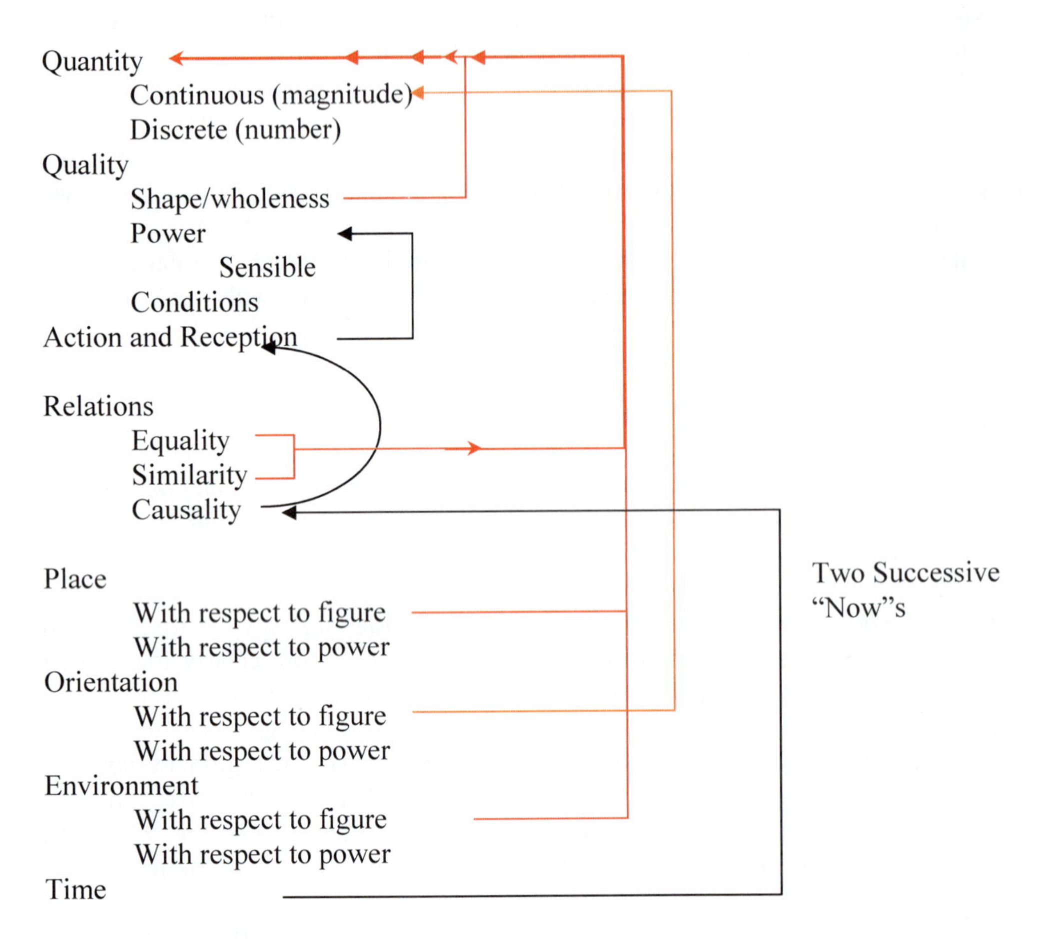

Figure 2: Some correlative relations of the categories that cannot be left behind as one abstracts from the bottom upward.

Appendix II

Outline of all the Sciences

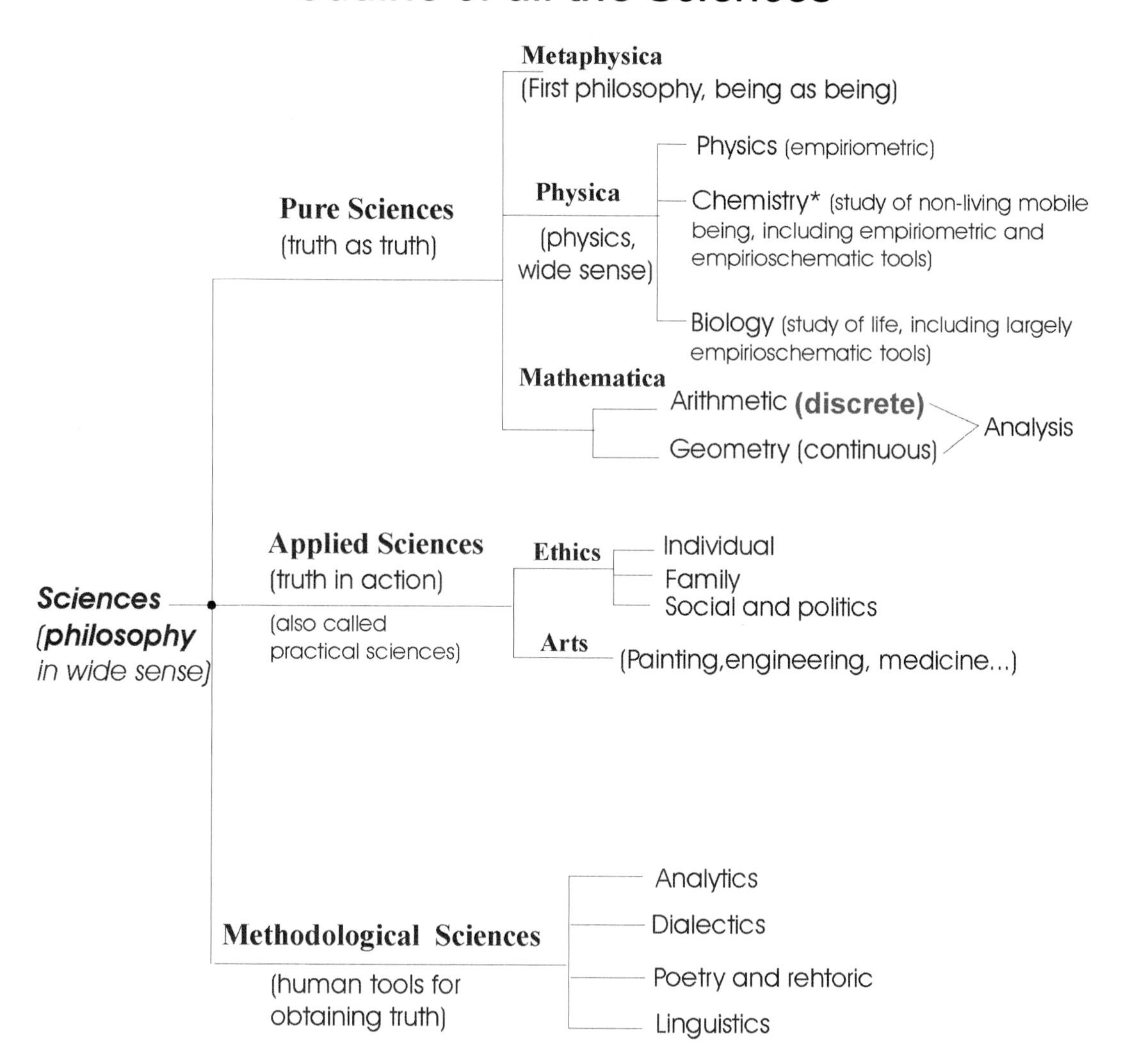

*Chemistry is usually discussed in reference to a particular environment. Thus, one can speak of the chemistry on the earth or on the Jovian moon, Io.

Appendix III: Mars Fact Sheet

Mars/Earth Comparison

Bulk parameters

(Mars/Earth)	Mars	Earth	Ratio
Mass (10^{24} kg)	0.64185	5.9736	0.107
Volume (10^{10} km^3)	16.318	108.321	0.151
Equatorial radius (km)	3397	6378.1	0.533
Polar radius (km)	3375	6356.8	0.531
Volumetric mean radius (km)	3390	6371.0	0.532
Core radius (km)	1700	3485	0.488
Ellipticity (Flattening)	0.00648	0.00335	1.93
Mean density (kg/m^3)	3933	5515	0.713
Surface gravity (m/s^2)	3.71	9.80	0.379
Surface acceleration (m/s^2)	3.69	9.78	0.377
Escape velocity (km/s)	5.03	11.19	0.450
GM (x 10^6 km^3/s^2)	0.04283	0.3986	0.107
Bond albedo	0.250	0.306	0.817
Visual geometric albedo	0.150	0.367	0.409
Visual magnitude V(1,0)	-1.52	-3.86	-
Solar irradiance (W/m^2)	589.2	1367.6	0.431
Black-body temperature (K)	210.1	254.3	0.826
Topographic range (km)	30	20	1.500
Moment of inertia (I/MR^2)	0.366	0.3308	1.106
J_2 (x 10^{-6})	1960.45	1082.63	1.811
Number of natural satellites	2	1	
Planetary ring system	No	No	

Orbital parameters

(Mars/Earth)	Mars	Earth	Ratio
Semimajor axis (10^6 km)	227.92	149.60	1.524
Sidereal orbit period (days)	686.980	365.256	1.881
Tropical orbit period (days)	686.973	365.242	1.881
Perihelion (10^6 km)	206.62	147.09	1.405
Aphelion (10^6 km)	249.23	152.10	1.639
Synodic period (days)	779.94	-	-
Mean orbital velocity (km/s)	24.13	29.78	0.810
Max. orbital velocity (km/s)	26.50	30.29	0.875
Min. orbital velocity (km/s)	21.97	29.29	0.750
Orbit inclination (deg)	1.850	0.000	-
Orbit eccentricity	0.0935	0.0167	5.599
Sidereal rotation period (hrs)	24.6229	23.9345	1.029
Length of day (hrs)	24.6597	24.0000	1.027
Obliquity to orbit (deg) *(tilt of axis)*	25.19	23.45	1.074

Mars Observational Parameters

Discoverer: Unknown
Discovery Date: Prehistoric

Distance from Earth	
Minimum (10^6 km)	55.7
Maximum (10^6 km)	401.3
Apparent diameter from Earth	
Maximum (seconds of arc)	25.1
Minimum (seconds of arc)	3.5
Mean values at opposition from Earth	
Distance from Earth (10^6 km)	78.39
Apparent diameter (seconds of arc)	17.9
Apparent visual magnitude	-2.0
Maximum apparent visual magnitude	-2.91

Mars Mean Orbital Elements (J2000)

Semimajor axis (AU)	1.52366231
Orbital eccentricity	0.09341233
Orbital inclination (deg)	1.85061
Longitude of ascending node (deg)	49.57854
Longitude of perihelion (deg)	336.04084
Mean Longitude (deg)	355.45332

North Pole of Rotation

Right Ascension: 317.681 - 0.108T
Declination : 52.886 - 0.061T
Reference Date : 12:00 UT 1 Jan 2000 (JD 2451545.0)
T = Julian centuries from reference date

Martian Atmosphere

Surface pressure: 6.36 mb at mean radius (variable from 4.0 to 8.7 mb depending on season)
[6.9 mb to 9 mb (Viking 1 Lander site)]
Surface density: ~0.020 kg/m^3
Scale height: 11.1 km
Total mass of atmosphere: ~2.5 x 10^{16} kg
Average temperature: ~210 K (-63 C)
Diurnal temperature range: 184 K to 242 K (-89 to -31 C) (Viking 1 Lander site)
Wind speeds: 2-7 m/s (summer), 5-10 m/s (fall), 17-30 m/s (dust storm) (Viking Lander sites)
Mean molecular weight: 43.34 g/mole
Atmospheric composition (by volume):
Major : Carbon Dioxide (CO_2) - 95.32% ; Nitrogen (N_2) - 2.7%
Argon (Ar) - 1.6%; Oxygen (O_2) - 0.13%;
Carbon Monoxide (CO) - 0.08%
Minor(ppm) : Water (H_2O) - 210; Nitrogen Oxide (NO) - 100;
Neon (Ne) - 2.5;
Hydrogen-Deuterium-Oxygen (HDO) - 0.85;
Krypton (Kr) - 0.3; Xenon (Xe) - 0.08

Satellites of Mars

	Phobos	Deimos
Semi-major axis* (km)	9378	23459
Sidereal orbit period (days)	0.31891	1.26244
Sidereal rotation period (days)	0.31891	1.26244
Orbital inclination (deg)	1.08	1.79
Orbital eccentricity	0.0151	0.0005
Major axis radius (km)	13.4	7.5
Median axis radius (km)	11.2	6.1
Minor axis radius (km)	9.2	5.2
Mass (10^{15} kg)	10.6	2.4
Mean density (kg/m^3)	1900	1750
Geometric albedo	0.07	0.08
Visual magnitude V(1,0)	+11.8	+12.89
Apparent visual magnitude (V_0)	11.3	12.40

*Mean orbital distance from the center of Mars.

If no sub- or superscripts appear on this page - for example, if the "Mass" is given in units of "(1024 kg)" - you may want to check the notes on the sub- and superscripts at http://www.spds.nasa.gov/planetary/factsheet/fact_notes.html.

Procession of Mars Equinox 175,000 earth years (cf. wikipedia)
*Procession of Earth Equinox 26,000 (*25,800) earth years
Tilt of earth's axis: 23 degrees 27 arcminutes See above Obliquity.

Highest point on surface Olympus Mons (about 24 km above surrounding lava plains)

Atmospheric components 95% carbon dioxide, 3% nitrogen, 1.6% argon Surface materials basaltic rock and altered materials[23]

[23] Most of information culled from NASA's Dr. David R. Williams' information sheet (dave.williams@gsfc.nasa.gov, Greenbelt, MD 20771 +1-301-286-1258).

Appendix IV

Common Physical Constants

Speed of light	c	$3.00 \times 10^{8}\ m/s$
Gravitational constant	G	$6.67 \times 10^{-11}\ N \cdot m^2/kg^2$
Avogadro constant	N_A	$6.02 \times 10^{23}\ mol^{-1}$
Universal gas constant	R	$8.31\ J/mol \cdot K$
Mass-energy relation	c^2	$8.99 \times 10^{16}\ J/kg$ (931.5 MeV/u)
Coulomb's law constant	k_e	$8.99 \times 10^{9}\ Nm^2/C^2$
Planck constant	h	$6.63 \times 10^{-34}\ J \cdot s$ ($4.14 \times 10^{-15}\ eV\,s$)
Boltzmann constant	k	$1.38 \times 10^{-23}\ J/K$ ($8.62 \times 10^{-5}\ V/K$)
Elementary charge	e	$1.60 \times 10^{-19}\ C$
Electron mass	m_e	$9.11 \times 10^{-31}\ kg$
Proton mass	m_p	$1.67 \times 10^{-27}\ kg$
Bohr radius	r_B	$5.29 \times 10^{-11}\ m$

Common Physical Properties

Air (dry, at 20°C and 1 atm)	
Density	$1.21\ kg/m^3$
Specific heat molar at constant pressure	$1010\ J/kg \cdot K$
Ratio of molar specific heats	1.40
Speed of sound	$343\ m/s$
Electrical breakdown strength	$3 \times 10^{6}\ V/m$
Effective molar mass	$0.0289\ kg/mol$
Water	
Density	$1000\ kg/m^3$
Speed of sound	$1460\ m/s$
Specific heat at constant pressure	$4190\ J/kg \cdot K$
Heat of fusion (0°C)	$333\ kJ/kg$
Heat of vaporization (100°C)	$2260\ kJ/kg$
Molar mass	$0.0180\ kg/mol$
Earth	
Mass	$5.98 \times 10^{24}\ kg$
Mean radius	$6.37 \times 10^{6}\ m$
Free-fall acceleration at the Earth's surface	$9.81\ m/s^2$
Standard atmosphere	$1.01 \times 10^{5}\ Pa$
Period of satellite at 100-km altitude	86.3 *min*
Radius of the geosynchronous orbit	42,200 *km*
Escape speed	$11.2\ km/s$
Magnetic dipole moment	$8.0 \times 10^{22}\ A \cdot m^2$
Mean electric field at surface	150 *V/m, down*

Astronomical Distances

From Earth to:

Moon	$3.82 \times 10^{8}\ m$
Sun	$1.50 \times 10^{11}\ m$
Nearest star	$4.04 \times 10^{16}\ m$
Galactic center	$2.2 \times 10^{20}\ m$
Andromeda galaxy	$2.1 \times 10^{22}\ m$
Edge of the observable universe	$\sim 10^{26}\ m$

Common Conversion Factors

Mass and Density

$1\ kg = 1000\ g = 6.02 \times 10^{26}\ u$

$1\ slug = 14.6\ kg$

$1\ u = 1.66 \times 10^{-27}\ kg$

$1\ kg/m^3 = 10^{-3}\ g/cm^3$

Length and Volume

$1\ m = 100\ \text{cm} = 39.4\ in. = 3.28\ ft$

$1\ mi = 1.61\ km = 5280\ ft$

$1\ in. = 2.54\ cm$

$1\ nm = 10^{-9} m = 10 Å$

$1\ light\text{-}year = 9.46 \times 10^{15}\ m$

$1\ m^3 = 1000\ L = 35.3\ ft^3 = 264\ gal$

Time

$1\ d = 86{,}400\ s$

$1\ y = 365\ ¼\ d = 3.16 \times 10^{7} s$

Angular Measure

$1\ rad = 57.3° = 0.159\ rev$

$\pi\ rad = 180° = ½\ rev$

Speed

$1\ m/s = 3.28\ ft/s = 2.24\ mi/h$

$1\ km/h = 0.621\ mi/h = 0.278\ m/s$

Force and Pressure

$1\ N = 10^{5}\ dyne = 0.225\ lb$

$1\ lb = 4.45\ N$

$1\ Pa = 1\ N/m2 = 10\ dyne/cm^2$
$= 1.45 \times 10^{-4}\ lb/in.^2$

$1\ atm = 1.01 \times 10^{5}\ Pa = 14.7\ lb/in.^2$
$= 76\ cm\text{-Hg}$

Energy and Power

$1\ J = 10^{7}\ erg = 0.239\ cal = 0.738\ ft - lb$

$1\ kW\text{-}h = 3.6 \times 10^{6} J$

$1\ cal = 4.19\ J$

$1\ eV = 1.60 \times 10^{-19} J$

$1\ horsepower = 746\ W = 550\ ft \cdot lb/s$

Index

A

B

C

D

E

F

G

H

I

J

K

L

M

N

O

P

Q

R

S

T

U

V

W